Computational Methods for Electromagnetic and Optical Systems

SECOND EDITION

OPTICAL SCIENCE AND ENGINEERING

Founding Editor
Brian J. Thompson
University of Rochester
Rochester, New York

RECENTLY PUBLISHED

Computational Methods for Electromagnetic and Optical Systems, Second Edition, *John M. Jarem, and Partha P. Banerjee*
Optical Methods of Measurement: Wholefield Techniques, Second Edition, *Rajpal S. Sirohi*
Optoelectronics: Infrared-Visible-Ultraviolet Devices and Applications, Second Edition, *edited by Dave Birtalan and William Nunley*
Photoacoustic Imaging and Spectroscopy, *edited by Lihong V. Wang*
Polarimetric Radar Imaging: From Basics to Applications, *Jong-Sen Lee and Eric Pottier*
Near-Earth Laser Communications, *edited by Hamid Hemmati*
Laser Safety: Tools and Training, *edited by Ken Barat*
Slow Light: Science and Applications, *edited by Jacob B. Khurgin and Rodney S. Tucker*
Dynamic Laser Speckle and Applications, *edited by Hector J. Rabal and Roberto A. Braga Jr.*
Biochemical Applications of Nonlinear Optical Spectroscopy, *edited by Vladislav Yakovlev*
Tunable Laser Applications, Second Edition, *edited by F. J. Duarte*
Optical and Photonic MEMS Devices: Design, Fabrication and Control, *edited by Ai-Qun Liu*
The Nature of Light: What Is a Photon?, *edited by Chandrasekhar Roychoudhuri, A. F. Kracklauer, and Katherine Creath*
Introduction to Nonimaging Optics, *Julio Chaves*
Introduction to Organic Electronic and Optoelectronic Materials and Devices, *edited by Sam-Shajing Sun and Larry R. Dalton*
Fiber Optic Sensors, Second Edition, *edited by Shizhuo Yin, Paul B. Ruffin, and Francis T. S. Yu*
Terahertz Spectroscopy: Principles and Applications, *edited by Susan L. Dexheimer*
Photonic Signal Processing: Techniques and Applications, *Le Nguyen Binh*
Smart CMOS Image Sensors and Applications, *Jun Ohta*
Organic Field-Effect Transistors, *Zhenan Bao and Jason Locklin*
Coarse Wavelength Division Multiplexing: Technologies and Applications, *edited by Hans Joerg Thiele and Marcus Nebeling*
Microlithography: Science and Technology, Second Edition, *edited by Kazuaki Suzuki and Bruce W. Smith*
Physical Properties and Data of Optical Materials, *Moriaki Wakaki, Keiei Kudo, and Takehisa Shibuya*
Microwave Photonics, *edited by Chi H. Lee*
Photonics: Principles and Practices, *Abdul Al-Azzawi*
Lens Design, Fourth Edition, *Milton Laikin*
Gas Lasers, *edited by Masamori Endo and Robert F. Walker*
Optical Waveguides: From Theory to Applied Technologies, *edited by Maria L. Calvo and Vasudevan Lakshiminarayanan*

Computational Methods for Electromagnetic and Optical Systems

SECOND EDITION

John M. Jarem

Partha P. Banerjee

CRC Press is an imprint of the
Taylor & Francis Group, an **informa** business

CRC Press
Taylor & Francis Group
6000 Broken Sound Parkway NW, Suite 300
Boca Raton, FL 33487-2742

First issued in paperback 2017

ISBN 13: 978-1-138-11629-0 (pbk)
ISBN 13: 978-1-4398-0422-3 (hbk)

Library of Congress Cataloging-in-Publication Data

Jarem, John M., 1948-
Computational methods for electromagnetic and optical systems, second edition / John M. Jarem, Partha P. Banerjee. -- 2nd ed.
p. cm. -- (Optical science and engineering ; 149)
Summary: "This text introduces and examines a variety of spectral computational techniques - including k-space theory, Floquet theory and beam propagation - that are used to analyze electromagnetic and optical problems. The book also presents a solution to Maxwell's equations from a set of first order coupled partial differential equations"-- Provided by publisher.
Includes bibliographical references and index.
ISBN 978-1-4398-0422-3 (hardback)
1. Electromagnetism--Mathematics. 2. Electromagnetism--Industrial applications. 3. Optics--Mathematics. 4. Optics--Industrial applications. I. Banerjee, Partha P. II. Title. III. Series.

QC760.J47 2011
537.01'51--dc22 2010045338

Visit the Taylor & Francis Web site at
http://www.taylorandfrancis.com

and the CRC Press Web site at
http://www.crcpress.com

I (John M. Jarem) would like to dedicate the book to Margaret and Francis Shea, and my parents, Sarah and John Jarem.

I (Partha P. Banerjee) would like to dedicate the book to my mom and dad, Prativa and Abani Banerjee.

Contents

Preface

Finding exact solutions of problems in electromagnetics (EM) and optics has become an increasingly important area of research. The analysis and design of modern applications in optics and those in traditional EM demand increasingly similar numerical computations due to a reduction of feature sizes in optics. In EM, a large amount of research concentrates on numerical analysis techniques such as the method of moments, finite element analysis, and finite difference analysis. In the field of optics (a part of EM), a large amount of research has been conducted on beam propagation methods (BPMs), analysis of thin and thick diffraction gratings, photorefractive (PR) materials and PR-induced diffraction gratings, chiral materials, and photonic band gap (PBG) structures. Also, over the last few years, extensive research in microwaves, terahertz, and optics has been conducted in the area of metamaterials, i.e., materials that show unusual properties such as negative refraction, amplification of nonpropagating waves, contradirectionality of power flow and phase velocities, etc.

From the late 1970s to the present, an extremely important technique called the state variable method has been developed by various researchers for the analysis of planar diffraction gratings; this technique is known as the rigorous coupled wave analysis (RCWA). It is based on expanding Maxwell's equations in periodic media in a set of Floquet harmonics, from which the unknown expansion variable is arranged in state variable form by which all unknowns of the system may be solved. For planar diffraction gratings, this technique has proved to be very effective, providing a fast, accurate solution and involving only a small matrix and eigenvalue equation to find the solution. A variant of the RCWA, called the transfer matrix method (TMM), is often used in the analysis of periodic layered structures, such as PBGs.

In the area of control theory and applications, the state variable method has been widely applied and in fact forms the very foundation for modern control. In EM, the state variable method, although a powerful analysis tool, has seen much less application. When used, it is applied in conjunction with other methods (e.g., the spectral domain method, transmission ladder techniques, *k*-space analysis techniques, and the spectral matrix method) and is rarely listed as a state variable method. In this book, we illustrate the application of the state variable method to various problems in EM and optics, with special emphasis on the analysis of diffraction gratings using RCWA.

This book aims to introduce students and researchers to a variety of spectral computational techniques including *k*-space theory, Floquet theory, TMM, and BPM, which are then used to analyze a variety of EM and optical systems. Examples include analysis of radiation through isotropic and anisotropic material slabs, planar diffraction gratings in isotropic and anisotropic media, propagation through inhomogeneous linear and nonlinear media, periodic layered structures including positive index material (PIM)–negative index material (NIM) stacks, radiation and scattering from three-dimensionally inhomogeneous cylindrical and spherical structures, and diffraction from PR materials. Many of the above cases are analyzed using a full-field vector approach to solve Maxwell's equations in anisotropic media where a standard wave equation approach is intractable. Efficient and rapidly convergent numerical algorithms are presented. The problems in this book have been analyzed using computational tools such as Fortran, MATLAB®, COMSOL®, and RSoft.

Since the publication of the first edition, several interesting new developments, such as metamaterials, nanostructures, photonic band gap materials, chiral materials, biomaterials, etc., have taken place in the fields of EM and optics. The second edition introduces these novel materials and outlines EM analysis and wave propagation through these materials. To this end, we have added sections to existing chapters, rearranged the chapters, and added new chapters. For instance, scalar wave propagation and analysis techniques now appear in Chapter 2, with new applications to induced transmission and reflection gratings in PR materials, analyzed using BPM, both unidirectional and bidirectional. In Chapter 3, we introduce concepts of EM as applied to chiral materials, metamaterials (NIMs), and photonic band gap structures. In this regard, the concepts of Poynting vector and stored energy, and energy, group, and phase velocities are clearly developed. In Chapter 4, we provide a concise review of k-space state variable formulation, including applications to anisotropic planar systems. In Chapter 5, we present the full-field rigorous coupled wave analysis of planar diffraction gratings, including application to H-mode, E-mode, crossed gratings; single- and multilayer diffraction grating analysis; and diffraction from anisotropic gratings. In Chapter 6, spectral techniques and RCWA are applied to the analysis of dynamic wave-mixing in PR materials with induced transmission and reflection gratings. In Chapters 7 and 8, we use the RCWA algorithm to analyze cylindrical and spherical systems using circular, bipolar cylindrical, and spherical coordinates. Finally, in Chapter 9, we present several RCWA computational case studies involving scattering from spatially inhomogeneous eccentric circular cylinders, solved in bipolar coordinates. In many examples throughout this book, we apply the complex Poynting theorem or the forward-scattering (optical) theorem to validate numerical solutions by verifying power conservation.

This book is intended primarily for senior and graduate students in electrical engineering, optical engineering, and physics. It should be useful for researchers in optics specializing in holography, gratings, nonlinear optics, PR materials, metamaterials, chiral media, etc., and researchers in EM working on antennas, propagation and scattering theory, or EM numerical methods. It should also be of interest to the military, industry, and academia, and to all interested in finding exact solutions to various types of EM problems. The book should be ideal for either classroom adoption or as an ancillary reference in graduate level courses like numerical methods in EM, diffractive optics, and EM scattering theory to name a few.

We would also like to acknowledge the support and encouragement of our wives, Elizabeth Jarem and Noriko Banerjee, and our parents and families during the writing of this book. PPB acknowledges the assistance of Drs. G. Nehmetallah and R. Aylo with proofreading, and with the solutions for some of the problems.

For MATLAB® and Simulink® product information, please contact:

The MathWorks, Inc.
3 Apple Hill Drive
Natick, MA, 01760-2098 USA
Tel: 508-647-7000
Fax: 508-647-7001
E-mail: info@mathworks.com
Web: www.mathworks.com

Authors

John M. Jarem received his BSEE, MSEE, and PhD from Drexel University, Philadelphia, Pennsylvania, in 1971, 1972, and 1975, respectively. His research interests include optical and microwave diffraction theory, numerical electromagnetics, and Green's function theory. Jarem has been a professor of electrical and computer engineering at the University of Alabama in Huntsville, Huntsville, Alabama, since 1987. Prior to this, he was a professor of electrical engineering in the Department of Electrical Engineering at the University of Texas at El Paso (1986–1987), an associate professor of electrical engineering at the University of Texas at El Paso (1981–1986), and an assistant professor of electrical engineering at the University of Petroleum and Minerals, Dhahran, Saudi Arabia (1975–1981). Jarem is the coauthor of one book and one book chapter, and is the author/coauthor of 51 refereed journal articles and 36 conference papers. He has been the PhD thesis advisor for four PhD students and the master's thesis advisor for 18 master's students. He is also a senior member of the IEEE.

Partha P. Banerjee is a professor of electro-optics and electrical and computer engineering (ECE) at the University of Dayton, where he was chair of the ECE department from 2000 to 2005. Prior to that, he was with the ECE department at the University of Alabama in Huntsville from 1991 to 2000 and at Syracuse University from 1984 to 2001. He received his BTech from the Indian Institute of Technology in 1979, and his MS and PhD from the University of Iowa in 1980 and 1983, respectively. His research interests include optical processing, nonlinear optics, photorefractive materials, and acousto-optics. He has authored/coauthored four books, several book chapters, over 100 refereed journal articles, and many conference papers. He is fellow of the Optical Society of America (OSA) and of the Society of Photo-optical Instrumentation Engineers (SPIE), and is a senior member of the Institute of Electrical and Electronics Engineers (IEEE). He received the National Science Foundation (NSF) Presidential Young Investigator award in 1987.

1 Mathematical Preliminaries

1.1 INTRODUCTION

Popular t-shirts advertising Maxwell's equations do not go beyond merely stating them. In this book, we enter into a little more depth and solve these equations for analyzing various electromagnetic (EM) and optical problems, e.g., diffraction gratings, radiation and scattering from dielectric objects, photorefractive materials, metamaterials, and chiral materials. The emphasis on finding the solutions in our text concerns the use of Fourier and state variable analyses, and beam propagation methods. In this chapter, we briefly review some mathematical techniques pertinent to the analyses presented in later chapters.

1.2 FOURIER SERIES AND ITS PROPERTIES

It is easy to show that a set of exponential functions $\{\exp jnKx\}$, $n = 0, \pm1, \pm2$, etc., where $j = \sqrt{-1}$, is orthogonal over an interval $(x_0, x_0 + 2\pi/K)$ for any value of x_0. The orthogonality can be demonstrated by considering the integral

$$I = \int_{x_0}^{x_0+2\pi/K} \exp(jnKx)\exp(-jmKx)dx \equiv \left(\frac{2\pi}{K}\right)\delta_{m,n} \tag{1.1}$$

where $\delta_{m,n}$ is the Kronecker delta function:

$$\delta_{m,n} = \begin{cases} 1, & m = n \\ 0, & m \neq n \end{cases} \tag{1.2}$$

Using this, a function $f(x)$ can be expanded in a Fourier series over an interval $(x_0, x_0 + 2\pi/K)$ as

$$f(x) = \sum_{n=-\infty}^{\infty} F_n \exp(jnKx) \tag{1.3}$$

Multiplying Equation 1.3 by $\exp(-jmKx)$, integrating over the interval $(x_0, x_0 + 2\pi/K)$, interchanging the summation and the integral, and using Equations 1.1 and 1.2, we get

$$\int_{x_0}^{x_0+2\pi/K} f(x)\exp(-jmKx)dx = \sum_{n=-\infty}^{\infty} F_n \int_{x_0}^{x_0+2\pi/K} \exp(jnKx)\exp(-jmKx)dx$$

$$= \sum_{n=-\infty}^{\infty} F_n\left[\left(\frac{2\pi}{K}\right)\delta_{m,n}\right] = \left(\frac{2\pi}{K}\right)F_m$$

Now, replacing m by n,

$$F_n = \left(\frac{K}{2\pi}\right) \int_{x_0}^{x_0+2\pi/K} f(x)\exp(-jnKx)dx \tag{1.4}$$

Note that if a function $f_e(x)$ is defined as $f_e(x) = f(x)\exp(-j\alpha x)$, where α is a constant, then over the interval $(x_0, x_0+2\pi/K)$, it can be written as

$$f_e(x) = \sum_{n=-\infty}^{\infty} F_n \exp(-jk_{xn}x), \quad k_{xn} = \alpha - nK \tag{1.5}$$

Li refers to this expansion in Equation 1.5 as a pseudo-Fourier series of $f_e(x)$ [1].

If two functions $f(x)$ and $g(x)$ having Fourier series expansions

$$f(x) = \sum_{n=-\infty}^{\infty} F_n \exp(jnKx), \quad g(x) = \sum_{n=-\infty}^{\infty} G_n \exp(jnKx) \tag{1.6}$$

over the same interval are multiplied, the product function $h(x)$ has a Fourier series expansion

$$h(x) = \sum_{n=-\infty}^{\infty} H_n \exp(jnKx) \tag{1.7}$$

over the same interval. We can find the Fourier coefficients of $h(x)$ in the following way:

$$\begin{aligned} h(x) = f(x)g(x) &= \sum_{n=-\infty}^{\infty} F_n \exp(jnKx) \sum_{m=-\infty}^{\infty} G_m \exp(jmKx) \\ &= \sum_{n=-\infty}^{\infty} \sum_{m=-\infty}^{\infty} F_n G_m \exp(j(n+m)Kx) = \sum_{l=-\infty}^{\infty} \sum_{m=-\infty}^{\infty} F_{l-m} G_m \exp(jlKx) \\ &\equiv \sum_{l=-\infty}^{\infty} H_l \exp(jlKx) \end{aligned} \tag{1.8}$$

The limits on l are $-\infty$ to $+\infty$ since $l = m+n$, and m and n each have limits $-\infty$ to $+\infty$. Hence the Fourier coefficients H_l of $h(x)$ can be expressed as

$$H_l = \sum_{m=-\infty}^{\infty} F_{l-m} G_m \tag{1.9}$$

Equation 1.7 is sometimes referred to as the *Laurent rule* [1]. To be more precise, Equations 1.8 and 1.9 should be understood in the following sense [1]:

$$h(x) = \lim_{N\to\infty} \sum_{l=-N}^{N} H_l \exp(jlKx) = \lim_{L\to\infty} \sum_{l=-L}^{L} \left(\lim_{M\to\infty} \sum_{m=-M}^{M} F_{l-m} G_m \right) \exp(jlKx) \tag{1.10}$$

The above equation, in the way it is written, emphasizes two important points. First, the two limits L and M are independent of each other, and the inner limit has to be taken first. Second, the upper and lower bounds in each sum should tend to infinity simultaneously [1].

1.3 FOURIER TRANSFORM

The one-dimensional temporal *Fourier transform* of a square-integrable function $f(t)$ is given as [2]

$$\hat{f}(\omega) = \int_{-\infty}^{\infty} f(t)\exp(-j\omega t)dt \tag{1.11}$$

The inverse Fourier transform is

$$f(t) = \frac{1}{2\pi}\int_{-\infty}^{\infty} \hat{f}(\omega)\exp(+j\omega t)d\omega \tag{1.12}$$

The one-dimensional spatial Fourier transform of a square-integrable function $f(x)$ is given as [2]

$$\tilde{f}(k_x) = \int_{-\infty}^{\infty} f(x)\exp(jk_x x)dx \tag{1.13}$$

The inverse Fourier transform is

$$f(x) = \frac{1}{2\pi}\int_{-\infty}^{\infty} \tilde{f}(k_x)\exp(-jk_x x)dx \tag{1.14}$$

The definitions for the forward and backward transforms are consistent with the engineering convention for a traveling wave, as explained in [2]. If $f(x)$ denotes a phasor EM field quantity, multiplication by $\exp j\omega t$ gives a collection or spectrum of forward traveling plane waves.

The two-dimensional extensions of Equations 1.13 and 1.14 are

$$\tilde{f}(k_x, k_y) = \int_{-\infty}^{\infty}\int_{-\infty}^{\infty} f(x,y)\exp(jk_x x + jk_y y)dx\,dy \tag{1.15}$$

$$f(x,y) = \frac{1}{(2\pi)^2}\int_{-\infty}^{\infty}\int_{-\infty}^{\infty} \tilde{f}(k_x, k_y)\exp(-jk_x x - jk_y y)dx\,dy \tag{1.16}$$

In many EM applications, the function $f(x, y)$ represents the transverse profile of an EM wave at a plane z. Hence in (1.15) and (1.16), $f(x, y)$ and $\tilde{f}(k_x, k_y)$ have z as a parameter. For instance (1.16) becomes

$$f(x,y,z) = \frac{1}{(2\pi)^2}\int_{-\infty}^{\infty}\int_{-\infty}^{\infty} \tilde{f}(k_x, k_y; z)\exp(-jk_x x - jk_y y)dx\,dy \tag{1.17}$$

The usefulness of this transform lies in the fact that when substituted into Maxwell's equations, one can reduce the set of three-dimensional PDEs to a set of one-dimensional differential equations (ODEs) for the spectral amplitudes $\tilde{f}(k_x, k_y; z)$.

1.4 HANKEL TRANSFORM

In the special case where the function $f(x,y)$ in Equation 1.15 is circularly symmetric, i.e., $f(x,y) = g(\rho)$, $\rho = \sqrt{x^2 + y^2}$, we can introduce new variables to change from rectangular to polar coordinates, in both the spatial and the spatial frequency domains:

$$x = \rho\cos\phi, \quad y = \rho\sin\phi; \quad k_x = k_\rho \cos\phi', \quad k_y = k_\rho \sin\phi' \tag{1.18}$$

Substituting (1.18) into (1.15), we get after some algebra:

$$\tilde{g}(k_\rho, \phi') = \int_0^\infty \int_0^{2\pi} g(\rho)\exp\{jk_\rho\rho\cos(\phi - \phi')\}\rho \, d\rho \, d\phi \tag{1.19}$$

It turns out that the result of the integration w.r.t. ϕ is independent of ϕ' since [3]

$$J_0(k_\rho\rho) = \frac{1}{2\pi}\int_0^{2\pi} \exp\{jk_\rho\rho\cos(\phi - \phi')\} d\phi \tag{1.20}$$

Substituting (1.20) into (1.19),

$$\tilde{g}(k_\rho) = \frac{1}{2\pi}\int_0^\infty g(\rho)J_0(k_\rho\rho)\rho \, d\rho \tag{1.21}$$

Equation 1.21 is called the *Fourier–Bessel transform* or the *Hankel transform* of the circularly symmetric function $g(\rho)$ [4]. The inverse transform is given by

$$g(\rho) = \frac{1}{2\pi}\int_0^\infty \tilde{g}(k_\rho)J_0(k_\rho\rho)k_\rho \, dk_\rho \tag{1.22}$$

1.5 DISCRETE FOURIER TRANSFORM

Given a discrete function $f(n\Lambda)$, $n = 0,\ldots N-1$, a corresponding periodic function $f_p(n\Lambda)$ with period $N\Lambda$ can be formed as [5]

$$f_p(n\Lambda) = \sum_{r=-\infty}^{\infty} f(n\Lambda + rN\Lambda) \tag{1.23}$$

The discrete function $f(n\Lambda)$ may be formed by the discrete values of a continuous function $f(x)$ evaluated at the points $x = n\Lambda$.

The *discrete Fourier Transform* (DFT) of $f_p(n\Lambda)$ is defined as

$$\tilde{f}_p(mK) = \sum_{n=0}^{N-1} f_p(n\Lambda)\exp(jmnK\Lambda), \quad K = \frac{2\pi}{N\Lambda} \tag{1.24}$$

The inverse DFT is defined as

$$f_p(n\Lambda) = \frac{1}{N}\sum_{n=0}^{N-1} \tilde{f}_p(mK)\exp(-jmnK\Lambda) \tag{1.25}$$

For properties of the DFT, e.g., linearity, symmetry, and periodicity, as well as relationship to the *z*-transform, the Fourier transform, and the Fourier series, the readers are referred to any standard book on digital signal processing [5].

For the purposes of this book, the DFT is a way of numerically approximating the continuous Fourier transform of a function. The DFT is of interest because it can be efficiently and rapidly evaluated by using standard Fast Fourier Transform (FFT) packages. The direct connection between the continuous Fourier transform and the DFT is given in the following. For a function $f(x)$ and its continuous Fourier transform $\tilde{f}(k_x)$,

$$\tilde{f}_p(mK) \cong \frac{1}{\Lambda}\tilde{f}(mK), \quad \frac{|mK| < \pi}{\Lambda} \tag{1.26}$$

In Equation 1.26, $\tilde{f}_p(mK)$ is defined, as in Equation 1.24, to be the DFT of $f_p(n\Lambda)$. The equality holds for the fictitious case when the function is both space and spatial frequency limited.

1.6 REVIEW OF EIGENANALYSIS

A large amount of the computations in this book are based on determining the eigenvalues and eigenvectors of a matrix $\underline{\underline{A}}$. Therefore, this section will briefly review the methods and techniques that are associated with numerically solving this problem [6,7]. The matrix $\underline{\underline{A}}$, which is a square matrix in general, transforms a column vector $\underline{x}$ into a matrix $\underline{y} = \underline{\underline{A}}\,\underline{x}$. We are interested in those vectors x that transform into themselves and satisfy

$$\underline{\underline{A}}\,\underline{x} = q\,\underline{x} \tag{1.27}$$

These column vectors are called the *eigenvectors* of the system. The values of q which satisfy (1.27) are known as the *eigenvalues*, the characteristic values, or the latent roots of the matrix $\underline{\underline{A}}$. Equation 1.27 may be written as a linear set of equations as

$$\begin{aligned}
&(a_{11} - q)x_1 + a_{12}x_2 + a_{13}x_3 + \cdots + a_{1n}x_n = 0 \\
&a_{12}x_1 + (a_{22} - q)x_2 + a_{23}x_3 + \cdots + a_{2n}x_n = 0 \\
&\vdots \\
&a_{n1}x_1 + a_{n2}x_2 + a_{n3}x_3 + \cdots + (a_{nn} - q)x_n = 0
\end{aligned} \tag{1.28}$$

A nontrivial solution exists for the above equations if and only if

$$P(q) \equiv \det(q\underline{I} - \underline{A}) = 0 \tag{1.29}$$

where $\underline{I}$ is the identity matrix. The result of Equation 1.29 is an nth-order polynomial called the *characteristic equation* or eigenvalue equation. The equation is given by

$$P(q) = q^n + a_1 q^{n-1} + a_2 q^{n-2} + \cdots + a_{n-1} q + a_n \tag{1.30}$$

The roots of this equation are the eigenvalues of the matrix $\underline{A}$. When the roots are all unequal to one another, the roots or eigenvalues are called distinct. When the eigenvalue occurs m times, the eigenvalue is a repeated value of order m. When the root has a real and nonzero imaginary part, the roots occur in complex conjugate pairs. In factored form, Equation 1.30 may be written as

$$P(q) = (q - q_1)(q - q_2) \cdots (q - q_n) \tag{1.31}$$

The coefficients of the eigenvalue equation may be found directly from the matrix $\underline{A}$. For instance, setting q to zero in (1.31), we find

$$P(0) = a_n = \det(-\underline{A}) = (-1)^n \det(\underline{A}) = (-1)^n q_1 q_2 \cdots q_n \tag{1.32}$$

and thus from Equation 1.32,

$$\det(\underline{A}) = q_1 q_2 \cdots q_n \tag{1.33}$$

The coefficient a_1 may be found by expanding the factored characteristic equation and comparing the polynomial coefficients of the resulting equation. For example, if $n = 2$,

$$P(q) = q^2 + a_1 q + a_2 = (q - q_1)(q - q_2) = q^2 - (q_1 + q_2)q + q_1 q_2 \tag{1.34}$$

and thus

$$a_1 = -(q_1 + q_2) \tag{1.35}$$

after equating coefficients. For general n,

$$a_1 = -(q_1 + q_2 + \cdots + q_n) \tag{1.36}$$

If the determinant is expanded, we also find that the determinant is the negative sum of diagonal coefficients, that is,

$$a_1 = -(a_{11} + a_{22} + \cdots + a_{nn}) \tag{1.37}$$

The quantity in parentheses is an important quantity and is called the *trace* of $\underline{A}$:

$$\mathrm{Tr}(\underline{A}) = a_{11} + a_{22} + \cdots + a_{nn} = q_1 + q_2 + \cdots + q_n \tag{1.38}$$

Let $T_k = \text{Tr}(\underline{A}^k)$. Then a useful formula for the coefficient a_n of the characteristic equation is

$$\begin{aligned}
a_1 &= -T_1 \\
a_2 &= -\frac{1}{2}(a_1T_1 + T_2) \\
a_3 &= -\frac{1}{3}(a_2T_1 + a_1T_2 + T_3) \\
&\vdots \\
a_n &= -\frac{1}{n}(a_{n-1}T_1 + a_{n-2}T_2 + \cdots + a_1T_{n-1} + T_n)
\end{aligned} \tag{1.39}$$

For the case when the roots of $P(q)$ are distinct, a nontrivial vector $\underline{x}_i$ can be found for each root which satisfies

$$\left(q_i\underline{I} - \underline{A}\right)\underline{x}_i = 0 \tag{1.40}$$

The matrix formed of the columns of $\underline{x}_i$ is called the *modal matrix* $\underline{M}$. The name modal matrix comes from control theory where a dynamical system can be decomposed into dynamic modes of operation. For EM diffraction grating problems and also for EM problems that use *k*-space techniques, the EM field solutions associated with a state variable analysis may be decoupled into spatial mode solutions. These modes are analogous to the dynamical modes of operation encountered in control systems.

If the eigenvalues are distinct, which is mainly the case under consideration in this text, the modal matrix is nonsingular and therefore its inverse exists. Letting $\underline{M}$ be the modal matrix, we may write

$$\underline{M}\underline{Q} = \underline{A}\underline{M} \tag{1.41}$$

where $\underline{Q}$ is a diagonal matrix holding the eigenvalues q_i on the diagonal. It can be shown that the inverse of $\underline{M}$ exists, hence, Equation 1.41 gives

$$\underline{Q} = \underline{M}^{-1}\underline{A}\underline{M} \tag{1.42}$$

If $\underline{Q}$ is squared, we have

$$\underline{Q}^2 = (\underline{M}^{-1}\underline{A}\underline{M})(\underline{M}^{-1}\underline{A}\underline{M}) = (\underline{M}^{-1}\underline{A}^2\underline{M}) \tag{1.43}$$

and if we further pre- and post-multiply by $\underline{M}$ and $\underline{M}^{-1}$, respectively, we have

$$\underline{A}^2 = \underline{M}\underline{Q}^2\underline{M}^{-1} \tag{1.44}$$

Similarly, if $\underline{A}$ is raised to the *p*th power, we have

$$\underline{A}^p = \underline{M}\underline{Q}^p\underline{M}^{-1} \tag{1.45}$$

where $\underline{Q}^p$ is the diagonal matrix formed by raising each eigenvalue q_i to the pth power. A matrix polynomial $N(\underline{A})$ can be conveniently evaluated as

$$N(\underline{A}) = \underline{M}N(\underline{Q})\underline{M}^{-1} \tag{1.46}$$

where linear combinations of powers of $\underline{A}$ as given by Equation 1.45 have been used. $N(\underline{Q})$ is the diagonal matrix formed by placing in each diagonal matrix entry the polynomial $N(q_i)$. Thus the modal matrix provides a convenient way to quickly and accurately evaluate powers and polynomials of the matrix $\underline{A}$.

In this text, we will be greatly concerned with calculating the exponential function of the matrix $\underline{A}$. The exponential function of the matrix $\underline{A}$, namely $\exp(\underline{A})$, is defined as [7]

$$\exp(\underline{A}) = \underline{I} + \underline{A} + \left(\frac{1}{2}\right)(\underline{A})^2 + \cdots + \frac{1}{k!}(\underline{A})^k + \cdots \tag{1.47}$$

which is the same infinite series expansion as that used to define the exponential function $\exp(a)$.

We now review two important aids that help in the solution and evaluation of exponential matrix and in fact any function of the matrix $\underline{A}$. These are called the *Cayley–Hamilton theorem* and the *Cayley–Hamilton technique*. These aids will be presented only for the cases of matrices with distinct eigenvalues.

The first theorem to be reviewed is the Cayley–Hamilton theorem. If we have a polynomial $N(q) = q^n + c_1 q^{n-1} + \cdots + c_{n-1} q + c_n$, then using Equation 1.46 we have

$$N(\underline{A}) = \underline{M}\begin{bmatrix} N(q_1) & 0 & 0 \\ 0 & N(q_2) & \\ 0 & & N(q_3) \end{bmatrix}\underline{M}^{-1} \tag{1.48}$$

where $\underline{M}$ is the modal matrix. If the polynomial $N(q)$ is chosen to the characteristic equation, that is, $N(q) = P(q)$, then $N(q_i) = P(q_i) = 0$; $i = 1, 2, \ldots, n$ and thus

$$P(\underline{A}) = \underline{M}\begin{bmatrix} 0 & 0 & 0 & \\ 0 & 0 & 0 & \\ 0 & 0 & 0 & \\ & & & \cdots \end{bmatrix}\underline{M}^{-1} = 0 \tag{1.49}$$

We thus see that the matrix $\underline{A}$ satisfies it own characteristic equation.

Another important aid to evaluating a function of a matrix, where the function is analytic over a given range of interest, is provided by the Cayley–Hamilton technique [7]. We first consider the case where the analytic function is a polynomial of higher degree than the characteristic polynomial $P(q)$ of order n. Let the polynomial be $N(q)$. We consider the case where the roots (or eigenvalues) of $P(q)$ are distinct. In this case,

$$\frac{N(q)}{P(q)} = Q(q) + \frac{R(q)}{P(q)} \tag{1.50}$$

where $Q(q)$ is a polynomial, and $R(q)$ is a polynomial of order $n - 1$ or less.

Multiplying by $P(q)$, we have

$$N(q) = Q(q)P(q) + R(q) \tag{1.51}$$

If q is an eigenvalue or root of $P(q)$, then $P(q) = 0$ and $N(q) = R(q)$. If we substitute $\underline{A}$ for q in Equation 1.51, we have

$$N(\underline{A}) = Q(\underline{A})\underline{P}(\underline{A}) + R(\underline{A}) \tag{1.52}$$

Since by the Cayley–Hamilton theorem the matrix $P(\underline{A}) = 0$, it follows that

$$N(\underline{A}) = R(\underline{A}) \tag{1.53}$$

Thus a higher order polynomial matrix may be represented and evaluated using an n–1 polynomial expression.

Consider next the case where the matrix function is a general analytic function over a region of interest, for example, $F(\underline{A}) = \exp(\underline{A})$. In this case, $F(q)$ may be expanded in an infinite power series over the analytic region of interest. As in the case when $F(q)$ was a polynomial, $F(q)$ may be written as

$$F(q) = Q(q)P(q) + R(q) \tag{1.54}$$

where $R(q)$ is a polynomial of order $n - 1$ given by

$$R(q) = \alpha_0 + \alpha_1 q + \alpha_2 q^2 + \cdots + \alpha_{n-1} q^{n-1} \tag{1.55}$$

Let $q = q_1, q_2, \ldots, q_n$ be the distinct roots of $P(q_i) = 0$, $i = 1, \ldots, n$. We have, after evaluating Equation 1.54,

$$\begin{aligned} F(q_1) &= R(q_1) \\ F(q_2) &= R(q_2) \\ &\vdots \\ F(q_n) &= R(q_n) \end{aligned} \tag{1.56}$$

This defines a set of $n \times n$ linear equations from which the coefficients α_i, $i = 1, \ldots, n$ may be determined. At this point, we would like to show that the function $Q(q)$ is analytic. To do this, we write $Q(q)$ as

$$Q(q) = \frac{F(q) - R(q)}{P(q)} \tag{1.57}$$

In this expression, we note that over the region of interest the numerator and denominator of Equation 1.57 have the same zeros. Since in Equation 1.57 all functions $F(q)$, $Q(q)$, $P(q)$, and $R(q)$ are analytic over the range where $F(q)$ is, we may replace q by the matrix $\underline{A}$. We therefore have

$$F(\underline{A}) = Q(\underline{A})P(\underline{A}) + R(\underline{A}) \tag{1.58}$$

Since by the Cayley–Hamilton theorem $P(\underline{A}) = 0$, we have

$$F(\underline{A}) = R(\underline{A}) \tag{1.59}$$

Thus we have shown that the analytic matrix function $F(\underline{A})$ may be evaluated by using a polynomial matrix expression of order $n - 1$ as given by $R(\underline{A})$ in Equation 1.55.

PROBLEMS

1.1 Derive the wave equation for the electric and magnetic fields starting from Maxwell's equations in a homogeneous, isotropic, source-free region (see Chapter 3). How does this change if the material is anisotropic?

1.2 Find from first principles the Fourier series coefficients for a periodic square wave $s(x)$ of unit amplitude and 50% duty cycle. Now find the Fourier series coefficients of $s^2(x)$: (a) from first principles and (b) using the Laurent rule. Plot $s^2(x)$ versus x by employing the Fourier series coefficients you found using (b). Use 5, 10, and 100 Fourier coefficients. Describe the general trend(s).

1.3 Find the two-dimensional Fourier transform of a rectangle (*rect*) function of unit height and width a in each dimension.

1.4 Show that the two-dimensional Fourier transform of a Gaussian function of width w is another Gaussian function. Functions like this are called self-Fourier transformable. Find its width in the spatial frequency domain. Can you think of any other functions that are self-Fourier transformable?

1.5 Find the Hankel transform of (a) a circular function defined as $g(\rho) = \mathrm{circ}(\rho/\rho_0)$, which has a value of 1 within a circle of radius ρ_0 and is 0 otherwise; (b) a Gaussian function given by $g(\rho) = \exp{-(\rho/\rho_0)^2}$.

1.6 Find the DFT of a square wave function using a software of your choice. Comment on the nature of the spectrum numerically computed as the width of the square wave changes.

1.7 Find $\sin \underline{\underline{A}}$, where $\underline{\underline{A}}$ is a matrix given by $\begin{pmatrix} 1 & 20 & 0 \\ -1 & 7 & 1 \\ 3 & 0 & -2 \end{pmatrix}$, using the Cayley–Hamilton theorem [7].

REFERENCES

1. L. Li, Use of Fourier series in the analysis of discontinuous periodic structures, *J. Opt. Soc. Am. A* 15, 1808–1816, 1996.
2. P.P. Banerjee and T.-C. Poon, *Principles of Applied Optics*, Irwin, New York, 1991.
3. M. Abramowitz and I. Stegun, *Handbook of Mathematical Functions*, Dover, New York, 1965.
4. I. Sneddon, *The Use of Integral Transforms*, McGraw-Hill, New York, 1972.
5. A. Antoniou, *Digital Filters: Analysis and Design*, McGraw-Hill, New York, 1979.
6. P.M. Deruso, R.J. Roy, and C.M. Close, *State Variables for Engineers*, John Wiley & Sons, Inc., New York, 1967.
7. L.A. Pipes and L.R. Harvill, *Advanced Mathematics for Engineers and Scientists*, McGraw-Hill, New York, 1970.

2 Scalar EM Beam Propagation in Inhomogeneous Media

2.1 INTRODUCTION

In the previous chapter, we reviewed some of the mathematical preliminaries that will be useful later on in the text. In this chapter, we discuss some of the basic concepts of scalar wave propagation, and discuss an important numerical method, called the beam propagation method (BPM), to study propagation in linear media and in media with induced nonlinearities. Furthermore, we also discuss propagation through induced gratings, both transmission and reflection type, in order to assess energy coupling between participating waves. Finally, we introduce readers to an important characterization method, called the *z*-scan method, which is often used to determine the focal length of an induced lens.

2.2 TRANSFER FUNCTION FOR PROPAGATION

For simplicity, we consider the *scalar wave equation*

$$\frac{\partial^2 E}{\partial t^2} - v^2 \nabla^2 E = 0 \tag{2.1}$$

and substitute

$$E(x, y, z, t) = \text{Re}\{E_e(x, y, z)\exp[j(\omega_0 t - k_0 z)]\} \tag{2.2}$$

with $\omega_0/k_0 = v$. The quantity E_e is related to the phasor E_p according to

$$E_p(x, y, z) = E_e(x, y, z)\exp(-jk_0 z) \tag{2.3}$$

and we will use one or the other notation according to convenience. Substituting Equation 2.2 into Equation 2.1 and assuming that E_e is a slowly varying function of z (the direction of propagation) in the sense that $\left|\partial^2 E_e/\partial z^2\right| / \left|\partial E_e/\partial z\right| \ll k_0$, we obtain the paraxial wave equation [1]

$$2jk_0 \frac{\partial E_e}{\partial z} = \nabla_\perp^2 E_e \tag{2.4}$$

where $\nabla_\perp^2$ denotes the transverse Laplacian. Equation 2.4 describes the propagation of the envelope $E_e(x, y, z)$ starting from the initial profile $E_e\big|_{z=0} = E_{e0}(x, y)$.

Equation 2.4 can be solved readily using Fourier transform techniques. Assuming E_e to be Fourier transformable, we can employ the definition of the Fourier transform as in Equation 1.15

$$\tilde{E}_e(k_x, k_y; z) = \Im_{x,y}\{E_e(x, y, z)\} = \int_{-\infty}^{\infty} E_e(x, y, z)\exp(jk_x x + jk_y y)dx\,dy \tag{2.5}$$

and its properties to transform Equation 2.4 into the ordinary differential equation (ODE)

$$\frac{d\tilde{E}_e}{dz} = \frac{j}{2k_0}(k_x^2 + k_y^2)\tilde{E}_e \tag{2.6}$$

We can easily solve Equation 2.6 to give

$$\tilde{E}_e(k_x, k_y; z) = \tilde{E}_{e0}(k_x, k_y)\exp\left\{\frac{j(k_x^2 + k_y^2)z}{2k_0}\right\} \tag{2.7}$$

where $\tilde{E}_{e0}(k_x, k_y)$ is the Fourier transform of $E_{e0}(x, y)$. We can interpret Equation 2.7 in the following way: Consider a linear system with an input spectrum of $\tilde{E}_{e0}(k_x, k_y)$ at $z = 0$ where the output spectrum at z is given by $\tilde{E}_e(k_x, k_y; z)$. The spatial frequency response of the system, which we will call the *paraxial transfer function for propagation* is then given by

$$\frac{\tilde{E}_e}{\tilde{E}_{e0}} \equiv H(k_x, k_y; z) = \exp\left\{\frac{j(k_x^2 + k_y^2)z}{2k_0}\right\} \tag{2.8}$$

As we will show later, in the split-step BPM, we model propagational diffraction by means of the transfer function for propagation derived above. For more exact calculations, the nonparaxial transfer function can be used. This may be derived starting from the nonparaxial wave equation, but will not be presented here for the sake of simplicity.

Incidentally, the inverse Fourier transform of the transfer function for propagation yields the *impulse response for propagation.* Starting from the paraxial transfer function for propagation which resembles a complex Gaussian, the inverse Fourier transform is a complex Gaussian as well, and has the form

$$h(x, y, z) = \frac{jk_0}{2\pi z}\exp\left[\frac{-j(x^2 + y^2)k_0}{2z}\right] \tag{2.9}$$

This, when convolved with the initial beam profile, yields the profile of the diffracted beam in the spatial domain directly. This convolution integral is in fact the Fresnel diffraction formula.

2.3 SPLIT-STEP BEAM PROPAGATION METHOD

If we wish to consider propagation in a material where the propagation constant or equivalently the refractive index is a function of position, either due to profiling of the material itself (such as a graded index fiber or a grating) or due to induced effects such as third order nonlinearities, the paraxial wave equation changes to

$$\frac{\partial E_e}{\partial z} = \frac{1}{2jk_0}\nabla_\perp^2 E_e - j\Delta n k_0 E_e \tag{2.10}$$

The quantity Δn is the change in the refractive index over the ambient refractive index $n_0 = c/v$, where c is the velocity of light in vacuum. Equation 2.10 is a modification of Equation 2.4 and can be derived from the scalar wave equation when the propagation constant or equivalently the velocity of the wave is a function of (x, y, z) explicitly as in gratings or fibers, or implicitly such as through the intensity dependent refractive index. An alternative, though heuristic, way to justify the presence of the additional term on the RHS of Equation 2.10 is to note that in the absence of diffraction (the first term on the RHS), the solution of the equation is of the form $E_e \propto \exp[-j\Delta n k_0 z]$, which explicitly shows the additional phase change due to propagation in the perturbed refractive index.

The paraxial propagation equation (2.10) is a partial differential equation (PDE) that does not always lend itself to analytical solutions, except for some very special cases involving special spatial variations of Δn, or when as in nonlinear optics, one looks for particular soliton solution of the resulting nonlinear PDE using exact integration or inverse scattering methods. Numerical approaches are often sought for to analyze beam (and pulse) propagation in complex systems such as optical fibers, volume diffraction gratings, Kerr and photorefractive (PR) media, etc. A large number of numerical methods can be used for this purpose. The pseudospectral methods are often favored over finite difference methods due to their speed advantage. The split-step BPM is an example of a pseudospectral method.

To understand the philosophy behind the BPM, it is useful to rewrite Equation 2.10 in the form [2,3]

$$\frac{\partial E_e}{\partial z} = (\hat{D} + \hat{S})E_e \tag{2.11}$$

where $\hat{D}$, $\hat{S}$ are a linear differential operator and a space dependent or nonlinear operator respectively (see, for instance, the structure of Equation 2.10). Thus, in general, the solution of Equation 2.11 can be symbolically written as

$$E_e(x, y, z + \Delta z) = \exp[(\hat{D} + \hat{S})\Delta z]E_e(x, y, z) \tag{2.12}$$

if $\hat{D}$, $\hat{S}$ are assumed to be z-independent. Now for two noncommuting operators $\hat{D}$, $\hat{S}$,

$$\exp(\hat{D}\Delta z)\exp(\hat{S}\Delta z) = \exp\left(\hat{D}\Delta z + \hat{S}\Delta z + \left(\frac{1}{2}\right)[\hat{D}, \hat{S}](\Delta z)^2 + \cdots\right) \tag{2.13}$$

according to the Baker–Hausdorff formula [2], where $[\hat{D}, \hat{S}]$ represents the commutation of $\hat{D}$, $\hat{S}$. Thus, up to second order in Δz,

$$\exp(\hat{D}\Delta z + \hat{S}\Delta z) \cong \exp(\hat{D}\Delta z)\exp(\hat{S}\Delta z) \tag{2.14}$$

which implies that in Equation 2.13 the diffraction and the inhomogeneous operators can be treated independent of each other.

The action of the first operator on the RHS of Equation 2.14 is better understood in the spectral domain. Note that this is the propagation operator that takes into account the effect of diffraction between planes z and $z + \Delta z$. Propagation is readily handled in the spectral or spatial frequency domain using the transfer function for propagation written in Equation 2.8 with z replaced by Δz. The second operator describes the effect of propagation in the absence of diffraction and in the presence of medium inhomogeneities, either intrinsic or induced, and is incorporated in the spatial

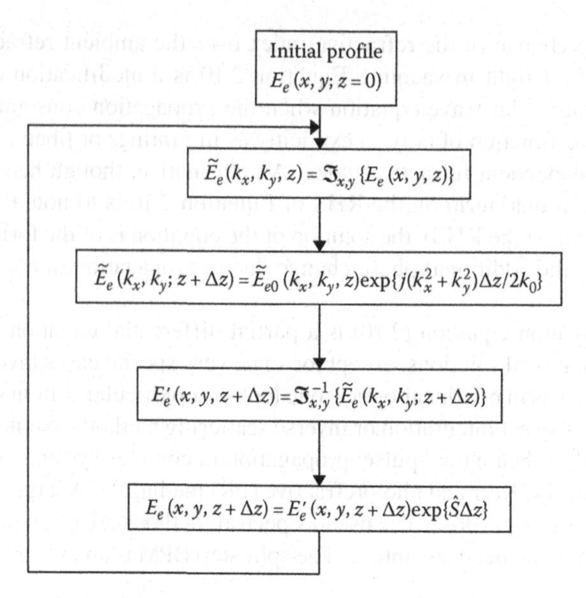

FIGURE 2.1 Flow diagram for the BPM split-step method.

domain. A schematic block diagram of the BPM method in its simplest form is shown in Figure 2.1. There are other modifications to the simple scheme, viz., the symmetrized split-step Fourier method, and the leap-frog techniques; these are discussed in detail elsewhere [2].

2.4 BEAM PROPAGATION IN LINEAR MEDIA

In this section, we will illustrate various cases where the BPM can be used to analyze propagation in inhomogeneous media. While most of the examples will be connected with beam propagation, we must point out to readers that the method can be used to analyze pulse propagation as well, simply by replacing z in Equation 2.11 with t (time), and making the linear spatial transverse differential operator a similar differential operator in z. With this modification, Equation 2.11 can model the propagation of one dimensional longitudinal pulse through an optical fiber with arbitrary group velocity dispersion. For details, we refer the readers to Agrawal [2].

2.4.1 Linear Free-Space Beam Propagation

The propagation of Gaussian beams through free space can be readily analyzed analytically using the transfer function of propagation. For a traditional scalar (or x- or y-polarized) Gaussian beam, $E_e(x, y; z = 0)$ is a Gaussian function of the form

$$E_e(x, y; 0) = E_0 \exp[-\alpha(x^2 + y^2)] \tag{2.15}$$

with a Fourier transform

$$\tilde{E}_e(k_x, k_y; 0) = E_0\left(\frac{\pi}{\alpha}\right)\exp\left[-\frac{(k_x^2 + k_y^2)}{4\alpha}\right] \tag{2.16}$$

where α may be complex. Hence, using Equation 2.8,

$$\begin{aligned}\tilde{E}_e(k_x,k_y;z) &= \tilde{E}_e(k_x,k_y;0)H(k_x,k_y;z)\\ &= E_0\left(\frac{\pi}{\alpha}\right)\exp\left[-\frac{(k_x^2+k_y^2)}{4\alpha}\right]\exp\left[\frac{j(k_x^2+k_y^2)z}{2k_0}\right]\\ &\equiv E_0\left(\frac{\pi}{\alpha}\right)\exp\left[\frac{j(k_x^2+k_y^2)q}{2k_0}\right]\end{aligned} \tag{2.17}$$

whose inverse transform is a Gaussian in transverse spatial dimensions:

$$E_e(x,y;z) = \frac{jk_0}{2\pi q}E_0\left(\frac{\pi}{\alpha}\right)\exp\left[-\frac{jk_0(x^2+y^2)}{2q}\right] \tag{2.18}$$

Equation 2.17 defines the well-known q-parameter of the Gaussian beam:

$$q = z + \frac{jk_0}{2\alpha} \tag{2.19}$$

If the initial Gaussian beam has waist w_0 and plane wavefronts, $\alpha = 1/w_0^2$ and $q = z + jz_R$, where $z_R = k_0 w_0^2/2$ is commonly referred to as the *Rayleigh range* of the Gaussian beam. Upon simplifying Equation 2.18 using Equation 2.19, we get the standard expression for the diffracted Gaussian beam profile as

$$E_e(x,y;z) = \frac{w_0}{w(z)}E_0\exp\left[-\frac{(x^2+y^2)}{w^2(z)}\right]\exp\left[-\frac{jk_0(x^2+y^2)}{2R(z)}\right]e^{-j\phi(z)} \tag{2.20}$$

where

$$w^2(z) = w_0^2\left[1+\left(\frac{z}{z_R}\right)^2\right], \quad R(z) = \frac{(z^2+z_R^2)}{z}, \quad \phi(z) = -\tan^{-1}\left(\frac{z}{z_R}\right) \tag{2.21}$$

Incidentally, if a Gaussian beam with initially plane wavefronts passes through a (thin) lens of focal length f, the latter introduces a quadratic phase curvature which can be accommodated by multiplying the initial Gaussian with a complex exponential $\exp[jk_0(x^2+y^2)/2f]$. In this case, the scalar optical field immediately behind the lens ($z = 0^+$) can be written in the form as in Equation 2.15 above, but where α is redefined as

$$\alpha = \frac{1}{w_0^2} - \frac{jk_0}{2f} = \left(\frac{1}{w_0^2}\right)\left(\frac{1-jz_R}{f}\right) \tag{2.22}$$

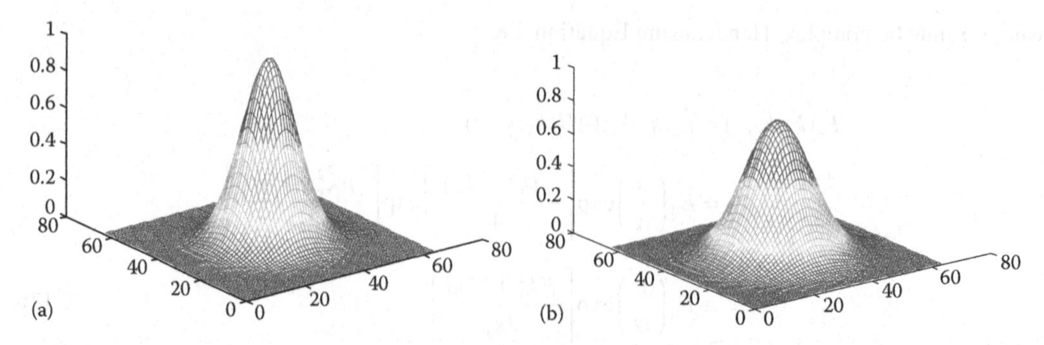

FIGURE 2.2 Diffraction of a Gaussian beam during free-space propagation: (a) profile at $z = 0$ (plane wavefronts assumed) and (b) profile at $z = z_R$, where z_R is the Rayleigh range of the original Gaussian beam.

The paraxial propagation of a Gaussian beam through an optical system involving translation and lensing can be equivalently studied in terms of the variation of the q-parameter. For instance, due to propagation in air through a distance z alone, the change in the q-parameter can be readily found from Equation 2.17 as

$$q_{new} = q_{old} + z \tag{2.23a}$$

On the other hand, if a Gaussian beam is incident on a lens of focal length f, the change in the q-parameter can be readily found from Equation 2.18 and the phase transformation by the lens given above as

$$\frac{1}{q_{new}} = \frac{1}{q_{old}} - \frac{1}{f} \tag{2.23b}$$

Equation 2.23a and b will be later used to analyze propagation of Gaussian beams through media which have a distributed lensing effect, such as a nonlinear medium.

We now demonstrate the application of BPM to Gaussian beam propagation through free space. In this case, the inhomogeneous operator is zero, and propagation from a plane $z = 0$ to arbitrary z can indeed be performed in one step in this case. However, in the example we provide, we use the split-step method to convince readers that the result is identical to what one would get if the propagation was covered in one step. In Figure 2.2, we show the profile of a diffracted Gaussian beam after propagation through free space, and the results agree with the physical intuition of increased width and decreased on-axis amplitude during propagation as well as the analytical results in Equations 2.20 and 2.21.

2.4.2 Propagation of Gaussian Beam through Graded Index Medium

A *graded index medium* has a refractive index variation of the form

$$n = n_0 + n^{(2)}(x^2 + y^2) \tag{2.24}$$

where n_0 denotes the intrinsic refractive index of the medium, and $n^{(2)}$ is a measure of the gradation in the refractive index.

In this case, the operator $\hat{S}$ becomes

$$\hat{S} = -jk_0 n^{(2)}(x^2 + y^2) \tag{2.25}$$

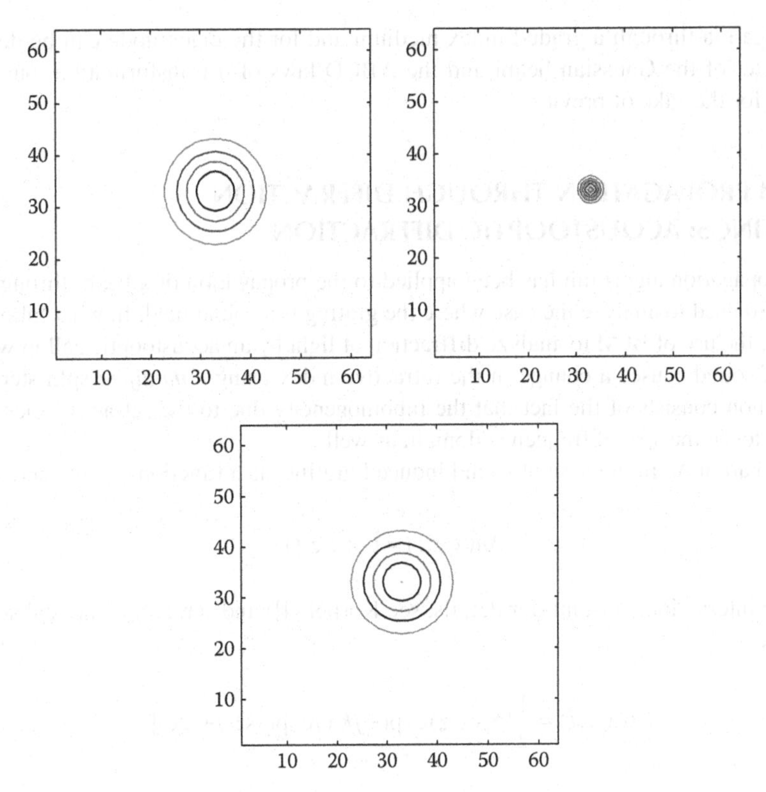

FIGURE 2.3 Contour plots showing periodic focusing of an initial Gaussian profile.

Propagation of a Gaussian beam in a medium with a graded index profile is shown in Figure 2.3. The contour plots show the initial (Gaussian) beam profile, the beam profile where the initial Gaussian attains its minimum waist during propagation before returning back to its original shape again, due to periodic focusing by the graded index distribution. Note that there exists a specific eigenmode (a Gaussian of a specific width, related to the refractive index gradient) for which the beam propagates through the material without a change in shape as a result of a balance between the diffraction of the beam and the guiding due to the parabolic gradient index profile. The contour plot of such a beam is shown in Figure 2.4. Analytical expressions for the Gaussian beam

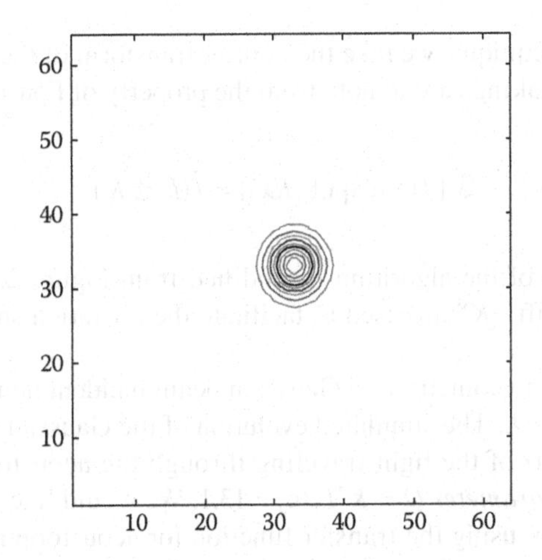

FIGURE 2.4 Fundamental mode in a graded index fiber.

during propagation through a graded index medium and for the eigenmode can be derived using the q-parameter of the Gaussian beam and the ABCD laws of q-transformation, but will not be pursued here for the sake of brevity.

2.5 BEAM PROPAGATION THROUGH DIFFRACTION GRATINGS: ACOUSTOOPTIC DIFFRACTION

The beam propagation algorithm has been applied to the propagation of a beam through a grating, and can be also used to analyze the case where the grating is a sound field. In what follows, we give an example of the use of BPM to analyze diffraction of light by an acoustooptic cell in which a traveling wave of sound causes a change in the refractive index using a *modified* split-step technique. The modification consists of the fact that the inhomogeneity due to the refractive index grating is accomodated for in the spatial frequency domain as well.

The perturbation Δn in the case of sound induced gratings is a function of time and space:

$$\Delta n(x,z,t) = Cs(x,z,t) \tag{2.26}$$

where C is an interaction constant (for details, see Korpel [4]) and $s(x, z, t)$ is the real sound amplitude given by,

$$s(x,z,t) = \frac{1}{2}[S_e(x,z)\exp(-jKx)\exp(j\Omega t) + \text{c.c.}] \tag{2.27}$$

where S_e is the complex amplitude of the sound field that interacts with the light beam and is traveling in the x direction, and c.c. denotes the complex conjugate.

The quantities K and Ω are the propagation constant and the angular frequency, respectively, of the sound field. Following Refs. [4,5], a snapshot of the sound field is used at $t = 0$, so that using 2.26 and 2.27,

$$\begin{aligned}\exp(\hat{S}\Delta z) &= \exp(-jk_0\Delta n\Delta z) \approx 1 - jk_0\Delta n\Delta z\\ &= 1 - \left(\frac{1}{2}\right) jk_0\Delta zC[S_e(x,z)\exp(-jKx) + S_e{}^*(x,z)\exp(+jKx)]\end{aligned} \tag{2.28}$$

In the modified split-step technique, we take the Fourier transform of the above operator operating on the optical field $E_e(x, z)$, taking care to note from the property of Fourier transforms that

$$\Im_x[f(x)\exp(\pm jKx)] = \tilde{f}(k_x \pm K) \tag{2.29}$$

The main propagation loop of the algorithm is modified from Figure 2.1 and is shown in Figure 2.5. The boxes marked "Shift $\pm K$" are used to facilitate the operation shown in 229 in the spatial frequency domain.

Figure 2.6 shows problem geometry of a Gaussian beam incident nominally at Bragg angle on a sound column of width $z = L$. The simulated evolution of the Gaussian beam is shown in Figure 2.7. The peak phase delay α of the light traveling through the acoustooptic cell is taken equal to π, and the *Klein–Cook parameter* $Q = K^2L/k_0 = 13.1$. We would like to point out that the same answers could be derived by using the transfer function for acoustooptic interaction, as given in Refs. [4,6].

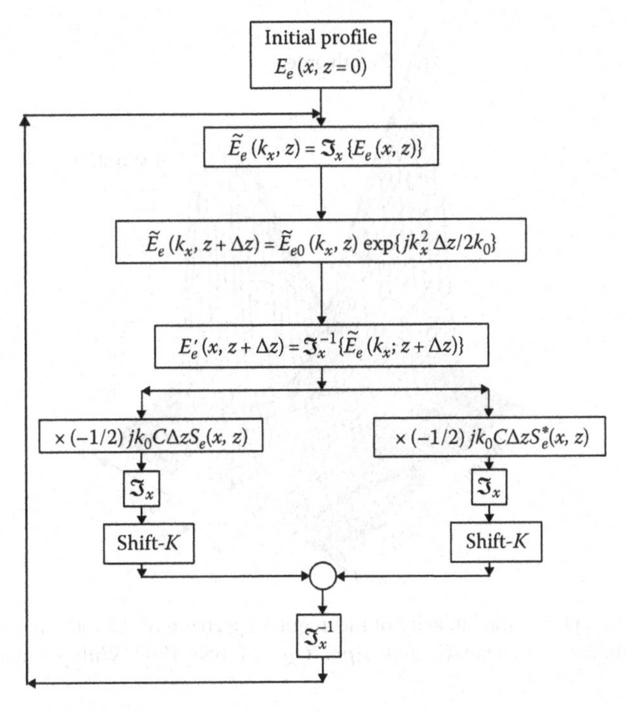

FIGURE 2.5 Flow diagram for the modified split-step technique to analyze acoustooptic interaction.

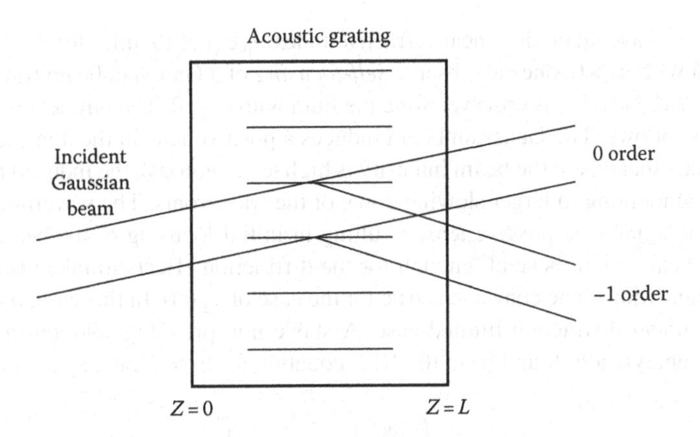

FIGURE 2.6 Geometry of acoustooptic interaction with a Gaussian beam at nominal Bragg incidence.

2.6 BEAM PROPAGATION IN KERR-TYPE NONLINEAR MEDIA

2.6.1 Nonlinear Schrodinger Equation

The nonlinear propagation of beams through a cubically nonlinear material is modeled by the nonlinear PDE also called the *nonlinear Schrodinger* (NLS) equation [7]

$$2\,jk_0\frac{\partial E_e}{\partial z} = \nabla_\perp^2 E_e + 2n_2k_0^2\left|E_e\right|^2 E_e \tag{2.30}$$

where n_2 is the *nonlinear refractive index coefficient* defined by the functional dependence of the total refractive index n on the intensity [7]:

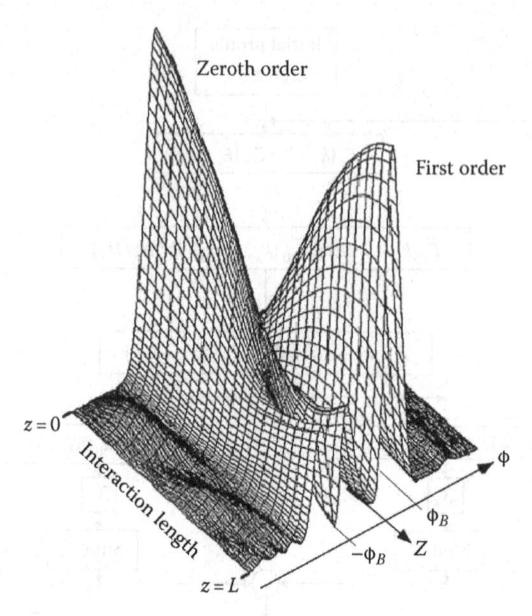

FIGURE 2.7 Simulation plot of the intensity of the angular spectrum of the total field at different positions along interaction length. (From Venzke, C. et al., *Appl. Opt.*, 31, 656, 1992. With permission.)

$$n = n_0 + \delta n\left(\left|E_e\right|^2\right) = n_0 + n_2\left|E_e\right|^2 \tag{2.31}$$

In writing (2.30), we have taken the linear refractive index n_0 equal to unity for the sake of simplicity. For a medium with $n_2 > 0$, one can observe *self-focusing* of a Gaussian beam traveling through a medium, while *self-defocusing* is observed for a medium with $n_2 < 0$. The physical reasoning behind self-focusing is as follows. The Gaussian beam induces a positive lens in the nonlinear material for $n_2 > 0$ due to the fact that where the beam intensity is high (e.g., on-axis), the induced refractive index is higher as well, amounting to larger slowing down of the wavefronts. The wavefronts are therefore bent similar to the action of a positive lens, resulting in initial focusing of the beam. This process continues till the beam width is small enough for the diffraction effects to take over, leading to an increase in the beam width. The converse is true for the case of $n_2 < 0$. In this case, the beam spreads more than in the linear diffraction limited case. A stable nonspreading solution in one transverse dimension can be analytically found from the NLS equation for $n_2 > 0$ and has the form

$$\left|E_e(x;z)\right| = \left(\frac{8\kappa}{n_2 k_0}\right)^{1/2} \operatorname{sech}\frac{x}{(2\kappa k_0)^{1/2}} \tag{2.32}$$

where κ is a free parameter. The phase of E_e in the nonspreading solution is linearly proportional to the propagation distance z.

As discussed above, self-focusing results in increase in the on-axis intensity and the narrowing of the beam width. For powers above a certain critical power P_c [7,8], the beam may theoretically collapse with an intensity so high that it can either cause breakdown in the material, triggering some other physical effects, such as saturation of the index of refraction or failure of the assumptions about slowly varying amplitude and paraxial approximation. Zakharov and Shabat [9] have pointed out that if the nonlinearity is strong enough, this results in higher index of refraction toward the center of the beam and on-axis rays undergo total internal reflection and are thereby trapped. As shown above, nonlinearity can balance diffraction of a beam in one dimension, resulting in the formation of first order spatial solitons, see Equation 2.32. Also, if the nonlinear effect is higher than diffraction, periodic focusing occurs, or may result in higher order solitons. This may not be the case in two or three dimensions where spatial collapse may occur.

In normalized form, and assuming $n_2 > 0$, the NLS equation in Equation 2.30 can be rewritten as

$$j\frac{\partial \tilde{u}_e}{\partial z} + \nabla_\perp^2 \tilde{u}_e + \left|\tilde{u}_e^2\right| \tilde{u}_e = 0 \tag{2.33}$$

In a system with cylindrical symmetry, Equation 2.33 becomes

$$j\frac{\partial \tilde{u}_e}{\partial z} + \frac{1}{r}\frac{\partial \tilde{u}_e}{\partial r} + \frac{\partial^2 \tilde{u}_e}{\partial r^2} + \left|\tilde{u}_e^2\right| \tilde{u}_e = 0 \tag{2.34}$$

The NLS equation as written in Equation 2.33 can also be modified to model pulse propagation through a nonlinear fiber and in the presence of group velocity dispersion. This is possible due to the fact that the interchange $x \to t$ in the NLS equation (2.33) with a suitable coefficient in front of the second order derivative term, signifying (anomalous) material dispersion and subsequent renormalization transforms the equation to one that can model the propagation of pulses in time τ along a fiber [2]:

$$j\frac{\partial \tilde{u}_e}{\partial z} + \frac{1}{2}\frac{\partial^2 \tilde{u}_e}{\partial \tau^2} + \gamma' \left|\tilde{u}_e\right|^2 \tilde{u}_e = 0 \tag{2.35}$$

Analogous to Equation 2.32, the *first order soliton* solution of (2.35) can be expressed as

$$\left|\tilde{u}_e(\tau; z)\right| = K \operatorname{sech}\left(K\sqrt{\gamma'}\tau\right) \tag{2.36}$$

where K is a free parameter. This profile is called a *temporal soliton* and can be regarded as a nonlinear eigenmode of the NLS system. The propagation of an initial profile $\tilde{u}_e(\tau; z = 0) = \operatorname{sech}\left(\sqrt{\gamma'}\tau\right)$ with $\gamma' = 1$ using the BPM to model the NLS equation as in Equation 2.35 is shown in Figure 2.8. The linear term in Equation 2.35 can be handled in the temporal frequency domain by using Fourier transforms, similar to the case of propagational diffraction. Note that the nonlinear operator is

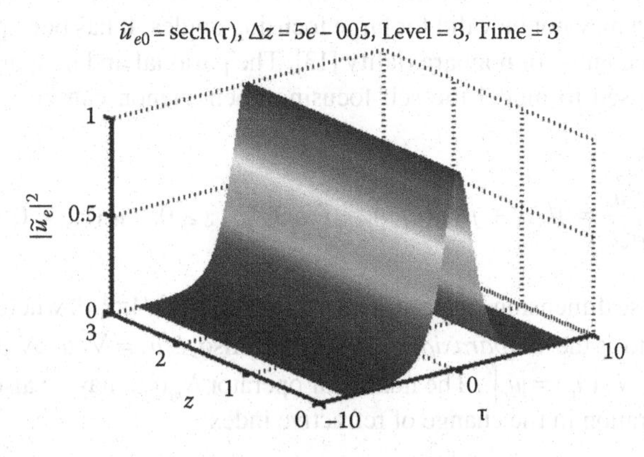

FIGURE 2.8 Evolution of a first order 1-D soliton. (From Nehmetallah, G. and Banerjee, P.P., *Nonlinear Optics and Applications*, H.A. Abdeldayem and D.O. Frazier, eds., Research Signpost, Trivandrum, India, 2007. With permission.)

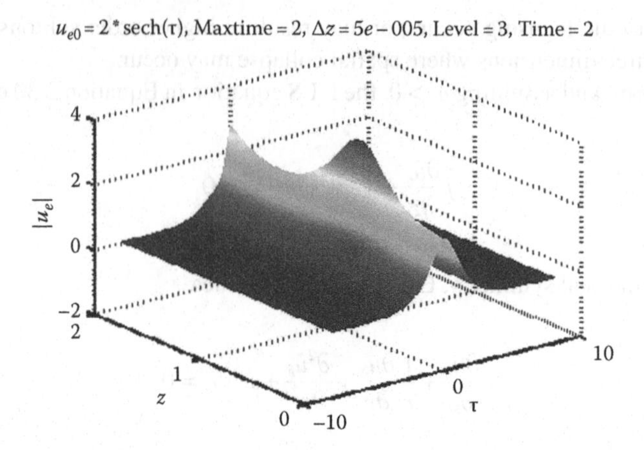

FIGURE 2.9 Evolution of a second order 1-D soliton. (From Nehmetallah, G. and Banerjee, P.P., *Nonlinear Optics and Applications*, H.A. Abdeldayem and D.O. Frazier, eds., Research Signpost, Trivandrum, India, 2007. With permission.)

$\exp\{\hat{S}\Delta z\} = \exp -j\gamma'|\tilde{u}_e|^2 \Delta z$. As expected, the "pulse" remains unchanged with propagation. A similar result is obtained if we program the propagation of an initial beam profile and use Equation 2.30 in one transverse dimension (x). The result is a *spatial soliton*.

The *second order soliton* input $\tilde{u}_e(\tau, 0) = 2\,\mathrm{sech}(\tau)$ and its evolution in time is shown in Figure 2.9. The split-step technique has also been applied to analyze propagation of profiles in two transverse dimensions [11], and also to analyze propagation of optical fields that are pulsed in time and have a spatial profile in the transverse dimension [12].

2.6.2 Simulation of Self-Focusing Using Adaptive Fourier and Fourier–Hankel Transform Methods

The NLS equation for beams using the paraxial approximation has been previously derived in Section 2.6.1 as

$$j\frac{\partial \tilde{u}_e}{\partial z} + \nabla_\perp^2 \tilde{u}_e + |\tilde{u}_e^2|\tilde{u}_e = 0 \tag{2.37}$$

During the last stages of self-focusing, the assumptions about slowly varying amplitude and the paraxial approximation may not be valid for large focusing angles. It has been proposed that there is no singularity if one accounts for nonparaxiality [13]. The paraxial and nonparaxial NLS equations which are classically used to model the self-focusing phenomenon can be written in the general operator form:

$$\frac{\partial \tilde{u}_e}{\partial z} - j\varepsilon\frac{\partial^2 \tilde{u}_e}{\partial z^2} = jL_r\tilde{u}_e + jN_{nl}(\tilde{u}_e)\tilde{u}_e, \quad \vec{r} \in \Re^D, \quad z \geq 0; \quad \tilde{u}_e(r, z=0) = \tilde{u}_{e0}(r) \tag{2.38}$$

where D is the transverse dimension in space. The parameter $\varepsilon = (\lambda/4\pi r_0)^2$, where r_0 is the initial beam radius is often referred to as the *nonparaxiality parameter*. Also, $L_r\tilde{u}_e = \nabla_\perp^2 \tilde{u}_e$, $N_{nl}(\tilde{u}_e) = |\tilde{u}_e|^{2\sigma}$. For most cases, $\sigma = 1$ and hence $N_{nl}(\tilde{u}_e) = |\tilde{u}_e|^2$. The nonlinear operator $N_{nl}(\tilde{u}_e)$ may be also conveniently modified to reflect any saturation in the change of refractive index:

$$\delta n\left(|\tilde{u}_e|^2\right) = n_2|\tilde{u}_e|^2 + n_4|\tilde{u}_e|^4, \quad n_4 < 0 \tag{2.39}$$

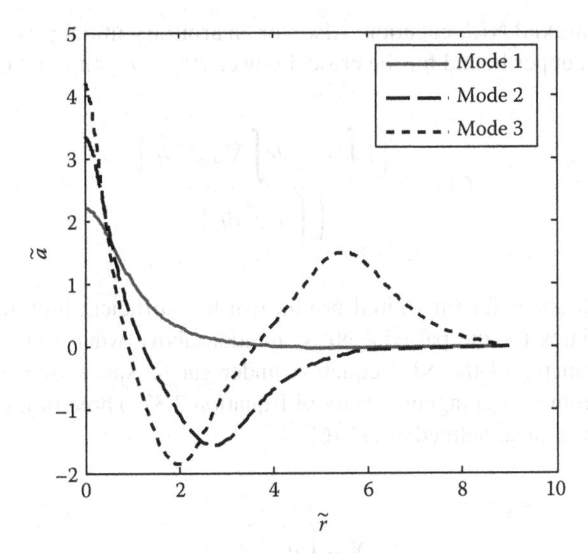

FIGURE 2.10 Townes soliton profiles for the first three "modes."

Incidentally, for the paraxial case, two transverse dimensions and assuming radial symmetry, and $\sigma = 1$, a special solution of (2.38), called the Townes soliton [7], can be found as the solution of the normalized differential equation

$$\nabla_{\perp}^{2} R(r) - R + R^{3} = 0, \quad R'(0) = 0, \quad R(\infty) = 0 \tag{2.40}$$

Typical solutions for three different initial conditions, called *modes*, are plotted in Figure 2.10. These modes, named $\tilde{a}(\tilde{r})$, are characterized by the fact that the solutions tend to zero for $\tilde{r} \to \infty$. The solutions of Equation 2.40 are very sensitive to initial conditions: for other initial conditions, the solutions do not converge to zero as $r \to \infty$.

The basic concept behind adaptive numerical algorithms is rather simple: imagine that a Gaussian beam is spreading during linear propagation due to diffraction. If a numerical solution is being computed, it is clear that after some distance of propagation, the size of the beam will become comparable to the total transverse grid size, which will result in errors in the numerical computation, such as aliasing if fast Fourier transform (FFT) techniques are used. The problem can be alleviated if the transverse profile is re-sampled using a coarser grid size. The step size for propagation can also be increased, based on the presumption that no sudden changes in the beam would occur at distances larger than the Rayleigh range. The converse should be true for beams that are focusing: the transverse grid size should be made finer and the propagational step size smaller.

In the case of the NLS equation, for the case $D = 2$, and assuming radial symmetry, McLaughlin et al. [14] predicted similarity solutions of the form

$$\tilde{u}_e(r,z) \propto \frac{1}{(z_r - z)^{1/2}} F\left(\frac{r}{(z_r - z)^{1/2}}\right) \tag{2.41}$$

where F is an arbitrary function. We can use this knowledge to adaptively vary the longitudinal stepping Δz and the transverse grid size for the beam.

A few more technical details must be mentioned before commencing the discussion on our adaptive numerical techniques. Weinstein [15] has shown that if the initial beam power P_0 is less than the lower bound for the critical power for blowup $P_c^{lb} = (\lambda^2/4\pi n_0 n_2)N_c$, where $N_c = \int |R|^2 \, r\,dr = 1.86225$, there is no collapse of the paraxial NLS equation. Also, for an arbitrary initial profile $\tilde{u}_{e0} \neq R$, Fibich [13] proved that there is an upper bound for the critical power $P_c^{ub} = (\lambda^2/4\pi n_0 n_2)\, G(\tilde{u}_{e0})$ where

$$G(\tilde{u}_{e0}) = \frac{\left(2\int |\tilde{u}_{e0}|^2\, dr \int |\nabla \tilde{u}_{e0}|^2\, dr\right)}{\left(\int |\tilde{u}_{e0}|^4\, dr\right)}$$

for which blowup will occur for this initial profile if it has sufficient high power. There are few known integral invariants for the paraxial NLS equation above when $\sigma D \geq 2$. These invariants are based on the symmetry of the NLS equation under gauge, space, and time transformations, and may be derived from a Lagrangian density of Equation 2.37. Three of these invariants are the Hamiltonian and the variance, defined as [14,16]

$$N = \int |\tilde{u}_e|^2\, dr \tag{2.42a}$$

$$H = \int \left(|\nabla \tilde{u}_e|^2 - \frac{1}{2} |\tilde{u}_e|^4 \right) dr \tag{2.42b}$$

$$\frac{1}{8}\frac{d^2}{dz^2} \int |r|^2 |\tilde{u}_e|^2\, dr = H - \frac{D-2}{4} \int |\tilde{u}_e|^4 dr \tag{2.42c}$$

where the term $|\nabla \tilde{u}_e|^2$ results from diffraction, and $|\tilde{u}_e|^4$ from the nonlinear effect.

Note that from Equation 2.42c blowup occurs only if $D \geq 2$. The "variance" is given as $V(z) = \int |r|^2 |\tilde{u}_e|^2 dr = 4Hz^2 + (dV(0)/dz)z + V(0)$ for $D = 2$ [17]. Thus, for $H < 0$, the function $V(z)$ vanishes at a distance $z_r = [V_0/-4H_0]^{1/2} > 0$. A sufficient condition for blowup is $H < 0$, i.e., when the nonlinear effect is stronger than diffraction, the beam self-focuses and collapse occurs at a distance $z_c \leq z_r$.

For an input Gaussian of the form $\tilde{u}_{e0} = ce^{-r^2/2r_0^2}$, $z_r = \left(1/(pN_c/2-1)\right)^{1/2}$, $z_c = 0.317(p-1)^{-0.6346}$ for $r_0 = 1$, and $z_c = 0.1585(p-1)^{-0.6346}$ for $r_0 = 1/\sqrt{2}$, where $p = N_0/N_c$ and $N_0 = c^2 r_0^D$ [18]. Note that the condition $H = 0$, which leads to P_c^{ub}, leads to an overestimate of the actual critical power.

We now outline our numerical adaptive spectral technique called the *adaptive split-step fast Hankel transform* (AFHTSS) used to track the solution of the NLS equation for $\sigma = 1$, $D = 2$ and variable ε. Our scheme is based on the combination of the standard split-step fast Fourier transform (SSFFT) and the Hankel transform, which exploits the cylindrical symmetry of the problem. This enhances the computation time and precision appreciably. In addition, we also use the concepts from the similarity solution developed by McLaughlin et al. [14] and apply them to our split-step spectral method mentioned above, so that the grid transverse spatial range and the longitudinal spatial step are adaptively updated. As its name indicates, we use the Hankel transform (see Chapter 1) instead of the usual Fourier transform, relying on algorithms already developed in the literature [19]. In cylindrical coordinates, Equation 2.38 becomes

$$\frac{\partial \tilde{u}_e}{\partial z} - j\varepsilon \frac{\partial^2 \tilde{u}_e}{\partial z^2} = j\left(\frac{\partial^2}{\partial r^2} + \frac{1}{r}\frac{\partial}{\partial r}\right)\tilde{u}_e + j|\tilde{u}_e|^2\, \tilde{u}_e; \quad \tilde{u}_e(r, z=0) = \tilde{u}_{e0}(r) = c\exp{-ar^2} \tag{2.43}$$

We use the definition of the lth order Hankel Transform pair [19,20]:

$$\Psi(\rho,z)=2\pi\int_0^\infty \psi(r,z)J_l(2\pi r\rho)rdr,\quad \psi(r,z)=2\pi\int_0^\infty \Psi(\rho,z)J_l(2\pi r\rho)\rho d\rho \tag{2.44a}$$

and the property:

$$j\left(\frac{\partial^2}{\partial r^2}+\frac{1}{r}\frac{\partial}{\partial r}\right)\psi(r,z)\xrightarrow{\text{HT}}-j4\pi^2\rho^2\Psi(\rho,z) \tag{2.44b}$$

The AFHTSS algorithm in Figure 2.11 resembles the symmetrized Fourier split-step technique, where we change the longitudinal spatial step $\Delta z \propto (1/c(z_1)^2 - 1/c(z_2)^2) \approx 1/c(z_1)^2$ when $c(z_2) \gg c(z_1)$, adaptively using McLaughlin's similarity formula (2.41) and the grid spatial range $\Delta r_{\max} \propto 1/c(z)$ in order to track the varying amplitude of the focusing beam when $\varepsilon = 0$.

There are several numerical approaches for implementing the Hankel transform [19,21]. The importance of Siegman's method [19] resides in the fact that, depending on the parameters, one can employ a non-uniform sampling that is denser near the focusing region, which has advantages over uniform sampling. Yu et al.'s method [21] is based on the expansion of the function and its transform by a zero order Bessel series that can be written as

$$\hat{\Psi}(m)=\sum_{n=1}^{N}C_{mn}\hat{\psi}(n),\quad \hat{\psi}(n)=\sum_{m=1}^{N}C_{nm}\hat{\Psi}(m) \tag{2.45a}$$

$$C_{mn}=\frac{2}{S}J_0\left(\frac{j_n j_m}{S}\right)\left|J_1^{-1}(j_n)\right|\left|J_1^{-1}(j_m)\right| \tag{2.45b}$$

$$\hat{\psi}(n)=\psi\left(\frac{j_n}{2\pi R_2}\right)\left|J_1^{-1}(j_n)\right|R_1,\quad \hat{\Psi}(n)=\Psi\left(\frac{j_n}{2\pi R_1}\right)\left|J_1^{-1}(j_n)\right|R_2 \tag{2.45c}$$

The j_n's are the positive roots of the zero order Bessel function J_0, J_1 is the first order Bessel function, and R_1, R_2 are the spatial and transform ranges respectively, with $S = 2R_1R_2$. For $r_n \in r \geq R_1$ and $\rho_m \in \rho \geq R_2$, we have $\psi(r_n) = \Psi(\rho_m) = 0$, where $r_n = j_n/2\pi R_2$ and $\rho_m = j_m/2\pi R_1$.

We now show sample simulation results using the AFHTSS method that uses Yu et al.'s Hankel transform technique [21], as well as a novel adaptive version of a split-step fast fourier transform (AFFTSS) technique. For the test function $\psi_0 = 4e^{-r^2/2}$, self-focusing and collapse is expected at $z_c = 0.1487$ for $\varepsilon = 0$. Note that $z_r = 0.288$, based on the study above, which is obviously overestimated. Figure 2.12 shows the maximal focusing as a function of grid size $h = \Delta r$, which proves the convergence of our method to the numerical focusing point when Δr decreases by varying the S parameter defined above [22]. Although the trend is similar, this is a considerable improvement over the convergence test results in Fibich and Ilan [23]. Figure 2.13 shows the growth of the on-axis intensity using AFHTSS technique for the test function above [22]. Also, in this figure, we take $S = 2\pi R_1 R_2 = 2\pi \times 2000$ (corresponding to approximately 4000 cylindrical samples), which permits computation for the paraxial case till $z = z_c = 0.14817$ corresponding to the expected $z_c = 0.1487$ written above. Figure 2.14 shows the corresponding AFFTSS technique for 1024 × 1024 grid where we get $z = z_c = 0.1581$ [22]. This implies that the critical distance z_c is a little bit over estimated and the maximum intensity reached is also less than the AFHTSS method. It is instructive to note that with increase in sampling points, the AFFTSS approaches the results from AFHTSS, but on the

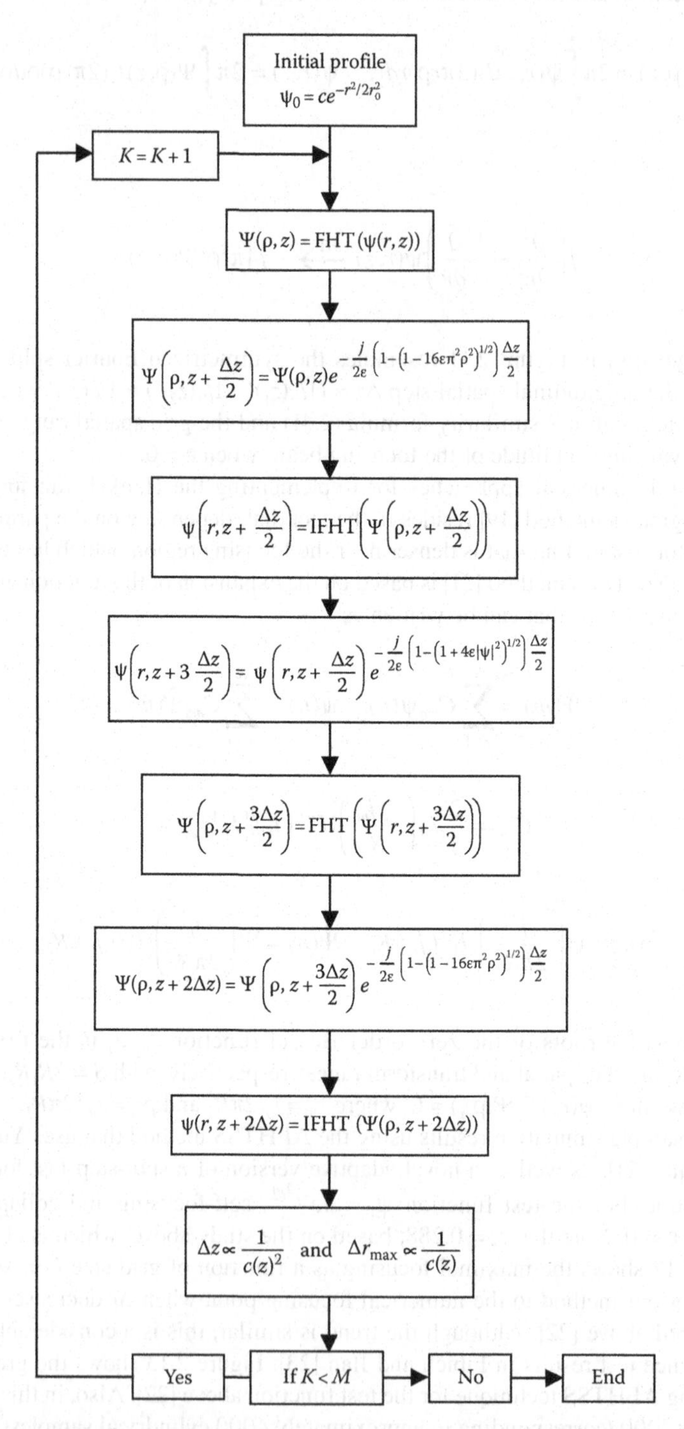

FIGURE 2.11 The AFHTSS algorithm, a symmetrized version of the split-step FFT using Hankel transform instead, and using adaptive longitudinal stepping and transverse grid management. FHT: fast Hankel transform, IFHT: inverse fast Hankel transform.

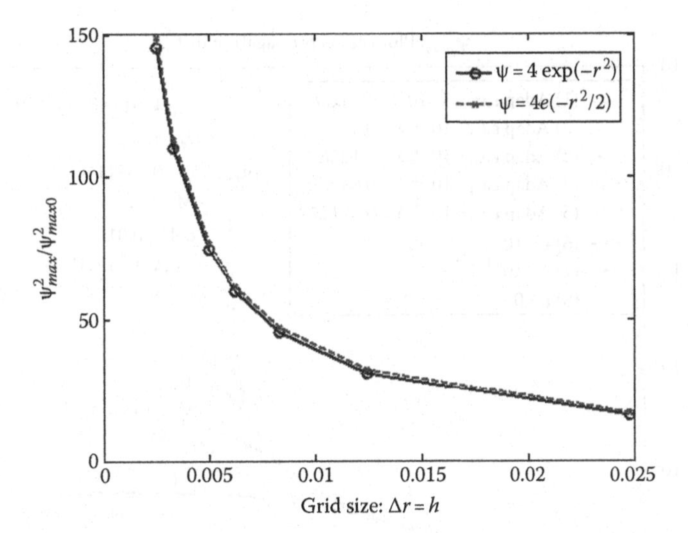

FIGURE 2.12 Maximal focusing as a function of grid size. (From Banerjee, P.P. et al., *Opt. Commun.*, 249, 293, 2005. With permission.)

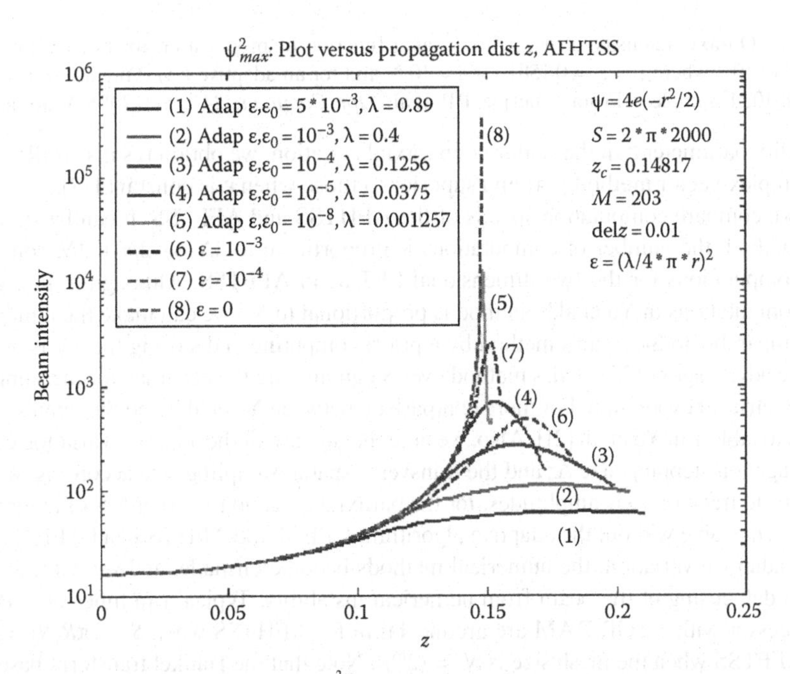

FIGURE 2.13 On-axis intensity of $\psi_0 = 4e^{-r^2/2}$ as a function of propagation for fixed values of ε ranging from 10^{-2} to 10^{-8} where $z = z_c = 0.1481$ for $\varepsilon = 10^{-8}$, and for an adaptive ε varying as $\varepsilon = (\lambda/4\pi r)^2$, using AFHTSS with $S = 2\pi R_1 R_2 = 2\pi \times 2000$ (4000 cylindrical samples). (From Banerjee, P.P. et al., *Opt. Commun.*, 249, 293, 2005. With permission.)

expense of calculation time. Also we note that changing the number of samples in both techniques does not affect the value of the critical distance drastically, but the maximum intensity reached at that point will be less or more depending on the number of samples, which is in agreement with the convergence test mentioned above. We stress that in all computations using AFHTSS and AFFTSS, energy is always conserved. Consequently, AFHTSS permits us to track peak intensities higher, faster, and more accurate than what is achievable by AFFTSS. Also, by using adaptive

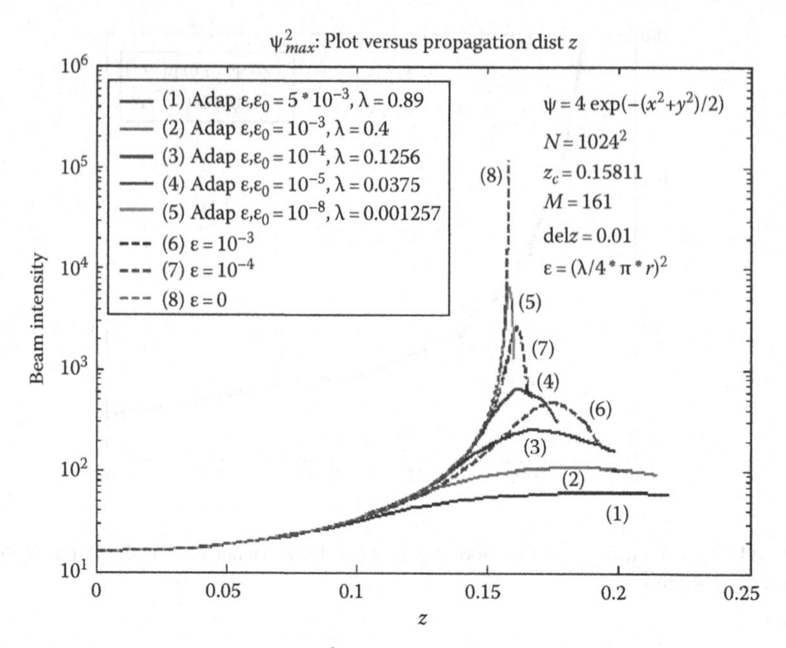

FIGURE 2.14 On-axis intensity of $\psi_0 = 4e^{-r^2/2}$ as a function of propagation for fixed values of ε ranging from 10^{-2} to 10^{-8} where $z = z_c = 0.1581$ for $\varepsilon = 10^{-8}$, and for an adaptive ε varying as $\varepsilon = (\lambda/4\pi r)^2$, using AFFTSS with 1024^2 samples. (From Banerjee, P.P. et al., *Opt. Commun.*, 249, 293, 2005. With permission.)

non-paraxiality parameter ε in the scalar nonparaxial equation, we obtain results similar to those of the more complex vector method [24] and superior to those when ε is constant [13].

Finally, we compare computation speeds of the AFHTSS and AFFTSS. It can be shown that for Siegman's method, the number of computations is proportional to $4N\log_2 2N + 2N$, compared with $2N^2\log_2 N$ computations for the two dimensional FFT as in AFFTSS. Although at first glance, the number of computations in Yu et al.'s method is proportional to N^2, we can make the number of computations comparable to Siegman's method by a priori computing and storing the zeros of the Bessel function. The advantages of Yu et al.'s method over Siegman's are the accuracy for the sampled points and a simple retrieval expression. For more comparison between Yu et al.'s and Siegman's method, we refer readers to table 1 in Yu et al. [21]. Also, we note that the use of the adaptive variation of the longitudinal propagation stepping size Δz and the transverse spatial sampling size according to $1/c^2(z)$ and $1/c(z)$ allow us to track on-axis amplitudes, for the paraxial case, up to two orders of magnitude more than what is achievable without the adaptive algorithm, for both the AFHTSS and AFFTSS methods. Without the adaptive variation, the numerical methods become unstable, and we witness oscillatory focusing and defocusing of the beam from numerical instability. Typical run times on a Pentium IV 2.4 GHz processor with a 2 GB RAM are around 1 min for AFHTSS when $S = 2\pi R_1 R_2 = 2\pi\times 2000$, 10 min for AFFTSS when the mesh size is $N^2 = (2^{10})^2$. Note that the Hankel transform based method, which exploits the cylindrical symmetry, is one dimensional, and therefore is expected to be faster than other two dimensional FFT based numerical methods.

2.7 BEAM PROPAGATION AND COUPLING IN PHOTOREFRACTIVE MEDIA

2.7.1 Basic Photorefractive Physics

In this section, a model for beam propagation through a nonlinear PR material that takes into account inhomogeneous induced refractive index changes due to the nonlinearity is first developed. In some cases a focused Gaussian beam asymmetrically distorts due to passage through the nonlinear material.

The PR effect has been used in a wide variety of applications, viz., image processing, optical interconnections, optical data storage, optical limiters, and self-pumped phase conjugators [25]. When a PR material is illuminated by a light beam or by a fringe pattern generated by the interference of two light beams, photoexcited carriers are redistributed in the volume of the crystal [25]. This sets up a space-charge field which, through the linear electro-optic effect, gives rise to a refractive index profile and hence a phase hologram.

The phenomenon of PR beam fanning, where the incident light beam is deflected and/or distorted when it passes through a high-gain PR crystal, has been observed in $BaTiO_3$, $LiNbO_3$, and SBN [26–28]. One of the ways this has been explained is through the fact that a symmetric beam may create an asymmetric refractive index profile, leading to beam distortion, or what we will call *deterministic beam fanning* (DBF) in the far field [29]. This analysis has been done for a thin sample, meaning one where diffraction of the beam is neglected during its travel through the PR material, and by using a linearized theory to determine the induced refractive index profile. We have recently extended the linearized approach to the case of a thick sample, and have included the transient effects, and are in the process of determining the effects of transient DBF when a reading beam is used to illuminate a previously stored hologram in the PR material [30].

Another school of thought is that beam "fanning" results from light scattering from the random distribution of space charges in the PR material. However, a larger contribution to *random beam fanning* (RBF) is the so-called amplified noise [31], which may arise from the couplings between the plane-wave components scattered from crystal defects.

In what follows, we examine steady state DBF in a diffusion-dominated PR material by deriving a closed form expression for the induced refractive index change from the nonlinearly coupled Kukhtarev equations. We also assess the role of propagational diffraction in DBF by determining the similarities and differences between the thin and thick sample models.

It must be stated that the simplified model for the induced refractive index described in the following sub-section can be also used to analyze two beam coupling and energy exchange in PR materials. When two approximately co-propagating beams are incident on the PR material, they give rise to *induced transmission gratings* which facilitate the energy exchange. On the other hand when two approximately contra-propagating beams are incident on the PR material, they give rise to *induced reflection gratings* which also facilitate the energy exchange. This energy exchange occurs over and above the beam fanning described above. While the first model described below for induced refractive index is based on carrier diffusion, other effects such as the *photovoltaic* (PV) effect in materials such as $LiNbO_3$ can also contribute to induced transmission and reflection gratings, as will be seen later.

2.7.2 Induced Transmission Gratings

It can be shown that the coupled set of simplified Kukhtarev equations [25] (see Chapter 6 for details) for a diffusion-dominated PR material can be decoupled in the steady state to yield an ODE for the space charge electric field [29]. In denormalized form, we can express this electric field $\vec{E}_s(x, y, z)$ approximately as

$$E_s \approx \frac{k_B T}{e} \frac{\vec{\nabla} I}{\beta / s + I} \tag{2.46}$$

where

e is the electronic charge

k_B is the Boltzmann constant

T is the temperature

s is the ionization cross section per unit photon energy

β is the thermal generation rate

$I(x, y, z)$ denotes the intensity distribution along x, y at a position z in the PR material

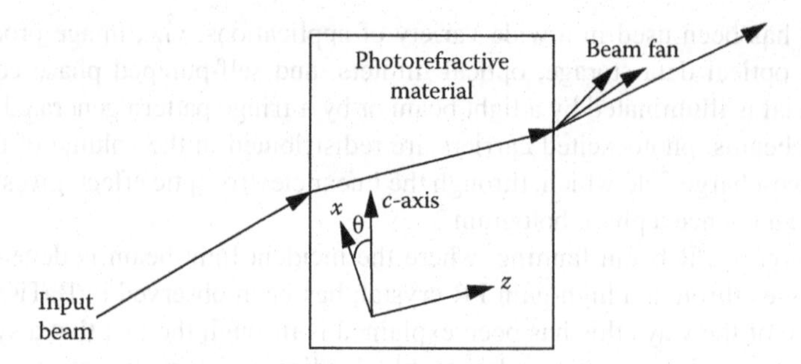

FIGURE 2.15 Geometry for analysis of DBF.

Now, this electrostatic field induces a refractive index change $\Delta n_{ext}(x,y,z)$ for extraordinary polarized (say along x, see Figure 2.15) *plane waves* of light in the PR material, assuming that $BaTiO_3$ in this example, through the linear electro-optic effect, is given by

$$\Delta n(x,y,z,\theta) = E_{sx}(x,y,z)f(\theta)$$

$$f(\theta) = -\frac{1}{2}n_e^3(\theta)\cos\theta(r_{13}\sin\theta + r_{33}\cos^2\theta + 2r_{42}\sin^2\theta) \tag{2.47}$$

$$n_e^2(\theta) = \left(\sin^2\frac{\theta}{n_0^2} + \cos^2\frac{\theta}{n_e^2}\right)^{-1}$$

where n_0, n_e are the linear ordinary and extraordinary refractive indices, and r_{ij} is the linear contracted electro-optic coefficient [25].

The angle θ in Equation 2.47 is defined in Figure 2.15. Note that $f(\theta)$ is a slowly varying function of θ over the spectral content of the optical field. It can be readily shown that, in general, propagation through the PR material under the slowly varying envelope approximation may be modeled by means of the PDE [29]

$$\frac{\partial E_e}{\partial z} = -jk_0\Delta nE_e - j\left[\frac{1}{2n_e(\theta)k_0}\right]\nabla_\perp^2 E_e$$

$$\Delta n(x,\theta) = \Im_x^{-1}\left[\Im_x[E_{sx}(x)]f\left(\frac{k_x}{n_e(\theta)k_0}+\theta\right)\right]$$

$$E_{sx}(x) = \frac{k_BT}{e}\frac{\partial|E_e(x)|^2/\partial x}{\eta\beta/s + |E_e(x)|^2} \tag{2.48}$$

For values of θ around 40°, a symmetric beam could induce an asymmetric refractive index profile, leading to beam bending and DBF in the far field. However, for some other value of θ, for instance 90°, our theory predicts symmetric beam shaping, in agreement with the findings of Segev et al. [31]. In this respect, the nature of the optical nonlinearity in a PR material is more involved as compared to that in a nonlinear Kerr-type material. We point out that in a Kerr-type material for instance, only an asymmetric beam profile can cause beam bending, as reported in [32], while a symmetric beam undergoes self-focusing or defocusing.

In what follows, we first provide results for the far-field beam profiles by assuming the PR material to be a thin sample, in the sense that we neglect the effects of propagational diffraction through the material. A Gaussian input

$$E_e(x,y,0) = (I_0\eta)^{1/2} \exp\left[-\frac{(x^2+y^2)}{W^2}\right] \tag{2.49}$$

with $I_0 = 2P/\pi W^2$, where I_0 denotes the on-axis intensity and P is the beam power, is phase modulated due to the induced refractive index profile. The resulting output field is $E_e(x,y,L) = E_e(x,y,0)\exp(-jk_0\Delta n(x)L)$, where L is the thickness of the PR material. Such a phase modulation results in a shift of the far-field pattern with respect to the axis (z) of propagation of the optical beam, and in the appearance of asymmetric sidelobes, the so-called fanning of the beam. Numerical simulations for $BaTiO_3$ with parameters $n_0 = 2.488$, $n_e = 2.434$, $r_{42} = 1640\,\text{pm/V}$, $r_{13} = 8\,\text{pm/V}$, $r_{33} = 28\,\text{pm/V}$, $s = 2.6 \times 10^{-5}\,\text{m}^2/\text{J}$, $\beta = 2/\text{s}$, $T = 298\,\text{K}$ [29] and $L = 1\,\text{cm}$ and using an incident wavelength of 514.5 nm shows a monotonic increase in the shift of the far-field main lobe from the z axis with increase in I_0 (implying either an increase in power P or a decrease in width W). In Figure 2.16a and b, k_x is the spatial frequency variable corresponding to x, and is related to the far-field coordinate x_f by $k_x = k_0x_f/d$, d being the distance of propagation from the exit of the crystal to the far field [29]. However, the amount of DBF (defined by the relative amount of power in the sidelobes) varies nonmonotonically with intensity, initially increasing as the intensity is increased from low levels to attain a maximum, and then decreasing with further increase in intensity.

Note that our results are different from those of Feinberg [26], in that the latter, based on a linearized *two-beam coupling* (TBC) theory that neglects coupling of the angular plane wave components of the Gaussian with any background illumination, yields $\vec{E}_s \propto \vec{\nabla}I/I_0$, where I_0 is the quiescent intensity (to be compared with our Equation 2.48). For a Gaussian intensity profile, the locations of the extrema of E in Feinberg's formulation are fixed w.r.t. to the incident profile, and hence may be shown to predict a monotonic increase in DBF with a decrease in W. In our nonlinear formulation, however, for decreasing W, the extrema of E move out with respect to the incident profile, so that the profile essentially sees a linear induced refractive index for sufficiently small W resulting in reduced DBF.

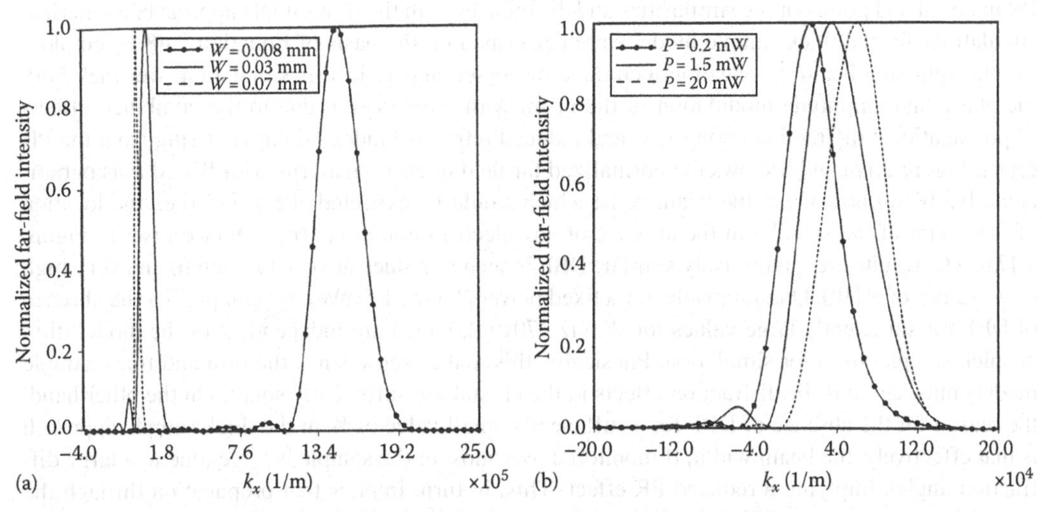

FIGURE 2.16 Normalized far-field intensity profiles for the thin-sample model: (a) $P = 1.5\,\text{mW}$ and (b) $W = 40\,\mu\text{m}$. (From Banerjee, P.P. and Misra, R.M., *Opt. Commun.*, 100, 166, 1993. With permission.)

Before comparing the thin sample results with the findings for the thick sample case, we will, at this point, provide a simple alternate explanation for the observed behavior of DBF when monitored as a function of the intensity. Our explanation is based on the examination of the spectrum of the phase modulation $\exp(-jk_0\Delta n(x)L)$. The far-field pattern is the convolution of the above spectrum with that of the input profile. Since $\Delta n(x)$ is an odd function of x (see Equations 2.47 and 2.48), it can be expanded in a power series of the form $ax^3 - bx$, where a and b are given by

$$b = \frac{4f(\theta)k_BT/eW^2}{\beta/sI_0+1}, \quad a = \frac{(2b/W^2)(\beta/sI_0)}{\beta/sI_0+1} \tag{2.50}$$

Note that the coefficients of this expansion hold for all values of the ratio β/sI_0. The spectrum $H(k_x)$ of $\exp(-ik_0\Delta n(x)L)$ is then

$$H(k_x) = \frac{2\pi}{(3a)^{1/3}} Ai \frac{k_x - bk_0L}{(3a)^{1/3}} \tag{2.51}$$

Once again, k_x has the same implication as in the discussion on Figure 2.16. We comment that if d is replaced by f, where f is the focal length of a lens at the exit plane of the crystal, k_x, and hence x_f, would be representative of the spatial coordinate on the back focal plane of the lens. Ai[.] is the *Airy function* [33]. The ith zero, α_i, of $H(k_x)$ is related to the ith zero, γ_i (<0) of Ai[.] by $\alpha_i = bk_0L + (3a)^{1/3}\gamma_i$. It then follows that the spatial extent of the Airy pattern for $k_x < bk_0L$, up to say the ith zero, and normalized by the spectral width $2/W$ of the incident Gaussian profile, varies nonmonotonically with I_0. The shift in the Airy pattern, bk_0L, however, increases with an increase in I_0. For large I_0, it can be shown that the shift is proportional to $1/W^2$, in agreement with the trend in Figure 2.16a. The resulting far-field pattern which is the convolution of the Gaussian spectrum and the Airy pattern generally exhibits decreased DBF when the Airy pattern has a (denormalized) width much smaller than that of the Gaussian spectrum (which may occur, for instance, for both small and large W). This is in agreement with our numerical simulations in Figure 2.16. Appreciable DBF occurs in the region where the normalized spectral width is greater than unity. As an example, for $P = 1.5$ mW, maximum beam fanning, defined by the maximum of the ratio of the peak value of the sidelobe and that of the mainlobe, occurs when $W = 30\,\mu$m. Details can be found in Misra and Banerjee [29].

We will now present the results for the far-field beam profiles using a thick sample model for the PR material and point out the similarities and differences with the thin sample approach. Numerical simulations for the thick sample model were performed on the basis of Equation 2.48 by employing the split-step beam propagation technique discussed above. In this simulation, we track both the phase and amplitude modulation of the beam within the crystal due to the combined effects of propagational diffraction (along x, y) and induced refractive index (along x) arising from the PR effect. Figure 2.17a and b shows the normalized far-field intensity patterns with W and P as parameters. By W, we now mean the beam waist which would be expected at $z = L/2$ (i.e., the location of the center of the sample) in the absence of any electro-optic effect ($r_{ij} = 0$) (see inset in Figure 2.17a). The results are qualitatively similar: DBF is seen to reduce at very low (high) and very high (low) values of P (W). Quantitatively, for a fixed power P (viz., 1.5 mW), we can predict the absence of DBF for sufficiently large values for W (viz., 70 μm), which are independent of the model (thin or thick sample) used for simulation. Physically, this makes sense since the thin and thick sample models must agree if the diffraction effects in the crystal are sufficiently small. On the other hand, the reason for the absence of DBF for a sufficiently small value of W in the thick sample approach is that effectively, the beam width, if monitored over most of the sample is large (due to a large diffraction angle), implying a reduced PR effect. This, in turn, implies that propagation through the crystal is predominantly diffraction limited. For small W, the thick sample model therefore is more accurate than the corresponding thin sample model for the same value of W since the latter model

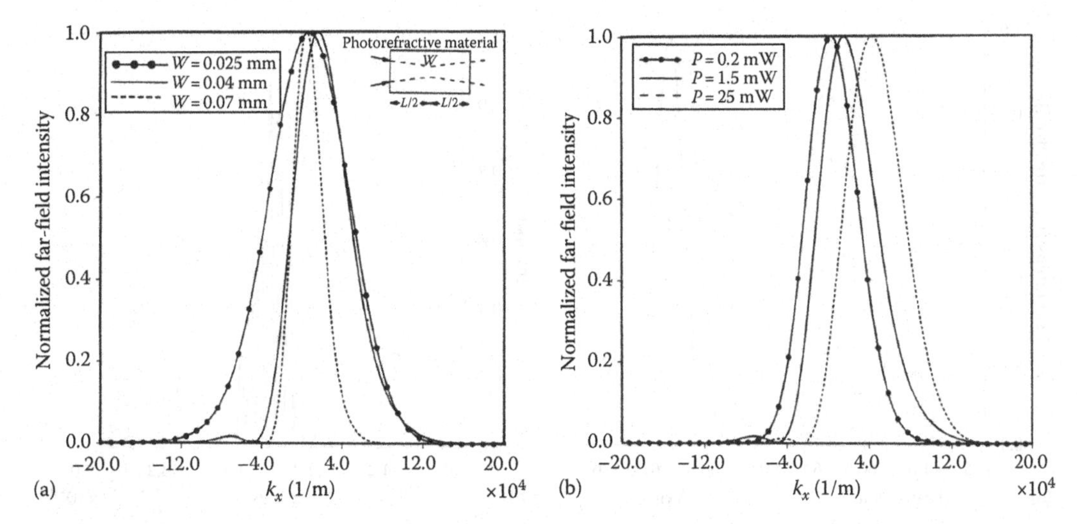

FIGURE 2.17 Normalized far-field profiles for the thick sample model: (a) $P = 1.5\,\text{mW}$ and (b) $W = 40\,\mu\text{m}$. (From Banerjee, P.P. and Misra, R.M., *Opt. Commun.*, 100, 166, 1993. With permission.)

overestimates the amount of cumulative PR effect. For the thick sample model, for the same value of P as above, we see negligible DBF for W less than 25 µm. On the other hand, the thin sample model predicts a value of W less than 5 µm for negligible beam fanning. The reason for the disappearance of DBF in the thin sample approach has been presented above using the Airy function argument and the movement of the extrema of E w.r.t. the incident optical field. Maximum DBF for $P = 1.5\,\text{mW}$ occurs for $W = 40\,\mu\text{m}$, in close agreement with the thin sample computations and the Airy function approach. However, the shift in the position of the mainlobe in the thick sample model is much smaller as compared to the thin sample case due to the effective decrease in the PR effect for a small waist size, as explained above. Referring to Figure 2.17a, we note that for $W = 40\,\mu\text{m}$, $P = 1.5\,\text{mW}$, and $f = 10\,\text{cm}$, the spatial shift in the back focal plane of a lens of focal length f located at the exit plane of the PR material is about 0.2 mm. We would like to comment that for the above parameters, DBF was also numerically observed at the exit face of the thick PR sample.

Thus far we have analyzed the propagation of a single focused Gaussian beam in a diffusion-dominated PR medium. When two beams are incident on such a medium with a small angle between each other, the induced refractive index profile is responsible for energy exchange between the two beams, a phenomenon referred to as TBC. This energy exchange occurs due to the phase shift between the intensity interference pattern and the induced refractive index pattern [25]. We can effectively study the interaction and the resulting energy exchange between two focused Gaussian beams incident on the material numerically using the split-step method. The problem geometry is shown in Figure 2.18. The two Gaussian beams are focused in the center of the PR material and the angle between them is 2θ. The Gaussian beams are expressed in terms of their q-parameters (see Section 2.4) at the entry face of the material. The split-step algorithm is used to determine the interaction and energy exchange between the two beams. The induced refractive index Δn is used to construct the operator representing the induced inhomogeneity in the material. The results on two

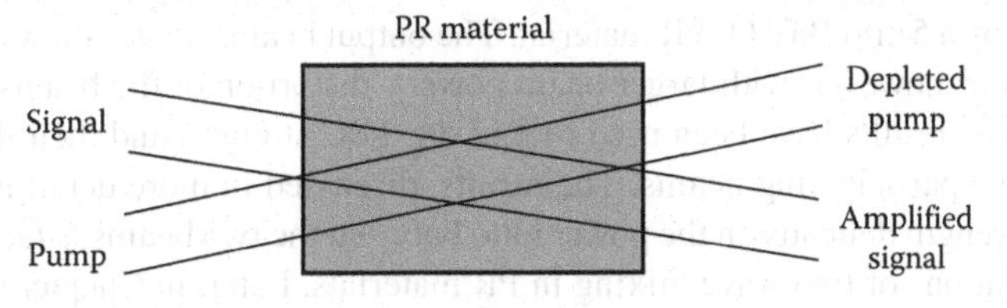

FIGURE 2.18 Geometry for TBC in a diffusion-dominated PR material.

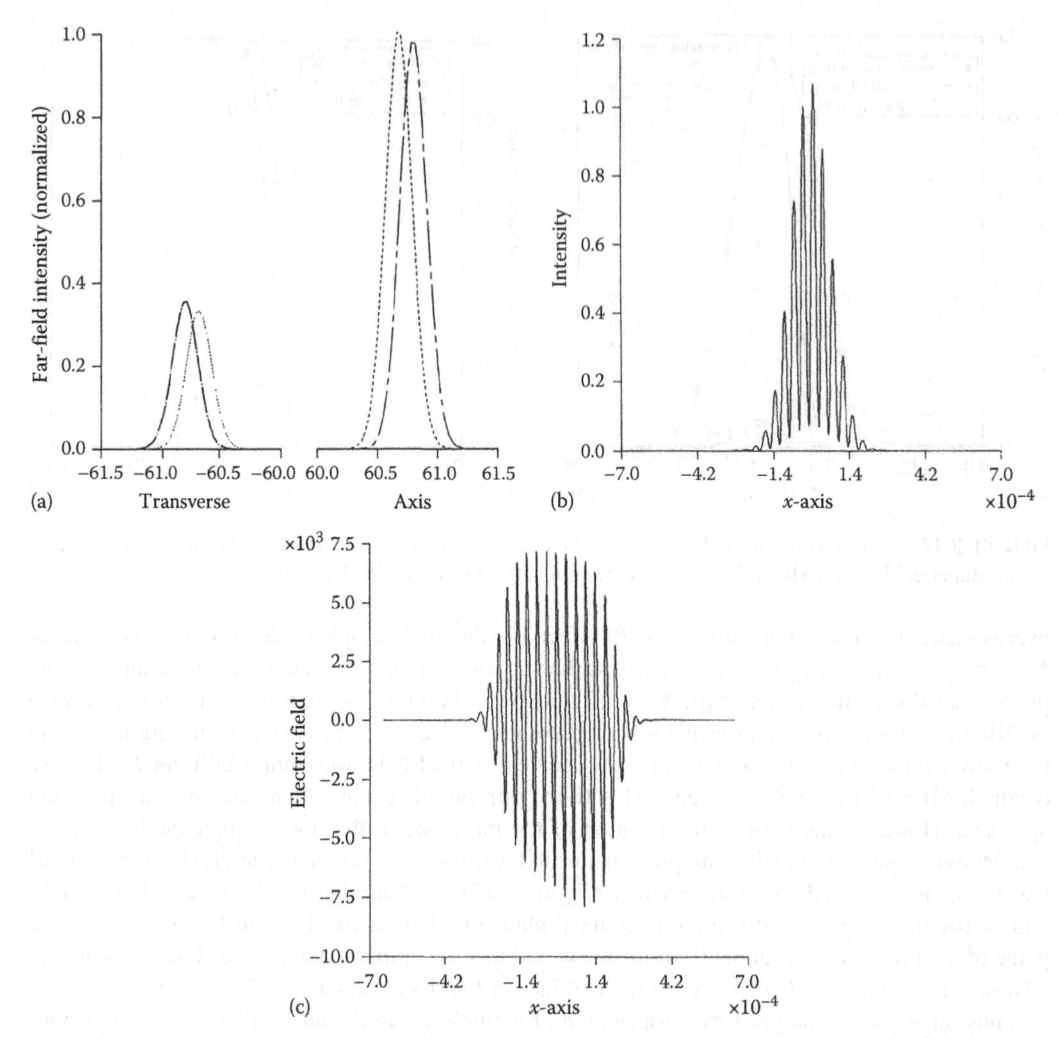

FIGURE 2.19 (a) Dotted and dashed lines are respectively the far-field signal and pump intensities with the absence of any PR material and chain dots and chain dashes represent the resulting far-field intensities after the beams have propagated through a 5 mm $BaTiO_3$ sample. Incident beams are focused to the center ($z = L/2$) of the PR crystal and the waist of each beam at wavelength 0.632 μm is 100 μm. Signal to pump ratio is 3 and semi-angle of crossing θ is 0.5°. (b) Interference pattern at center ($z = L/2$) of the PR crystal for the beams described in (a). (c) Space-charge field (V/m) at the center ($z = L/2$) of the crystal for the beams of (a). (From Ratnam, K. and Banerjee, P.P., *Opt. Commun.*, 107, 522, 1994. With permission.)

wave mixing are shown in Figure 2.19. The dot-dashed lines show the far-field intensity profiles of the two Gaussian beams in the absence of the PR material. The dashed lines show the beams after energy transfer due to the induced refractive index. The initial pump to signal power is 3. The peak intensity of the pump and signal beams are 63 and 21 W/cm^2, respectively, before the interaction. The beams are coupled by a 5 mm $BaTiO_3$ PR material. The output beams do not show any effect of beam fanning at this power; however, with larger beam powers, distortion of the beams due to beam fanning is observed. The results have been used to find the TBC strength and their dependence on the intensities of the two participating beams. The results, discussed in more detail in Ref. [34], depict that the coupling strength depends on the power ratio between the two beams, a fact that is ignored in perturbation calculations of two-wave mixing in PR materials. Later, in Chapter 6, we will analyze this effect in more detail with participating plane waves and using rigorous coupled wave theory.

2.7.3 Induced Reflection Gratings and Bidirectional Beam Propagation Method

Contrary to transmission gratings which are induced in a PR medium by two waves nominally traveling at a small angle with each other, reflection gratings are formed by two nearly contra-propagating waves. In a PR material such as lithium niobate ($LiNbO_3$), this can be simply formed due to the interference between an incident beam (typically traveling down the *c*-axis) and its Fresnel reflection from the back surface of the crystal due to the linear refractive index mismatch. The period of the reflection grating is therefore much smaller than that of the transmission grating, determined by the wavelength of the interacting waves (or beams). Energy exchange between the forward and backward traveling beams can give rise to depletion of the forward traveling beam (pump) and amplification of the backward traveling beam (signal), and has applications in optical limiting [35,36]. In what follows, we outline the use of the BPM, suitably modified to include forward and backward propagation, to determine the energy exchange during self-pumped TBC in a reflection grating geometry. It turns out that a PR material like $LiNbO_3$ has contributions to its PR effect from diffusion as well as the PV effect, where the latter gives rise to an induced refractive index profile in phase with the intensity profile, further complicating the analysis.

The bi-directionality of the simulation (opposing directions of the pump and signal beams) is handled by treating the two counter-propagating directions in sequence and then using an iterative shooting method to find the converged solution. The schematic of the beam propagation algorithm is shown in Figure 2.20. The dark lines on the left and right sides of Figure 2.20 represent the front and rear surfaces of the PR material, respectively, and the crystal *c*-axis is perpendicular to these faces. In this work, we only consider the case of $LiNbO_3$, which is a uniaxial crystal. For uniaxial crystals with light propagation along the polar axis, only one transverse polarization vector is required to be defined. The crystal longitudinal propagation direction is split into n steps (shown by the dashed lines). The incident pump beam is defined with a Gaussian amplitude profile and the e^{-1} spot size characterizes the focusing condition. Other transverse beam profiles may be substituted for the Gaussian, to suit specific experimental parameters.

In typical practical applications, the pump beam is usually chosen to be focused at the front surface of the crystal. Focus at any point inside the crystal is accomplished by doing beam propagation in a crystal through the desired length, taking the complex conjugate of the field and using that as the initial field. The amplitude of the incident optical field E_{in} is adjusted to use any desired power for the pump beam. During simulation, the field retains both amplitude and

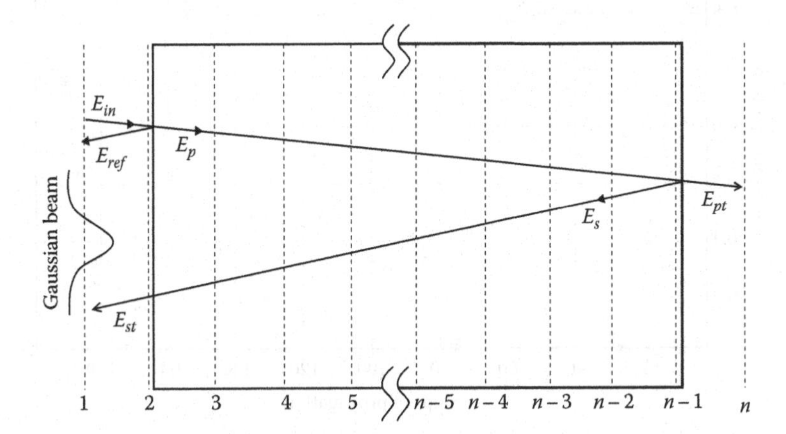

FIGURE 2.20 A schematic representation of the origin and path of the pump and signal beams. The signal beam (E_s) is generated by the Fresnel reflection of the pump beam (E_p) at the rear surface of the crystal. Both beams are nominally perpendicular to the boundaries. The paths are shown with angles to help distinguish the beams.

phase information at every propagation step. The paraxial approximation used in this split-step method allows us to simulate the propagation of beams with spot sizes as small as 3 μm without significant error [37].

As the beam enters the crystal (at step 2), part of it is reflected (E_{ref}). The transmitted beam (E_p) is propagated to the rear boundary, where part of this beam is transmitted (E_{pt}) and remainder is reflected (E_s) (at step $n - 1$). At each propagation step, the local value of the pump field ($E_p(i)$) is saved in a buffer. The signal beam, E_s, is then propagated in the opposite direction to the pump beam, until it travels through the front face of the crystal (E_{st}). At each step the signal beam field $E_s(i)$ is also saved in a buffer. The $E_s(i)$ buffer is then added to the previously buffered pump beam field $E_p(i)$ to produce the interference array. At the ith longitudinal step, the intensity of the interference pattern is then given by $|E_p(i) + E_s(i)|^2$. Since the fields are defined as transverse arrays, the interference at every ith longitudinal step also has a transverse profile. This transverse profile of the interference, however, is computationally very costly, since for each step this has to be saved in the buffer. Moreover, since we are simulating an interference pattern, the step size along the longitudinal direction must be substantially smaller than the wavelength.

It is determined that for a crystal of thickness of the order of the wavelength (in the material), the number of points n has to be at least 40 to produce a smooth interference pattern. For contra-directional TBC in $LiNbO_3$, it is the intensity modulation in the longitudinal direction that is of greatest importance, since the electro-optic coefficient is weak perpendicular to the c-axis. Hence, we have employed an alternate approximate method of simulating the interference with less computational load. We have saved only the peak of the field (instead of the whole transverse array) in the buffer. This results in an interference pattern without a transverse profile. This approximation does not result in any significance difference in our simulation, but has improved the required CPU time by two orders of magnitude.

Figure 2.21 shows the peak of the interference at every step in the longitudinal direction (each point of the plot represent the peak of the transverse profile at that step) for a thin $LiNbO_3$ crystal. The length of the simulated crystal equals two wavelengths inside the crystal, where the free-space wavelength $\lambda_{\bar{0}}$ is taken to be 532 nm, and the ordinary refractive index (n_0) of $LiNbO_3$ is used. Because there is no absorption loss and hardly any beam diffraction in this short distance, the interference has constant amplitude. For large crystal lengths, and for smaller beam spot sizes focused at the entrance of the crystal, the beam amplitude changes with propagation due to diffraction, and consequently results in a gradually decaying interference pattern.

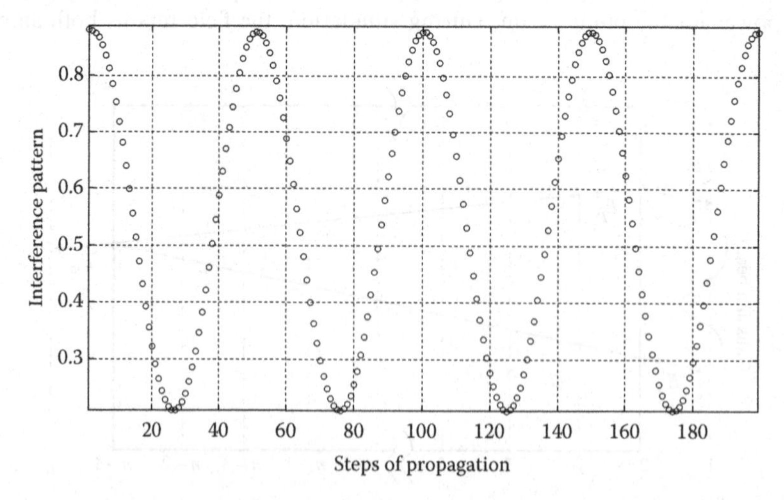

FIGURE 2.21 Interference pattern in the longitudinal direction for a $LiNbO_3$ crystal. Crystal length is equal to two wavelengths (532 nm) of light (inside). The number of longitudinal steps (n) per wavelength is 100. Spot size is 5 μm.

Following the linear cycle, the PR effect is now simulated through a nonlinear cycle of beam propagation. In this cycle, the buffered modulated intensity can be used to determine the space-charge field. The change in index then can then be determined by considering the electro-optic effect. For qualitative simulation, the following simplified general expressions are used [38,39]:

$$n = n_0 + \Delta n \tag{2.52}$$

$$\Delta n \propto E_s \tag{2.53}$$

$$E_s(\text{diffusion}) = C_{df} \cdot \nabla I(x,y,z) \tag{2.54}$$

$$E_s(\text{photovoltaic}) = C_{pv} \cdot I(x,y,z) \tag{2.55}$$

where n_0 is the bulk index of refraction of the crystal, and Δn is the change in index due to the space-charge field.

Note that in writing (2.54), we have approximated Equation 2.46 for the case where the "dark intensity" which may include background illumination is much larger than the intensity interference pattern from the participating optical waves or beams involved in coupling ($\beta/s \gg I$). Furthermore, in $LiNbO_3$, there is a substantial PV effect, which provides an additional contribution to the space charge field, as shown through Equation 2.55. The total space-charge field is approximately the sum of the contributions from diffusion of charge carriers and the PV effect.

The nonlinear cycle proceeds in a similar manner to the linear cycle. The buffered interference pattern intensity profile from the linear cycle is used to estimate the modified local refractive index variations through Equations 2.52 through 2.55. The refractive index change at each space-step in the calculations is treated as a dielectric boundary. As the pump beam E_p propagates through the index modulated crystal, the power of the beam is diminished at every step due to reflection losses at the dielectric boundaries, and the local values of E_p are stored in a buffer. Each grating reflected beam from the pump (pumpscatter) is propagated all the way out of the crystal. This grating scattered light is added to the buffered signal array. Each grating reflected beam from the signal (going toward the back surface of the crystal) is propagated all the way out of the crystal and added to the buffered pump array. As a first order approximation, it is assumed that scattered light from the pump and signal beams do not go through additional scattering during the propagation through the grating.

At the end of this first nonlinear cycle, the interference pattern is slightly modified. Subsequent iterations of this nonlinear cycle continue to modify the interference pattern (and the modulated index). When the difference between two successive interference patterns is within a chosen tolerance value (we have used 5%), convergence is assumed and the TBC efficiency is determined. All of the scattered beams that exit through the front and rear faces of the crystal are summed ($E_{p(scatter)}$ and $E_{s(scatter)}$ respectively). Throughout the crystal, the value of the pump buffer and signal buffer reflects the converged value of the pump and signal after loss or gain. In the case of pure charge diffusion, for the *c*-axis oriented along the direction of pump propagation, constructive interference occurs for the backscattered light, while the transmitted light suffers destructive interference. This interference occurs due to the phase difference between the interference and grating patterns, and the phase change in the scattered beams due to the reflections from the grating. The constructive and destructive interference directions are reversed when the crystal *c*-axis is reversed. Because of the destructive and constructive interference, the pump beam and the signal beam losses and gains power in the direction of respective propagation.

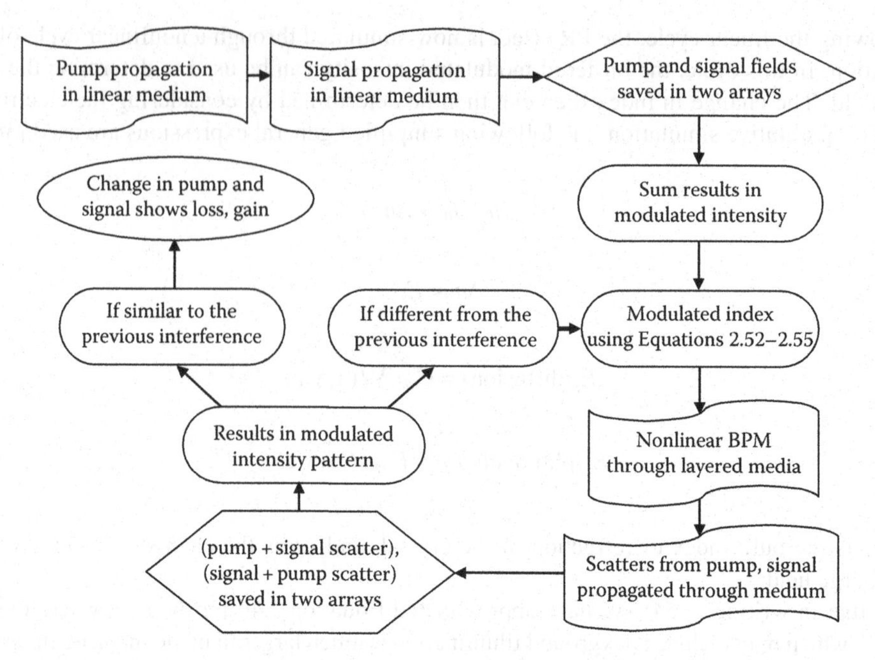

FIGURE 2.22 Flowchart showing the novel algorithm for simulation of self-pumped contra-directional TBC for a Gaussian beam.

This algorithm for self-pumped contra-directional TBC is summarized in the flowchart of Figure 2.22.

As stated earlier, a grating can only cause power coupling between two counter-propagating plane waves if the grating is phase shifted (ideally, by 90°) with respect to the interference produced by the two waves. For a grating that is in phase with the interference pattern, there can be no beam coupling. Extending on this established fact, we assume that for a weakly focusing beam, similar beam coupling behavior should be observed. We simulate a beam-coupling where the grating is 90° phase shifted with the interference (Figure 2.23), and in a crystal that is 2 wavelengths long. Following Equation 2.50, we have chosen C_{df} such that the change in index is 0.027. Figure 2.24 shows the

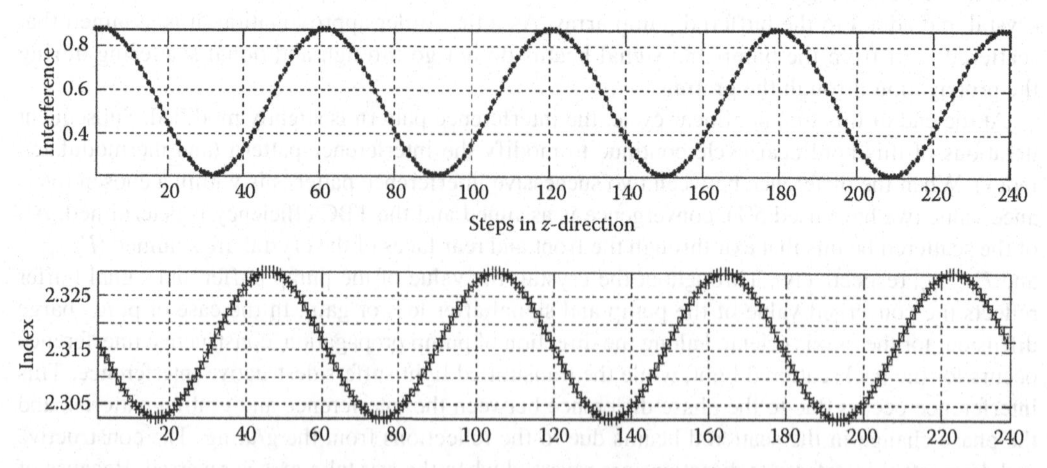

FIGURE 2.23 Interference pattern (top) and resulting 90° phase shifted index modulation (bottom) for pure diffusion alone. Crystal length is 2 wavelengths inside. Power inside is 1 unit and change in index is 0.027. (From Saleh, M.A., Self-pumped Gaussian beam coupling and stimulated backscatter due to reflection gratings in a photorefractive material, PhD dissertation, University of Dayton, Dayton, OH, 2007.)

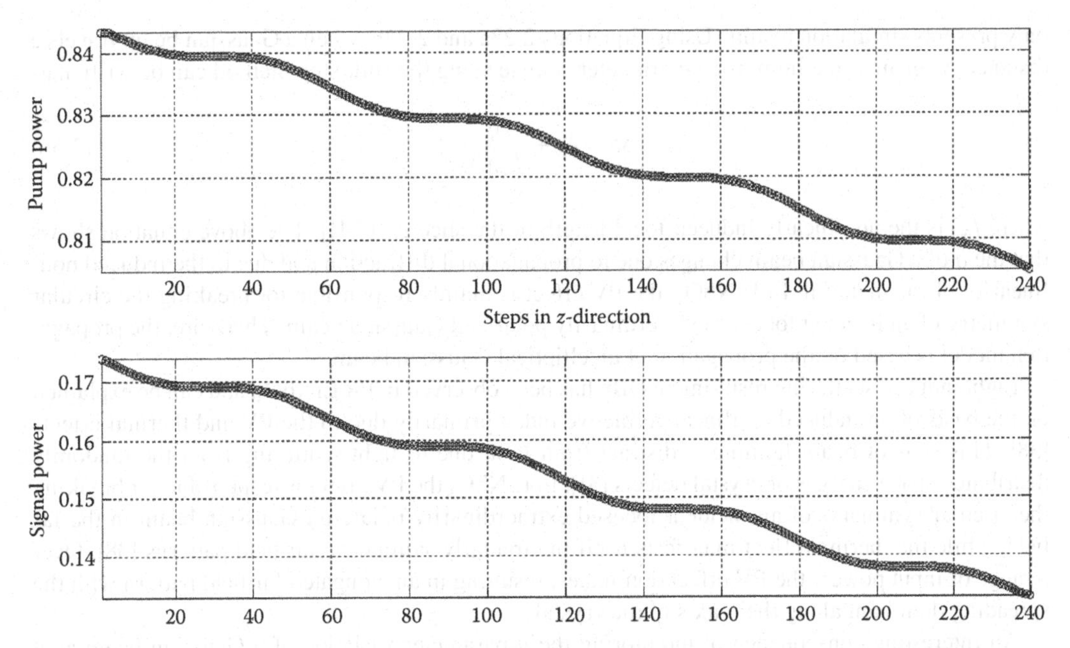

FIGURE 2.24 Energy exchange between the pump and the signal beam due to the modulated index from pure diffusion. Pump power is reduced by about 4% and signal power gains equally. Oscillation in power is artifact, as explained in the text. (From Saleh, M.A., Self-pumped Gaussian beam coupling and stimulated backscatter due to reflection gratings in a photorefractive material, PhD dissertation, University of Dayton, Dayton, OH, 2007.)

energy exchange between the pump and the signal beam due to the out-of-phase induced refractive index. Scattered light from the signal beam destructively interferes with the pump beam to diminish the pump power. Conversely, scattered light from the pump beam constructively interferes with the signal beam to increase the signal power. This is a distinctive feature of energy exchange due to beam coupling. In this case of simulation, pump power is reduced by about 4% and signal power gains equally. The observed oscillation in the power in Figure 2.24 is an artifact. The oscillation comes from the difficulty of calculating the power of (pump + signal scatters) or (signal + pump scatter) at any given point inside the nonlinear medium. The actual coupling is calculated by accounting for the amplified signal or diminished pumps outside the crystal, where the index is constant.

The presence of PV field usually has a negative effect on the phase of the index modulation. However, the PV field may change the magnitude of Δn to overcome this negative effect. It has been shown, both theoretically and experimentally, that a larger PV field may actually enhance the beam coupling [40]. This can also be verified using a constant value for C_{df} and changing the value of C_{pv} in our model Equations 2.52 through 2.55.

2.8 *z*-SCAN METHOD

The previous examples illustrated the use of the split-step method in calculating the beam profiles during diffraction in space or during propagation through a guided (externally or internally induced) medium. If a Gaussian beam is however assumed, the split-step method can be reformulated in terms of a differential equation that shows the evolution of the Gaussian beam's parameters, e.g., width, during propagation. The ensuing equation can be exactly solved in some cases, e.g., for a Kerr-type material, and is therefore physically more transparent than the results obtained using the split-step method. The differential equation for the parameter(s) may not be simpler to solve than the split step, however, having an analytical solution (Gaussian beam) adds a tremendous insight into the actual propagation of the wave through the material, whereas the split-step method only presents simulation results. Using Equations 2.23a and 2.23b, when a Gaussian beam travels a distance Δz in an n_2 medium, the q-parameter change using the split-step method can be written as

$$\Delta q = \Delta z + \frac{q^2}{f_{ind}(\Delta z)} \tag{2.56}$$

where f_{ind} is the nonlinearly induced focal length of the slice Δz [7,41]. The above equation shows that the q of a Gaussian beam changes due to propagational diffraction and due to the induced nonlinearity of the material. In $LiNbO_3$, the PV effect is mainly responsible for breaking the circular symmetry of an incident focused extraordinarily polarized Gaussian beam. Therefore, the propagation model is based on the propagation of an elliptical Gaussian beam.

Light induced scattering resulting in DBF has been observed in PR $LiNbO_3$, and can be explained on the basis of an induced nonlinear refractive index primarily due to the PV and thermal effects [38]. This type of beam fanning is distinct from RBF due to light scattering from the randomly distributed space charges or crystal defects [31]. In $LiNbO_3$ the PV effect is responsible for breaking the circular symmetry of an incident focused extraordinarily polarized Gaussian beam in the far field, while the thermal effect manifests itself in circularly symmetric far-field patterns [38]. Over a range of input powers the PV effect dominates, resulting in an elongated far-field pattern with the spreading dominant along the c-axis of the crystal.

An interesting consequence of monitoring the q-parameter variation of a Gaussian beam as it propagates through a nonlinear material is the fact one can thereby estimate the amount of nonlinearity in the material. Conventional methods of estimating the sign and magnitude of the optical nonlinearity in materials include the z-scan technique where the far-field on-axis transmittance is monitored as a function of the scan distance about the back focal plane of an external lens [41–43], as shown in Figure 2.25.

We point out that sometimes the z-scan method, however, may be rather cumbersome since it involves physically scanning the material, leading to our development of a simpler technique where the longitudinal position of the sample is not changed. Instead the beam ellipticity is monitored as a function of the incident beam power P, while testing materials with induced inhomogeneous nonlinearities, e.g., PR $LiNbO_3$ [44]. Another disadvantage of the z-scan is that monitoring the on-axis intensity may be difficult due to aberrations, optical misalignments, sample imperfections, refractive index mismatch, and non-parallelism of the entry and exit faces of the material. The imperfections can give rise to fine interference patterns within the far-field intensity profile. These problems have been observed during z-scan measurements of $LiNbO_3$, leading us to develop the P-scan technique as an attractive and simple alternative [44].

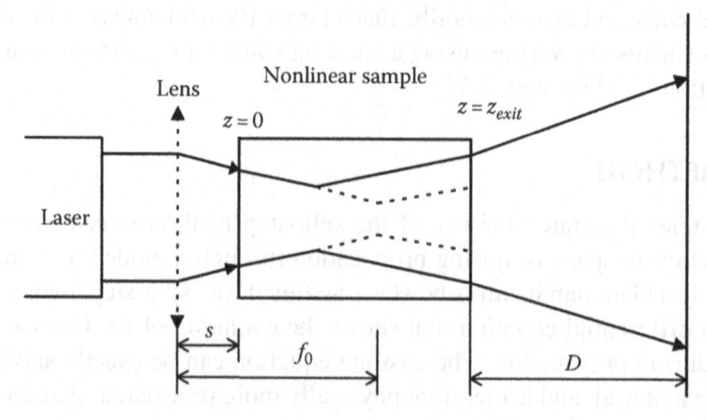

FIGURE 2.25 z-scan setup for a thick sample. The thick lines represent the path of the "rays," described as the locus of the $1/e$ points of the Gaussian beam. The thin lines show the ray path in the absence of the medium. Circular symmetry of the Gaussian beam is assumed throughout the sample.

In what follows, we develop the theoretical model for determining the nonlinear refractive index of PR $LiNbO_3$ that uses an appropriate model for beam propagation through a nonlinear material. The model takes into account inhomogeneous induced refractive index changes due to the optical nonlinearity. For the case of $LiNbO_3$, induced refractive index changes are primarily due to PV contributions over the range of powers used. The model is based on the evolution of beam widths of an incident circularly symmetric Gaussian beam focused by a lens onto the material in order to reduce RBF. The calculations closely follow the analysis for the z-scan determination of nonlinearities in a *thick* sample of a nonlinear material [41,44]. Under certain approximations, the model reduces to that used by Song et al. to study anisotropic light-induced scattering and "position dispersion" in PR materials [43]. Since we consider a "thick" sample, i.e., a sample whose thickness is much larger than the Rayleigh range of the focused Gaussian beam, diffraction effects become important and cannot be neglected. Therefore we determine the beam shape as it leaves the nonlinear sample and then calculate the beam profile after it has propagated some distance outside the medium. The information about the effective n_2 is contained in the nature of this profile. In general, the magnitude and sign of the nonlinearity can be determined from the beam profile variation as the sample position is varied about the back focal length of the external lens. The nonlinearity depends on the acceptor-to-donor concentration ratio N_A/N_D, which in turn determines the far-field diffraction pattern. Conversely, measurements of the far-field pattern can be used to calculate N_A/N_D and used as a tool for characterizing different $LiNbO_3$ samples.

2.8.1 Model for Beam Propagation through PR Lithium Niobate

Assume an incident Gaussian beam in the form

$$E_e(x,y,z) = a(z)\exp\left(-\frac{x^2}{w_x^2}\right)\exp\left(-\frac{y^2}{w_y^2}\right) \tag{2.57}$$

Extending 2.56, for an elliptical Gaussian beam, the following relationships holds:

$$\Delta q_x = \Delta z + \frac{q_x^2}{f_{ind_x}}, \quad \Delta q_y = \Delta z + \frac{q_y^2}{f_{ind_y}} \tag{2.58}$$

Since

$$n = n_e + n_2|E_e|^2 \approx n_e - 2n_2a^2(z)\left(\frac{x^2}{w_x^2} + \frac{y^2}{w_y^2}\right) \tag{2.59}$$

where

n_2 is the effective nonlinear refractive index coefficient
n_e is the linear refractive index
E_e is the optical field

We can compute the phase change upon nonlinear propagation through a section Δz of the sample and thereby determine the induced focal length. As expected, these focal lengths are inversely proportional to Δz and can be expressed as

$$f_{ind_x} = \frac{n_e w_x^2}{4n_{2_x}a^2(z)\Delta z}, \quad f_{ind_y} = \frac{n_e w_y^2}{4n_{2_y}a^2(z)\Delta z} \tag{2.60}$$

Substituting (2.60) into (2.58) and taking the limit as $\Delta z \to 0$ we obtain the system of equations

$$\frac{dq_x}{dz} = 1 + \frac{4n_{2_x} a(z) q_x^2}{n_e w_x^2}$$
$$\frac{dq_y}{dz} = 1 + \frac{4n_{2_y} a(z) q_y^2}{n_e w_y^2} \tag{2.61}$$

Using the well-known relationship: $(1/q) = (1/R) + j(\lambda/n_e \pi w^2)$ where R is a radius of Gaussian beam curvature $(1/R) = (1/w)(dw/dz)$, and λ is the wavelength in vacuum, we obtain

$$\frac{1}{R_x^2}\frac{dR_x}{dz} = \frac{n_e^2 \pi^2 w_x^4 - \lambda^2 R_x^2}{(n_e \pi w_x^2 R_x)^2} - \frac{4n_{2_x} a^2}{n_e w_x^2}$$
$$\frac{1}{R_y^2}\frac{dR_y}{dz} = \frac{n_e^2 \pi^2 w_y^4 - \lambda^2 R_y^2}{(n_e \pi w_y^2 R_y)^2} - \frac{4n_{2_y} a^2}{n_e w_y^2} \tag{2.62}$$

$$\frac{d^2 w_x}{dz^2} = \frac{\lambda^2}{n_e^2 \pi^2 w_x^3} - \frac{4n_{2_x} a^2}{n_e w_x}$$
$$\frac{d^2 w_y}{dz^2} = \frac{\lambda^2}{n_e^2 \pi^2 w_y^3} - \frac{4n_{2_y} a^2}{n_e w_y} \tag{2.63}$$

Taking into account the relationship for the beam's power, $P = (\pi/2\eta)a^2(z)w_x(z)w_y(z)$ where η is the characteristic impedance of the material, which is conserved, we finally have the system of equations describing the Gaussian beam propagation in a thick $LiNbO_3$ crystal:

$$\frac{d^2 w_x}{dz^2} = \frac{\lambda^2}{n_e^2 \pi^2 w_x^3} - \frac{8n_{2_x} P\eta}{\pi n_e w_x^2 w_y}$$
$$\frac{d^2 w_y}{dz^2} = \frac{\lambda^2}{n_e^2 \pi^2 w_y^3} - \frac{8n_{2_y} P\eta}{\pi n_e w_y^2 w_x} \tag{2.64}$$

Assuming $n_{2_x} \gg n_{2_y}$ (true for PR lithium niobate) the variation of the widths w_x and w_y of an elliptic Gaussian beam propagating through a thick $LiNbO_3$ sample as shown in the z-scan setup of Figure 2.25 can be modeled by the coupled differential equations:

$$\frac{d^2 w_x}{dz^2} = \frac{\lambda^2}{n_e^2 \pi^2 w_x^3} - \frac{8n_2 \eta P}{\pi n_e w_x^2 w_y}$$
$$\frac{d^2 w_y}{dz^2} = \frac{\lambda^2}{n_e^2 \pi^2 w_y^3} \tag{2.65}$$

The case when $n_{2_x} = n_{2_y}$ has been studied in [41] by employing the q-transformation approach to find the widths of a circular Gaussian beam in a nonlinear medium in the presence of diffraction. Equation 2.65 assume that the nonlinearity is highly inhomogeneous and only affects the width along the x-axis (which coincides with the c-axis of our crystals) due to the large electron mobility along that axis [25]. The effective n_2 can be written as [41]

$$n_2 \cong -\frac{1}{2} n_e^3 r_{33} \frac{k\alpha\gamma_R N_A}{\mu e \beta N_D} \tag{2.66}$$

where

r_{33} is the electro-optic coefficient
k is the PV constant
α is the absorption coefficient
γ_R is the recombination constant
μ is the mobility
e is the electron charge
β is the thermal generation rate

In the above equation, we have made the assumption $\beta \gg sI$, where s is the ionization cross-section per quantum of light and I is the optical intensity.

2.8.2 z-Scan: Analytical Results, Simulations, and Sample Experiments

In this sub-section, we present analytical and numerical simulation results and compare them with sample experiments using PR $LiNbO_3$. If the Gaussian beam incident on the sample is assumed to have planar wavefronts and waist w_0 (approximately at the back focus of the lens), then

$$w_y^2(z) = w_0^2\left(1+\frac{z^2}{z_{R_y}^2}\right); \quad z_{R_y} = \frac{n_e \pi w_0^2}{\lambda_0} \tag{2.67}$$

For a sample length L assumed to be much larger than the Rayleigh ranges z_{R_y} and z_{R_x} along z for the elliptic beam, the evolution of w_x can be approximated as

$$w_x^2(z) = w_0^2\left(1+\frac{z^2}{z_{R_x}^2}\right); \quad z_{R_x} = \frac{n_e \pi w_0^2}{\lambda_0}\left(1+\frac{4 n_e n_2 \eta \pi P}{\lambda_0^2}\right) \tag{2.68}$$

It is clear that in the x-direction, the beam spread is more than that in the linear diffraction-limited case when $n_2 < 0$ and less when $n_2 > 0$. As seen from relation 2.63 the nonlinearity does not affect the beam width along the y-direction, which leads to elliptic beam cross-section profile at the exit of the crystal and, in general, in the far field.

For more general geometry, where the incident beam does not have a planar wavefront, we have solved Equations 2.65 numerically. Figure 2.26 shows typical z-scan graphs plotted for four different values of power for the initially circularly symmetric Gaussian beam. In the calculations, we have used the following parameters: crystal width L = 10 mm, lens focal length f_0 = 10 cm, λ_0 = 514 nm, initial beam width w_0 = 1.0 mm, n_e = 2.20, $n_2 = -1.4\times10^{-12}$ m²/V², P = 1 mW, crystal exit plane to observation plane distance D = 1 m. A simple explanation of the behavior in the limiting case (s much smaller or larger than f_0) seen in Figure 2.26 can be given by referring to Figure 2.25. When the distance s, lens-to-sample separation, is much smaller than the lens focal length f_0, the incident beam is weakly focused and therefore the beam widths lie close to their linear values leading to semi-linear diffraction-limited propagation. When s is much larger than f_0, the incident beam is weakly diverging and the overall nonlinear effect is small that, in turn, leads to semi-linear diffraction-limited propagation. If $s \sim f_0$, the incident beam is highly focused and therefore the nonlinear effect is large. In this region, as s decreases, the normalized intensity decreases from its linear value, passes through a minimum, and then reaches its maximum before approaching its linear value again. The overall negative slope (between the peak and the valley) of the z-scan confirms the net negative nonlinearity of the sample.

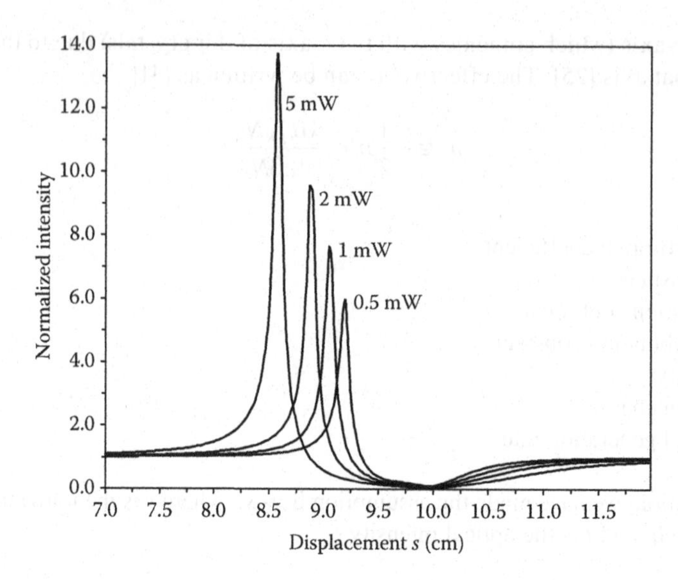

FIGURE 2.26 Typical z-scan graph drawn by solving Equations 2.65 and propagating the Gaussian beam a distance D behind the sample.

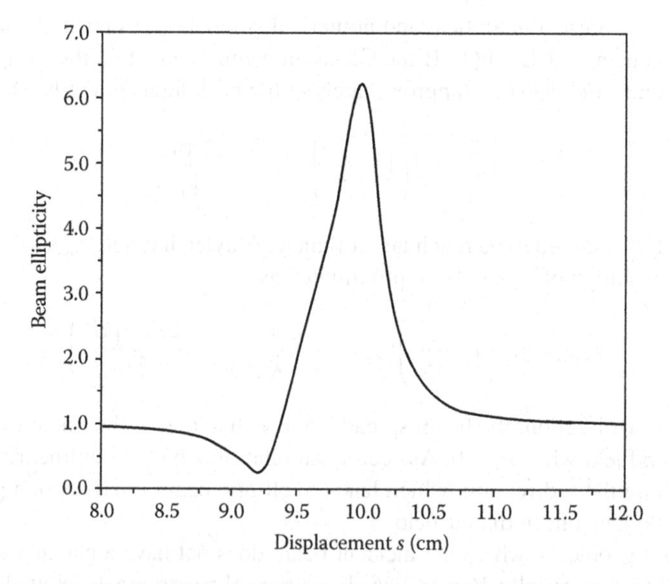

FIGURE 2.27 Plot of ellipticity as function of displacement s for parameters same as in Figure 2.26 but for $P = 0.2$ mW. (From Banerjee, P.P. et al., *J. Opt. Soc. Am. B*, 15, 2446, 1998. With permission.)

Figure 2.27 depicts ellipticity w_x/w_y in the far field versus displacement s drawn for the same set of parameters as that used to draw Figure 2.26, but for $P = 0.2$ mW. We have done a series of sample experiments and compared results. It turns out that the on-axis intensity measurement of far-field patterns may lead to significant errors due to fine structures in the pattern as seen in Figure 2.28 (obtained using a $LiNbO_3$ crystal doped with Fe). We have used this crystal for all experimentation to validate our theory, unless otherwise stated. Possible reasons for this include

1. Interference patterns stemming from single-beam holography [45]
2. Interference patterns from optical misalignment
3. Light diffraction and scattering on crystal defects
4. Interference patterns from nonparallel crystal edges

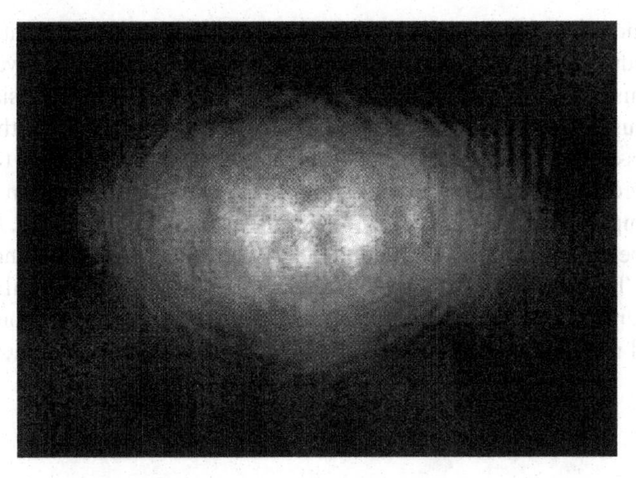

FIGURE 2.28 Typical beam pattern at $D = 0.5\,\text{m}$ for $P = 0.05\,\text{mW}$, $f_0 = 20\,\text{cm}$, and $s = 19.5\,\text{cm}$ for Fe doped $LiNbO_3$ crystal. (From Banerjee, P.P. et al., *J. Opt. Soc. Am. B*, 15, 2446, 1998. With permission.)

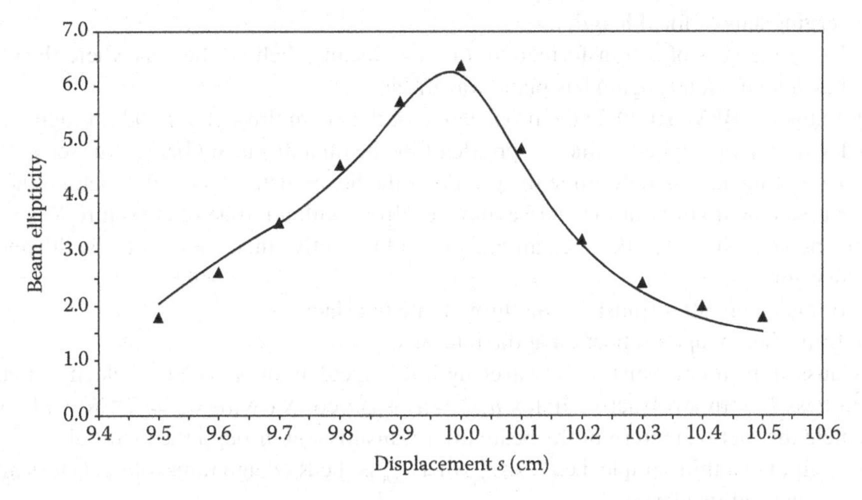

FIGURE 2.29 Experimental (points) and theoretical (line) variation of the beam ellipticity on the observation plane as a function of scan distance. Here, $P = 0.2\,\text{mW}$, $D = 0.5\,\text{m}$, $f_0 = 10\,\text{cm}$. Upon comparison, $n_2 = -1.4 \times 10^{-12}\,\text{m}^2/\text{V}^2$. (From Banerjee, P.P. et al., *J. Opt. Soc. Am. B*, 15, 2446, 1998. With permission.)

Note that the pattern is approximately symmetric (along x and y). This symmetry arises because the refractive index changes that are due to PV (and thermal effects) are symmetric and because there is little contribution from diffusion. Experimental results based on the measurement of ellipticity, as shown in Figure 2.29, show the same trend as the theoretical predictions superposed on the same figure. The ellipticity was calculated from experimental observations by first determining the extent w_x, w_y of the bright or gray region along x and y, respectively, from pictures such as Figure 2.28 and taking the ratio of the two. Note that Figure 2.29 is in fact a blowup of Figure 2.27 over the interval 9.5–10.5 cm. The theoretical graph in Figure 2.29 was drawn after examining the experimental results shown in the same figure and choosing that value n_2 for the analytical graph that minimizes the sum of the differences between the experimental points and the corresponding theoretical data.

As a final note, we would like to point out that each time the crystal was displaced along the longitudinal direction for a fresh z-scan ellipticity measurement, we also made a transverse movement of the crystal in order to make sure that we were starting out from a virgin location in the crystal for each data point. In other words, we always started out from an initially unexposed region of the crystal and exposed it to the incident illumination until steady state was achieved.

In summary, a model for beam propagation through a nonlinear material that takes into account inhomogeneous induced refractive index changes due to the nonlinearity was developed. The theory based on this model can be used to analyze the propagation of Gaussian beams through PR $LiNbO_3$. A focused Gaussian beam of circular cross section incident on the sample emerges as an elliptic Gaussian after interaction in this material. As stated earlier, the simpler *P*-scan method can be used to evaluate the effective nonlinearities (resulting from the PV effect) of lithium niobate samples doped with different materials such as Fe, Co, Cr, Rh, Mn, etc. The value of the nonlinear coefficient can then be used to determine the acceptor-to-donor ratio of dopants in the PR samples. This method can be used to characterize any optically nonlinear material that has an induced intensity dependent refractive index. We would like to point out that this method is very general and in principle may be applied to any nonlinear electromagnetic material and at any frequency.

PROBLEMS

2.1 Assume a Gaussian beam in air with plane wavefronts and waist w_0 at a distance d_0 from a converging lens of focal length f.

(a) Using the laws of q-transformation, find the distance behind the lens where the Gaussian beam focuses, i.e., again has plane wavefronts.

(b) Using the BPM, simulate the propagation of the beam through air and through a lens.

(c) By setting $d_0 = f$, determine the profile of the beam a distance f behind the lens.

(d) By setting $d_0 = 2f$, determine the profile of the beam distances f and $2f$ behind the lens.

2.2 A Gaussian beam of width w and having wavefront with a radius of curvature R is normally incident on the interface between air and glass of refractive index n. Find the width and radius of curvature

(a) Immediately after transmission through the interface

(b) Immediately upon reflection at the interface

2.3 A Gaussian beam of waist w_0 of wavelength λ is incident on a slice of dielectric material of thickness L with a refractive index $n(x) = n_0 + \Delta n \cos Kx$ with $w_0 \gg 2\pi/K$. Calculate the far-field diffraction pattern of the beam after transmission through the material

(a) Assuming a thin sample, i.e., $L \ll z_R$ where z_R is the Rayleigh range of the Gaussian beam, and normal incidence

(b) Assuming a thick sample $L > z_R$ and with $K^2L\lambda \gg 1$, and incidence at Bragg angle given by $\phi_B = \sin^{-1}(K\lambda/4\pi)$

2.4 Use the split-step beam propagation technique to analyze propagation along z of a one-dimensional Gaussian beam of $w_0 = 100\lambda_0$ (λ_0 is the free-space wavelength) incident onto a grating made using a material of quiescent refractive index n_0. The grating has a thickness of $L = 100\lambda_0$ with a refractive index profile $n(x) = n_0 + \Delta n \,\text{sgn}\,(\cos Kx)$, $K = 2\pi/\Lambda$, $\Lambda = 5\lambda_0$, where sgn denotes the signum function. Assume $n_0 = 1.5$ and $\Delta n = 0.00015$. Calculate the profile at the exit plane of the grating and in the far field. Repeat the problem for the case where the thickness of the grating is $L = 1000\lambda_0$ and characterize the differences between the two cases.

2.5 Analyze the propagation of a Gaussian beam of waist $w_0 = 100\lambda_0$ through a material of thickness $L = 100\lambda_0$ having a refractive index profile $n(x) = n_0 + \alpha(x/w_0)$, $|x| < 5w_0$. Let $n_0 = 1.5$, $\alpha = 0.015$. Determine the far-field intensity profile. You may use analytical techniques and/or the BPM.

2.6 A Gaussian beam of waist $w_0 = 100\lambda_0$ symmetric about $x = 0$ is incident from air onto a nonlinear material slab of thickness $L = 100\lambda_0$ and of refractive index $n(x) = n_0 + n_2I(x)$, $n_0 = 1$ where $I(x)$ is the intensity of the Gaussian beam. Assume that a knife edge is present at $z = 0$. Use the split-step method to determine the far-field profile. At $z = 0^-$, assume $n_2I(0) = 10^{-4}$.

2.7 The paraxial NLS equation can be written as

$$j\frac{\partial u}{\partial z}+\frac{1}{r^{D-1}}\frac{\partial}{\partial r}\left(r^{D-1}\frac{\partial u}{\partial r}\right)+f\left(|u|^2\right)u=0$$

where D represents the dimension of the problem. For the case $D = 2$ (cylindrical symmetry) and $f(|u|^2) = |u|^2$ (a) use the Hankel transform technique to numerically plot representative beam on-axis amplitudes during propagation all the way to near the self-focusing point for an initial profile $u_0 = 4e^{-r^2/2}$; (b) repeat part (a) for the case when, as in the text, a fixed and adaptive nonparaxiality parameter has been included.

2.8 (a) Plot the on-axis amplitudes as a function of propagation distance for the case of the paraxial NLS equation with cylindrical symmetry but for an initial profile $u_0 = 4e^{-r^2}$. (b) Repeat part (a) for the case when, as in the text, a fixed and adaptive nonparaxiality parameter has been included [22].

2.9 In the paraxial NLS equation

$$j\frac{\partial u}{\partial z}+\frac{1}{r^{D-1}}\frac{\partial}{\partial r}\left(r^{D-1}\frac{\partial u}{\partial r}\right)+f\left(|u|^2\right)u=0$$

setting $D=3$ implies spherical symmetry. Assume $f(|u|^2) = \left(|u|^2/(1+\mu|u|^2)\right)$. Use a change of variable $u = r^{-l}v$; $l = 1/2$ to change the radial operator from spherical to cylindrical coordinates. Thereafter, by using a suitable initial condition, sketch typical profiles of the spherically symmetric shapes that are stable during propagation. For hints and details of the Hankel transform to be used, readers are referred to Nehmetallah and Banerjee [46].

2.10 Derive representative z-scan graphs for the case where the (thin) lens to be characterized is a linear lens with a fixed focal length f which is independent of the intensity of the light, but may depend on other parameters such as applied voltage across the sample lens as in an electro-optic lens. In this case, show that the z-derivative of the on-axis intensity is similar to the traditional z-scan signature of a nonlinear induced lens.

2.11 Derive representative z-scan graphs for the case when the sample under test has an induced refractive index as in a diffusion-dominated PR material, and given by $\Delta n \propto \partial I/\partial x$, assuming one transverse dimension. Assume thin sample, for simplicity. Show that the nature of the z-scan graph is an even function of the displacement from the back focal plane of the external lens. For hints and details, readers are referred to Noginov et al. [47].

2.12 If the induced refractive index of a material is proportional to the gradient of the intensity, show how this effect may be used in image processing applications such as edge enhancement of an image. For hints, readers are referred to Banerjee et al. [48].

REFERENCES

1. P.P. Banerjee and T.C. Poon, *Principles of Applied Optics*, CRC Press, Boca Raton, FL, 1991.
2. G.P. Agrawal, *Nonlinear Fiber Optics*, Academic, London, 1989.
3. R.H. Hardin and F.D. Tappert, Applications of the split-step Fourier method to the numerical solution of nonlinear and variable coefficient wave equations, *SIAM* 15, 423, 1973.
4. A. Korpel, *Acousto-Optics*, 2nd edn., Dekker, Amsterdam, 1997.
5. C. Venzke, A. Korpel, and D. Mehrl, Improved space-marching algorithm for strong acousto-optic interaction of arbitrary fields, *Appl. Opt.* 31, 656–654, 1992.

6. P.P. Banerjee and C.-W. Tarn, A Fourier transform approach to acoustooptic interactions in the presence of propagational diffraction, *Acustica* 74, 181–191, 1991.
7. P.P. Banerjee, *Nonlinear Optics*, Dekker, New York, 2004.
8. G. Fibich and A. Gaeta, Critical power for self-focusing in bulk optical media and in hollow waveguides, *Opt. Lett.* 25, 335–337, 2000.
9. V.E. Zakharov and D. Shabat, Exact theory of two-dimensional self-focusing and one-dimensional self-modulation of waves in nonlinear media, *Sov. Phys. JETP* 34, 62–69, 1972.
10. G. Nehmetallah and P.P. Banerjee, Study of soliton stabilization in ($D=1$) dimensions using novel analytical and numerical techniques, in *Nonlinear Optics and Applications*, H.A. Abdeldayem and D.O. Frazier, eds., Research Signpost, Trivandrum, India, 2007.
11. A. Korpel, K.E. Lonngren, P.P. Banerjee, H.K. Sim, and M.R. Chatterjee, Split-step-type angular plane-wave spectrum method for the study of self-refractive effects in nonlinear wave propagation, *J. Opt. Soc. Am. B* 3, 885–890, 1986.
12. H.K. Sim, A. Korpel, K.E. Lonngren, and P.P. Banerjee, Simulation of two-dimensional nonlinear envelope pulse dynamics by a two-step spatiotemporal angular spectrum method, *J. Opt. Soc. Am. B* 5, 1900–1909, 1988.
13. G. Fibich, Small beam nonparaxiality arrests self-focusing of optical beams, *Phys. Rev. Lett.* 76, 4356–4359, 1996.
14. D.W. McLaughlin, G.C. Papanicolaou, C. Sulem, and P. L. Sulem, The focusing singularity of the cubic Schrödinger equation, *Phys. Rev. A* 34, 1200–1210, 1986.
15. M.I. Weinstein, Nonlinear Schrodinger equation and sharp interpolation estimates, *Comm. Math. Phys.* 87, 567–576, 1983.
16. R.T. Glassey, On the blowing up of solutions to the Cauchy problem for nonlinear Schrodinger equations, *J. Math Phys.* 18, 1794–1797, 1977.
17. P. Sulem, C. Sulem, and A. Patera, Numerical solution of the singular solutions to the 2-dimensional cubic Schrodinger equation, *Commun. Pure Appl. Math.* 37, 755–778, 1984.
18. G. Fibich and G.C. Papanicolaou, Self-focusing in the perturbed and unperturbed nonlinear Schrodinger equation in critical dimension, *SIAM J. Appl. Math.* 60, 183–240, 1999.
19. A.E. Siegman, Quasi-fast Hankel transform, *Opt. Lett.*, 1, 13–15, 1977.
20. L.C. Andrews and B.K. Shivamoggi, *Integral Transforms for Engineers*, SPIE, Bellingham, WA, 1998.
21. L. Yu, M. Huang, M. Chen, W. Chen, W. Huang, and Z. Zhu, Quasi-discrete Hankel transform, *Opt. Lett.* 23, 409–411, 1998.
22. P.P. Banerjee, G. Nehmetallah, and M.R. Chatterjee, Numerical modeling of cylindrically symmetric nonlinear self-focusing using an adaptive fast Hankel split-step method, *Opt. Commun.* 249, 293–300, 2005.
23. G. Fibich and B. Ilan, Discretization effects in the nonlinear Schrodinger equation, *Appl. Num. Math.* 44, 63–75, 2003.
24. S. Chi and Q. Guo, Vector theory of self-focusing of an optical beam in Kerr media, *Opt. Lett.* 20, 1598–1600, 1995.
25. P. Yeh, *Introduction to Photorefractive Nonlinear Optics*, Wiley, New York, 1993.
26. J. Feinberg, Asymmetric self-defocusing of an optical beam from the photorefractive effect, *J. Opt. Soc. Am.* 72, 46–51, 1982.
27. J.J. Liu, P.P. Banerjee, and Q.W. Song, Role of diffusive, photovoltaic, and thermal effects in beam fanning in $LiNbO_3$, *J. Opt. Soc. Am. B* 11, 1688–1693, 1994.
28. V.V. Voronov, I.R. Dorosh, Yu.S. Kuz'minov, and N.V. Tkachenko, Photoinduced light scattering in cerium-doped barium strontium niobate crystals, *Sov. J. Quantum Electron.* 10, 1346–1349, 1980.
29. P.P. Banerjee and R.M. Misra, Dependence of photorefractive beam fanning on beam parameters, *Opt. Commun.* 100, 166–172, 1993.
30. P.P. Banerjee and J.-J. Liu, Perturbational analysis of steady state and transient beam fanning in thin and thick photorefractive media, *J. Opt. Soc. Am. B* 10, 1417–1423, 1993.
31. M. Segev, Y. Ophir, and B. Fischer, Nonlinear mutli two-wave mixing. The fanning process and its bleaching in photorefractive media, *Opt. Commun.* 77, 265–274, 1990.
32. G.A. Swartzlander and A.E. Kaplan, Self-deflection of laser beams in a thin nonlinear film, *J. Opt. Soc. Am. B* 5, 765–768, 1988.
33. M. Abramowitz and I.A. Stegun, *Handbook of Mathematical Functions*, Dover, New York, 1970.
34. K. Ratnam and P.P. Banerjee, Nonlinear theory of two-beam coupling in a photorefractive material, *Opt. Commun.* 107, 522–530, 1994.

35. M. Cronin-Golomb and A. Yariv, Optical limiter using photorefractive nonlinearities, *J. Appl. Phys.* 57, 4906–4910, 1985.
36. S.E. Bialkowski, Application of $BaTiO_3$ beam-fanning optical limiter as an adaptive spatial filter for signal enhancement in pulsed infrared laser-excited photothermal spectroscopy, *Opt. Lett.* 14, 1020–1022, 1989.
37. S. Guha, Validity of the paraxial approximation in the focal region of a small *f*-number lens, *Opt. Lett.* 26, 1598–1600, 2001.
38. J.-J. Liu and P.P. Banerjee, Role of diffusive, photovoltaic and thermal fields in beam fanning in LiNbO3, *J. Opt. Soc. Am. B* 11, 1688–1693, 1994.
39. M.A. Saleh, Self-pumped Gaussian beam coupling and stimulated backscatter due to reflection gratings in a photorefractive material, PhD dissertation, University of Dayton, Dayton, OH, 2007.
40. G. Cook, C.J. Finman, and D.C. Jones, Photovoltaic contribution to cunterpropagating two-beam coupling in photorefractive lithium niobate, *Opt. Commun.* 192, 393–398, 2001.
41. P.P. Banerjee, R.M. Misra, and M. Maghraoui, Theoretical and experimental studies of propagation of beams through a finite sample of a cubically nonlinear material, *JOSA B* 8, 1072–1080, 1991.
42. M. Sheik-Bahae, A.A. Said, and E.W. Van Stryland, High-sensitivity, single-beam n_2 measurements, *Opt. Lett.* 14, 955–957, 1989.
43. Q.W. Song, C.-P. Zhang, and P.J. Talbot, Anisotropic light-induced scattering and "position dispersion" in $KNbO_3$:Fe crystal, *Opt. Commun.* 98, 269–273, 1993.
44. P.P. Banerjee, A. Danilieko, T. Hudson, and D. McMillen, P-scan analysis of induced inhomogeneous optical nonlinearities, *J. Opt. Soc. Am. B* 15, 2446–2456, 1998.
45. N. Kukhtarev, G. Dovgalenko, G. Duree, G. Salamo, E. Sharp, B. Wechler, and M. Klein, Single beam polarization holographic grating recording, *Phys. Rev. Lett.* 71, 4330–4332, 1993.
46. G. Nehmetallah and P.P. Banerjee, Numerical modeling of spatio-temporal solitons using an adaptive spherical Forurier Bessel split-step method, *Opt. Commun.* 257, 197–205, 2006.
47. M. Noginov, S.W. Helzer, G.B. Loutts, P.P. Banerjee, M. Morrisey, and Y. Kim, Study of photorefraction response and diffraction efficiency in Mn:YAlO3 crystals, *J. Opt. Soc. Am. B* 20, 1233–1241, 2003.
48. P.P. Banerjee, H.-L. Yu, D. Gregory, and N. Kukhtarev, Phase conjugation, edge detection and image broadcasting using two-beam coupling in photorefractive potassium niobate, *Opt. Photonics Tech. Lett.* 28, 89–92, 1996.

3 EM Wave Propagation in Linear Media

3.1 INTRODUCTION

In the previous chapter, we discussed various aspects of scalar wave propagation, including numerical techniques to model scalar wave propagation. Most importantly, we discussed the unidirectional and bidirectional beam propagation method and applied it to various cases of determining beam propagation in homogeneous and inhomogeneous media, including media where the induced refractive index is a function of the intensity of the beam. In this chapter, we start from Maxwell's equations and analyze plane wave solutions of electromagnetic (EM) waves in various media. We show that the constitutive relations that relate the EM field variables in the frequency domain play an important part in the behavior of EM waves. Particularly, we bring out the distinction between phase, energy, and group velocities, and show the characteristics of wave propagation in negative index media, i.e., media where the phase and group velocities may be counterdirected due to possible negative values of the permittivity and permeability over a certain frequency range. We also discuss wave propagation through chiral media, where the constitutive relations allow for coupling between the electric and magnetic fields. Finally, we also derive the transfer matrix approach to EM propagation through a layer of a material, which can be described by arbitrary dispersive permittivity and permeability, and show how this technique can be effectively used to model EM plane wave propagation through inhomogeneous materials.

3.2 MAXWELL'S EQUATIONS

In the study of EM and optics, we are concerned with four vector quantities called electromagnetic fields: the *electric field strength* $\vec{E}$ (V/m); the *electric flux density* $\vec{D}$ (C/m^2); the *magnetic field strength* $\vec{H}$ (A/m); and the *magnetic flux density* $\vec{B}$ (Wb/m^2). The fundamental theory of EM fields is based on *Maxwell's equations*. In differential form, these are expressed as [1]

$$\vec{\nabla}\cdot\vec{D}=\rho \tag{3.1}$$

$$\vec{\nabla}\cdot\vec{B}=0 \tag{3.2}$$

$$\vec{\nabla}\times\vec{E}=\frac{\partial\vec{B}}{\partial t} \tag{3.3}$$

$$\vec{\nabla}\times\vec{H}=\vec{J}=\vec{J}_i+\vec{J}_d=\vec{J}_i+\frac{\partial\vec{D}}{\partial t} \tag{3.4}$$

where

$\vec{J}$ is the *current density* (A/m^2)

ρ denotes the electric *charge density* (C/m^3)

$\vec{J}_i$ and ρ are the sources generating the electromagnetic fields

$\vec{J}_i$ is called the conduction current density

We can summarize the physical interpretation of Maxwell's equations as follows: Equation 3.1 is the differential representation of *Gauss's law for electric fields.* To convert this to an integral form, which is more physically transparent, we integrate Equation 3.1 over a volume V bounded by a surface S and use the *divergence theorem* (or *Gauss's theorem*),

$$\int_V \vec{\nabla}\cdot\vec{D}dV = \oint_S \vec{D}\cdot d\vec{S} \tag{3.5}$$

to get

$$\oint_S \vec{D}\cdot d\vec{S} = \int_V \rho dV \tag{3.6}$$

This states that the electric flux *flowing out of a surface S enclosing V equals the total charge enclosed in the volume.*

Equation 3.2 is the *magnetic analog* of Equation 3.1 and can be converted to an integral form similar to Equation 3.6 by using the divergence theorem once again:

$$\oint_S \vec{B}\cdot d\vec{S} = 0 \tag{3.7}$$

The RHSs of Equations 3.2 and 3.7 are zero because, in the classical sense, magnetic monopoles do not exist. Thus, the *magnetic flux is always conserved.*

Equation 3.3 enunciates *Faraday's law of induction.* To convert this to an integral form, we integrate over an open surface S bounded by a line C and use *Stokes' theorem,*

$$\int_S (\vec{\nabla}\times\vec{E})\cdot d\vec{S} = \int_C \vec{E}\cdot d\ell \tag{3.8}$$

to get

$$\int_C \vec{E}\cdot d\ell = -\int_S \left(\frac{\partial\vec{B}}{\partial t}\right)\cdot d\vec{S} \tag{3.9}$$

This states that the *electromotive force* (emf) *induced in a loop is equal to the time rate of change of the magnetic flux passing through the area of the loop.* The emf is induced in a sense such that it opposed the variation of the magnetic field, as indicated by the minus sign in Equation 3.9; this is known as *Lenz's law.*

Analogously, the integral form of Equation 3.4 reads

$$\int_C \vec{H}\cdot d\ell = \int_S \left(\frac{\partial\vec{D}}{\partial t}\right)\cdot d\vec{S} + \int_S \vec{J}_i\cdot d\vec{S} \tag{3.10}$$

which states that the *line integral of $\vec{H}$ around a closed loop C equals the total current (conduction and displacement) passing through the surface of the loop.* When first formulated by Ampere, Equations 3.4 and 3.10 only had the conduction current term $\vec{J}_i$ on the RHS. Maxwell proposed the addition of the displacement current term $\partial\vec{D}/\partial t$ to include the effect of currents flowing through, for instance, a capacitor.

For a given current and charge density distribution, note that there are four equations (Equations 3.1 through 3.4) and, at first sight, four unknowns that need to be determined to solve a given electromagnetic problem. As such, the problem appears well posed. However, a closer examination reveals that Equations 3.3 and 3.4, which are vector equations, are really equivalent to six scalar equations. Also, by virtue of the *continuity equation*,

$$\vec{\nabla} \cdot \vec{J}_i + \frac{\partial \rho}{\partial t} = 0 \tag{3.11}$$

Equation 3.1 is not independent of Equation 3.4 and, similarly, Equation 3.2 is a consequence of Equation 3.3. We can verify this by taking the divergence on both sides of Equations 3.3 and 3.4 and by using the continuity equation (3.11) and a vector relation

$$\vec{\nabla} \cdot (\vec{\nabla} \times \vec{A}) = 0 \tag{3.12}$$

to simplify. The upshot of this discussion is that, strictly speaking, there are 6 independent scalar equations and 12 unknowns (viz., the x, y, and z components of $\vec{E}$, $\vec{D}$, $\vec{H}$, and $\vec{B}$ to solve for). The six more scalar equations required are provided by the *constitutive relations*, which relate $\vec{E}$, $\vec{D}$, $\vec{H}$, and $\vec{B}$.

3.3 CONSTITUTIVE RELATIONS: FREQUENCY DEPENDENCE AND CHIRALITY

3.3.1 Constitutive Relations and Frequency Dependence

For a *linear, homogeneous, isotropic* medium, the constitutive relations are commonly given as

$$\hat{\vec{D}} = \hat{\varepsilon}\hat{\vec{E}}, \quad \hat{\vec{B}} = \hat{\mu}\hat{\vec{H}} \tag{3.13}$$

where $\hat{\varepsilon}$, $\hat{\mu}$ are scalars and denote the permittivity (F/m) and the permeability (H/m) of the medium, respectively. Note that the hats on top of the variables denote the fact that these are functions of temporal frequency. For instance, $\hat{\vec{E}}(\bar{r},\omega)$ represents the temporal Fourier transform of $\vec{E}(\bar{r},t)$, which is the electric field in Maxwell's equations, all of which are in the time domain. Also, $\hat{\varepsilon}(\omega)$, $\hat{\mu}(\omega)$ denote the frequency-dependent permittivity and permeability of the medium. In the special case where $\hat{\varepsilon}(\omega)$, $\hat{\mu}(\omega)$ are constants w.r.t. frequency, Equation 3.13 revert to $\vec{D} = \varepsilon\vec{E}$, $\vec{B} = \mu\vec{H}$, where ε, μ now represent the permittivity and permeability constants (with respect to frequency).

A medium is *linear* if its properties do not depend on the amplitude of the fields in the medium. It is *homogeneous* if its properties are not functions of space. The medium is, furthermore, *isotropic* if its properties are the same in all directions from any given point.

Returning our focus to linear, homogeneous, isotropic media, constants worth remembering are the values of ε and μ for free space or vacuum: $\varepsilon_0 = (1/36\pi) \times 10^{-9}$ F/m and $\mu_0 = 4\pi \times 10^{-7}$ H/m. For *dielectrics*, the value of ε is greater than that of ε_0, and a material part characterized by a *dipole moment density* or *polarization* $\vec{P}$ (C/m^2). $\vec{P}$ is related to the electric field $\vec{E}$ as

$$\hat{\vec{P}} = \hat{\chi}_e \varepsilon_0 \hat{\vec{E}} \tag{3.14}$$

where $\hat{\chi}_e$ is the *electric susceptibility* and indicates the ability of the electric dipoles in the dielectric to align themselves with the electric field. The $\vec{\hat{D}}$ field is the sum of $\varepsilon_0\vec{\hat{E}}$ and $\vec{\hat{P}}$:

$$\vec{\hat{D}} = \varepsilon_0\vec{\hat{E}} + \vec{\hat{P}} = \varepsilon_0(1+\hat{\chi}_e)\vec{\hat{E}} \equiv \varepsilon_0\hat{\varepsilon}_r\vec{\hat{E}} \tag{3.15}$$

where $\hat{\varepsilon}_r$ is the relative permittivity, so that

$$\hat{\varepsilon} = \varepsilon_0(1+\hat{\chi}_e) \tag{3.16}$$

Similarly, for magnetic materials, $\hat{\mu}$ is greater than μ_0 and is given by the magnetic susceptibility $\hat{\chi}_m$:

$$\hat{\mu} = \mu_0(1+\hat{\chi}_m) \tag{3.17}$$

We would like to point out that if the permittivity and permeability are functions of frequency, so are the electric and magnetic susceptibilities, and the constitutive relations are valid in the frequency domain. If they are frequency independent, the constitutive relationships have the same form in the time and frequency domains. In this case, the hats on top of ε and μ can be dropped for the sake of notational convenience.

An example of a frequency-dependent (relative) permittivity is provided by the Drude model [2]:

$$\frac{\hat{\varepsilon}(\omega)}{\varepsilon_0} = 1 - \frac{\omega_p^2}{\omega(\omega + j\gamma)} = 1 - \frac{\omega_p^2}{\omega^2+\gamma^2} + j\frac{\gamma\omega_p^2}{\omega(\omega^2+\gamma^2)} = \hat{\varepsilon}_r' - j\hat{\varepsilon}_r'' \tag{3.18}$$

For silver, $\omega_p = 21.75 \times 10^{14}$ rad/s and $\gamma = 4.35 \times 10^{12}$ rad/s, and the frequency dependence or *dispersion* of the real and imaginary parts of the permittivity is shown in Figure 3.1.

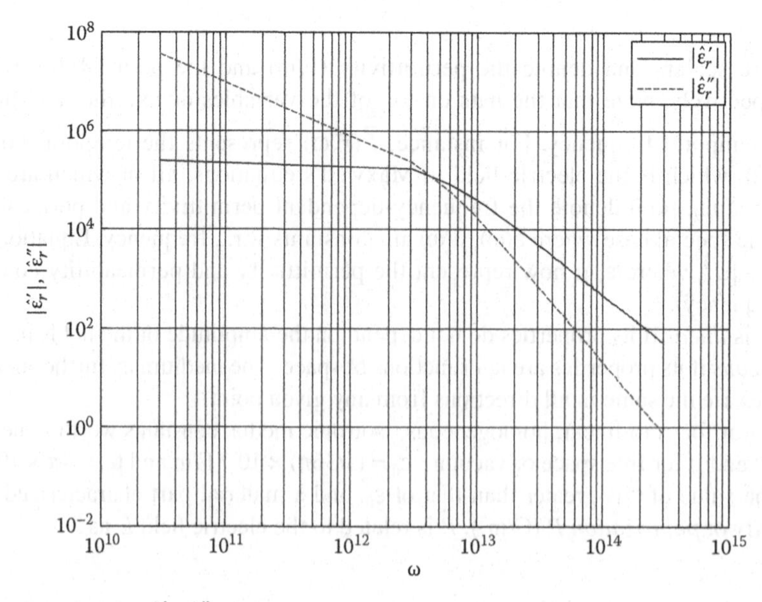

FIGURE 3.1 Permittivity $|\hat{\varepsilon}_r'|$, $|\hat{\varepsilon}_r''|$ of silver versus frequency with $\omega_p = 21.75 \times 10^{14}$ rad/s and $\gamma = 4.35 \times 10^{12}$ rad/s.

Consider a material that has an electric permittivity $\hat{\varepsilon}(\omega)$ and a magnetic permeability $\hat{\mu}(\omega)$, where both are functions of the frequency ω. The *refractive index* is defined as

$$\hat{n}^2(\omega) = \frac{\hat{\varepsilon}(\omega)\hat{\mu}(\omega)}{(\varepsilon_0\mu_0)} \tag{3.19}$$

which in the (inverse) time domain is equivalent to

$$n^2(t) = \frac{[\varepsilon(t) * \mu(t)]}{(\varepsilon_0\mu_0)} \tag{3.20}$$

where "*" denotes convolution. Assuming that the susceptibilities $\hat{\chi}_e(\omega)$ and $\hat{\chi}_m(\omega)$ are two complex analytic functions in the upper half of the complex plane, it follows according to the Kramers–Kronig (K–K) theorem that $\varepsilon(t)$ and $\mu(t)$ are two causal functions. Knowing that the convolution of two causal functions is also a causal function [3], it also follows that the square of the refractive index, $n^2(t)$, should also be a causal function. More on dispersion relations follows in later sections of this chapter.

3.3.2 Constitutive Relations for Chiral Media

In certain other kinds of linear materials, the constitutive relations as written in Equation 3.13 may be modified to [4]

$$\hat{\vec{D}} = \hat{\varepsilon}\hat{\vec{E}} - j\hat{\kappa}\sqrt{\mu_0\varepsilon_0}\,\hat{\vec{H}}, \quad \hat{\vec{B}} = j\hat{\kappa}\sqrt{\mu_0\varepsilon_0}\,\hat{\vec{E}} + \hat{\mu}\hat{\vec{H}} \tag{3.21}$$

These are the dispersive constitutive relationships in a reciprocal *chiral* medium. Again, frequency independence of chirality, permittivity, and permeability would suggest that the constitutive relations (3.21) are valid in the time domain. For the simple nondispersive case, the dimensionless chirality parameter κ is given by [5]

$$\kappa = N\mu_0\omega_0\beta c \tag{3.22}$$

where

β [C^2m^3/J] is the *optical rotatory parameter* of molecule, defined as the difference between the orientationally averaged polarizabilities for the left-handed and right-handed polarizations of light

N is the concentration of chiral molecules

This is related to the macroscopically observed rotation φ [rad/m] according to [5]

$$\varphi = \frac{1}{3}(n^2+2)\omega_0^2 N\mu_0\beta \tag{3.23}$$

Combining (3.22) and (3.23),

$$\kappa = \frac{3\varphi c}{(n^2+2)\omega_0} = \frac{3\varphi\lambda_0}{(n^2+2)} \tag{3.24}$$

3.4 PLANE WAVE PROPAGATION THROUGH LINEAR HOMOGENEOUS ISOTROPIC MEDIA

We can derive the wave equation describing the propagation of the electric and magnetic fields. By taking the curl of both sides of Equation 3.3 and assuming frequency-independent constitutive relations, we get

$$\vec{\nabla}\times\vec{\nabla}\times\vec{E} = -\mu\frac{\partial(\vec{\nabla}\times\vec{H})}{\partial t} \tag{3.25}$$

Now, employing Equation 3.4, Equation 3.25 becomes

$$\vec{\nabla}\times\vec{\nabla}\times\vec{E} = -\mu\varepsilon\frac{\partial^2\vec{E}}{\partial t^2} - \mu\frac{\partial\vec{J}_i}{\partial t} \tag{3.26}$$

Then, by using the vector relationship

$$\vec{\nabla}\times\vec{\nabla}\times\vec{E} = \vec{\nabla}(\vec{\nabla}\cdot\vec{E}) - \nabla^2\vec{E} \tag{3.27}$$

we get

$$\nabla^2\vec{E} - \mu\varepsilon\frac{\partial^2\vec{E}}{\partial t^2} = \mu\frac{\partial\vec{J}_i}{\partial t} + \vec{\nabla}(\vec{\nabla}\cdot\vec{E}) \tag{3.28}$$

If we now assume the permittivity to be space independent as well, then using the first of Maxwell's equations (Equation 3.1) in (3.28), we finally obtain

$$\nabla^2\vec{E} - \mu\varepsilon\frac{\partial^2\vec{E}}{\partial t^2} = \mu\frac{\partial\vec{J}_i}{\partial t} + \left(\frac{1}{\varepsilon}\right)\vec{\nabla}\rho \tag{3.29}$$

which is a wave equation having source terms on the RHS. In fact, Equation 3.29, being a vector equation, is really equivalent to three scalar equations, one for every component of E. Expressions for the Laplacian (∇^2) operator in Cartesian (x, y, z), cylindrical (R, ϕ, z), and spherical (R, θ, ϕ) coordinates can be found in standard EM books.

In space free of all sources ($\vec{J}_i = 0$, $\rho = 0$), Equation 3.29 reduces to the *homogeneous wave equation* for the electric field:

$$\nabla^2\vec{E} - \mu\varepsilon\frac{\partial^2\vec{E}}{\partial t^2} = 0 \tag{3.30}$$

A similar equation may be derived for the magnetic field.

We caution readers that the ∇^2 operator must be applied only after decomposing Equations 3.29 and 3.30 into scalar equations for three orthogonal components. However, for the rectangular coordinate case only, these scalar equations may be recombined and interpreted as the Laplacian ∇^2_{rect} acting on the total vector. A scalar version of Equation 3.30 has been used in Chapter 2 to develop the paraxial wave equation and the beam propagation method.

In the case of time harmonic fields, the electric field can be written as

$$\vec{E} = \text{Re}\,[\vec{E}_p \exp j\omega_0 t] \tag{3.31}$$

where ω_0 represents the carrier frequency. Substituting (3.31) into (3.30) results in the Helmholtz equation

$$\nabla^2\vec{E} + \omega_0^2\mu\varepsilon\vec{E} = 0 \tag{3.32}$$

3.4.1 Dispersive Media

As discussed above, dispersion can arise from the frequency dependence of the permittivity and the permeability (and the chirality parameter in the case of chiral materials). This type of dispersion will be termed *material dispersion.* Dispersion can also arise from boundary conditions imposed on the EM wave, viz., through boundaries such as in the case of metallic waveguides. This type of dispersion will be termed *topological dispersion.* In what follows, we will provide a simplified analysis of plane wave propagation through media that have frequency dependence of the permittivity and the permeability. In particular, we will assume propagation around a carrier frequency ω_0 and examine wave propagation when there are excursions Ω around the carrier due to modulation. This approach also leads to the concept of phase, group, and energy (or signal) velocities, and their dependence on frequency, to be discussed later. Dispersion relations can also be specified in terms of the variation of the frequency ω w.r.t. the propagation constant k of the wave.

The four vector EM fields $\vec{D}(\bar{r},t)$, $\vec{B}(\bar{r},t)$, $\vec{E}(\bar{r},t)$, $\vec{H}(\bar{r},t)$ in Maxwell's equations are time varying functions. However, as is true for all constitutive relations, the ones relating $\hat{\vec{D}}(\bar{r},\omega)$, $\hat{\vec{B}}(\bar{r},\omega)$, $\hat{\vec{E}}(\bar{r},\omega)$, $\hat{\vec{H}}(\bar{r},\omega)$ are valid in the frequency domain:

$$\hat{\vec{D}}(\omega) = \hat{\varepsilon}(\omega)\hat{\vec{E}}(\omega) \tag{3.33}$$

$$\hat{\vec{B}}(\omega) = \hat{\mu}(\omega)\hat{\vec{H}}(\omega) \tag{3.34}$$

where $\hat{\varepsilon}(\omega)$ and $\hat{\mu}(\omega)$ are the (scalar) frequency-dependent electric permittivity, and magnetic permeability parameters, often about a carrier frequency ω_0.

In order to incorporate the constitutive relations into Maxwell's equations, we need to express the vector fields $\vec{D}(t)$, $\vec{B}(t)$, $\vec{E}(t)$, $\vec{H}(t)$ in terms of slowly time varying phasor fields as

$$\begin{bmatrix} \vec{D}(\bar{r},t) \\ \vec{B}(\bar{r},t) \\ \vec{E}(\bar{r},t) \\ \vec{H}(\bar{r},t) \end{bmatrix} = \mathrm{Re}\left\{ \begin{bmatrix} \vec{D}_p(\bar{r},t) \\ \vec{B}_p(\bar{r},t) \\ \vec{E}_p(\bar{r},t) \\ \vec{H}_p(\bar{r},t) \end{bmatrix} e^{j\omega_0 t} \right\} \tag{3.35}$$

where ω_0 is the carrier frequency. Remembering that the Fourier transform of the complex exponential is $\Im_t\{\exp \pm j\omega_0 t\} = \delta(\omega \mp \omega_0)$, and upon using the convolution property, the Fourier transforms of the above variables, which are also space dependent, are

$$\begin{bmatrix} \hat{\vec{D}}(\omega) \\ \hat{\vec{B}}(\omega) \\ \hat{\vec{E}}(\omega) \\ \hat{\vec{H}}(\omega) \end{bmatrix} = \frac{1}{2}\left\{ \begin{bmatrix} \hat{\vec{D}}_p(\omega-\omega_0) + \hat{\vec{D}}_p^*(-\omega-\omega_0) \\ \hat{\vec{B}}_p(\omega-\omega_0) + \hat{\vec{B}}_p^*(-\omega-\omega_0) \\ \hat{\vec{E}}_p(\omega-\omega_0) + \hat{\vec{E}}_p^*(-\omega-\omega_0) \\ \hat{\vec{H}}_p(\omega-\omega_0) + \hat{\vec{H}}_p^*(-\omega-\omega_0) \end{bmatrix} \right\} \tag{3.36}$$

where "*" denotes the complex conjugate.

Now, the material parameters appearing in the constitutive relations are specified in the frequency domain, and similarly must possess a time domain counterpart. Accordingly, we express them as

$$\begin{bmatrix}\varepsilon(t)\\ \mu(t)\end{bmatrix}=\left\{\begin{bmatrix}\varepsilon_p(t)\\ \mu_p(t)\end{bmatrix}e^{j\omega_0 t}+\text{c.c}\right\} \tag{3.37}$$

with their Fourier transforms

$$\begin{bmatrix}\hat{\varepsilon}(\omega)\\ \hat{\mu}(\omega)\end{bmatrix}=\begin{bmatrix}\hat{\varepsilon}_p(\omega-\omega_0)+\hat{\varepsilon}_p^*(-\omega-\omega_0)\\ \hat{\mu}_p(\omega-\omega_0)+\hat{\mu}_p^*(-\omega-\omega_0)\end{bmatrix} \tag{3.38}$$

Equation 3.38 represents two shifted spectra of the permittivity and permeability, respectively, around $\omega=\pm\omega_0$.

As mentioned earlier, all the above material parameters, viz., $\varepsilon(t)$, $\mu(t)$, are causal functions of time, and as such, the real and imaginary parts of their Fourier transform form a Hilbert transform pair, which is also referred to as the Kramers–Kronig (K–K) relationship.

Dispersion of the above material parameters in the frequency domain can be expressed as a series expansion of the shifted spectrum around $\Omega\equiv\omega-\omega_0$. To *first order*, this leads to, after replacing $\Omega\rightarrow\omega-\omega_0$,

$$\begin{bmatrix}\hat{\varepsilon}_p(\Omega)\\ \hat{\mu}_p(\Omega)\end{bmatrix}=\begin{bmatrix}\hat{\varepsilon}_p(\omega-\omega_0)\\ \hat{\mu}_p(\omega-\omega_0)\end{bmatrix}\approx\begin{bmatrix}\hat{\varepsilon}_{p0}+\Omega\hat{\varepsilon}'_{p0}\\ \hat{\mu}_{p0}+\Omega\hat{\mu}'_{p0}\end{bmatrix}=\begin{bmatrix}\hat{\varepsilon}_p(\omega_0)+(\omega-\omega_0)\left(\dfrac{\partial\hat{\varepsilon}_p}{\partial\omega}\right)_{\omega_0}\\ \hat{\mu}_p(\omega_0)+(\omega-\omega_0)\left(\dfrac{\partial\hat{\mu}_p}{\partial\omega}\right)_{\omega_0}\end{bmatrix} \tag{3.39}$$

where we have written $\Omega=\omega-\omega_0$ to represent excursion around the carrier frequency ω_0. Higher order expansions, not covered in this chapter, lead to second derivatives of ε, μ which can be related to (normal or anomalous) group velocity dispersion.

We note that while (3.39) is strictly an expansion around ω_0, its coefficients are constants evaluated at ω_0. In some literature [6,7], the dispersive behavior described by (3.39) is sometimes represented through the ad hoc replacement $\hat{\varepsilon}(\omega)\rightarrow\partial(\omega\hat{\varepsilon})/\partial\omega$. As is easily seen, this replacement can be written as $\hat{\varepsilon}+\omega\partial\hat{\varepsilon}/\partial\omega$, which is neither a series expansion around $\omega=0$ nor one around $\omega=\omega_0$, since it does not possess constant coefficients evaluated around the center of expansion.

Using the constitutive relation for $\hat{D}$ (as in (3.33)), the displacement field may be expressed in terms of the phasor electric and magnetic fields as

$$\hat{D}=\frac{1}{2}\left[\hat{E}_p(\omega-\omega_0)\hat{\varepsilon}_p(\omega-\omega_0)+\hat{E}_p^*(-\omega-\omega_0)\hat{\varepsilon}_p^*(-\omega-\omega_0)\right] \tag{3.40}$$

where it has been assumed that all fields and material parameters are bandlimited, thus eliminating cross-coupling between positive and negative frequencies. The derivation of (3.41) is qualitatively illustrated by the spectral characteristics shown in Figure 3.2. Note that each of the quantities $\hat{E}$, $\hat{\varepsilon}$, $\hat{D}$ has been assumed to be bandlimited, with spectral profiles around a carrier ω_0. Thus, the spectral product for $\hat{D}$ according to (3.33) will contain the direct products at $\pm\omega_0$, while the cross-product terms vanish.

When $\hat{\varepsilon}_p$ is a constant in frequency ($=\varepsilon$), we find from inverse Fourier transform (3.40) that the displacement field in the time domain reduces to the familiar relation $D(t)=\varepsilon E(t)$. This validates the choice for the expressions of ε, μ in (3.37).

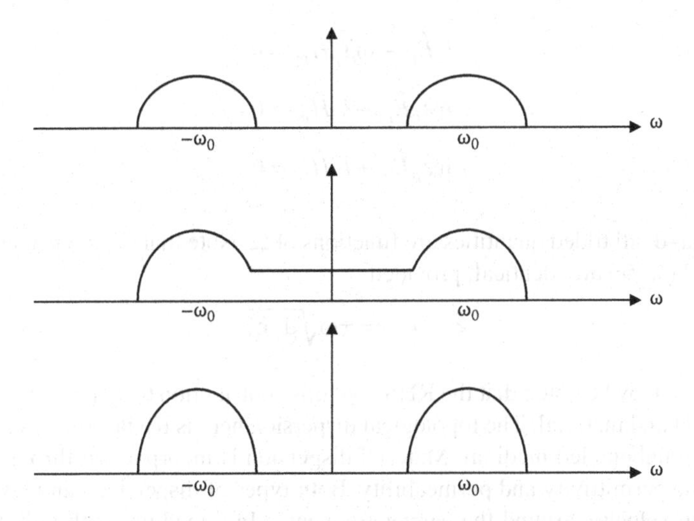

FIGURE 3.2 Bandlimited electric field, permittivity, and bandlimited displacement spectra around a carrier frequency ω_0. Both $\pm\omega_0$ have to be considered due to the Fourier transform operation. There is no cross-coupling between the upper sideband (USB) (lower sideband [LSB]) of the electric field and the LSB (USB) of the permittivity in the product leading to the spectrum of the displacement due to the bandlimited nature of the electric field spectrum.

From Equations 3.36, 3.39, and 3.40,

$$\hat{D}_p(\Omega) = \hat{E}_p(\Omega)\hat{\varepsilon}_p(\Omega) \approx \hat{E}_p(\Omega)(\hat{\varepsilon}_{p0} + \Omega\hat{\varepsilon}'_{p0}) \tag{3.41}$$

The corresponding slowly time varying displacement phasor is obtained from (3.41) as

$$D_p(t) \approx \left(\hat{\varepsilon}_{p0} - j\hat{\varepsilon}'_{p0} \frac{\partial}{\partial t} \right) E_p(t) \tag{3.42}$$

The corresponding relation for the slowly varying flux density phasor $B_p(t)$ may be found similarly. Note that in the absence of dispersion, the above result reduces to the expected constitutive relationship between the electric displacement and the electric field.

In order to determine possible propagation constants and the corresponding field solutions, we start from Maxwell's curl equations. For instance, from $\vec{\nabla} \times \vec{E} = -\partial\vec{B}/\partial t$, Fourier transforming and using the constitutive relations, we get for the phasor fields ($\omega > 0$) for the relation

$$\vec{k} \times \hat{\vec{E}}_p(\Omega) = \omega\hat{\mu}_p(\Omega)\hat{\vec{H}}_p(\Omega) \tag{3.43}$$

In deriving Equation 3.43, it has also been assumed that the phasor fields have a space dependence of the form $\exp -j\vec{k}\cdot\vec{r}$, so that $\vec{\nabla} \to -j\vec{k}$. Similarly, from the other Maxwell curl equation,

$$\vec{k} \times \hat{\vec{H}}_p(\Omega) = \omega\hat{\varepsilon}_p(\Omega)\hat{\vec{E}}_p(\Omega) \tag{3.44}$$

The solutions for the above fields are obtained more readily by assuming a wave vector $\vec{k}(\Omega) = k_z(\Omega)$ $\hat{a}_z$ pointed in the z-direction (which is arbitrary, and hence general, in an unbounded medium). This approach simplifies the resulting characteristic matrix to a 4×4 instead of a 6×6. The following set of homogeneous equations for the field components is then obtained from Equations 3.43 and 3.44:

$$k_z\hat{E}_{py} + \omega\hat{\mu}_p\hat{H}_{px} = 0 \tag{3.45a}$$

$$k_z\hat{E}_{px} - \omega\tilde{\mu}_p\hat{H}_{py} = 0 \tag{3.45b}$$

$$\omega\hat{\varepsilon}_p\hat{E}_{px} - k_z\hat{H}_{py} = 0 \tag{3.45c}$$

$$\omega\hat{\varepsilon}_p\hat{E}_{py} + k_z\hat{H}_{px} = 0 \tag{3.45d}$$

In Equation 3.45a–d, all tilded quantities are functions of Ω. Note that (3.45a) is identical to (3.45d), while (3.45b) and (3.45c) are identical, provided

$$k_z = k_{z1,2} = \pm\omega\sqrt{\hat{\mu}_p\hat{\varepsilon}_p} \tag{3.46}$$

In Equation 3.46, it may be noted that the RHS contains contribution to dispersion from two mechanisms: topological and material. The topological dispersion here is trivial, since we are considering propagation in an unbounded medium. Material dispersion is incorporated through the frequency dependence of the permittivity and permeability. Both types of dispersion can play a role in determining the group velocity, around the carrier frequency. In a medium with only topological dispersion, and no material dispersion, excursion of the generalized frequency ω around the carrier frequency ω_0 should be performed to determine the group velocity. This justifies why we have to retain the generalized frequency ω to begin with, in the slowly varying phasor form of Maxwell's equations (3.43 and 3.44). It also turns out that if instead of writing ω one used ω_0, the expression for the group velocity does not converge to the standard nondispersive result in the absence of material dispersion.

As stated earlier, general dispersion relations can be described as the variation of ω with k, i.e., $\omega = \omega(k)$, or conversely as the variation of k with ω, i.e., $k = k(\omega)$. The inverse Fourier transform of $k = k(\omega)$ will be denoted as $\breve{k}(t) = \Im_t^{-1}[k(\omega)]$. Dispersion relations can be complex as well. The complexity may arise, for instance, from the complex nature of $\hat{\mu}_p$, $\hat{\varepsilon}_p$. From linear systems theory, $k(\omega)$ can be perceived as the "propagation frequency response" to an excitation at an angular frequency ω. Accordingly, in the inverse or time domain, $\breve{k}(t)$ should correspond to a "propagation impulse response" [8]. The apparent nonphysicality of such a concept is akin to the apparent nonphysicality of negative frequencies in Fourier transforms, which however must be incorporated into mathematical analyses. Hence, suppose that

$$k(\omega) = k_r(\omega) - jk_i(\omega) \tag{3.47}$$

then a generic (scalar) wavefunction E (denoting the *real* electric field) propagating in the dispersive environment can be constituted from a collection of propagating plane waves:

$$\begin{aligned} E(z,t) &= \frac{1}{2\pi}\int_{-\infty}^{\infty}\hat{E}(\omega)\exp\left[j(\omega t - k(\omega)z)\right]d\omega \\ &= \frac{1}{2\pi}\int_{-\infty}^{\infty}\hat{E}(\omega)\exp\left[j(\omega t - k_r(\omega)z + jk_i(\omega)z)\right]d\omega \end{aligned} \tag{3.48a}$$

Hence,

$$\begin{aligned} E^*(z,t) &= \frac{1}{2\pi}\int_{-\infty}^{\infty}\hat{E}^*(\omega)\exp\left[-j(\omega t - k_r(\omega)z)\right]\exp\left[-k_i(\omega)z\right]d\omega \\ &= \frac{1}{2\pi}\int_{-\infty}^{\infty}\hat{E}^*(-\omega)\exp\left[j(\omega t + k_r(-\omega)z)\right]\exp\left[-k_i(-\omega)z\right]d\omega \end{aligned} \tag{3.48b}$$

through the replacement $\omega \to -\omega$. Since $E(z, t)$ is real, $E(z, t) = E^*(z, t)$, so that upon comparing (3.48a) and (3.48b), it follows that

$$k_r(-\omega) = -k_r(\omega), \quad k_i(-\omega) = k_i(\omega) \tag{3.49}$$

This means that $k_r(\omega)$ must be an odd function of ω, while $k_i(\omega)$ must be an even function of ω [8]. In acoustics, $k_i(\omega)$ is often referred to as the attenuation constant $\alpha(\omega)$.

3.4.2 Chiral Media

Using the constitutive relations in (3.21) and following the steps as in the last subsection, it can be shown that the analog of Equation 3.45a–d becomes

$$j\hat{\alpha}\hat{E}_{px} + k_z\hat{E}_{py} + \omega\hat{\mu}\hat{H}_{px} = 0 \tag{3.50a}$$

$$k_z\hat{E}_{px} - j\hat{\alpha}\hat{E}_{py} - \omega\hat{\mu}\hat{H}_{py} = 0 \tag{3.50b}$$

$$\omega\hat{\varepsilon}\hat{E}_{px} - j\hat{\alpha}\hat{H}_{px} - k_z\hat{H}_{py} = 0 \tag{3.50c}$$

$$\omega\hat{\varepsilon}\hat{E}_{py} + k_z\hat{H}_{px} - j\hat{\alpha}\hat{H}_{py} = 0 \tag{3.50d}$$

where the parameter $\hat{\alpha} = \omega_0\hat{\kappa}\sqrt{\mu_0\varepsilon_0}$, which has the dimension m^{-1}, can be termed the *chiral wavenumber.*

Finding nontrivial solutions for the homogeneous Equation 3.50a–d requires that the determinant of the coefficient matrix must vanish. This leads to the well-known condition for the wavenumber in the chiral medium [9]:

$$k_{z1} = +\omega\hat{\kappa}\sqrt{\mu_0\varepsilon_0} + \omega\sqrt{\hat{\mu}\hat{\varepsilon}} \tag{3.51a}$$

$$k_{z2} = +\omega\hat{\kappa}\sqrt{\mu_0\varepsilon_0} - \omega\sqrt{\hat{\mu}\hat{\varepsilon}} \tag{3.51b}$$

$$k_{z3} = -\omega\hat{\kappa}\sqrt{\mu_0\varepsilon_0} + \omega\sqrt{\hat{\mu}\hat{\varepsilon}} \tag{3.51c}$$

$$k_{z4} = -\omega\hat{\kappa}\sqrt{\mu_0\varepsilon_0} - \omega\sqrt{\hat{\mu}\hat{\varepsilon}} \tag{3.51d}$$

which indicates a set of four possible values of the wavenumber that satisfy the nontrivial field solutions. We may note that the k_z values depend on the chirality parameter $\hat{\kappa}$.

We next assume that the field component $\hat{E}_{px}$ is known. From the homogeneous Equation 3.50a–d, assuming the known $\hat{E}_{px}$ value, the field solutions are obtained after some algebra as follows [9]:

$$\hat{E}_{px} \text{ arbitrary} \tag{3.52a}$$

$$\hat{E}_{py} = \frac{-\omega^2\hat{\mu}\hat{\varepsilon} + \hat{\alpha}^2 + k_z^2}{2j\hat{\alpha}k_z}\hat{E}_{px} \tag{3.52b}$$

$$\hat{H}_{px} = \frac{-k_z^2 + \omega^2\hat{\mu}\hat{\varepsilon} + \hat{\alpha}^2}{2j\hat{\alpha}\omega\hat{\mu}}\hat{E}_{px} \tag{3.52c}$$

$$\hat{H}_{py} = \frac{-\hat{\alpha}^2 + k_z^2 + \omega^2\hat{\mu}\hat{\varepsilon}}{2\omega\hat{\mu}k_z}\hat{E}_{px} \tag{3.52d}$$

An important observation may be made here in view of the above solutions. Note that these solutions are true for an unbounded medium (and may be readily shown to satisfy the Maxwell curl equations), and are general in the sense that the z-directed wave vector is essentially arbitrary in an unbounded medium. We may also observe that the transverse electric (TE) field components have a phase difference of $\pm\pi/2$, but are unequal in amplitude. Such a condition arises from the superposition of a left- and a right-circular waves that have unequal amplitudes. Hence, the total electric field in the chiral medium (regardless of positive or negative index) is *elliptically polarized* with a phase difference of $\pm\pi/2$, and is made up of two circular polarizations (left circular polarization [LCP] and right circular polarization [RCP]).

3.5 POWER FLOW, STORED ENERGY, ENERGY VELOCITY, GROUP VELOCITY, AND PHASE VELOCITY

In this section, we will calculate the power flow, stored energy, and three characteristic velocities, viz., the phase velocity, group velocity, and energy (or signal) velocity pertinent to propagation in a dispersive medium [10]. For simplicity, we consider only material dispersion and a non-chiral medium. Therefore, it suffices to consider only one type of linear polarization, viz., $\hat{E}_{px}$, for the electric field (and correspondingly $\hat{H}_{py}$ for the magnetic field). The case of a chiral medium is more complicated and has been treated in Ref. [9].

By choosing, say, the positive root for k_z in (3.46), we obtain from (3.45c)

$$\hat{H}_{py} = \sqrt{\frac{\hat{\varepsilon}_p}{\hat{\mu}_p}}\hat{E}_{px} \tag{3.53}$$

which, upon expanding $\hat{\varepsilon}_p$ and $\hat{\mu}_p$ to the first power in Ω per Equation 3.39, leads to

$$\hat{H}_{py} = \sqrt{\frac{\hat{\varepsilon}_{p0}}{\hat{\mu}_{p0}}}\left[1+\frac{\Omega}{2}\left(\frac{\hat{\varepsilon}'_{p0}}{\hat{\varepsilon}_{p0}} - \frac{\hat{\mu}'_{p0}}{\hat{\mu}_{p0}}\right)\right]\hat{E}_{px} \tag{3.54}$$

Writing $\hat{H}_{py}$ as $\hat{H}_{py} = (A + B\Omega)\,\hat{E}_{px}$, we obtain the coefficients A and B as

$$A = \sqrt{\frac{\hat{\varepsilon}_{p0}}{\hat{\mu}_{p0}}}, \quad B = \frac{1}{2}\sqrt{\frac{\hat{\varepsilon}_{p0}}{\hat{\mu}_{p0}}}\left(\frac{\hat{\varepsilon}'_{p0}}{\hat{\varepsilon}_{p0}} - \frac{\hat{\mu}'_{p0}}{\hat{\mu}_{p0}}\right) \tag{3.55}$$

Inverse Fourier transforming (3.54) leads to

$$H_{py}(t) = AE_{px}(t) - jB\frac{\partial E_{px}}{\partial t} \tag{3.56}$$

Using (3.56), we then have for the slowly time varying *average Poynting vector* (averaged over a time period $2\pi/\omega_0$) as

$$S_z = \frac{1}{2}E_{px}(t)H^*_{py}(t) = \frac{1}{2}AE^2_{px} + j\frac{1}{4}B\frac{\partial}{\partial t}\left(E^2_{px}\right) \tag{3.57}$$

and its associated spectrum becomes

$$\hat{S}_z(\Omega) = \left(\frac{1}{2}A - \frac{1}{4}B\Omega\right)\Im\left[E_{px}^2\right] \tag{3.58}$$

We will now compute the same time average of the *stored electric* and *magnetic energies* given by the definitions

$$w_e(t) = \frac{1}{4}D_p^*(t)\cdot E_p(t), \quad w_m(t) = \frac{1}{4}B_p(t)\cdot H_p^*(t) \tag{3.59}$$

Using (3.42),

$$w_e(t) = \frac{1}{4}\left[\left(\hat{\varepsilon}_{p0} - j\hat{\varepsilon}'_{p0}\frac{\partial}{\partial t}\right)E_{px}^2\right] \tag{3.60}$$

assuming $E_{px}(t)$ to be real once again. Now, simplifying, taking the Fourier transform, and upon retaining only up to the first power of Ω, we obtain

$$\hat{w}_e(\Omega) = \frac{1}{4}\left[\hat{\varepsilon}_{p0} + \hat{\varepsilon}'_{p0}\Omega\right]\Im_t[E_{px}^2(t)] \tag{3.61}$$

Similarly,

$$\hat{w}_m(\Omega) = \frac{1}{4}A^2\left[\hat{\mu}_{p0} + \hat{\mu}'_{p0}\Omega\right]\Im_t[E_{px}^2(t)] \tag{3.62}$$

The *total stored energy* can then be written as

$$\hat{w}_{t1}(\Omega) = \frac{1}{4}\left\{\left[2\hat{\varepsilon}_{p0} + \left(\frac{\hat{\varepsilon}'_{p0} + \hat{\varepsilon}_{p0}\hat{\mu}'_{p0}}{\hat{\mu}_{p0}}\right)\Omega\right]\right\}\Im_t[E_{px}^2(t)] \tag{3.63}$$

Consequently, the z-directed *energy velocity* for k_{z1} is obtained using (3.61) and (3.63) as

$$\begin{aligned} v_{e1}(\Omega) &= \frac{\hat{S}_{z1}(\Omega)}{\hat{w}_{t1}(\Omega)}\hat{a}_z \\ &\approx \frac{1}{\left[\left\{\sqrt{\hat{\varepsilon}_{p0}\hat{\mu}_{p0}}\right\} + \left\{\frac{1}{4}\sqrt{\hat{\varepsilon}_{p0}\hat{\mu}_{p0}}\left(\frac{\hat{\varepsilon}'_{p0}}{\hat{\varepsilon}_{p0}} - \frac{\hat{\mu}'_{p0}}{\hat{\mu}_{p0}}\right) + \frac{1}{2}\sqrt{\hat{\varepsilon}_{p0}\hat{\mu}_{p0}}\left(\frac{\hat{\varepsilon}'_{p0}}{\hat{\varepsilon}_{p0}} + \frac{\hat{\mu}'_{p0}}{\hat{\mu}_{p0}}\right)\right\}\Omega\right]} \end{aligned} \tag{3.64}$$

The z-directed *group velocity* can be found using the following relationship:

$$v_{g1} = \frac{1}{\partial k_{z1}/\partial\omega} = \frac{1}{\partial k_{z1}/\partial\Omega} \tag{3.65}$$

where k_{z1} defined explicitly in terms of the sideband Ω becomes

$$k_{z1} = (\omega_0 + \Omega)\sqrt{\hat{\mu}_p\,\hat{\varepsilon}_p} \tag{3.66}$$

and where $\hat{\mu}_p$, $\hat{\varepsilon}_p$ are described through Equation 3.39. Finally, the z-directed *phase velocity* can be directly found using the definition

$$v_{p1}(\Omega) = \frac{1}{k_{z1}/\omega} \tag{3.67}$$

The following points pertain to discussion on comparison between the three velocities derived above:

1. All three velocities have explicit and implicit dependence on the carrier frequency ω_0. Implicit dependence is manifested through values of the permittivity and permeability, and their derivatives, at the carrier frequency.
2. In the nondispersive limit (primed quantities are zero), all three velocities are equal. Note that all three velocities depend on the values of the permittivity $\hat{\varepsilon}_{p0}$ and permeability $\hat{\mu}_{p0}$ centered at the carrier frequency, and hence they all implicitly depend on the carrier frequency ω_0.
3. The group velocity has an explicit dependence on the carrier frequency ω_0 even when excursions (Ω) around the carrier frequency are not present. The reason for this is as follows. Note that in expressing $k_{z1,2}$ in (3.46), we have incorporated the general frequency $\omega(= \omega_0 + \Omega)$ instead of simply the carrier ω_0, because taking the derivative of k_z w.r.t. ω en route to finding the group velocity using ω_0 instead of ω leads to a result that in the nondispersive limit does not match the expected classical result. On the other hand, restoring the term ω in k_z leads to the correct result for the nondispersive case. An interesting outcome of this approach is the persistence of ω_0 in the expression for group velocity under dispersion, as may be readily verified. This additional contribution is a result of the material dispersion of the permittivity and permeability in the neighborhood of the carrier ω_0, reflected through the operation of taking the partial derivative of the propagation constant w.r.t. the frequency ω.
4. Furthermore, the group velocity also has an explicit dependence on the carrier frequency ω_0 in the coefficient of the Ω term when excursions (Ω) around the carrier frequency are present. The reason is clear from the discussion in (2) above.
5. In the absence of modulation, both the energy and phase velocities are found to be explicitly independent of the carrier frequency ω_0 and are equal to each other. For the case of the phase velocity, nondependence on the carrier frequency ω_0 is due to the fact that the electric and magnetic field components, expressed in terms of $\hat{E}_{px}$, are independent of ω_0. For the case of the energy velocity, its nondependence on the carrier frequency ω_0 is due to the fact that the frequency variable, which appears both in the numerator and the denominator, simply drops out. The equality between the two velocities in the absence of modulation intuitively makes sense since in this case, the energy is transported essentially through the carrier alone, and hence the carrier propagation velocity (or phase velocity) must match the energy transport velocity.
6. In the presence of dispersion, and in the presence of modulation, all three velocities are, in principle, different from each other.

If the above analysis is performed for chiral materials, the calculations become much more involved. For instance, for the value of the propagation constant $k_{z3} = -\omega\hat{\kappa}\sqrt{\mu_0\varepsilon_0} + \omega\sqrt{\hat{\mu}\hat{\varepsilon}}$ from Equation 3.51c, the calculation of the average Poynting vector yields, after considerable algebra [9],

$$\hat{S}_{av3} \approx \left\{\frac{-A_{23} + B_{23}\Omega}{2}\right\}\Im_t[E_{px}^2(t)]\hat{a}_z \tag{3.68}$$

where

$$A_{23} = -\sqrt{\frac{\hat{\varepsilon}_{p0}}{\hat{\mu}_{p0}}}, \quad B_{23} = -\sqrt{\frac{\hat{\varepsilon}_{p0}}{\hat{\mu}_{p0}}}\left(\frac{\hat{\varepsilon}'_{p0}}{\hat{\varepsilon}_{p0}} - \frac{\hat{\mu}'_{p0}}{\hat{\mu}_{p0}}\right)\Big/2 \tag{3.69}$$

denote the first order expansion coefficients of $\hat{H}_{px}$ in terms of $\hat{E}_{px}$ pertaining to the above propagation constant.

An especially interesting situation arises when, for the simple case of no modulation ($\Omega = 0$, $\omega = \omega_0$), it is observed that the Poynting vector may be contradirectional to the propagation constant when $\hat{\kappa}_{p0} > \sqrt{\hat{\mu}_{p0}\hat{\varepsilon}_{p0}}\Big/\sqrt{\mu_0\varepsilon_0} = \sqrt{\hat{\mu}_{r0}\hat{\varepsilon}_{r0}}$ [4]. The Poynting vector is directed along the positive z-axis, while the propagation vector is directed along the negative z-axis. Contradirectionality of the Poynting and propagation vectors is one of the necessary conditions for wave propagation in a *negative index medium* (NIM), and will be examined in more detail in the following sections. It turns out, however, that it is difficult to find such materials [11]; however, the condition for contradirectionality of the group and phase velocities can be achieved over a range of frequencies, provided there is the right dispersion of the permittivity, permeability, and chirality [9]. Traditionally, however, NIMs are realizable through negative values of the real parts of the permittivity and permeability, which can occur in fabricated *metamaterials* over a certain frequency range.

3.6 METAMATERIALS AND NEGATIVE INDEX MEDIA

If the permittivity and permeability of a material have negative values (for their real parts), then the refractive index becomes negative [12]. This is because, simplistically speaking,

$$n \propto \varepsilon^{1/2}\mu^{1/2} = |\varepsilon|^{1/2}e^{j\pi/2}|\mu|^{1/2}e^{j\pi/2} = |\varepsilon|^{1/2}j|\mu|^{1/2}j = -|\varepsilon|^{1/2}|\mu|^{1/2} \tag{3.70}$$

However, as mentioned earlier, both ε and μ are in general (complex) functions of frequency, and can take negative values (for their real parts) over a certain frequency range. Thus the refractive index also has a negative value over a certain frequency range. It must be noted that the characteristic impedance of the material, even when the permittivity and permeability are negative and real, is positive and given by

$$\eta = \mu^{1/2}\varepsilon^{-1/2} = |\varepsilon|^{-1/2}e^{-j\pi/2}|\mu|^{1/2}e^{j\pi/2} = \frac{|\mu|^{1/2}}{|\varepsilon|^{1/2}} \tag{3.71}$$

A consequence of propagation in an NIM is the contradirection of phase and group velocities [12,13]. The usual reasoning for this is that the energy velocity, which is defined as the ratio of the Poynting vector and the stored energy, is also equal to the group velocity; thus the group velocity has the same sign as the Poynting vector. Hence, when there is opposition between the Poynting vector and the phase velocity, there should be contradirection between the group velocity and the phase velocity. The sign difference between the Poynting vector and the phase velocity comes from the fact that the latter depends on the refractive index, which is negative, while the former depends on the characteristic impedance, which is still positive.

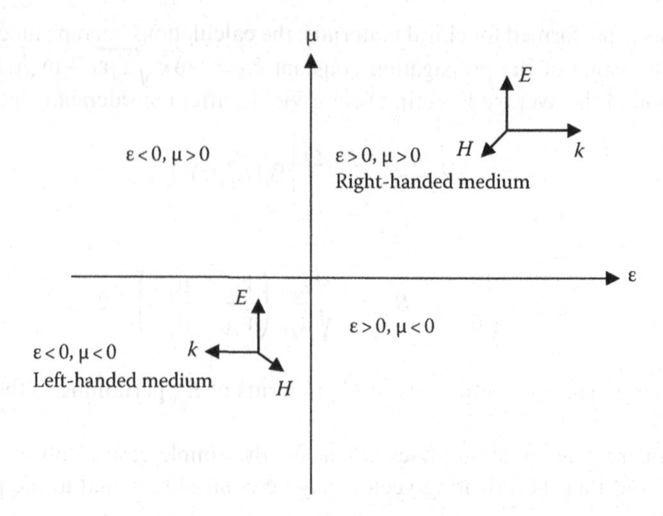

FIGURE 3.3 ε and μ space showing regions for right-handed and left-handed media.

From a more rigorous EM perspective, it can be argued from Maxwell's equations (3.1 through 3.4) along with the constitutive relations for the simple nondispersive case that in NIMs, the wave vector, the electric field, and the magnetic field form a left-handed system; hence the name *left-handed material* (LHM). Landau and Lifshitz [14] showed that the direction of the energy flow determined by the Poynting vector $\vec{S}$ is independent of the signs and values of ε and μ. The Poynting vector is always directed away from the source of radiation [15]. But the wave vector is directed toward the source for LHM and away from the source for a positive index material (PIM), also called *right-handed material* (RHM). We can say that the phase wave fronts run backward in LHM.

Figure 3.3 shows the ε and μ space [12]. Materials in the first quadrant with $\varepsilon > 0$ and $\mu > 0$ allow the propagation of wave and are called RHMs. Materials in the second and fourth quadrants have either $\varepsilon < 0$ or $\mu < 0$. EM waves cannot propagate inside them and evanescent waves will occur. For the third quadrant, $\varepsilon < 0$ and $\mu < 0$. LHMs with $\varepsilon < 0$ and $\mu < 0$, which fall in the third quadrant, also allow propagating waves.

One of the most interesting properties of LHMs is *negative refraction*. Negative refraction of a wave propagating from one medium to another is said to be observed when the wave in the second medium appears at the same side of the normal to the interface as in the first medium. Veselago [16] has shown that Snell's law remains valid, as long as we choose the negative sign for the index of refraction in LHMs. Figure 3.4 shows the refraction of EM plane wave incident from vacuum onto a denser EM medium with its real part of the effective refractive index being positive (a) or negative (b) [17]. Note that the angle of the reflected beam remains unaffected by the LHM.

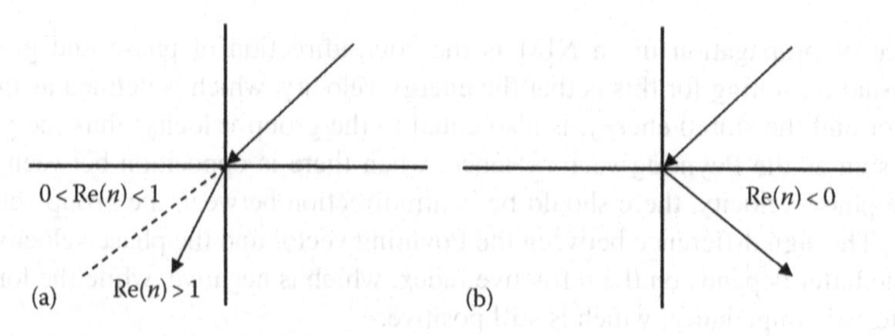

FIGURE 3.4 Refraction in (a) RHM and (b) LHM.

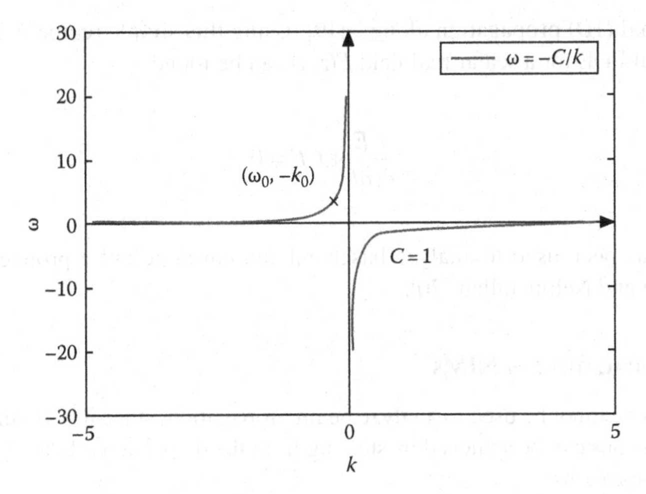

FIGURE 3.5 Dispersion relation of the form $\omega = -C/k$, where $C = 1$.

In what follows, we will describe continuous wave (CW) propagation in an NIM based on a hypothetical dispersion relation, which demonstrates positive group velocity and negative phase velocity, and is simple to analyze. The simplest model dispersion relation that can be used for an NIM can be expressed in the form

$$\omega = W(k) = -\frac{C}{k}, \quad C > 0 \tag{3.72}$$

which is an odd function of k, and is plotted schematically in Figure 3.5. For

$$\omega = \omega_0 > 0, \quad k = -k_0^- = -\frac{C}{\omega_0} < 0 \tag{3.73}$$

so that from (3.72) and (3.73),

$$v_p(\omega_0) = \left.\frac{\omega}{k}\right|_{\omega_0} = -\frac{\omega_0^2}{C} < 0 \tag{3.74}$$

while

$$v_g(\omega_0) = \left.\frac{d\omega}{dk}\right|_{\omega_0} = \left.\frac{C}{k^2}\right|_{\omega_0} = \frac{\omega_0^2}{C} > 0 \tag{3.75}$$

The negative superscript has been inserted on k_0 in Equation 3.73 to remind readers that it pertains to an NIM. The phase velocity is equal in magnitude and opposite in sign to the group velocity for the chosen dispersion relation. Note that the dispersion relation above is consistent with the propagation characteristics of a lumped circuit transmission line in the high pass configuration, where the series element is a capacitor instead of an inductor, and the shunt element is an inductor instead of a capacitor [18]. It has been shown that this model of the transmission line behaves as a waveguide filled with an NIM [12].

Given a certain dispersion relation, the corresponding partial differential equation (PDE) for a generic wavefunction $E(z, t)$ (which can represent, for instance, a component of the electric field in EM or optics) can be found by substituting ω and k by the operators

$$\omega \to -j\frac{\partial}{\partial t}, \quad k \to j\frac{\partial}{\partial z} \tag{3.76}$$

for one-dimensional (1D) propagation along z [19]. Using this simple recipe in Equation 3.72, the (1 + 1) dimensional PDE for a scalar real field $E(z, t)$ can be found as

$$\frac{\partial^2 E}{\partial z \partial t} + CE = 0 \tag{3.77}$$

The PDE (3.77) has been used to analyze baseband and envelope pulse propagation in NIMs, as shown in Banerjee and Nehmetallah [20].

3.6.1 Beam Propagation in NIMs

Equation 3.77 above cannot be used to analyze beam propagation, since it is 1D in space. The extension to multidimensions can be achieved by starting from the dispersion relation (3.72) and rewriting the propagation constant as

$$k = \sqrt{k_x^2 + k_y^2 + k_z^2} \approx k_z + \frac{1}{2}\left(k_x^2 + k_y^2\right)k_z^{-1} \equiv -C\omega^{-1} \tag{3.78}$$

or, equivalently,

$$\omega\left[k_z^2 + \frac{1}{2}\left(k_x^2 + k_y^2\right)\right] \equiv -Ck_z \tag{3.79}$$

Now replacing ω, k_x, k_y, k_z by their respective operators [19],

$$\omega \to -j\frac{\partial}{\partial t}, \quad k_x \to j\frac{\partial}{\partial x}, \quad k_y \to j\frac{\partial}{\partial y}, \quad k_z \to j\frac{\partial}{\partial z} \tag{3.80}$$

and operating on the wavefunction $E(x, y, z, t)$, we derive the PDE [20]

$$\frac{\partial^3 E}{\partial z^2 \partial t} + \frac{1}{2}\frac{\partial^3 E}{\partial x^2 \partial t} + \frac{1}{2}\frac{\partial^3 E}{\partial y^2 \partial t} + C\frac{\partial E}{\partial z} = 0 \tag{3.81}$$

Now, writing

$$E(x, y, z, t) = \mathrm{Re}\left\{E_e(x, y, z)\exp\left[j\left(\omega_0 t + k_0^- z\right)\right]\right\}, \quad k_0^- = \frac{C}{\omega_0} \tag{3.82}$$

in (3.81) and equating the coefficients of $\exp\left[j(\omega_0 t + k_0^- z)\right]$, we obtain the paraxial wave equation for beam propagation in a negative index medium as [20]

$$\frac{\partial E_e}{\partial z} = j\frac{\omega_0}{2C}\left(\frac{\partial^2 E_e}{\partial x^2} + \frac{\partial^2 E_e}{\partial y^2}\right) \tag{3.83}$$

Proceeding in exactly the same way as that for propagation in a conventional *positive index medium* (PIM), one can define the *transfer function for propagation* for NIM by taking the Fourier transform of (3.84) w.r.t. x, y and solving the ordinary differential equation for the spectrum

$$\tilde{E}_e(k_x,k_y,z) = \Im_{x,y}\{E_e(x,y,z)\} = \int_{-\infty}^{\infty}\int_{-\infty}^{\infty} E_e(x,y,z)\exp[j(k_x x + k_y y)]dx\,dy \tag{3.84}$$

This yields

$$H_{-}(k_x,k_y,z) = \frac{\tilde{E}_e(k_x,k_y,z)}{\tilde{E}_e(k_x,k_y,z=0)} = \exp\left[-j\left(\frac{\omega_0}{2C}\right)\left(k_x^2+k_y^2\right)z\right] \tag{3.85}$$

which is in agreement with the findings of Tassin et al. [15].

We would like to point out that the important difference between the transfer function for propagation in an NIM as compared to a conventional medium is that the sign on the argument of the exponential is negative for the case of the NIM. Recall that for a conventional positive index material (see Equation 2.8),

$$H(k_x,k_y,z) = \frac{\tilde{E}_e(k_x,k_y,z)}{\tilde{E}_e(k_x,k_y,z=0)} = \exp\left[j\frac{(k_x^2+k_y^2)z}{2k_0}\right] \tag{3.86}$$

where k_0 denotes the propagation constant in a PIM. The corresponding *impulse response* for propagation in an NIM is given by the inverse transform of Equation 3.85:

$$h_{-}(x,y,z) = -j\frac{C}{2\pi\omega_0 z}\exp\left[j\frac{C}{2\omega_0 z}(x^2+y^2)\right] \tag{3.87}$$

It is interesting to examine the implications of the negative sign in the argument of the exponential of the transfer function for the NIM. Note that for a positive index material, the transfer function (3.86) can be used to explain the spreading of a Gaussian beam due to diffraction. As shown in Chapter 2, a Gaussian beam initially of waist w_0 acquires positive (diverging) phase curvature and spreads after traveling a distance z_0 in a conventional PIM. The complex envelope of the Gaussian can be expressed as (Equation 2.18)

$$E_e = j\frac{k_0 w_0^2}{2q(z_0)}E_0\exp\left[\frac{-jk_0(x^2+y^2)}{2q(z_0)}\right],\quad q(z_0) = z_0 + j\frac{k_0 w_0^2}{2} \tag{3.88}$$

and can be taken as the general expression for a Gaussian with increasing width and diverging wave fronts. If such a complex Gaussian is introduced into an NIM at $z = 0$, we can use the transfer function to determine its shape after propagation through an arbitrary distance z in the medium. Note that corresponding to (3.88),

$$\tilde{E}_e(k_x,k_y,z=0) = \pi w_0^2 E_0\exp\left[j\frac{\left(k_x^2+k_y^2\right)q(z_0)}{2k_0}\right] \tag{3.89}$$

Using (3.85) and the definition of $q(z_0)$ from (3.88), we find that after a distance of propagation

$$z = z_f = \frac{Cz_0}{\omega_0 k_0} \tag{3.90}$$

the spectrum of the beam will be purely real, which corresponds to the spectrum of a Gaussian with plane wave fronts and of waist $w_f = w_0$, under paraxial approximation, indicating focusing of the Gaussian beam. This proves that the NIM can act as a focusing medium for a diverging Gaussian beam. The wave picture developed here can be corroborated with a corresponding ray matrix approach to Gaussian beam propagation in a negative index medium, as developed in Banerjee and Nehmetallah [20].

Alternatively, consider a Gaussian with initially plane wave fronts at $z = 0$ in an NIM:

$$E_e(x, y, z = 0) = \exp\left[-\frac{(x^2 + y^2)}{w_0^2}\right] \tag{3.91}$$

Then, $\tilde{E}_e(k_x, k_y, z = 0) = \pi w_0^2 \exp\left[-\left(k_x^2 + k_y^2\right) w_0^2/4\right]$ and using (3.85),

$$\tilde{E}_e(k_x, k_y, z) = \pi w_0^2 \exp\left[-\frac{j\left(k_x^2 + k_y^2\right) q_- \omega_0}{2C}\right] \tag{3.92}$$

where we have defined a q-parameter of the Gaussian beam propagating in an NIM as

$$q_-(z) = z - j\frac{Cw_0^2}{2\omega_0} \tag{3.93}$$

Note that the definition of the q of the Gaussian beam in the negative index medium is consistent with the usual definition of q in a positive index medium. In the latter case, the q is defined (see Equation 3.88) so as to have the expression for the spectrum of the Gaussian to be of the same form as that of the transfer function for propagation, with z replaced by q. Comparison of Equation 3.92 with Equation 3.85 shows that the same is true for the negative index medium with q defined as in (3.93). Furthermore, as in the positive index medium, the second term in Equation 3.93 contains the information about the propagation constant in the medium, which is negative in the case of a negative index medium. The inverse transform of Equation 3.92 yields, analogous to Equation 3.88,

$$E_e(x, y, z) = -j\frac{Cw_0^2}{2\omega_0 q_-} \exp\left[j\frac{C}{2\omega_0 q_-}(x^2 + y^2)\right] \tag{3.94}$$

Upon substituting Equation 3.93 into Equation 3.94 and separating into magnitude and phase, it is possible to determine variation of the width and radius of curvature of the Gaussian during propagation in the negative index medium, along with the on-axis amplitude and the Guoy phase. In fact, the exponential part in Equation 3.94 gives the information about the width and radius of curvature, and can be re-expressed as

$$\exp\left[j\frac{C}{2\omega_0 q_-}(x^2 + y^2)\right] = \exp\left\{\frac{-C^2 w_0^2/2\omega_0^2}{2\left[z^2 + \left(Cw_0^2/2\omega_0\right)^2\right]}(x^2 + y^2)\right\} \exp\left\{j\frac{(C/\omega_0)z}{2\left[z^2 + \left(Cw_0^2/2\omega_0\right)^2\right]}(x^2 + y^2)\right\} \tag{3.95}$$

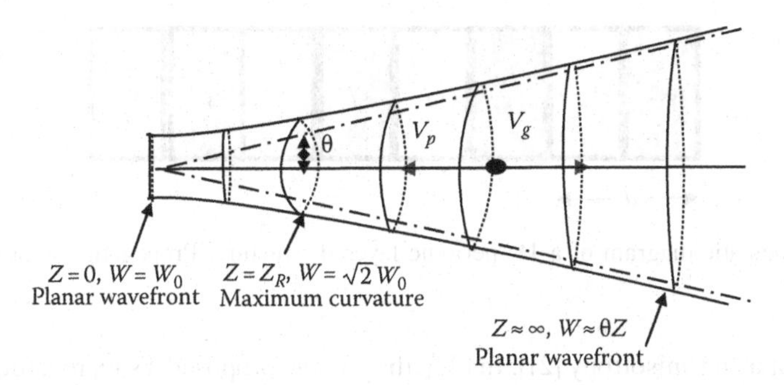

FIGURE 3.6 Changes in wave front radius with propagation distance. Solid line is when $n < 0$, dotted line is when $n > 0$.

Note that the width $w(z)$ monotonically increases in this case, just like propagation in a conventional positive index medium. The radius of curvature $R(z)$, on the other hand, has a sign opposite to that for a positive index medium. This means that forward propagation gives converging wave fronts. With propagation, the phase fronts move from the right to the left, as shown in Figure 3.6. The variations of $w(z)$ and $R(z)$ are given as

$$w^2(z) = \frac{2\left[z^2 + \left(Cw_0^2/2\omega_0\right)^2\right]}{C^2 w_0^2/2\omega_0^2}, \quad R(z) = -\frac{\left[z^2 + \left(Cw_0^2/2\omega_0\right)^2\right]}{z} \tag{3.96}$$

3.7 PROPAGATION THROUGH PHOTONIC BAND GAP STRUCTURES: THE TRANSFER MATRIX METHOD

In Chapter 2, we have examined forward and backward propagations of light due to induced reflection gratings in nonlinear materials, such as plasmon resonance (PR) materials. In particular, we have seen how two-wave mixing is possible due to induced reflection gratings in these materials. The key point in two-wave mixing is that for the induced reflection grating, the grating period is exactly one half of the wavelength of the contrapropagating light waves in the medium. Another way of stating this is to say that the grating vector length is twice the propagation constant of the light in the medium. In this case, there is perfect phase matching between the forward and the backward traveling waves, which gives rise to substantial reflected light.

Now consider the following scenario: assume that the grating is built into the material a priori, i.e., it is not induced due to the material response. In this case, one can have the liberty of making gratings of any arbitrary period or wavevector in the material. Hence, if the fundamental spatial frequency of the grating is exactly equal to twice the propagation constant of the contrapropagating light waves, one can expect maximum reflection of an incident light wave. However, if this is not the case, the interaction is no longer phase matched and the reflection is not as high. In other words, given a certain grating period, there is a band of frequencies for which appreciable amount of light is reflected back, leading to minimal transmission. This implies that if the frequency of light is changed, there may be a certain stop band, or *band gap*, in the transmission. Such periodic structures are referred to as *photonic crystals* or *photonic band gap structures*.

Photonic crystals consist of periodically modulated dielectric constants, as shown in Figure 3.7. Photonic crystal structures may be designed to possess interesting optical properties, particularly when the photonic band structure of the material is highly anisotropic [21]. It has been shown that diffraction effects of photonic crystals can produce negative refraction [22]. There are two kinds of negative refraction possible in photonic crystals. One is from the left-handed behavior described above. The other negative refraction is realized without negative index but by high-order Bragg scattering or anisotropy. Xiao has proposed a structure consisting of concentric silicon photonic crystal

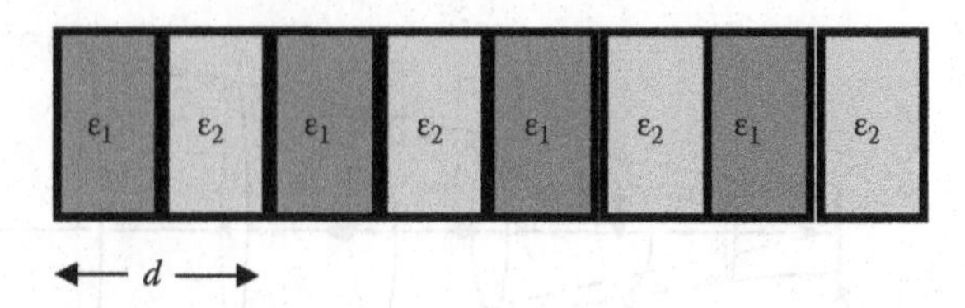

FIGURE 3.7 Schematic diagram of a 1D periodic layered structure. Propagation is along the horizontal direction ($\pm z$).

layer to bend light using anisotropy [21]. In [22], the authors proposed a square lattice of Si cylinders surrounded by nonlinear liquid crystal (NLC) to achieve negative refraction. Applying an external field allows one to change the refracted angle.

The standard approach used for analyzing stacks or layers of various refractive indices is the *transfer matrix method* (TMM). We first summarize the derivation of the transfer matrix from the first principles using Maxwell's equations and the constitutive equations:

$$\vec{\nabla} \times \vec{E} = \frac{\partial \vec{B}}{\partial t} \tag{3.97a}$$

$$\vec{\nabla} \times \vec{H} = \frac{\partial \vec{D}}{\partial t} \tag{3.97b}$$

$$\hat{\vec{D}} = \hat{\varepsilon}\hat{\vec{E}} \tag{3.97c}$$

$$\hat{\vec{B}} = \hat{\mu}\hat{\vec{H}} \tag{3.97d}$$

Consider a CW EM wave at frequency ω_0 traveling at an angle θ w.r.t. the z-axis (horizontal direction) in a layer in the periodical layered structure as shown in Figure 3.8. Suppose that the wave vector $\vec{k}(\omega)$ lies in the y–z plane, where y is the vertical direction. For the TE case, assume that the electric field $\vec{E} = \mathrm{Re}[\vec{E}_p e^{j\omega_0 t}]$ is in the x direction (out of the plane of the paper), and hence $E_{py} = E_{pz} = 0$. Using (3.97a), it follows that $H_{px} = 0$. Also,

$$\frac{\partial H_{pz}}{\partial y} - \frac{\partial H_{py}}{\partial z} = j\omega_0 \varepsilon E_{px} \tag{3.98a}$$

$$\frac{\partial E_{px}}{\partial z} + j\omega_0 \mu H_{py} = 0 \tag{3.98b}$$

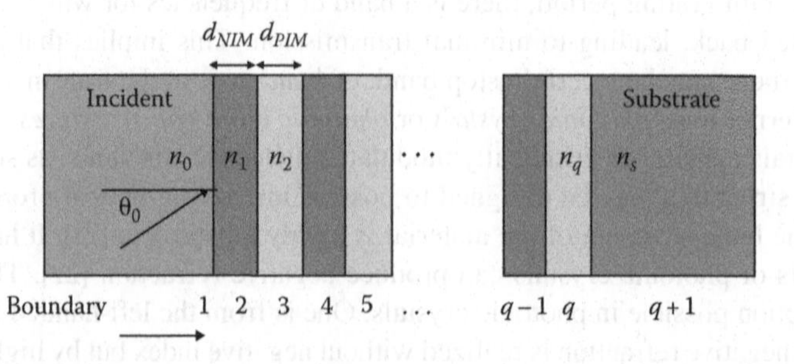

FIGURE 3.8 Schematic of 1D periodic structure composed of alternating refractive index materials.

$$\frac{\partial E_{px}}{\partial y} - j\omega_0 \mu H_{pz} = 0 \tag{3.98c}$$

In Equations 3.98a–c, ε, μ refer to values of the permittivity and permeability at the frequency ω_0. Eliminating H_{py}, H_{pz}, we get from (3.98a–c)

$$\frac{\partial^2 E_{px}}{\partial y^2} + \frac{\partial^2 E_{px}}{\partial z^2} + \varepsilon_r \mu_r k_0^2 E_{px} = 0 \tag{3.99}$$

where we assumed that $d\mu/dz = 0$ and $d\varepsilon/dz = 0$ (true within each layer) and where k_0 is the propagation constant in vacuum. Substituting $E_{px} = E_{px}(z)\exp - j(k_0 \alpha y)$ in (3.99) with $\alpha^2 = \left(\varepsilon_r \mu_r / k_0^2\right)\sin^2\theta$, we get

$$\frac{d^2 E_{px}(z)}{dz^2} + k_0^2 \varepsilon_r \mu_r \cos^2\theta E_{px}(z) = 0 \tag{3.100}$$

In a similar way, by writing $H_{py} = H_{py}(z)\exp - j(k_0 \alpha y)$, solving (3.100) and using (3.98b),

$$E_{px}(z) = A\cos(\beta z) + B\sin(\beta z),$$

$$H_{py}(z) = -\frac{\beta}{j\omega_0 \mu_0 \mu_r}\left[B\cos(\beta z) - A\sin(\beta z)\right] \tag{3.101}$$

where $\beta = k_0\sqrt{\varepsilon_r}\sqrt{\mu_r}\cos(\theta)$. Now, upon replacing z by $z + \Delta z$ in (3.101) and after simple algebra,

$$E_{px}(z+\Delta z) = \cos(\beta\Delta z)E_{px}(z) - \frac{j\omega_0 \mu_0 \mu_r}{\beta}\sin(\beta\Delta z)H_{py}(z)$$

$$H_{py}(z+\Delta z) = \frac{\beta}{j\omega_0 \mu_0 \mu_r}\sin(\beta\Delta z)E_{px}(z) + \cos(\beta\Delta z)H_{py}(z) \tag{3.102}$$

Equivalently, Equations 3.102 can be written as

$$\begin{bmatrix} E_{px}(z+\Delta z) \\ H_{py}(z+\Delta z) \end{bmatrix} = M \begin{bmatrix} E_{px}(z) \\ H_{py}(z) \end{bmatrix} \tag{3.103}$$

with

$$M = \begin{pmatrix} \cos(\beta\Delta z) & -\frac{j}{P}\sin(\beta\Delta z) \\ -jP\sin(\beta\Delta z) & \cos(\beta\Delta z) \end{pmatrix} \tag{3.104}$$

where M is called the *transfer matrix*. In Equation 3.104, $P = (\beta/\omega_0 \mu_0 \mu_r) = (\cos\theta/\eta_0)\left(1/\left(\sqrt{\mu_r}/\sqrt{\varepsilon_r}\right)\right) = (\cos\theta/\eta)$, ε_r is the relative permittivity, μ_r is the relative permeability which can be the function of frequency, Δz is the layer width d_i, $\eta = \eta_0\left(\sqrt{\mu_r}/\sqrt{\varepsilon_r}\right)$, and $\eta_0 = \sqrt{\mu_0/\varepsilon_0}$ is the impedance of vacuum.

In general, for the multilayered structure shown in Figure 3.8, it can be shown that the tangential component of E and H at the rth interface is related to the tangential component of E and H of the $(m + 1)$th interface through the transfer matrix M above, and can be rewritten in the form

$$M_m = \begin{pmatrix} \cos(\delta_m) & -j\eta_m \sin(\delta_m) \\ \dfrac{-j}{\eta_m}\sin(\delta_m) & \cos(\delta_m) \end{pmatrix} \tag{3.105}$$

where

$\delta_m = (2\pi n_m d_m (\cos\theta_m)/\lambda$

$n_m = \varepsilon_{rm}^{1/2}\mu_{rm}^{1/2}$

$\eta_m = (\eta_0 \mu_{rm}^{1/2}\varepsilon_{rm}^{-1/2})/\cos\theta_m$ for s-polarized (TE) waves and $\eta_m = \eta_0\mu_{rm}^{1/2}\varepsilon_m^{-1/2}\cos\theta_m$ for p-polarized (transverse magnetic [TM]) waves

θ_0 is the angle of incidence

θ_m is given by Snell's law $n_0 \sin\theta_0 = n_m \sin\theta_m = n_s \sin\theta_s$

Then for the case of q layers, the characteristic matrix is simply the product of the individual matrices

$$\begin{bmatrix} B \\ C \end{bmatrix} = \left\{ \prod_{m=1}^{q} \begin{pmatrix} \cos\delta_m & j\eta_m \sin\delta_m \\ \dfrac{j}{\eta_m}\sin\delta_m & \cos\delta_m \end{pmatrix} \right\} \begin{bmatrix} 1 \\ \dfrac{1}{\eta_s} \end{bmatrix} \tag{3.106}$$

where $B = E_1/E_{q+1}$, $C = H_1/E_{q+1}$. The net irradiance at the exit of the assembly (entering the substrate) is given by $I_q = (1/2)\mathrm{Re}(1/\eta_s^*)E_qE_q^*$, and that at the entrance to the assembly (immediately inside layer 1) is given by $I_1 = (1/2)\mathrm{Re}(BC^*)E_qE_q^*$. Let the incident irradiance be denoted by I_i, then $I_1 = (1-R)I_i = (1/2)\mathrm{Re}(BC^*)E_qE_q^*$. Hence

$$I_i = \frac{\mathrm{Re}(BC^*)E_qE_q^*}{2(1-R)} \tag{3.107}$$

where

$$R = \left(\frac{B-\eta_0 C}{B+\eta_0 C}\right)\left(\frac{B-\eta_0 C}{B+\eta_0 C}\right)^* \tag{3.108}$$

The transmittance T into the substrate is defined as

$$T = \frac{I_q}{I_i} = \frac{\mathrm{Re}(1/\eta_s)(1-R)}{\mathrm{Re}(BC^*)} = \frac{4\eta_0\,\mathrm{Re}(1/\eta_s)}{(B+\eta_0 C)(B+\eta_0 C)^*} \tag{3.109}$$

The absorbance A in the multilayered structure is related to R and T by $R + T + A = 1$. Hence the absorbance is written as

$$A = (1-R)\left[1 - \frac{\mathrm{Re}(1/\eta_s)}{\mathrm{Re}(BC^*)}\right] = \frac{4\eta_0\,\mathrm{Re}(BC^* - 1/\eta_s)}{(B+\eta_0 C)(B+\eta_0 C)^*} \tag{3.110}$$

3.7.1 Periodic PIM–NIM Structures

As an example, we assume a multilayer structure composed of seven periods of NIM/PIM. It has been shown that in a 1D stack of alternating PIM and NIM layers, a volume averaged effective refractive index ($<n>$) equal to zero gives rise to a new type of band gap [23]. Commonly called the *zero-*$<n>$ *band gap*, it has different properties than the conventional *Bragg band gap*. For instance, unlike the conventional Bragg band gap where the center frequency is a function of the lattice constant or periodicity of the structure, the center frequency for the zero-$< n >$ band gap remains unchanged when the layer lengths of the PIM and the NIM are scaled by the same factor. The simple reason for this is that if a material has a net zero refractive index, the reflection coefficient is equal to ±1, irrespective of the refractive index n_0 of the incident medium. The zero-$< n >$ gap has been demonstrated experimentally by Yuan et al. [24]. The transmittivity of the stack has been measured and the existence of zero-$< n >$ gap has been confirmed.

For simplicity, we choose the PIM to be a nondispersive dielectric. The index of refraction of PIM is taken to be 1.5. Assume that the dispersion properties of the NIM are described by *Lorentz characteristics* for both $\hat{\varepsilon}_r$ and $\hat{\mu}_r$ as

$$\hat{\varepsilon}_r(\omega) = 1 + \frac{\omega_{pe}^2}{\omega_{1e}^2 - \omega^2 + j\gamma_e\omega} \tag{3.111}$$

$$\hat{\mu}_r(\omega) = 1 + \frac{\omega_{pm}^2}{\omega_{1m}^2 - \omega^2 + j\gamma_m\omega} \tag{3.112}$$

where

ω_{pe} is the electric plasma frequency
ω_{1e} is the electric resonance frequency
γ_e is the electric damping constant
ω_{pm} is the magnetic plasma frequency
ω_{1m} is the magnetic resonance frequency
γ_m is the magnetic damping constant [25]

For this example, we take $\omega_{pe} = 1.1543 \times 10^{11}$ rad/s, $\omega_{1e} = \omega_{1m} = 2\pi \times 5 \times 10^6$ rad/s, $\omega_{pm} = 1.6324 \times 10^{11}$ rad/s, and $\gamma_e = 2 \times \gamma_m = 2\pi \times 6 \times 10^6$ rad/s [25]. The real parts of $\hat{\varepsilon}_r$ and $\hat{\mu}_r$ are negative for frequencies $f < 18.5$ GHz. Assume that the thickness of each "cell" composed of an NIM layer and a PIM layer is $d = d_{PIM} + d_{NIM} = 0.01$ m. Figure 3.9 shows the transmittance and reflectance for a multilayer structure composed of 14 layers (7 cells), where $d_{PIM} = d_{NIM} = 0.005$ m. Normal incidence is assumed in all simulations in this example. The zero-$< n >$ band gap is approximately between 12.25 and 15.5 GHz, while a neighboring Bragg band gap is approximately between 8 and 10 GHz. Figure 3.10 shows the variation of the transmittance with frequency for three values of the "duty cycle" of the NIM layer ($D = d_{NIM}/d$). For different duty cycles, the zero-$< n >$ band gap occurs around different center frequencies with all other parameters kept unchanged. For $D = 0.5$, the center frequency is 13.75 GHz. For smaller duty cycle ($D = 0.3$), the center frequency for the zero-$< n >$ band gap is lower, viz., at 10.2 GHz, while for a higher duty cycle ($D = 0.7$), the center frequency is 16.32 GHz. This is because for lower (higher) frequency, the refractive index of the NIM, calculated from the permittivity and permeability relations, is more (less) negative; hence, a smaller (larger) layer thickness d_{NIM} is needed to attain zero-$< n >$. We remark that other band gaps seen in Figure 3.10 for $D = 0.3$, 0.7 correspond to Bragg gaps.

3.7.2 EM Propagation in Complex Structures

Metals are often regarded as an excellent source of relatively free electrons. When EM radiation is incident on a metal, the electrons oscillate at a characteristic frequency called the *plasma frequency* ω_p. Light of frequency below the plasma frequency is reflected, while that above the

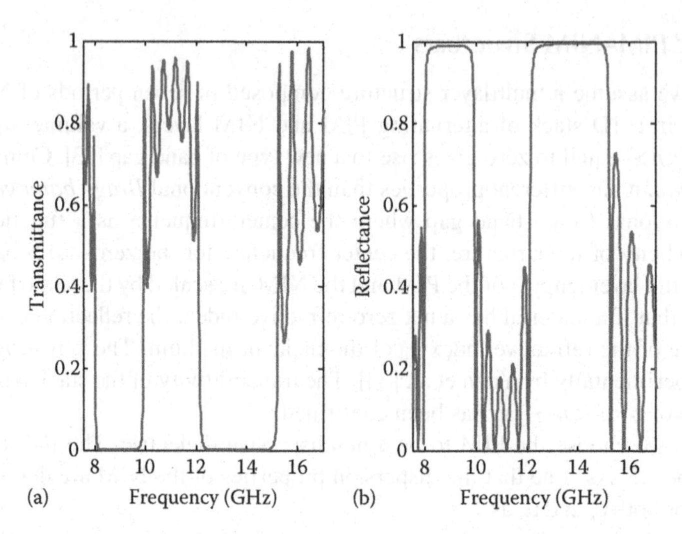

FIGURE 3.9 (a) Transmittance and (b) reflectance of a 14 layers stack with $n_{PIM} = 1.5$, n_{NIM} given by Equations 3.111 and 3.112, where $\omega_{pe} = 1.1543 \times 10^{11}$ rad/s, $\omega_{1e} = \omega_{1m} = 2\pi \times 5 \times 10^6$ rad/s, $\omega_{pm} = 1.6324 \times 10^{11}$ rad/s, $\gamma_e = 2 \times \gamma_m = 2\pi \times 6 \times 10^6$ rad/s, and $d_{PIM} = d_{NIM} = 0.005$ m. Normal incidence is assumed. The Bragg gap is between 8 and 10 GHz, while the zero-$< n >$ band gap is approximately between 12.25 and 15.5 GHz. (From Aylo, R. et al., *SPIE* 7604, 760412-1, 2010. With permission.)

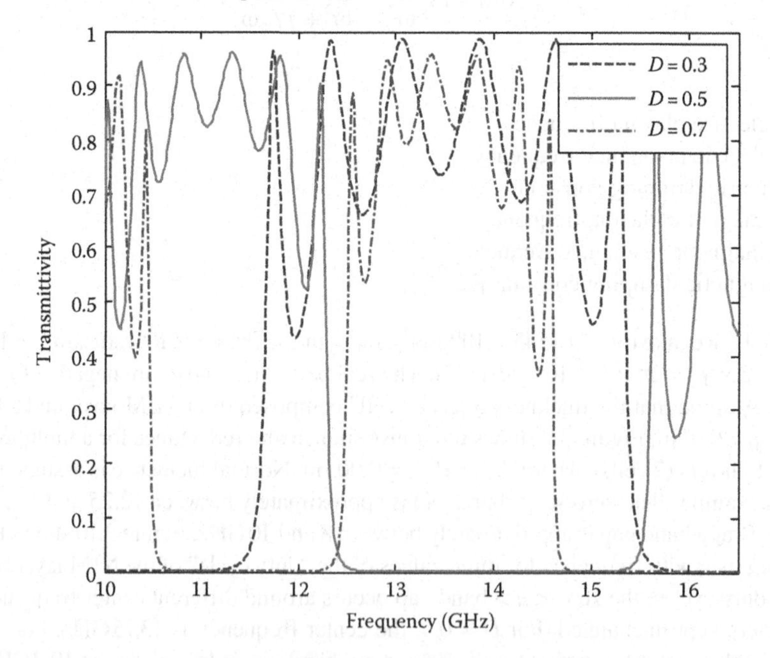

FIGURE 3.10 Variation of transmittance with frequency for three different values of NIM layer thickness fraction. All other parameters are as in Figure 3.9. (From Aylo, R. et al., *SPIE* 7604, 760412-1, 2010. With permission.)

plasma frequency is transmitted. *Surface plasmons* occur at the interface of a material with a positive permittivity and a material with a negative permittivity, such as a metal. The standard model for permittivity for a metal is the Drude model given as [2]

$$\hat{\varepsilon}_m = \varepsilon_0\left(1 - \frac{\omega_p^2}{\omega^2 - j\omega\gamma}\right) \tag{3.113}$$

which is similar to the dispersion relation given in (3.111).

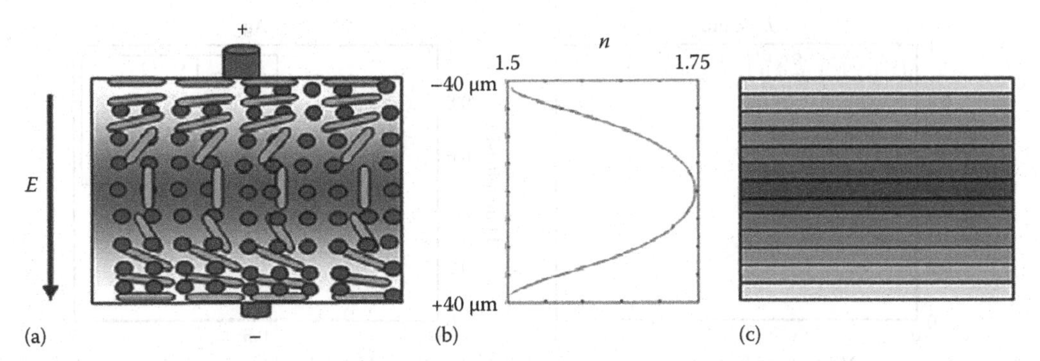

FIGURE 3.11 (a) Electric field applied on a NDLCC—the bars represent the directors in the LC, which has approximately a parabolic orientation profile; (b) index profile of the LC due to the electric field; (c) equivalent setup showing a layered stack with each layer's index of refraction according to the profile in (b).

In recent years, *plasmonics* has been used in a variety of sensor applications (see, for instance, Dhawan et al. [27]). Many of these utilize periodic arrays of apertures with subwavelength dimensions and periodicity [27]. Others use fiber-optic *surface plasmon resonance* (SPR) sensors of different chemical, physical, and biochemical parameters [28]. Localized SPRs of metallic nanoparticles such as gold, silver, copper, or other suitable nanoshells with appropriate sensory coatings have also been employed for sensing of chemical and biological molecules [29]. There are reported operations that use nanoparticles, in particular those dispersed, for instance, in cholesteric liquid crystals (LCs) [30].

In what follows, we will show only the results of transmission through a complex structure comprising a nematic LC dispersed with plasmonic nanoparticles, as shown in Figure 3.11a, to bring out the power of the TMM. Since the index of refraction profile of the virgin LC with different voltages applied across the LC structure changes along the propagation direction as shown in Figure 3.11b, we can consider the setup to be equivalent to a layered stack as shown in Figure 3.11c. In the presence of nanoparticles, each layer in the stack, therefore, has different refractive indices for the host in which these nanoparticles are embedded. The presence of nanoparticles in each layer further modifies the refractive index, which can be determined using a rather complicated theory of Mie scattering. This leads to what is termed different *effective* refractive indices for each layer. The overall transmission is thus determined by the change in these effective refractive indices, which is also therefore a function of the applied voltage. If an EM wave is applied on the nanoparticle dispersed LC cell (NDLCC) with center frequency around the PR of the nanoparticles, the effective index of refraction due to the PR varies significantly in that region, hence resulting in a noticeable change in transmittivity for two different electrostatic fields as compared to the case when there are no nanoparticles.

Figure 3.12a shows the transmittivities of a LC cell without nanoparticles at visible frequencies for two different electrostatic fields $V = 1$ V and $V = 2$ V. Figure 3.12b shows the transmittivities of an LC cell with Ag nanoparticles at visible frequencies for two different electrostatic fields $V = 1$ V and $V = 2$ V. Figure 3.12c shows the transmittivities of an LC cell without nanoparticles at far infrared (FIR) frequencies for two different electrostatic fields $V = 1$ V and $V = 2$ V. Finally, Figure 3.12d shows the transmittivities of an LC cell with Ge nanoparticles at FIR frequencies for two different electrostatic fields $V = 1$ V and $V = 2$ V. The quasiperiodic behavior of Figure 3.12a and c is due to the fact that the optical path length of the LC in terms of wavelength changes with frequency. However, the variation in Figure 3.12b and d is due to the existence of a zero transmission band gap, similar to that observed in layered positive index–negative index media, as illustrated in the previous example.

The calculations have been performed using finite element method (FEM) in COMSOL® to model the LC equations to derive the refractive index variation, MATLAB® to determine the effective refractive indices of each layer based on Mie scattering, and RSoft to implement the TMM approach and determine the overall transmission graphs as a function of frequency.

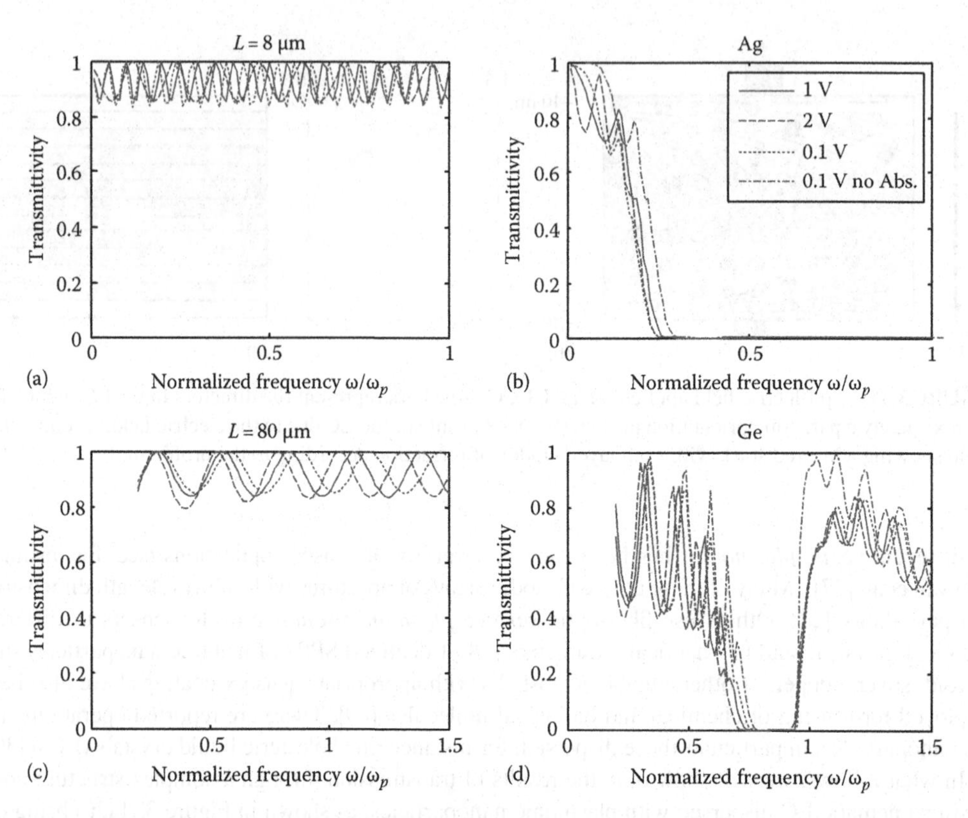

FIGURE 3.12 Transmittivity of a LC cell for two different electrostatic fields $V = 1$ V (solid line) and $V = 2$ V (dashed line), along with $V = 0.1$ V (dotted line) and $V = 0.1$ V neglecting absorption (dash-dot line): (a) without nanoparticles at optical frequencies; (b) with Ag nanoparticles at visible frequencies; $\varepsilon_{rLC} \approx 1$, filling fraction $f = 0.1$, nanoparticle radius $r = 20 \times 10^{-9}$ m, $\omega_p = 2\pi \times 2.18 \times 10^{15}$ rad/s, $\gamma = 2\pi \times 4.35 \times 10^{12}$/s, $\mu_r = 1$; (c) without nanoparticles at FIR frequencies; and (d) with Ge nanoparticles at FIR frequencies; $\varepsilon_r = 15.8$, $f = 0.38$, $r = 2.25 \times 10^{-6}$, $\omega_p = 26.7 \times 10^{12}$, $\varepsilon_r = 15.8$, $\gamma = \omega_p/100$, and $\mu = 1$. LC cell thickness is 8 μm for optical frequencies, and 80 μm for FIR.

PROBLEMS

3.1 For time harmonic uniform plane waves in a linear isotropic homogeneous medium, the $\vec{E}, \vec{H}$ fields vary according to $\exp j(\omega_0 t - \vec{k}_0 \cdot \vec{R})$. Show that in this case, Maxwell's equations can be expressed as $\vec{k}_0 \cdot \vec{E} = 0$; $\vec{k}_0 \cdot \vec{H} = 0$; $\vec{k}_0 \times \vec{E} = \omega_0 \mu \vec{H}$; $\vec{k}_0 \times \vec{H} = -\omega_0 \varepsilon \vec{E}$.

3.2 Assume that a function $f(t)$, which is causal, has a Fourier transform $\hat{f}(\omega) = \hat{f}_r(\omega) - j\hat{f}_i(\omega)$. Show that the real and imaginary parts of the Fourier transform are related through

$$\hat{f}_r(\omega) = \frac{1}{\pi}\int_{-\infty}^{\infty} \frac{\hat{f}_i(\omega')}{\omega - \omega'} d\omega'; \quad \hat{f}_i(\omega) = \frac{1}{\pi}\int_{-\infty}^{\infty} \frac{\hat{f}_r(\omega')}{\omega - \omega'} d\omega'$$

Such integrals are called the Hilbert transform, and the two equations above are called the Kramers–Kronig relationship.

3.3 Assume that a complex propagation constant is given by

$$k_{\pm}(\omega) = k_{r\pm}(\omega) - jk_{i\pm}(\omega) \equiv \pm\left(\frac{\omega_0}{v}\right)\left[\frac{(\omega/\omega_0)}{1+(\omega/\omega_0)^2} + j\frac{1}{1+(\omega/\omega_0)^2}\right]$$

(a) Find $\breve{k}_{\pm}(t) \equiv \Im_t^{-1}\left[k_{\pm}(\omega)\right]$.

(b) Using the dispersion relation, derive the PDE for the wave propagation using the operator formalism for k and ω.

3.4 Show that the real and imaginary parts of the electric susceptibility (assuming nonmagnetic material) in terms of real and imaginary parts of the propagation constant are given by $\chi_r(\omega) = \left(k_r^2 - k_i^2\right)(c/\omega)^2 - 1$, $\chi_i(\omega) = (2k_r k_i)(c/\omega)^2$, where c denotes the velocity of light in free space. Hence find $\chi_{r,i}(\omega)$ for the dispersion relation given in Problem 3.3.

3.5 Find the phase, group, and energy velocities for the case of a nonmagnetic material where the permittivity is given by the Drude model

$$\frac{\hat{\varepsilon}(\omega)}{\varepsilon_0} = 1 - \frac{\omega_p^2}{\omega(\omega + j\gamma)} = 1 - \frac{\omega_p^2}{\omega^2 + \gamma^2} + j\frac{\gamma\omega_p^2}{\omega(\omega^2 + \gamma^2)} = \hat{\varepsilon}_r - j\hat{\varepsilon}_i$$

What is the significance of the complex values for the velocities?

3.6 Show that the transmission matrix relating the position and angle of the optical ray in a negative index medium is given by $T = \begin{pmatrix} 1 & z \\ 0 & 1 \end{pmatrix}$ while the refraction matrix from air to a negative index medium is given by

$$\tilde{R} = \begin{pmatrix} 1 & 0 \\ -\dfrac{|n|+1}{|n|}\dfrac{1}{R} & -\dfrac{1}{|n|} \end{pmatrix}$$

where R is the radius of curvature of the interface.

3.7 The q-parameter of a Gaussian beam (see Chapter 2) transforms according to the relation $q' = (Aq + B)/(Cq + D)$ in terms of the ABCD parameters of the propagating medium. A Gaussian beam with plane wave fronts and waist w_0 propagates a distance z_0 to the plane interface between air and a negative index medium. Determine the distance in the negative index medium where the Gaussian beam focuses [20].

3.8 From the result of Problem 3.3 for $k\_(\omega)$, find the PDE describing envelope propagation in an NIM. Solve this numerically to find the evolution of an initial Gaussian pulse in time after a certain distance of propagation [8]. How would you determine numerically that the medium is indeed an NIM?

3.9 Consider a system composed of 14 alternating layers of a nondispersive positive index material A (permittivity $\varepsilon_r = 1.5$, and permeability $\mu_r = 1$) and a material B with dispersive permittivity characteristics given by the Drude model $\hat{\varepsilon}(\omega)/\varepsilon_0 = 1 - \left(\omega_p^2/\omega(\omega + j\gamma)\right)$, with $\omega_p = 2\pi \times 8.13\,\text{THz}$; $\gamma = \omega_p/100$. Assume the thicknesses of the layers to be $d_A = d_B = 0.02125\,\text{mm}$. Plot the transmittivity of the stack as a function of frequency.

3.10 Consider an alternating stack ($d_A = d_B = 1\,\text{cm}$) of a PIM (air) and a dispersive NIM with the permittivity and permeability given by

$$\hat{\varepsilon}_r(f) = 1 + \frac{5^2}{0.9^2 - f^2} + \frac{10^2}{11.5^2 - f^2}, \quad \hat{\mu}_r(f) = 1 + \frac{3^2}{0.902^2 - f^2}$$

where f is the frequency measured in gigahertz. Plot the permittivity and permeability as a function of the frequency. Also, using the TMM, plot the transmittance for normal incidence for different numbers of layer pairs, namely $N = 4, 25, \ldots, \infty$. From your plots, find the zero $<n>$ gap and the Bragg gap [31].

3.11 Repeat Problem 3.10 by changing the angle of incidence and for both TE and TM cases. Does the location of the zero <n> gap and the Bragg gap change with changing angle of incidence? Give a physical reason for your answer [31].

3.12 Redo the plots of Figure 3.9 to study the effect of structure scaling by, say, ±10%. Comment on the widths and positions of the zero <n> gap and the Bragg gap. Discuss possible use of this setup as a sensor [26].

REFERENCES

1. D. Cheng, *Fundamentals of Engineering Electromagnetics*, Prentice-Hall, Upper Saddle River, NJ, 1993.
2. M.A. Ordal, L.L. Long, R.J. Bell, S.E. Bell, R.R. Bell, R.W. Alexander, Jr., and C. A. Ward, Optical properties of the metals Al, Co, Cu, Au, Fe, Pb, Ni, Pd, Pt, Ag, Ti, and W in the infrared and far infrared, *Appl. Opt.* 22, 1099–1120, 1983.
3. R.J. Beerends, H.G. terMorsche, J.C. van denBerg, and E.M. deVries, *Fourier and Laplace Transforms*, Cambridge University Press, Cambridge, U.K., 2003.
4. T. Mackay and A. Lakhtakia, Plane waves with negative phase velocity in Faraday chiral mediums, *Phys. Rev. E* 69, 026602.1–026602.9, 2004.
5. A. Baev, M. Samoc, P.N. Prasad, M. Krykunov, and J. Autschbach, A quantum chemical approach to the design of chiral negative index materials, *Opt. Exp.* 15, 5730–5741, 2007.
6. H. Haus, *Waves and Fields in Optoelectronics*, Prenctice-Hall, New York, 1983.
7. A. Bers, Note on group velocity and energy propagation, *Am. J. Phys.* 68, 482–485, 2000.
8. P.P. Banerjee, R. Aylo, and G. Nehmetallah, Baseband and envelope propagation in media modeled by a class of complex dispersion relations, *J. Opt. Soc. Am. B* 25, 990–994, 2008.
9. P.P. Banerjee and M.R. Chatterjee, Negative index in the presence of chirality and material dispersion, *J. Opt. Soc. Am. B* 26, 194–202, 2009.
10. J.A. Stratton, *Electromagnetic Theory*, IEEE Press, Piscataway, NJ, 2007.
11. R.C. Qiu and I.-T. Lu, Guided waves in chial optical fibers, *J. Opt. Soc. Am. A* 11, 3212–3219, 1994.
12. S.A. Ramakrishna, Physics of negative refractive index materials, *Rep. Prog. Phys.* 68, 449–521, 2005.
13. J.B. Pendry, Negative refraction makes a perfect lens, *Phys. Rev. Lett.* 85, 3966–3969, 2000.
14. L.D. Landau and E.M. Lifshitz, *Electrodynamics of Continuous Media*, Pergamon Press, New York, 1960.
15. P. Tassin, G. Van der Sande, and I. Veretennicoff, Left-handed materials: The key to sub-wavelength resolution? *Proc. Symp. IEEE/LEOS Benelux*, 2004, pp. 41–44.
16. V.G. Veselago, The electrodynamics of substances with simultaneously negative value of ε and μ, *Sov. Phys. USP* 10, 509–514, 1968.
17. Z. Jakšić, N. Dalarsson, and M. Maksimović, Negative refractive index metamaterials: Principles and applications, *Microwave Rev.* 12, 36–49, 2006.
18. K.E. Lonngren and S.V. Savov, *Fundamentals of Electromagnetics with MATLAB*, Scitech, Raleigh, NC, 2005.
19. A. Korpel and P.P. Banerjee, A heuristic guide to nonlinear dispersive wave equations and soliton-type solutions, *Proc. IEEE* 72, 1109–1130, 1984.
20. P.P. Banerjee and G. Nehmetallah, Linear and nonlinear propagation in negative index materials, *J. Opt. Soc. Am. B* 23, 2348–2355, 2006.
21. D. Xiao and H.T. Johnson, Approximate optical cloaking in an axisymmetric silicon photonic crystal structure, *Opt. Lett.* 33, 860–862, 2008.
22. Y.Y. Wang and L.W. Chen, Tunable negative refraction photonic crystals achieved by liquid crystals, *Opt. Expr.* 14, 10580–10587, 2006.
23. J. Li, L. Zhou, and P. Sheng, Photonic band gap from a stack of positive and negative index materials, *Phys. Rev. Lett.* 90, 083901.1–083901.4, 2003.
24. Y. Yuan, L. Ran, J. Huangfu, and H. Chen, Experimental verification of zero order bandgap in a layered stack of left-handed and right-handed materials, *Opt. Expr.* 14, 2220–2227, 2006.
25. M. Feise, I. Shadrivov, and Y. Kivshar, Tunable transmission and bistability in left-handed band-gap structures, *Appl. Phys. Lett.* 85, 1451–1453, 2004.
26. R. Aylo, P.P. Banerjee, A.K. Ghosh, and P. Verma, Design of metamaterial based sensors for pressure measurement, *SPIE* 7604, 760412-1–760412-8, 2010.

27. A. Dhawan, M.D. Gerhold, and J.F. Muth, Plasmonic structures based on subwavelength apertures for chemical and biological sensing applications, *IEEE Sensors J.* 8, 942–950, 2008.
28. W. Yuan, H. Ho, C. Wong, S. Kong, and C. Lin, Surface plasmon resonance biosensor incorporated in a Michelson interferometer with enhanced sensitivity, *IEEE Sensors J.* 7, 70–73, 2007.
29. J.J. Mock, D.R. Smith, and S. Schultz, Local refractive index dependence of plasmon resonance spectra from individual nanoparticles, *Nano Lett.* 3, 485–491, 2003.
30. M. Mitov, C. Bourgerette, and F. deGuerville, Fingerprint patterning of solid nanoparticles embedded in a cholesteric liquid crystal, *J. Phys.: Condens. Matter* 16, S1981–S1988, 2004.
31. R. Aylo, P.P. Banerjee, and G. Nehmetallah, Perturbed multilayer structures of positive and negative index materials, *J. Opt. Soc. Am. B* 27, 599–604, 2010.

4 Spectral State Variable Formulation for Planar Systems

4.1 INTRODUCTION

A problem that is extremely important in optics, microwave theory, antenna theory, and electromagnetics (EMs) in general [1–34] is the way radiation is transmitted, reflected, refracted, and propagates through two-dimensional, infinite homogeneous material layer systems. This problem has been studied for a wide variety of different material layers, e.g., isotropic dielectric materials, isotropic permeable materials, anisotropic dielectric and permeable materials, and bianisotropic materials. It has also been studied when a wide variety of different types of EM source radiation is incident on, or is present in, a layer of the planar system, e.g., incident plane wave, dipole source, line source, Gaussian beam, antenna source, waveguide-flange system, and microstrip line source strip. The synthesis and design of isotropic, planar multilayer optical systems have also received considerable attention [11–13].

In carrying out EM studies of these types of systems, a very powerful tool for analysis [1–10] is provided by one- and two-dimensional Fourier transform theory (also called *k*-space theory). This theory is a powerful tool because it allows virtually any time reduced EM source in any layer to be represented as a sum of plane waves whose propagation through the layers of the system can be analyzed in several manageable, tractable ways. Thus by using two-dimensional Fourier transform theory, one can study: (1) how individual plane spectral components propagate through the overall EM system, (2) what strength spectral components are excited by the source in the system, and (3) the overall spatial response of the system at any given point in the system by adding up (using superposition) the different spectral components.

The determination of the EM fields and their propagation, reflection, transmission, and scattering from isotropic, anisotropic, and bianisotropic planar layered media has been a topic that has received wide attention for a long time. References [2,8] give a very complete review and description of the topic of reflection from planar isotropic single and multilayers. A topic that has received less attention but still has been studied by a number of researchers is the problem of determining the radiation and scattering when sources and external incident fields (plane waves, Gaussian beam, etc.) excite EM fields in an anisotropic or bianisotropic planar, multilayer system. The anisotropic and bianisotropic EM scattering problem is considerably more difficult to analyze than the isotropic case because the anisotropic or bianisotropic, constitutive material parameters couple the field components together, creating from Maxwell's equations a much more complicated system than arises in the isotropic case. In most isotropic propagation problems, the typical approach, based on Maxwell's equations, is to decouple the components from one another and then derive a second-order, partial differential wave equation from which the solution to the EM problem may be obtained. For most anisotropic and bianisotropic scattering problems, this procedure is very intractable. Attempting this procedure for most anisotropic or bianisotropic systems would lead to fourth-, sixth-, or eighth-order partial differential equations, which would be quite intractable to solve. For anisotropic and bianisotropic materials, an alternate procedure that has been developed for transversely homogeneous, planar layers is to Fourier transform all EM field quantities with respect to the transverse coordinate(s) and then algebraically manipulate the reduced Fourier transformed field variable equations into a standard state variable form. Eigenanalysis of these first-order state variable equations yields the propagation constants and propagation modes of the system.

In this procedure, the two longitudinal field components are expressed in terms of the four transverse field components and then substituted into Maxwell's equations to reduce the system to a 4×4 state variable form. Expressing the longitudinal fields in terms of the transverse fields is very useful as it allows simple boundary matching of the tangential field components from one layer interface to another. The eigenanalysis method is also known as the exponential-matrix method, discussed in Chapter 1.

As discussed above, in this chapter, an alternate approach to determining the EM fields is presented. This method [18–29] consists of (1) replacing first-order, transverse derivative operators with terms proportional to their wave numbers ($\partial F/\partial x \propto -jk_xF$, $\partial F/\partial z \propto -jk_zF$), (2) writing out the six field component equations (these equations will contain first-order, longitudinal derivative operator terms $\partial F/\partial y$), (3) manipulating these equations in such a way as to eliminate the longitudinal electric field component E_y and longitudinal magnetic field component H_y (this reduces the number of curl equations from six to four), and finally (4) putting the four remaining equations into a standard, 4×4, first-order, state variable matrix equation form. The four transverse components E_x, E_z, H_x, H_z form the components of the 4×1 state variable column matrix. As shown in Section 4.4, this procedure provides a straightforward method of analyzing anisotropic or bianisotropic material layers whenever oblique, and arbitrarily polarized plane wave radiation is incident on the material layers.

This 4×4, state variable matrix procedure has been first implemented by Teitler and Henvis [19] and perhaps others who have reduced Maxwell's equations in an anisotropic layer to a set of four first-order linear differential equations and then, assuming an exponential form of solution, have solved for the normal or eigenmodes which describes propagation in the layer. The method is further developed by Berreman [20], who, starting from Maxwell's six component equations, puts the general anisotropic equations in a 4×4 form (where the 4×1 column vector contained the two tangential electric field components and two tangential magnetic field components) and then solves, using matrix techniques, for the four eigenvectors and eigenvalues of the system. Berreman [20] has studied several anisotropic material examples, including propagation in an orthorhombic crystal, propagation in optically active materials (described by the Drude model), and propagation involving Faraday rotation based on Born's model. Berreman [20] has also considered the state variable method as applied to determine propagation in media which is anisotropic and longitudinally periodic. Lin-Chung and Teitler [21], Krowne [22], and Morgan et al. [23] have used the 4×4 matrix method of Berreman [20] to study propagation of plane waves in stratified or multilayer anisotropic media. Weiss and Gaylord [24] have used the Berreman method to study stratified or multilayer resonators and optical filters (Fabry-Perot/Solc filter) composed of anisotropic materials. Two papers by Yang [25,26] study the important problem of formulating the EM state variable equations in such a way that efficient numerical solution of the equations arises.

Dispersion in anisotropic and birefringent materials and properties of the EM field propagation in these materials have also been studied by many other researchers. Yeh [27] has studied EM propagation in layer birefringent media. Alexopoulos and Uslenghi [28] studies reflection and transmission with arbitrarily graded parameters. Graglia et al. [29] study dispersion relations for bianisotropic materials and its symmetry properties. The book by Lindell et al. [6] also quotes many papers that have studied propagation in bianisotropic materials.

Another area where the k-space state variable analysis is useful is in the problem of characterizing radiation from antennas, dipoles, and metallic structures in millimeter and microwave integrated circuits (MMICs). Several papers [30–34] have studied the problem of determining the radiation from arbitrarily oriented electric and magnetic dipoles embedded in anisotropic planar layers. Tsalamengas and Uzunoglu [32] have studied the problem of determining the EM fields of an electric dipole in the presence of a general anisotropic layer backed by a ground plane. Their method consists of Fourier transforming all EM fields in the transverse coordinates, casting the Fourier transformed differential equations into the form of a first-order matrix differential equation, and after solving this, matching EM boundary conditions at the half-space anisotropic layer interface, to determine all fields of the system. An interesting feature of the Tsalamengas and Uzunoglu [32] method is that they have defined auxiliary vector components (the electric field and magnetic field were resolved into components parallel and perpendicular to the planar interfaces) that allow them to construct a matrix solution where the ground plane boundary condition is built into their matrix solution. This simplifies the problem to matching of the boundary conditions at the half-space anisotropic layer boundary. Tsalamengas and Uzunoglu [32] have solved several numerical examples including radiation from a dipole when uniaxial materials, ferrites, or magneto-plasmas comprise the anisotropic layer. The method differs from other methods in that the fundamental matrix differential equation is for a 2×2 matrix rather than the usual column matrices used by almost all other researchers.

Krowne [34] has used Fourier transform theory and the 4×4 matrix formalism of Berreman [20] to study propagation in layered, completely general bianisotropic media and to study Green's functions in bianisotropic media. In Krowne's [34] analysis, in addition to determining the modes of propagation in all bianisotropic layers, the effect of arbitrary electric and magnetic surface currents located at the interfaces of the bianisotropic layers is also studied. In their analysis, the surface current sources are delta source functions in the spatial domain and, therefore, planar sources in the Fourier *k*-space transform domain. Golden and Stewart [35] and Jarem [16] have used two-dimensional, Fourier *k*-space theory to study waveguide slot radiation through an inhomogeneous plasma layer.

Tang [31] has studied the EM fields in anisotropic media due to dipole sources using Sommerfield integrals and a transverse electric and transverse magnetic decomposition of the fields of the system. Ali and Mahmoud [30] have also studied dipole radiation in stratified anisotropic materials using a 3×3 state variable matrix technique.

In addition to the state variable analysis, a second theme that will be developed in this chapter is the use of the complex Poynting theorem as an information aid to the computation of the EM fields of the system. First, the complex Poynting theorem will be used as a cross-check of the numerical calculations themselves. The use of this theorem over a given region of space, regardless of whether the region contains lossy (gain) material or not, must show equality between the power radiated out of the region and the power dissipated and energy stored in the region. This is a more stringent and useful test than the more standard test of checking conservation of power from one layer to another layer. Checking power conservation from one layer to another is a conclusive test as long as the materials inside the layers are nonlossy. It is inconclusive if the layers inside are lossy since in the lossy case the power transmitted out of a given region will necessarily be less than the power transmitted into the given region since some power must be dissipated as heat in the lossy layer. The complex Poynting theorem, on the other hand, accounts not only for all power transmitted into and out of a given region but also for all power dissipated and energy stored in the region. In a given computation, if the surface and volume integrals of the complex Poynting theorem do not agree closely, some degree of numerical error has been made in the computation. If the agreement is too poor, most likely a significant computational error has been made somewhere in the calculations and it is most likely that the computations cannot be trusted.

A second way that the complex Poynting theorem is an aid to EM field analysis is it can give insight into the way that energy is stored and power is dissipated in a given region of space. Often in making EM field plots, the plots of the individual field components either electric or magnetic can be deceptive since, for example, the fields can appear large but in reality may be standing waves, which are actually transmitting very little real power into a system. Plots of the energy stored and power dissipated then give great insight into how EM radiation is actually interacting with a material at a given place in space.

In what follows, both the state variable method (in conjunction with k-space analysis) and the complex Poynting theorem will be applied to study a wide variety of different EM planar reflection and transmission problems. Section 4.2 considers one of the simplest possible cases, namely, when a normally incident plane wave impinges on an isotropic, lossy material slab. Section 4.3 studies the case when an oblique, incident plane wave impinges on an anisotropic layer. Section 4.4 considers cases when EM sources which are not plane waves impinge on an anisotropic layer. In this section, k-space theory is used to decompose the EM source into a plane wave Fourier spectrum from which a tractable analysis can be carried out. In particular, the cases of a waveguide flange system that radiates into an anisotropic, lossy layer are considered. The expression for the wave slot admittance is developed. The two-dimensional state variable formulation that applies to the study of EM fields in bianisotropic materials is given as a homework problem at the end of this chapter.

Overall, in this chapter, only cases of homogeneous, single layer material slabs are considered. Only a single layer analysis has been carried out in order to make the analysis as simple and clear as possible. Extension to multilayer analysis is straightforward. Later chapters use multilayer analyses extensively. The multilayer analysis is described thoroughly in these chapters.

4.2 STATE VARIABLE ANALYSIS OF AN ISOTROPIC LAYER

4.2.1 INTRODUCTION

In this section, we study one of the simplest EM state variable problems, namely, the problem of determining the EM fields that result when a plane wave propagates with normal incidence in an isotropic, lossy dielectric slab ($\tilde{\varepsilon}_2 = \tilde{\varepsilon}' - j\tilde{\varepsilon}''$, $\tilde{\mu}_2 = \tilde{\mu}' - j\tilde{\mu}''$) (see Figure 4.1). Three cases are studied, namely, (1) a plane wave is normally incident on the slab, (2) a plane wave is normally incident on the slab backed by a perfect conductor, and (3) the EM fields are excited by an electric or magnetic current source. These cases are solved by the state variable method. Because of the fact that all eigenvectors or eigenmodes of the state variable system can be solved in closed form, these examples show in a simple manner the principles and properties of the state variable formalism that apply to much more complicated problems (anisotropic planar slabs, diffraction grating, etc.).

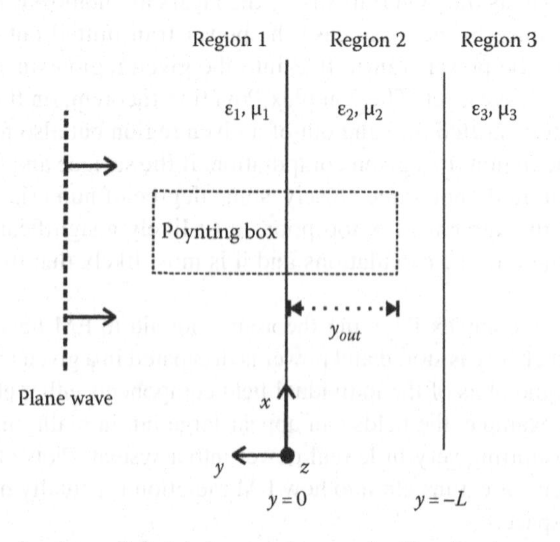

FIGURE 4.1 Geometry of a planar dielectric layer and a complex Poynting box.

4.2.2 Analysis

To begin the analysis in this section, we assume that all propagation is at normal incidence and that the EM fields of the system in Regions 1, 2, and 3 in a $(\tilde{x}, \tilde{y}, \tilde{z})$ coordinate system are given by

$$\vec{E}_\ell = E_{x\ell}(\tilde{y})\hat{a}_x$$

$$\vec{H}_\ell = H_{z\ell}(\tilde{y})\hat{a}_z \quad \ell = 1, 2, 3 \tag{4.1}$$

where ℓ denotes the region number.

From Maxwell's equations assuming source-free regions

$$\tilde{\nabla}\times\vec{E}_\ell = -j\omega\tilde{\mu}_\ell\vec{H}_\ell$$

$$\tilde{\nabla}\times\vec{H}_\ell = j\omega\tilde{\varepsilon}_\ell\vec{E}_\ell \tag{4.2}$$

we find substituting Equation 4.1 that

$$\frac{\partial E_{x\ell}}{\partial\tilde{y}} = j\omega\tilde{\mu}_\ell H_{z\ell}$$

$$\frac{\partial H_{x\ell}}{\partial\tilde{y}} = j\omega\tilde{\varepsilon}_\ell E_{x\ell} \tag{4.3}$$

It is convenient to make the above equations dimensionless. We introduce the state variables $E_{x\ell} = S_{x\ell}$ and $H_{z\ell} = U_{z\ell}/\eta_0$, and after letting $y = k_0\tilde{y}$, $k_0 = \omega\sqrt{\mu_0\varepsilon_0} = 2\pi/\lambda$, $\omega = 2\pi f$, $\varepsilon_\ell = \tilde{\varepsilon}_\ell/\varepsilon_0 = \varepsilon'_\ell - j\varepsilon''_\ell$, $\mu_\ell \equiv \tilde{\mu}_\ell/\mu_0 = \mu'_\ell - j\mu''_\ell$, $\eta_0 = \sqrt{\mu_0/\varepsilon_0} = 377\,\Omega$, we find

$$\frac{\partial S_{x\ell}}{\partial y} = j\mu_\ell U_{z\ell}$$

$$\frac{\partial U_{x\ell}}{\partial y} = j\varepsilon_\ell S_{z\ell} \tag{4.4}$$

In Equations 4.3 and 4.4, ε_0 and μ_0 represent the permittivity (F/m) and permeability (H/m) of free space, respectively; η_0 is intrinsic impedance of free space (Ω); f is the excitation frequency (H) of the system; and λ is the free space wavelength (m) of the system.

Letting

$$\underline{V} = \begin{bmatrix} S_{x\ell} \\ U_{z\ell} \end{bmatrix}, \quad \underline{\underline{A}} = \begin{bmatrix} 0 & j\mu_\ell \\ j\varepsilon_\ell & 0 \end{bmatrix} \tag{4.5}$$

(and dropping the ℓ subscript for the moment), we may write Equation 4.5 in the general state variable from

$$\frac{\partial\underline{V}(y)}{\partial y} = \underline{\underline{A}}\,\underline{V}(y) \tag{4.6}$$

Equation 4.6 may be solved by determining the eigenvalues and eigenvectors of the matrix $\underline{\underline{A}}$ according to the equation

$$\underline{\underline{A}}\,\underline{V} = q\underline{V} \tag{4.7}$$

From this, the general solution of Equation 4.6 is then given by

$$\underline{V} = \sum_{n=1}^{N} C_n \underline{V}_n e^{q_n y} \tag{4.8}$$

where $N = 2$; $\underline{V}_n$ and q_n, $n = 1,2$, are eigenvectors and eigenvalues of the matrix $\underline{\underline{A}}$; and C_n are general constants. We may demonstrate that $\underline{V}_n e^{q_n y}$ is a solution of Equation 4.8 by direct substitution. We have for $n = 1,2$,

$$\frac{d}{dy}(\underline{V}_n e^{q_n y}) = \underline{V}_n \frac{d}{dy} e^{q_n y} = q_n \underline{V}_n e^{q_n y}$$

But $q_n \underline{V}_n = \underline{\underline{A}}\underline{V}_n$; hence

$$\frac{d}{dy}(\underline{V}_n e^{q_n y}) = \underline{\underline{A}}(\underline{V}_n e^{q_n y}) \tag{4.9}$$

which is the original equation. Superposition of the distinct modes of $\underline{V}_n e^{q_n y}$ then gives the full EM solution.

The eigenvalues of q_n, $n = 1, 2$, of $\underline{\underline{A}}$ in Equation 4.7 satisfy

$$\det[\underline{\underline{A}} - q\underline{\underline{I}}] = (-q)^2 - (j^2 \mu\varepsilon) = 0$$

or

$$q^2 + \mu\varepsilon = 0 \tag{4.10}$$

Let $\gamma \equiv \alpha + j\beta \equiv q_1$, $\alpha > 0$, $\beta > 0$ (α, β are real numbers) be the forward traveling mode in the $\ell = 1,2,3$ region. Substituting in Equation 4.10, we have

$$(\alpha + j\beta)^2 + [\mu' - j\mu''][\varepsilon' - j\varepsilon''] = 0 \tag{4.11}$$

After performing algebra, it is found

$$\alpha = \left\{\frac{1}{2}\left[-r' + [r'^2 + r''^2]^{1/2}\right]\right\}^{1/2}$$

$$\beta = \left\{\frac{1}{2}\left[r' + [r'^2 + r''^2]^{1/2}\right]\right\}^{1/2} \tag{4.12}$$

where $\mu\varepsilon = [\mu'\varepsilon' - \mu''\varepsilon''] - j[\mu''\varepsilon' + \mu'\varepsilon''] = r' - jr''$. Usually, $r'' > 0$.

We note that $q_n = \alpha + j\beta$, $n = 1$, corresponds to a forward traveling wave and $q_n = -(\alpha + j\beta)$, $n = 2$, corresponds to a backward traveling wave in all regions of the system. We also note that these solutions obey proper boundary conditions in all regions. For example, in Region 3, we have for the forward traveling wave ($n = 1$); for the exponential part of the EM wave, $E_x \propto \exp(\alpha y) \rightarrow 0$ as $y \rightarrow -\infty$ when $\alpha > 0$; and for the oscillatory part of the wave, $E_x(y,t)\alpha\cos(\beta y + \omega t)$, which indicates a wave traveling to right since the phase velocity $v_\varphi = -(\omega/\beta) < 0$. A similar analysis in Region 1 shows that the second eigenvalue $q_2 = -(\alpha + j\beta)$ corresponds to a backward traveling wave.

The eigenvectors $\underline{V}_1 = [S_{x1}, U_{z1}]^T$ and $\underline{V}_2 = [S_{x2}, U_{z2}]^T$ may be determined from Equation 4.7 after substitution of the eigenvalue q_n, $n = 1,2$, into Equation 4.10. For the forward traveling wave in any of the three regions, we have

$$q_1 = \alpha + j\beta = \gamma$$

$$0 = \begin{bmatrix} -q_1 & j\mu \\ j\varepsilon & -q_1 \end{bmatrix} \begin{bmatrix} S_{x1} \\ U_{z1} \end{bmatrix} = 0 \tag{4.13}$$

Because q_1 is an eigenvalue, the two equations of Equation 4.13 are linearly dependent. We have $-q_1 S_{x1} + j\mu U_{z1} = 0$ or $U_{z1} = (-jq_1/\mu)S_{x1}$. Letting $S_{x1} = 1$, the forward traveling eigenvector is $\underline{V}_1 = [1, -j\gamma/\mu]^T$, where T denotes the matrix transpose. Substituting the backward traveling wave with $q_2 = (\alpha + j\beta) = -\gamma$, the backward traveling eigenvector corresponding to q_2 is $\underline{V}_2 = [1, j\gamma/\mu]^T$.

The electric field associated with the eigenmodes q_n, $n = 1,2$, is given in Regions $\ell = 1,2,3$ as

$$\vec{E}_n^{(\ell)} = S_{xn\ell}\, e^{q_{n\ell} y}\, \hat{a}_x$$

$$\vec{H}_n^{(\ell)} = \frac{1}{\eta_0} U_{zn\ell} e^{q_{n\ell} y}\, \hat{a}_z \tag{4.14a}$$

where

$$S_{xn\ell} = 1, \quad n = 1,2$$

$$U_{z1\ell} = \frac{-j\gamma_\ell}{\mu_\ell}, \quad U_{z2\ell} = \frac{j\gamma_\ell}{\mu_\ell}$$

$$\gamma_\ell = \alpha_\ell + j\beta_\ell \tag{4.14b}$$

Since the medium is linear, a superposition over the modes in Equation 4.14 gives the total field in any region. The total electric and magnetic fields that can exist in Regions 1, 2, and 3 are given by

$$\vec{E}^{(\ell)} = \sum_{n=1}^{2} C_{n\ell} \vec{E}_n^{(\ell)} \tag{4.15a}$$

$$\vec{H}^{(\ell)} = \sum_{n=1}^{2} C_{n\ell} \vec{H}_n^{(\ell)} \tag{4.15b}$$

where $C_{n\ell}$ are general complex coefficients that need to be determined from boundary conditions.

As a cross-check of the solution, we note that for any region (suppressing the ℓ subscript or superscript),

$$H_z = \frac{1}{j\omega\tilde{\mu}} \frac{\partial E_x}{\partial \tilde{y}} = \frac{1}{\eta_0}\left(\frac{-j}{\mu}\right)\frac{\partial E_x}{\partial y} \tag{4.16}$$

From Equation 4.15a, we note

$$E_x = C_1 \exp(\gamma y) + C_2 \exp(-\gamma y)$$

Substituting this in Equation 4.16, we have

$$H_z = \frac{1}{\eta_0}\left(-\frac{j}{\mu}\right)[C_1\gamma \exp(\gamma y) - C_2\gamma \exp(-\gamma y)] \tag{4.17}$$

which is the same solution as Equation 4.15b when the eigenvectors of Equation 4.14 are used.

In addition to the field amplitudes of the electric and magnetic fields, another important quantity to calculate is the time-averaged power that passes through any layer that is parallel to the material interface. This is explained in detail in the next subsection.

4.2.3 Complex Poynting Theorem

The previous subsection has presented the EM field solution for a normally incident plane wave on a uniform, isotropic, lossy material layer. An important numerical consideration in all computations is the accuracy with which the numerical computations have been formed. A relatively simple test of the computation, which applies only when the slab is lossless, is provided by calculating the power incident on the slab, calculating the sum of the power that is transmitted and reflected from the slab, and then calculating the difference of these two sums to compute the error in the numerical solution. However, as just mentioned, this test applies only when the layer is lossless. When the layer is lossy, the power reflected and transmitted does not equal the incident power since some of the power is absorbed as heat inside the material layer. In the case when the layer is lossy, one may test numerical accuracy results by using the complex Poynting theorem. The purpose of this section is to present the complex Poynting theorem (Harrington [3]) as it applies to the lossy material slab and also to test the numerical accuracy of the EM field solutions, which will be studied in Section 4.3.2.

For an isotropic material, the complex Poynting theorem states that the time-averaged power delivered at a point P contained in a volume $\Delta\tilde{V} \to 0$ by the electric and magnetic current sources $\vec{J}_i$ and $\vec{M}_i$ should be balanced by the sum of (1) the time-averaged power $\bar{P}_f$ radiated over the surface $\Delta\tilde{S}$ enclosing the volume $\Delta\tilde{V}$, (2) the electric power $\bar{P}_{DE}$ and magnetic power $\bar{P}_{DM}$ dissipated over the volume $\Delta\tilde{V}$, and (3) $(j2\omega)$ times the difference between the time-averaged magnetic energy $\bar{W}_M$ stored in $\Delta\tilde{V}$ and the time-averaged electric energy $\bar{W}_E$ stored in $\Delta\tilde{V}$, where $\omega = 2\pi f$ (rad) is angular frequency and f is the frequency in Hertz.

Mathematically, the complex Poynting theorem for a general anisotropic material is given by [3]

$$\left(\frac{1}{2}\right)\iint_{\tilde{S}} \vec{E}\times\vec{H}^{*}\cdot\hat{a}_n d\tilde{S} + \left(\frac{1}{2}\right)\iiint_{\tilde{V}} (\vec{E}\cdot\vec{J}^{t*} + \vec{H}^{*}\cdot\vec{M}^{t})d\tilde{V} = 0 \tag{4.18}$$

where $\vec{J}^t$ is a general electric displacement, conduction, and source current term and $\vec{M}^t$ is the generalized magnetic current.

Mathematically, these currents are given by

$$\vec{J}^t = j\omega(\underline{\underline{\varepsilon}}' - j\underline{\underline{\varepsilon}}'')\vec{E} + \vec{J}^i \tag{4.19}$$

$$\vec{M}^t = j\omega(\underline{\underline{\tilde{\mu}}}' - j\underline{\underline{\tilde{\mu}}}'')\vec{H} + \vec{M}^i \tag{4.20}$$

where $\vec{J}^i$ and $\vec{M}^i$ are impressed source terms, and we have assumed that the permittivity and permeability are complex anisotropic quantities. After some algebra, we get from (4.18)

$$\bar{P}_s = \bar{P}_f + \bar{P}_{DE} + \bar{P}_{DM} + j(-\bar{P}_{WE} + \bar{P}_{WM}) \tag{4.21}$$

where

$$\bar{P}_s = -\frac{1}{2}\iiint_{\tilde{V}}\left[\vec{E}\cdot\vec{J}^{i*} + \vec{H}^*\cdot\vec{M}^i\right]d\tilde{V} \quad \text{(source power)}$$

$$\bar{P}_f = \frac{1}{2}\oiint_{\tilde{S}}\vec{E}\times\vec{H}^*\cdot\hat{a}_n d\tilde{S} \quad \text{(net outward power flow)}$$

$$\bar{P}_{WE} = 2\omega\bar{W}_E = 2\omega\left\{\frac{1}{4}\iiint_{\tilde{V}}\vec{E}\cdot\left[\underline{\underline{\tilde{\varepsilon}}}'\vec{E}\right]^* d\tilde{V}\right\} \quad \text{(proportional to stored electric energy)}$$

$$\bar{P}_{WM} = 2\omega\bar{W}_M = 2\omega\left\{\frac{1}{4}\iiint_{\tilde{V}}\vec{H}^*\cdot\left[\underline{\underline{\tilde{\mu}}}'\vec{H}\right]d\tilde{V}\right\} \quad \text{(proportional to stored magnetic energy)}$$

$$\bar{P}_{DE} = \frac{\omega}{2}\iiint_{\tilde{V}}\vec{E}\cdot\left[\underline{\underline{\tilde{\varepsilon}}}''\vec{E}\right]^* d\tilde{V} \quad \text{(electric power dissipated)}$$

$$\bar{P}_{DM} = \frac{\omega}{2}\iiint_{V}\vec{H}^*\cdot\left[\underline{\underline{\tilde{\mu}}}''\vec{H}\right]d\tilde{V} \quad \text{(magnetic power dissipated)} \tag{4.22}$$

For present applications, we will consider a Poynting box as shown in Figure 4.1. This box is assumed to have end faces which have a cross section $\Delta\tilde{S}$ and are assumed to be parallel to the interfaces of the slab. For this box, we first note that in the power flow integral $\bar{P}_f$ the integral over the lateral portion of the box (the portion between the end faces of the box) is zero. This follows since there is no variation in the EM fields or power flow in the x- and z-directions. Thus, the power flow integral can be written as a sum of the power flow as calculated over the two end faces of the box:

$$\bar{P}_f = -\bar{P}_{IN} + \bar{P}_{OUT} \tag{4.23}$$

where

$$\bar{P}_{IN} = \frac{1}{2}\iint_{\Delta\tilde{S}}\vec{E}\times\vec{H}^*\Big|_{\tilde{y}=\tilde{y}^+}\cdot(-\hat{a}_y)d\tilde{S} \tag{4.24}$$

$$\bar{P}_{OUT} = \frac{1}{2}\iint_{\Delta\tilde{S}}\vec{E}\times\vec{H}^*\Big|_{\tilde{y}=\tilde{y}^-}\cdot(-\hat{a}_y)d\tilde{S} \tag{4.25}$$

The minus sign in Equation 4.25 is a result of the fact that the outward normal on the $\tilde{y}^+$ end cap is $\hat{a}_y$. Using Equations 4.21–4.25, we find that the complex Poynting theorem for the present problem may be written as

$$\bar{P}_{IN} = -\bar{P}_s + \bar{P}_{OUT} + \bar{P}_{DE} + \bar{P}_{DM} + j(-\bar{P}_{WE} + \bar{P}_{WM}) \tag{4.26}$$

It is convenient to express the above power and energy integrals in dimensionless coordinates $x = k_0\tilde{x}$, etc. and to normalize the complex Poynting theorem equations by an amount of power $P_{INC}^{FS} = [\Delta\tilde{S}/(2\eta_0)](E_0'^2/\eta_1)$ (Watts), where $\eta_0 = \sqrt{\tilde{\mu}_0/\tilde{\varepsilon}_0} = 377\,\Omega$, $\eta_1 = \sqrt{\tilde{\mu}_1/\tilde{\varepsilon}_1}/\eta_0$ (dimensionless), and $(E_0'^2/\eta_1) = 1.0$ (V^2/m^2). With this normalization and also carrying out all integrals in Equations 4.22, 4.24, and 4.25, each term in (4.26) can be written as

$$P_{IN} = -P_s + P_{OUT} + P_{DE} + P_{DM} + j(-P_{WE} + P_{WM}) \tag{4.27}$$

$$P_{WE} = \frac{\eta_1}{E_0'^2}\int_{\ell_y} \vec{E}\cdot\underline{\underline{\varepsilon}}'\vec{E}^* dy \quad \text{(dimensionless)}$$

$$P_{WM} = \frac{\eta_1}{E_0'^2}\int_{\ell_y} (\eta_0\vec{H}^*)\cdot\underline{\underline{\mu}}'(\eta_0\vec{H})dy \quad \text{(dimensionless)}$$

$$P_{DE} = \frac{\eta_1}{E_0'^2}\int_{\ell_y} \vec{E}\cdot\underline{\underline{\varepsilon}}''\vec{E}^* dy \quad \text{(dimensionless)}$$

$$P_{DM} = \frac{\eta_1}{E_0'^2}\int_{\ell_y} (\eta_0\vec{H}^*)\cdot\underline{\underline{\mu}}''(\eta_0\vec{H})dy \quad \text{(dimensionless)}$$

$$P_{OUT} = \frac{\eta_1}{E_0'^2}\left[\vec{E}\times(\eta_0\vec{H}^*)\right]\Bigg|_{y=y^-}\cdot(-\hat{a}_y) \quad \text{(dimensionless)}$$

$$P_{IN} = \frac{\eta_1}{E_0'^2}\left[\vec{E}\times(\eta_0\vec{H}^*)\right]_{y=y^+}\cdot(-\hat{a}_y) \quad \text{(dimensionless)}$$

$$P_s = \frac{\eta_1}{E_0'^2}k_0^{-1}\eta_0\int_{\ell_y}\left[\vec{E}\cdot\vec{J}_i^* + \vec{H}^*\cdot\vec{M}_i\right]dy \quad \text{(dimensionless)}$$

where $\underline{\underline{\varepsilon}} = \underline{\underline{\varepsilon}}' - j\underline{\underline{\varepsilon}}''$ and $\underline{\underline{\mu}} = \underline{\underline{\mu}}' - j\underline{\underline{\mu}}''$ represent relative permittivity and permeability, respectively. Substitution of the field solutions as obtained through the state variable technique into the above one-dimensional integrals gives the various power terms that make up the complex Poynting theorem. Because all permittivity and permeability tensor elements are constant and because all EM field solutions in the equations are exponentials, we note that all the one-dimensional power integrals may be carried out in closed form. This is important for checking numerical error since estimates of the error using these formulas do not depend on the accuracy of the numerical integration.

4.2.4 State Variable Analysis of an Isotropic Layer in Free Space

In this subsection, we consider the case when a plane wave from $y = \infty$ is normally incident as a dielectric slab. In this case, the C_{11} and C_{23} coefficients are known (see Equation 4.15), with $C_{11} = E_0$, where E_0 is the incident amplitude (V/m), and $C_{23} = 0$ also since there is no reflected wave from Region 3. As coefficient C_{21} represents the complex amplitude of the reflected field in Region 1, we let $C_{21} = R$, and since the coefficient C_{13} represents the complex amplitude of the transmitted fields in Region 3, we let $C_{13} = T$. Using these coefficients, the fields in Regions 1, 2, and 3 are given by (see Figure 4.1)

Region 1

$$E_x^{(1)} = E_0 \exp(\gamma_1 y) + R\exp(-\gamma_1 y)$$

$$H_z^{(1)} = \frac{j\gamma_1}{\eta_0\mu_1}\left[-E_0 \exp(\gamma_1 y) + R\exp(-\gamma_1 y)\right] \tag{4.28}$$

Region 2

$$E_x^{(2)} = C_{12}\exp(\gamma_2 y) + C_{22}\exp(-\gamma_2 y)$$

$$H_z^{(2)} = \frac{1}{\eta_0}\frac{j\gamma_2}{\mu_2}\left[-C_{12}\exp(\gamma_2 y) + C_{22}\exp(-\gamma_2 y)\right] \tag{4.29}$$

Region 3

$$E_x^{(3)} = T\exp(\gamma_3 (y+L))$$

$$H_z^{(3)} = \frac{1}{\eta_0}\frac{j\gamma_3}{\mu_3}\left[-T\exp(\gamma_3 (y+L))\right] \tag{4.30}$$

The $E_x^{(3)}$ and $H_z^{(3)}$ fields have been written with an $\exp(\gamma_3(y + L))$ in order to refer the phase of T coefficient to the $y = -L$ boundary.

The boundary conditions require that the tangential electric and magnetic fields match at $y = 0, -L$. Matching of the tangential electric and magnetic fields at $y = 0$ and $y = -L$ leads to four equations in four unknowns from which the EM fields in all regions can be determined. It is convenient to use the electric field equations at the boundaries to eliminate the unknowns in exterior Regions 1 and 3, thus reducing the number of equations from four to two. Doing so, it is found that

$$\frac{2\gamma_1}{\mu_1}E_0 = a_{11}C_{12} + a_{12}C_{22}$$

$$0 = a_{21}C_{12} + a_{22}C_{22} \tag{4.31}$$

where

$$a_{11} = \frac{\gamma_2}{\mu_2} + \frac{\gamma_1}{\mu_1}, \quad a_{12} = \frac{\gamma_1}{\mu_1} - \frac{\gamma_2}{\mu_2}$$

$$a_{21} = \left[-\frac{\gamma_2}{\mu_2} + \frac{\gamma_3}{\mu_3} \right] \exp(-\gamma_2 L), \quad a_{22} = \left[\frac{\gamma_2}{\mu_2} + \frac{\gamma_3}{\mu_3} \right] \exp(\gamma_2 L) \tag{4.32}$$

also

$$R = -E_0 + C_{12} + C_{22}$$

$$T = C_{12} \exp(-\gamma_2 L) + C_{22} \exp(\gamma_2 L) \tag{4.33}$$

Inversion of the 2 × 2 as given by Equation 4.31 then determines the unknown coefficients C_{12} and C_{22} of the system.

We now apply the complex Poynting theorem of Equation 4.27 to the normal incident plane wave case being studied in this section. We assume that the Poynting box has its left face 0.5λ from the Region 1–2 interface, i.e., $\tilde{y}_+ = \tilde{y}_{in} = 0.5\lambda$, and has its right face at $\tilde{y}_- = -\tilde{y}_{out}$, $\tilde{y}_{out} \geq 0$. For the present analysis, there are no sources in the layer, so $P_s = 0$. Substituting we find that the complex Poynting theorem is given by

$$P_{IN} = P_{OUT} + P_{DE} + P_{DM} + j(-P_{WE} + P_{WM}) \equiv P_{BOX} \tag{4.34}$$

where

$$P_{DE} = P_{DE1} + P_{DE2} + P_{DE3}$$

$$P_{DE1} = P_{DE3} = 0$$

$$P_{DE2} = \varepsilon_2'' \int_{y_{2-}}^{0} \left| c_{12} \exp(\gamma_2 y) + c_{22} \exp(-\gamma_2 y) \right|^2 dy$$

where

$$y_{2-} = \begin{cases} -y_{out}, & -y_{out} > -L \\ -L, & -y_{out} < -L \end{cases}$$

$$P_{DM} = P_{DM1} + P_{DM2} + P_{DM3}$$

$$P_{DM1} = P_{DM3} = 0$$

$$P_{DM2} = \mu_2'' \left| \frac{\gamma_2}{\mu_2} \right|^2 \int_{y_{2-}}^{0} \left| c_{12} \exp(\gamma_2 y) - c_{22} \exp(-\gamma_2 y) \right|^2 dy$$

$$P_{WE} = P_{WE1} + P_{WE2} + P_{WE3}$$

$$P_{WE1} = \varepsilon_1' \int_{0}^{y_{in}} \left| E_0 \exp(\gamma_1 y) + R \exp(-\gamma_1 y) \right|^2 dy$$

$$P_{WE2} = \varepsilon_2' \int_{y_{2-}}^{0} |c_{12}\exp(\gamma_2 y) + c_{22}\exp(-\gamma_2 y)|^2 \, dy$$

$$P_{WE3} = \varepsilon_3' \int_{y_{3-}}^{-L} |T\exp(\gamma_3(y+L)|^2 \, dy$$

where

$$y_{3-} = \begin{cases} -L, & -y_{out} > -L \\ -y_{out}, & -y_{out} < -L \end{cases}$$

$$P_{WM} = P_{WM1} + P_{WM2} + P_{WM3}$$

$$P_{WM1} = \mu_1' \left|\frac{\gamma_1}{\mu_1}\right|^2 \int_{0}^{y_{in}} |E_0\exp(\gamma_1 y) - R\exp(-\gamma_1 y)|^2 \, dy$$

$$P_{WM2} = \mu_2' \left|\frac{\gamma_2}{\mu_2}\right|^2 \int_{y_{2-}}^{0} |c_{12}\exp(\gamma_2 y) - c_{22}\exp(-\gamma_2 y)|^2 \, dy$$

$$P_{WM3} = \mu_3' \left|\frac{\gamma_3}{\mu_3}\right|^2 \int_{y_{3-}}^{-L} |T\exp(\gamma_3(y+L))|^2 \, dy$$

$$P_{IN} = j\left(\frac{\gamma_1}{\mu_1}\right)^* [E_0\exp(\gamma_1 y_{in}) + R\exp(-\gamma_1 y_{in})] \times [E_0\exp(\gamma_1 y_{in}) - R\exp(-\gamma_1 y_{in})]^*$$

$$P_{OUT} = j\left(\frac{\gamma_2}{\mu_2}\right)^* [c_{12}\exp(-\gamma_2 y_{out}) + c_{22}\exp(\gamma_2 y_{out})] \times [c_{12}\exp(-\gamma_2 y_{out}) - c_{22}\exp(\gamma_2 y_{out})]^*$$

when $-y_{out} > -L$

$$P_{OUT} = j\left(\frac{\gamma_3}{\mu_3}\right)^* |T\exp(-\gamma_3(y_{out}+L))|^2$$

when $-y_{out} < -L$. In these equations, R is the reflection coefficient in Region 1, T is the transmission coefficient in Region 3, and c_{12} and c_{22} are wave coefficients in Region 2. The expressions for P_{WE3} and P_{WM3} have been chosen in such a way that when $-y_{out} > -L$ (that is, $-y_{out}$ is in Region 2), the lower limit y_{3-} equals the upper limit and P_{WE3} and P_{WM3} are zero as they should be.

The conservation theorem as given by Equation 4.34 states (1) that the sum of Re(P_{OUT}) and $P_D = P_{DE} + P_{DM}$ ($P_D = P_{DE} + P_{DM}$ is real and nonnegative), which by definition equals Re(P_{BOX}) should equal Re(P_{IN}), and (2) that the sum of Imag(P_{OUT}) and the energy–power difference $-P_{WE} + P_{WM}$, which by definition equals Imag(P_{BOX}) should equal the sum of Imag(P_{IN}).

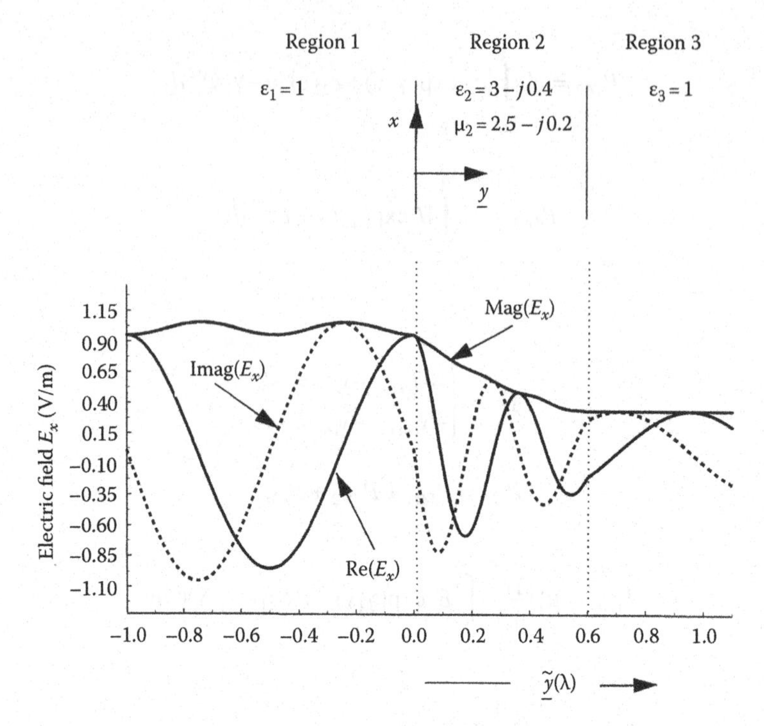

FIGURE 4.2 E_x electric field (magnitude, real and imaginary parts) plotted versus the distance $\underline{\tilde{y}}$ from the incident side interface is shown.

As a numerical example for the normal incidence case, we assume the layer thickness is $\tilde{L} = 0.6\lambda$, free space bounds the layer in Regions 1 and 3, and the slab has a lossy permittivity given by $\varepsilon_2 = 3 - j0.4$ and relative permeability $\mu_2 = 2.5 - j0.2$. Figures 4.2 through 4.4 show plots of the EM fields and different power terms associated with the present example. Figure 4.2 shows the E_x electric field (magnitude, real and imaginary parts) plotted versus the distance $\underline{\tilde{y}} = -\tilde{y}$

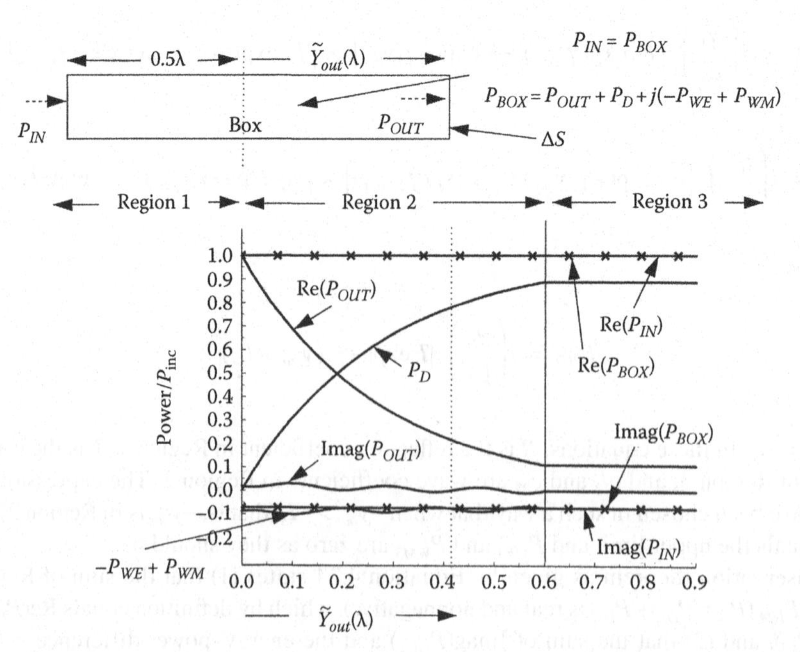

FIGURE 4.3 Plots of the real and imaginary parts of P_{IN} and P_{BOX} as a function of the distance $\tilde{Y}_{out}$.

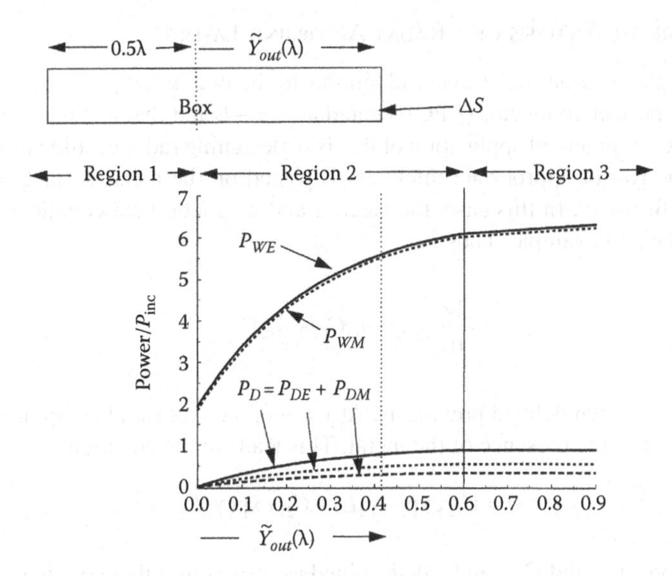

FIGURE 4.4 Plots of the electric energy term, magnetic energy term, power stored and power dissipated in the Poynting box versus the distance $\tilde{Y}_{out}$.

from the incident side interface. In observing the real and imaginary plots of E_x, one notices that the standing wave wavelength of the E_x electric field is greatly shortened in Region 2 as opposed to Region 1. This is due to the greater magnitude of the material constants $|\varepsilon_2| = |3 - j0.4|$ and $|\mu_2| = |2.5 - j0.2|$ in Region 2 as opposed to those in Region 1, namely, $|\varepsilon_1| = 1$ and $|\mu_1| = 1$. In observing the plots of Figure 4.2, one also notices that the continuity of the E_x is numerically obeyed as expected. In Figure 4.2, one also notices that the presence of the lossy layer causes a standing wave in Region 1 with a standing wave ratio (SWR)

$$\text{SWR} = \frac{|E_{xMAX}|}{|E_{xMIN}|} = \frac{|E_0 + R|}{|E_0 - R|} \cong 1.2$$

This means that the lossy layer represents a fairly matched load to the normally incident plane wave. In Region 2 of Figure 4.2, it is observed that the $|E_x|$ is attenuated to about 30% as the EM wave is multiply reflected in the lossy layer.

In Figure 4.3, plots of the real and imaginary parts of P_{IN} and P_{BOX} are made as a function of the distance $\tilde{Y}_{out}$, the distance that the Poynting Box extends to the right of the Region 1–2 interface. As can be seen from Figure 4.3, the complex Poynting theorem is obeyed to a high degree of accuracy as the real and imaginary parts of P_{IN} ("solid line") and P_{BOX} ("cross") agree very closely. One also observes that as the distance $\tilde{Y}_{out}$ increases, the power dissipated P_D increases and the Re(P_{OUT}) decreases and that both change in such a way as to leave the sum constant and equal to Re(P_{IN}). Also plotted in Figure 4.3 are the Imag(P_{OUT}) and the energy difference term $-P_{WE} + P_{WM}$. One observes from these plots that the Imag(P_{OUT}) and $-P_{WE} + P_{WM}$ vary sinusoidally in Region 2 and that the nonconstant portion of these curves are out of phase with one another by 180°. Thus the sum of Imag(P_{OUT}) and $-P_{WE} + P_{WM}$ is a constant equal to Imag(P_{IN}). Thus the imaginary part of the power is exchanged periodically between Imag(P_{OUT}) and $-P_{WE} + P_{WM}$ in such a way as to keep the Imag(P_{IN}) a constant throughout the system. Figure 4.4 shows plots of the electric and magnetic energy and power stored and dissipated in the Poynting box versus the distance $\tilde{Y}_{out}$. As can be seen from Figure 4.4, the electric and magnetic stored energy terms P_{WE} and P_{WM} are nearly equal to each other.

4.2.5 State Variable Analysis of a Radar Absorbing Layer

As a second example, assume that a material similar to the one in the previous example is placed against an electric perfect conductor (EPC) located at $y = -L$ and that a plane wave from $y = \infty$ is incident on the layer. A practical application of this is in designing radar evading aircraft (see Figure 4.3), where such a layer of appropriate thickness is pasted on the metal surface of the aircraft to minimize radar reflectivity. In this case, the electric and magnetic field equations at $\tilde{y} = 0$ are the same as those in the first example. Thus

$$\frac{2\gamma_1}{\mu_1} E_0 = a_{11}C_{12} + a_{12}C_{22} \tag{4.35}$$

where a_{11} and a_{12} have been defined previously. At $\tilde{y} = -\tilde{L}$, the tangential component of the electric field must vanish due to the presence of the metal. This leads to the equation

$$0 = C_{12}\exp(-\gamma_2 L) + C_{22}\exp(\gamma_2 L) \tag{4.36}$$

From these equations, C_{12} and C_{22} can be determined as well as all other coefficients in the system.

Figure 4.5 shows the Re(E_x), Imag(E_x), and $|E_x|$ electric fields plotted versus the distance $\tilde{y}$ from the Region 1–2 interface, using the material parameter values of Section 4.2.4. As can be seen from Figure 4.5, the presence of the EPC in Region 3 causes a larger SWR than was observed when a free space occupied Region 3. One also notices that the presence of the EPC causes more internal reflection within the slab layer, Region 2, as can be seen by the increased ripple or decaying SWR pattern displayed by the $|E_x|$ plot. Figure 4.6 shows the various normalized power terms associated with complex Poynting theorem of Equation 4.34. Figure 4.6 uses the same geometry as Figure 4.3. The only difference between Figures 4.3 and 4.6 is that an EPC is in Region 3 of Figure 4.6, whereas free space was in Region 3 of Figure 4.3. As can be seen in Figure 4.6, in Figure 4.3 we observe that the complex Poynting theorem is obeyed to a high degree of accuracy since the real and imaginary parts of P_{IN} ("solid line") and P_{OUT} ("cross") agree very closely with each other. We also

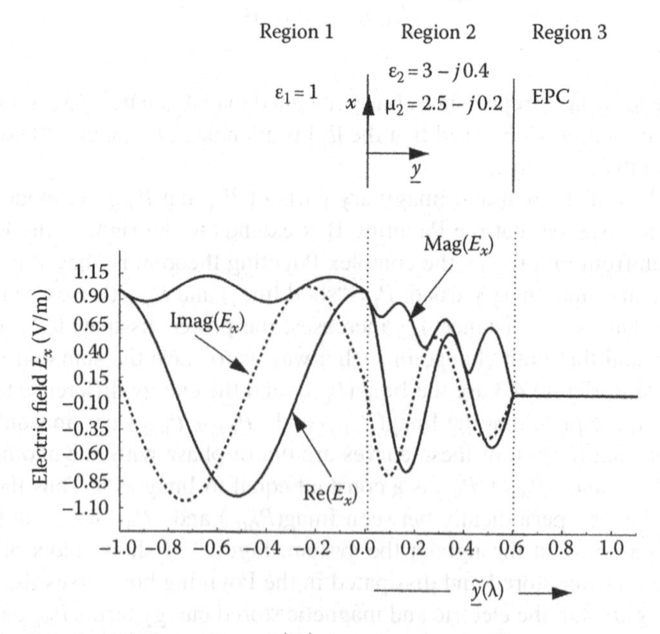

FIGURE 4.5 Plots of the Re(E_x), Imag(E_x), and $|E_x|$ plotted versus the distance $\tilde{y}$.

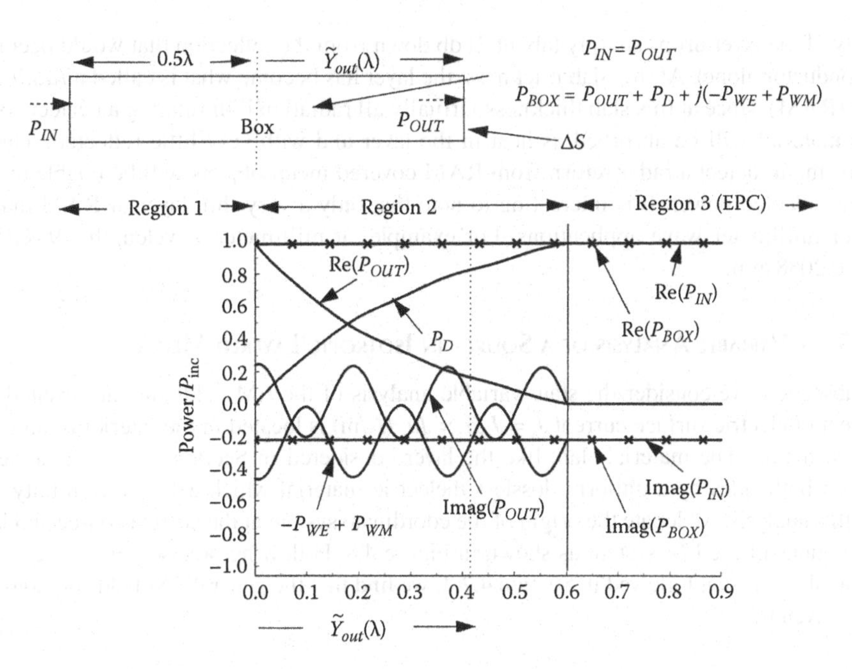

FIGURE 4.6 Plots of the various normalized power terms associated with complex Poynting theorem of Equation 4.34. This figure uses the same geometry as Figure 4.3.

notice from Figure 4.6 that a higher oscillation of $-P_{WE} + P_{WM}$ and Imag(P_{OUT}) occurs than did in Figure 4.2. This higher internal reflection in the slab is caused by the high reflectivity of the EPC at the Region 2–3 interface.

Figure 4.7 shows a plot of normalized reflected power (reflected power/incident power [db]) of a uniform slab that results when a plane wave is normally on the slab. Region 3 is an EPC and in Region 2, $\varepsilon_2 = 7 - j3.5$ and $\mu_2 = 2.5 - j0.2$. In this figure, the normalized reflected power is plotted versus the slab length $\tilde{L}$. As can be seen from Figure 4.7, at a slab thickness of $\tilde{L} = 0.066\lambda$, the

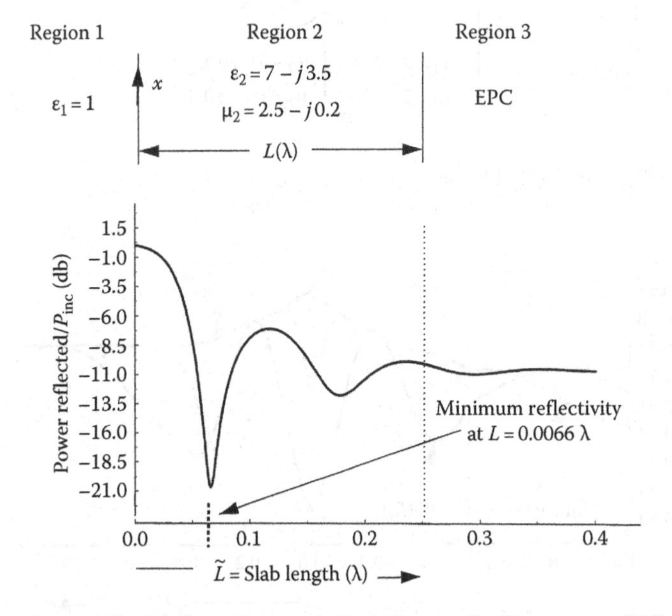

FIGURE 4.7 Plots of normalized reflected power (reflected power/incident power, [db]) for the case where Region 3 is an EPC and Region 2 has $\varepsilon_2 = 7 - j3.5$ and $\mu_2 = 2.5 - j0.2$.

reflectivity of the layer drops sharply (about 21 db down from the reflection that would occur from a perfect conductor alone). At this slab thickness, the layer has become what is called a *radar absorbing layer* (RAM), since at this slab thickness virtually all radiation illuminating a perfect conductor with this material will be absorbed as heat in the layer and with very little reflected. Thus radar systems trying to detect a radar return from RAM covered metal objects will be unable to observe significant reflected power. It is interesting to note that only a very thin layer of RAM material is needed for millimeter wave applications. For example, at millimeter wavelengths (94 GHz), $\tilde{L} = 0.066\lambda = 0.2088\,\text{mm}$.

4.2.6 State Variable Analysis of a Source in Isotropic Layered Media

In this subsection, we consider the state variable analysis of the EM fields that are excited when a planar sheet of electric surface current $\vec{J}_S = J_{sx}\hat{a}_x = J\hat{a}_x$ (A/m) is located in the interior of an isotropic, two-layered media. The material slab, like the layer considered in Section 4.2.2, is assumed to be bounded on both sides by a uniform, lossless dielectric material which extends to infinity on each side. For this analysis, we locate the origin of the coordinate system at the current source and label the different regions of the EM system as shown in Figure 4.8. Following precisely the same state variable EM analysis as that followed in Section 4.2.2, we find that the general EM field solutions in each region are given by

Region 1′

$$E_x^{(1')} = C_{11'}\exp(\gamma_{1'}(y - L_+)) + C_{21'}\exp(-\gamma_{1'}(y - L_+)), \quad C_{11'} = 0 \tag{4.37a}$$

$$U_z^{(1')} = \eta_0 H_z^{(1')} = \frac{j\gamma_{1'}}{\mu_{1'}} C_{21'}\exp(-\gamma_{1'}(y - L_+)) \tag{4.37b}$$

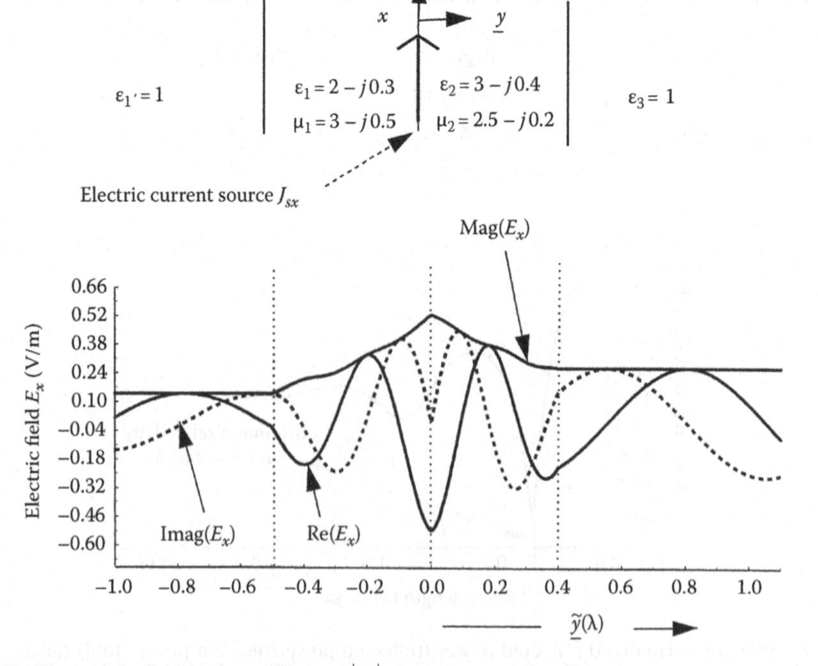

FIGURE 4.8 Plots of the Re(E_x), Imag(E_x), and $|E_x|$ plotted versus the distance $\tilde{y}$.

Region 1

$$E_x^{(1)} = C_{11}\exp(\gamma_1 y) + C_{21}\exp(-\gamma_1 y) \tag{4.38a}$$

$$U_z^{(1)} = \eta_0 H_z^{(1)} = \frac{-j\gamma_1}{\mu_1}\left[C_{11}\exp(\gamma_1 y) - C_{21}\exp(-\gamma_1 y)\right] \tag{4.38b}$$

Region 2

$$E_x^{(2)} = C_{12}\exp(\gamma_2 y) + C_{22}\exp(-\gamma_2 y) \tag{4.39a}$$

$$U_z^{(2)} = \eta_0 H_z^{(2)} = \frac{-j\gamma_2}{\mu_2}\left[C_{12}\exp(\gamma_2 y) - C_{22}\exp(-\gamma_2 y)\right] \tag{4.39b}$$

Region 3

$$E_x^{(3)} = C_{13}\exp(\gamma_3(y + L_-)) \tag{4.40a}$$

$$U_z^{(3)} = \eta_0 H_z^{(3)} = \frac{-j\gamma_3}{\mu_3}C_{13}\exp(\gamma_3(y + L_-)) \tag{4.40b}$$

The total layer thickness is $L = L_+ + L_-$, where $L_+ \geq 0$ and $L_- \geq 0$.

Matching the tangential electric and magnetic fields at the Region 1–1′ interface and eliminating the $C_{21'}$ coefficient, it is found that

$$C_{11} = \alpha C_{21}, \quad \alpha = \frac{(-\gamma_{1'}/\mu_{1'}) + (\gamma_1/\mu_1)}{(\gamma_{1'}/\mu_{1'}) + (\gamma_1/\mu_1)}\exp(-2\gamma_1 L_+) \tag{4.41}$$

Matching the tangential electric and magnetic fields at the Region 2–3 interface and eliminating the C_{13} coefficient, it is found that

$$C_{12} = \beta C_{22}, \quad \beta = \frac{(\gamma_3/\mu_3) + (\gamma_2/\mu_2)}{(-\gamma_3/\mu_3) + (\gamma_2/\mu_2)}\exp(2\gamma_2 L_-) \tag{4.42}$$

To proceed further, we match EM boundary conditions at the Region 1–2 boundary $y = 0$. These boundary conditions are given by

$$\eta_0 H_z^{(1)} - \eta_0 H_z^{(2)} = U_z^{(1)} - U_z^{(2)} = \eta_0 J \tag{4.43a}$$

$$E_x^{(1)} - E_x^{(2)} = 0 \tag{4.43b}$$

In the present problem, because an electric current source is present at the Region 1–2 boundary, the tangential magnetic field given by Equation 4.43a is discontinuous at $y = 0$. Performing algebra,

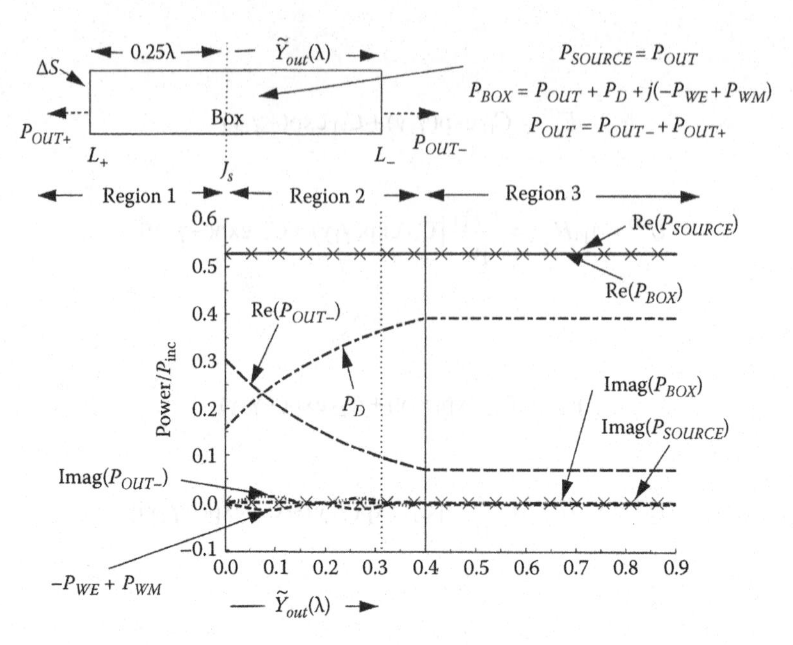

FIGURE 4.9 Plots of different power terms that make up the complex Poynting theorem of Equation 4.45 plotted versus the distance $\tilde{Y}_{OUT}$.

it is found that the following equations result, from which the unknown coefficients of the system may be found:

$$\frac{-j\gamma_1}{\mu_1}[\alpha-1]C_{21}+\frac{j\gamma_2}{\mu_2}[\beta-1]C_{22}=\eta_0 J \tag{4.44a}$$

$$[\alpha+1]C_{21}-[\beta+1]C_{22}=0 \tag{4.44b}$$

To give a numerical example of the EM fields and complex Poynting results, we assume that the material slab (Region 2) has the parameters $\varepsilon_1 = 2 - j0.3$, $\mu_1 = 3 - j0.5$, $\varepsilon_2 = 3 - j0.4$, $\mu_2 = 2.5 - j0.2$, $\tilde{L}_- = 0.4\lambda$, and $\tilde{L}_+ = 0.5\lambda$, and assume that Regions 1 and 3 are free space. In this example, we further assume that the Poynting box is the same one described in Section 4.2 except that its leftmost face is located $\tilde{y}_{OUT+} = 0.25\lambda$ to the left of the Region 1–2 interface (the source is located at the Region 1–2 interface at $\tilde{y} = 0$), and the rightmost face is located at $\tilde{y} = -\tilde{y}_{OUT}$, $\tilde{y}_{OUT} \geq 0$ from the Region 1–2 interface (see Figure 4.9). For the present source problem, the complex Poynting theorem is given by

$$P_S = P_{OUT+} + P_{OUT-} + P_{DE} + P_{DM} + j(-P_{WE} + P_{WM}) \equiv P_{BOX} \tag{4.45}$$

where

$$P_S = -\eta_0 \vec{E}\cdot\vec{J}_s^*\Big|_{\tilde{y}=0} = -\eta_0 E_x J^*\Big|_{\tilde{y}=0} \tag{4.46}$$

the electric field $E_x\big|_{\tilde{y}=0} = (C_{11}+C_{21})$ is continuous at $\tilde{y}=0$ and also from Equation 4.44a,

$$\eta_0 J = \frac{-j\gamma_1}{\mu_1}(C_{11}-C_{21})+\frac{j\gamma_2}{\mu_2}(C_{12}-C_{22}) \tag{4.47}$$

Thus

$$P_s = -(C_{11}+C_{21})\left[\frac{-j\gamma_1}{\mu_1}(C_{11}-C_{21})+\frac{j\gamma_2}{\mu_2}(C_{12}-C_{22})\right]^* \quad (4.48)$$

The terms P_{OUT+} and P_{OUT-} are given by

$$P_{OUT+} = -j\left(\frac{\gamma_1}{\mu_1}\right)^*\left[C_{11}\exp(\gamma_1 y_{OUT+})+C_{21}\exp(-\gamma_1 y_{OUT+})\right]\times\left[C_{11}\exp(\gamma_1 y_{OUT+})-C_{21}\exp(-\gamma_1 y_{OUT+})\right]^* \quad (4.49a)$$

$$\tilde{y}_{OUT+} = 0.25\lambda$$

$$P_{OUT-} = j\left(\frac{\gamma_2}{\mu_2}\right)^*\left[C_{12}\exp(-\gamma_2 y_{OUT})+C_{22}\exp(\gamma_2 y_{OUT})\right]\times\left[C_{12}\exp(-\gamma_2 y_{OUT})-C_{22}\exp(\gamma_2 y_{OUT})\right]^* \quad (4.49b)$$

when $0 \le y_{OUT} \le L_-$,

$$P_{OUT-} = j\left(\frac{\gamma_3}{\mu_3}\right)^*\left|C_{13}\exp(-\gamma_3(y_{OUT}+L_-))\right|^2 \quad (4.49c)$$

when $-y_{OUT} < -L_-$. The other terms in Equation 4.49c are given in Equation 4.34.

Figure 4.8 shows the Re(E_x), Imag(E_x), and $|E_x|$ electric fields plotted versus the distance $\tilde{y}$ (in units of λ) from the Region 1–2 interface. As can be seen from Figure 4.8, the presence of the electric current source in a lossy medium causes the electric field to be the greatest at the source location and to attenuate as distance from the source increases. Because the regions are different to the left and right of the source, the fields are not symmetric about the source location. In observing Figure 4.8, one notices that the Re(E_x), Imag(E_x), and $|E_x|$ electric fields are all continuous at the different interfaces as they must be to satisfy EM boundary conditions. Figure 4.9 shows different power terms that make up the complex Poynting theorem of Equation 4.45 plotted versus the distance $\tilde{y}_{OUT}$. As can be seen from Figure 4.9, the real and imaginary parts of P_S ("cross") and P_{BOX} ("solid line") agree with each other to a high degree of accuracy, thus showing that the complex Poynting theorem is being obeyed numerically for the present example. One also observes that as the distance $\tilde{y}_{OUT}$ increases, the power dissipated P_D increases and the Re(P_{OUT-}) decreases, and that both change in such a way as to leave the sum constant and equal to Re(P_S). Also plotted in Figure 4.9 is the Imag(P_{OUT-}) and the energy–power difference $-P_{WE} + P_{WM}$. One observes from these plots that the Imag(P_{OUT-}) and the energy–power difference $-P_{WE} + P_{WM}$ vary sinusoidally in Region 2 and that the nonconstant portion of these curves are out of phase with one another. Thus the sum of Imag(P_{OUT}) and $-P_{WE} + P_{WM}$ is a constant equal to Imag(P_S). Thus the imaginary part of the power is exchanged periodically between Imag(P_{OUT}) and $-P_{WE} + P_{WM}$ in such a way as to keep the Imag(P_S) a constant throughout the system. Despite the fact that the EM fields were excited by an electric current source in Figure 4.9 rather than a plane wave as in Figure 4.3, the complex Poynting numerical results in the two figures are very similar.

4.3 STATE VARIABLE ANALYSIS OF AN ANISOTROPIC LAYER

4.3.1 Introduction

Thus far we have discussed several examples of EM scattering from isotropic layers. Another interesting problem is EM scattering from anisotropic media, e.g., crystals and ionosphere. This section differs from the previous sections in two ways: the media is anisotropic and couples the field components into one another and also, the EM fields are obliquely incident on the dielectric slab at an angle θ_I. The analysis [18–29] is based on the state variable analysis similar to that in the previous section, and gives a reasonably straightforward and direct solution to the problem. We note that a traditional second-order wave equation analysis would lead to a fairly intractable equation set to solve due to the anisotropic coupling of the fields.

We assume that the plane wave is polarized with its electric field in the plane of incidence of the EM wave. The dielectric slab is assumed to be characterized by a lossy, anisotropic relative dielectric permittivity tensor, where ε_{xx}, ε_{xy}, ε_{yx}, ε_{yy}, ε_{zz} are nonzero and the other tensor elements are zero. The geometry is the shown in Figure 4.10. The slab relative permeability is assumed to be isotropic and lossy, and characterized by $\mu = \mu' - j\mu''$. The basic analysis to be carried out is to solve Maxwell's equations on the incident side (Region 1), in the slab region (Region 2), and on the transmitted side (Region 3), and then from these solutions match EM boundary conditions at the interfaces of the dielectric slab.

4.3.2 Basic Equations

A state variable analysis will be used to determine the EM fields in the dielectric slab region. We begin by specifying the EM fields in Regions 1 and 3 of the system, assumed to be nonmagnetic ($\mu_1 = \mu_3 = 1$). The EM fields in Region 1 are given by

$$\begin{aligned} E_x^{(1)} = S_x^{(1)}(y)\exp(-jk_x x) &= \frac{k_{y1}}{\varepsilon_1}\left[E_0 \exp(jk_{y1}y) - R\exp(-jk_{y1}y)\right]\exp(-jk_x x) \\ &= \left[C_{x11}\exp(\gamma_{11}y) + C_{x21}\exp(\gamma_{21}y)\right]\exp(-jk_x x) \end{aligned} \tag{4.50}$$

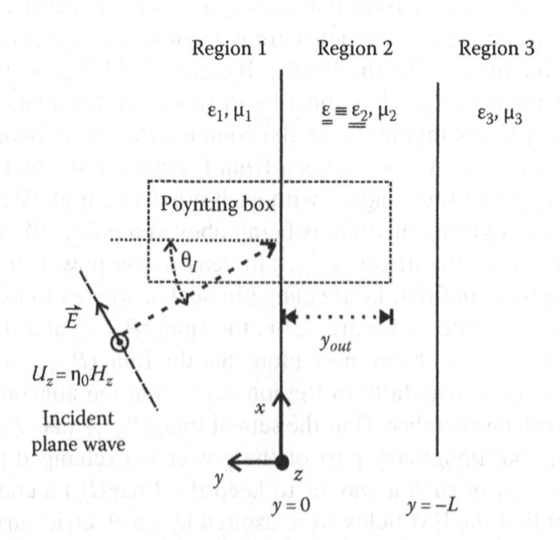

FIGURE 4.10 Geometry of a planar dielectric layer and a complex Poynting box is shown. A plane wave parallel polarization is obliquely incident on the layer $U_z = \eta_0 H_z$.

$$E_y^{(1)} = S_y^{(1)}(y)\exp(-jk_x x) = \frac{k_x}{\varepsilon_1}\left[E_0 \exp(jk_{y1}y) + R\exp(-jk_{y1}y)\right]\exp(-jk_x x)$$

$$= \left[C_{y11}\exp(\gamma_{11}y) + C_{y21}\exp(\gamma_{21}y)\right]\exp(-jk_x x) \tag{4.51}$$

$$\eta_0 H_{z1} = U_z^{(1)}(y)\exp(-jk_x x) = \left[E_0\exp(jk_{y1}y) + R\exp(-jk_{y1}y)\right]\exp(-jk_x x)$$

$$= \left[C_{z11}\exp(\gamma_{11}y) + C_{z21}\exp(\gamma_{21}y)\right]\exp(-jk_x x) \tag{4.52}$$

where

$x = k_0\tilde{x}$, $y = k_0\tilde{y}$, $z = k_0\tilde{z}$, $k_0 = 2\pi/\lambda$, $k_x = \sqrt{\varepsilon_1}\sin(\theta_I)$, $k_{y1} = \sqrt{\varepsilon_1 - k_x^2}$, and $\eta_0 = 377\ \Omega$
E_0 is the incident plane wave amplitude
λ is the free space wavelength in meters
ε_1 is the relative permittivity of Region 1

The EM fields in Region 3 consist only of a transmitted wave and are given by

$$E_x^{(3)} = S_x^{(3)}(y)\exp(-jk_x x) = \frac{k_{y3}}{\varepsilon_3}\left[T\exp(jk_{y3}(y+L))\right]\exp(-jk_x x)$$

$$= \left[C_{x13}\exp(\gamma_{13}y) + C_{x23}\exp(\gamma_{23}y)\right]\exp(-jk_x x) \tag{4.53}$$

$$E_y^{(3)} = S_y^{(3)}(y)\exp(-jk_x x) = \frac{k_x}{\varepsilon_3}\left[T\exp(jk_{y3}(y+L))\right]\exp(-jk_x x)$$

$$= \left[C_{y13}\exp(\gamma_{13}y) + C_{y23}\exp(\gamma_{23}y)\right]\exp(-jk_x x) \tag{4.54}$$

$$\eta_0 H_z^{(3)} = U_z^{(3)}(y)\exp(-jk_x x) = \left[T\exp(jk_{y3}(y+L))\right]\exp(-jk_x x)$$

$$= \left[C_{z13}\exp(\gamma_{13}y) + C_{z23}\exp(\gamma_{23}y)\right]\exp(-jk_x x) \tag{4.55}$$

where

$k_{y3} = \sqrt{\varepsilon_3 - k_x^2}$
T is the transmitted plane wave amplitude
ε_3 is the relative permittivity of Region 3
$C_{x23} = C_{y23} = C_{z23} = 0$

In the anisotropic, dielectric slab region, Maxwell's equations are given by

$$\nabla\times\vec{E} = -j\underline{\mu}(\eta_0\vec{H})$$

$$\nabla\times(\eta_0\vec{H}) = j\underline{\varepsilon}(\vec{E}) \tag{4.56}$$

where we assume that $\underline{\mu}$ is a dimensionless, magnetic permeability matrix, assumed to be diagonal. The x component of $\underline{\varepsilon}\,\vec{E}$, for example, is given by $\varepsilon_{xx}E_x + \varepsilon_{xy}E_y + \varepsilon_{xz}E_z$. In order that the EM fields of Regions 1 and 3 phase match with the EM fields of Region 2 for all x, it is necessary that the EM fields of Region 2 all be proportional to $\exp(-jk_x x)$. (This factor follows from the application of the separation of variable method to Maxwell's equations). Using this fact the electric and magnetic fields in Region 2 may be expressed as

$$\vec{E} = (S_x(y)\hat{a}_x + S_y(y)\hat{a}_y + S_z(y)\hat{a}_z)\exp(-jk_x x)$$

$$\eta_0\vec{H} = (U_x(y)\hat{a}_x + U_y(y)\hat{a}_y + U_z(y)\hat{a}_z)\exp(-jk_x x) \quad (4.57)$$

Using the fact that the only nonzero EM field components in Region 1 are the E_x, E_y, and H_z, a small amount of analysis shows that in Equation 4.56 a complete field solution can be found taking only S_x, S_y, and U_z to be nonzero with $S_z = U_x = U_y = 0$. Substituting Equation 4.57 in 4.56, taking appropriate derivatives with respect to x, it is found that the following equations result:

$$-jk_x S_y - \frac{\partial S_x}{\partial y} = -j\mu_{zz}U_z \quad (4.58)$$

$$\frac{\partial U_z}{\partial y} = j\varepsilon_{xx}S_x + j\varepsilon_{xy}S_y \quad (4.59)$$

$$jk_x U_z = j\varepsilon_{yx}S_x + j\varepsilon_{yy}S_y \quad (4.60)$$

To proceed further, it is possible to eliminate the longitudinal electric field component and express the equations in terms of the S_x and U_z components alone. Although other components could be eliminated, the S_y is the best since the remaining equations involve variables that are transverse or parallel to the layer interfaces. This is very useful since these variables may be used to match tangential EM boundary conditions directly. The S_y component is given by (from Equation 4.60)

$$S_y = -\frac{\varepsilon_{yx}}{\varepsilon_{yy}}S_x + \frac{k_x}{\varepsilon_{yy}}U_z \quad (4.61)$$

Substituting (4.61) into (4.58 and 4.59),

$$\frac{\partial S_x}{\partial y} = j\left[k_x\frac{\varepsilon_{yx}}{\varepsilon_{yy}}\right]S_x + j\left[\mu_{zz} - \frac{k_x^2}{\varepsilon_{yy}}\right]U_z \quad (4.62)$$

$$\frac{\partial U_z}{\partial y} = j\left[\varepsilon_{xx} - \frac{\varepsilon_{xy}\varepsilon_{yx}}{\varepsilon_{yy}}\right]S_x + j\left[k_x\frac{\varepsilon_{xy}}{\varepsilon_{yy}}\right]U_z \quad (4.63)$$

The above equations are in state variable form and can be rewritten as

$$\frac{\partial \underline{V}}{\partial y} = \underline{\underline{A}}\,\underline{V} \quad (4.64)$$

where

$$a_{11} = j\left[k_x\frac{\varepsilon_{yx}}{\varepsilon_{yy}}\right],\quad a_{12} = j\left[\mu_{zz} - \frac{k_x^2}{\varepsilon_{yy}}\right] \quad (4.65)$$

$$a_{21} = j\left[\varepsilon_{xx} - \frac{\varepsilon_{xy}\varepsilon_{yx}}{\varepsilon_{yy}}\right], \quad a_{22} = j\left[k_x\frac{\varepsilon_{xy}}{\varepsilon_{yy}}\right] \tag{4.66}$$

where $\underline{V} = [S_x, U_z]^T$.

The basic solution method is to find the eigenvalues and eigenvectors of the state variable matrix $\underline{\underline{A}}$, form a full field solution from these eigensolutions, and then match boundary conditions to find the final solution. The general eigenvector solution is given by

$$\underline{V} = \underline{V}_n \exp(q_n y) \tag{4.67}$$

where q_n and $\underline{V}_n = [S_{xn}, U_{zn}]^t$ are eigenvalues and eigenvectors of $\underline{\underline{A}}$ and satisfy

$$\underline{\underline{A}}\,\underline{V}_n = q_n \underline{V}_n, \quad n = 1,2 \tag{4.68}$$

Because $\underline{\underline{A}}$ is only a 2 × 2 matrix, it is possible to find the eigenvalues and eigenvectors of the system in closed form. The quantities q_n and $\underline{V}_n$ are given by

$$\begin{bmatrix} a_{11} - q_n & a_{12} \\ a_{21} & a_{22} - q_n \end{bmatrix}\begin{bmatrix} S_{xn} \\ U_{zn} \end{bmatrix} = 0 \tag{4.69}$$

For this to have nontrivial solutions,

$$\det\left(\begin{bmatrix} a_{11} - q_n & a_{12} \\ a_{21} & a_{22} - q_n \end{bmatrix}\right) = (a_{11} - q_n)(a_{22} - q_n) - a_{12}a_{21} = 0 \tag{4.70}$$

Using the quadratic equation to solve for q_n, we find

$$q_n = 0.5(a_{11} + a_{22}) \pm 0.5\left[a_{11}^2 - 2a_{11}a_{22} + 4a_{12}a_{21} + a_{22}^2\right]^{1/2}, \quad n = 1,2 \tag{4.71}$$

Letting $S_{xn} = 1$, $n = 1, 2$, it is found that the eigenvectors are given by

$$\underline{V}_n = \left[1, \frac{(q_n - a_{11})}{a_{12}}\right]^t \tag{4.72}$$

The longitudinal eigenvector components S_{yn} are given by, using Equation 4.61,

$$S_{yn} = -\frac{\varepsilon_{yx}}{\varepsilon_{yy}} S_{yn} + \frac{k_x}{\varepsilon_{yy}} U_{zn}, \quad n = 1, 2 \tag{4.73}$$

Using these eigenvalues and eigenvectors, it is found that the EM fields in Region 2 are given by

$$\begin{aligned} E_x^{(2)} = S_x^{(2)}(y)\exp(-jk_x x) &= [C_1 S_{x1} \exp(q_1 y) + C_2 S_{x2} \exp(q_2 y)]\exp(-jk_x x) \\ &\equiv \left[C_{x12} \exp(\gamma_{12} y) + C_{x22} \exp(\gamma_{22} y)\right]\exp(-jk_x x) \end{aligned} \tag{4.74}$$

$$E_y^{(2)} = S_y^{(2)}(y)\exp(-jk_x x) = [C_1 S_{y1}\exp(q_1 y) + C_2 S_{y2}\exp(q_2 y)]\exp(-jk_x x)$$
$$\equiv \left[C_{y12}\exp(\gamma_{12}y) + C_{y22}\exp(\gamma_{22}y)\right]\exp(-jk_x x) \tag{4.75}$$

$$\eta_0 H_z^{(2)} = U_z^{(2)}(y)\exp(-jk_x x) = \left[C_1 U_{z1}\exp(q_1 y) + C_2 U_{z2}\exp(q_2 y)\right]\exp(-jk_x x)$$
$$\equiv \left[C_{z12}\exp(\gamma_{12}y) + C_{z22}\exp(\gamma_{22}y)\right]\exp(-jk_x x) \tag{4.76}$$

In these equations, C_1 and C_2 are field coefficients that are yet to be determined.

To proceed further, it is necessary to determine the unknown coefficients of the field solution in Regions 1–3. In this case, the unknown coefficients are R, T, C_1, and C_2. In the present problem, the boundary conditions require that the tangential electric field (the E_x field) and the tangential magnetic field (H_z) must be equal at the two slab interfaces. Thus in this analysis there are four boundary condition equations from which the four unknown constants of the system may be determined. Matching boundary conditions at the Region 1–2 interface, we find

$$\frac{k_{y1}}{\varepsilon_1}\left[E_0 - R\right] = C_1 S_{x1} + C_2 S_{x2} \tag{4.77}$$

$$E_0 + R = C_1 U_{z1} + C_2 U_{z2} \tag{4.78}$$

$$\frac{k_{y3}}{\varepsilon_3} T = C_1 S_{x1}\exp(-q_1 L) + C_2 S_{x2}\exp(-q_2 L) \tag{4.79}$$

$$T = C_1 U_{z1}\exp(-q_1 L) + C_2 U_{z2}\exp(-q_2 L) \tag{4.80}$$

By substituting R and T from Equations 4.77 and 4.80 in 4.78 and 4.79, the 4 × 4 system may be reduced to the following 2 × 2 set of equations:

$$\frac{2k_{y1}}{\varepsilon_1}E_0 = \left[\frac{k_{y1}}{\varepsilon_1}U_{z1} + S_{x1}\right]C_1 + \left[\frac{k_{y1}}{\varepsilon_1}U_{z2} + S_{x2}\right]C_2 \tag{4.81}$$

$$0 = \exp(-q_1 L)\left[-\frac{k_{y3}}{\varepsilon_3}U_{z1} + S_{x1}\right]C_1 + \exp(-q_2 L)\left[-\frac{k_{y3}}{\varepsilon_3}U_{z2} + S_{x2}\right]C_2 \tag{4.82}$$

The C_1 and C_2 may found from the above in closed form. Using Equations 4.77 and 4.80, the other coefficients may be found.

4.3.3 Numerical Results

This section will be concerned with presenting a numerical example of EM reflection and transmission from an anisotropic layer when an obliquely incident plane wave impinges on the layer. In this example, Regions 1 and are free space, and Region 2 is a material slab with a thickness $\tilde{L} = 0.6\lambda$, and having material parameters $\varepsilon_{xx} = \varepsilon_{yy} = 2.25 - j0.3$, $\varepsilon_{xy} = \varepsilon_{yx} = 0.75 - j0.1$. We assume the permeability to be isotropic but lossy, with $\mu_{zz} = \mu_2 = 2.5 - j0.2$. The incident plane wave (incident amplitude $E_0 = 1$ [V/m], electric field polarization in the plane of incidence) is assumed to have an angle of incidence $\theta_I = 25°$. Figure 4.11 shows plots of the magnitudes of the E_x, E_y, and $U_z = \eta_0 H_z$ EM fields in Regions 1–3 as a function of $\tilde{y}$ (units of λ) ($\tilde{y}$ shows location of the

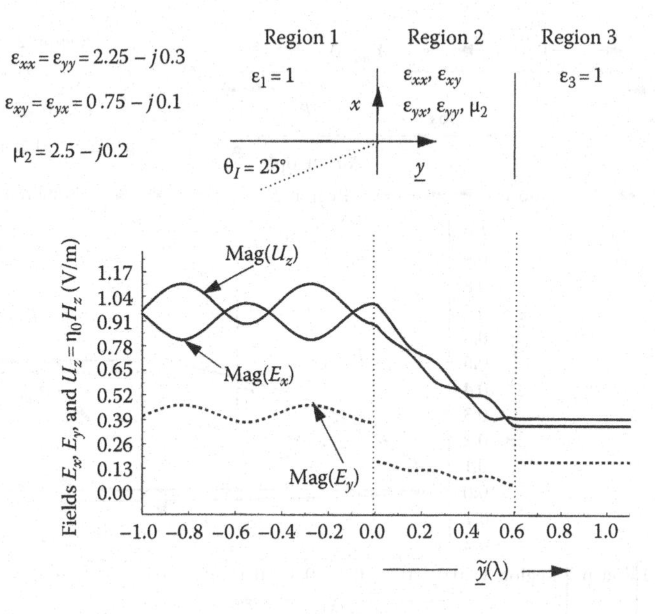

FIGURE 4.11 Plots of the magnitudes of the E_x, E_y, and $U_z = \eta_0 H_z$ EM fields in Regions 1–3 as a function of $\underline{y} = -y$, which is the location of the field relative to the incidence side of the Region 1–2 interface (see Figure 4.10), are shown.

field relative to the incidence side of the Region 1–2 interface [see Figure 4.10]). As can be seen from Figure 4.11, the material slab represents a mismatched medium to the incident wave and thus the incident and reflected waves interfere in Region 1 forming a standing wave pattern. In Region 2, because the layer is lossy, one also observes that all three EM field magnitudes attenuate as the distance from the incident side increases. In Region 2, an SWR pattern is also observed in addition to the attenuation which has already been mentioned. The SWR pattern is caused by the multiple internal reflections which occur within the slab. In Region 3, only a forward traveling transmitted wave is excited, thus the EM field amplitude is constant in this region. One also notices from Figure 4.11 that the tangential electric field (E_x) and tangential magnetic field ($U_z = \eta_0 H_z$) are continuous as they should be and that the normal electric field (E_y) is discontinuous as it should be.

Figure 4.12 shows plots of normalized, dissipated power that results when the complex Poynting theorem of Section 4.2 is used to study the example of this section. In this figure, the Poynting box has been chosen to extend a half wavelength into Region 1 (see Figure 4.12), to extend a variable distance $\tilde{Y}_{out}$ (units of λ) into Region 2 when $\tilde{Y}_{out} \leq 0.6\lambda$, and into Region 3 when $\tilde{Y}_{out} > 0.6\lambda$. In this figure, P_{dexx}, P_{dexy}, etc., are given by the integrals $P_{dexx} = \int S_x \varepsilon''_{xx} S_x^* dy$, $P_{dexy} = \int S_x \varepsilon''_{xy} S_y^* dy$, etc., and $P_{DE} = P_{dexx} + P_{dexy} + P_{deyx} + P_{deyy}$. Also, $P_{DM} = P_{dmzz} = \int U_z \mu''_{zz} U_z^* dy$. As can be seen from Figure 4.12, the dissipated electric and magnetic powers P_{DE} and P_{DM} are zero at $\tilde{Y}_{out} = 0$ and increase in a monotonic fashion until $\tilde{Y}_{out} = 0.6\lambda$, where they become constant for $\tilde{Y}_{out} > 0.6\lambda$. This is exactly to be expected since the only loss in the system is in Region 2, where $0 \leq \tilde{Y}_{out} \leq 0.6\lambda$. We also note that the integrals P_{dexy} and P_{deyx} are complex and satisfy $P_{dexy} = P_{deyx}^*$, as expected. Thus $P_{dexy} + P_{deyx} = 2$ Re(P_{dexy}). The integrals P_{dexx} and P_{deyy} are purely real, thus the electric dissipation integral P_{DE} is purely real. Note, as can be seen from Figure 4.12, that although the total electric dissipation integral is positive, the cross-term contribution given by $P_{dexy} + P_{deyx} = 2$ Re(P_{dexy}) is negative. This is interesting as one would usually associate only positive values with typical power dissipation terms.

Figure 4.13 shows plots of normalized energy–power terms as results from Equations 4.21 through 4.27 using the example of this section. In this figure, as in the previous one, the Poynting

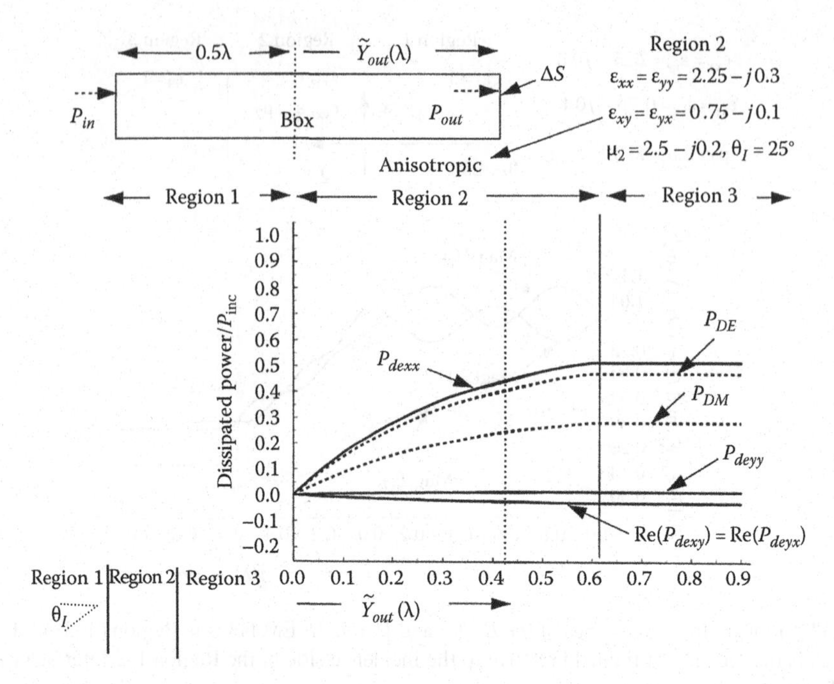

FIGURE 4.12 Plots of normalized, dissipated power that results when the complex Poynting theorem is used to study the numerical example specified in Section 4.3.3.

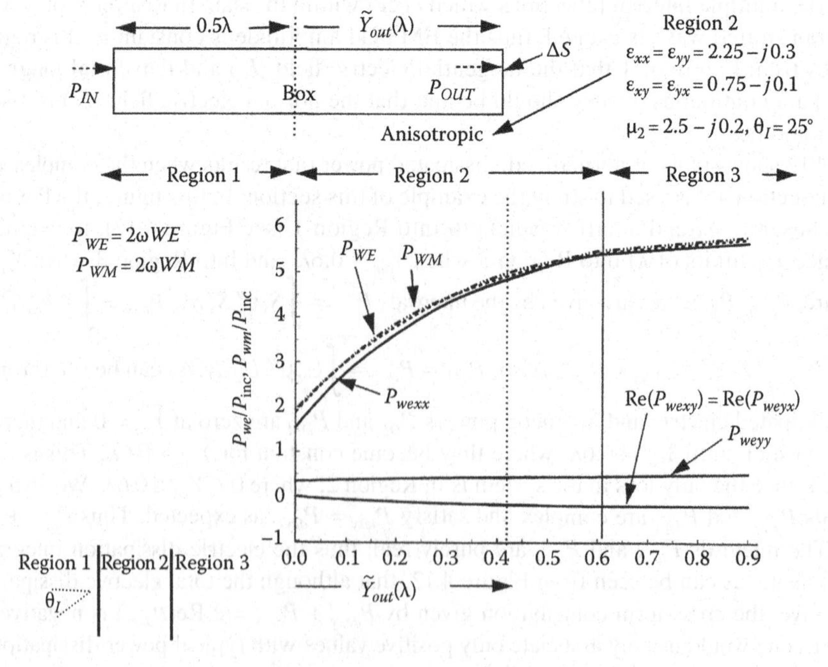

FIGURE 4.13 Plots of normalized energy–power terms using the numerical example specified in Section 4.3.3.

box has been chosen to extend a half wavelength into Region 1 (see Figure 4.13), to extend a variable distance $\tilde{Y}_{out}$ into Region 2 when $\tilde{Y}_{out} \le 0.6\lambda$, and into Region 3 when $\tilde{Y}_{out} > 0.6\lambda$. In this figure, P_{wexx}, P_{wexy}, etc., are given by the integrals $P_{wexx} = \int S_x \varepsilon'_{xx} S_x^* dy$, $P_{wexy} = \int S_x \varepsilon'_{xy} S_y^* dy$, etc., and $P_{WE} = P_{wexx} + P_{wexy} + P_{weyx} + P_{weyy}$. Also, $P_{WM} = P_{wmzz} = \int U_z \mu'_{zz} U_z^* dy$. As can be seen from Figure 4.13, the stored electric and magnetic energy–powers P_{WE} are nonzero at $\tilde{Y}_{out} = 0$ and increase in a monotonic fashion thereafter. As in the case of the dissipation power integrals, we note that the integrals P_{wexy} and P_{weyx} are complex and satisfy $P_{wexy} = P_{weyx}^*$. Thus $P_{wexy} + P_{weyx} = 2\,\text{Re}(P_{wexy})$. The integrals P_{wexx} and P_{weyy} are purely real, thus the electric energy–power integral P_{WE} is purely real. Note, as can be seen from Figure 4.13, that although the total electric energy–power integral is positive, the cross-term contribution given by $P_{wexy} + P_{weyx} = 2\,\text{Re}(P_{wexy})$ is negative.

Figure 4.14 shows plots of the real and imaginary parts of the complex Poynting theorem terms as results from Equations 4.21 through 4.27, using the same Poynting box as in Figures 4.12 and 4.13. In this figure, since we are testing the numerical accuracy of the computation formulas, we let $P_{BOX} = P_{OUT} + P_{DE} + P_{DM} + j(-P_{WE} + P_{WM})$ and compare P_{IN} and P_{BOX}. As can be seen from Figure 4.14, the real and imaginary parts of P_{IN} ("cross") and P_{BOX} ("solid line") are numerically indistinguishable from one another, showing that the numerical computations have been carried out accurately. Figure 4.14 also shows plots of $\text{Re}(P_{OUT})$ which decreases as $\tilde{Y}_{out}$ increases, and $P_D = P_{DE} + P_{DM}$ (P_D is purely real) which increases as $\tilde{Y}_{out}$ increases. As can be seen from Figure 4.14, the sum of these two quantities, namely $\text{Re}(P_{OUT}) + P_D$, adds to $\text{Re}(P_{IN})$ which is constant as $\tilde{Y}_{out}$ increases. It makes sense that the $\text{Re}(P_{OUT})$ decreases as $\tilde{Y}_{out}$ increases since the EM fields are increasingly dissipated in power loss as $\tilde{Y}_{out}$ increases. Figure 4.14 shows plots of $\text{Imag}(P_{OUT})$ and the energy difference term $-P_{WE} + P_{WM}$. As can be seen from Figure 4.14 within Region 2, the two terms are oscillatory, with the oscillatory terms out of phase with one another by 180°. When the two terms are added together, they add to a constant value which equals as it should be to $\text{Imag}(P_{OUT}) = \text{Imag}(P_{IN})$. The complex Poynting results of this section are similar to those of Section 4.2.

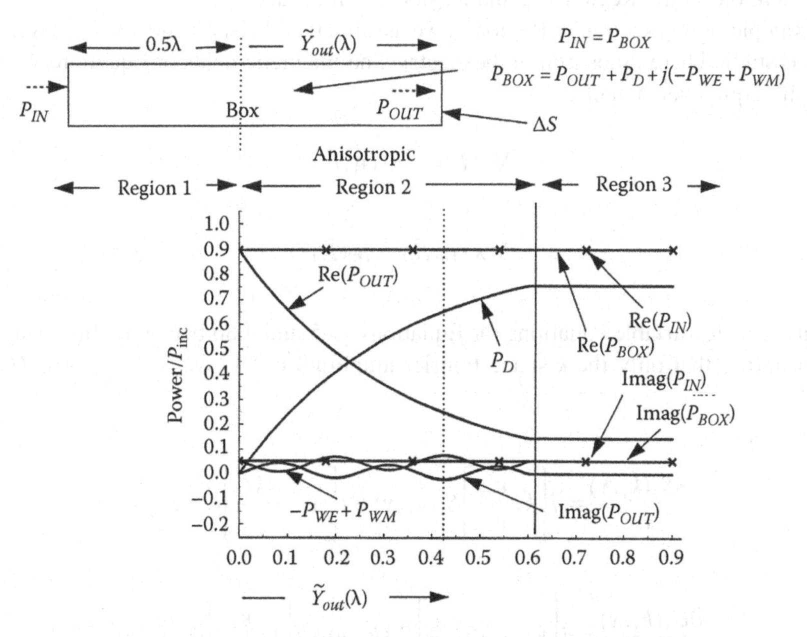

FIGURE 4.14 Plots of the real and imaginary parts of the complex Poynting theorem terms are shown.

4.4 ONE-DIMENSIONAL k-SPACE STATE VARIABLE SOLUTION

4.4.1 INTRODUCTION

In this section, we apply the state variable method to solve problems where the EM field profiles vary in one transverse dimension and are incident on, in general, anisotropic slab. The anisotropic slab is assumed to be bounded by either a homogeneous lossless half space or a perfect electric or magnetic conductor. Examples of this type of problem are a one-dimensional Gaussian beam incident on the material slab, an electric or magnetic line source incident on the slab (or located within the slab), or a slot radiating from a ground plane located adjacent to the material slab. In this section, we assume that the EM fields vary in the x- and y-directions and are constant in the z-direction.

4.4.2 k-SPACE FORMULATION

To begin the analysis, we expand the EM fields in Regions 1–3 in a one-dimensional Fourier transform [1–8] (also called a k-space expansion) and substitute these fields in Maxwell's equations. As in other sections, all coordinates are normalized as $x = k_0\tilde{x}$, $y = k_0\tilde{y}$, etc. We have

$$\vec{E}(x,y) = \int_{-\infty}^{\infty} \vec{S}(k_x, y)\exp(-j\psi)\,dk_x \tag{4.83}$$

$$\eta_0\vec{H}(x,y) = \int_{-\infty}^{\infty} \vec{U}(k_x, y)\exp(-j\psi)\,dk_x \tag{4.84}$$

where $\psi = k_x x$, and $\vec{E}$ and $\vec{H}$ may be the EM fields in any of the regions under consideration. Our objective is to find the EM field solutions in Regions 1–3 of space and then to match appropriate EM boundary conditions at the Region 1–2 and Region 2–3 interfaces.

For an example in this section in Region 2, we assume that the same anisotropic layer studied in Section 4.3 is studied here. Substituting the electric and magnetic fields of Equations 4.83 and 4.84 into Maxwell's equations, namely,

$$\nabla \times \vec{E} = -j\underline{\underline{\mu}}(\eta_0\vec{H}) \tag{4.85}$$

$$\nabla \times (\eta_0\vec{H}) = j\underline{\underline{\varepsilon}}(\vec{E}) \tag{4.86}$$

we find that the state variable equations for Equations 4.85 and 4.86 become, after using Equation 4.57 and assuming that only the k-space Fourier amplitudes $S_x(k_x, y)$, $S_y(k_x, y)$, and $U_z(k_x, y)$ are nonzero,

$$\frac{\partial S_x(k_x,y)}{\partial y} = j\left[k_x\frac{\varepsilon_{yx}}{\varepsilon_{yy}}\right]S_x(k_x,y) + j\left[\mu_{zz} - \frac{k_x^2}{\varepsilon_{yy}}\right]U_z(k_x,y) \tag{4.87}$$

$$\frac{\partial U_z(k_x,y)}{\partial y} = j\left[\varepsilon_{xx} - \frac{\varepsilon_{xy}\varepsilon_{yx}}{\varepsilon_{yy}}\right]S_x(k_x,y) + j\left[k_x\frac{\varepsilon_{xy}}{\varepsilon_{yy}}\right]U_z(k_x,y) \tag{4.88}$$

The eigenvalue solution of Equations 4.87 and 4.88 for $S_x(k_x, y)$ and $S_y(k_x, y)$ for each k_x, when substituted in the Fourier integrals of Equations 4.83 and 4.84, respectively, then represents a general solution for the EM field in Region 2.

4.4.3 Ground Plane Slot Waveguide System

As a specific example of the theory of this section, we consider the problem of a slot parallel plate waveguide radiating from an infinite ground plane through an anisotropic material slab into a homogeneous half space. Figure 4.15 shows the geometry of the system. We initially assume that the EM fields inside the slot waveguide consist only of an incident and reflected TEM waveguide mode whose incident amplitude is E_0 (V/m) and whose reflected amplitude is R_0 (V/m) to be determined. The material parameters in the slot are taken to be lossless, isotropic, and characterized by relative parameters ε_3 and μ_3. We assume that the material layer (Region 2) has a finite thickness L and assume that the only nonzero, lossy, relative material parameters in the slab are ε_{xx}, ε_{xy}, ε_{yx}, ε_{yy}, and $\mu_{xx} = \mu_{yy} = \mu_{zz} = \mu = \mu' - j\mu''$. The infinite half space is assumed to have lossless material parameters ε_1 and μ_1. Assuming only a TEM wave in Region 3, we find that the EM fields in the waveguide slot referring to Figure 4.15 are given by

$$E_{xI} = E_0 \exp(-jk_3(y+L)) \tag{4.89}$$

$$H_{zI} = \left(-\frac{E_0}{\eta_3}\right)\exp(-jk_3(y+L)) \tag{4.90}$$

$$E_{xR} = R_0 \exp(jk_3(y+L)) \tag{4.91}$$

$$H_{zR} = \left(\frac{R_0}{\eta_3}\right)\exp(jk_3(y+L)) \tag{4.92}$$

$$E_x^{(3)} = E_{xI} + E_{xR} \tag{4.93}$$

$$H_z^{(3)} = H_{zI} + H_{zR} \tag{4.94}$$

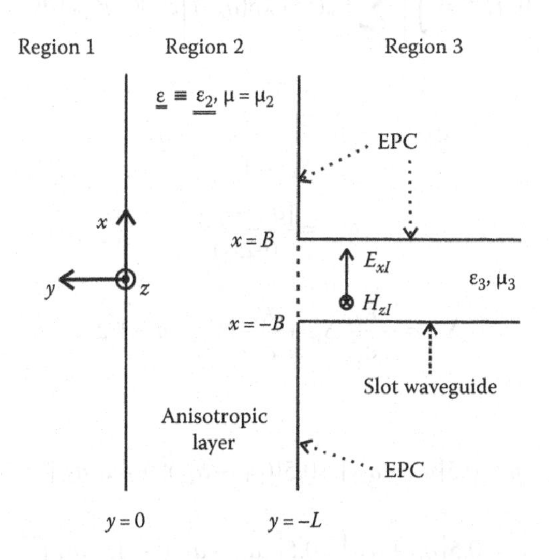

FIGURE 4.15 The geometry of the ground plane slot waveguide system.

for $|x| \le 2B$ and zero elsewhere in Region 3. In Equations 4.89 through 4.94, $\eta_0 = 377\ \Omega$, $\eta_3 = \sqrt{\mu_3/\varepsilon_3}$, $k_3 = \sqrt{\mu_3\varepsilon_3}$, $B = k_0\tilde{B}$, and $\tilde{B}$ (m) is the waveguide slot half width. Since the EM fields are independent of the z-direction, it turns out that the only nonzero field components in all regions of space are the E_x, E_y, and H_z components. The general state variable equations in matrix form are from Equations 4.87 and 4.88:

$$\frac{\partial \underline{V}}{\partial y} = \underline{A}\underline{V}, \quad \underline{A} = \begin{bmatrix} a_{11} & a_{12} \\ a_{21} & a_{22} \end{bmatrix} \tag{4.95}$$

where

$$a_{11} = j\left[k_x \frac{\varepsilon_{yx}}{\varepsilon_{yy}}\right], \quad a_{12} = j\left[\mu_{zz} - \frac{k_x^2}{\varepsilon_{yy}}\right] \tag{4.96}$$

$$a_{21} = j\left[\varepsilon_{xx} - \frac{\varepsilon_{xy}\varepsilon_{yx}}{\varepsilon_{yy}}\right], \quad a_{22} = j\left[k_x \frac{\varepsilon_{xy}}{\varepsilon_{yy}}\right] \tag{4.97}$$

and where $\underline{V} = [S_x, U_z]^t$. These are in fact the same equations as were studied in Section 4.3 except that here S_x, U_z represent k-space Fourier amplitudes rather than the spatial EM field components studied in Section 4.3. The general solution to Equation 4.95 in Region 2 is

$$E_x^{(2)} = \int_{-\infty}^{\infty}\left[\sum_{n=1}^{2} C_n S_{xn} \exp(q_n y)\right]\exp(-jk_x x)dk_x \tag{4.98}$$

$$E_y^{(2)} = \int_{-\infty}^{\infty}\left[\sum_{n=1}^{2} C_n S_{yn} \exp(q_n y)\right]\exp(-jk_x x)dk_x \tag{4.99}$$

$$\eta_0 H_z^{(2)} = \int_{-\infty}^{\infty}\left[\sum_{n=1}^{2} C_n U_{zn} \exp(q_n y)\right]\exp(-jk_x x)\,dk_x \tag{4.100}$$

where

$$S_{xn} = 1 \tag{4.101}$$

$$U_{zn} = \frac{[a_{11} - q_n]}{(-a_{12})} \tag{4.102}$$

$$S_{yn} = -\frac{\varepsilon_{yx}}{\varepsilon_{yy}} S_{yn} + \frac{k_x}{\varepsilon_{yy}} U_{zn}, \quad n = 1, 2 \tag{4.103}$$

and where

$$q_1 = 0.5[a_{11} + a_{22}] + 0.5[(a_{11} - a_{22})^2 + 4a_{12}a_{21}]^{1/2} \tag{4.104}$$

$$q_2 = 0.5\left[a_{11} + a_{22}\right] - 0.5\left[(a_{11} - a_{22})^2 + 4a_{12}a_{21}\right]^{1/2} \tag{4.105}$$

From Maxwell's equations and including the boundary condition that only an outgoing wave can propagate in Region 1 ($y > 0$), the EM fields in Region 1 are given by

$$E_x^{(1)} = \int_{-\infty}^{\infty} \left[-\frac{k_{y1}}{\varepsilon_1} U_z^{(1)}(k_x) \right] \exp(-jk_x x - jk_{y1} y) dk_x \tag{4.106}$$

$$E_y^{(1)} = \int_{-\infty}^{\infty} \left[\frac{k_x}{\varepsilon_1} U_z^{(1)}(k_x) \right] \exp(-jk_x x - jk_{y1} y) dk_x \tag{4.107}$$

$$\eta_0 H_z^{(1)} = \int_{-\infty}^{\infty} U_z^{(1)}(k_x) \exp(-jk_x x - jk_{y1} y) dk_x \tag{4.108}$$

where

$$k_{y1} = \begin{cases} \left[\mu_1\varepsilon_1 - k_x^2\right]^{1/2}, & \mu_1\varepsilon_1 - k_x^2 \geq 0 \\ -j\left[\mu_1\varepsilon_1 - k_x^2\right]^{1/2}, & \mu_1\varepsilon_1 - k_x^2 < 0 \end{cases} \tag{4.109}$$

The minus sign of k_{y1} (or branch of k_{y1}) was chosen on the physical grounds that the integrals converge as $y \to \infty$ when $|k_x| > \sqrt{\mu_1\varepsilon_1}$.

To proceed further, it is necessary to match EM boundary conditions at the Region 1–2 interfaces and Region 2–3 interfaces. To facilitate the Region 2–3 EM boundary matching, it is convenient to represent and replace the waveguide aperture slot with an equivalent magnetic surface current $\vec{M}_s$ which is backed by an electrical perfect conductor. The boundary condition equation to determine the equivalent magnetic surface current $\vec{M}_s$ backed by an infinite ground plane is

$$\hat{a}_y \times \left(\vec{E}^{(2)}\Big|_{y=-L^+} - \vec{E}^{(3)}\Big|_{y=-L^-} \right) = -\vec{M}_s \tag{4.110}$$

where

$$\vec{E}^{(3)}\Big|_{y=-L^-} = 0 \tag{4.111}$$

since the magnetic surface current is assumed to be backed by an infinite ground plane and where

$$\vec{E}_2\Big|_{y=-L^+} = E_A(x)\,\mathrm{rect}\left(\frac{x}{2B}\right)\hat{a}_x \tag{4.112}$$

where

$$\mathrm{rect}\left(\frac{x}{2B}\right) = \begin{cases} 1, & |x| < B \\ 0, & |x| > B \end{cases} \tag{4.113}$$

$E_A(x)$ represents the x-component of the electric field in the aperture. Using Equation 4.112, it is found that the equivalent magnetic surface current is given by

$$\vec{M}_s = \hat{a}_z E_A(x)\, U_{step}(x) = \hat{a}_z \int_{-\infty}^{\infty} M(k_x)\exp(-jk_x x)\,dk_x \tag{4.114}$$

The last part of Equation 4.114 expresses $\vec{M}_s$ in k-space. For the present problem, the aperture electric field is given by Equation 4.112 evaluated at $y = -L_-$. Thus in this equation, E_A is a constant and is given by $E_A = E_0 + R_0$. Using this value of E_A, it is found from Fourier inversion that

$$M(k_x) = \frac{BE_A}{\pi}\left[\frac{\sin(k_x B)}{(k_x B)}\right] \tag{4.115}$$

Matching the tangential electric field (E_x-component) and the tangential magnetic field (H_z-component) at the Region 1–2 interface, matching the tangential electric field (E_x-component) at $y = -L^+$ (Region 2) to the magnetic surface current $\vec{M}_s$, and recognizing that the Fourier amplitudes of all the k-space integrals must equal each other for all values of k_x, we find the following equations:

$$-\frac{k_{y1}}{\varepsilon_1} U_z^{(1)}(k_x) = \sum_{n=1}^{2} C_n S_{xn} \tag{4.116}$$

$$U_z^{(1)}(k_x) = \sum_{n=1}^{2} C_n U_{zn} \tag{4.117}$$

$$\sum_{n=1}^{2} C_n S_{xn} \exp(-q_n L) = M(k_x) \tag{4.118}$$

If we eliminate $U_z^{(1)}(k_x)$ from Equations 4.116 through 4.117, we are left with a 2×2 set of equations from which to determine C_1 and C_2 in terms of $M(k_x)$. We find

$$C_1 = \frac{-T_2 M(k_x)}{T_1 \exp(-q_2 L) - T_2 \exp(-q_1 L)} \tag{4.119}$$

$$C_2 = \frac{T_2 M(k_x)}{T_1 \exp(-q_2 L) - T_2 \exp(-q_1 L)} \tag{4.120}$$

where

$$T_n = a_{12} - \frac{k_{y1}}{\varepsilon_1}\left[a_{11} - q_n\right], \quad n = 1,2 \tag{4.121}$$

The last boundary condition to be imposed is that the tangential magnetic field at $y = -L^+$ (Region 2) match the tangential magnetic field at $y = -L^-$ (Region 3, inside the waveguide aperture). We have

$$\eta_0 H_z^{(2)}\Big|_{y=-L^+} = \eta_0 H_z^{(3)}\Big|_{y=-L^-}, \quad |x| \le 2B \tag{4.122}$$

In this section, we will enforce this boundary condition by averaging Equation 4.122 over the width of the waveguide slot $|x| \le 2B$. Integrating over $|x| \le 2B$ and dividing by $2B$, we have

$$\frac{1}{2B}\int_{-B}^{B} \eta_0 H_z^{(2)}\Big|_{y=-L^+} dx = \frac{1}{2B}\int_{-B}^{B} \eta_0 H_z^{(3)}\Big|_{y=-L^-} dx \tag{4.123}$$

The right-hand side of Equation 4.123 integrates after using Equation 4.94 to

$$\frac{1}{2B}\int_{-B}^{B} \eta_0 H_z^{(3)}\Big|_{y=-L^-} dx = -\frac{1}{\eta_3}[E_0 - R_0] \tag{4.124}$$

Thus

$$-\frac{1}{\eta_3}[E_0 - R_0] = \frac{1}{2B}\int_{-B}^{B} \eta_0 H_z^{(2)}\Big|_{y=-L^-} dx \tag{4.125}$$

When only TEM waves propagate in a parallel plate waveguide, the parallel plate waveguide forms a two-conductor transmission line system. An important quantity associated with the transmission line system is a quantity called the transmission line admittance, which for the present case at location y on the line $y \le -L$ is defined as

$$\tilde{Y}(y) = -\frac{H_z^{(3)}(y)}{E_x^{(3)}(y)} \tag{4.126}$$

and for the present case using Equations 4.93 and 4.94 is given by

$$\tilde{Y}(y) = \frac{1}{\eta_0\eta_3}\frac{E_0\exp(-jk_3(y+L)) - R_0\exp(-jk_3(y+L))}{E_0\exp(-jk_3(y+L)) + R_0\exp(-jk_3(y+L))} \tag{4.127}$$

This quantity is useful for transmission lines because once a transmission line admittance load, call it $\tilde{Y}_{LOAD}$, is specified at a given point on the line, it is possible to find a relation between the incident wave amplitude E_0 (assumed known) and the reflected wave amplitude R_0. With E_0 assumed known and R_0 known from Equation 4.127, the fields everywhere on the line may then be determined using Equations 4.89 through 4.94.

In the present problem, we define a transmission line load admittance to be located at the waveguide aperture at $y = -L$. In this case, we find, calling the transmission line load admittance $\tilde{Y}_{A\ell}$ (in units of Ω^{-1} (or mho), the subscript A refers to aperture) that

$$\tilde{Y}_{A\ell} = -\frac{H_z^{(3)}\Big|_{y=-L^-}}{E_x^{(3)}\Big|_{y=-L^-}} = -\frac{H_z^{(3)}\Big|_{y=-L^-}}{E_A} = -\left[\frac{-1}{\eta_0\eta_3}\frac{E_0 - R_0}{E_A}\right] \tag{4.128}$$

If we replace $-(1/\eta_3)[E_0 - R_0]$ by $(1/2B)\int_{-B}^{B} \eta_0 H_z^{(2)}\Big|_{y=-L^-} dx$ using Equation 4.125, we find

$$\tilde{Y}_{A\ell} = -\frac{1}{\eta_0}\frac{1}{E_A}\left\{\frac{1}{2B}\int_{-B}^{B} \eta_0 H_z^{(2)}\Big|_{y=-L^+} dx\right\} \tag{4.129}$$

Defining a normalized aperture load admittance, we have

$$Y_{A\ell} = \eta_0 \tilde{Y}_{A\ell} = -\frac{(1/2B)\int_{-B}^{B} \eta_0 H_z^{(2)}\Big|_{y=-L^+} dx}{E_A} \tag{4.130}$$

If we substitute the EM field solution for the magnetic field in Region 2 into Equation 4.130, interchange the "dx" and "dk_x" integrals in the numerator of Equation 4.130, and cancel the common constant E_A in the numerator ($\eta_0 H_z^{(2)}$ depends on $M(k_x)$ which is proportional to E_A, see Equations 4.100, 4.115, 4.119, and 4.120) and denominator of Equation 4.130, we find the following expression for the normalized aperture load admittance:

$$Y_{A\ell} = \int_{-\infty}^{\infty} Y(k_x)\,dk_x \tag{4.131}$$

where

$$Y(k_x) = \frac{B}{\pi}\left[\frac{T_2 U_{z1}\exp(-q_1 L) - T_1 U_{z2}\exp(-q_2 L)}{T_1\exp(-q_2 L) - T_2\exp(-q_1 L)}\right]\left(\frac{\sin(k_x B)}{k_x B}\right)^2 \tag{4.132}$$

We remind readers that in the above equation, the quantity in square brackets is a complicated function of k_x, and U_{zn}, $n = 1, 2$, are eigenvector components associated with the magnetic field in Region 2. Once the integral in Equation 4.131 is carried out, $Y_{A\ell}$ is known and then a relationship between E_0 and R_0 can be found through the equation

$$Y_{A\ell} = \frac{1}{\eta_3}\frac{E_0 - R_0}{E_0 + R_0} \tag{4.133}$$

If E_0 is assumed to be known, then the normalized reflection coefficient of the system is

$$r \equiv \frac{R_0}{E_0} = \frac{(1/\eta_3) - Y_{A\ell}}{(1/\eta_3) + Y_{A\ell}} \tag{4.134}$$

In computing the integral as given in Equation 4.134, care must be used in carrying out the integral near the points where $k_x = \pm k_1$ when $k_1 - \delta \le |k_x| \le k_1$ (this interval is in the visible region) and $k_1 + \delta \ge |k_x| \ge k_1$ (this interval is in the invisible region), where δ is a small number say on the order of $k_1/4$ or possibly less. The reason for this is that the function in square brackets in the integrand of $Y_{A\ell}$ integral may be discontinuous (or even singular) near the points $k_x = \pm k_1$, and thus significant numerical error could occur if a very fine numerical integration grid was not used around these points. In this section, Chebyshev quadrature (using the formulas $k_x = k_1\cos(u)$, $0 \le u \le \pi$ in the visible region and $k_x = k_1\cosh(u)$, $0 \le u \le \infty$ in the invisible region) was employed to integrate the $Y_{A\ell}$ integral. These formulas provide a very dense grid near $k_x = \pm k_1$ and thus provide an accurate integration of the $Y_{A\ell}$ integral.

Harrington [3, p. 183, eqs (4-104–4-105)] defines an aperture admittance for the present slot radiator problem through the Parseval power relation

$$\tilde{Y}_A = \frac{\tilde{P}^*}{|V|^2} \tag{4.135}$$

where $V = 2\tilde{B}E_A$, $E_A = 1$ (V/m) and where

$$\tilde{P} = -\int_{-\infty}^{\infty}\left[E_x^{(2)}\Big|_{\tilde{y}=-\tilde{L}^+}\right]\left[H_z^{(2)}\Big|_{\tilde{y}=-\tilde{L}^+}\right]^* d\tilde{x} = -\frac{1}{2\pi}\int_{-\infty}^{\infty}\bar{E}_x(\tilde{k}_x)\bar{H}_z^*(\tilde{k}_x)d\tilde{k}_x \tag{4.136}$$

where $\bar{E}_x(\tilde{k}_x)$ and $\bar{H}_z(\tilde{k}_x)$ are the Fourier amplitudes (or k-space pattern space factors) of the of $E_x^{(2)}$ electric field and the $H_z^{(2)}$ magnetic field, respectively. $\tilde{P}$ has units of (W/m) = (VA/m), so $\tilde{Y}_A$ has units of $(\Omega\ \mathrm{m})^{-1}$ (or mho/m). Substituting the EM field solutions derived earlier in Equation 4.136, it is found that the aperture admittance $\tilde{Y}_A$ as defined by Equation 4.136 is very closely related to the transmission line load admittance expression $\tilde{Y}_{A\ell}$. It is related by the equation

$$\tilde{Y}_A = \frac{\tilde{Y}_{A\ell}}{(2\tilde{B})} \tag{4.137}$$

where $2\tilde{B}$ is the width of the slot.

We note that in calculating the $Y_{A\ell}$ integral using Equation 4.131 in the limit as $L \to 0$, the exponential terms in Equation 4.132 approach unity, and it is found after a small amount of algebra that

$$Y_{A\ell} = \eta_0\,\tilde{Y}_{A\ell} = \int_{-\infty}^{\infty}\frac{B\varepsilon_1}{\pi k_{y1}}\left[\frac{\sin(k_xB)}{k_xB}\right]^2 dk_x \tag{4.138}$$

which is an expression for the aperture load admittance of a slot waveguide radiating into a homogeneous lossless half space. If one substitutes $\tilde{Y}_{A\ell}$ as given by Equation 4.138 in the aperture admittance expression as given by Equation 4.137, one derives the same expression as derived by Harrington [3, p. 183, eqs (4-104–4-105)] for a ground plane slot radiating into a lossless half space.

Another quantity of interest is the power that is radiated as one moves infinitely far away from the radiating slot. The Poynting vector at a location $x = \rho\cos(\varphi_c)$, $y = \rho\sin(\varphi_c)$, $\rho \to \infty$ is given by

$$\vec{S} = \frac{1}{2}\mathrm{Re}(\vec{E}^{(1)}\times\vec{H}^{(1)*}) = \frac{1}{2}\frac{\eta_1}{\eta_0}\left|U_z^{(1)}\right|^2\hat{a}_\rho \tag{4.139}$$

where

$$U_z^{(1)} = \int_{-\infty}^{\infty} A(k_x)\exp(-jk_xx - jk_{y1}y)\,dk_x \tag{4.140}$$

and where

$$A(k_x) = -\varepsilon_1\left(\frac{BE_A}{\pi}\right)\frac{[-T_2+T_1]}{k_{y1}\left[T_1\exp(-q_2L)-T_2\exp(-q_1L)\right]}\left(\frac{\sin(k_xB)}{k_xB}\right) \tag{4.141}$$

We note in passing that Equation 4.140 for $U_z^{(1)}$ is identical to that given by Ishimaru [4, chapter 14] when one (1) lets the dielectric layer to be isotropic, (2) lets the slot waveguide width $2\tilde{B}$ to approach zero while holding the voltage potential difference between the parallel plate conductors constant, and (3) makes the correct geometric association between Ishimaru's analysis and the present one.

Ishimaru [4] shows by using the method of steepest descent that as $\rho \to \infty$, the integral in Equation 4.140 asymptotically approaches the value

$$U_z^{(1)} = F(k_1 \sin(\varphi_c))\left[\frac{2}{k_1\rho}\right]^{1/2} \exp\left(-jk_1\rho + \frac{j\pi}{4}\right) \tag{4.142}$$

where

$$F(k_1 \sin(\varphi_c)) \equiv [k_1 \cos(\varphi_c)]A(k_1 \sin(\varphi_c)) = -\varepsilon_1\left(\frac{BE_A}{\pi}\right)\frac{[-T_2 + T_1]}{[T_1 \exp(-q_2L) - T_2 \exp(-q_1L)]}\left(\frac{\sin(k_xB)}{k_xB}\right) \tag{4.143}$$

and where $k_1 \sin(\varphi_c)$ and $k_1 \cos(\varphi_c)$ have been substituted for k_x and k_{y1}, respectively. To describe the radiation from the waveguide aperture and material slab system in the far field ($\rho \to \infty$), we plot the normalized radiation intensity which is defined here as the radiation intensity $\rho S_\rho(\rho, \varphi_c)$, $\rho \to \infty$ divided by the total radiation intensity integrated from $\varphi_c = -\pi/2$ to $\varphi_c = \pi/2$. This quantity is called the "Directive Gain" $D(\varphi_c)$. Applying this definition and using Equations 4.142 and 4.143 after canceling common factors, we find

$$D(\varphi_c) = \pi \frac{|F(k_1 \sin(\varphi_c))|^2}{\int_{-\pi/2}^{\pi/2} |F(k_1 \sin(\varphi_c))|^2 \, d\varphi_c} \tag{4.144}$$

4.4.4 Ground Plane Slot Waveguide System, Numerical Results

As a numerical example of the radiation through a waveguide slot radiating through the anisotropic layer under study, we consider the layer formed when

$$\varepsilon_1 = 1, \quad \mu_1 = 1, \quad \mu_2 = \mu = 1.2 - j2.6, \quad \underline{\underline{\varepsilon}}_2 = \begin{bmatrix} \varepsilon_{xx} & \varepsilon_{xy} & 0 \\ \varepsilon_{yx} & \varepsilon_{yy} & 0 \\ 0 & 0 & \varepsilon_{zz} \end{bmatrix} \tag{4.145}$$

where $\varepsilon_{xx} = 2$, $\varepsilon_{xy} = 0.3$, $\varepsilon_{yx} = 0.9 - j0.2$, and $\varepsilon_{yy} = 2.1$. The value of ε_{zz} is immaterial to the present analysis and is not specified here. For all calculations in this section, the slot width has been taken to be $2\tilde{B} = 0.6\lambda$.

Figure 4.16 shows a plot of the $Y(k_x)$ aperture admittance integrand when the layer thickness has been taken to be $\tilde{L} = 0.6\lambda$. As can be seen from Figure 4.16 for the values used in the present example, the integrand converges fairly rapidly for values of $|k_x| \geq 5k_1 = 5$. An inspection of Equation 4.132 for $Y(k_x)$ shows that when k_x is large, the integrand approaches $1/k_x^3$ and thus is guaranteed to converge. In an inspection of Figure 4.16, one also finds that the integrand $Y(k_x)$ is not symmetric with respect to the k_x variable. This is a result of the slot radiating through an anisotropic rather than an isotropic medium. For the present example, the boundary of the visible and invisible radiation range is at $k_x = k_1 = \pm 1$. One observes from Figure 4.16 the effect that the discontinuous k_{y1} function

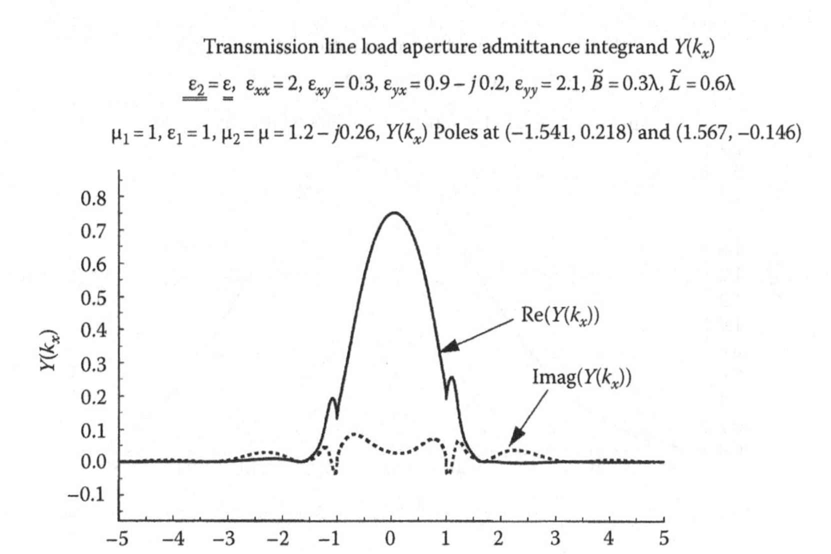

FIGURE 4.16 A plot of the $Y(k_x)$ aperture admittance integrand.

of Equation 4.109 has on the $Y(k_x)$ integrand in the k_x regions near $k_x = k_1 = \pm 1$. Figure 4.16 also lists values of the two lowest magnitude poles that were associated with the $Y(k_x)$ integrand. The two pole locations in the complex k_x plane ($k_{xp1} = -1.541 + j0.218$ and $k_{xp2} = 1.567 - j0.146$) were nonsymmetric because of the anisotropy of the material slab. The values of the poles were listed as they influence the real k_x integration when the k_x integration variable passes close to the poles' location.

Figure 4.17 shows a plot of the $Y_{A\ell}$ aperture load admittance as a function of the layer thickness $\tilde{L}$. At a value of $\tilde{L} = 0$, the layer does not exist and the waveguide aperture radiates into free space. As $\tilde{L}$ increases, the real and imaginary parts of the aperture admittance are oscillatory up to a value of about $\tilde{L} = \lambda$, where it starts to approach a constant value.

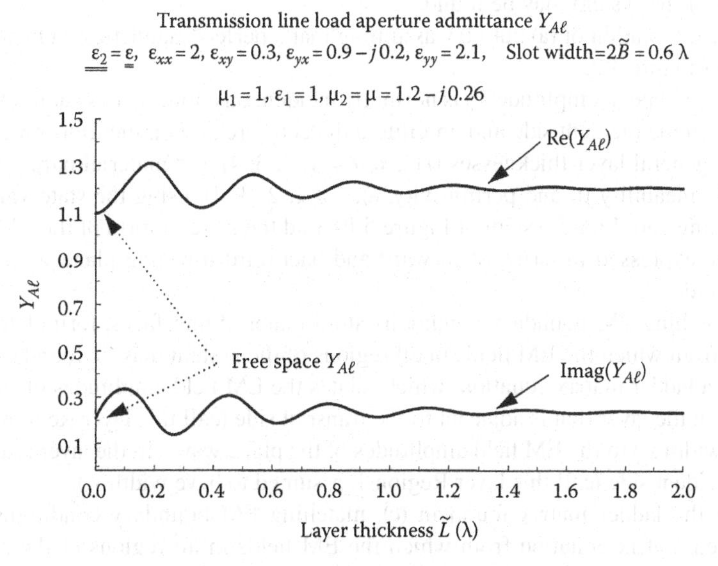

FIGURE 4.17 A plot of the $Y_{A\ell}$ aperture load admittance as a function of the layer thickness $\tilde{L}$.

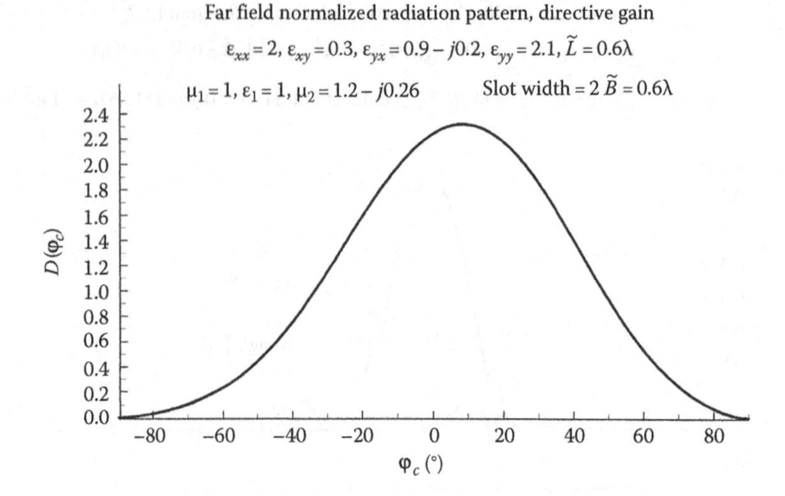

FIGURE 4.18 A plot of the directive gain as a function of the angle φ_c.

Figure 4.18 shows a plot of the directive gain as a function of the angle φ_c. One observes from this figure that the radiation pattern is concentrated in a 90° angle around the broadside direction and one also observes that the radiation pattern is asymmetric in the angle φ_c, with the peak radiation value occurring at about angle $\varphi_c = 10°$. The asymmetry is caused by the fact that the slot has radiated through an anisotropic material slab.

PROBLEMS

4.1 Using the wave equation for the electric field, write down solutions for the electric field that correspond to the three regions shown in Figure 4.1. Assume normal incidence from Region 1. Show that your results are the same as the state variable solutions of Section 4.2.

4.2 (a) Assume that Region 3 shown in Figure 4.2 is a perfect electrical conductor and further assume that a plane wave, located in Region 1 of peak amplitude E_0 (V/m), is normally incident on Region 2. Using the state variable method, formulate equations from which the EM fields in all regions of the system may be found.
(b) Repeat the solution of (a) but now assuming that a perfect magnetic conductor is located in Region 3 of Figure 4.2.

4.3 A plane wave of peak amplitude E_0 is normally incident on a four-layer system located between free space on the incident side and an infinite dielectric region on the transmit side.

(a) Using general layer thicknesses (i.e., s_ℓ, $\ell = 1, 2, 3, 4$) and material parameters (i.e., relative permeability μ_ℓ and permittivity, ε_ℓ, $\ell = 1, 2, 3, 4$), using the state variable method, and using coordinate system of Figure 5.19, find the general form of the EM fields in each region, expressed in terms of forward and backward traveling plane waves of unknown amplitude.

(b) By matching EM boundary conditions at all material interfaces, formulate matrix equations from which the EM fields in all regions of the system may be found.

(c) Form a ladder matrix equation, which relates the EM field amplitudes of the plane waves of (b) in the layer that is adjacent to the transmit side (call this layer Region 4, assumed to have width s_4) to the EM field amplitudes of the plane waves in the layer that is adjacent to the incident side (call this layer Region 1, assumed to have width s_1).

(d) Using the ladder matrix found in (c), matching EM boundary conditions, formulate a reduced matrix equation from which the EM fields in all regions of the system may be found.

(e) Picking specific layer thicknesses (i.e., s_ℓ, $\ell = 1, 2, 3, 4$) and lossless layer material parameters (i.e., relative permeability μ_ℓ and permittivity, ε_ℓ, $\ell = 1, 2, 3, 4$) of your choice, solve numerically all equations formulated in (a)–(d). Present and compare numerical results of (b) and (d).
(f) Repeat (e) using lossy layer material parameters.
(g) Comment on the advantages and disadvantages of the solutions as determined by the methods described in (b) and with the solution described in (d) and (e).

4.4 Referring to Problem 4.3 and the geometry described therein, this problem concerns calculating the real time-averaged power,

$$S_{AV} \equiv \frac{1}{T}\int_0^T \vec{E}(\omega,t)\times\vec{H}(\omega,t)\cdot(-\hat{a}_y)dt = \left\langle \vec{S}(\omega,t)(-\hat{a}_y)\right\rangle_T = \frac{1}{2\eta_0}\mathrm{Re}(\vec{E}(\omega)\times\vec{U}(\omega)^*\cdot(-\hat{a}_y))$$

where $\omega = 2\pi/T$, T is the temporal period and $\vec{E}(\omega, t)$ and $\vec{H}(\omega, t)$ represent the real, time varying EM fields of the system.
(a) Using the general eigensolutions of Problem 4.3, give a general expression for the real time-averaged power that is transmitted in the $-\hat{a}_y$ direction.
(b) Using the formula of (a), numerically calculate S_{AV} for the numerical case used in your Problem 4.3e solution (this was the lossless case) in the regions exterior to the multilayer system. Verify numerically that the incident power equals the sum of the power reflected and transmitted from the multilayer system, thus verifying that conservation of power holds for the system.

4.5 A plane wave of E_0 is obliquely incident on a four-layer system located between free space on the incident side and an infinite dielectric region on the transmit side. Using the coordinate system of Figure 5.19 (assuming the vertical direction in this figure to be the x-direction), the incident plane wave is given by $\vec{E} = E_0\exp(-j\vec{k}_I\cdot\vec{r})\hat{a}_z$, where $\vec{k}_I = (\mu_I\varepsilon_I)^{1/2}\left[x\sin(\theta_I)\hat{a}_x - y\cos(\theta_I)\hat{a}_y\right]$, where μ_I, ε_I are the relative permeability and permittivity, respectively, of the incident region and θ_I is the incident angle of the plane wave.
(a) Using general layer thicknesses (i.e., s_ℓ, $\ell = 1, 2, 3, 4$) and material parameters (i.e., relative permeability μ_ℓ and permittivity, ε_ℓ, $\ell = 1, 2, 3, 4$), write Maxwell's equations in component form in any region for the problem under consideration.
(b) Put the resulting equations of (a) in state variable form for the EM field components that are parallel to the interfaces of the multilayer system and solve for the eigenfunctions associated with the state variable equations.
(c) Using the eigenfunction solutions of (b), find the general form of the EM fields in each region, expressed in terms of forward and backward, obliquely, traveling plane waves of unknown amplitude.
(d) Assuming that the transmit region has material parameters μ_T, ε_T, by matching EM boundary conditions at all material interfaces, formulate matrix equations from which the EM fields in all regions of the system may be found.
(e) Form a ladder matrix equation, which relates the EM field amplitudes of the plane waves of (b) in the layer that is adjacent to the transmit side (call this layer Region 4, assumed to have width s_4) to the EM field amplitudes of the plane waves in the layer that is adjacent to the incident side (call this layer Region 1, assumed to have width s_1).
(f) Using the ladder matrix found in (e), matching EM boundary conditions, formulate a reduced matrix equation from which the EM fields in all regions of the system may be found.
(g) Picking specific layer thicknesses (i.e., s_ℓ, $\ell = 1, 2, 3, 4$) and lossless layer material parameters (i.e., relative permeability μ_ℓ and permittivity, ε_ℓ, $\ell = 1, 2, 3, 4$) of your choice, solve numerically all equations formulated in (a)–(f). Present numerical results of (d), (e), and (f).
(h) For your numerical example, verify that the conservation of power holds in the system.

4.6 This problem concerns using the state variable method to determine the EM fields that result when a plane wave, possessing arbitrary polarization, is obliquely incident on a uniform bianisotropic layer located between free space on the incident side and an infinite dielectric region on the transmit side. The EM fields of the bianisotropic system are defined by the relations of Ref. [6], namely, $\vec{D} = \underline{\tilde{\varepsilon}} \cdot \vec{E} + \underline{\tilde{\xi}} \cdot \vec{H}$ and $\vec{B} = \underline{\tilde{\zeta}} \cdot \vec{E} + \underline{\tilde{\mu}} \cdot \vec{H}$, where $\underline{\tilde{\varepsilon}}$, $\underline{\tilde{\xi}}$, $\underline{\tilde{\zeta}}$, $\underline{\tilde{\mu}}$ are lossy, in general, nonzero complex constant dyadic quantities. Figure 4.10 shows the general layer system geometry assuming that an arbitrarily polarized plane wave is obliquely incident on the system and that $\underline{\tilde{\varepsilon}}$, $\underline{\tilde{\xi}}$, $\underline{\tilde{\zeta}}$, $\underline{\tilde{\mu}}$ characterize the Region 2 material parameters of the system. The analysis of this problem is to be carried by formulating state variable and associated matrix equations to solve Maxwell's equations and then using this formulation to solve a specific numerical example. Complete the following parts to perform the just stated analysis.

(a) Assume that the oblique incident plane wave is given mathematically by

$$\vec{E}_{INC} = \vec{E}_I \exp(-j\psi_I)$$

$$\eta_0 \vec{H}_{INC} = \eta_0 \vec{H}_I \exp(-j\psi_I)$$

$$\psi_I = \vec{k}_I \cdot \vec{r} = k_x x - k_{y1} y + k_z z$$

where

$$\vec{k}_I = k_x \hat{a}_x - k_{y1} \hat{a}_y + k_z \hat{a}_z, \quad \vec{r} = x\hat{a}_x + y\hat{a}_y + z\hat{a}_z$$

$$k_{y1} = \left[\varepsilon_1 \mu_1 - k_x^2 - k_z^2\right] \neq 0$$

Assume that the wave vector values k_x and k_z are known and given, and that the incident plane wave's polarization is specified by known, given values of $E_{xI} = S_{xI}$ and $E_{zI} = S_{zI}$. From Maxwell's equations and the assumed known value of $\vec{k}_I$, find the other field components of the incident plane wave. Find the form of the reflected and transmitted plane waves that propagate in Regions 1 and 3, respectively.

(b) In Region 2, write out Maxwell's curl equations in dimensionless form letting $x = \tilde{k}_0 \tilde{x}$, etc., $\tilde{k}_0 = 2\pi/\lambda$, $\vec{U} = \eta_0 \vec{H}$ using dimensionless dyadics defined by

$$\underline{\varepsilon} = \underline{\tilde{\varepsilon}}/\varepsilon_0 = \underline{\varepsilon}' - j\underline{\varepsilon}'', \quad \underline{\mu} = \underline{\tilde{\mu}}/\mu_0 = \underline{\mu}' - j\underline{\mu}''$$

$$\underline{a} = \underline{a}' - j\underline{a}'' = \sqrt{\mu_0 \varepsilon_0}\ \underline{\tilde{\zeta}} = \frac{k_0}{\omega} \underline{\tilde{\zeta}}, \quad \underline{b} = \underline{b}' - j\underline{b}'' = \sqrt{\mu_0 \varepsilon_0}\ \underline{\tilde{\xi}} = \frac{k_0}{\omega} \underline{\tilde{\xi}}$$

(c) Let all EM field components in the Region 2 material layer be proportional to the factor $\exp(-j\psi)$, where $\psi \equiv \tilde{k}_x \tilde{x} + \tilde{k}_z \tilde{z} = k_x x + k_z z$ (since an incident plane wave possessing this factor is incident on the layer, and phase matching must occur at the interfaces of the slab). Substitute these EM field expressions into Maxwell's normalized equations, and find and write the resulting equations.

(d) From the resulting six equations of (c), simplify these equations to a set of four equations involving only the transverse electric and magnetic field components (i.e., x-, z-components) by expressing the longitudinal electric and magnetic fields in terms of the transverse electric and magnetic field components. Put the resulting four equations in the state variable form

$$\frac{\partial \underline{V}}{\partial y} = \begin{bmatrix} A_{11} & A_{12} & A_{13} & A_{14} \\ A_{21} & A_{22} & A_{23} & A_{24} \\ A_{31} & A_{32} & A_{33} & A_{34} \\ A_{41} & A_{42} & A_{43} & A_{44} \end{bmatrix} \underline{V} = \underline{\underline{A}}\underline{V}$$

Why, possibly, is it useful to reduce the set of six equations to a set of four equations involving only tangential field components?

(e) Solve the state variable equation of (d) for the eigenfunctions of Region 2 and express the general EM field solution of Region 2 in terms of these eigenfunctions.

(f) By matching EM boundary conditions, using the Regions 1, 2, and 3 solutions of (a) and (e), formulate the final matrix equation from which all of the unknowns of the system may be determined.

(g) Using the formulation described in (a)–(f), solve the following numerical example assuming: in Region 1, $\varepsilon_1 = 1.3$ and $\mu_1 = 1.8$, the incident plane wave has the amplitude $S_{xI} = 1$ (V/m), $S_{zI} = 0.9$ (V/m), $k_x = \sin(\theta_I)\cos(\phi_I)$, and $k_z = \sin(\theta_I)\sin(\phi_I)$, where $\theta_I = 35°$, $\phi_I = 65°$; in Region 3, $\varepsilon_3 = 1.9$ and $\mu_3 = 2.7$; in Region 2, we take the layer thickness $\tilde{L} = 0.6\lambda$, and we consider a complicated numerical example where all material parameters of $\underline{\underline{\varepsilon}}$, $\underline{\underline{\mu}}$, $\underline{\underline{a}}$, $\underline{\underline{b}}$ are taken to be nonzero and are given by

$$\underline{\underline{a}} = \begin{bmatrix} -0.3+0.2j & -0.15 & -0.2+0.2j \\ -0.1+0.05j & -0.3 & -0.6+0.65j \\ -0.05 & -0.1+0.1j & -0.25 \end{bmatrix}, \quad \underline{\underline{b}} = \begin{bmatrix} -0.1 & -0.05+0.05j & -0.3 \\ -0.1+0.1j & -0.01 & -0.01 \\ -0.05 & -0.04+0.08j & -0.14 \end{bmatrix}$$

$$\underline{\underline{\varepsilon}} = \begin{bmatrix} 1.3-0.2j & 0.3-0.1j & 0.33-0.07j \\ 0.1 & 2.0 & 0.01 \\ 0.02 & 0.01 & 3.0 \end{bmatrix}, \quad \underline{\underline{\mu}} = \begin{bmatrix} 0.1 & 0.01 & 1.0-0.4j \\ -0.15j & 2.0-0.3j & 0.013 \\ 0.011 & 0.012 & 1.3-0.2j \end{bmatrix}$$

(h) Make sample plots of the EM fields in Regions 1, 2, and 3.

4.7 This problem concerns, using a multilayer k-space algorithm, the study of EM radiation from a parallel plate waveguide (see Figure 4.15).

(a) Re-solve the example of Section 4.4.4 using the k-space single layer algorithm described in the text.

(b) Divide the single layer of Figure 4.15, referred to in (a), into a discrete number of layers and solve for the general form of the EM fields there. Using these solutions and by matching EM boundary conditions at all interfaces, develop a cascaded, multilayer ladder algorithm that relates all EM fields at one interface to other interfaces in the system. Formulate and solve a system matrix from which all of the unknowns of the system may be found. Your solution to this part should almost exactly equal that of (a).

(c) Use the algorithm in (b) to solve the case where instead of each layer having the same material parameters, let the material parameters be different from one another. The solution of this problem represents how k-space theory can be used to solve radiation from an inhomogeneous layer, aperture, waveguide system. Golden and Stewart [35] and Jarem [16] have used two-dimensional, Fourier k-space theory to study reentry vehicle, rectangular waveguide slot radiation through an inhomogeneous plasma layer sheath.

REFERENCES

1. D.R. Rhodes, *Synthesis of Planar Aperture Antenna Sources*, Oxford University Press, London, U.K., 1974.
2. H.A. MacLeod, *Thin-Film Optical Filters*, Macmillan, New York, 1986.
3. R.F. Harrington, *Time Harmonic Fields*, McGraw-Hill Book Company, New York, 1961.
4. A. Ishimaru, *Electromagnetic Wave Propagation, Radiation and Scattering*, Prentice Hall, Upper Saddle River, NJ, 1991.
5. L.B. Felson and N. Marcuvitz, *Radiation and Scattering of Waves*, Prentice Hall, Upper Saddle River, NJ, 1973.
6. I.V. Lindell, A.H. Sihvola, S.A. Tretyakov, and A.J. Vitanen, *Electromagnetic Waves in Chiral and Bi-isotropic Media*, Artech House, Norwood, MA, 1994.
7. J. Galejs, *Antennas in Inhomogeneous Media,* Pergamon Press, Oxford, U.K., 1969.
8. J.R. Wait, *Electromagnetic Waves in Stratified Media*, Pergamon Press, Oxford, U.K., 1970.
9. S.V. Marshall, R.E. DuBroff, and G.G. Skitek, *Electromagnetic Concepts and Applications*, 4th edn., Prentice Hall, Englewood Cliffs, NJ, 1996.
10. P.P. Banerjee and T.C. Poon, *Principles of Applied Optics*, Aksen Associates, Homewood, IL, 1991.
11. P. Baumeister, Utilization of Kard's equations to suppress the high frequency reflectance bands of periodic multilayers, *Appl. Opt.* 24, 2687–2689, 1985.
12. J.A. Dobrowolski and D. Lowe, Optical thin film synthesis based on the use of Fourier transforms, *Appl. Opt.* 17, 3039–3050, 1978.
13. W.H. Southwell, Coating design using very thin high- and low-index layers, *Appl. Opt.* 24, 457–459, 1985.
14. J.M. Jarem, The minimum quality factor of a rectangular antenna aperture, *Arab. J. Sci. Eng.* 7(1), 27–32, 1982.
15. J.M. Jarem, Method-of-moments solution of a parallel-plate waveguide aperture system, *J. Appl. Phys.* 59(10), 3566–3570, 1986.
16. J.M. Jarem, The input impedance and antenna characteristics of a cavity-backed plasma covered ground plane antenna, *IEEE Trans. Antennas Propag.* AP-34(2), 262–267, 1986.
17. J.M. Jarem and F.T. To, A K-space method of moments solution for the aperture electromagnetic fields of a circular cylindrical waveguide radiating into an anisotropic dielectric half space, *IEEE Trans. Antennas Propag.* AP-37, 187–193, 1989.
18. D.A. Holmes and D.L. Feucht, Electromagnetic wave propagation in birefringent multilayers, *J. Opt. Soc. Am.* 56, 1763–1769, 1966.
19. S. Teitler and B.W. Henvis, Refraction in stratified, anisotropic media, *J. Opt. Soc. Am.* 56, 830–834, 1970.
20. D.W. Berreman, Optics in stratified and anisotropic media, *J. Opt. Soc. Am.* 62, 502–510, 1972.
21. P.J. Lin-Chung and S. Teitler, 4 × 4 matrix formalisms for optics in stratified anisotropic media, *J. Opt. Soc. Am. A* 1, 703–705, 1984.
22. C.M. Krowne, Fourier transformed matrix method of finding propagation characteristics of complex anisotropic layer media, *IEEE Trans. Microwave Theory Tech.* MTT-32, 1617–1625, 1984.
23. M.A. Morgan, D.L. Fisher, and E.A. Milne, Electromagnetic scattering by stratified inhomogeneous anisotropic media, *IEEE Trans. Antennas Propag.* AP-35, 191–197, 1987.
24. R.S. Weiss and T.K. Gaylord, Electromagnetic transmission and reflection characteristics of anisotropic multilayered structures, *J. Opt. Soc. Am. A* 4, 1720–1740, 1987.
25. H.Y.D. Yang, A spectral recursive transformation method for electromagnetic waves in generalized anisotropic layered media, *IEEE Trans. Antennas Propag.* AP-45, 520–526, 1997.
26. H.Y.D. Yang, A numerical method of evaluating electromagnetic fields in a generalized anisotropic layered media, *IEEE Trans. Microwave Theory Tech.* MTT-43, 1626–1628, 1995.
27. P. Yeh, Electromagnetic propagation in birefringent layered media, *J. Opt. Soc. Am.* 69, 742–756, 1979.
28. N.G. Alexopoulos and P.L.E. Uslenghi, Reflection and transmission with arbitrarily graded, *J. Opt. Soc. Am.* 71, 1508–1512, 1981.
29. R.D. Graglia, P.L.E. Uslenghi, and R.E. Zich, Dispersion relation for bianisotropic materials and its symmetry properties, *IEEE Trans. Antennas Propag.* AP-39, 83–90, 1991.
30. S.M. Ali and S.F. Mahmoud, Electromagnetic fields of buried sources in stratified anisotropic media, *IEEE Trans. Antennas Propag.* AP-37, 671–678, 1979.
31. C.M. Tang, Electromagnetic fields due to dipole antennas embedded in stratified anisotropic media, *IEEE Trans. Antennas Propag.* AP-27, 665–670, 1979.

32. J.L. Tsalamengas and N.K. Uzunoglu, Radiation from a dipole in the proximity of a general anisotropic grounded layer, *IEEE Trans. Antennas Propag.* AP-33(2), 165–176, 1985.
33. J.L. Tsalamengas, Electromagnetic fields of elementary dipole antennas embedded in stratified inhomogeneous anisotropic media, *IEEE Trans. Antennas Propag.* AP-37, 399–403, 1989.
34. C.M. Krowne, Determination of the Green's function in the spectral domain using a chain matrix method: Application to radiators or resonators immersed in a complex anisotropic layered medium, *IEEE Trans. Antennas Propag.* AP-34, 247–253, 1986.
35. K.E. Golden and G.E. Stewart, Self and mutual admittance of rectangular slot antennas in the presence of an inhomogeneous plasma layer, *IEEE Trans. Antennas Propag.* AP-17(6), 763–771, 1969.

5 Planar Diffraction Gratings

5.1 INTRODUCTION

In the past 40 years, the study and use of periodic structures and diffraction gratings have become increasingly important. Diffraction gratings have been constructed for applications in the frequency ranges of microwaves, millimeter waves, far infrared, infrared, optics, and x-rays. Diffraction gratings occur in applications such as holography; memory storage; spectroscopy; phase conjugation; photorefractives; image reconstruction; optical computing; transducers; integrated optics; microwave phased arrays; acousto-optics; interdigitated, voltage-controlled, liquid crystal displays, and many other areas. Petit (editor) [1], Gaylord and Moharam [2], Solymar and Cooke [3], and Maystre (editor) [4] give extensive reviews of the applications of diffraction gratings. Chapter 7 cites many references that study diffraction gratings in photorefractive materials.

We will give a brief description and overview of the physical makeup of diffraction gratings. Diffraction gratings have been manufactured and constructed in many different forms and types. Two main classifications of diffraction gratings are those that are *metallic* and those that are *dielectric*. Metallic gratings have grooves that are etched or cut from a flat metal surface. These grooves may be rectangular or triangular in shape. Triangular grooves are referred to as *blazed* gratings. Metallic gratings are operated in the reflection mode as the diffracted waves are reflected from the metal surface. Metallic gratings are also examples of *surface relief* gratings as the rectangular or triangular groove shape of the grating is cut from the flat metal surface.

Dielectric gratings are constructed of dielectric materials that are transparent to the electromagnetic radiation which impinges on it. Dielectric gratings may be classified into two major types: dielectric gratings that are surface relief gratings and dielectric gratings that are *volume* gratings. Surface relief dielectric gratings tend to have a large periodic modulation but small thickness, whereas volume dielectric gratings tend to have a small periodic modulation but large thickness. The large modulation of the surface relief grating occurs because the grating material from which the grating is constructed has a large difference in index of refraction compared to the medium that is adjacent to the grating. Dielectric gratings may be operated either in the *transmission* or in the *reflection* mode. Transmission gratings have periods on the order of a few wavelengths with the grating vector parallel to the grating surface, whereas reflection gratings have periods on the order of a half wavelength and grating vectors perpendicular to grating surface. Gratings that are neither exactly parallel nor perpendicular to the grating surface are referred to as *slanted* gratings.

Scattering from diffraction gratings depends strongly on three main factors, namely, the type and strength of the periodic variation of the index of refraction which exists in the grating, the type of material (anisotropic or isotropic, nonlossy or lossy) the grating is made from, and the type of EM wave which is incident on the grating. We will now briefly discuss these three factors.

The periodic variation of the index of refraction that induces diffraction when a grating is illuminated may consist of many different forms. The periodic variation may be one dimensional or two dimensional, in which case it is referred to as crossed grating, or the grating may consist of two superimposed one-dimensional gratings. In addition to the index varying in one or two dimensions, the periodic variation of the index of refraction may vary longitudinally throughout the grating. A sinusoidal surface relief grating and a triangular blaze grating that has air as an interface are examples of this type of variation. A surface relief grating is an example of a *longitudinally inhomogeneous* because at a plane where the groove is deeper, more material will be included in the duty cycle of the grating than at a plane that is closer to the homogeneous air half space.

The type of material that makes up the grating may be isotropic and nonlossy, such as glass; may be anisotropic, such as calcite or $LiNbO_3$ (lithium niobate); or may be either weakly lossy (e.g., $BaTiO_3$) or strongly lossy. Lossy gratings attenuate the diffracted waves as they propagate through the system. In anisotropic materials, the anisotropy tends to couple the polarization states of the incident wave in the medium and induce new polarization states in the system. In anisotropic systems, the diffracted waves consist of ordinary and extraordinary waves that are coupled together through the grating vector.

The type of EM radiation that is incident on the grating strongly influences the diffraction that results from the grating. The EM radiation may consist of either a plane wave or a collection or spectrum of plane waves (e.g., a Gaussian beam). Furthermore, each of these types of waves may be incident on the grating at an oblique angle and possess an arbitrary polarization. Later in this chapter, we will show examples of H-mode (magnetic field in plane of incidence) and E-mode polarization (electric field in plane of incidence) states that may be used to illuminate a diffraction grating. Particularly for anisotropic gratings, the type of incident wave and its polarization determine strongly how the EM wave will couple and diffract from the grating.

Many mathematical analyses and numerical algorithms have been developed in order that the diffraction that occurs from planar gratings may be predicted. Some of the main diffraction grating methods and algorithms are (1) *coupled wave analysis* [5–9], (2) *rigorous coupled wave analysis* (RCWA) [2,10–51], (3) *coupled mode theory* [52–59] (Refs. [55–57] have been referred to as the *Australian* method), (4) the *differential method* [1,60–63], (5) the *integral method* [64], (6) *the finite difference method* [65–67], (8) the *boundary element method* [68], (7) the *unimoment method* [69], and (9) other methods [70,71] which are either closely related or variations of the methods listed above. References [72–74] list papers that have studied energy and power conservation in electromagnetic and electromagnetic diffraction grating systems.

Concerning the first three methods, we mention that recently within the last few years, several researchers [22,23] have been concerned with the problem of improving the convergence or increasing the stability (that is, allowing the analysis of thicker grating structures that have increased grating strength) of the coupled mode and coupled wave algorithms and also with the problem of understanding in the first place, for certain polarizations and material types, why the coupled mode and coupled wave algorithms are unstable and why they do not converge well.

Just about all the above-mentioned algorithms solve the EM grating diffraction problem in three basic steps: one must (1) express the EM fields outside the diffraction grating region as *Rayleigh series* of propagating and evanescent plane waves whose amplitudes are unknown and yet to be determined (the series is transversely periodic with the period equal to grating period of the periodic structure), (2) by an appropriate method, find a general solution of Maxwell's equations in the diffraction grating region, and (3) match EM boundary conditions at the diffraction grating and homogeneous grating interfaces to determine all the unknown coefficients of the diffraction grating system. Most of the methods differ in the way that Maxwell's equations are solved in the diffraction grating region. We will now give a brief description of all of these algorithms. This chapter will primarily focus on the rigorous coupled wave approach. The reader may refer to the references for further details on the other methods.

We will now give a brief description of the above-mentioned algorithms. The description here, in order to simplify the discussion and description, is assumed to apply only to longitudinally homogeneous gratings. When using *coupled wave analysis* [5–9] and *RCWA* [10–51], Maxwell's equations in the diffraction grating region are solved by expanding the periodic dielectric in the diffraction grating region in a Fourier series, expanding the EM fields in the diffraction grating region in a set of Floquet harmonics whose amplitudes are functions of the longitudinal coordinate, and after substituting these expansions in Maxwell's equations, organizing the resulting equations into a *state variable* form where eigensolutions to the state variable system can be found. Coupled wave analysis [5–9] differs from RCWA [10–51] in that in coupled wave analysis, only a very few *Floquet harmonics* are used (two or three), whereas in RCWA, the analysis is made nearly exact by including as many Floquet harmonics as necessary until convergence of the solution is obtained. Typical state variable matrix sizes in RCWA method may range from 10×10 to 100×100.

In *coupled mode theory* algorithms, the transverse periodic region of the grating is divided into homogeneous subregions, and wave equation solutions in the homogeneous subregions (which are linear combinations of sinusoids proportional to a longitudinal propagation factor $\exp(-\gamma z)$ where z is the longitudinal coordinate) are EM boundary matched to the adjacent homogeneous subregions. After imposing the boundary condition that the overall EM solution across the grating period repeat itself in every grating period, one derives a nonstandard eigenvalue equation whose multiple roots thus determine the propagation constant γ of the modes which can propagate in the system. The propagation constant γ can of course be purely imaginary (nonevanescent), purely real (evanescent or attenuating), or complex if the medium is lossy, propagating with attenuation. By summing the forward and backward modes in the diffraction grating region, a complete solution of Maxwell's equations in the grating region is found. This method is particularly useful for lamellar gratings or step gratings where there are just two, or a few uniform layers within one grating period. This method is called a coupled mode approach because it is based on determining the propagating modes of the system. In the special one-dimensional case, when the grating period is bounded by perfect conductors and the overall grating region is uniform, the method reduces to the well-known problem of determining the propagating modes in a parallel plate waveguide. We would like to caution readers that the algorithm names, coupled *mode* analysis and coupled *wave* analysis, have been sometimes used interchangeably in the literature.

The *differential method* [1,60–63] is designed primarily for solving for diffraction from surface relief gratings. This method is based on solving Maxwell's equations in the diffraction grating region, by (1) defining a function $y = f(x)$ which specifies the shape of the surface relief grating in the diffraction grating region (that is, over the range $0 < y < L$, where L is the grating thickness), (2) expanding the dielectric permittivity function $\varepsilon(x, y)$ in a Fourier series over the grating period, (3) expanding the EM fields of the system in a Fourier series with the series amplitudes expressed as a function of the longitudinal coordinate y, (4) substituting this $\varepsilon(x, y)$ and the expanded EM fields either in the wave equation resulting from Maxwell's equations or into Maxwell's equations directly, (5) organizing the system of Fourier series amplitudes into a state variable form (with first-order derivatives of the system being taken with respect to the coordinate y), and (6) solving the state variable system using differential equation shooting methods. Petit [1] gives a very detailed description and survey of this method and its application to metallic and dielectric surface relief gratings.

The *integral method* [64] which is particularly useful for metallic surface relief gratings is based on four basic steps: (1) deriving a periodic Green's function which describes the way that an electrical surface current radiates from one point on the metal surface to an arbitrary point in space, (2) using this Green's function, writing an electric field integral which represents the way that the grating current radiates to an arbitrary point in space, (3) summing the electric field integral of step (2) and the incident electric plane wave field together, and (4) setting the total tangential electric field at the grating surface to zero to thus form an integral equation from which the surface current of the grating can be determined. This formulation is very similar to that used to solve for surface currents on an antenna or on a metallic scatterer.

The *finite difference method* [65–67] of determining diffraction from a grating is based on solving Maxwell's equation in the diffraction grating region by dividing the diffraction grating region over one period into a large grid, and then approximating the spatial partial derivatives of Maxwell's equations by using finite difference. In *the finite element method* [68], the diffraction grating region is divided into cells, and the field variables over a cell are expanded as boundary element functions. By substituting these as boundary element expansions in Maxwell's equations, a large system matrix is formed from which the fields of the system are determined. In both the finite difference and finite element methods, the solutions found in the diffraction grating region are matched to the plane wave Rayleigh expansion exterior to the diffraction grating system. The *unimoment method* [69] determines, by using either finite difference or finite elements, sets of special expansion functions in the diffraction grating that satisfy the wave equation and can be used to expand the unknown fields of the overall system.

Other methods and applications of diffraction gratings (including diffraction analysis of interdigitated, voltage-controlled, liquid crystal displays are listed in Refs. [70–95].

In the previous paragraphs, we have given a brief overview of available methods to solve diffraction grating problems. In what follows, we will concentrate on analyzing several different diffraction grating structures using the RCWA method [2,10–51]. The RCWA technique is relatively simple and straightforward to use, provides rapid convergence in many cases, and can apply equally well to thick or surface relief gratings. Hence it has become a popular method to use for solving diffraction grating problems. Section 5.2 will study two formulations of the RCWA algorithm when an H-mode polarized wave is incident on the grating. Section 5.3 will study the RCWA algorithm when an E-mode polarized wave is incident on the grating. In both Sections 5.2 and 5.3, the *complex Poynting theorem* [72–74,96–101] will be used to check convergence of the solutions and to study *evanescent power* in the grating. Section 5.4 will study a multilayer RCWA algorithm for the case when an E-mode polarized wave is incident on the grating. Section 5.5 will study diffraction from a crossed diffraction grating when a general oblique incident plane wave illuminates the grating.

5.2 H-MODE PLANAR DIFFRACTION GRATING ANALYSIS

In this section, we are interested in using the RCWA algorithm to study the diffraction case that occurs when a plane wave is incident on the planar grating shown in Figure 5.1. The diffraction grating is assumed to have its grating vector specified by $\vec{K} = \tilde{K}\hat{a}_x$, where $\tilde{K} = 2\pi/\tilde{\Lambda}$ and $\tilde{\Lambda}$ is the grating period or grating wavelength. In this case, the electric field is assumed to be polarized perpendicular to the plane of incidence as $\vec{E} = E_z\hat{a}_z$. In this section, two RCWA formulations will be presented. In the first formulation (given in Section 5.2.1), the state variable equations will be derived directly from Maxwell's equations, whereas in the following section, Maxwell's equations will be reduced to a second-order wave equation, and then placed in a state variable form. The complex Poynting theorem [72–74,96–101] using the solutions found from the full-field RCWA algorithm will be used to calculate the real and reactive power of the diffraction grating system and thus validate the overall analysis.

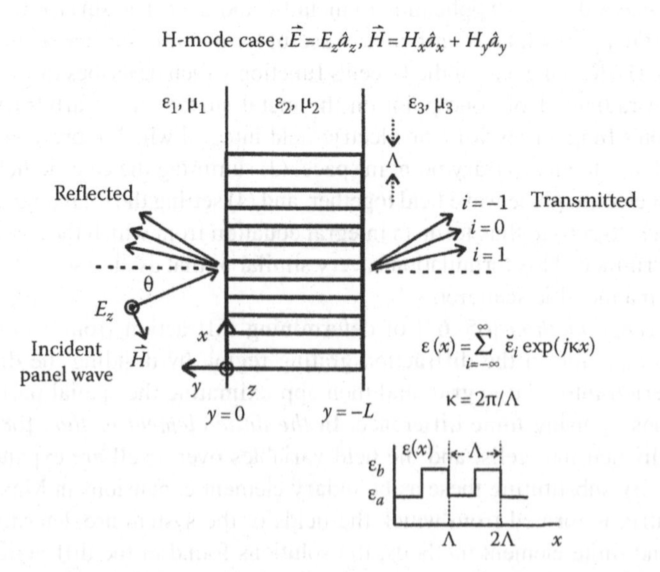

FIGURE 5.1 H-mode problem geometry assuming an arbitrary dielectric permittivity profile occupies Region 2. The inset shows one possible profile.

5.2.1 Full-Field Formulation

The basic overall RCWA [16–24] approach that will be used to study diffraction in this section will be to solve Maxwell's equations in Regions 1–3, and then using the general solutions, match electromagnetic boundary to determine the specific EM fields in each region. The EM field solutions in Regions 1 and 3 after solving Maxwell's equations in homogeneous space consist of an infinite set of propagating and evanescent, reflected and transmitted plane waves. The EM field solution in Region 2 is determined by expanding the electric and magnetic fields in a set of periodic or Floquet harmonics (the periodicity of the Floquet harmonics equals that of the diffraction grating), substituting these expansions into Maxwell's equations, and from the resulting equations, developing a set of state variable equations from whose solution the general EM fields of Region 2 are found.

We begin the analysis by determining the general EM field solution of Region 2, the diffractive grating region. Using normalized coordinates where $x = k_0\tilde{x}$, $y = k_0\tilde{y}$, and $z = k_0\tilde{z}$, where $k_0 = 2\pi/\lambda$ and λ (m) is the free space wavelength, Maxwell's normalized equations in Region 2 are given by

$$\nabla \times \vec{E} = -j\mu(\eta_0\vec{H}) \tag{5.1}$$

$$\nabla \times (\eta_0\vec{H}) = j\varepsilon\vec{E} \tag{5.2}$$

where $\eta_0 = \sqrt{\mu_0/\varepsilon_0} = 377\,\Omega$ is the intrinsic impedance of free space; $\mu = \tilde{\mu}/\mu_0$ is the relative permeability of Region 2, where μ_0 is the permeability of free space; and $\varepsilon = \tilde{\varepsilon}/\varepsilon_0 \equiv \varepsilon' - j\varepsilon''$ is the relative permittivity of Region 2, where ε_0 is the permittivity of free space. We expand the electric and magnetic field as

$$\vec{E} = \sum_{i=-\infty}^{\infty} S_{zi}(y)\exp(-jk_{xi}x)\hat{a}_z \tag{5.3}$$

$$\vec{U} = \eta_0\vec{H} = \sum_{i=-\infty}^{\infty}\left[U_{xi}(y)\hat{a}_x + U_{yi}(y)\hat{a}_y\right]\exp(-jk_{xi}x) \tag{5.4}$$

$$k_{xi} = k_{x0} - iK_x, \quad k_{x0} = \sqrt{\mu_1\varepsilon_1}\sin(\theta), \quad K_x = \frac{2\pi}{\Lambda}, \quad \Lambda = k_0\tilde{\Lambda}$$

Substituting, we have

$$\nabla \times \vec{E} = \sum_{i=-\infty}^{\infty}\left[\frac{\partial S_{zi}}{\partial y}\hat{a}_x + jk_{xi}S_{zi}\hat{a}_y\right]\exp(-jk_{xi}x) = -j\mu(\eta_0\vec{H}) = -j\mu\sum_{i=-\infty}^{\infty}\left[U_{xi}\,\hat{a}_x + U_{yi}\,\hat{a}_y\right]\exp(-jk_{xi}x) \tag{5.5}$$

$$\nabla \times \vec{U} = \sum_{i=-\infty}^{\infty}\left[-jk_{xi}U_{yi} - \frac{\partial U_{xi}}{\partial y}\right]\exp(-jk_{xi}x)\hat{a}_z = j\varepsilon(x)E_z\hat{a}_z = j\varepsilon(x)\sum_{i=-\infty}^{\infty} S_{zi}\exp(-jk_{xi}x)\hat{a}_z \tag{5.6}$$

The term $\varepsilon(x)E_z$ may be written as

$$\varepsilon(x)E_z = \left[\sum_{i''=-\infty}^{\infty} \breve{\varepsilon}_{i''} e^{ji''K_x x}\right]\left[\sum_{i'=-\infty}^{\infty} S_{zi''} e^{-j(k_{x0}-i'K_x)x}\right] \tag{5.7}$$

or after combining sums, we find that

$$\varepsilon(x)\, E_z = \sum_{i''=-\infty}^{\infty}\sum_{i'=-\infty}^{\infty} \breve{\varepsilon}_{i''} S_{zi'} e^{-jk_{x0}x} e^{j(i'+i'')K_x x} \tag{5.8}$$

At this point, we will make a substitution and let $i = i' + i''$, or $i'' = i - i'$. We notice in the i'' summation that when $i'' = -\infty$, $i = -\infty$ for a fixed finite i'. Thus in making the substitution of $i = i' + i''$, $\sum_{i''=-\infty}^{\infty}$ may be replaced by the sum $\sum_{i=-\infty}^{\infty}$. Carrying out the substitution $i'' = i - i'$, we find

$$\varepsilon(x)E_z = \sum_{i=-\infty}^{\infty}\left[\sum_{i'=-\infty}^{\infty} \breve{\varepsilon}_{i-i'} S_{zi'}\right] e^{-j(k_{x0}-iK_x)x} \tag{5.9}$$

Using $k_{xi} = k_{x0} - iK_x$, we find

$$\varepsilon(x)E_z = \sum_{i=-\infty}^{\infty}\left[\sum_{i'=-\infty}^{\infty} \breve{\varepsilon}_{i-i'} S_{zi'}\right] e^{-jk_{xi}x} \tag{5.10}$$

Coupling of the different i orders occurs through the term

$$\frac{\partial S_{zi}}{\partial y} = -j\mu U_{xi}$$

$$jk_{xi} S_{zi} = -j\mu U_{yi}$$

$$-jk_{xi}U_{yi} - \frac{\partial U_{xi}}{\partial y} = j\sum_{i'=-\infty}^{\infty} \breve{\varepsilon}_{i-i'} S_{zi'} \tag{5.11}$$

It is useful to introduce column and square matrices and put the preceding equations into a state variable form. Let $\underline{U_x} = [U_{xi}]$, $\underline{U_y} = [U_{yi}]$, $\underline{S_z} = [S_{zi}]$, and $i = -\infty, \ldots, \infty$ and let $\underline{\underline{\varepsilon}} = [\varepsilon_{i,i'}] = [\breve{\varepsilon}_{i-i'}]$, $\underline{\underline{K_x}} = [k_{xi}\delta_{i,i'}]$, $\underline{\underline{I}} = [\delta_{i,i'}]$, and $(i, i') = -\infty, \ldots, \infty$ be square matrices. $\delta_{i,i'}$ is the Kronecker delta and $\underline{\underline{I}}$ is the identity matrix. We find

$$\frac{\partial \underline{S_z}}{\partial y} = -j\mu \underline{U_x}$$

$$j\underline{\underline{K_x}}\,\underline{S_z} = -j\mu \underline{U_y}$$

$$-j\underline{\underline{K_x}}\,\underline{U_y} - \frac{\partial \underline{U_x}}{\partial y} = j\underline{\underline{\varepsilon}}\,\underline{S_z} \tag{5.12}$$

We eliminate the longitudinal vector component $\underline{U}_y$ and find that

$$-j\underline{\underline{K}}_x\underline{U}_y - \frac{\partial \underline{U}_x}{\partial y} = -j\underline{\underline{K}}_x\left(\frac{-1}{\mu}\underline{\underline{K}}_x\underline{S}_z\right) - \frac{\partial \underline{U}_x}{\partial y} = j\underline{\underline{\varepsilon}}\underline{S}_z \tag{5.13}$$

Rearranging, we find the following state variable form:

$$\frac{\partial \underline{S}_z}{\partial y} = \underline{\underline{0}}\underline{S}_z - j\mu\underline{\underline{I}}\,\underline{U}_x$$

$$\frac{\partial \underline{U}_x}{\partial y} = j\left(\frac{1}{\mu}\underline{\underline{K}}_x\underline{\underline{K}}_x - \underline{\underline{\varepsilon}}\right)\underline{S}_z + \underline{\underline{0}}\underline{U}_x \tag{5.14}$$

These equations may be put in a state variable form if we introduce the super matrices

$$\underline{V}^e = \begin{bmatrix} \underline{S}_z \\ \underline{U}_x \end{bmatrix}$$

$$\underline{\underline{A}} = j\begin{bmatrix} \underline{\underline{0}} & -\mu\underline{\underline{I}} \\ (\underline{\underline{K}}_x\underline{\underline{K}}_x/\mu - \underline{\underline{\varepsilon}}) & \underline{\underline{0}} \end{bmatrix} \tag{5.15}$$

We then have

$$\frac{\partial \underline{V}^e(y)}{\partial y} = \underline{\underline{A}}\underline{V}^e(y) \tag{5.16}$$

These equations may be solved numerically by truncating the matrices $\underline{\underline{A}}$ and $\underline{V}$, and using state variable techniques to solve the resulting equation. The truncation may be carried out by keeping mode orders whose magnitude is not greater than M_T, that is, keeping modal terms where $(i, i') = -M_T, \ldots, -1, 0, 1, \ldots, M_T$. Making the truncation, we find $\underline{V^e(y)}$ is a column matrix of size $N_T = 2(2M_T + 1)$ and $\underline{\underline{A}}$ is a constant matrix of size $N_T \times N_T$. Equation 5.16 when truncated to size N_T may be solved by finding the eigenvector and eigenvalues of the constant coefficient matrices $\underline{\underline{A}}$ as was done in Chapter 2. Let q_n and $\underline{V}_n$ be the eigenvalues and eigenvector of the matrix $\underline{\underline{A}}$. We have

$$\underline{\underline{A}}\underline{V}_n = q_n\underline{V}_n \tag{5.17}$$

The general solution for the electromagnetic field in the grating region may be found from the state variable solution. The electric field associated with the nth eigenvector mode is given by

$$\vec{E}_n^e = \left\{\sum_{i=-M_T}^{M_T}\left[S_{zin}\hat{a}_z\right]\exp(-jk_{xi}x)\right\}\exp(q_n y) \tag{5.18}$$

where S_{zin}, $i = -M_T, \ldots, M_T$ correspond to the electric field part of the eigenvector $\underline{V}_n$. The magnetic field associated with the nth eigenvector mode similarly is given by

$$\vec{U}_n^e = \eta_0\vec{H}_n^e = \left\{\sum_{i=-M_T}^{M_T}\left[U_{xin}\hat{a}_x + U_{yin}\hat{a}_y\right]\exp(-jk_{xi}x)\right\}\exp(q_n y) \tag{5.19}$$

where U_{xin}, $i = -M_T, \ldots M_T$ corresponds to the magnetic field part of the eigenvector $\underline{V}_n$ and U_{yin} is found from Equation 5.11. Summing over the individual eigenmodes, we find

$$\vec{E}^{(2)} = \sum_{n=1}^{N_T} C_n \vec{E}_n^e = \sum_{i=-M_T}^{M_T} \left\{ \sum_{n=1}^{N_T} C_n [S_{zin} \hat{a}_z] \exp(q_n y) \right\} \exp(-jk_{xi} x) \tag{5.20}$$

$$\vec{U}^{(2)} = \eta_0 \vec{H}^{(2)} = \sum_{n=1}^{N_T} C_n \vec{U}_n^e = \sum_{i=-M_T}^{M_T} \left\{ \sum_{n=1}^{N_T} C_n [U_{xin} \hat{a}_x + U_{yin} \hat{a}_y] \exp(q_n y) \right\} \exp(-jk_{xi} x) \tag{5.21}$$

where $\underline{U}_{yn} = -(1/\mu)\underline{\underline{K}}_x \underline{S}_{zn}$.

Equations 5.20 and 5.21 represent $N_T = 2(2M_T + 1)$ forward and backward traveling, propagating and nonpropagating eigenmodes which when summed together given the general electromagnetic field solution in Region 2, the grating region.

An important problem that remains is to determine the N_T coefficients C_n of the Equations 5.20 and 5.21. Up to this point, we have specified the general form of the diffracted fields in the grating region. The EM fields on the incident side of the diffraction grating (Region 1 of Figure 5.1) and on the transmission side of the diffraction grating (Region 3 of Figure 5.1) consist of an infinite number of propagating and nonpropagating plane waves whose tangential wave vectors are given by k_{xi}, $i = -\infty, \ldots, -1, 0, 1, \ldots, \infty$. The EM fields in Region 1 consists of a single, incident H-mode polarized wave making an angle θ with the y-axis, and an infinite number of reflected propagating and evanescent H-mode polarized plane waves. The tangential incident field in Region 1 is given by

$$E_{zinc}^{(1)} = E_0 \exp(-jk_{xi} x + jk_{y1} y) \delta_{i,0} \tag{5.22}$$

$$H_{xinc}^{(1)} = -\frac{k_{y1i}}{\eta_0} e^{-jk_{xi}x + jk_{y1i}y} E_0 \delta_{i,0} \tag{5.23}$$

where $n_1 = \sqrt{\mu_1 \varepsilon_1}$ is the index of refraction

$$E_{zref}^{(1)} = \sum_{i=-\infty}^{\infty} r_i e^{-jk_{xi}x - jk_{y1}y} \tag{5.24}$$

$$H_{xref}^{(1)} = \frac{1}{\eta_0} \sum_{i=-\infty}^{\infty} k_{y1i} r_i e^{-jk_{xi}x - jk_{y1i}y} \tag{5.25}$$

where

$$k_{y1i} = \begin{cases} [n_1^2 - k_{xi}^2]^{1/2}, & n_1 > k_{xi} \\ -j[k_{xi}^2 - n_1^2]^{1/2}, & k_{xi} > n_1 \end{cases} \tag{5.26}$$

as $y = +\infty$, we note for $k_{xi} > n_1$

$$e^{-j\left[-j\left[k_{xi}^2 - n_1^2\right]^{1/2}\right]y} = e^{-\left[\left[k_{xi}^2 - n_1^2\right]^{1/2}\right]y} \to 0 \tag{5.27}$$

$y \to \infty$ and thus the evanescent fields meet proper boundary conditions physically as $y \to \infty$. The total tangential fields in Region 1 are given by

$$E_z^{(1)} = E_{zinc}^{(1)} + E_{zref}^{(1)} \tag{5.28}$$

$$U_x^{(1)} = \eta_0 H_x^{(1)} = \eta_0 (H_{xinc}^{(1)} + H_{xref}^{(1)}) \tag{5.29}$$

In the transmitted region $y < -L$, the tangential electric and magnetic fields are given by

$$E_z^{(3)} = \sum_{i=-\infty}^{\infty} t_i e^{-jk_{xi}x + jk_{y3i}(y+L)} \tag{5.30}$$

$$U_x^{(3)} = \eta_0 H_x^{(3)} = -\sum_{i=-\infty}^{\infty} k_{y3i} t_i e^{-jk_{xi}x + jk_{y3i}(y+L)} \tag{5.31}$$

where

$$k_{y3i} = \begin{cases} [n_3^2 - k_{xi}^2]^{1/2}, & n_3 > k_{xi} \\ -j[k_{xi}^2 - n_3^2]^{1/2}, & k_{xi} > n_3 \end{cases} \tag{5.32}$$

as $y \to -\infty$, we note that for $k_{xi} > n_3$,

$$e^{j\left[-j\left[k_{xi}^2 - n_3^2\right]^{1/2}\right](y+L)} = e^{\left[\left[k_{xi}^2 - n_3^2\right]^{1/2}\right](y+L)} \to 0 \tag{5.33}$$

as $y \to -\infty$. We thus see that the evanescent fields in Region 3 meet the proper boundary condition as $y \to -\infty$.

Now that the EM fields have been defined in Regions 1–3, the next step is to match boundary conditions at the interfaces $y = 0$ and $-L$. At the $y = 0$ interface, we have

$$E_z^{(1)}\Big|_{y=0^+} = E_z^{(2)}\Big|_{y=0^-} \tag{5.34}$$

$$H_x^{(1)}\Big|_{y=0^+} = H_x^{(2)}\Big|_{y=0^-} \tag{5.35}$$

Substituting Equations 5.33 and 5.34 and keeping orders of $|i| \le M_T$, we find

$$\sum_{i=-M_T}^{M_T} \{E_0 \delta_{i,0} + r_i\} e^{-jk_{xi}x} = \sum_{i=-M_T}^{M_T} \left\{ \sum_{n=1}^{N_T} C_n S_{zin} \right\} e^{-jk_{xi}x} \tag{5.36}$$

$$\sum_{i=-M_T}^{M_T} \{-k_{y1i}\, \delta_{i,0} E_0 + k_{y1i} r_i\} e^{-jk_{xi}x} = \sum_{i=-M_T}^{M_T} \left\{ \sum_{n=1}^{N_T} C_n U_{xin} \right\} e^{-jk_{xi}x} \tag{5.37}$$

At the $y = -L$ boundary, we have

$$E_z^{(2)}\Big|_{y=-L^+} = E_z^{(3)}\Big|_{y=L^-} \tag{5.38}$$

$$H_x^{(2)}\Big|_{y=-L^+} = H_x^{(3)}\Big|_{y=L^-} \tag{5.39}$$

$$\sum_{i=-M_T}^{M_T}\left\{\sum_{n=1}^{N_T} C_n S_{zin} e^{-q_n L}\right\} e^{-jk_{xi}x} = \sum_{i=-M_T}^{M_T} \{t_i\} e^{-jk_{xi}x} \tag{5.40}$$

$$\sum_{i=-M_T}^{M_T}\left\{\sum_{n=1}^{N_T} C_n U_{xin} e^{-q_n L}\right\} e^{-jk_{xi}x} = \sum_{i=-M_T}^{M_T} \{-k_{y3i} t_i\} e^{-jk_{xi}x} \tag{5.41}$$

In Equations 5.37 and 5.38 and Equations 5.40 and 5.41 in order for the left- and right-hand side (RHS) of the equations to agree, it is necessary for the Fourier coefficients of $e^{-jk_{xi}x}$ to agree for each Floquet harmonic $e^{-jk_{xi}x}$. Thus for the unknown coefficients r_i, C_n, and t_i, we have the following equations:

$$E_0\delta_{i,0} + r_i = \sum_{n=1}^{N_T} C_n S_{zin} \tag{5.42}$$

$$-k_{y10}\delta_{i,0}E_0 + k_{y1i}r_i = \sum_{n=1}^{N_T} C_n U_{xin} \tag{5.43}$$

$$\sum_{n=1}^{N_T} C_n S_{zin} e^{-q_n L} = t_i \tag{5.44}$$

$$\sum_{n=1}^{N_T} C_n U_{xin} e^{-q_n L} = -k_{y3i} t_i \tag{5.45}$$

for $i = -M_T, \ldots, M_T$. We notice in Equations 5.42 and 5.44 that the r_i and t_i variables can be eliminated. These equations may be simplified by substituting r_i and t_i of Equations 5.42 and 5.44, respectively, into Equations 5.43 and 5.45. We have

$$-k_{y10}\delta_{i,0}E_0 + k_{y1i}\left[-E_0\delta_{i,0} + \sum_{n=1}^{N_T} C_n S_{zin}\right] = \sum_{n=1}^{N_T} C_n U_{xin} \tag{5.46}$$

$$\sum_{n=1}^{N_T} C_n U_{xin} e^{-q_n L} = -k_{y3i}\left[\sum_{n=1}^{N_T} C_n S_{zin} e^{-q_n L}\right] \tag{5.47}$$

or altogether

$$\sum_{n=1}^{N_T} C_n\{k_{y1i}S_{zin} - U_{xin}\} = 2E_0 k_{y10}\delta_{i,0} \tag{5.48}$$

$$\sum_{n=1}^{N_T} C_n\{e^{-q_n L}[U_{xin} + k_{y3i}S_{zin}]\} = 0 \tag{5.49}$$

where $i = -M_T, \ldots, M_T$.

The above equations constitute a set of $N_T = 2(2M_T + 1)$ equations for the N_T unknown coefficients C_n. Power is excited in the diffraction grating system through the $2E_0 k_{y10}\delta_{i,0}$ term in the RHS of Equation 5.48. Once the C_n are determined, the r_i and t_i can be found from Equations 5.42 and 5.44.

5.2.2 Differential Equation Method

A different way of analyzing the diffraction from a grating in the H-mode case under consideration is to eliminate the magnetic field from Maxwell's equations directly and then analyze the second-order partial differential for the electric field that results. In the analysis to be presented, it will be assumed that dielectric permittivity is a sinusoidal one. In this section, we will follow the formulation of Moharam and Gaylord's [16] original paper but use the geometry of Figure 5.1. We refer to this formulation as a wave equation formulation as it is based on placing the wave equation in a state variable form and proceeding with the solution from that point.

To start the analysis, we assume that $\vec{E} = E_z(x,y)\hat{a}_z$ in all regions and that all fields are independent of z. In Region 2 in normalized coordinates $x = k_0\tilde{x}$, $y = k_0\tilde{y}$, and $z = k_0\tilde{z}$, we have

$$\nabla \times \vec{E} = -j\mu(\eta_0\vec{H}) \tag{5.50}$$

$$\nabla \times (\eta_0\vec{H}) = j\varepsilon(x)\vec{E} \tag{5.51}$$

$$\nabla \times \nabla \times \vec{E} = -j\mu\nabla \times (\eta_0\vec{H}) = \mu\varepsilon(x)\vec{E} \tag{5.52}$$

$$\nabla \times \nabla \times \vec{E} = \nabla\nabla \cdot \vec{E} - \nabla^2\vec{E} \tag{5.53}$$

$$\nabla \cdot \vec{E} = \frac{\partial E_z}{\partial z} = 0 \tag{5.54}$$

Therefore, we have

$$\nabla^2 E_z + \mu\varepsilon(x)E_z = 0 \tag{5.55}$$

or since $(\partial^2 E_z/\partial z^2) = 0$, we have

$$\frac{\partial^2}{\partial x^2}E_z + \frac{\partial^2 E_z}{\partial y^2} + \mu\varepsilon(x)E_z = 0 \tag{5.56}$$

For the present analysis, we will let $\varepsilon(x) = \varepsilon_2 + \Delta\varepsilon \cos Kx$ and take $\mu = 1$. K is a normalized wave number

$$K = \frac{\tilde{K}}{k_0} = \frac{2\pi}{k_0\tilde{\Lambda}} = \frac{2\pi}{\Lambda} \tag{5.57}$$

and $\Lambda = k_0\tilde{\Lambda}$ is the normalized grating period of the system. To start the analysis, we expand the electric field of Region 2, namely E_z, in the Floquet harmonic series

$$E_z = \sum_{i=-\infty}^{\infty} S_i(y)\exp(j\psi_i) \tag{5.58}$$

$$\xi_2 = \left(\frac{\tilde{k}_2}{k_0}\right)\cos\theta' = \sqrt{\varepsilon_2}\cos\theta', \quad \psi_i = -\beta_i x + \xi_2 y \tag{5.59}$$

$$\beta_i = \sqrt{\varepsilon_1}\sin\theta - iK \tag{5.60}$$

where $i = \ldots, -1, 0, 1, \ldots$ and θ' is the angle of light refracted into the dielectric grating. $S_i(y)$ are Floquet modal amplitudes which need to be determined. Differentiating E_z with respect to y and x, we find that

$$\frac{\partial^2}{\partial y^2}E_z = \sum_{i=-\infty}^{\infty}\left[\frac{\partial^2}{\partial y^2}S_i(y) + 2j\xi_2\frac{\partial}{\partial y}S_i(y) - \xi_2^2 S_i(y)\right]\exp(j\psi_i) \tag{5.61}$$

$$\frac{\partial^2}{\partial x^2}E_z = \sum_i\left[-\beta_i^2 S_i(y)\right]\exp(j\psi) \tag{5.62}$$

The term

$$\varepsilon(x)E_z = \left\{\varepsilon_2 + \frac{\Delta\varepsilon}{2}[\exp(-jKx) + \exp(jKx)]\right\}\sum_{i=-\infty}^{\infty} S_i(y)\exp(-j\beta_i x + j\xi_2 y) \tag{5.63}$$

equals

$$\varepsilon(x)E_z = \varepsilon_2\sum_{i=-\infty}^{\infty} S_i(y)\exp(j\psi_i) + \frac{\Delta\varepsilon}{2}\sum_{i=-\infty}^{\infty} S_i(y)\exp(-j(\beta_i + K)x + j\xi_2 y)$$

$$+\frac{\Delta\varepsilon}{2}\sum_{i=-\infty}^{\infty} S_i(y)\exp(-j(\beta_i + K)x + j\xi_2 y) \tag{5.64}$$

The terms in the exponential factors may be manipulated to give

$$\beta_i + K = \left(\sqrt{\varepsilon_1}\sin\theta - iK\right) + K \tag{5.65}$$

$$\beta_i + K = \left(\sqrt{\varepsilon_1}\sin\theta - (i-1)K\right) = \beta_{i-1} \tag{5.66}$$

Similarly,

$$\beta_i - K = \beta_{i+1} \tag{5.67}$$

The second term of Equation 5.64 may be rewritten as

$$T_2 \equiv \frac{\Delta\varepsilon}{2}\sum_{i=-\infty}^{\infty} S_i(y)\exp(-j(\beta_i + K)x + j\xi_2 y) = \frac{\Delta\varepsilon}{2}\sum_{i=-\infty}^{\infty} S_i(y)\exp(-j\beta_{i-1}x + j\xi_2 y) \tag{5.68}$$

In this RHS summation, we will make a substitution and $i' = i - 1$. Doing this, we have

$$T_2 = \frac{\Delta\varepsilon}{2}\sum_{i'=-\infty}^{\infty} S_{i'+1}(y)\exp(-j\beta_{i'}x + j\xi_2 y) \tag{5.69}$$

Similarly, the third term of Equation 5.64 may be written as

$$T_3 = \frac{\Delta\varepsilon}{2}\sum_{i=-\infty}^{\infty} S_i(y)\exp(-j(\beta_i - K)x + j\xi_2 y) = \frac{\Delta\varepsilon}{2}\sum_{i'=-\infty}^{\infty} S_{i'-1}(z)\exp(-j\beta_{i'}x + j\xi_2 y) \tag{5.70}$$

Substituting T_2 and T_3 into Equation 5.64 and using i, instead of i', in summation, we find

$$\varepsilon(x)E_z = \sum_{i=-\infty}^{\infty}\left[\varepsilon_2 S_i(y) + \frac{\Delta\varepsilon}{2}S_{i+1}(y) + \frac{\Delta\varepsilon}{2}S_{i-1}(y)\right]\exp(j\psi_i) \tag{5.71}$$

Substituting into the original differential equation for E_z, we find

$$0 = \sum_{i=-\infty}^{\infty}\left\{\frac{\partial^2}{\partial y^2}S_i(y) + 2j\xi_2\frac{\partial}{\partial y}S_i(y) - \xi_2^2 S_i(y) - \beta_i^2 S_i(y) + \varepsilon_2 S_i(y) + \frac{\Delta\varepsilon}{2}S_{i+1}(y) + \frac{\Delta\varepsilon}{2}S_{i-1}(y)\right\}\exp(j\psi_i) \tag{5.72}$$

The only way that the above equation can be zero for all values of x and y is if the curly bracketed expression is zero. Thus Equation 5.72 describes a series of coupled modal amplitude equations to determine the EM fields of the system. At this point, it is useful to introduce scaled coordinates into analysis. We let

$$u = -\frac{j\pi\Delta\varepsilon}{2\lambda\sqrt{\varepsilon_2}}\tilde{y} = -\frac{j\pi\Delta\varepsilon}{2\lambda\sqrt{\varepsilon_2}k_0}y = -\frac{j\Delta\varepsilon}{4\sqrt{\varepsilon_2}}y = -j\kappa y \tag{5.73}$$

Substituting the above scaling into Equation 5.72, we find after algebra that

$$\frac{-\Delta\varepsilon}{8\varepsilon_2}\frac{d^2S_i}{du^2}+\cos\theta'\frac{dS_i}{du}-\rho i[i-B]S_i+S_{i+1}+S_{i-1}=0 \tag{5.74}$$

where

$$B=\frac{2\tilde{\Lambda}\sqrt{\varepsilon_2}}{\lambda}\sin\theta'=\frac{2\tilde{\Lambda}\sqrt{\varepsilon_1}}{\lambda}\sin\theta \tag{5.75}$$

and

$$\rho=\frac{2\lambda^2}{\tilde{\Lambda}^2\Delta\varepsilon} \tag{5.76}$$

and the last equation of B follows from Snell's law

$$\sqrt{\varepsilon_2}\sin\theta'=\sqrt{\varepsilon_1}\sin\theta \tag{5.77}$$

If we further let $a = (8\varepsilon_2/\Delta\varepsilon)$, $b_i = -\rho i(i - B)a$, $c = a\cos\theta'$, we may rewrite the equation as

$$\frac{-1}{a}\frac{d^2S_i}{du^2}+\frac{c}{a}\frac{dS_i}{du}+\frac{b_i}{a}S_i+S_{i+1}+S_{i-1}=0 \tag{5.78}$$

Equation 5.74 is a second-order coupled differential equation. It may be put into the form of a first-order state variable equation, if the following new variables are defined. Let

$$S_{1i}=S_i(u) \tag{5.79}$$

$$S_{2i}=\frac{dS_i(u)}{du} \tag{5.80}$$

Making these substitutions, we find that the second-order Equation 5.74 may be written as

$$\frac{dS_{1i}}{du}=S_{2i} \tag{5.81}$$

$$\frac{dS_{2i}}{du}=aS_{1i+1}+b_iS_{1i}+aS_{1i-1}+cS_{2i} \tag{5.82}$$

If we differentiate Equation 5.80 with respect to u, we find that

$$\frac{d^2S_{1i}}{du^2}=\frac{dS_{2i}}{du}=aS_{1i+1}+b_iS_{1i}+aS_{1i-1}+cS_{2i} \tag{5.83}$$

Dividing Equation 5.82 by a, transferring the second derivative term to the RHS, and substituting the original definitions of S_{1i} and S_{2i}, we have

$$\frac{-1}{a}\frac{d^2}{du^2}S_i + S_{i+1} + \frac{b_i}{a}S_i + S_{i-1} + \frac{c}{a}\frac{dS_i}{du} = 0 \tag{5.84}$$

This is identical to Equation 5.83, thus showing that Equation 5.83 is the correct first-order state variable form of Equation 5.78.

The full matrix form for Equations 5.80 and 5.81 when written out when truncated for $M_T = 2$ is

$$\frac{d}{du}\begin{bmatrix}\underline{S}_1\\ \underline{S}_2\end{bmatrix} = \begin{bmatrix}\underline{\underline{A}}_{11} & \underline{\underline{A}}_{12}\\ \underline{\underline{A}}_{21} & \underline{\underline{A}}_{22}\end{bmatrix}\begin{bmatrix}\underline{S}_1\\ \underline{S}_2\end{bmatrix} \tag{5.85}$$

where

$$\underline{S}_1 = \begin{bmatrix}S_{1,-2} & S_{1,-1} & S_{1,0} & S_{1,1} & S_{1,2}\end{bmatrix}^t \tag{5.86}$$

$$\underline{S}_2 = \begin{bmatrix}S_{2,-2} & S_{2,-1} & S_{2,0} & S_{2,1} & S_{2,2}\end{bmatrix}^t \tag{5.87}$$

$$\underline{\underline{A}}_{11} = [0]_{5\times5} \tag{5.88}$$

$$\underline{\underline{A}}_{12} = [\underline{\underline{I}}]_{5\times5} = [\delta_{ii'}]_{5\times5} \tag{5.89}$$

$$\underline{\underline{A}}_{21} = \begin{bmatrix}b_{-2} & a & 0 & 0 & 0\\ a & b_{-1} & a & 0 & 0\\ 0 & a & b_0 & a & 0\\ 0 & 0 & a & b_1 & a\\ 0 & 0 & 0 & a & b_2\end{bmatrix}_{5\times5} \tag{5.90}$$

$$\underline{\underline{A}}_{22} = [c\delta_{ii'}]_{5\times5} \tag{5.91}$$

$$\delta_{i,i'} = \begin{cases}1, & i = i'\\ 0, & i \neq i'\end{cases} \tag{5.92}$$

If we let $\underline{V} = [\underline{S}_1\ \underline{S}_2]^t$ and

$$\underline{\underline{A}} = \begin{bmatrix}\underline{\underline{A}}_{11} & \underline{\underline{A}}_{12}\\ \underline{\underline{A}}_{21} & \underline{\underline{A}}_{22}\end{bmatrix}_{10\times10} = \begin{bmatrix}a_{ii'}\end{bmatrix}_{10\times10} \tag{5.93}$$

$a_{ii'}(i,i') = (1,\ldots,10)$ represent the individual matrix elements of the overall matrix $\underline{\underline{A}}$. Using the just defined matrices, Equation 5.85 may be written in full-state variable form as

$$\frac{d}{du}\underline{V} = \underline{\underline{A}}\,\underline{V} \tag{5.94}$$

If we let $\underline{V}_n$ and q_n be the eigenvalues and eigenvectors of the matrix $\underline{\underline{A}}$, truncated at values of $i = -M_T,\ldots,-1,0,1,\ldots,M_T$, then we find that the solution for $S_i(u)$ and $S_{1i}(u)$ is given by

$$S_i(u) = S_{1i}(u) = \sum_{n=1}^{N_T} C_n w_{in} \exp(q_n u) \tag{5.95}$$

where $N_T = 2(2M_T + 1)$ and w_{in} represents the ith row of the nth eigenvector $(\underline{S}_1)_n$. The electric field E_z is given by Equation 5.58 with $S_i(u)$ substituted. We have

$$E_z = \sum_{i=-M_T}^{M_T} \left\{ \exp[-j(\beta_i x - \xi_2 y)] \sum_{n=1}^{N_T} C_n w_{in} \exp[-jq_n \kappa y] \right\} \tag{5.96}$$

where, as already defined, $\kappa = (\Delta\varepsilon/4\sqrt{\varepsilon_2})$. To proceed further, it is necessary to find the magnetic field associated with E_z. Using Maxwell's equations, the tangential magnetic field H_x is found from

$$H_x = -\frac{1}{j\eta_0}\frac{\partial E_z}{\partial y} \tag{5.97}$$

Altogether tangential electromagnetic fields in Region 2, the diffraction gating region, are given by (including now the Region 2 subscript label)

$$E_{z2} = \sum_{i=-M_T}^{M_T} \sum_{n=1}^{N_T} C_n w_{in} \exp\{-j[\beta_i x - (\xi_2 - q_n \kappa) y]\} \tag{5.98}$$

$$U_{x2} = \eta_0 H_{x2} = \sum_{i=-M_T}^{M_T} \sum_{n=1}^{N_T} C_n w_{in} [(q_n \kappa - \xi_2)] \exp\{-j[\beta_i x - (\xi_2 - q_n \kappa) y]\} \tag{5.99}$$

The differential equation method provides an alternate state variable representation from which to obtain the electromagnetic fields of Region 2. Although the state variable representations of Sections 5.2.1 and 5.2.2 are exactly equal as $M_T \to \infty$, the two representations give different solutions for finite M_T. Thus a comparison of the two methods for different values of M_T gives a good measure of how well both representations are converging.

The final matrix equations for C_n of this section may be found by matching the tangential electromagnetic fields as given in Section 5.3 by Equations 5.28 through 5.32 for Regions 1 and 3 with the EM field solutions of Region 2 that have been just derived. The final matrix equations from which C_n result are

$$2k_{y10} E_0 \delta_{i,0} = \sum_n C_n w_{in} \left[k_{y1i} + \xi_2 - q_n \kappa \right] \tag{5.100}$$

$$0 = \sum_n C_n e_n w_{in} \left[-k_{y3i} + \xi_2 - q_n \kappa \right] \tag{5.101}$$

where

$$e_n = \exp\left[j(q_n\kappa - \xi_2)L\right] \tag{5.102}$$

The reflection and transmission coefficients are given by

$$r_i = \sum_n C_n w_{in} - E_0\delta_{i,0} \tag{5.103}$$

$$t_i = \sum_n C_n w_{in} e_n \tag{5.104}$$

These equations have been presented in Ref. [16].

5.2.3 Numerical Results

In this section, we will present numerical examples of the diffraction as results from RCWA and will present complex Poynting theorem power balance results using the RCWA method. The examples to be presented consist of a RCWA study of a cosine diffraction grating (lossless and lossy bulk dielectric cases) and a RCWA study of a square or step profile diffraction grating. Both gratings, consistent with the theory of this section, are assumed to be homogeneous in the longitudinal direction. These two gratings have been chosen because the cosine grating is relatively smooth, containing low spatial frequencies $i = -1, 0, 1$, whereas the square wave or step profile contains sharp dielectric discontinuities at dielectric steps and thus possesses a high spectral content $i = -\infty, \ldots, -1, 0, 1, \ldots, \infty$. The complex Poynting theorem power–energy balance analysis is based on the formulation presented in Chapter 4. The complex Poynting theorem numerical results are performed using the Poynting box shown in Figure 5.15. Details of the calculation are given in Section 5.3.

We begin presenting results for the cosine grating. The solid line plots in Figure 5.2 show the transmitted diffraction efficiencies $DE_T(\%)$ for five orders $i = -2, -1, 0, 1, 2$ as calculated by the

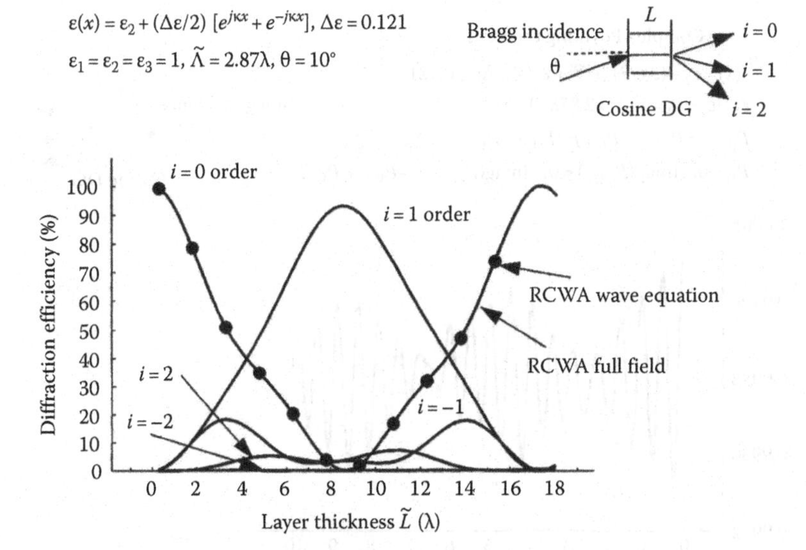

FIGURE 5.2 Transmitted diffraction efficiencies $DE_T(\%)$ for five orders $i = -2, -1, 0, 1, 2$ as calculated by Equations 5.48 and 5.49 using a lossless cosine grating. Refer to transmitted diffraction efficiency of Moharam and Gaylord, Ref. [16].

full-field method (see Section 5.2.1, Equations 5.48 and 5.49) using the lossless cosine grating as specified in Figure 5.2 inset and heading. These plots show $DE_T(\%)$ versus the layer length $\tilde{L}$ (in units of free space wavelength λ). As can be seen from Figure 5.2, as the layer length $\tilde{L}$ increases from 0λ to 9λ because θ is at the Bragg angle (implying Bragg incidence), power is primarily diffracted from the $i = 0$ order into the $i = 1$ order, with a small amount of power being diffracted into the other orders $i = -2, -1, 2$. For larger values of $\tilde{L}$, $9\lambda - 18\lambda$ power is diffracted from the $i = 1$ order into the $i = 0$ order, with a small amount of power being diffracted into the other orders $i = -2, -1, 2$. This cycle is repeated over a long range of $\tilde{L}$ values. Because the bulk regions had matched permittivities, the reflected diffractions were small and have not been plotted. Also shown in Figure 5.2 is the $DE_T(\%)$ as calculated by a differential equation, state variable method described in Section 5.2.2, and derived originally in Ref. [16] (*dots*, $i = 0$). In this analysis, Maxwell's equations are reduced to a second-order differential equation for the electric field, and this differential equation is put in a state variable form. The state variable form that results is different from the one which has been presented in Section 5.2.1, although as $M_T \rightarrow \infty$, the two methods are mathematically equivalent. As can be seen from Figure 5.2, a comparison of the $i = 0$ order plots (typical of all orders) shows that virtually identical results occur from the use of the two methods.

Figures 5.3 and 5.4 show plots of the real and imaginary parts of the normalized complex power P_{IN} and $P_{BOX} \equiv P_{OUT}$ of the complex Poynting theorem (Figure 5.15 shows the Poynting box used). For more detail on the application of the Poynting theorem to gratings, see the next subsection. As mentioned earlier, this case represents a lossless diffraction grating, bulk dielectric case. In these plots, the complex power is plotted versus the layer length $\tilde{L}$. As can be seen from Figures 5.3 and 5.4, excellent agreement in both plots is obtained from the calculation. Figure 5.5 shows a plot of the electric and magnetic energies P_{WE} and P_{WM}, versus layer length $\tilde{L}$ that results for the example under consideration. As can be seen from Figure 5.5, the electric and magnetic energies are very nearly equal to one another, and in a $\tilde{L} = 1\lambda$ size slab, the electric and magnetic energies P_{WE} and P_{WM} are much larger than the peak magnitude energy difference between the two energies.

Figure 5.6 shows the Imag(P_{BOX}) versus layer length $\tilde{L}$ when the electromagnetic fields are computed using $M_T = 3$ and $M_T = 6$. As can be seen from Figure 5.6, extremely good convergence is observed using the two different truncation sizes.

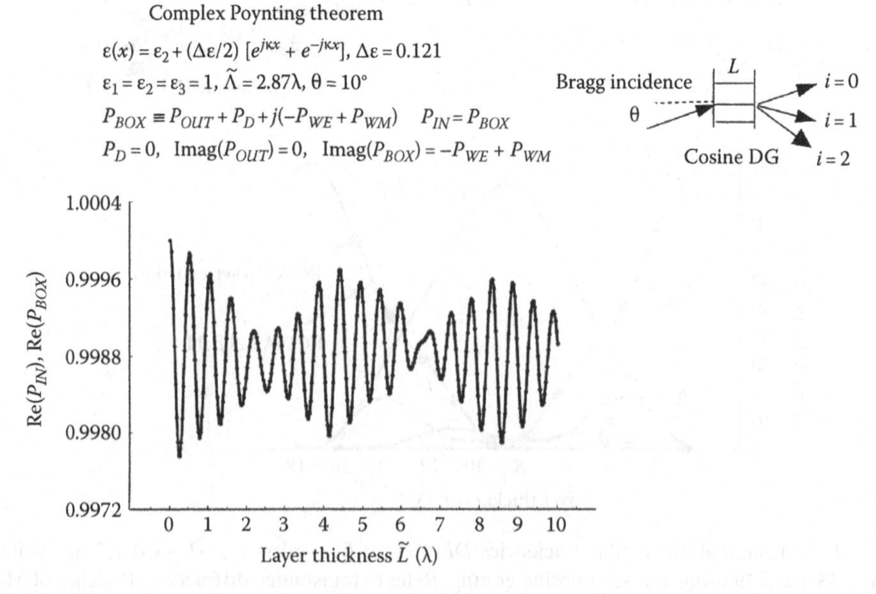

FIGURE 5.3 The real part of the normalized complex power P_{IN} and P_{BOX} of the complex Poynting theorem.

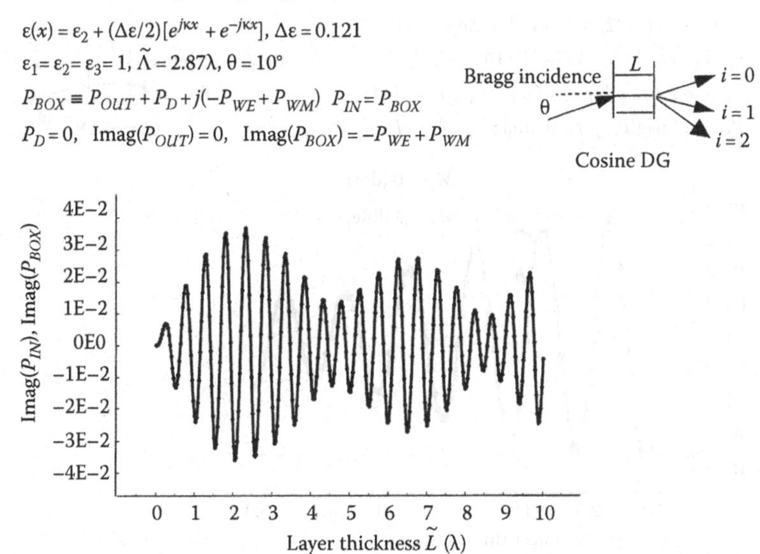

FIGURE 5.4 The imaginary part of the normalized complex power P_{IN} and P_{BOX} of the complex Poynting theorem.

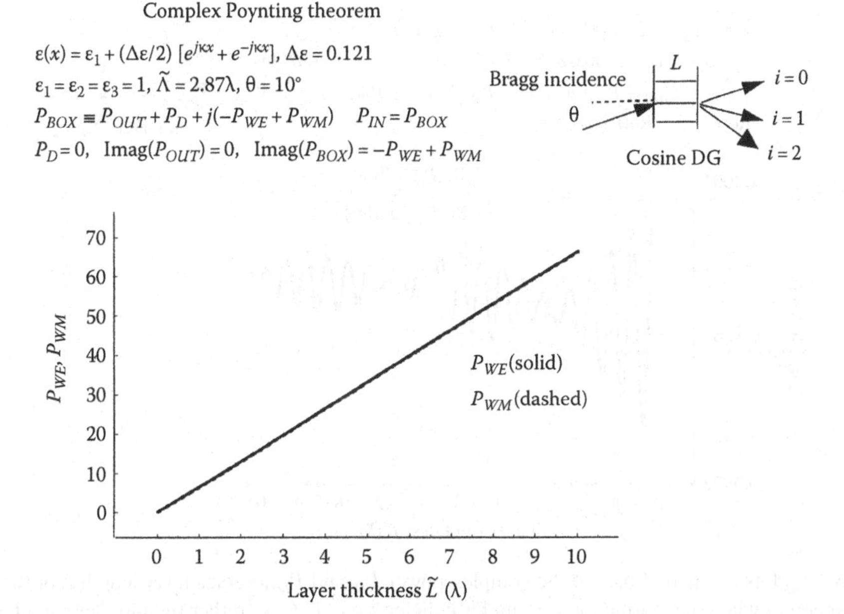

FIGURE 5.5 Plots of the electric and magnetic energies P_{WE} and P_{WM} versus layer length are shown.

Figures 5.7 and 5.8 show plots of the real and imaginary parts of the complex power P_{IN} and $P_{BOX} \equiv P_{IN} = P_{OUT} + P_{DE} + P_{DM} + j(-P_{WE} + P_{WM})$ (see Equation 5.163a), versus layer length $\tilde{L}$ when the diffraction grating bulk dielectric ε_2 is lossy rather than lossless, and has a value of $\varepsilon_2 = 1 - j0.02$. In this figure, one again observes extremely good agreement between the real and imaginary parts of P_{IN} and P_{BOX}, again showing that the complex Poynting theorem is obeyed to a high degree of accuracy. A comparison of Figures 5.3 and 5.4 (lossless case) with Figures 5.7 and 5.8 (lossy case) shows a very clear difference in the shapes of the real and imaginary parts of P_{IN} and P_{BOX} which is being

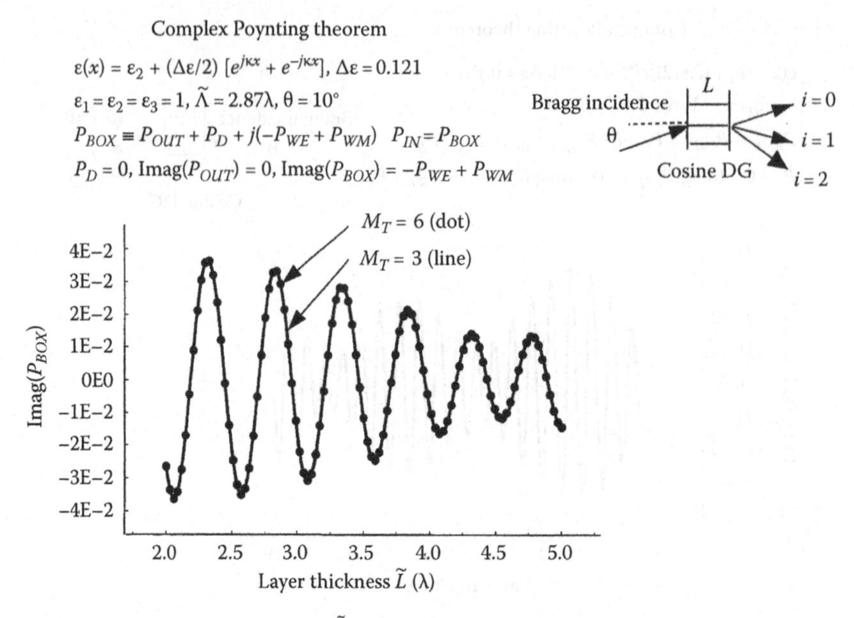

FIGURE 5.6 Imag(P_{BOX}) versus layer length $\tilde{L}$ when the electromagnetic fields are computed using $M_T = 3$ and 6.

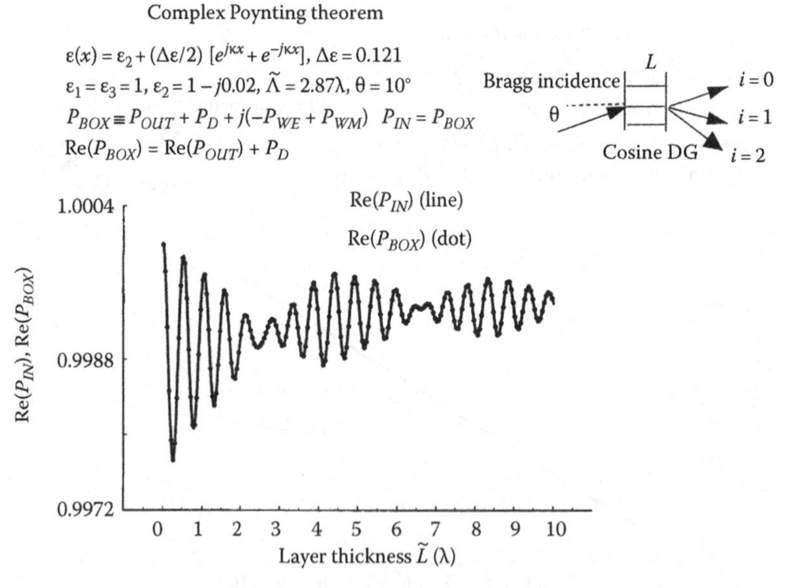

FIGURE 5.7 Plots of the real part of the complex power P_{IN} and P_{BOX} versus layer length $\tilde{L}$ of the complex Poynting theorem when the diffraction grating bulk dielectric ε_2 is lossy rather than lossless and has a value of $\varepsilon_2 = 1 - j0.02$.

computed in the four figures. In the lossy case as $\tilde{L}$ increases, the envelope of the oscillations of P_{IN} and P_{BOX} damps out as $\tilde{L}$ increases, whereas in the lossless case the envelope maintains a longitudinal periodic shape. The damping of the envelope with increasing $\tilde{L}$ in the lossy case is expected, since as the layer length increases, the EM fields in the system attenuate near the exit side of the diffraction grating due to the losses in the grating. When the diffraction grating becomes sufficiently long, the EM fields at the exit side approach zero; therefore, P_{IN} and P_{BOX} become independent of $\tilde{L}$, and thus there is no oscillation.

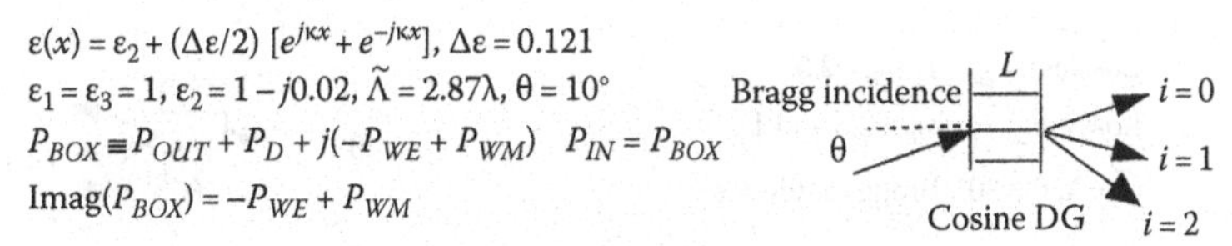

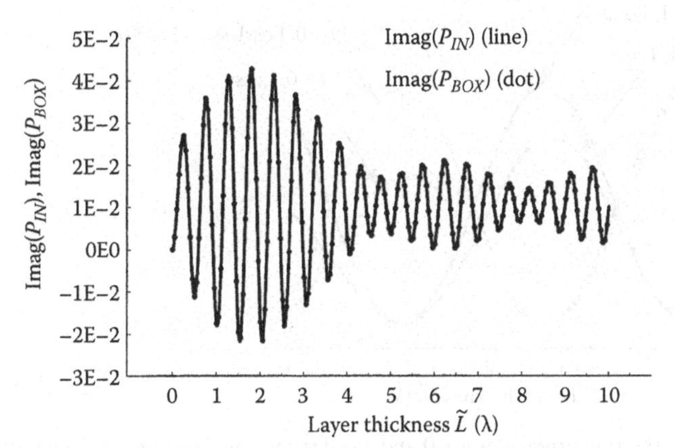

FIGURE 5.8 Plots of the imaginary part of the complex power P_{IN} and P_{BOX} versus layer length $\tilde{L}$ of the complex Poynting theorem when the diffraction grating bulk dielectric ε_2 is lossy rather than lossless and has a value of $\varepsilon_2 = 1 - j0.02$.

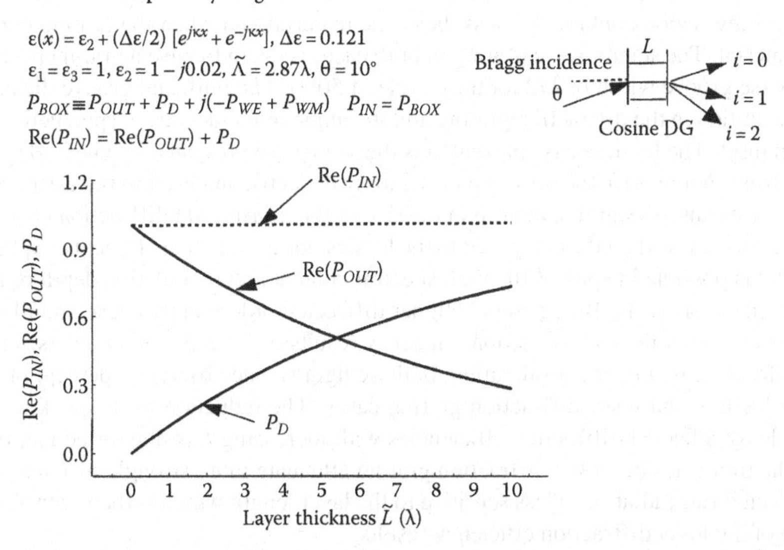

FIGURE 5.9 Plot of the power dissipated P_D, $\mathrm{Re}(P_{IN})$, and $\mathrm{Re}(P_{OUT})$ versus layer length $\tilde{L}$ is shown. The ripple in the $\mathrm{Re}(P_{IN})$ observed in Figure 5.7 is not observed here because of the scale of the plot of this figure.

Figure 5.9 shows a plot of the power dissipated P_D, $\mathrm{Re}(P_{IN})$, and $\mathrm{Re}(P_{OUT})$ versus layer length $\tilde{L}$. The ripple in the $\mathrm{Re}(P_{IN})$ observed in Figure 5.7 is not observed here because of the scale of the Figure 5.9 plot. As can be seen from Figure 5.9, the sum of power radiated out of the Poynting box and the power dissipated is balanced by the real power radiated into the box as one would physically expect.

Figure 5.10 shows the transmitted diffraction efficiency (i = 0 and 1 orders) versus layer length $\tilde{L}$ that arises when a plane wave is incident on a square wave or step profile dielectric grating.

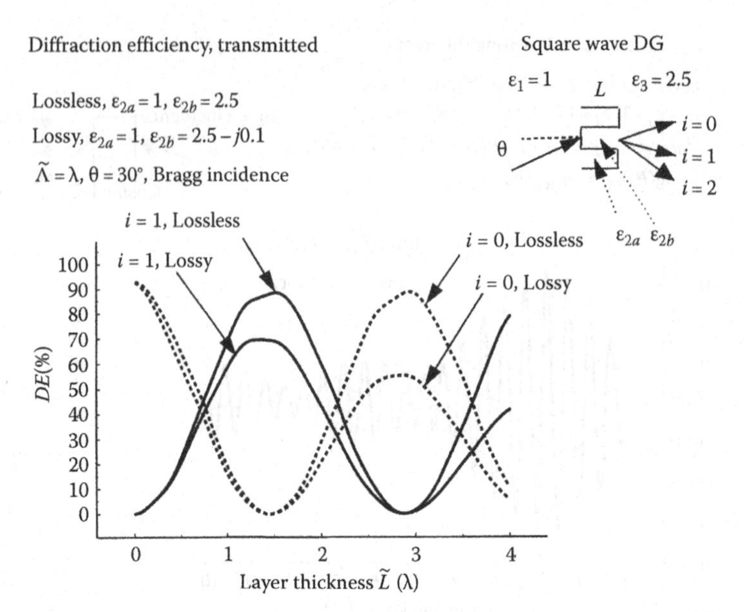

FIGURE 5.10 Transmitted diffraction efficiency ($i = 0$ and 1 orders) versus layer length $\tilde{L}$ that arises when a plane wave is incident on a square wave or step profile dielectric grating is shown. The diffraction efficiency results are presented for a lossless and a lossy case. The square wave grating in both cases is taken to have a grating period of $\Lambda = \lambda$ and a transverse groove width of $\lambda/2$ (or duty cycle of 50%).

In the present figure, diffraction efficiency results are presented for two cases, namely, when the diffraction grating region contains lossless dielectric material and when the grating contains lossy dielectric material. The square wave grating in both cases is taken to have a grating period of $\Lambda = \lambda$ and a transverse groove width of $\lambda/2$ (or duty cycle of 50%). The bulk and groove dielectric values and their orientation in the diffraction grating and the angle of incidence are specified in the Figure 5.10 title and inset. The lossless case presented is the same case presented by Gaylord et al. [19]. As can be seen from Figure 5.10, the presence of the lossy dielectric material in the diffraction grating for the lossy case causes a significant drop in the size of the transmitted diffraction efficiency as the layer length $\tilde{L}$ increases when the compared to the lossless case. Note that efficient coupling between 0 and 1 orders is possible in spite of the high spectral content and modulation depth of the grating, as long as incidence is at the Bragg angle. Higher diffracted orders in this case are all evanescent. Figure 5.11 shows the reflected diffraction efficiency results ($i = 0$ and 1 orders) versus layer length $\tilde{L}$ that arises for the case under consideration. In these figures, one observes a perceptible difference between the lossless and lossy diffraction grating cases. The reduction in the peak-to-peak envelopes in the lossy reflected diffraction efficiencies with increasing $\tilde{L}$ is due to the fact that the EM fields near the transmit side of the diffraction grating attenuate more strongly as $\tilde{L}$ becomes larger, and thus reflected EM radiation is less sensitive to the layer length which is then seen as a reduction in the ripple of the lossy diffraction efficiency results.

Figures 5.12 and 5.13 show the real and imaginary parts of P_{IN} and P_{BOX} versus $\tilde{L}$ for the lossy square wave diffraction case under study. In the lossy case, the ripple envelope slowly decays as layer length $\tilde{L}$ increases. The reduction in ripple envelope for this lossy case is due to the attenuation of electromagnetic fields near the transmit side as has been described earlier.

5.2.4 Diffraction Grating Mirror

Another important EM case that may be studied using a RCWA consists of the determining the EM fields that are diffracted when an H-mode polarized plane wave is incident on a diffraction grating that is backed by a mirror (also called a short circuit plate). Thus Region 3 is an electrical perfect conductor

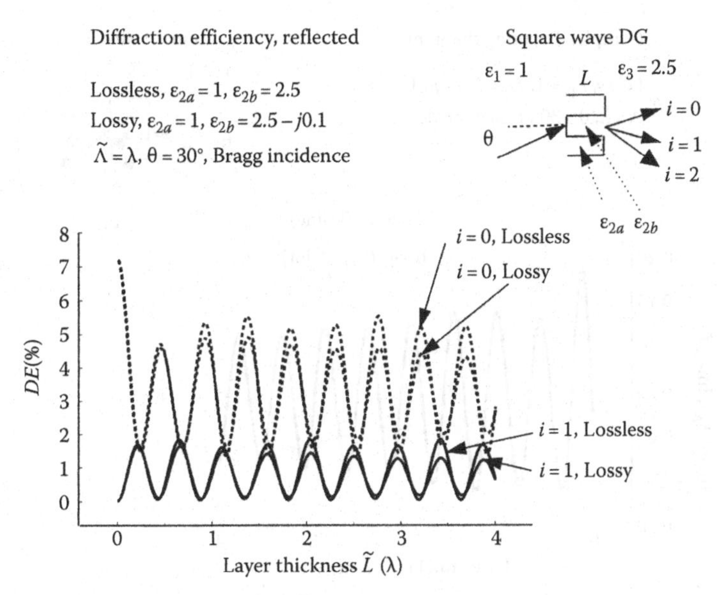

FIGURE 5.11 Reflected diffraction efficiency results ($i = 0$ and 1 orders) versus layer length $\tilde{L}$ which arises for the case under consideration is shown.

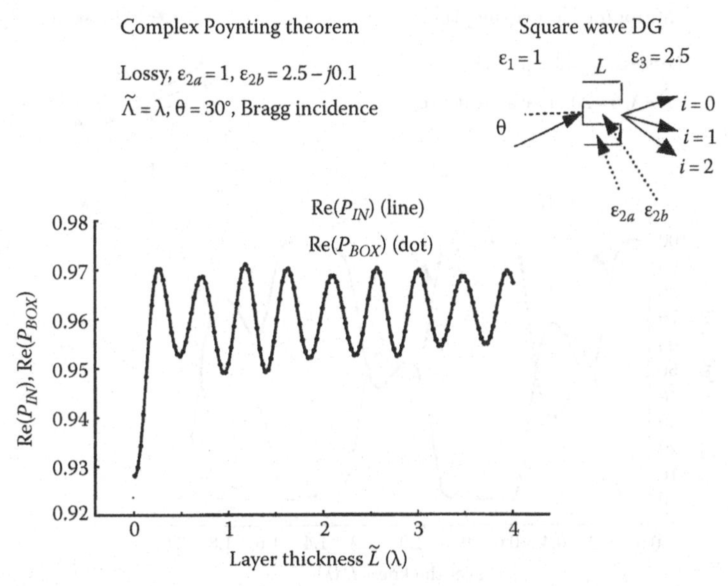

FIGURE 5.12 The real part of P_{IN} and P_{BOX} versus $\tilde{L}$ for the lossy square wave diffraction cases which were studied in Figures 5.10 and 5.11.

rather than a dielectric material (see the inset of Figure 5.14). The analysis for this case is identical to that presented in Section 5.2.1 except that inside Region 3 ($y < -L$), the EM fields are taken to be zero, and at Region 2–3 interface ($y = -L$), the EM fields are required to meet the well-known boundary condition that the tangential electric fields are zero. Mathematically, for the present H-mode polarization case, this requires

$$E_z^{(2)}(x, y, z)\Big|_{y=-L^+} = 0 \tag{5.105}$$

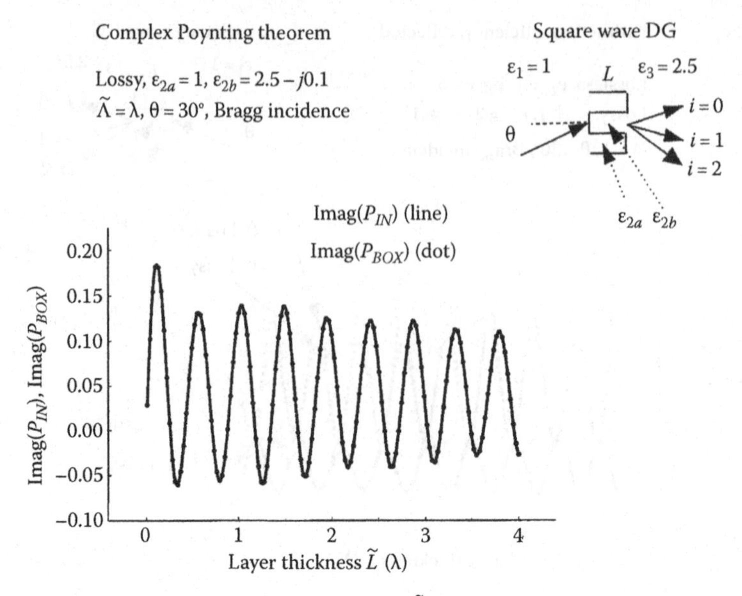

FIGURE 5.13 The imaginary part of P_{IN} and P_{BOX} versus $\tilde{L}$ for the lossy square wave diffraction cases which were studied in Figures 5.10 and 5.11.

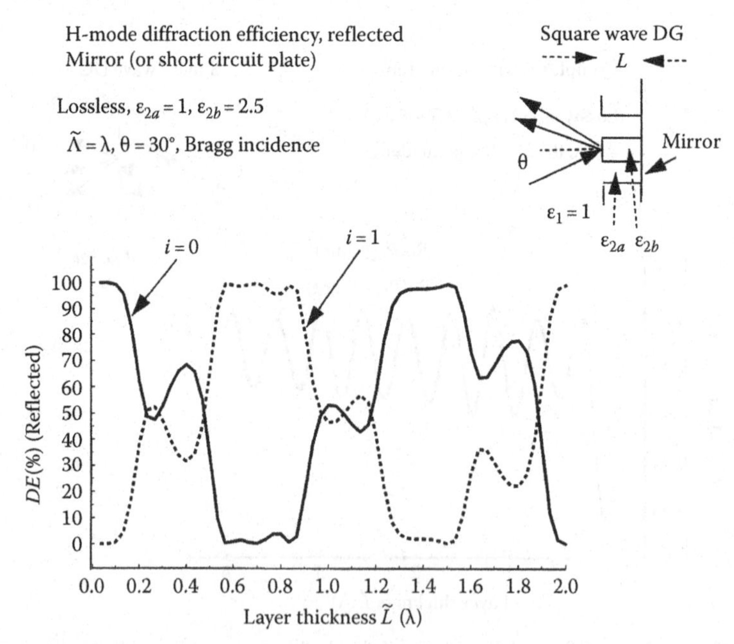

FIGURE 5.14 Reflected diffraction efficiency ($i = 0$ and 1 orders) versus layer length $\tilde{L}$.

If this boundary condition is imposed, it is found that the overall matrix equation that must be solved to determine the EM fields of the grating mirror system is

$$\sum_{n=1}^{N_T} C_n\{k_{y1i}S_{zin} - U_{xin}\} = 2E_0 k_{y10}\delta_{i,0} \tag{5.106}$$

$$\sum_{n=1}^{N_T} C_n\{e^{-q_n L}S_{zin}\} = 0 \tag{5.107}$$

Equation 5.106 is identical to Equation 5.48 as it should be since it results from matching EM fields at the Region 1–2 interface, and the general form of the unknown EM fields in the two regions is the same whether the mirror is present or not. The second matrix equation (Equation 5.107) is quite different from the second transmission grating matrix equation, Equation 5.49. Equation 5.107 was determined by imposing boundary condition that the tangential electric field at a perfect conductor boundary is zero, whereas Equation 5.49 was determined by matching the electric and magnetic fields at $y = -L$, and then eliminating the electric field unknown coefficients. Substitution of the C_n coefficients of Region 2 into Equation 5.42 then allows for the determination of the r_i reflection coefficients of the system. Incident and reflected power are given by the same formulas as already given for the transmission grating analysis.

Figure 5.14 shows the reflected diffraction efficiency ($i = 0$ and 1 orders) versus layer length $\tilde{L}$ which arises when a plane wave is incident on a square wave or step profile dielectric grating that is backed by a mirror (or short circuit plate). In Figure 5.14, the same lossless square grating that was studied in Figure 5.10 is analyzed. The square wave grating was taken to have a grating period of $\Lambda = \lambda$ and a transverse groove width of $\lambda/2$ (or duty cycle of 50%). The bulk and groove dielectric values and their orientation in the diffraction grating and the angle of incidence are specified in the Figure 5.14 caption and inset. For the present case, for the angle of incidence used, it turns out that the $i = 0, 1$ orders are the only orders that are reflected, diffracted propagating plane waves. All the other orders are evanescent. The value of $M_T = 6$ was used to calculate the data of Figure 5.14. As can be seen from Figure 5.14, power for small grating thickness is diffracted from the $i = 0$ order into the $i = 1$ ($0 \le \tilde{L} \le 0.6\lambda$). As the thickness increases, however, power is transferred back to $i = 0$ order from the $i = 1$ ($1\lambda \le \tilde{L} \le 1.6\lambda$). This cycle is repeated for larger values of $\tilde{L}$. In observing the $i = 0, 1$ plots, its very interesting to note that the transfer of power between the $i = 0, 1$ orders is not periodic with increasing $\tilde{L}$, but irregular and unpredictable. The nonperiodicity is undoubtedly due to the interaction of the evanescent and propagating waves that result from the matrix solution. Conservation of incident and reflected power $\left(\sum_i [DE_{Ri} + DE_{Ti}] = 1,\ DE_{Ti} = 0\right)$ was observed to a high degree of accuracy.

5.3 APPLICATION OF RCWA AND THE COMPLEX POYNTING THEOREM TO E-MODE PLANAR DIFFRACTION GRATING ANALYSIS

In the previous section, RCWA was used to study the H-mode polarization as it diffracts from isotropic diffraction gratings. In many real-life applications, it is necessary to study diffraction from anisotropic gratings, e.g., photorefractive materials (discussed in detail in Chapter 7). In this section, RCWA and the complex Poynting theorem [96–101] will be used to study, respectively, the EM fields and power flow and energy storage when a plane wave (E-mode polarization) is scattered from a general lossy and anisotropic diffraction grating. Full calculation of the diffraction efficiency; electromagnetic energy (electric and magnetic); and real, reactive, dissipative, and evanescent power of the grating will be made. In this section, several numerical examples involving a step profile will be studied.

The grating in Figure 5.15a is assumed to have its grating vector specified by $\vec{K} = \tilde{K}\hat{a}_x$, where $\tilde{K} = 2\pi/\tilde{\Lambda}$ and $\tilde{\Lambda}$ is the grating period or grating wavelength. In this case, the magnetic field is assumed to be polarized perpendicular to the plane of incidence as $\vec{H} = H_z\hat{a}_z$. In the present study, the complex Poynting theorem will be applied to a Poynting box whose length extends over the grating region $\tilde{L}$, whose width extends over a grating period $\tilde{\Lambda}$, and whose thickness is $\Delta\tilde{z}$ (the electromagnetic fields do not vary in the z-direction; thus the thickness of the Poynting box is immaterial to the Poynting power calculation). In Section 5.3.1, we will briefly summarize the E-mode RCWA equations for anisotropic diffraction gratings. In Section 5.3.2, the pertinent equations for the power budget as results from the complex Poynting theorem will be presented. In Section 5.3.3, illustrative examples will be given for anisotropic media where the permittivity tensor is either Hermitian or arbitrary.

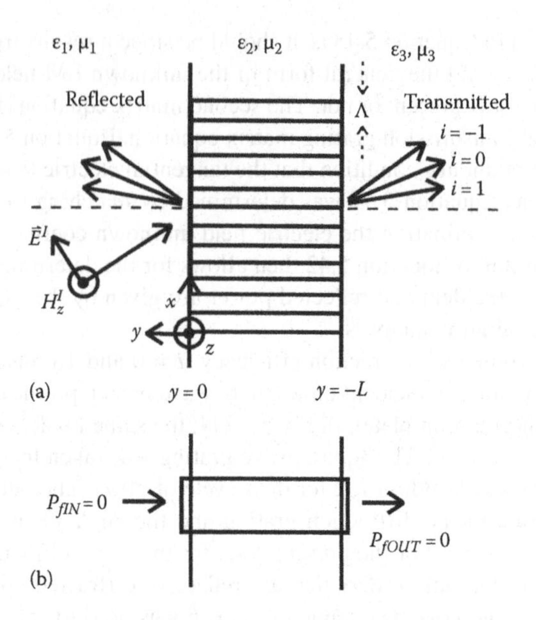

FIGURE 5.15 (a) The geometry of the E-mode diffraction grating system is shown. (b) The complex Poynting box used for calculations is shown. (From Jarem, J.M. and Banerjee, P.P., Application of the complex Poynting theorem to diffraction gratings, *J. Opt. Soc. Am. A*, 16, 819, 1999. With permission of Optical Society of America.)

In much of the existing diffraction grating literature, power conservation is verified by calculating the time-averaged real power which is transmitted and reflected from a lossless grating and then verifying that the sum of these powers equals the power incident on the grating. Computing the power budget using only the time-averaged real power has two large limitations associated with it. First, it cannot be used to verify power conservation for the very common case of lossy gratings, since in this case some power is dissipated as heat, and thus the transmitted and reflected powers will not equal the incident power. A second limitation of computing the power budget using only the time-averaged real power is the fact that information about the reactive power, the evanescent fields, the electric energy, magnetic energy, and power dissipated within the grating is left underdetermined and therefore unknown. All of these quantities contain important information about the nature and behavior of the grating. In the area of near-field optics, considerable attention has been paid to evanescent waves since these carry information about the diffracting or scattering object. Specifically, evanescent wave monitoring has applications in the area of submicron microscopy. Evanescent waves may also be excited from sharp discontinuities in the grating, e.g., corners and blaze tips [95]. A power budget approach that can study energy and power, both real and reactive, during diffraction from such gratings, is incorporated in the framework of the complex Poynting theorem. Botten et al. [56] consider the problem of energy balance in isotropic, lossy gratings when both E- and H-mode polarized incidence plane waves impinge on the grating.

Energy flow distributions, generation of plasmon surface waves, and absorption of EM energy by metallic sinusoidal gratings have been studied by Popov et al. [96–99] for shallow and deep gratings. The nature of the Poynting vector in a dielectric sinusoidal grating under total internal reflection has been studied by Shore et al. [100].

Our discussion of the Poynting vector here and in Jarem and Banerjee [101] is fundamentally different from that of Popov et al. [96–99] and Shore et al. [100]. In the work of Popov et al. [96–99] and Shore et al. [100], they were concerned with the problem of studying the spatial variation of the Poynting vector (and energy density) on a point-to-point basis over a region of space close to the diffraction grating surface. The point of their work was to relate local variation of the Poynting vector to the diffraction that occurred from the grating. They studied the physical mechanisms of blazing and anti-blazing and its relation to Poynting vector. In this section, we focus on the Poynting vector power which has been averaged transversely over a diffraction grating period and relate this averaged Poynting power to the power dissipated, transmitted, and reflected from the grating [101]. We apply the complex Poynting theorem for EM incidence on periodic diffraction gratings of arbitrary profile and made of anisotropic, lossy materials. We explicitly show also that the energy dissipated in the grating can result from both imaginary and real parts of the permittivity and permeability for the case of anisotropic, nonreciprocal grating media.

5.3.1 E-Mode RCWA Formulation

We begin the analysis by determining the general EM field solution of Region 2, the diffractive grating region. Using normalized coordinates where $x = k_0\tilde{x}$, $y = k_0\tilde{y}$, and $z = k_0\tilde{z}$, where $k_0 = 2\pi/\lambda$ and λ (m) is the free space wavelength, we find that Maxwell's normalized equations in Region 2 are given by

$$\nabla \times \vec{E} = -j\mu(\eta_0\vec{H}) \tag{5.108}$$

$$\nabla \times (\eta_0\vec{H}) = j\underline{\varepsilon}\vec{E} \tag{5.109}$$

where $\eta_0 = \sqrt{\mu_0/\varepsilon_0} = 377\,\Omega$ is the intrinsic impedance of free space; $\mu = \tilde{\mu}/\mu_0$ is the relative permeability of Region 2, where μ_0 is the permeability of free space; and $\underline{\varepsilon} = \underline{\tilde{\varepsilon}}/\varepsilon_0 \equiv \underline{\varepsilon}' - j\underline{\varepsilon}''$ is the relative, tensor permittivity of Region 2, where ε_0 is the permittivity of free space. In this section, we consider the important case when the relative permittivity tensor is anisotropic and has the specific form

$$\underline{\varepsilon} = \begin{bmatrix} \varepsilon_{xx} & \varepsilon_{xy} & 0 \\ \varepsilon_{yx} & \varepsilon_{yy} & 0 \\ 0 & 0 & \varepsilon_{zz} \end{bmatrix} \tag{5.110}$$

We expand the electric and magnetic field as

$$\vec{E} = \sum_{i=-\infty}^{\infty} [S_{xi}(y)\hat{a}_x + S_{yi}(y)\hat{a}_y] \exp(-jk_{xi}x) \tag{5.111}$$

$$\vec{U} \equiv \eta_0\vec{H} = \sum_{i=-\infty}^{\infty} U_{zi}(y)\exp(-jk_{xi}x)\hat{a}_z \tag{5.112}$$

$$k_{xi} = k_{x0} - iK_x, \quad k_{x0} = \sqrt{\mu_1\varepsilon_1}\sin(\theta), \quad K_x = 2\pi/\Lambda, \quad \Lambda = k_0\tilde{\Lambda} \tag{5.113}$$

where

θ is the angle of incidence
$\tilde{\Lambda}$ is the grating wavelength

Letting $\varepsilon(x)$ represent any of the elements of the tensor $\underline{\varepsilon}$ of Equation 5.110, we also expand those permittivity elements as

$$\varepsilon(x) = \sum_{i=-\infty}^{\infty} \breve{\varepsilon}_i \exp(jiK_x x) \tag{5.114}$$

where $\breve{\varepsilon}_i$ represents the Fourier coefficients of $\varepsilon(x)$. Substituting Equations 5.111 through 5.114 in Maxwell's equations; taking the relative permeability of Region 2 to be $\mu = 1$; introducing column and square matrices, namely, $\underline{S_x} = [S_{xi}]$, $\underline{S_y} = [S_{yi}]$, $\underline{U_z} = [U_{zi}]$, $i=-\infty,\ldots,\infty$, $\underline{\varepsilon_{xx}} = [\varepsilon_{xx_{i,i'}}] = [\breve{\varepsilon}_{xx_{i-i'}}]$, $\underline{\varepsilon_{xy}} = [\varepsilon_{xy_{i,i'}}] = [\breve{\varepsilon}_{xy_{i-i'}}]$, $\underline{\varepsilon_{yx}} = [\varepsilon_{yx_{i,i'}}] = [\breve{\varepsilon}_{yx_{i-i'}}]$, and $\underline{\varepsilon_{yy}} = [\varepsilon_{yy_{i,i'}}] = [\breve{\varepsilon}_{yy_{i-i'}}]$ (here the *underbar* denotes a square (i,i') matrix), $\underline{K_x} = [k_{xi}\delta_{i,i'}]$, $\underline{I} = [\delta_{i,i'}]$, $(i,i') = -\infty,\ldots\infty$, be square matrices. $\delta_{i,i'}$ is the Kronecker delta and $\underline{I}$ is the identity matrix; eliminating $\underline{S_y}$ using the equation

$$\underline{S_y} = \underline{\varepsilon_{yy}^{-1}}\left(\underline{K_x}\ \underline{U_z} - \underline{\varepsilon_{yx}}\ \underline{S_x}\right) \tag{5.115}$$

and rearranging terms, we find the following state variable form

$$\frac{\partial \underline{V^e(y)}}{\partial y} = \underline{A}\,\underline{V^e(y)} \tag{5.116}$$

where

$$\underline{V^e} = \begin{bmatrix} \underline{S_x} \\ \underline{U_z} \end{bmatrix}, \quad \underline{A} = \begin{bmatrix} \underline{a_{11}} & \underline{a_{12}} \\ \underline{a_{21}} & \underline{a_{22}} \end{bmatrix} \tag{5.117}$$

where

$$\underline{a_{11}} = j\left(\underline{K_x}\,\underline{\varepsilon_{yy}^{-1}}\ \underline{\varepsilon_{yx}}\right), \quad \underline{a_{12}} = j\left(-\underline{K_x}\,\underline{\varepsilon_{yy}^{-1}}\ \underline{K_x} + \underline{I}\right) \tag{5.118}$$

$$\underline{a_{21}} = j\left(\underline{\varepsilon_{xx}} - \underline{\varepsilon_{xy}}\ \underline{\varepsilon_{yy}^{-1}}\ \underline{\varepsilon_{yx}}\right), \quad \underline{a_{22}} = j\left(\underline{\varepsilon_{xy}}\,\underline{\varepsilon_{yy}^{-1}}\,\underline{K_x}\right) \tag{5.119}$$

where the superscript -1 in these equations denotes the matrix inverse. The above equations have been found by expressing each of the product terms $\varepsilon_{xx}(x)E_x(x,y)$, $\varepsilon_{xy}(x)E_y(x,y)$, etc., in a convolution form when the Fourier series expansions of $\underline{\varepsilon(x)}$ and $\vec{E}(x,y)$ are substituted in each of the product terms making up $\underline{\varepsilon(x)}\vec{E}$ and collecting coefficients on common i orders.

Let q_n and $\underline{V}_n$ be the eigenvalues and eigenvectors of the matrix $\underline{A}$ after truncation. Summing over the individual eigenmodes, we find that the overall electric and magnetic fields in Region 2 are given by

$$\vec{E}^{(2)} = \sum_{n=1}^{N_T} C_n \vec{E}_n^e = \sum_{i=-M_T}^{M_T}\left\{\sum_{n=1}^{N_T} C_n [S_{xin}\hat{a}_x + S_{yin}\hat{a}_y]\exp(q_n y)\right\}\exp(-jk_{xi}x) \tag{5.120}$$

$$\vec{U}^{(2)} = \eta_0 \vec{H}^{(2)} = \sum_{n=1}^{N_T} C_n \vec{U}_n^e = \sum_{i=-M_T}^{M_T}\left\{\sum_{n=1}^{N_T} C_n [U_{zin}\,\hat{a}_z]\exp(q_n y)\right\}\exp(-jk_{xi}x) \tag{5.121}$$

where $N_T = 2(2M_T + 1)$. Equations 5.120 and 5.121 represent the sum of $N_T = 2(2M_T + 1)$ forward and backward traveling, propagating and nonpropagating eigenmodes $\vec{E}_n^e$ and $\vec{U}_n^e \equiv \eta_0 \vec{H}_n^e$, which gives the general electromagnetic field solution in Region 2, the diffraction grating region.

An important problem that remains is to determine the N_T coefficients C_n of the Equations 5.120 and 5.121. Up to this point, we have specified the general form of the diffracted fields in the grating region. The EM fields on the incident side of the diffraction grating (Region 1 of Figure 5.15a) and on the transmission side of the diffraction grating (Region 3 of Figure 5.15a) consist of an infinite number of propagating and nonpropagating plane waves whose tangential wave vectors are given by k_{xi}, $i = -\infty, \ldots, -1, 0, 1, \ldots, \infty$. The electromagnetic fields in Region 1 consist of a single E-mode polarized, incident plane wave, and an infinite number of reflected propagating and evanescent plane waves. The total electric and magnetic fields in Regions 1 and 3 after summing the incident and reflected fields are given by

Region 1

$$U_z^{(1)} = \eta_0 H_z^{(1)} = \sum_{i=-\infty}^{\infty} \left[E_0 \delta_{i,0} \exp(jk_{y1i}y) + r_i \exp(-jk_{y1i}y) \right] \exp(-jk_{xi}x) \tag{5.122}$$

$$E_x^{(1)} = \frac{1}{\varepsilon_1} \sum_{i=-\infty}^{\infty} \left[k_{y1i} \left(E_0\, \delta_{i,0} \exp(jk_{y1i}y) - r_i \exp(-jk_{y1i}y) \right) \right] \exp(-jk_{xi}x) \tag{5.123}$$

$$E_y^{(1)} = \frac{1}{\varepsilon_1} \sum_{i=-\infty}^{\infty} \left[k_{xi} \left(E_0\, \delta_{i,0} \exp(jk_{y1i}y) + r_i \exp(-jk_{y1i}y) \right) \right] \exp(-jk_{xi}x) \tag{5.124}$$

Region 3

$$U_z^{(3)} = \eta_0 H_z^{(3)} = \sum_{i=-\infty}^{\infty} \left[t_i \exp(jk_{y3i}(y+L)) \right] \exp(-jk_{xi}x) \tag{5.125}$$

$$E_x^{(3)} = \frac{1}{\varepsilon_3} \sum_{i=-\infty}^{\infty} \left[k_{y3i} t_i \exp(jk_{y3i}(y+L)) \right] \exp(-jk_{xi}x) \tag{5.126}$$

$$E_y^{(3)} = \frac{1}{\varepsilon_3} \sum_{i=-\infty}^{\infty} \left[k_{xi} t_i \exp(jk_{y3i}(y+L)) \right] \exp(-jk_{xi}x) \tag{5.127}$$

where

$$k_{yri} = \begin{cases} [\mu_r \varepsilon_r - k_{xi}^2]^{1/2}, & \sqrt{\mu_r \varepsilon_r} > k_{xi} \\ -j[k_{xi}^2 - \mu_r \varepsilon_r]^{1/2}, & k_{xi} > \sqrt{\mu_r \varepsilon_r} \end{cases} \qquad r = 1,3 \tag{5.128}$$

Now that the electromagnetic fields have been defined in Regions 1–3, the next step is to match boundary conditions at the interfaces $y = 0$ and $-L$. Matching the tangential components at the grating interfaces, we have the final matrix equation

$$\sum_{n=1}^{N_T} C_n \left\{ \frac{k_{y1i}}{\varepsilon_1} U_{zin} + S_{xin} \right\} = \frac{2 E_0 k_{y1i}}{\varepsilon_1} \delta_{i0} \tag{5.129}$$

$$\sum_{n=1}^{N_T} C_n \left\{ \exp(-q_n L) \left[S_{xin} - \frac{k_{y3i}}{\varepsilon_3} U_{zin} \right] \right\} = 0 \tag{5.130}$$

where $i = -M_T, \ldots, M_T$.

The above equations constitute a set of $N_T = 2(2M_T + 1)$ equations for the N_T unknown coefficients C_n. Power is excited in the diffraction grating system through the RHS term of Equation 5.129.

5.3.2 Complex Poynting Theorem

We will use the complex Poynting theorem to study the power transmitted into and from the diffraction grating which is under consideration. In the present case, we will choose the Poynting box to have a width Λ (normalized grating period length) in the x-direction, a thickness Δz in the z-direction (the diffraction grating is z-independent, thus the Poynting box dimension may be chosen to have an arbitrary value in this direction), and have its front face located at $y = 0^+$ (in Region 1, just in front of the Region 1–2 interface) and its back face located at $y = -L^-$ (in Region 3, just behind the Region 2–3 interface). Figure 5.15b shows the Poynting box. We assume that no sources are present in Region 2. With these assumptions, we find that the complex Poynting theorem is given by

$$P^u_{fIN} = P^u_{fOUT} + P^u_{DE} + P^u_{DM} + j(-P^u_{WE} + P^u_{WM}) \tag{5.131}$$

where

$$P^u_{fIN} = \iint_{\Delta S} \vec{E}^{(1)} \times \vec{U}^{(1)*}\Big|_{y=0^+} \cdot (-\hat{a}_y)\, dS \tag{5.132}$$

$$P^u_{fOUT} = \iint_{\Delta S} \vec{E}^{(3)} \times \vec{U}^{(3)*}\Big|_{y=-L^-} \cdot (-\hat{a}_y)\, dS \tag{5.133}$$

$$P^u_{DE} = \iiint_{\Delta V} \vec{E}^{(2)} \cdot \left[\underline{\varepsilon}'' \vec{E}^{(2)} \right]^* dV \tag{5.134}$$

$$P^u_{DM} = \iiint_{\Delta V} \vec{U}^{(2)} \cdot \left[\underline{\mu}'' \vec{U}^{(2)} \right]^* dV \tag{5.135}$$

$$P^u_{WE} = \iiint_{\Delta V} \vec{E}^{(2)} \cdot \left[\underline{\varepsilon}' \vec{E}^{(2)} \right]^* dV \tag{5.136}$$

$$P^u_{WM} = \iiint_{\Delta V} \vec{U}^{(2)} \cdot \left[\underline{\mu}' \vec{U}^{(2)} \right]^* dV \tag{5.137}$$

and where

$$dS = k_0^2\, d\tilde{S},\ dV = k_0^3\, d\tilde{V}.$$

In Equations 5.131 through 5.137, $P_{fIN}^{u}(V^2/m^2)$ is proportional to the complex power which is radiated into the diffraction grating (it is the sum of the incident power, reflected power, and interaction power between the incident and reflected power), and P_{DE}^{u} and P_{DM}^{u} are proportional to the electric and magnetic dissipated power loss, while P_{WE}^{u} and P_{WM}^{u} are proportional to the reactive powers proportional to the electric and magnetic energies in the case when the grating material is isotropic. In the general anisotropic case, all of these four quantities can be complex. Hence, for instance, energy loss can result from both the imaginary and real parts of $\underline{\varepsilon}$ and μ as in a non-Hermitian medium. The superscript "u" in Equations 5.131 through 5.137 refers to "unnormalized." These power terms will be later normalized to the incident plane wave power. We will now be concerned with evaluating these equations for the E-mode plane wave polarization case under consideration.

5.3.2.1 Sample Calculation of P_{WE}^{u}

We illustrate the evaluation of the integrals in Equations 5.131 through 5.137 by calculating the P_{WE}^{u} integral of Equation 5.136. Substituting the electric field of Region 2, we find that the dot product inside the integral is

$$\vec{E}^{(2)}\cdot\underline{\varepsilon}'\vec{E}^{(2)*}=E_x^{(2)}\varepsilon'_{xx}(x)E_x^{(2)*}+E_x^{(2)}\varepsilon'_{xy}(x)E_y^{(2)*}+E_y^{(2)}\varepsilon'_{yx}(x)E_x^{(2)*}+E_y^{(2)}\varepsilon'_{yy}(x)E_y^{(2)*} \tag{5.138}$$

Each of the four terms in Equation 5.138 must be substituted in Equation 5.136 and the subsequent volume energy integrals must be evaluated. The analysis consists of substituting the Fourier series expansions of the electric field quantities and dielectric tensor quantities into the energy volume integral, interchanging sum and integral expressions, carrying out all exponential integrals exactly and in closed form, and finally simplifying all summations. Letting

$$V(x,y)=\sum_{i',n'}C_{n'}V_{i',n'}\exp(q_{n'}y)\exp(-jk_{xi'}x) \tag{5.139}$$

represent the electric field $E_x(x,y)$ or $E_y(x,y)$, letting

$$\varepsilon(x)=\sum_{i'''}\breve{\varepsilon}'_{i'''}\exp\left[ji'''K_x x\right] \tag{5.140}$$

represent $\varepsilon'_{xx}(x)$, $\varepsilon'_{xy}(x)$, $\varepsilon'_{yx}(x)$, or $\varepsilon'_{yy}(x)$, and letting

$$W(x,y)=\sum_{i'',n''}C_{n''}W_{i'',n''}\exp(q_{n''}y)\exp(-jk_{xi''}x) \tag{5.141}$$

represent the electric field $E_x(x,y)$ or $E_y(x,y)$, we find that any of the four terms of the unnormalized energy volume integral P_{WE}^{u} may be expressed in the general form

$$P=\iiint_{\Delta V}V(x,y)\varepsilon(x)W(x,y)^{*}dV=\iiint_{\Delta V}\left\{\sum_{i',n'}C_{n'}V_{i'n'}\exp\left[q_{n'}y\right]\exp\left[-jk_{xi'}x\right]\right\}$$
$$\times\left\{\sum_{i'''}\breve{\varepsilon}_{i'''}\exp\left[ji'''K_x x\right]\right\}\left\{\sum_{i'',n''}C_{n''}W_{i''n''}\exp\left[q_{n''}y\right]\exp\left[-jk_{xi''}x\right]\right\}^{*}dx\,dy\,dz \tag{5.142}$$

The integrals are independent of the z-coordinate. We find after some algebra that

$$\iiint_{\Delta V} V(x,y)\varepsilon(x)W(x,y)^* dV = \Delta z\Lambda \sum_{n',n''} C_{n'}C_{n''}^* I_{yn',n''} \sum_{i',i''} \breve{\varepsilon}_{i''-i'} V_{i'n'} W_{i''n''}^* \tag{5.143}$$

where

$$I_{yn',n''} = \int_{-L^-}^{0^+} \exp[(q_{n'} + q_{n''}^*)y]dy \tag{5.144}$$

Substitution of the four terms of Equation 5.138, with each term simplified according to Equation 5.143, produces a closed form expression (that is, all integrations have been carried out exactly) from which normalized energy volume integral P_{WE}^u may be evaluated. The evaluation of the P_{DE}^u is identical to the analysis of the P_{WE}^u integral except that the lossy relative permittivity $\underline{\varepsilon''(x)}$ tensor is used rather than $\underline{\varepsilon'(x)}$.

5.3.2.2 Other Poynting Theorem Integrals

The evaluation of the magnetic volume integrals P_{WM}^u and P_{DM}^u is identical to that of P_{WE}^u and P_{DE}^u except that the magnetic field is used rather than the electric field. In the present case under analysis, the permeability $\mu = \mu' - j\mu''$ is uniform in the x-direction and its Fourier representation is

$$\mu(x) \equiv \mu = \sum_{i=-\infty}^{\infty} \breve{\mu}_i e^{ji\kappa x} \tag{5.145}$$

where

$$\breve{\mu}_i = (\mu' - j\mu'')\delta_{i,0} \tag{5.146}$$

Following the analysis used to determine P_{WE}^u and evaluating the discrete Kronecker delta found, we find that

$$P_{WM}^u = \Delta z\Lambda\mu' \sum_{n',n''} C_{n'}C_{n''}^* I_{yn',n''} \sum_i \left[U_{zin'}^{(2)} U_{zin''}^{(2)*}\right] \tag{5.147}$$

$$P_{DM}^u = \Delta z\Lambda\mu'' \sum_{n',n''} C_{n'}C_{n''}^* I_{yn',n''} \sum_i \left[U_{zin'}^{(2)} U_{zin''}^{(2)*}\right] \tag{5.148}$$

5.3.2.3 Simplification of Results and Normalization

Substituting the electromagnetic fields of Region 1 into Equation 5.132, integrating over the cross section $\Delta z\Lambda$, and performing algebra, we find that the power radiated into the Poynting box at $y = 0^+$ is given by

$$P_{fIN}^u = \frac{\Delta z\Lambda}{\varepsilon_1} \sum_i k_{y1i} \left[E_0\delta_{i,0} - r_i\right]\left[E_0\delta_{i,0} + r_i\right]^* \tag{5.149}$$

It is convenient to normalize the power terms of Equation 5.131 to the incident power radiated into the Poynting box. Integration of the incident power P_{inc}^{u} over the cross section $\Delta z\Lambda$ at $y = 0^{+}$ for the present E-mode polarization case, after noting that the integral is purely real, gives the incident power as

$$P_{inc}^{u} = \frac{\Delta z\Lambda}{\varepsilon_1}\left|E_0\right|^2 k_{y10} \tag{5.150}$$

If we normalize the power of Equation 5.149 to the incident power, we find that

$$P_{IN} \equiv \frac{P_{fIN}^{u}}{P_{inc}^{u}} = \frac{\sum_i k_{y1i}\left[E_0\delta_{i,0} - r_i\right]\left[E_0\delta_{i,0} + r_i\right]^*}{\left[k_{y10}\left|E_0\right|^2\right]} \tag{5.151}$$

Of interest are the powers that are reflected and transmitted from the diffracting grating at $y = 0^{+}$ and $-L^{-}$, respectively, and the relation of that these powers have to the power P_{fIN} which is radiated into the Poynting box $y = 0^{+}$. The unnormalized reflected and transmitted powers are given by the expressions

$$P_{ref}^{u} = \iint_{\Delta S} \vec{E}_{ref}^{(1)} \times \vec{U}_{ref}^{(1)*}\Big|_{y=0^+} \cdot \hat{a}_y \, dS = \frac{\Delta z\Lambda}{\varepsilon_1}\sum_i k_{y1i}\, r_i\, r_i^* \tag{5.152}$$

$$P_{fOUT}^{u} \equiv P_{trans}^{u} = \iint_{\Delta S} \vec{E}^{(3)} \times \vec{U}^{(3)*}\Big|_{y=-L^-} \cdot (-\hat{a}_y)\, dS = \frac{\Delta z\Lambda}{\varepsilon_3}\sum_i k_{y3i} t_i t_i^* \tag{5.153}$$

To find a relation between P_{fIN}^{u} and P_{ref}^{u}, we take $E_0 = |E_0|\exp(j0°)$ and note that k_{y1i} is purely real for $i = 0$, and we analyze the summation in the numerator of the RHS of Equation 5.151, namely,

$$T \equiv \sum_i k_{y1i}\left[E_0\delta_{i,0} - r_i\right]\left[E_0\delta_{i,0} + r_i\right]^* \tag{5.154}$$

After expanding the product term of Equation 5.154 in square brackets and after separating the $i = 0$ term from the $i \neq 0$, we find that

$$T = \sum_i k_{y1i}\left[E_0^2\,\delta_{i,0} + (-r_i + r_i^*)E_0\delta_{i,0} - r_i\, r_i^*\right] \tag{5.155}$$

$$T = k_{y10}\left[E_0^2 + (-2j\,\mathrm{Imag}(r_0)\,E_0 - r_0\, r_0^*\right] - \sum_{i,i\neq 0} k_{y1i} r_i\, r_i^* \tag{5.156}$$

$$T = k_{y10}E_0^2 - 2jk_{y10}\,\mathrm{Imag}(r_0)\,E_0 - \sum_i k_{y1i} r_i\, r_i^* \tag{5.157}$$

Thus

$$P_{fIN}^{u} = \frac{\Delta z\Lambda}{\varepsilon_1}\left[\sum_i k_{y1i}\, E_0 E_0^*\,\delta_{i,0}\right] - \frac{\Delta z\Lambda}{\varepsilon_1}\sum_i\left[k_{y1i} r_i\, r_i^*\right] - \frac{\Delta z\Lambda}{\varepsilon_1}\left[2jk_{y10}\,\mathrm{Imag}(r_0)\,E_0\right] \tag{5.158}$$

The first and second summation terms of Equation 5.158 represent the unnormalized incident and reflected power at $y = 0^+$. The third term is an interaction term between the incident and reflected EM wave. We have

$$P_{fIN}^u = P_{inc}^u - P_{ref}^u - \frac{\Delta z \Lambda}{\varepsilon_1}\left[2jk_{y10}\,\text{Imag}(r_0)E_0\right] \tag{5.159}$$

We now substitute P_{fIN}^u of Equation 5.159 into left-hand side (LHS) of Equation 5.131. We find that

$$P_{inc}^u - P_{ref}^u - \frac{\Delta z \Lambda}{\varepsilon_1}\left[2jk_{y10}\,\text{Imag}(r_0)E_0\right] = P_{fOUT}^u + P_{DE}^u + P_{DM}^u + j(-P_{WE}^u + P_{WM}^u) \tag{5.160}$$

Transposing the reflected power term and the interaction power term to the RHS of Equation 5.160, we find

$$P_{inc}^u = P_{ref}^u + \frac{\Delta z \Lambda}{\varepsilon_1}\left[2jk_{y10}\,\text{Imag}(r_0)E_0\right] + P_{fOUT}^u + P_{DE}^u + P_{DM}^u + j(-P_{WE}^u + P_{WM}^u) \tag{5.161}$$

Defining the normalized power terms (taking $E_0 = |E_0|\exp(j0^\circ)$)

$$P_{inc} \equiv \frac{P_{inc}^u}{P_{inc}^u} = 1$$

$$P_{ref} = \frac{P_{ref}^u}{P_{inc}^u} = \frac{1}{k_{y10}}\sum_i k_{y1i} r_i r_i^*$$

$$P_{OUT} \equiv P_{trans} = \frac{P_{fOUT}^u}{P_{inc}^u} = \frac{\varepsilon_1}{\varepsilon_3}\frac{1}{k_{y10}}\sum_i k_{y3i} t_i t_i^*$$

$$P_{DE} \equiv \frac{P_{DE}^u}{P_{inc}^u}, \quad P_{DM} \equiv \frac{P_{DM}^u}{P_{inc}^u}$$

$$P_{WE} \equiv \frac{P_{WE}^u}{P_{inc}^u}, \quad P_{WM} \equiv \frac{P_{WM}^u}{P_{inc}^u} \tag{5.162}$$

we now find that the complex Poynting theorem of Equation 5.131 after division of all terms by P_{inc}^u may be written in normalized form as

$$P_{IN} = P_{OUT} + P_{DE} + P_{DM} + j(-P_{WE} + P_{WM}) \equiv P_{BOX} \tag{5.163a}$$

or in terms of the reflected and transmitted powers as

$$1 = P_{inc} = P_{ref} + 2j\,\text{Imag}(r_0) + P_{trans} + P_{DE} + P_{DM} + j(-P_{WE} + P_{WM}) \tag{5.163b}$$

Equation 5.163a and b represent the main complex Poynting theorem conservation relation which relates the input, incident, reflected, and transmitted powers to the dissipated power and stored energy of the system.

By taking the real and imaginary parts of Equation 5.163b, we may derive useful relations from which the numerical accuracy of the diffraction analysis may be checked and we may also derive useful expressions into which numerical insight of the diffraction process may be gained. Taking the real part of Equation 5.163b, we find that

$$1 = P_{inc} = \mathrm{Re}(P_{ref}) + \mathrm{Re}(P_{trans}) + \mathrm{Re}(P_{DE}) + \mathrm{Re}(P_{DM}) + \mathrm{Re}\left[j(-P_{WE} + P_{WM})\right] \quad (5.164)$$

Taking the imaginary part of Equation 5.163b, we find that

$$0 = \mathrm{Imag}(P_{ref}) + \mathrm{Imag}(P_{trans}) + 2\,\mathrm{Imag}(r_0) + \mathrm{Imag}(P_{DE}) + \mathrm{Imag}(P_{DM}) + \mathrm{Imag}\left[j(-P_{WE} + P_{WM})\right] \quad (5.165)$$

We remind the reader that from its definition that the quantities P_{DE}, P_{DM}, P_{WE}, and P_{WM} in the general, anisotropic case are not purely real and thus taking the real and imaginary parts as specified in Equation 5.165 is necessary as shown.

From Equations 5.164 and 5.165, we will now define three useful relations from which numerical plots can be made and which give insight into the diffraction process. We will now give the first relation. From Equation 5.164, if we transpose the $\mathrm{Re}[j(-P_{WE} + P_{WM})]$ term, we have

$$\mathrm{Re}\left[j(-P_{WE} + P_{WM})\right] = P_{inc} - \mathrm{Re}(P_{ref}) - \mathrm{Re}(P_{trans}) - \mathrm{Re}(P_{DE}) - \mathrm{Re}(P_{DM}) \quad (5.166)$$

Letting the LHS of Equation 5.166 be

$$P_{diffR}^{WEM} \equiv \mathrm{Re}\left[j(-P_{WE} + P_{WM})\right] \quad (5.167)$$

and letting the RHS of Equation 5.166 be

$$P_{diffR} \equiv P_{inc} - \mathrm{Re}(P_{ref}) - \mathrm{Re}(P_{trans}) - \mathrm{Re}(P_{DE}) - \mathrm{Re}(P_{DM}) \quad (5.168)$$

we have from inspecting Equation 5.166

$$P_{diffR}^{WEM} = P_{diffR} \quad (5.169)$$

Equations for P_{diffR}^{WEM} and P_{diffR} are in general useful quantities to calculate. When the medium is reciprocal, P_{WE}, P_{WM}, P_{DE}, and P_{DM} are all purely real quantities; therefore, from Equation 5.168, $P_{diffR}^{WEM} = 0$ and $P_{diffR} = 0$, and thus Equation 5.168 represents a conservation relation stating that the incident power should equal the sum of the transmitted, reflected and dissipated powers. When the medium is anisotropic, P_{WE} and P_{WM} are in general complex and thus P_{diffR}^{WEM} is not necessarily zero. In this case, P_{diffR}^{WEM} (which should equal P_{diffR}) gives a sense of to what degree the anisotropy of the medium will effect the EM field numerical results. The computation of P_{diffR}^{WEM} and P_{diffR} is also useful in this case as a cross-check of the numerical calculation. It is useful since both terms are computed from EM field quantities located in different regions of space. Numerically, if P_{diffR}^{WEM} and P_{diffR} are not equal or almost equal to each other, the numerical computation has to be in error. A second useful relation is given by Equation 5.165. We define

$$P_{diffI}^{WEM} \equiv \mathrm{Imag}\left[j(-P_{WE} + P_{WM})\right] \quad (5.170)$$

$$P_{diffI} \equiv -\mathrm{Imag}(P_{ref} + P_{trans} + P_{DE} + P_{DM}) - 2\,\mathrm{Imag}(r_0) \quad (5.171)$$

Proper balance of the complex Poynting theorem requires

$$P_{diffl}^{WEM} = P_{diffl} \tag{5.172}$$

We will now define a third useful relation that results from Equations 5.164 and 5.165. Transposing the terms $\text{Imag}(P_{ref})$ and $\text{Imag}(P_{trans})$ to the LHS of Equation 5.165 and multiplying by −1, we find that

$$\text{Imag}(P_{ref}) + \text{Imag}(P_{trans}) = -2\,\text{Imag}(r_0) - \text{Imag}(P_{DE}) - \text{Imag}(P_{DM}) - \text{Imag}\left[j(-P_{WE} + P_{WM})\right] \tag{5.173}$$

Letting

$$P_{evan} \equiv \text{Imag}(P_{ref}) + \text{Imag}(P_{trans}) \tag{5.174}$$

and letting

$$P_{evan}^{diff} \equiv -2\,\text{Imag}(r_0) - \text{Imag}(P_{DE}) - \text{Imag}(P_{DM}) - \text{Imag}\left[j(-P_{WE} + P_{WM})\right] \tag{5.175}$$

we have from inspecting Equations 5.173 through 5.175

$$P_{evan} = P_{evan}^{diff} \tag{5.176}$$

Equation 5.176 is a useful quantity to calculate as it gives a measure of the evanescent power stored in the diffraction. The fact that it gives the evanescent power may be seen from the equations for the reflected power in Region 1 and the transmitted power in Region 3. For the reflected power in Region 1, we have

$$P_{ref} = \frac{P_{ref}^{u}}{P_{inc}^{u}} = \frac{1}{k_{y10}} \sum_{i} k_{y1i} r_i r_i^* \tag{5.177}$$

where from Equation 5.128

$$k_{y1i} = \begin{cases} [\mu_1\varepsilon_1 - k_{xi}^2]^{1/2}, & \sqrt{\mu_1\varepsilon_1} > k_{xi} \\ -j[k_{xi}^2 - \mu_1\varepsilon_1]^{1/2}, & k_{xi} > \sqrt{\mu_1\varepsilon_1} \end{cases} \tag{5.178}$$

In these equations, we note that k_{y10} is real and positive. Because the term $r_i r_i^*$ is purely real, the reflected power P_{ref} is purely imaginary only when k_{y1i} is purely imaginary. Thus we see that $\text{Imag}(P_{ref}) \neq 0$ is only nonzero whenever $k_{y1i} = -j[k_{xi}^2 - \mu_1\varepsilon_1]^{1/2}$, $k_{xi} > \sqrt{\mu_1\varepsilon_1}$. This occurs only for the case when the space harmonics that are evanescent. The transmitted power is evanescent for $\text{Imag}(P_{trans}) \neq 0$ the same reason as the reflected power.

Relations 5.174 through 5.176, like Equations 5.167 through 5.169, are useful quantities to calculate for two reasons. First, because they give the evanescent power they give a measure of how much power and energy is stored in non propagating EM waves near the diffraction grating interfaces. The larger the evanescent power and energy, the larger and rougher are the diffraction grating interfaces relative to the bounding regions. A second reason the computation of equations is useful is that they provide an excellent cross-check of the numerical diffraction solution. P_{evan} and P_{evan}^{diff} are computed from EM terms that exist in different regions of the EM system. Thus equality or very close numerical equality of P_{evan} and P_{evan}^{diff} help show that the computations are being made correctly. We also note that the sum of evanescent powers tends to be small relative to the other diffraction power terms in the system. Thus the sum of evanescent powers tends to be a fairly sensitive test of the EM algorithm.

The type of power budget analysis presented here can be applied to virtually any type of diffraction grating and any kind of polarization for the incident wave, and may be extended to multilayer grating structures in a straightforward way. For example, results for H-mode incidence on an isotropic grating have already been presented in the previous section.

5.3.3 Numerical Results

In this section, we will present numerical examples of the diffraction as results from RCWA and the complex Poynting theorem that have been discussed previously for the E-mode polarization case under consideration. The examples to be presented consist of a RCWA study of a square or step profile diffraction grating when several different isotropic and anisotropic materials make up the step profile. The step grating is assumed to be homogeneous in the longitudinal direction and to contain sharp dielectric discontinuities at dielectric steps and thus possesses a high spectral content $i = -\infty, \ldots, -1, 0, 1, \ldots, \infty$. Thus, studying isotropic and anisotropic materials with relatively short grating periods $\tilde{\Lambda} = \lambda$ tests the RCWA and the complex Poynting theorem in a fairly severe way.

We begin presenting results for the isotropic step grating. Figure 5.16a shows the transmitted diffraction efficiencies $DE_T(\%)$ for the $i = 0, 1$ orders when a lossless and lossy grating is present. The parameters of the grating are given in the Figure 5.16 insets. The plots show $DE_T(\%)$ versus the layer length $\tilde{L}$ (in units of free space wavelength λ). The lossless grating example of Figure 5.16a has been studied in Ref. [52] and the lossless diffraction efficiency results of Figure 5.16a are identical to the results of Ref. [52]. As can be seen from Figure 5.16a, as the layer length $\tilde{L}$ increases from 0λ to 2λ because θ is at the Bragg angle, power is diffracted from the $i = 0$ order into the $i = 1$ order. For larger values of $\tilde{L}$, 2λ to 4λ, power is diffracted from the $i = 0$ order to the $i = 1$ order. This cycle of blazing and anti-blazing is repeated over a long range of $\tilde{L}$ values. See Refs. [96–99] for an insightful discussion of blazing and anti-blazing and its relation to the Poynting vector. One also observes that in the case that ε_{2b} is lossy, the diffraction efficiency of both the $i = 0$ and 1 orders is attenuated as one would expect in a lossy material.

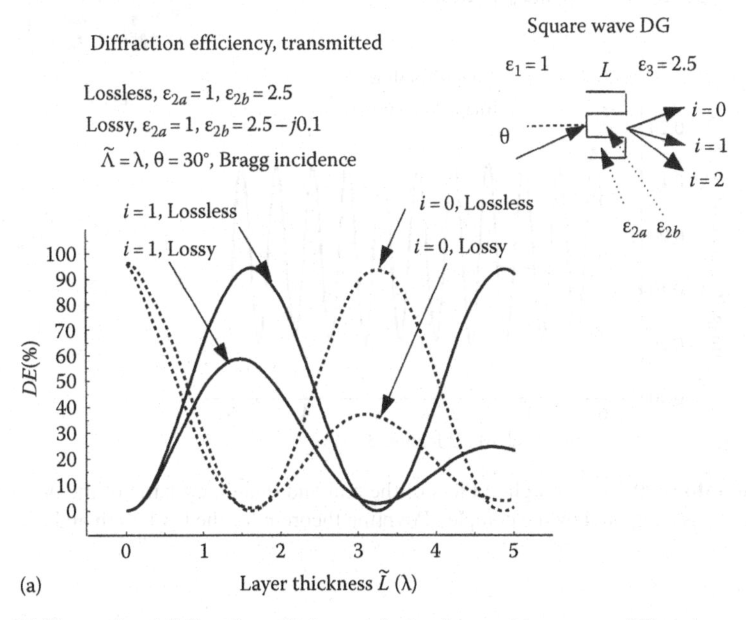

FIGURE 5.16 (a) Transmitted diffraction efficiency of a lossless and lossy step diffraction grating is shown.

(continued)

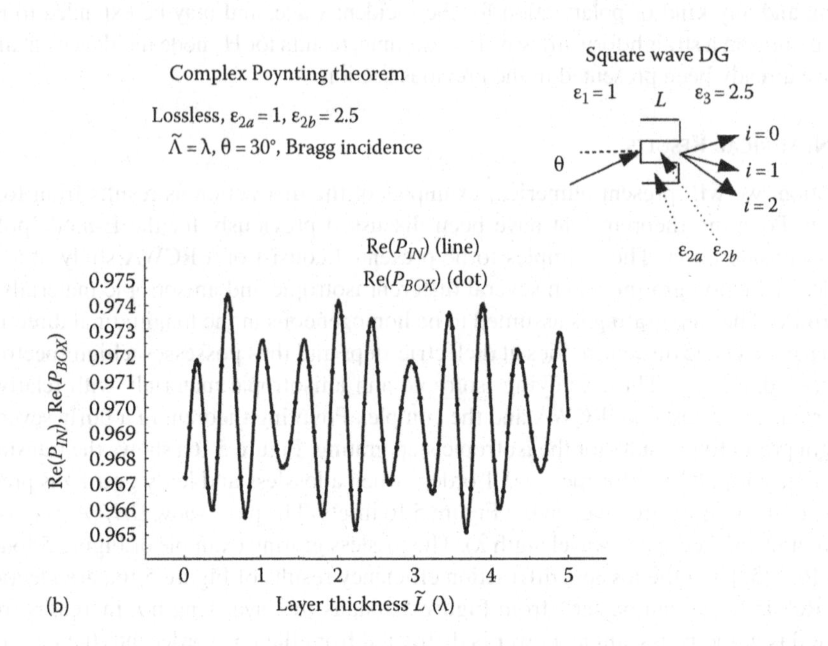

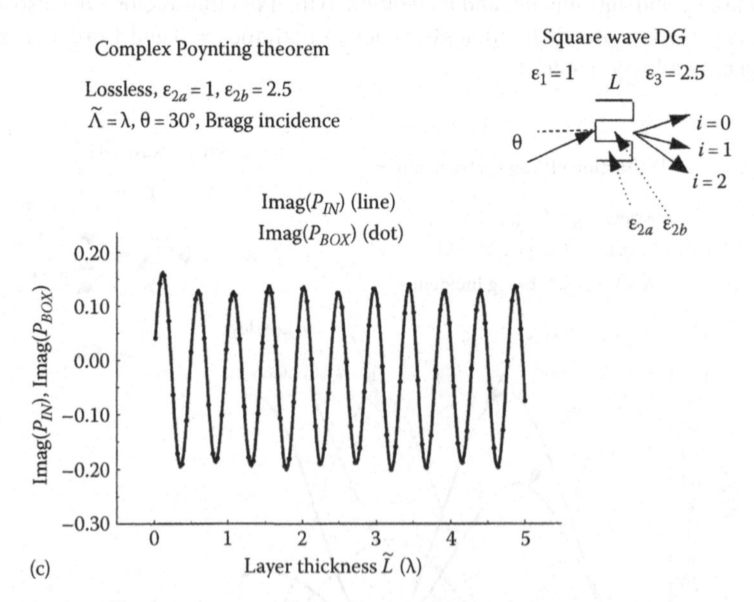

FIGURE 5.16 (continued) (b through e) Plots of the real and imaginary parts of the normalized complex power P_{IN} and P_{BOX} as computed by the complex Poynting theorem for the lossless (b and c)

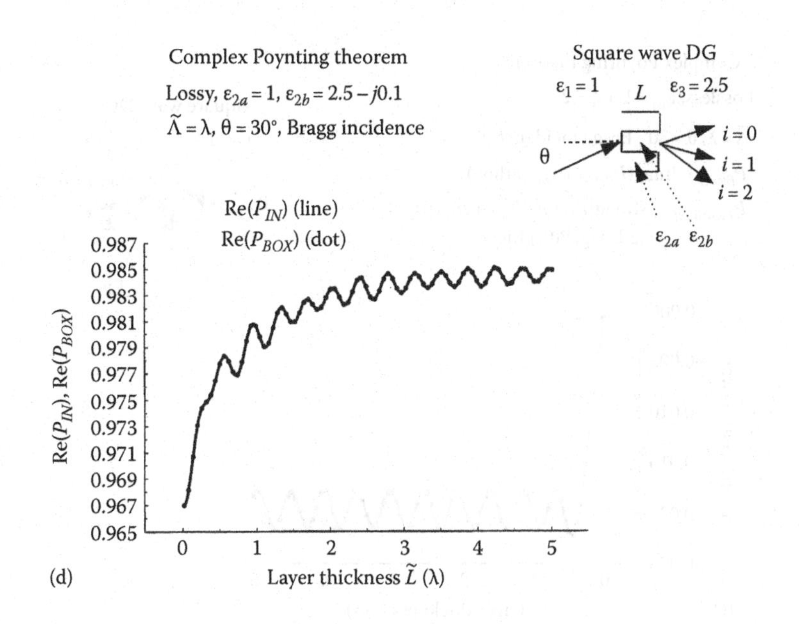

(d)

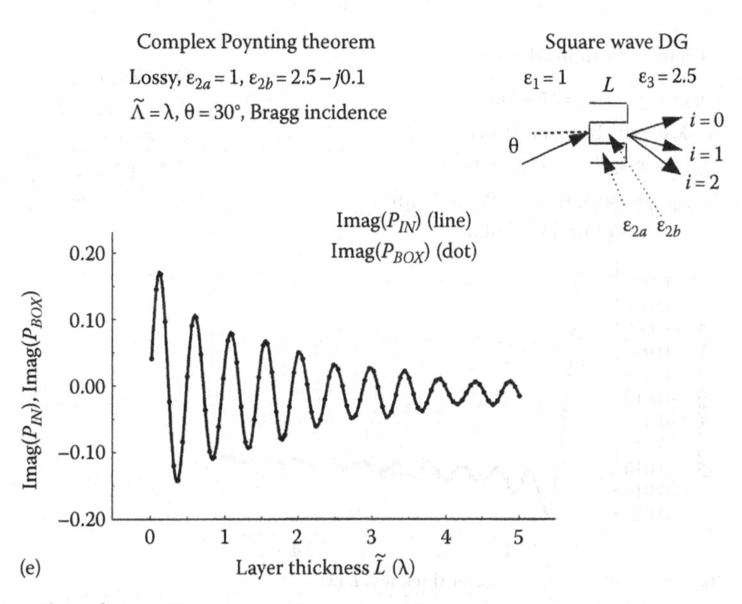

(e)

FIGURE 5.16 (continued) and lossy cases (d and e) are shown.

(continued)

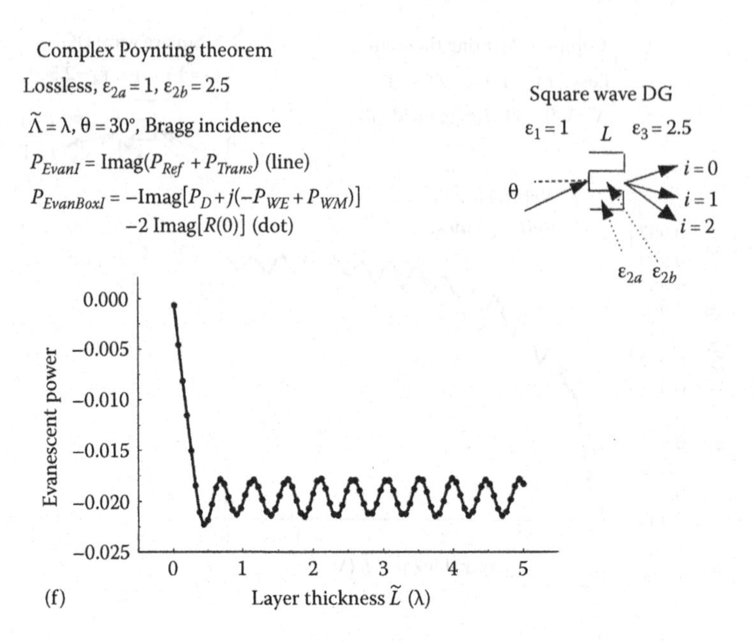

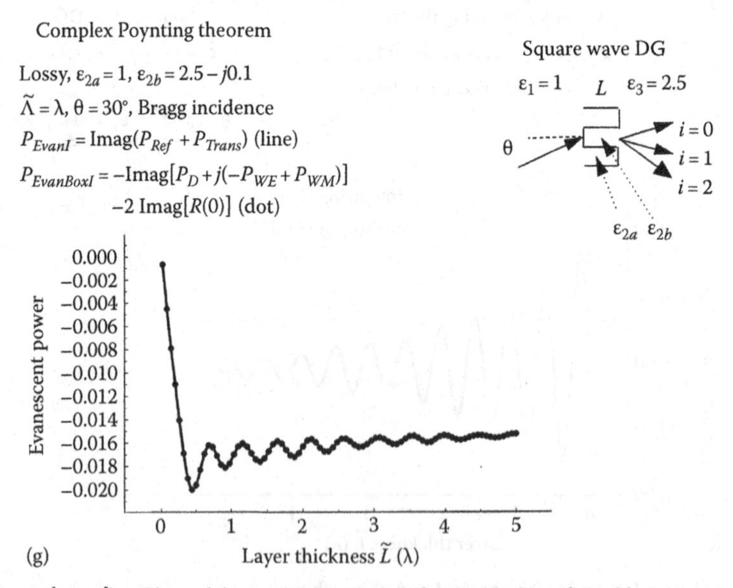

FIGURE 5.16 (continued) Plots of the evanescent power for the lossless (f) and lossy cases (g) are shown. (From Jarem, J.M. and Banerjee, P.P., Application of the complex Poynting theorem to diffraction gratings, *J. Opt. Soc. Am. A*, 16, 819, 1999. With permission of Optical Society of America.)

Figure 5.16b through e show plots of the real and imaginary parts of the normalized complex power P_{IN} and P_{BOX} of the complex Poynting theorem for the lossless (Figure 5.16b and c) and lossy cases (Figure 5.16d and e) as specified by Equation 5.163a and b. As can be seen from these four plots, the complex Poynting theorem is obeyed to a high degree of accuracy as evidenced by the close fit between the data for P_{IN} (*solid line*) and the data for P_{BOX} (*dots*). In comparing the lossless and lossy cases (using Equation 5.163a and b), one also notices a significant difference in the complex Poynting results for these cases. In the lossy case, as the layer becomes larger, the oscillations of the power results decrease. This is because in the lossy case, less EM power is reflected from the boundary and thus interferes less with forward traveling waves than in the case when the medium is lossless. Figure 5.16f and g shows plots of the evanescent power as computed by Equations 5.174 through 5.176. As can be seen in these figures, excellent agreement from the complex Poynting theorem is observed in both the lossy and nonlossy cases. As discussed earlier, because the grating width is on the order of a wavelength, a certain amount of energy is stored in the evanescent fields of Regions 1 and 3. The comparison of the lossy and nonlossy different material cases shows a definite difference in the evanescent field quantities of the system in the two cases.

Figure 5.17a and b shows plots of the real and imaginary parts, respectively, of the normalized complex power as computed by Equation 5.163a and b for the lossless case when a general, nonreciprocal, anisotropic material occupies Region 2a of the step diffractive region as shown in the inset of Figure 5.17a and b. Region 2b of the step was chosen to have $\varepsilon_{2b} = 2.5$ and Region 2a of the step was chosen to have $\varepsilon_{2ayy} = 1.5\varepsilon_{2axx}$, $\varepsilon_{2axx} = 1 - j0.1$, $\varepsilon_{2axy} = 0.2\varepsilon_{2axx}$, and $\varepsilon_{2ayx} = 0$. The example being considered is non-Hermitian since $\varepsilon_{2axy} \neq \varepsilon_{2ayx}^*$. This situation may be encountered in, for instance, materials with stimulated Raman scattering, a process whose governing susceptibility does not exhibit overall permutation symmetry [94]. As can be seen from these figures, the complex Poynting theorem is obeyed to a high degree of accuracy. Figure 5.17c shows a plot of the evanescent power as calculated by Equations 5.174 through 5.176 for the nonreciprocal, anisotropic case under consideration.

Figure 5.17d and e shows plots of P_{diffR}^{WEM} and P_{diffR} (calculated in Equations 5.167 through 5.169) and plots of P_{diffI}^{WEM} and P_{diffI} (calculated in Equations 5.170 through 5.172) for the same anisotropic

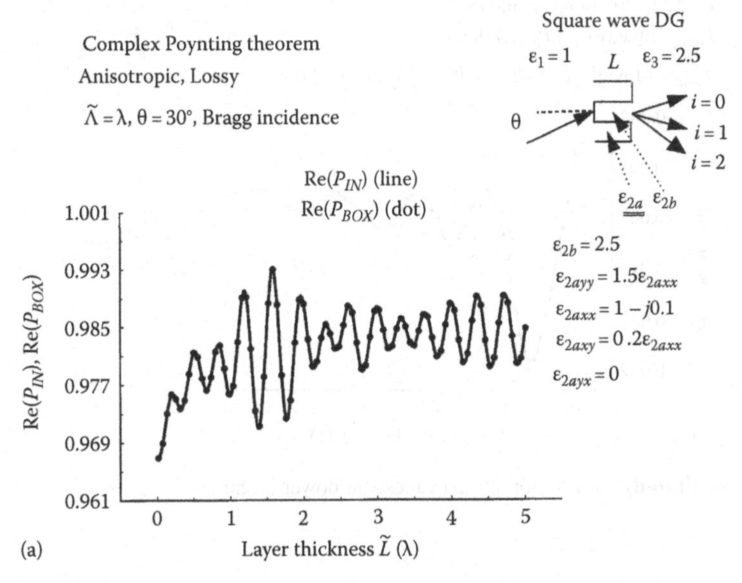

FIGURE 5.17 (a,b) Plots of the real and imaginary parts, respectively, of the normalized complex power P_{IN} and P_{BOX} as computed by the complex Poynting theorem are shown.

(continued)

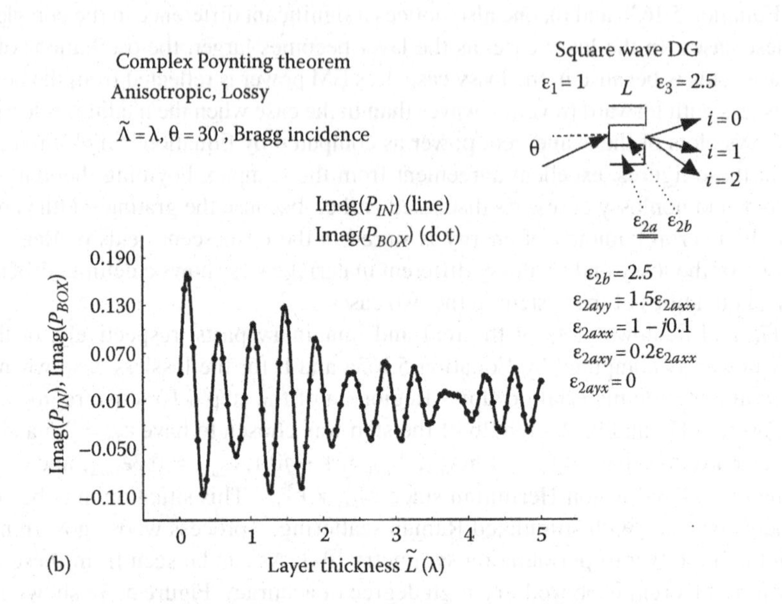

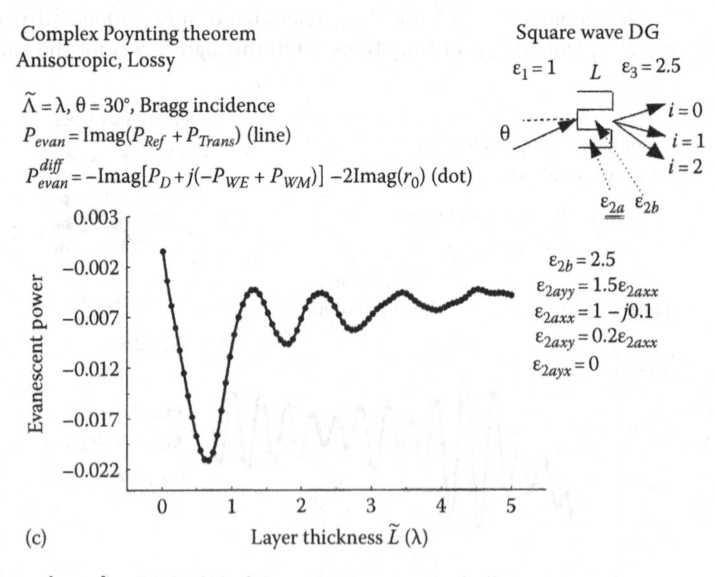

FIGURE 5.17 (continued) (c) A plot of the evanescent power is shown.

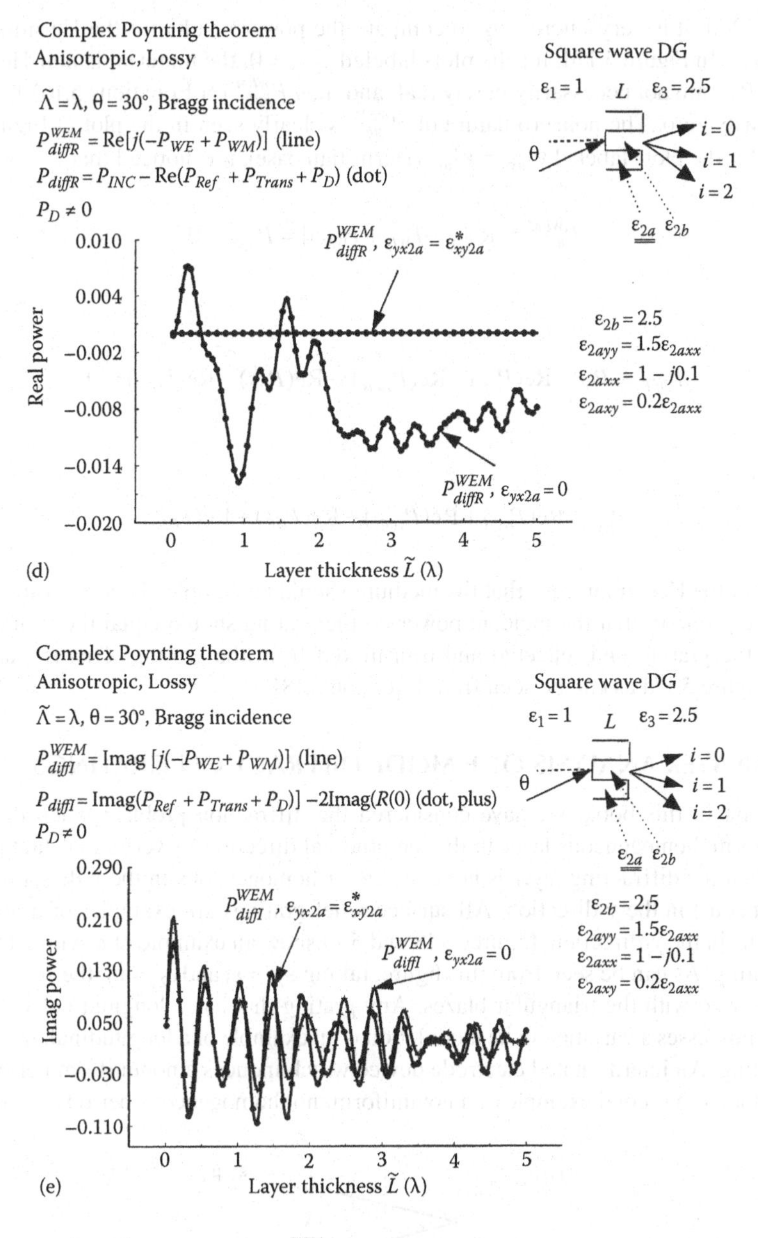

FIGURE 5.17 (continued) (d) Plots of P_{diffR}^{WEM} and P_{diffR} for Hermitian and non-Hermitian step diffraction gratings are shown. (e) Plots of P_{diffI}^{WEM} and P_{diffI} for Hermitian and non-Hermitian step diffraction gratings are shown. (From Jarem, J.M. and Banerjee, P.P., Application of the complex Poynting theorem to diffraction gratings, *J. Opt. Soc. Am. A*, 16, 819, 1999. With permission of Optical Society of America.)

case as was considered in Figure 5.17a through c (plots labeled $\varepsilon_{2ayx} = 0$). Also shown in these figures are plots made for the case when all the permittivity elements are the same as in Figure 5.17a through c except that instead of taking $\varepsilon_{2ayx} = 0$, ε_{2axy} has been taken to be $\varepsilon_{2axy} = \varepsilon_{2ayx}^*$ and thus the medium is Hermitian. These plots are labeled $\varepsilon_{2axy} = \varepsilon_{2ayx}^*$ in Figure 5.17d and e. As can be seen from all the plots shown in Figure 5.17d and e, Equations 5.169 and 5.172 (namely, $P_{diffR}^{WEM} = P_{diffR}$ and $P_{diffI}^{WEM} = P_{diffI}$) are obeyed to a very high degree of accuracy as can be seen by the close agreement between *line* and *dot* as displayed in the figures.

In Figure 5.17d, it is very interesting to compare the power results for the Hermitian and non-Hermitian cases. In Figure 5.17d, for the plots labeled $\varepsilon_{2ayx} = 0$, the material is non-Hermitian, and thus P_{WE} and P_{WM} are not necessarily purely real, and thus P_{diffR}^{WEM} (of Equations 5.167 through 5.169) is not necessarily zero. The nonzero nature of P_{diffR}^{WEM} is clearly seen in the plot of Figure 5.17d. On the contrary, for the plots labeled $\varepsilon_{2axy} = \varepsilon_{2ayx}^*$ (Hermitian case), it is noticed that

$$P_{diffR}^{WEM} = \mathrm{Re}[j(-P_{WE} + P_{WM})] = P_{diffR} = 0 \tag{5.179}$$

or

$$P_{diffR} \equiv P_{inc} - \mathrm{Re}(P_{ref}) - \mathrm{Re}(P_{trans}) - \mathrm{Re}(P_{DE}) - \mathrm{Re}(P_{DM}) = 0 \tag{5.180}$$

or

$$P_{inc} = \mathrm{Re}(P_{ref}) + \mathrm{Re}(P_{trans}) + \mathrm{Re}(P_{DE}) + \mathrm{Re}(P_{DM}) \tag{5.181}$$

It is expected in the Hermitian case that the medium should be completely conservative and should show the basic property that the incident power on the grating should equal the sum of the power dissipated in the grating and reflected and transmitted from the grating. This is exactly what is observed in Figure 5.17d as can be seen from Equation 5.181.

5.4 MULTILAYER ANALYSIS OF E-MODE DIFFRACTION GRATINGS

Up to this point in this book, we have considered the diffraction problem when the diffracting layer is a uniform, homogeneous layer in the longitudinal direction. A very important problem that remains is when the diffracting layer is not constant or homogeneous in the y-direction, but varies (is inhomogeneous) in the y-direction. All surface relief gratings are examples of gratings that are inhomogeneous in the y-direction. Figures 5.18 and 5.19 show an example of a symmetric blaze surface relief grating. As can be seen from this figure, taking $\varepsilon_{2a} = \varepsilon_1$ and $\varepsilon_{2b} = \varepsilon_2$, the surface of Region 3 is cut or grooved with the triangular blazes. Any grating that has a longitudinal variation in the bulk index or possesses a variation in its modulation is an example of a longitudinal, inhomogeneous dielectric grating. An interdigitated electrode device which induces a nonuniform periodic index of refraction change is a second example of a nonuniform, nonhomogeneous periodic grating [75–80].

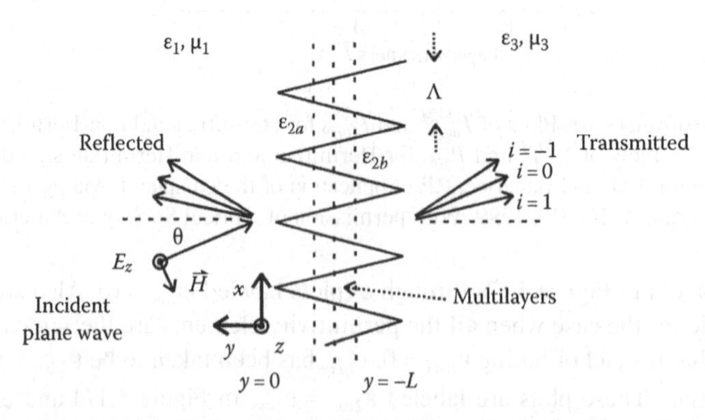

FIGURE 5.18 H-mode symmetric blaze grating with multilayers is shown. The polarization of an E-mode grating is shown in Figure 5.15.

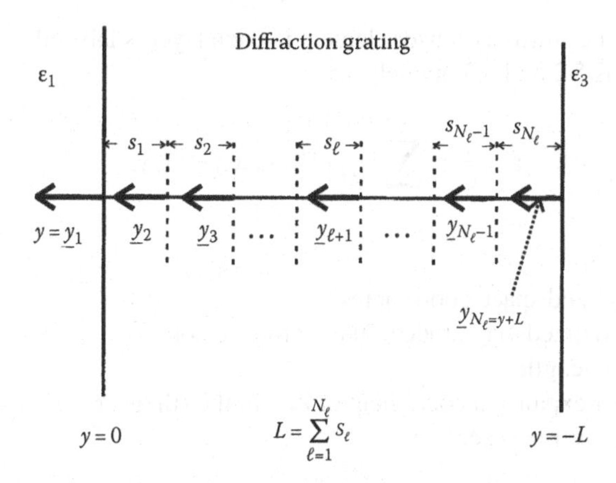

FIGURE 5.19 Local coordinate systems of the multilayers of Figure 5.18 is shown.

The basic theory that will be used in this section to analyze longitudinally inhomogeneous gratings will be to divide the grating into a set of thin layers (sufficiently thin that very little longitudinal variation occurs inside the grating thin layer) and then in each thin layer; (1) find the Fourier coefficients that correspond to that layer, (2) formulate a set of state variable equations within the thin layer as has been done in earlier sections, and (3) boundary match the state variable solutions at all thin layer interfaces and at all regions exterior to the grating to find an EM solution to the inhomogeneous grating problem. This procedure has been used by numerous researchers to study longitudinally inhomogeneous diffraction gratings [18,19,21–23,28,31,33,53,63] and other EM systems.

In observing Figure 5.18, one also notices why the grating must be considered inhomogeneous in the y-direction. The *dotted* lines of Figure 5.18 show examples of the thin layers that may be used to analyze the grating. In observing these dotted lines, for the example presented, one notices that if the thin layer is chosen close to the incident side of the grating, then that little material of Region 3 is included in the thin layer, and the Fourier series representing that thin layer will nearly be that of Region 1 alone. On the other hand, if the thin layer is chosen close to the transmit side of the grating, then most of the material in the thin layer is that of Region 3 alone and the Fourier series representing this thin layer will nearly be that of Region 3. Clearly, from this discussion and observing the thin layer in Figure 5.18, one can see that the Fourier series in the thin layer near the incident side are significantly different from those at the transmit side. Thus from this discussion, the RCWA method must reflect the change in Fourier series coefficients as one changes y position in the grating.

In the following sections, we will carry out the multilayer analysis for the E-mode case studied in Section 5.3. The analysis for the H-mode case is similar to that for the E-mode case, and only the main equations will be presented.

5.4.1 E-Mode Formulation

To begin the analysis of longitudinal inhomogeneous gratings, we divide the inhomogeneous region into N_ℓ regions as shown in Figure 5.19. If the width of the lth layer is s_ℓ, then the overall layer thickness is given by $L = \sum_{\ell=1}^{N_\ell} s_\ell$. Each layer is assigned a local coordinate system y_ℓ with its local origin as shown in Figure 5.19. The first layer has the coordinate $\underline{y}_1$ and the last layer to the right is $\underline{y}_{N_\ell}$. Enough layers N_ℓ are used so that the grating inhomogeneity is nearly constant in each layer.

The periodic dielectric permittivity tensor element in each layer is labeled $\tilde{\varepsilon}_{ss'\ell}$ and is expanded the same way as in Sections 5.2 and 5.3, namely, as

$$\tilde{\varepsilon}_{ss'\ell} = \varepsilon_0 \sum_{i=-\infty}^{\infty} \breve{\varepsilon}_{ss'i\ell} e^{jiK_x x} (s, s') = (x, y, z) \tag{5.182}$$

where

x, y, and z are normalized space coordinates
$K_x = 2\pi/\Lambda$ is the normalized magnitude of the grating vector
Λ is the grating wave length
$\breve{\varepsilon}_{ss'i\ell}$ is the ith Fourier expansion coefficient of $\tilde{\varepsilon}_{ss'\ell}$ in the ℓth layer
ε_0 is the permittivity of free space

Using a local coordinate system is an important feature of the analysis.

The first step is to determine the full EM fields in the ℓth layer of the grating. This analysis has been discussed already in Section 5.3 since the grating is assumed to be uniform in each layer. Thus the EM fields from Maxwell's equations in Region ℓ after truncation to order $|i| \le M_T$ are given by $\vec{H} = H_z \hat{a}_z$ (all regions)

$$U_{z\ell} = \eta_0 H_{z\ell} = \sum_{i=-M_T}^{M_T} \left\{ \sum_{n=1}^{N_T} C_{n\ell} U_{zin\ell} e^{q_{n\ell} \underline{y}_\ell} \right\} e^{-jk_{xi}x} \tag{5.183}$$

$$E_{x\ell} = \sum_{i=-M_T}^{M_T} \left\{ \sum_{n=1}^{N_T} C_{n\ell} S_{xin\ell} e^{q_{n\ell} \underline{y}_\ell} \right\} e^{-jk_{xi}x} \tag{5.184}$$

In these equations, $k_{xi} = k_{x0} - iK_x$, $N_T = 2(2M_T + 1)$, $q_{n\ell}$ and $\underline{V}_\ell = [\underline{S}_{x\ell}^T, \underline{U}_{z\ell}^T]^T$ ($\underline{S}_{x\ell} = [\underline{S}_{xi\ell}]$, $\underline{U}_{z\ell} = [\underline{U}_{zi\ell}]$) are the eigenvalues and eigenvectors of the ℓth region. The eigenvector $\underline{V}_{n\ell}(y) = \underline{V}_n e^{q_{n\ell} \underline{y}}$ is assumed to satisfy the eigenvalue equation

$$\frac{\partial \underline{V}_{n\ell}}{\partial \underline{y}} = \underline{\underline{A}}_\ell \underline{V}_{n\ell} \tag{5.185}$$

where the matrix $\underline{\underline{A}}_\ell$ is given by the general form of Equations 5.117 through 5.119, with the $\breve{\varepsilon}_{ss'i\ell}$ of Equation 5.182 used to define $\underline{\underline{A}}_\ell$ in each thin layer.

The EM fields in Regions 1 and 3 are the same as in the uniform case and are given in Section 5.3, Equations 5.122 through 5.127. The analysis proceeds by matching the tangential electric field (E_x in this case) and tangential magnetic field (H_z in this case) at every boundary interface, namely, Region 1: *Layer* $\ell = 1$ interface; *Layer* $\ell = 2$: *Layer* $\ell = 3$ interface; …; *Layer* $N_\ell - 1$: *Layer* N_ℓ interface; *Layer* N_ℓ: Region 3 interface. The E_x and H_z fields at Region 1: *Layer* $\ell = 1$ interface are

$$E_x^{(1)}\Big|_{y=0^+} = E_{x1}\Big|_{\underline{y}_1=0^-}, \quad H_z^{(1)}\Big|_{y=0^+} = H_{z1}\Big|_{\underline{y}_1=0^-} \tag{5.186}$$

Substituting, we have

$$\sum_{i=-M_T}^{M_T} \frac{k_{y1i}}{\varepsilon_1} (E_0 \delta_{i,0} - R_i) e^{-jk_{xi}x} = \sum_{i=-M_T}^{M_T} \left\{ \sum_{n=1}^{N_T} C_{n1} S_{xin1} e^{q_{n1} \underline{y}_1} \right\} e^{-jk_{xi}x} \Bigg|_{\underline{y}_1=0^-} \tag{5.187a}$$

$$\frac{1}{\eta_0}\sum_{i=-M_T}^{M_T}(E_0\delta_{i,0}+R_i)e^{-jk_{xi}x}=\frac{1}{\eta_0}\sum_{i=-M_T}^{M_T}\left\{\sum_{n=1}^{N_T}C_{n1}U_{zin1}e^{q_{n1}\underline{y}_1}\right\}e^{-jk_{xi}x}\Bigg|_{\underline{y}_1=0^-} \tag{5.187b}$$

…

At the *Layer ℓ*: *Layer ℓ* + 1 interface, we have

$$E_{x,\ell}\Big|_{\underline{y}_\ell=-S_\ell^+}=E_{x,\ell+1}\Big|_{\underline{y}_{\ell+1}=0^-},\quad H_{z,\ell}\Big|_{\underline{y}_\ell=-S_\ell^+}=H_{z,\ell+1}\Big|_{\underline{y}_{\ell+1}=0^-} \tag{5.188}$$

$$\sum_{i=-M_T}^{M_T}\left\{\sum_{n=1}^{N_T}C_{n\ell}S_{xin\ell}e^{q_{n\ell}\underline{y}_\ell}\right\}e^{-jk_{xi}x}\Bigg|_{\underline{y}_\ell=-S_\ell^+}=\sum_{i=-M_T}^{M_T}\left\{\sum_{n=1}^{N_T}C_{n,\ell+1}S_{xin,\ell+1}e^{q_{n,\ell+1}\underline{y}_{\ell+1}}\right\}e^{-jk_{xi}x}\Bigg|_{\underline{y}_{\ell+1}=0^-} \tag{5.189a}$$

$$\frac{1}{\eta_0}\sum_{i=-M_T}^{M_T}\left\{\sum_{n=1}^{N_T}C_{n\ell}U_{zin\ell}e^{q_{n\ell}\underline{y}_\ell}\right\}e^{-jk_{xi}x}\Bigg|_{\underline{y}_\ell=-S_\ell^+}=\frac{1}{\eta_0}\sum_{i=-M_T}^{M_T}\left\{\sum_{n=1}^{N_T}C_{n,\ell+1}U_{zin,\ell+1}e^{q_{n,\ell+1}\underline{y}_{\ell+1}}\right\}e^{-jk_{xi}x}\Bigg|_{\underline{y}_{\ell+1}=0^-} \tag{5.189b}$$

At the *Layer* N_ℓ: Region 3 interface, we find that

$$\sum_{i=-M_T}^{M_T}\left\{\sum_{n=1}^{N_T}C_{nN_\ell}S_{xinN_\ell}\exp(q_{nN_\ell}\underline{y}_{N_\ell-1})\right\}e^{-jk_{xi}x}\Bigg|_{\underline{y}_{N_\ell-1}=-S_{N\ell}}=\sum_{i=-M_T}^{M_T}\frac{k_{y3i}}{\varepsilon_3}T_ie^{-jk_{xi}x} \tag{5.190a}$$

$$\frac{1}{\eta_0}\sum_{i=-M_T}^{M_T}\left\{\sum_{n=1}^{N_T}C_{nN_\ell}U_{zinN_\ell}\exp(q_{nN_\ell}\underline{y}_{N_\ell-1})\right\}e^{-jk_{xi}x}\Bigg|_{\underline{y}_{N_\ell-1}=-S_{N\ell}}=\frac{1}{\eta_0}\sum_{i=-M_T}^{M_T}T_ie^{-jk_{xi}x} \tag{5.190b}$$

Each of the Fourier coefficients of each exponential $e^{-jk_{xi}x}$ on the LHS and RHS of the above equations must be equal in order that the equations are all satisfied. Equating Fourier coefficients after evaluation of the y-dependent terms, we find that

$$\frac{k_{y1i}}{\varepsilon_1}(E_0\delta_{i,0}-R_i)=\sum_{n=1}^{N_T}C_{n1}S_{xin1} \tag{5.191a}$$

$$(E_0\delta_{i,0}+R_i)=\sum_{n=1}^{N_T}C_{n1}U_{zin1} \tag{5.191b}$$

$$\sum_{n=1}^{N_T}C_{n\ell}S_{xin\ell}e^{-q_{n\ell}S_\ell}=\sum_{n=1}^{N_T}C_{n,\ell+1}S_{xin,\ell+1}\quad \ell=1,\ldots,N_\ell-1 \tag{5.192a}$$

$$\sum_{n=1}^{N_T}C_{n\ell}U_{zin\ell}e^{-q_{n\ell}S_\ell}=\sum_{n=1}^{N_T}C_{n,\ell+1}U_{zin,\ell+1}\quad \ell=1,\ldots,N_\ell-1 \tag{5.192b}$$

$$\sum_{n=1}^{N_T} C_{n,N_\ell} S_{xinN_\ell} e^{(-q_{n,N_\ell} S_{N\ell})} = \frac{k_{y3i}}{\varepsilon_3} T_i \tag{5.193a}$$

$$\sum_{n=1}^{N_T} C_{n,N_\ell} U_{zinN_\ell} e^{(-q_{n,N_\ell} S_{N\ell})} = T_i \tag{5.193b}$$

Each equation assumes a range of i values from $i = -M_T, \ldots, M_T$, $N_T = 2(2M_T + 1)$. The above equations may be simplified. Equations 5.193a and 5.193b for $\ell = 1, \ldots, N_\ell - 1$ may be written, with $i = -M_T, \ldots, M_T$, as

$$\underline{\underline{K}}_\ell^- = \begin{bmatrix} [S_{xin\ell}] \\ [U_{zin\ell}] \end{bmatrix} e^{-q_{n\ell} S_\ell}, \quad \underline{\underline{K}}_{\ell+1}^+ = \begin{bmatrix} [S_{xin,\ell+1}] \\ [U_{zin,\ell+1}] \end{bmatrix} \tag{5.194}$$

$$\begin{aligned} \underline{\underline{K}}_1^- \underline{C}_1 &= \underline{\underline{K}}_2^+ \underline{C}_2 \\ \underline{\underline{K}}_2^- \underline{C}_2 &= \underline{\underline{K}}_3^+ \underline{C}_3 \\ &\vdots \\ \underline{\underline{K}}_{N_\ell-1}^- \underline{C}_{N_\ell-1} &= \underline{\underline{K}}_{N_\ell}^+ \underline{C}_{N_\ell} \end{aligned} \tag{5.195}$$

If we invert $\underline{\underline{K}}_\ell$, $\ell = 1, \ldots, N_\ell - 1$ in each of the above equations, we have

$$\begin{aligned} \underline{C}_1 &= (\underline{\underline{K}}_1^-)^{-1}(\underline{\underline{K}}_2^+)\underline{C}_2 \\ \underline{C}_2 &= (\underline{\underline{K}}_2^-)^{-1}(\underline{\underline{K}}_3^+)\underline{C}_3 \\ &\vdots \\ \underline{C}_{N_\ell-2} &= (\underline{\underline{K}}_{N_\ell-2}^-)^{-1}(\underline{\underline{K}}_{N_\ell}^+)\underline{C}_{N_\ell-1} \\ \underline{C}_{N_\ell-1} &= (\underline{\underline{K}}_{N_\ell-2}^-)^{-1}(\underline{\underline{K}}_{N_\ell}^+)\underline{C}_{N_\ell} \end{aligned} \tag{5.196}$$

Substituting, we find that

$$\underline{C}_1 = \{(\underline{\underline{K}}_1^-)^{-1}(\underline{\underline{K}}_2^+)\}\{(\underline{\underline{K}}_2^-)^{-1}(\underline{\underline{K}}_3^+)\} \cdots \{(\underline{\underline{K}}_{N_\ell-2}^-)^{-1}(\underline{\underline{K}}_{N_\ell}^+)\}\underline{C}_{N_\ell} \tag{5.197}$$

Letting the cascaded matrix be $\underline{\underline{M}}_{N_T \times N_T}$, we have

$$\underline{C}_1 = \underline{\underline{M}}\underline{C}_{N_\ell} \tag{5.198}$$

We may simplify Equation 5.191a by solving for R_i in Equation 5.191b and substituting in Equation 5.191a. We may also substitute T_i from Equation 5.193b into Equation 5.193a. Doing both operations, we find that

$$\frac{k_{y1i}}{\varepsilon_1}\left(E_0\delta_{i,0}+E_0\delta_{i,0}-\sum_{n=1}^{N_T}C_{n1}U_{zin1}\right)=\sum_{n=1}^{N_T}C_{n1}S_{xin1}$$

$$V_i\equiv\frac{2E_0k_{y1i}}{\varepsilon_1}\delta_{i,0}=\sum_{n=1}^{N_T}C_{n1}\left[U_{zin1}\frac{k_{y1i}}{\varepsilon_1}+S_{xin1}\right]$$

$$0=\sum_{n=1}^{N_T}C_{nN_\ell}\exp(-q_{n,N_\ell}s_{N_\ell})\left[-\frac{k_{y3i}}{\varepsilon_3}U_{zinN_\ell}+S_{xinN_\ell}\right] \tag{5.199}$$

We also have $k_{y10}=n_1\cos\theta_1$ so that

$$\frac{k_{y10}}{\varepsilon_1}=\frac{n_1\cos\theta_1}{n_1^2}=\frac{\cos\theta_1}{n_1} \tag{5.200}$$

so that

$$V_i=\frac{2E_0\cos\theta_1}{n_1}\delta_{i,0} \tag{5.201}$$

Letting

$$\underline{\underline{R}}^{(1)}=\left[U_{zin1}\frac{k_{y1i}}{\varepsilon_1}+S_{xin1}\right]_{(N_T/2)\times N_T}$$

$$\underline{\underline{I}}=[\delta_{i,i'}]_{N_T\times N_T},\quad \underline{\underline{0}}=[0]_{M_T\times N_T}$$

$$\underline{\underline{R}}^{(N_\ell)}=e^{-q_{nN_\ell}s_{N_\ell}}\left[-\frac{k_{y3i}}{\varepsilon_{31}}U_{zin1}+S_{xin1}\right]_{(N_T/2)\times N_T}\qquad i=-M_T,\ldots,0,\ldots,M_T,\quad n=1,\ldots,N_T$$

$$\underline{V}=[V_i\delta_{i,0}],\quad i=-M_T,\ldots,0,\ldots,M_T$$

$$\underline{V}^{ext}=\begin{bmatrix}\underline{V}\\0\\0\\0\end{bmatrix}_{2N_T},\quad \underline{C}=\begin{bmatrix}\underline{C}_1\\\underline{C}_{N_\ell}\end{bmatrix}_{2N_T}$$

$$\underline{\underline{A}}=\begin{bmatrix}\underline{\underline{R}}^{(1)}&\underline{\underline{0}}\\\underline{\underline{M}}&-\underline{\underline{I}}\\\underline{\underline{0}}&\underline{\underline{R}}^{(N_\ell)}\end{bmatrix}_{2N_T\times 2N_T} \tag{5.202}$$

We find the final matrix equation from which $\underline{C}_1$ and $\underline{C}_{N_\ell}$ can be found. It is given by

$$\underline{V}^{ext}=\underline{\underline{A}}\,\underline{C} \tag{5.203}$$

where $\underline{C}^T = \left[\underline{C}_1^T, \underline{C}_{N_\ell}^T\right]$ and T refers to the matrix transpose. Inversion of this equation gives $\underline{C}_1$ and $\underline{C}_{N_\ell}$. Power may be analyzed in the same way as in the single-layer E-mode case.

The analysis of the H-mode multilayer case (see Figure 5.18) is very similar to that of the E-mode case and the details are not given.

5.4.2 Numerical Results

This section will present some numerical examples of H-mode and E-mode diffraction efficiency as results from the multilayer analysis described in this section. Figure 5.20 shows transmitted and reflected diffraction efficiencies of a sinusoidal surface relief grating when an H-mode polarized plane wave is incident on the grating. The layer thickness is taken to be $L = k_0\tilde{L}$ and extend from the peak to trough of the grating. In this example, $\varepsilon_{2a} = \varepsilon_1 = 1$, $\varepsilon_{2b} = \varepsilon_3 = 2.5$, $\tilde{\Lambda} = \lambda$, $N_\ell = 5$, and $\theta = 30°$. In this example, the full-field formulation of Section 5.2.1 was used in each thin layer to calculate the EM fields of the diffraction grating system. As can be seen from the plots of Figure 5.20, diffraction power is mainly transferred from the zero-order diffracted power into the first order, with a small amount of diffracted power being reflected and transferred into higher order modes. This example was analyzed by Moharam and Gaylord [19, Fig. 4, p. 1388] using the multilayer RCWA method to determine the EM fields of the system. Comparing their figure to the one we present here, almost identical results came from the two formulations. In Figure 5.20, we mention that conservation of the real power was observed to a high degree of accuracy.

Figure 5.21 shows transmitted ($i = -1, 0, 1, 2$) and reflected ($i = 0, 1$) diffraction efficiencies of a sinusoidal surface relief grating when an E-mode polarized rather than an H-mode polarization plane wave is incident on the grating. The layer thickness is taken to be $L = k_0\tilde{L}$ and to extend from the peak to trough of the grating as in the previous figure. In this example, $\varepsilon_{2a} = \varepsilon_1 = 1$, $\varepsilon_{2b} = \varepsilon_3 = 4$, $\tilde{\Lambda} = \lambda$, $N_\ell = 5$, and $\theta = 30°$. In this example, the full-field formulation of Section 5.4 was used to calculate the EM fields of the diffraction grating system. As can be seen from the plots of Figure 5.21, diffraction power is mainly transferred from the zero-order diffracted power to the first order, with a small amount of diffracted power being reflected and transferred into higher order modes. This example has been analyzed by Yamakita et al. [53, Fig. 6, p. 156] who used a multilayer, coupled mode method to determine the EM fields of the system. In comparing their figure to the present one,

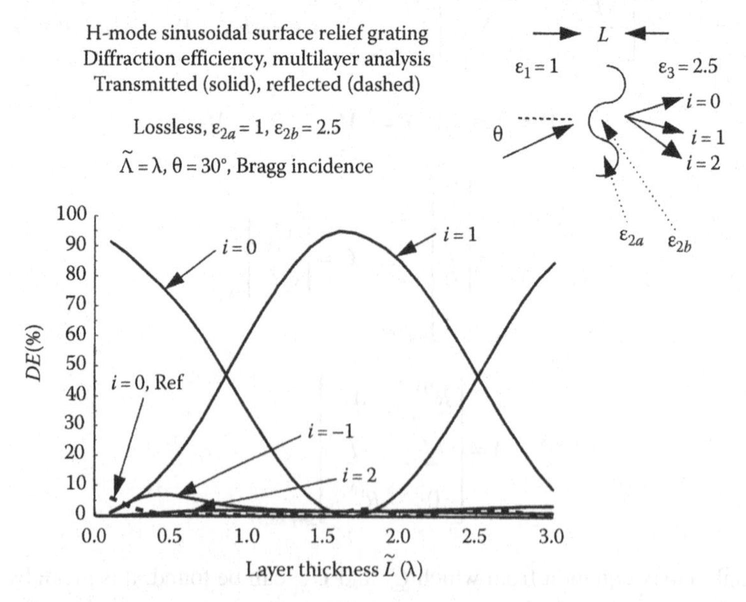

FIGURE 5.20 Transmitted and reflected diffraction efficiencies of a sinusoidal surface relief grating when an H-mode polarized plane wave is incident on the grating.

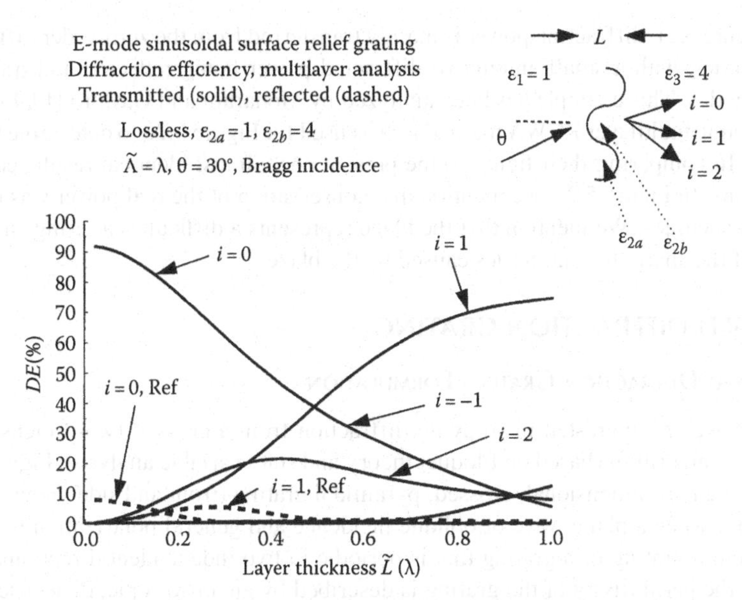

FIGURE 5.21 The transmitted ($i = -1, 0, 1, 2$) and reflected ($i = 0, 1$) diffraction efficiencies of a sinusoidal surface relief grating when an E-mode polarized rather than an H-mode polarization plane wave is incident on the grating is shown.

almost identical results came from the two formulations. In Yamakita et al.'s paper [53], the numerical value of the diffraction orders was opposite of that used in Ref. [19] or that used in this section. (That is, Yamakita's [53] $i = -1$ order is the same as the $i = 1$ order in this section.) In Figure 5.21, we mention that conservation of the real power was observed to a high degree of accuracy.

Figure 5.22 shows transmitted and reflected diffraction efficiencies of a symmetric blaze surface relief grating when an H-mode polarized plane wave is incident on the grating. The layer thickness is taken to be $L = k_0\tilde{L}$ and to extend from the peak to trough of the grating. In this example, $\varepsilon_{2a} = \varepsilon_1 = 1$, $\varepsilon_{2b} = \varepsilon_3 = 2.5$, $\tilde{\Lambda} = \lambda$, $N_\ell = 5$, and $\theta = 30°$. In this example, the full-field formulations of Sections 5.2.1 and 5.4 were used to calculate the EM fields of the diffraction grating system. As can be seen from

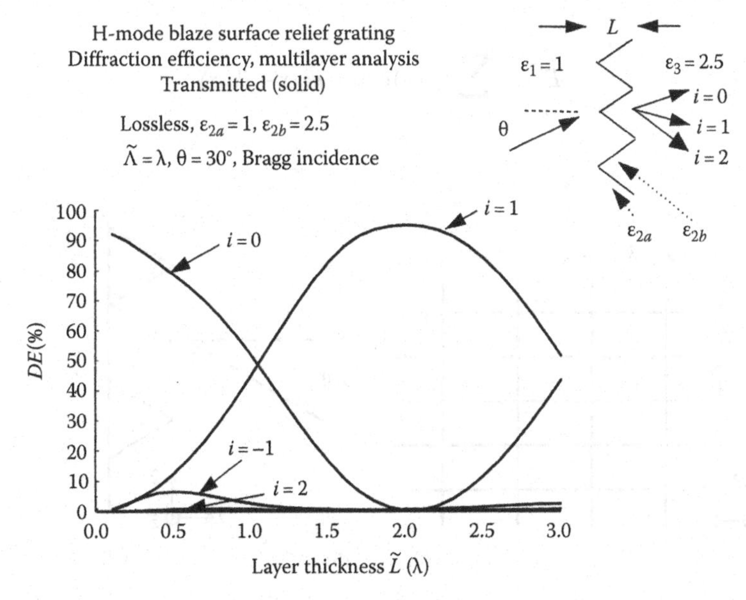

FIGURE 5.22 Transmitted and reflected diffraction efficiencies of a symmetric blaze surface relief grating when an H-mode polarized plane wave is incident on the grating.

the plots of Figure 5.21, diffraction power is mainly transferred from the zero-order diffracted power into the first order, with a small amount of diffracted power being reflected and transferred into higher order modes. This example has been analyzed by Moharam and Gaylord [19, Fig. 6, p. 1389] who used the same multilayer RCWA method as was used for Figure 5.20 to determine the EM fields of the system. In comparing their figure to the present one, almost identical results came from the two formulations. In Figure 5.22, we mention that conservation of the real power was observed to a high degree of accuracy. We mention that the blaze represents a difficult scattering case because of the presence of the sharp discontinuities caused by the blaze.

5.5 CROSSED DIFFRACTION GRATING

5.5.1 Crossed Diffraction Grating Formulation

In this section, we are interested in studying diffraction from a crossed two-dimensional grating using the RCWA algorithm (based on Floquet theory and state variable analysis). Figure 5.23 shows the geometry of a two-dimensional, crossed, pyramidal grating (front and side view). We are interested in the case when a plane wave of oblique incidence and general polarization is incident on a crossed diffraction grating or a grating that is periodic in two independent directions. We assume the case when the permittivity of the grating is described by an anisotropic, dielectric permittivity tensor and the permeability of the grating is that of homogeneous space, and that the permittivity tensor elements ε_{pq}, $(p,q) = (x,y,z)$ may be expressed as a sum of Floquet harmonics as

$$\tilde{\varepsilon}_{pq} = \varepsilon_0 \sum_{i=-\infty}^{\infty} \sum_{\underline{i}=-\infty}^{\infty} \breve{\varepsilon}_{pq,i\underline{i}} \exp j(iK_x x + \underline{i}K_z z) \tag{5.204}$$

where $K_x = (2\pi/\Lambda_x)$, $K_z = (2\pi/\Lambda_z)$, where Λ_x, Λ_z are normalized, and x and z are normalized grating periods given by $\Lambda_x = k_0\tilde{\Lambda}_x$, $\Lambda_z = k_0\tilde{\Lambda}_z$. The system is nonmagnetic. The analysis to be presented assumes Region 2 ($-L < y < 0$) homogenous which applies, for example, to a square wave grating (Figure 5.24). Longitudinally inhomogenous gratings, for example, the pyramidal grating of Figure 5.23 requires multilayer RCWA approach similar to that of Section 5.4.

To start the analysis, we expand the electric fields in Region 2 in a set of Floquet and space harmonics as

$$\vec{E}^{(2)} = \sum_{i,\underline{i}} \vec{S}_{i,\underline{i}}(y) \exp(-jk_{xi}x - jk_{z\underline{i}}z) \tag{5.205}$$

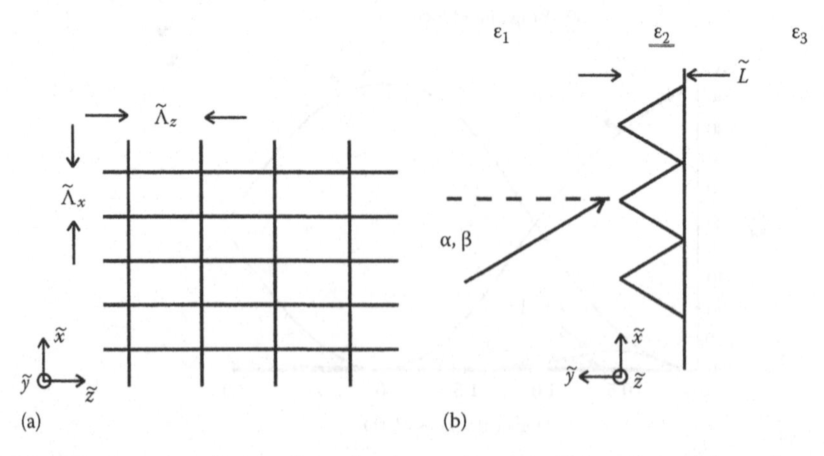

FIGURE 5.23 The geometry of a two-dimensional, crossed, pyramidal grating, (a) front view and (b) side view, is shown.

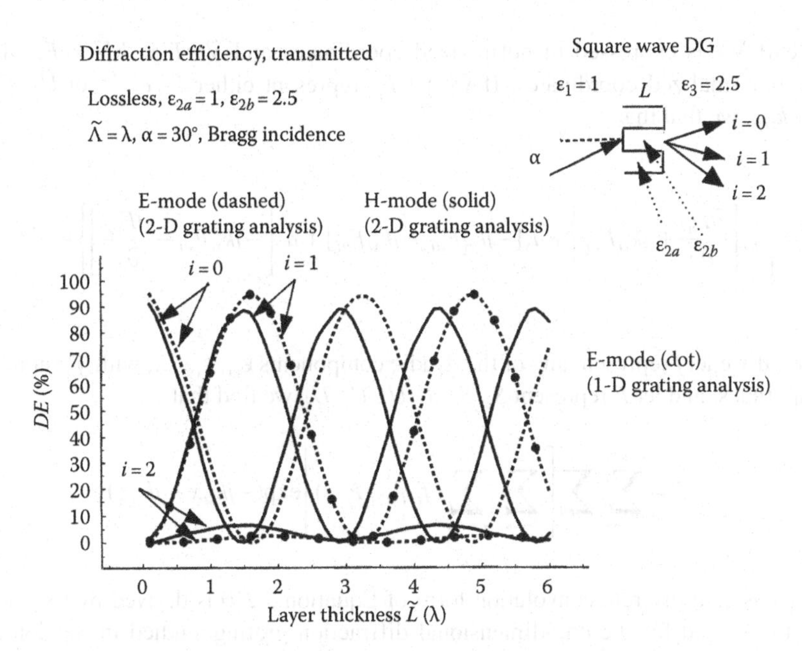

FIGURE 5.24 Diffraction efficiency results using the one-dimensional theory of Sections 5.2.1 and 5.3 are used to validate the two-dimensional theory, crossed diffraction grating theory of this section. Results here are identical to those of Ref. [52, Fig. 5, p. 242] where this example was first calculated using coupled mode theory.

where

$$k_{xi} = k_{x0} - iK_x, \quad i = -M_T, \ldots, 0, \ldots, M_T$$
$$k_{z\underline{i}} = k_{z0} - \underline{i}K_z, \quad \underline{i} = -\underline{M}_T, \ldots, 0, \ldots, \underline{M}_T \tag{5.206}$$

where

$$k_{x0} = n_1 \sin\alpha \sin\beta$$
$$k_{z0} = n_1 \sin\alpha \cos\beta \tag{5.207}$$

where $\tilde{k}_{x0} = k_0 k_{x0}$, $\tilde{k}_{z0} = k_0 k_{z0}$, $n_1 = \sqrt{\mu_1 \varepsilon_1}$, and α, β are the incident angles of the incoming plane wave. The quantity n_1 is the index of refraction of Region 1. The magnetic field $\vec{H}^{(2)}$ in Region 2 may also be expanded in Floquet harmonics and it is given by

$$\vec{H}^{(2)} = \frac{\vec{U}}{\eta_0} = \frac{1}{\eta_0} \sum_{i,\underline{i}} \vec{U}_{i,\underline{i}}(y) \exp(-jk_{xi}x - jk_{z\underline{i}}z) \tag{5.208a}$$

where $\eta_0 = \sqrt{\mu_0/\varepsilon_0}$. If $\vec{E}^{(2)}$ and $\vec{H}^{(2)}$ are substituted in Maxwell's equations, we find that

$$\nabla \times \vec{S} = -j\vec{U} \tag{5.208b}$$

$$\nabla \times \vec{U} = j\underline{\varepsilon} \cdot \vec{S} \tag{5.208c}$$

where the curl $\nabla\times$ is expressed in normalized coordinates and $\nabla\times\vec{F} = k_0\tilde{\nabla}\times\vec{F}$, where $\tilde{\nabla}\times$ is the curl in unnormalized coordinates. If we let $\vec{F}_{i,\underline{i}}$ represent either $\vec{S}_{i,\underline{i}}\,e^{-j\psi_{i,\underline{i}}}$ or $\vec{U}_{i,\underline{i}}\,e^{-j\psi_{i,\underline{i}}}$ where $\psi_{i,\underline{i}} = k_{xi}x + k_{z\underline{i}}z$, we find that

$$\nabla\times\vec{F}_{i,\underline{i}} = \left\{\hat{a}_x\left[\frac{\partial F_{zi\underline{i}}}{\partial y} + jk_{z\underline{i}}F_{yi\underline{i}}\right] + \hat{a}_y[-jk_{z\underline{i}}F_{xi\underline{i}} + jk_{x\underline{i}}F_{zi\underline{i}}] + \hat{a}_z\left[-jk_{xi}F_{yi\underline{i}} - \frac{\partial F_{xi\underline{i}}}{\partial y}\right]\right\}e^{-j\psi_{i,\underline{i}}} \tag{5.209}$$

Furthermore, if we let f represent any of the dyadic components $\varepsilon_{xx}, \varepsilon_{xy}, \ldots$, with $\breve{f}_{i\underline{i}}$ representing the Fourier amplitudes, and let F represent $S_x, S_y, S_z, U_x, U_y, U_z$, we find that

$$fF = \sum_{i=-\infty}^{\infty}\sum_{\underline{i}=-\infty}^{\infty}\left[\sum_{i'=-\infty}^{\infty}\sum_{\underline{i}'=-\infty}^{\infty}(\breve{f}_{i-i',\underline{i}-\underline{i}'}F_{i',\underline{i}'})\right]\exp(-jk_{xi}x - jk_{z\underline{i}}z) \tag{5.210}$$

The two-dimensional discrete convolution form of Equation 5.210 is derived by the same manipulations as those used for the one-dimensional diffraction grating studied in Section 5.2. When the Floquet expansions of $\vec{S}$, $\vec{U}$, and $\underline{\underline{\varepsilon}}$ are substituted in Maxwell's equations and common modal amplitudes of $\exp(-jk_{xi}x - jk_{z\underline{i}}z)$ are equated, the following set of coupled equations are found:

$$\frac{\partial S_{zi\underline{i}}}{\partial y} + jk_{z\underline{i}}S_{yi\underline{i}} = -jU_{xi\underline{i}} \tag{5.211a}$$

$$-jk_{z\underline{i}}S_{xi\underline{i}} + jk_{xi}S_{zi\underline{i}} = -jU_{yi\underline{i}} \tag{5.211b}$$

$$-jk_{x\underline{i}}S_{yi\underline{i}} - \frac{\partial S_{xi\underline{i}}}{\partial y} = -jU_{zi\underline{i}} \tag{5.211c}$$

$$\frac{\partial U_{zi\underline{i}}}{\partial y} + jk_{z\underline{i}}U_{yi\underline{i}} = j\sum_{i,\underline{i}'}[\breve{\varepsilon}_{xx,i-i',\underline{i}-\underline{i}'}S_{xi'\underline{i}} + \breve{\varepsilon}_{xy,i-i',\underline{i}-\underline{i}'}S_{yi'\underline{i}} + \breve{\varepsilon}_{xz,i-i',\underline{i}-\underline{i}'}S_{zi'\underline{i}}] \tag{5.211d}$$

$$-jk_{z\underline{i}}U_{xi\underline{i}} + jk_{xi}U_{zi\underline{i}} = j\sum_{i',\underline{i}'}[\breve{\varepsilon}_{yx,i-i',\underline{i}-\underline{i}'}S_{xi'\underline{i}} + \breve{\varepsilon}_{yy,i-i',\underline{i}-\underline{i}'}S_{yi'\underline{i}} + \breve{\varepsilon}_{yz,i-i',\underline{i}-\underline{i}'}S_{zi'\underline{i}}] \tag{5.211e}$$

$$-jk_{xi}U_{yi\underline{i}} - \frac{\partial U_{zi\underline{i}}}{\partial y} = j\sum_{i',\underline{i}'}[\breve{\varepsilon}_{zx,i-i',\underline{i}-\underline{i}'}S_{xi'\underline{i}} + \breve{\varepsilon}_{zy,i-i',\underline{i}-\underline{i}'}S_{yi'\underline{i}} + \breve{\varepsilon}_{zz,i-i',\underline{i}-\underline{i}'}S_{zi'\underline{i}}] \tag{5.211f}$$

To make further progress, we eliminate the longitudinal electric and magnetic field Floquet harmonic amplitudes $S_{yi\underline{i}}$, $U_{yi\underline{i}}$ which occur in Equation 5.211a–f, by expressing these amplitudes in terms of the tangential field components $S_{xi\underline{i}}$, $S_{zi\underline{i}}$, $U_{xi\underline{i}}$, $U_{zi\underline{i}}$. To do this, it is necessary to express Equation 5.211a–f in matrix form. We now define the column matrices $\underline{S}_q$, $\underline{U}_q$ $(q = x, y, z)$ defined by $\underline{S}_q = [S_{qi\underline{i}}]$ and $\underline{U}_q = [U_{qi\underline{i}}]$. Each of the numbers $i\ \underline{i}$ defines a position in the column matrix. We also define the square diagonal matrices $\underline{\underline{K}}_x$ and $\underline{\underline{K}}_z$, where these matrices represent the multipliers k_{xi} and $k_{z\underline{i}}$, respectively. These square matrices are defined by

$$\underline{\underline{K}}_x = [k_{xi}\,\delta_{i\underline{i},i'\underline{i}'}] \tag{5.212}$$

and

$$\underline{\underline{K}}_z = [k_{z\underline{i}}\,\delta_{i\underline{i},i'\underline{i}'}], \quad ((i\underline{i}),(i'\underline{i}')) = ((-M_T,-M_T),\ldots,(0,0),\ldots(M_T,M_T)) \tag{5.213}$$

We also define the square matrices $\underline{\underline{\breve{\varepsilon}}}_{pq}$, $(p,q) = (x,y,z)$

$$\underline{\underline{\breve{\varepsilon}}}_{pq} = [\breve{\varepsilon}_{pq\,i\underline{i},i'\underline{i}'}] = [\breve{\varepsilon}_{pq\;i-i',\underline{i}-\underline{i}'}] \tag{5.214}$$

where $\breve{\varepsilon}_{pqi\underline{i},i'\underline{i}'}$ is the $(i\underline{i}),(i'\underline{i}')$ matrix element of matrix $\underline{\underline{\breve{\varepsilon}}}_{pq}$ and $\breve{\varepsilon}_{pq,i\underline{i}}$ represents the Fourier coefficients associated with the dielectric tensor element $\varepsilon_{pq}(x,y)$.

With these matrix definitions, it is possible to express Equation 5.211a–f in matrix form. Doing so, we find that Equation 5.211a–f satisfy

$$\frac{\partial \underline{S}_z}{\partial y} + j\underline{\underline{K}}_z\underline{S}_y = -j\underline{U}_x \tag{5.215a}$$

$$-\underline{\underline{K}}_z\underline{S}_x + \underline{\underline{K}}_x\underline{S}_z = -\underline{U}_y \tag{5.215b}$$

$$-j\underline{\underline{K}}_x\underline{S}_y - \frac{\partial \underline{S}_z}{\partial y} = -j\underline{U}_z \tag{5.215c}$$

$$\frac{\partial \underline{U}_z}{\partial y} + j\underline{\underline{K}}_z\underline{U}_y = j[\underline{\underline{\varepsilon}}_{xx}\underline{S}_x + \underline{\underline{\varepsilon}}_{xy}\underline{S}_y + \underline{\underline{\varepsilon}}_{xz}\underline{S}_z] \tag{5.215d}$$

$$-j\underline{\underline{K}}_z\underline{U}_x + j\underline{\underline{K}}_x\underline{U}_z = j[\underline{\underline{\varepsilon}}_{yx}\underline{S}_x + \underline{\underline{\varepsilon}}_{yy}\underline{S}_y + \underline{\underline{\varepsilon}}_{yz}\underline{S}_z] \tag{5.215e}$$

$$-j\underline{\underline{K}}_x\underline{U}_y - \frac{\partial \underline{U}_x}{\partial y} = j[\underline{\underline{\varepsilon}}_{zx}\underline{S}_x + \underline{\underline{\varepsilon}}_{zy}\underline{S}_y + \underline{\underline{\varepsilon}}_{zz}\underline{S}_z] \tag{5.215f}$$

We may express $\underline{S}_y$ in terms of the tangential mode amplitudes $\underline{S}_x$, $\underline{S}_z$, $\underline{U}_x$, $\underline{U}_z$ if we express $\underline{\underline{\varepsilon}}_{yy}\underline{S}_y$ in terms of $\underline{S}_x$, $\underline{S}_z$, $\underline{U}_x$, $\underline{U}_z$ and then multiply the resulting equation by $\underline{\underline{\varepsilon}}_{yy}^{-1}$. Doing so, we find that the longitudinal modal amplitudes $\underline{S}_y$ and $\underline{U}_y$ are given by

$$\underline{S}_y = -\underline{\underline{\varepsilon}}_{yy}^{-1}\underline{\underline{K}}_z\underline{U}_x + \underline{\underline{\varepsilon}}_{yy}^{-1}\underline{\underline{K}}_x\underline{U}_z - \underline{\underline{\varepsilon}}_{yy}^{-1}\underline{\underline{\varepsilon}}_{yx}\underline{S}_x - \underline{\underline{\varepsilon}}_{yy}^{-1}\underline{\underline{\varepsilon}}_{yz}\underline{S}_z \tag{5.216a}$$

$$\underline{U}_y = \underline{\underline{K}}_z\underline{S}_x - \underline{\underline{K}}_x\underline{S}_z \tag{5.216b}$$

If we substitute $\underline{S}_y$ and $\underline{U}_y$, and perform algebra, we find that

$$\frac{\partial}{\partial y}\begin{bmatrix}\underline{S}_x\\ \underline{S}_z\\ \underline{U}_x\\ \underline{U}_z\end{bmatrix} = \begin{bmatrix}A_{11} & A_{12} & A_{13} & A_{14}\\ A_{21} & A_{22} & A_{23} & A_{24}\\ A_{31} & A_{32} & A_{33} & A_{34}\\ A_{41} & A_{42} & A_{43} & A_{44}\end{bmatrix}\begin{bmatrix}\underline{S}_x\\ \underline{S}_z\\ \underline{U}_x\\ \underline{U}_z\end{bmatrix} \tag{5.217}$$

where

$$\underline{\underline{A}}_{11} = j\underline{\underline{K}}_x \underline{\underline{\varepsilon}}_{yy}^{-1} \underline{\underline{\varepsilon}}_{yx}$$

$$\underline{\underline{A}}_{12} = j\underline{\underline{K}}_x \underline{\underline{\varepsilon}}_{yy}^{-1} \underline{\underline{\varepsilon}}_{yz}$$

$$\underline{\underline{A}}_{13} = j\underline{\underline{K}}_x \underline{\underline{\varepsilon}}_{yy}^{-1} \underline{\underline{K}}_z$$

$$\underline{\underline{A}}_{14} = j[\underline{\underline{I}} - \underline{\underline{K}}_x \underline{\underline{\varepsilon}}_{yy}^{-1} \underline{\underline{K}}_x]$$

$$\underline{\underline{A}}_{21} = j[\underline{\underline{K}}_z \underline{\underline{\varepsilon}}_{yy}^{-1} \underline{\underline{\varepsilon}}_{yx}]$$

$$\underline{\underline{A}}_{22} = j[\underline{\underline{K}}_z \underline{\underline{\varepsilon}}_{yy}^{-1} \underline{\underline{\varepsilon}}_{yz}]$$

$$\underline{\underline{A}}_{23} = j[\underline{\underline{K}}_z \underline{\underline{\varepsilon}}_{yy}^{-1} \underline{\underline{K}}_z - \underline{\underline{I}}]$$

$$\underline{\underline{A}}_{24} = j[-\underline{\underline{K}}_z \underline{\underline{\varepsilon}}_{yy}^{-1} \underline{\underline{K}}_x]$$

$$\underline{\underline{A}}_{31} = j[-\underline{\underline{K}}_x \underline{\underline{K}}_z - \underline{\underline{\varepsilon}}_{zx} + \underline{\underline{\varepsilon}}_{zy} \underline{\underline{\varepsilon}}_{yy}^{-1} \underline{\underline{\varepsilon}}_{yx}]$$

$$\underline{\underline{A}}_{32} = j[\underline{\underline{K}}_x^2 + \underline{\underline{\varepsilon}}_{zy} \underline{\underline{\varepsilon}}_{yy}^{-1} \underline{\underline{\varepsilon}}_{yz} - \underline{\underline{\varepsilon}}_{zz}]$$

$$\underline{\underline{A}}_{33} = j[\underline{\underline{\varepsilon}}_{zy} \underline{\underline{\varepsilon}}_{yy}^{-1} \underline{\underline{K}}_z]$$

$$\underline{\underline{A}}_{34} = j[-\underline{\underline{\varepsilon}}_{zy} \underline{\underline{\varepsilon}}_{yy}^{-1} \underline{\underline{K}}_x]$$

$$\underline{\underline{A}}_{41} = j[-\underline{\underline{K}}_z^2 + \underline{\underline{\varepsilon}}_{xx} - \underline{\underline{\varepsilon}}_{xy} \underline{\underline{\varepsilon}}_{yy}^{-1} \underline{\underline{\varepsilon}}_{yx}]$$

$$\underline{\underline{A}}_{42} = j[\underline{\underline{K}}_z \underline{\underline{K}}_x - \underline{\underline{\varepsilon}}_{xy} \underline{\underline{\varepsilon}}_{yy}^{-1} \underline{\underline{\varepsilon}}_{yz} + \underline{\underline{\varepsilon}}_{xz}]$$

$$\underline{\underline{A}}_{43} = j[-\underline{\underline{\varepsilon}}_{xy} \underline{\underline{\varepsilon}}_{yy}^{-1} \underline{\underline{K}}_z]$$

$$\underline{\underline{A}}_{44} = j[\underline{\underline{\varepsilon}}_{xy} \underline{\underline{\varepsilon}}_{yy}^{-1} \underline{\underline{K}}_x] \tag{5.218}$$

where $\underline{I}$ is the identity matrix. If we introduce the column matrix (T is the matrix transpose)

$$\underline{V} = [\underline{S}_x^T \; \underline{S}_z^T \; \underline{U}_x^T \; \underline{U}_z^T]^T \tag{5.219}$$

and the square matrix

$$\underline{\underline{A}} = [\underline{\underline{A}}_{rs}], \quad (r, s) = (1, 2, 3, 4) \tag{5.220}$$

we may express Equation 5.217 as

$$\frac{\partial \underline{V}}{\partial y} = \underline{\underline{A}}\,\underline{V} \tag{5.221}$$

The solution of this equation using the state variable method, as has been discussed previously, is given by

$$\underline{V}_n(y) = \sum_{n=1}^{N_T} V_n e^{q_n y} \tag{5.222}$$

where

$$\underline{\underline{A}}\underline{V}_n = q_n \underline{V}_n \tag{5.223a}$$

The quantities $\underline{V}_n$ and q_n are the eigenvectors and eigenvalues of the matrix $\underline{\underline{A}}$, and have dimension $N_T = 4(2M_T + 1)(2\underline{M}_T + 1)$. The electromagnetic fields in Region 2 are given by

$$E_q^{(2)}(x, y, z) = \sum_{i,\underline{i}} \left(\sum_{n=1}^{N_T} C_n S_{qi\underline{i}n} e^{q_n y} \right) e^{-j(k_{xi}x + k_{z\underline{i}}z)} \tag{5.223b}$$

and

$$H_q^{(2)}(x, y, z) = \frac{1}{\eta_0} \sum_{i,\underline{i}} \left(\sum_{n=1}^{N_T} C_n U_{qi\underline{i}n} e^{q_n y} \right) e^{-j(k_{xi}x + k_{z\underline{i}}z)}, \quad q = (x, y, z) \tag{5.224}$$

where $S_{xi\underline{i}n}, S_{zi\underline{i}n}, U_{xi\underline{i}n}, U_{zi\underline{i}n}$ are eigenvectors obtained from the eigenvector $\underline{V}_n$. The quantities $S_{yi\underline{i}n}, U_{yi\underline{i}n}$ are obtained from Equation 5.216a and b, using the known eigenvectors $S_{xi\underline{i}n}, S_{zi\underline{i}n}, U_{xi\underline{i}n}, U_{zi\underline{i}n}$.

Now that the EM fields have been determined in Region 2, the next step is to determine the EM fields in Region 1 (incident side) and Region 3 (transmit side) of the grating. The fields in Region 1 consist of an obliquely incident plane wave and an infinite number of Floquet harmonic reflected waves. Using the coordinates shown in Figure 5.23 and assuming that the incident plane wave has polarization (α and β are the incident angles of incoming wave),

$$\vec{E}^I = [E_\beta^I \hat{a}_\beta + E_\alpha^I \hat{a}_\alpha] \exp(-jk_{xi}x + jk_{y1i\underline{i}}y - jk_{z\underline{i}}z), \quad i = \underline{i} = 0 \tag{5.225}$$

where

$$k_{y1i\underline{i}} = \begin{cases} [\varepsilon_1 - k_{xi}^2 - k_{z\underline{i}}^2]^{1/2}, & k_{xi}^2 + k_{z\underline{i}}^2 \le \varepsilon_1 \\ -j[k_{xi}^2 + k_{z\underline{i}}^2 - \varepsilon_1]^{1/2}, & k_{xi}^2 + k_{z\underline{i}}^2 > \varepsilon_1 \end{cases} \tag{5.226}$$

Noting that

$$\hat{a}_\beta = -\sin\beta\, \hat{a}_z + \cos\beta\, \hat{a}_x \tag{5.227a}$$

$$\hat{a}_\alpha = \cos\alpha\cos\beta\, \hat{a}_z + \cos\alpha\sin\beta\, \hat{a}_x - \sin\alpha\, \hat{a}_y \tag{5.227b}$$

letting $\psi_{i\underline{i}}^I = k_{xi}x - k_{y1i\underline{i}}y + k_{z\underline{i}}z$, and substituting $\hat{a}_\beta$ and $\hat{a}_\alpha$, we find that

$$\vec{E}^I = \left\{ [\cos\beta E_\beta^I + \cos\alpha\sin\beta E_\alpha^I]\hat{a}_x + [-\sin\alpha E_\alpha^I]\hat{a}_y + [-\sin\beta E_\beta^I + \cos\alpha\cos\beta E_\alpha^I]\hat{a}_z \right\} \exp(-j\psi_{00}^I) \tag{5.228a}$$

$$\vec{E}^I = \sum_{i,\underline{i}} [E_{xi\underline{i}}^I \hat{a}_x + E_{yi\underline{i}}^I \hat{a}_y + E_{zi\underline{i}}^I \hat{a}_z] \exp(-j\psi_{i\underline{i}}^I)\delta_{i\underline{i},00} \tag{5.228b}$$

where $\delta_{i\underline{i},00} \equiv \delta_{i,0}\delta_{\underline{i},0}\delta_{\alpha,\beta}$ is the Kronecker delta. The incident magnetic field may be determined from the second Maxwell curl equation. We have

$$\vec{H}^I = \frac{1}{-j\eta_0}\nabla\times\vec{E}^I \tag{5.229a}$$

$$\vec{H}^I = \frac{1}{\eta_0}\left\{[-k_{y1i\underline{i}}\vec{E}^I_{zi,\underline{i}} - k_{z\underline{i}}\vec{E}^I_{yi\underline{i}}]\hat{a}_x + [k_{z\underline{i}}\vec{E}^I_{xi\underline{i}} - k_{xi}\vec{E}^I_{zi\underline{i}}]\hat{a}_y + [k_{xi}\vec{E}^I_{yi\underline{i}} + k_{y1i\underline{i}}\vec{E}^I_{xi,\underline{i}}]\hat{a}_z\right\} e^{-j\psi^I_{i\underline{i}}}\delta_{i\underline{i},00} \tag{5.229b}$$

$$\vec{H}^I = \sum_{i,\underline{i}}[H^I_{xi\underline{i}}\hat{a}_x + H^I_{yi\underline{i}}\hat{a}_y + H^I_{zi\underline{i}}\hat{a}_z]\exp(-j\psi^I_{i\underline{i}})\delta_{i\underline{i},00} \tag{5.229c}$$

The reflected EM field consists of an infinite number of propagating Floquet harmonic and evanescent backward traveling plane waves. The reflected EM electric field is given by

$$\vec{E}^R = \sum_{i,\underline{i}}\vec{E}^R_{i\underline{i}}\exp(-j\psi^R_{i\underline{i}}) \tag{5.230a}$$

$$\vec{H}^R = \sum_{i,\underline{i}}\vec{H}^R_{i\underline{i}}\exp(-j\psi^R_{i\underline{i}}) \tag{5.230b}$$

where

$$\vec{E}^R_{i\underline{i}} = [R_{xi\underline{i}}\hat{a}_x + R_{yi\underline{i}}\hat{a}_y + R_{zi\underline{i}}\hat{a}_z] \tag{5.230c}$$

$$\psi^R_{i\underline{i}} = k_{xi}x + k_{y1i\underline{i}}y + k_{z\underline{i}}z \tag{5.230d}$$

Notice that in Equation 5.226 for the case that $k_{y1i\underline{i}}$ is evanescent, $e^{-j(-j|k_{y1i\underline{i}}|y)} = e^{-|k_{y1i\underline{i}}|y} \to 0$ as $y \to \infty$. We thus see that for the evanescent plane wave wavenumber

$$k_{y1i\underline{i}} = \pm j[k_{xi}^2 + k_{z\underline{i}}^2 - \varepsilon_1]^{1/2},\ (k_{xi}^2 + k_{z\underline{i}}^2 > \varepsilon_1) \tag{5.231}$$

the minus is the correct root. This is the one used in Equation 5.226. The reflected magnetic field in Region 1 is given by

$$\vec{H}^R_{i\underline{i}} = [H^R_{xi\underline{i}}\hat{a}_x + H^R_{yi\underline{i}}\hat{a}_y + H^R_{zi\underline{i}}\hat{a}_z] = \frac{1}{\eta_0}\{[k_{y1i\underline{i}}R_{zi\underline{i}} - k_{z\underline{i}}R_{yi\underline{i}}]\hat{a}_x + [k_{z\underline{i}}R_{xi\underline{i}} - k_{xi}R_{zi\underline{i}}]\hat{a}_y + [k_{xi}R_{yi\underline{i}} - k_{y1i\underline{i}}R_{xi\underline{i}}]\hat{a}_z\} \tag{5.232}$$

In Equation 5.228b for $\vec{E}^I$ and Equation 5.230a for $\vec{E}^R$, the longitudinal y-electric field component E_y can be expressed in terms of the tangential electric fields E_x and E_z. Using the electric flux density equation in Region 1,

$$\nabla\cdot\vec{D} = \nabla\cdot\varepsilon_1\vec{E} = \varepsilon_1\nabla\cdot\vec{E} = 0,\quad \nabla\cdot\vec{E} = 0 \tag{5.233}$$

where $\vec{E}$ represents either $\vec{E}^I$ or $\vec{E}^R$. Using this equation, we have

$$\nabla\cdot\vec{E}^I = \nabla\cdot\left[\sum_{i,\underline{i}}\vec{E}^I_{i\underline{i}}e^{-j\psi^I_{i\underline{i}}}\delta_{i\underline{i},00}\right] = 0 \tag{5.234a}$$

$$-jk_{x0}E^I_{x00} + jk_{y100}E^I_{y00} - jk_{z0}E_{z00} = 0 \tag{5.234b}$$

$$E_{y00}^{I} = \frac{k_{x0}}{k_{y100}} E_{x00}^{I} + \frac{k_{z0}}{k_{y100}} E_{z00}^{I} \tag{5.234c}$$

and for $\vec{E}^R$

$$\nabla \cdot \vec{E}^R = 0 \tag{5.235a}$$

$$-jk_{xi}R_{xi\underline{i}} - jk_{y1i\underline{i}}R_{yi\underline{i}} - jk_{z\underline{i}}R_{zi\underline{i}} = 0 \tag{5.235b}$$

$$R_{yi\underline{i}} = -\frac{k_{xi}}{k_{y1i\underline{i}}} R_{xi\underline{i}} - \frac{k_{z\underline{i}}}{k_{y1i\underline{i}}} R_{zi\underline{i}} \tag{5.235c}$$

We thus see that $\vec{E}^I$ and $\vec{E}^R$ are known once they can be expressed entirely in terms of the E_{x00}^I, E_{z00}^I, $R_{xi\underline{i}}$, $R_{zi\underline{i}}$ coefficients.

The incident and reflected magnetic fields $\vec{H}^I$ and $\vec{H}^R$ may be expressed in terms of the tangential $\vec{E}^I$ and $\vec{E}^R$ fields. After substitution of E_{x00}^I, E_{y00}^I, E_{z00}^I in Equation 5.229b, we find that the tangential incident magnetic field amplitudes H_{x00}^I and H_{z00}^I are given by

$$H_{x00}^I = \frac{1}{\eta_0}\left[Y_{xx00}^I E_{x00}^I + Y_{xz00}^I E_{z00}^I\right] \tag{5.236a}$$

$$H_{z00}^I = \frac{1}{\eta_0}\left[Y_{zx00}^I E_{x00}^I + Y_{zz00}^I E_{z00}^I\right] \tag{5.236b}$$

where

$$\begin{aligned}
Y_{xx00}^I &= -\frac{k_{z0}k_{x0}}{k_{y100}} \\
Y_{xz00}^I &= -k_{y100} - \frac{k_{z0}^2}{k_{y100}} \\
Y_{zx00}^I &= k_{y100} + \frac{k_{x0}^2}{k_{y100}} \\
Y_{zz00}^I &= \frac{k_{x0}\,k_{z0}}{k_{y100}}
\end{aligned} \tag{5.237}$$

The quantities Y_{pq00}^I, $(p,q) = (x,z)$ may be considered the normalized surface admittances of the system. They are analogous to the surface aperture admittances used in k-space theory to analyze radiation from inhomogeneously covered, surface aperture antennas (see Chapter 4, [35]). The tangential magnetic field reflected modal amplitudes may also be expressed in terms of the tangential reflected electric field modal amplitudes $R_{xi\underline{i}}$ and $R_{zi\underline{i}}$ using Equation 5.235c. We have

$$H_{xi\underline{i}}^R = \frac{1}{\eta_0}\left[Y_{xxi\underline{i}}^R R_{xi\underline{i}} + Y_{xzi\underline{i}}^R R_{zi\underline{i}}\right] \tag{5.238a}$$

$$H_{z i \underline{i}}^{R}=\frac{1}{\eta_{0}}\left[Y_{z x i \underline{i}}^{R} R_{x i \underline{i}}+Y_{z z i \underline{i}}^{R} R_{z i \underline{i}}\right] \tag{5.238b}$$

$$\begin{aligned}
Y_{x x i \underline{i}}^{R} &= \frac{k_{z \underline{i}} k_{x i}}{k_{y 1 i \underline{i}}} \\
Y_{x z i \underline{i}}^{R} &= k_{y 1 i \underline{i}}+\frac{k_{z \underline{i}}^{2}}{k_{y 1 i \underline{i}}} \\
Y_{z x i \underline{i}}^{R} &= -\left[\frac{k_{x \underline{i}}^{2}}{k_{y 1 i \underline{i}}}+k_{y 1 i \underline{i}}\right] \\
Y_{z z i \underline{i}}^{R} &= -\frac{k_{x i} k_{z \underline{i}}}{k_{y 1 i \underline{i}}}
\end{aligned} \tag{5.239}$$

Overall, the EM fields in Region 1 are given by

$$\vec{E}^{(1)}=\sum_{i, \underline{i}}\left[\vec{E}_{i \underline{i}}^{I} \exp \left(-j \psi_{i \underline{i}}^{I}\right) \delta_{i \underline{i}, 00}+\vec{E}_{i \underline{i}}^{R} \exp \left(-j \psi_{i \underline{i}}^{R}\right)\right] \tag{5.240a}$$

$$\vec{H}^{(1)}=\sum_{i, \underline{i}}\left[\vec{H}_{i \underline{i}}^{I} \exp \left(-j \psi_{i \underline{i}}^{I}\right) \delta_{i \underline{i}, 00}+\vec{H}_{i \underline{i}}^{R} \exp \left(-j \psi_{i \underline{i}}^{R}\right)\right] \tag{5.240b}$$

The analysis for the EM fields in Region 3 on the transmit side is very similar to the analysis made in Region 1. In Region 3, the electric and magnetic fields consist of infinite number Floquet harmonic diffracted plane waves. The electric field in Region 3 is given by

$$\vec{E}^{(3)} \equiv \vec{E}^{T}=\sum_{i \underline{i}}\left[T_{x i \underline{i}} \hat{a}_{x}+T_{y i \underline{i}} \hat{a}_{y}+T_{z i \underline{i}} \hat{a}_{z}\right] \exp \left(-j \psi_{i \underline{i}}^{T}\right) \tag{5.241}$$

where

$$\psi_{i \underline{i}}^{T}=k_{x i} x-k_{y 3 i \underline{i}}(y+L)+k_{z \underline{i}} z \tag{5.242}$$

where

$$k_{y 3 i \underline{i}}=\begin{cases}\left[\varepsilon_{3}-k_{x i}^{2}-k_{z \underline{i}}^{2}\right]^{1 / 2}, & k_{x i}^{2}+k_{z \underline{i}}^{2} \leq \varepsilon_{3} \\ -j\left[k_{x i}^{2}+k_{z \underline{i}}^{2}-\varepsilon_{3}\right]^{1 / 2}, & k_{x i}^{2}+k_{z \underline{i}}^{2}>\varepsilon_{3}\end{cases} \tag{5.243}$$

Note that when the plane wave is evanescent ($k_{xi}^2+k_{z\underline{i}}^2>\varepsilon_3$), the exponent in (5.241) tends to zero as $y \rightarrow -\infty$. Note also that in (5.242), $\psi_{i\underline{i}}^T$ has been chosen so that $\psi_{i\underline{i}}^T\big|_{y=-L}=k_{xi}x+k_{z\underline{i}}z$, which simplifies boundary matching. The transmitted electric field satisfies the equation

$$0=\nabla \cdot \vec{D}^{T}=\varepsilon_{3} \nabla \cdot \vec{E}^{T}=\varepsilon_{3}\left[\sum_{i \underline{i}} \nabla \cdot\left(\vec{E}_{i \underline{i}}^{T} \exp \left(-j \psi_{i \underline{i}}^{T}\right)\right)\right] \tag{5.244}$$

Differentiating Equation 5.244, we find that

$$0 = -jk_{xi}T_{xi\underline{i}} + jk_{y3i\underline{i}}T_{yi\underline{i}} - jk_{z\underline{i}}T_{zi\underline{i}} \tag{5.245}$$

$$T_{yi\underline{i}} = \frac{k_{xi}}{k_{y3i\underline{i}}}T_{xi\underline{i}} + \frac{k_{z\underline{i}}}{k_{y3i\underline{i}}}T_{zi\underline{i}} \tag{5.246}$$

The magnetic field in Region 3 may be found from Maxwell's first curl equation. We have

$$\vec{H}^{(3)} \equiv \vec{H}^{(T)} = \left[\sum_{i\underline{i}}[H^T_{xi\underline{i}}\hat{a}_x + H^T_{yi\underline{i}}\hat{a}_y + H^T_{zi\underline{i}}\hat{a}_z]\exp(-j\psi^T_{i\underline{i}})\right] \tag{5.247}$$

$$\vec{H}^{(3)} = \frac{1}{\eta_0}\sum_{i\underline{i}}\{[-k_{y3i\underline{i}}T_{zi\underline{i}} - k_{z\underline{i}}T_{yi\underline{i}}]\hat{a}_x + [k_{z\underline{i}}T_{xi\underline{i}} - k_{xi}T_{zi\underline{i}}]\hat{a}_y + [k_{xi}T_{yi\underline{i}} + k_{y3i\underline{i}}T_{xi\underline{i}}]\hat{a}_z\}\exp(-j\psi^T_{i\underline{i}}) \tag{5.248}$$

Using Equation 5.246, $T_{yi\underline{i}}$ may be expressed in terms of $T_{xi\underline{i}}$ and $T_{zi\underline{i}}$. Thus it is possible to express all the magnetic field components in terms of $T_{xi\underline{i}}$ and $T_{zi\underline{i}}$. The tangential magnetic field modal amplitudes $H^T_{xi\underline{i}}$ and $H^T_{zi\underline{i}}$ are given by

$$H^T_{xi\underline{i}} = \frac{1}{\eta_0}[Y^T_{xxi\underline{i}}T_{xi\underline{i}} + Y^T_{xzi\underline{i}}T_{zi\underline{i}}]$$

$$H^T_{zi\underline{i}} = \frac{1}{\eta_0}[Y^T_{zxi\underline{i}}T_{xi\underline{i}} + Y^T_{zzi\underline{i}}T_{zi\underline{i}}] \tag{5.249}$$

where

$$Y^T_{xxi\underline{i}} = -\frac{k_{z\underline{i}}k_{xi}}{k_{y3i\underline{i}}}$$

$$Y^T_{xzi\underline{i}} = -k_{y3i\underline{i}} - \frac{k^2_{z\underline{i}}}{k_{y3i\underline{i}}}$$

$$Y^T_{zxi\underline{i}} = \frac{k^2_{xi}}{k_{y3i\underline{i}}} + k_{y3i\underline{i}} \tag{5.250}$$

$$Y^T_{zzi\underline{i}} = \frac{k_{xi}k_{z\underline{i}}}{k_{y3i\underline{i}}}$$

The next step in the analysis is to match the EM field solutions at the $y = 0$ and $-L$ interfaces and determine all the unknown constants of the system.

Now that the EM fields have been defined in Regions 1–3, the next step in the analysis is to match the tangential electric and magnetic fields at boundary plane $y = 0$ and $-L$. At $y = 0$, we have

$$E^{(1)}_{x,z}\Big|_{y=0^+} = E^{(2)}_{x,z}\Big|_{y=0^-}$$

$$H^{(1)}_{x,z}\Big|_{y=0^+} = H^{(2)}_{x,z}\Big|_{y=0^-} \tag{5.251}$$

Substituting the previous equations for the EM fields and evaluating at $y = 0$, we have

$$\sum_{i,\underline{i}}[E^I_{xi\underline{i}}\delta_{i\underline{i},00} + R_{xi\underline{i}}]\, e^{-jk_{xi}x-jk_{z\underline{i}}z} = \sum_{i,\underline{i}}\left[\sum_{n=1}^{N_T} C_n S_{xi\underline{i}n}\right] e^{-jk_{xi}x-jk_{z\underline{i}}z}$$

$$\sum_{i,\underline{i}}[E^I_{zi\underline{i}}\delta_{i\underline{i},00} + R_{zi\underline{i}}]\, e^{-jk_{xi}x-jk_{z\underline{i}}z} = \sum_{i,\underline{i}}\left[\sum_{n=1}^{N_T} C_n S_{zi\underline{i}n}\right] e^{-jk_{xi}x-jk_{z\underline{i}}z}$$

$$\frac{1}{\eta_0}\sum_{i,\underline{i}}\left[(Y^I_{xxi\underline{i}}E^I_{xi\underline{i}} + Y^I_{xzi\underline{i}}E^I_{zi\underline{i}})\delta_{i\underline{i},00} + (Y^R_{xxi\underline{i}}R_{xi\underline{i}} + Y^R_{xzi\underline{i}}R_{zi\underline{i}})\right] e^{-jk_{xi}x-jk_{z\underline{i}}z} = \frac{1}{\eta_0}\sum_{i,\underline{i}}\left[\sum_{n=0}^{N_T} C_n U_{xi\underline{i}n}\right] e^{-jk_{xi}x-jk_{z\underline{i}}z}$$

$$\frac{1}{\eta_0}\sum_{i,\underline{i}}\left[(Y^I_{zxi\underline{i}}E_{xi\underline{i}} + Y^I_{zzi\underline{i}}E^I_{zi\underline{i}})\delta_{i\underline{i},00} + (Y^R_{zxi\underline{i}}R_{xi\underline{i}} + Y^R_{zzi\underline{i}}R_{zi\underline{i}})\right] e^{-jk_{xi}x-jk_{z\underline{i}}z} = \frac{1}{\eta_0}\sum_{i,\underline{i}}\left[\sum_{n=0}^{N_T} C_n U_{zi\underline{i}n}\right] e^{-jk_{xi}x-k_{z\underline{i}}z} \quad (5.252)$$

At the $y = -L$ interface, we have

$$E^{(3)}_{x,z}\Big|_{y=-L^-} + E^{(2)}_{x,z}\Big|_{y=-L^+}$$

$$H^{(3)}_{x,z}\Big|_{y=-L^-} + H^{(2)}_{x,z}\Big|_{y=-L^+}$$

$$\sum_{i,\underline{i}} T_{xi\underline{i}} e^{-jk_{xi}x-jk_{z\underline{i}}z} = \sum_{i,\underline{i}}\left[\sum_{n=0}^{N_T} C_n S_{xi\underline{i}n} e^{-q_nL}\right] e^{-jk_{xi}x-jk_{z\underline{i}}z}$$

$$\sum_{i,\underline{i}} T_{zi\underline{i}} e^{-jk_{xi}x-jk_{z\underline{i}}z} = \sum_{i,\underline{i}}\left[\sum_{n=0}^{N_T} C_n S_{zi\underline{i}n} e^{-q_nL}\right] e^{-jk_{xi}x-k_{z\underline{i}}z}$$

$$\frac{1}{\eta_0}\sum_{i,\underline{i}}[Y^T_{xxi\underline{i}}T_{xi\underline{i}} + Y^T_{xzi\underline{i}}T_{zi\underline{i}}]\, e^{-jk_{xi}x-jk_{z\underline{i}}z} = \frac{1}{\eta_0}\sum_{i,\underline{i}}\left[\sum_{n=0}^{N_T} C_n U_{xi\underline{i}n} e^{q_nL}\right] e^{-jk_{xi}x-jk_{z\underline{i}}z}$$

$$\frac{1}{\eta_0}\sum_{i,\underline{i}}[Y^T_{zxi\underline{i}}T_{xi\underline{i}} + Y^T_{zzi\underline{i}}T_{zi\underline{i}}]\, e^{-jk_{xi}x-jk_{z\underline{i}}z} = \frac{1}{\eta_0}\sum_{i,\underline{i}}\left[\sum_{n=0}^{N_T} C_n U_{zi\underline{i}n} e^{q_nL}\right] e^{-jk_{xi}x-jk_{z\underline{i}}z} \quad (5.253)$$

Equating modal coefficients, we obtain the following set of equations:

$$E^I_{xi\underline{i}}\delta_{i\underline{i},00} + R_{xi\underline{i}} = \sum_{n=0}^{N_T} C_n S_{xi\underline{i}n} \quad (5.254)$$

$$E^I_{zi\underline{i}}\delta_{i\underline{i},00} + R_{zi\underline{i}} = \sum_{n=0}^{N_T} C_n S_{zi\underline{i}n} \tag{5.255}$$

$$\left[(Y^I_{xxi\underline{i}}E^I_{xi\underline{i}} + Y^I_{xzi\underline{i}}E^I_{zi\underline{i}})\delta_{i\underline{i},00} + (Y^R_{xxi\underline{i}}R_{xi\underline{i}} + Y^R_{xzi\underline{i}}R_{zi\underline{i}})\right] = \sum_{n=1}^{N_T} C_n U_{xi\underline{i}n} \tag{5.256}$$

$$\left[(Y^I_{zxi\underline{i}}E^I_{xi\underline{i}} + Y^I_{zzi\underline{i}}E^I_{zi\underline{i}})\delta_{i\underline{i},00} + (Y^R_{zxi\underline{i}}R_{xi\underline{i}} + Y^R_{zzi\underline{i}}R_{zi\underline{i}})\right] = \sum_{n=1}^{N_T} C_n U_{zi\underline{i}n} \tag{5.257}$$

We may eliminate $R_{xi\underline{i}}$ and $R_{zi\underline{i}}$ and determine equations for C_n alone. We have

$$(Y^I_{xxi\underline{i}}E^I_{xi\underline{i}} + Y^I_{xzi\underline{i}}E^I_{zi\underline{i}})\delta_{i\underline{i},00} + Y^R_{xxi\underline{i}}\left[-E^I_{zi\underline{i}}\delta_{i\underline{i},00} + \sum_{n=1}^{N_T} C_n S_{xi\underline{i}n}\right] + Y^R_{xzi\underline{i}}\left[-E^I_{zi\underline{i}}\delta_{i\underline{i},00} + \sum_{n=1}^{N_T} C_n S_{zi\underline{i}n}\right] = \sum_{n=1}^{N_T} C_n U_{xi\underline{i}n} \tag{5.258}$$

Collecting common terms, we have

$$\left[(Y^I_{xxi\underline{i}} - Y^R_{xxi\underline{i}})E^I_{xi\underline{i}} + (Y^I_{xzi\underline{i}} - Y^R_{xzi\underline{i}})E^I_{zi\underline{i}}\right]\delta_{i\underline{i},00} = \sum_{n=1}^{N_T} C_n[-Y^R_{xxi\underline{i}}S_{xi\underline{i}n} - Y^R_{xzi\underline{i}}S_{zi\underline{i}n} + U_{xi\underline{i}n}] \tag{5.259}$$

A similar analysis for the H_z, $y = 0$ equation shows that

$$\left[(Y^I_{zxi\underline{i}} - Y^R_{zxi\underline{i}})E^I_{xi\underline{i}} + (Y^I_{zzi\underline{i}} - Y^R_{zzi\underline{i}})E^I_{zi\underline{i}}\right]\delta_{i\underline{i},00} = \sum_{n=1}^{N_T} C_n\,[-Y^R_{zxi\underline{i}}S_{xi\underline{i}n} - Y^R_{zzi\underline{i}}S_{zi\underline{i}n} + U_{zi\underline{i}n}] \tag{5.260}$$

If the modal coefficients of the H_x and H_z, $y = -L$ equations are computed, we find that

$$T_{xi\underline{i}} = \sum_{n=1}^{N_T} C_n S_{xi\underline{i}n} \exp(-q_n L)$$

$$T_{zi\underline{i}} = \sum_{n=1}^{N_T} C_n S_{zi\underline{i}n} \exp(-q_n L)$$

$$Y^T_{xxi\underline{i}}T_{xi\underline{i}} + Y^T_{xzi\underline{i}}T_{zi\underline{i}} = \sum_{n=1}^{N_T} C_n U_{xi\underline{i}n} \exp(-q_n L) \tag{5.261a}$$

$$Y^T_{zxi\underline{i}}T_{xi\underline{i}} + Y^T_{zzi\underline{i}}T_{zi\underline{i}} = \sum_{n=1}^{N_T} C_n U_{zi\underline{i}n} \exp(-q_n L) \tag{5.261b}$$

If we substitute $T_{xi\underline{i}n}$ and $T_{zi\underline{i}n}$ into Equation 5.261a and b, we find

$$0 = \sum_{n=1}^{N_T} C_n \exp(-q_n L)\left\{-Y^T_{xxi\underline{i}} S_{xi\underline{i}n} - Y^T_{xzi\underline{i}} S_{zi\underline{i}n} + U_{xi\underline{i}n}\right\} \tag{5.262a}$$

$$0 = \sum_{n=1}^{N_T} C_n \exp(-q_n L)\left\{-Y^T_{zxi\underline{i}} S_{xi\underline{i}n} - Y^T_{zzi\underline{i}} S_{zi\underline{i}n} + U_{zi\underline{i}n}\right\} \tag{5.262b}$$

Equation 5.262a and b form a set of $N_T \times N_T$ equations from which all of the modal coefficients can be determined.

Another important quantity that needs to be studied is the power that is incident on the cross grating and the power that is reflected, diffracted, and transmitted from the grating. The power that is incident on the grating over one grating cell in the $(-\hat{a}_y)$ direction is given by

$$P^I = \frac{1}{2}\operatorname{Re}(P_c^I) \tag{5.263}$$

where P_c^I is given by

$$P_c^I = \int_{-\tilde{\Lambda}_z/2}^{\tilde{\Lambda}_z/2} \int_{-\tilde{\Lambda}_x/2}^{\tilde{\Lambda}_x/2} \vec{E}^I \times \vec{H}^{I*} \cdot (-\hat{a}_y\, d\tilde{x}\, d\tilde{z}) \tag{5.264}$$

Or, after being put in normalized form and carrying out the $-\hat{a}_y$, the dot product is given by

$$P_c^I = \frac{1}{k_0^2} \int_{-\Lambda_z/2}^{\Lambda_z/2} \int_{-\Lambda_x/2}^{\Lambda_x/2} \left[E^I_{z00} H^I_{x00} + E^I_{x00} H^{I*}_{z00}\right] dx\, dz \tag{5.265}$$

The quantity in square brackets is a constant. After integrating Equation 5.265 and substituting the incident modal admittances $Y^I_{xx00}, \ldots$, it is found that P_c^I is given by

$$P_c^I = \frac{\Lambda_x \Lambda_z}{k_0^2 \eta_0}\left\{E^I_{z00}\left[\theta Y^{I*}_{xx00} E^{I*}_{x00} + Y^{I*}_{xz00} E^{I*}_{z00}\right] - E^I_{x00}\left[Y^{I*}_{zx00} E^{I*}_{x00} + Y^{I*}_{zz00} E^{I*}_{z00}\right]\right\} \tag{5.266}$$

The quantities E^I_{x00} and E^I_{z00} are given in terms of the incident angles and polarizations by Equation 5.228a.

The reflected power from the crossed grating is given by

$$P^R = \frac{1}{2}\operatorname{Re}(P_c^R) \tag{5.267}$$

where

$$P_c^R = \frac{1}{k_0^2} \int_{-\Lambda_z/2}^{\Lambda_z/2} \int_{-\Lambda_x/2}^{\Lambda_x/2} \vec{E}^R \times \vec{H}^{R*} \cdot (\hat{a}_y \, dx \, dz) \tag{5.268}$$

$$P_c^R = \frac{1}{k_0^2} \int_{-\Lambda_z/2}^{\Lambda_z/2} \int_{-\Lambda_x/2}^{\Lambda_x/2} \left[E_z^R H_x^{R*} - E_x^R H_z^{R*} \right] dx \, dz \tag{5.269}$$

$$P_c^R = \frac{1}{k_0^2} \left(I_{zx}^R - I_{xz}^R \right) \tag{5.270}$$

where I_{zx}^R and I_{xz}^R refer to the first and second terms in Equation 5.269. If we substitute E_x^R and E_z^R into I_{zx}, we find after interchanging summation and integration

$$I_{zx} = \sum_{i\underline{i}} \sum_{i'\underline{i}'} R_{zi\underline{i}} \left[Y_{xxi'\underline{i}'}^{R*} R_{xi'\underline{i}'}^{*} + Y_{xzi'\underline{i}'}^{R*} R_{zi'\underline{i}'}^{*} \right] \cdot \left[\int_{-\Lambda_x/2}^{\Lambda_x/2} e^{-j(k_{xi} - k_{xi'})x} \, dx \right] \left[\int_{-\Lambda_z/2}^{\Lambda_z/2} e^{-j(k_{z\underline{i}} - k_{z\underline{i}'})z} \, dz \right] \tag{5.271}$$

The first integral (x-integral) equals $\Lambda_x \delta_{ii'}$ and the second integral equals $\Lambda_z \delta_{\underline{i}\underline{i}'}$, where $\delta_{ii'}$ is the Kronecker delta. Substituting these values in Equation 5.271, we find that

$$I_{zx} = \Lambda_x \Lambda_z \sum_{i,\underline{i}} R_{zi\underline{i}} \left[Y_{xxi'\underline{i}'}^{R*} R_{xi'\underline{i}'}^{*} + Y_{xzi'\underline{i}'}^{R*} R_{zi'\underline{i}'}^{*} \right] \tag{5.272}$$

Carrying out a similar analysis for I_{xz} and substituting the expressions into Equation 5.270, we find that

$$P_c^R = \sum_{i\underline{i}} P_{ci\underline{i}}^R = \frac{\Lambda_x \Lambda_z}{k_0^2 \eta_0} \sum_{i'\underline{i}'} \left\{ \left[R_{zi\underline{i}} (Y_{xxi\underline{i}}^{R*} R_{xi\underline{i}}^{*} + Y_{xzi\underline{i}}^{R*} R_{zi\underline{i}}^{*}) \right] - \left[R_{xi\underline{i}} (Y_{zxi\underline{i}}^{R*} R_{xi\underline{i}}^{*} + Y_{zzi\underline{i}}^{R*} R_{zi\underline{i}}^{*}) \right] \right\} \tag{5.273}$$

The power transmitted in the $-\hat{a}_y$ direction at $y = -L^-$ over one crossed $\Lambda_x \Lambda_z$ grating cell is given by

$$P^T = \frac{1}{2} \operatorname{Re}(P_c^T) \tag{5.274}$$

where

$$P_c^T = \int_{-\Lambda_z/2}^{\Lambda_z/2} \int_{-\Lambda_x/2}^{\Lambda_x/2} \vec{E}^T \times \vec{H}^{T*} \cdot (-\hat{a}_y) \, d\tilde{x} \, d\tilde{z} \tag{5.275}$$

$$P_c^T = \frac{1}{k_0^2} \int_{-\Lambda_z/2}^{\Lambda_z/2} \int_{-\Lambda_x/2}^{\Lambda_x/2} \left[-E_z^T H_x^{T*} + E_x^T H_z^{T*} \right] dx \, dz \tag{5.276}$$

Substituting the transmitted electric and magnetic fields into Equation 5.276 for P_c^T and carrying out an analysis similar to that used to determine P_c^R, we find that

$$P_c^T = \sum_{i\underline{i}} P_{ci\underline{i}}^T = \frac{\Lambda_x \Lambda_z}{k_0^2 \eta_0} \sum_{i'\underline{i}'} \left\{ -\left[T_{zi\underline{i}} (Y_{xxi\underline{i}}^{T*} T_{xi\underline{i}}^* + Y_{xzi\underline{i}}^{T*} T_{zi\underline{i}}^*) \right] + \left[T_{xi\underline{i}} (Y_{zxi\underline{i}}^{T*} T_{xi\underline{i}}^* + Y_{zzi\underline{i}}^{T*} T_{zi\underline{i}}^*) \right] \right\} \tag{5.277}$$

An important factor associated with the transmitted and reflected differential power is the diffraction efficiency of the $i,\underline{i}$th order. The diffraction efficiency of the reflected $i\underline{i}$th order is given by and defined by

$$D_{i\underline{i}}^R = \frac{\frac{\text{Re}}{2}\left(P_{ci\underline{i}}^R\right)}{\frac{\text{Re}}{2}\left(P_c^I\right)} \tag{5.278}$$

The diffraction efficiency of the transmitted $i,\underline{i}$th order is given by and defined by

$$D_{i\underline{i}}^T = \frac{\frac{\text{Re}}{2}\left(P_{ci\underline{i}}^T\right)}{\frac{\text{Re}}{2}\left(P_c^I\right)} \tag{5.279}$$

For a lossless crossed grating, the reflected and transmitted diffraction efficiencies obey the conservation of power relation

$$\sum_{i\underline{i}} \left(D_{i\underline{i}}^R + D_{i\underline{i}}^T\right) = 1 \tag{5.280}$$

5.5.2 Numerical Results

This section will present some numerical examples of the diffraction efficiency that results when an oblique plane wave is scattered or diffracted from a crossed or two-dimensional diffraction grating.

The example to be presented involves scattering from a one-dimensional square wave grating, where $\varepsilon_1 = \varepsilon_3 = 1$, $\varepsilon_2 = 2.5$, $\alpha = 30°$, $\tilde{\Lambda}_x = \lambda$, and $\tilde{\Lambda}_z = \infty$. This example has been previously studied for the H-mode case in Section 5.2.1, for the E-mode case in Section 5.3, and in the literature it has been first presented by Yamakita [52]. The purpose of using the more general crossed grating algorithm to study a one-dimensional case is to validate that in the limiting case, the operation of the RCWA crossed grating algorithm presented in this section can produce the same results as the one-dimensional RCWA algorithm. The *H-mode* square wave case was numerically studied by taking $\tilde{\Lambda}_z$ to have a large but not infinite value. $\tilde{\Lambda}_z$ in the algorithm was set to $\tilde{\Lambda}_z = 15\lambda$; $M_T = 6$; $\underline{M}_T = 0$; $\beta = 270°$; $\alpha = 30°$; and $E_\beta^I = E_0$, $E_\alpha^I = 0$. The nonzero relative dielectric permittivities were taken to be $\varepsilon_{xx}(x,z) = \varepsilon_{yy}(x,z) = \varepsilon_{zz}(x,z) = \sum_{i=-M_T}^{M_T} \breve{\varepsilon}_i \exp(jiK_x x)$, where $\breve{\varepsilon}_i$ are the Fourier coefficients of the square profile used in the square wave example of Section 3.3. The E-mode square wave case was studied using the same parameters as those used in the H-mode square wave case, except that the polarization was taken to be $E_\beta^I = 0$, $E_\alpha^I = -E_0$. Figure 5.24 shows the diffraction efficiency results using the one-dimensional theory of Sections 5.2.1 and 5.3, and shows the crossed diffraction grating theory results of this section. As can be seen from Figure 5.24, nearly identical diffraction efficiency results from the two algorithms.

The crossed diffraction grating theory has also been used to calculate the scattering from the H-mode cosine grating [16] which was described in Section 5.2.1. After setting the parameters of crossed grating algorithm to match those of the H-mode cosine grating, identical diffraction efficiency results were obtained for the one- and two-dimensional RCWA algorithms for this case also.

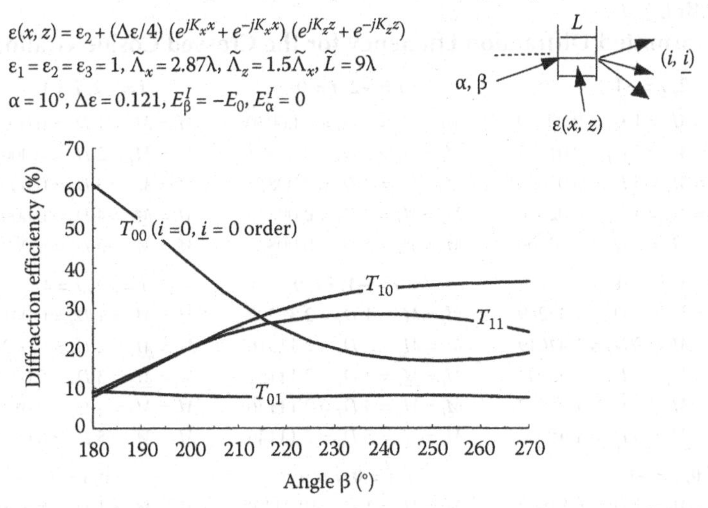

FIGURE 5.25 Diffraction efficiency data that result when crossed grating theory is applied to study scattering from a two-dimensional, crossed, cosine wave grating is shown.

Figure 5.25 shows the diffraction efficiency data that result when the crossed grating theory of this section is applied to study scattering from a two-dimensional, crossed, cosine wave grating, where $\varepsilon_1 = \varepsilon_2 = \varepsilon_3 = 1$, $\alpha = 10°$, $\tilde{\Lambda}_x = 2.8747\lambda$, and $\tilde{\Lambda}_z = 1.5\tilde{\Lambda}_x$, $\tilde{L} = 9\lambda$, $M_T = 3$, $\underline{M}_T = 3$, $180° \le \beta \le 270°$, $E_\beta^I = -1$, and $E_\alpha^I = 0$. The nonzero relative dielectric permittivities were taken to be

$$\varepsilon_{xx}(x,z) = \varepsilon_{yy}(x,z) = \varepsilon_{zz}(x,z) = \sum_{\underline{i}=-\underline{M}_T}^{\underline{M}_T} \sum_{i=-M_T}^{M_T} \breve{\varepsilon}_{i,\underline{i}} \exp\left[j(iK_x x + \underline{i}K_z z)\right]$$

$$\varepsilon_{rs}(x,z) = 0, \quad r \ne s, \quad (r,s) = (x,y,z) \tag{5.281}$$

In Equation 5.281, $\breve{\varepsilon}_{0,0} = \varepsilon_1$, $\breve{\varepsilon}_{1,1} = \breve{\varepsilon}_{-1,1} = \breve{\varepsilon}_{1,-1} = \breve{\varepsilon}_{-1,-1} = \Delta\varepsilon/4$, $\Delta\varepsilon = 0.121$ and all other Fourier coefficients $\breve{\varepsilon}_{i,\underline{i}}$ in Equation 5.281 are zero. In Figure 5.25, the transmitted diffraction efficiencies (denoted by $T_{i,\underline{i}}$) of the T_{00}, T_{01}, T_{10}, and T_{11} orders were plotted versus the azimuthal angle β which was varied over the range $180° \le \beta \le 270°$. As can be seen from the Figure 5.25 plot, changing the β-angle of incidence causes a very perceptible change in the diffraction efficiency which is observed from the grating. In making the Figure 5.25 plot, conservation of real power, Equation 5.280 was observed to a high degree of accuracy.

Table 5.1 shows the transmitted diffraction efficiency for the crossed cosine diffraction grating studied in Figure 5.25 (taking $\beta = 270°$) that results for fifteen orders (taking all pair combinations of $i = -2, -1, 0, 1, 2$ and $\underline{i} = -1, 0, 1$) when five different matrix truncations $M_T = \underline{M}_T = 1, 2, 3, 4, 5$ were used. (For the truncations where the $(i, \underline{i})$ pair order exceeds the truncation order [for example, when the pair $(i, \underline{i}) = (-2, 1)$ and exceeds, the truncation order is $M_T = \underline{M}_T = 1$], the diffraction efficiency is set to zero.) A striking and reassuring feature of the diffraction efficiencies displayed in Table 5.1 is the fact how rapidly the diffraction efficiency converges to a final value that does not change with increasing order. After the value of $M_T = \underline{M}_T = 3$, there is virtually no change in diffraction efficiency for any of the fifteen orders displayed.

The second diffraction efficiency example to be presented consists of the diffraction efficiency data that result when the crossed grating theory of this section is applied to study scattering

TABLE 5.1
Transmitted Diffraction Efficiency for the Crossed Cosine Grating

$i=-2, \underline{i}=-1$	$i=-2, \underline{i}=0$	$i=-2, \underline{i}=1$
$M_t=\underline{M}_T=1\ D_T=0.00000$	$M_t=\underline{M}_T=1\ D_T=0.00000$	$M_t=\underline{M}_T=1\ D_T=0.00000$
$M_t=\underline{M}_T=2\ D_T=0.01983$	$M_t=\underline{M}_T=2\ D_T=0.00881$	$M_t=\underline{M}_T=2\ D_T=0.00012$
$M_t=\underline{M}_T=3\ D_T=0.02066$	$M_t=\underline{M}_T=3\ D_T=0.00835$	$M_t=\underline{M}_T=3\ D_T=0.00014$
$M_t=\underline{M}_T=4\ D_T=0.02066$	$M_t=\underline{M}_T=4\ D_T=0.00835$	$M_t=\underline{M}_T=4\ D_T=0.00014$
$M_t=\underline{M}_T=5\ D_T=0.02066$	$M_t=\underline{M}_T=5\ D_T=0.00835$	$M_t=\underline{M}_T=5\ D_T=0.00014$
$i=-1, \underline{i}=-1$	$i=-1, \underline{i}=0$	$i=-1, \underline{i}=1$
$M_t=\underline{M}_T=1\ D_T=1.43269$	$M_t=\underline{M}_T=1\ D_T=2.39502$	$M_t=\underline{M}_T=1\ D_T=0.04159$
$M_t=\underline{M}_T=2\ D_T=1.44619$	$M_t=\underline{M}_T=2\ D_T=2.11516$	$M_t=\underline{M}_T=2\ D_T=0.08341$
$M_t=\underline{M}_T=3\ D_T=1.45093$	$M_t=\underline{M}_T=3\ D_T=2.11151$	$M_t=\underline{M}_T=3\ D_T=0.08284$
$M_t=\underline{M}_T=4\ D_T=1.45092$	$M_t=\underline{M}_T=4\ D_T=2.11146$	$M_t=\underline{M}_T=4\ D_T=0.08284$
$M_t=\underline{M}_T=5\ D_T=1.45092$	$M_t=\underline{M}_T=5\ D_T=2.11146$	$M_t=\underline{M}_T=5\ D_T=0.08284$
$i=0, \underline{i}=-1$	$i=0, \underline{i}=0$	$i=0, \underline{i}=1$
$M_t=\underline{M}_T=1\ D_T=6.81922$	$M_t=\underline{M}_T=1\ D_T=19.21145$	$M_t=\underline{M}_T=1\ D_T=8.52301$
$M_t=\underline{M}_T=2\ D_T=5.97634$	$M_t=\underline{M}_T=2\ D_T=18.76203$	$M_t=\underline{M}_T=2\ D_T=7.67222$
$M_t=\underline{M}_T=3\ D_T=5.94877$	$M_t=\underline{M}_T=3\ D_T=18.77696$	$M_t=\underline{M}_T=3\ D_T=7.65204$
$M_t=\underline{M}_T=4\ D_T=5.94865$	$M_t=\underline{M}_T=4\ D_T=18.77734$	$M_t=\underline{M}_T=4\ D_T=7.65192$
$M_t=\underline{M}_T=5\ D_T=5.94865$	$M_t=\underline{M}_T=5\ D_T=18.77734$	$M_t=\underline{M}_T=5\ D_T=7.65192$
$i=1, \underline{i}=-1$	$i=1, \underline{i}=0$	$i=1, \underline{i}=1$
$M_t=\underline{M}_T=1\ D_T=0.75138$	$M_t=\underline{M}_T=1\ D_T=36.76304$	$M_t=\underline{M}_T=1\ D_T=24.02316$
$M_t=\underline{M}_T=2\ D_T=1.14026$	$M_t=\underline{M}_T=2\ D_T=36.54341$	$M_t=\underline{M}_T=2\ D_T=24.00149$
$M_t=\underline{M}_T=3\ D_T=1.15368$	$M_t=\underline{M}_T=3\ D_T=36.57226$	$M_t=\underline{M}_T=3\ D_T=24.01952$
$M_t=\underline{M}_T=4\ D_T=1.15374$	$M_t=\underline{M}_T=4\ D_T=36.57190$	$M_t=\underline{M}_T=4\ D_T=24.01921$
$M_t=\underline{M}_T=5\ D_T=1.15374$	$M_t=\underline{M}_T=5\ D_T=36.57190$	$M_t=\underline{M}_T=5\ D_T=24.01921$
$i=2, \underline{i}=-1$	$i=2, \underline{i}=0$	$i=2, \underline{i}=1$
$M_t=\underline{M}_T=1\ D_T=0.00000$	$M_t=M_T=1\ D_T=0.00000$	$M_t=\underline{M}_T=1\ D_T=0.00000$
$M_t=\underline{M}_T=2\ D_T=0.01071$	$M_t=M_T=2\ D_T=0.41210$	$M_t=\underline{M}_T=2\ D_T=1.26739$
$M_t=\underline{M}_T=3\ D_T=0.01149$	$M_t=M_T=3\ D_T=0.39627$	$M_t=\underline{M}_T=3\ D_T=1.23198$
$M_t=\underline{M}_T=4\ D_T=0.01150$	$M_t=M_T=4\ D_T=0.39622$	$M_t=\underline{M}_T=4\ D_T=1.23244$
$M_t=\underline{M}_T=5\ D_T=0.01150$	$M_t=M_T=5\ D_T=0.39622$	$M_t=\underline{M}_T=5\ D_T=1.23245$

from a two-dimensional rectangular surface relief grating which is composed of isotropic dielectric material. The rectangular dielectric (shown in the Figure 5.26 inset) making up the surface relief grating in Region 2 was assumed to be centered in each two-dimensional grating period and have a width $2x_1 = \Lambda_x/2$, length of $2z_1 = \Lambda_z/2$, thickness $\tilde{L}$, and relative permittivity value of ε_3. The region surrounding the rectangular dielectric was assumed to have a dielectric value of ε_1. In Region 2, mathematically, the permittivity tensor of the surface relief grating is given by

$$\varepsilon_{xx}(x,y,z) = \varepsilon_{yy}(x,y,z) = \varepsilon_{zz}(x,y,z) = \varepsilon(x,y,z)$$

$$\varepsilon_{rs}(x,z) = 0, \quad r \neq s, \quad (r,s) = (x,y,z) \tag{5.282}$$

where

$$\varepsilon(x,y,z) = \begin{cases} \varepsilon_3, & |x| \le x_1,\ |z| \le z_1 \\ \varepsilon_1, & \text{elsewhere in the cell } |x| \le \Lambda_x,\ |z| \le \Lambda_z \end{cases} \tag{5.283}$$

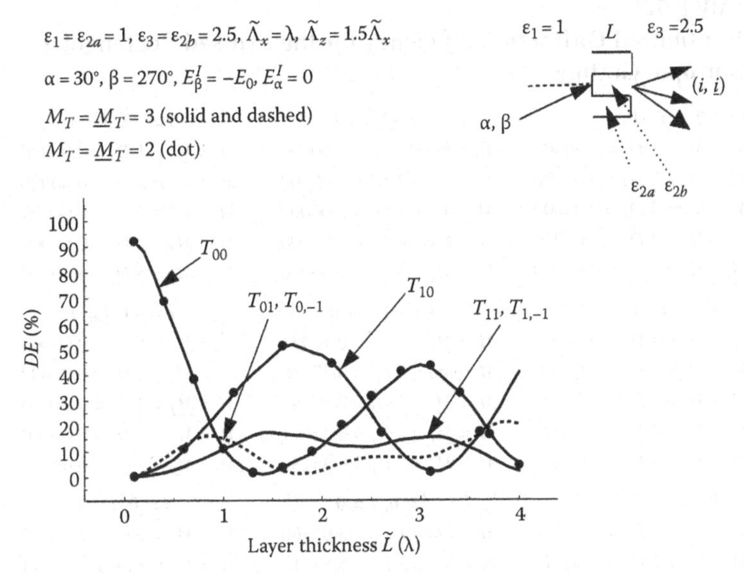

FIGURE 5.26 Transmitted diffraction efficiencies (*solid line*) of the $T_{i,\underline{i}}$ orders when i and $\underline{i}$ range from $(-1, 0, 1)$ and when the grating thickness is varied from $\tilde{L} = 0\lambda$ to $\tilde{L} = 4\lambda$ are shown.

and where $x_1 = \Lambda_x/4$ and $z_1 = \Lambda_z/4$ (please recall that $\Lambda_x = k_0\tilde{\Lambda}_x$, $k_0 = 2\pi/\lambda$, etc.). Figure 5.26 shows the transmitted diffraction efficiencies (*solid line*) of the $T_{i,\underline{i}}$ orders when i and $\underline{i}$ range from $(-1, 0, 1)$; when the grating thickness is varied from $\tilde{L} = 0\lambda$ to $\tilde{L} = 4\lambda$; when $\varepsilon_1 = 1$, $\varepsilon_3 = 2.5$, $\alpha = 30°$, $\tilde{\Lambda}_x = \lambda$, and $\tilde{\Lambda}_z = 1.5\tilde{\Lambda}_x$, $\beta = 270°$, $E^I_\beta = -E_0$, and $E^I_\alpha = 0$; and when $M_T = \underline{M}_T = 3$. As can be seen from Figure 5.26, EM power is diffracted out of the $T_{0,0}$ order (pump wave or incident wave) and is subsequently diffracted into the higher orders. Because of symmetry, the diffraction efficiencies of the $T_{1,1}$ and $T_{1,-1}$ orders were the same and the diffraction efficiencies of the $T_{0,1}$ and $T_{0,-1}$ orders were the same. By the same token, for $\beta = 0°, 90°, 180°$, we should observe similar symmetry in the diffracted orders. Figure 5.26 also shows the diffraction efficiency of the $T_{0,0}$ and $T_{1,0}$ orders (*dotted* curves) when the truncation was taken to be $M_T = \underline{M}_T = 2$ rather than $M_T = \underline{M}_T = 3$, as was done for the curves discussed earlier. As can be seen from the figure, very little diffraction efficiency difference exists between the two truncations.

Table 5.2 shows the transmitted diffraction efficiency for the crossed rectangular diffraction grating studied in Figure 5.26 (taking $\beta = 270°$ and $\tilde{L} = 1.7\lambda$) that results for fifteen orders (taking all pair combinations of $i = -2, -1, 0, 1, 2$ and $\underline{i} = -1, 0, 1$) when five different matrix truncations $M_T = \underline{M}_T = 1, 2, 3, 4, 5$ are used. (For the truncations where the $(i, \underline{i})$ pair order exceeds the truncation order [for example, when the pair $(i, \underline{i}) = (-2, 1)$ and exceeds, the truncation order is $M_T = \underline{M}_T = 1$], the diffraction efficiency is set to zero.) A striking and reassuring feature of the diffraction efficiencies displayed in Table 5.2, like that of Table 5.1, is the fact that the diffraction efficiency converges fairly rapidly to a final value that does not change with increasing order. Comparing Table 5.2 to Table 5.1, it is interesting to note that the convergence to a final value is slightly slower in Table 5.2 than in Table 5.1. This is believable since the grating studied in Table 5.2 is much smaller in size than the grating studied in Table 5.1, and also the grating studied in Table 5.2 has a much higher spatial spectral content than the grating studied in Table 5.1 (cosine grating). Both of these factors would cause a slower convergence with truncation order.

The third diffraction efficiency example to be presented consists of the diffraction efficiency data that result when the crossed grating theory of this section is applied to study scattering from a two-dimensional rectangular surface relief grating which contains anisotropic dielectric material.

TABLE 5.2
Transmitted Diffraction Efficiency for the Crossed Rectangular Isotropic Grating

$i=-2, \underline{i}=-1$	$i=-2, \underline{i}=0$	$i=-2, \underline{i}=1$
$M_t=\underline{M}_T=1$ $D_T=0.00000$	$M_t=\underline{M}_T=1$ $D_T=0.00000$	$M_t=\underline{M}_T=1$ $D_T=0.00000$
$M_t=\underline{M}_T=2$ $D_T=0.00000$	$M_t=\underline{M}_T=2$ $D_T=0.00000$	$M_t=\underline{M}_T=2$ $D_T=0.00000$
$M_t=\underline{M}_T=3$ $D_T=0.00000$	$M_t=\underline{M}_T=3$ $D_T=0.00000$	$M_t=\underline{M}_T=3$ $D_T=0.00000$
$M_t=\underline{M}_T=4$ $D_T=0.00000$	$M_t=\underline{M}_T=4$ $D_T=0.00000$	$M_t=\underline{M}_T=4$ $D_T=0.00000$
$M_t=\underline{M}_T=5$ $D_T=0.00000$	$M_t=\underline{M}_T=5$ $D_T=0.00000$	$M_t=\underline{M}_T=5$ $D_T=0.00000$
$i=-1, \underline{i}=-1$	**$i=-1, \underline{i}=0$**	**$i=-1, \underline{i}=1$**
$M_t=\underline{M}_T=1$ $D_T=0.00000$	$M_t=\underline{M}_T=1$ $D_T=0.10971$	$M_t=\underline{M}_T=1$ $D_T=0.00000$
$M_t=\underline{M}_T=2$ $D_T=0.00000$	$M_t=\underline{M}_T=2$ $D_T=0.37916$	$M_t=\underline{M}_T=2$ $D_T=0.00000$
$M_t=\underline{M}_T=3$ $D_T=0.00000$	$M_t=\underline{M}_T=3$ $D_T=0.35999$	$M_t=\underline{M}_T=3$ $D_T=0.00000$
$M_t=\underline{M}_T=4$ $D_T=0.00000$	$M_t=\underline{M}_T=4$ $D_T=0.36691$	$M_t=\underline{M}_T=4$ $D_T=0.00000$
$M_t=\underline{M}_T=5$ $D_T=0.00000$	$M_t=\underline{M}_T=5$ $D_T=0.36130$	$M_t=\underline{M}_T=5$ $D_T=0.00000$
$i=0, \underline{i}=-1$	**$i=0, \underline{i}=0$**	**$i=0, \underline{i}=1$**
$M_t=\underline{M}_T=1$ $D_T=1.44451$	$M_t=\underline{M}_T=1$ $D_T=8.01360$	$M_t=\underline{M}_T=1$ $D_T=1.44451$
$M_t=\underline{M}_T=2$ $D_T=1.40744$	$M_t=\underline{M}_T=2$ $D_T=5.88113$	$M_t=\underline{M}_T=2$ $D_T=1.40744$
$M_t=\underline{M}_T=3$ $D_T=1.47110$	$M_t=\underline{M}_T=3$ $D_T=5.37599$	$M_t=\underline{M}_T=3$ $D_T=1.47110$
$M_t=\underline{M}_T=4$ $D_T=1.45656$	$M_t=\underline{M}_T=4$ $D_T=5.38856$	$M_t=\underline{M}_T=4$ $D_T=1.45656$
$M_t=\underline{M}_T=5$ $D_T=1.49127$	$M_t=\underline{M}_T=5$ $D_T=5.34491$	$M_t=\underline{M}_T=5$ $D_T=1.49127$
$i=1, \underline{i}=-1$	**$i=1, \underline{i}=0$**	**$i=1, \underline{i}=1$**
$M_t=\underline{M}_T=1$ $D_T=9.56752$	$M_t=\underline{M}_T=1$ $D_T=65.76953$	$M_t=\underline{M}_T=1$ $D_T=9.56752$
$M_t=\underline{M}_T=2$ $D_T=15.86283$	$M_t=\underline{M}_T=2$ $D_T=52.30565$	$M_t=\underline{M}_T=2$ $D_T=15.86283$
$M_t=\underline{M}_T=3$ $D_T=16.30174$	$M_t=\underline{M}_T=3$ $D_T=51.79183$	$M_t=\underline{M}_T=3$ $D_T=16.30174$
$M_t=\underline{M}_T=4$ $D_T=16.50874$	$M_t=\underline{M}_T=4$ $D_T=51.42793$	$M_t=\underline{M}_T=4$ $D_T=16.50874$
$M_t=\underline{M}_T=5$ $D_T=16.57968$	$M_t=\underline{M}_T=5$ $D_T=51.30075$	$M_t=\underline{M}_T=5$ $D_T=16.57968$
$i=2, \underline{i}=-1$	**$i=2, \underline{i}=0$**	**$i=2, \underline{i}=1$**
$M_t=\underline{M}_T=1$ $D_T=0.00000$	$M_t=\underline{M}_T=1$ $D_T=0.00000$	$M_t=\underline{M}_T=1$ $D_T=0.00000$
$M_t=\underline{M}_T=2$ $D_T=0.00000$	$M_t=\underline{M}_T=2$ $D_T=1.78114$	$M_t=\underline{M}_T=2$ $D_T=0.00000$
$M_t=\underline{M}_T=3$ $D_T=0.00000$	$M_t=\underline{M}_T=3$ $D_T=1.84399$	$M_t=\underline{M}_T=3$ $D_T=0.00000$
$M_t=\underline{M}_T=4$ $D_T=0.00000$	$M_t=\underline{M}_T=4$ $D_T=1.83670$	$M_t=\underline{M}_T=4$ $D_T=0.00000$
$M_t=\underline{M}_T=5$ $D_T=0.00000$	$M_t=\underline{M}_T=5$ $D_T=1.82095$	$M_t=\underline{M}_T=5$ $D_T=0.00000$

In Region 2, we first need to express mathematically the permittivity tensor of the surface relief grating cell for this example. Let

$$f(x,z)=\begin{cases}\varepsilon_b, & |x|\le x_1,\ |z|\le z_1\\ \varepsilon_a, & \text{elsewhere in the cell } |x|\le\Lambda_x,\ |z|\le\Lambda_z\end{cases} \tag{5.284}$$

where

$$\varepsilon_b=\frac{(n_o^2+n_e^2)}{2} \tag{5.285a}$$

and also let

$$\varepsilon_{xxC} = \frac{\left[n_o^2 + \left(n_e^2 - n_o^2\right)C_x^2\right]}{\varepsilon_b}$$

$$\varepsilon_{yyC} = \frac{\left[n_o^2 + \left(n_e^2 - n_o^2\right)C_y^2\right]}{\varepsilon_b}$$

$$\varepsilon_{zzC} = \frac{\left[n_o^2 + \left(n_e^2 - n_o^2\right)C_z^2\right]}{\varepsilon_b}$$

$$\varepsilon_{xyC} = \varepsilon_{yxC} = \frac{\left[\left(n_e^2 - n_o^2\right)C_xC_y\right]}{\varepsilon_b}$$

$$\varepsilon_{xzC} = \varepsilon_{zxC} = \frac{\left[\left(n_e^2 - n_o^2\right)C_xC_z\right]}{\varepsilon_b}$$

$$\varepsilon_{yzC} = \varepsilon_{zyC} = \varepsilon_{zxC} \tag{5.285b}$$

$$C_x = \sin(\phi_c)\sin(\theta_c)$$

$$C_y = \cos(\phi_c)\sin(\theta_c)$$

$$C_z = \cos(\theta_c)$$

where $\phi_C = \theta_C = 45°$, $x_1 = \Lambda_x/4$, $z_1 = \Lambda_z/4$, $\varepsilon_1 = 1$, $\tilde{\Lambda}_x = \lambda$, and $\tilde{\Lambda}_z = 1.5\tilde{\Lambda}_x$.

Using these parameters and functions, we define the relative dielectric permittivity to be

$$\underline{\underline{\varepsilon}}(x,z) = \begin{bmatrix} \varepsilon_{xxC} & \varepsilon_{xyC} & \varepsilon_{xzC} \\ \varepsilon_{yxC} & \varepsilon_{yyC} & \varepsilon_{yzC} \\ \varepsilon_{zxC} & \varepsilon_{zyC} & \varepsilon_{zzC} \end{bmatrix} f(x,z) \tag{5.286}$$

Note that the diffraction grating in Region 2 is made up of two different anisotropic materials, as specified above. Figure 5.27 shows plots of the diffraction efficiency of the T_{10} order ($i = 1, \underline{i} = 0$) when the grating thickness is varied from $\tilde{L} = 0\lambda$ to 4λ; when $\varepsilon_3 = 2.5$, $\alpha = 30°$, $\tilde{\Lambda}_x = \lambda$, and $\tilde{\Lambda}_z = 1.5\tilde{\Lambda}_x$, $\beta = 270°$, $E_\beta^I = -E_0$, and $E_\alpha^I = 0$; when $M_T = \underline{M}_T = 2$; and when the values of the parameters n_o^2, n_e^2 were taken to be $n_o^2 = 2$, $n_e^2 = 3$ (curve marked T_{10a}), $n_o^2 = 2.4$, $n_e^2 = 2.6$ (curve marked T_{10b}), and $n_o^2 = 2.5$, $n_e^2 = 2.5$ (curve marked T_{10c}). As can be seen from Figure 5.27, considerable power is diffracted into the T_{10} order. Figure 5.27 also shows that as the grating is made more anisotropic (that is, by increasing the magnitude of the difference between n_o^2 and n_e^2), a more perceptible difference between the isotropic and anisotropic diffraction efficiencies occurs. It is interesting to note that the diffraction efficiencies of the isotropic and anisotropic cases (second and third examples, respectively) presently under consideration are similar to one another.

The fourth diffraction efficiency example to be presented consists of the diffraction efficiency data that result when the crossed grating theory of this section is applied to study scattering from a

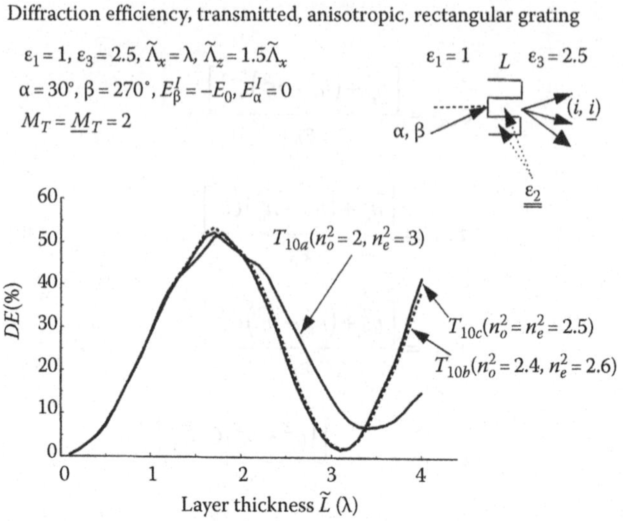

FIGURE 5.27 Plots of the diffraction efficiency of the T_{10} order ($i = 1$, $\underline{i} = 0$) when the grating thickness is varied from $\tilde{L} = 0\lambda$ to 4λ are shown.

rectangular, pyramidal surface relief grating which contains anisotropic dielectric material. Let the function $f(x, y, z)$ at any given value of y in the interval $-L \leq y \leq 0$ be defined as

$$f(x,y,z) = \begin{cases} \varepsilon_b, & |x| \leq x_1(y),\ |z| \leq z_1(y) \\ \varepsilon_a, & \text{elsewhere in } |x| \leq \dfrac{\Lambda_x}{2},\ |z| \leq \dfrac{\Lambda_z}{2} \end{cases} \tag{5.287}$$

where

$$x_1(y) = \frac{\sqrt{3}\,|y|}{4\,\Lambda_x}$$

$$z_1(y) = \frac{\sqrt{3}\,|y|}{4\,\Lambda_z}$$

Then the permittivity in the grating cell in Region 2 is given by

$$\underline{\varepsilon}(x,y,z) = \begin{bmatrix} \varepsilon_{xxC} & \varepsilon_{xyC} & \varepsilon_{xzC} \\ \varepsilon_{yxC} & \varepsilon_{yyC} & \varepsilon_{yzC} \\ \varepsilon_{zxC} & \varepsilon_{zyC} & \varepsilon_{zzC} \end{bmatrix} f(x,y,z) \tag{5.288}$$

where the parameters of Equation 5.288 are already given in Equation 5.285a and b. Because the grating is longitudinally inhomogeneous, a two-dimensional multilayer analysis based on the theory of Section 5.4 was used to calculate the diffraction efficiency.

Figure 5.28 shows plots of the diffraction efficiency of the T_{00}, T_{10}, T_{01}, and T_{11} orders when the grating thickness is varied from $\tilde{L} = 0\lambda$ to 2.5λ; $\varepsilon_3 = 2.5$, $\tilde{\Lambda}_x = \lambda$, and $\tilde{\Lambda}_z = 1.5\tilde{\Lambda}_x$; when $\alpha = 30°$, $\beta = 270°$, $E^I_\beta = -E_0$, and $E^I_\alpha = 0$; when $M_T = \underline{M}_T = 2$; when the values of the parameters n_o^2, n_e^2 were taken to be $n_o^2 = 2$, $n_e^2 = 3$; and when 10 layers ($N_\ell = 10$) were used to carry out the two-dimensional multilayer analysis. As can be seen from Figure 5.28 for the grating under study, power is diffracted out of the T_{00} order into higher orders. Conservation of power as specified by equations was observed to a high degree of accuracy. Table 5.3 shows the transmitted diffraction efficiency for the crossed rectangular, anisotropic diffraction grating studied in Figure 5.27 (taking $n_o^2 = 2$, $n_e^2 = 3$,

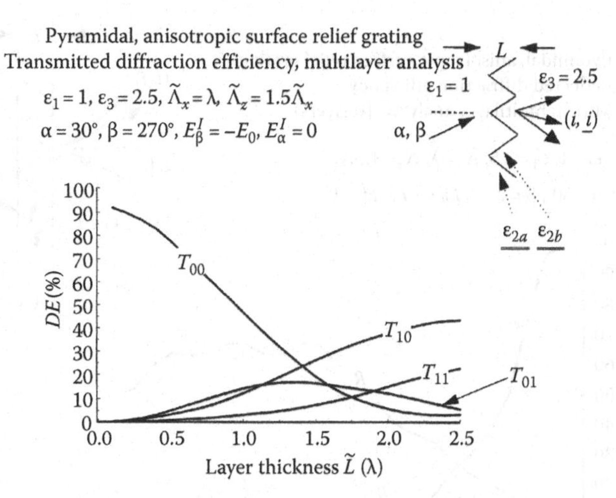

FIGURE 5.28 Plots of the diffraction efficiency of the T_{00}, T_{10}, T_{01}, and T_{11} orders when the grating thickness is varied from $\tilde{L} = 0\lambda$ to $\tilde{L} = 2.5\lambda$ are shown.

TABLE 5.3
Transmitted Diffraction Efficiency for the Crossed Rectangular Anisotropic Grating

$i=-2, \underline{i}=-1$	$i=-2, \underline{i}=0$	$i=-2, \underline{i}=1$
$M_t = \underline{M}_T = 1$ $D_T = 0.00000$	$M_t = \underline{M}_T = 1$ $D_T = 0.00000$	$M_t = \underline{M}_T = 1$ $D_T = 0.00000$
$M_t = \underline{M}_T = 2$ $D_T = 0.00000$	$M_t = \underline{M}_T = 2$ $D_T = 0.00000$	$M_t = \underline{M}_T = 2$ $D_T = 0.00000$
$M_t = \underline{M}_T = 3$ $D_T = 0.00000$	$M_t = \underline{M}_T = 3$ $D_T = 0.00000$	$M_t = \underline{M}_T = 3$ $D_T = 0.00000$
$M_t = \underline{M}_T = 4$ $D_T = 0.00000$	$M_t = \underline{M}_T = 4$ $D_T = 0.00000$	$M_t = \underline{M}_T = 4$ $D_T = 0.00000$
$M_t = \underline{M}_T = 5$ $D_T = 0.00000$	$M_t = \underline{M}_T = 5$ $D_T = 0.00000$	$M_t = \underline{M}_T = 5$ $D_T = 0.00000$
$i=-1, \underline{i}=-1$	**$i=-1, \underline{i}=0$**	**$i=-1, \underline{i}=1$**
$M_t = \underline{M}_T = 1$ $D_T = 0.00000$	$M_t = \underline{M}_T = 1$ $D_T = 0.02602$	$M_t = \underline{M}_T = 1$ $D_T = 0.00000$
$M_t = \underline{M}_T = 2$ $D_T = 0.00000$	$M_t = \underline{M}_T = 2$ $D_T = 0.14482$	$M_t = \underline{M}_T = 2$ $D_T = 0.00000$
$M_t = \underline{M}_T = 3$ $D_T = 0.00000$	$M_t = \underline{M}_T = 3$ $D_T = 0.13863$	$M_t = \underline{M}_T = 3$ $D_T = 0.00000$
$M_t = \underline{M}_T = 4$ $D_T = 0.00000$	$M_t = \underline{M}_T = 4$ $D_T = 0.14003$	$M_t = \underline{M}_T = 4$ $D_T = 0.00000$
$M_t = \underline{M}_T = 5$ $D_T = 0.00000$	$M_t = \underline{M}_T = 5$ $D_T = 0.13735$	$M_t = \underline{M}_T = 5$ $D_T = 0.00000$
$i=0, \underline{i}=-1$	**$i=0, \underline{i}=0$**	**$i=0, \underline{i}=1$**
$M_t = \underline{M}_T = 1$ $D_T = 2.12381$	$M_t = \underline{M}_T = 1$ $D_T = 1.70374$	$M_t = \underline{M}_T = 1$ $D_T = 3.81748$
$M_t = \underline{M}_T = 2$ $D_T = 2.28781$	$M_t = \underline{M}_T = 2$ $D_T = 3.21351$	$M_t = \underline{M}_T = 2$ $D_T = 2.52697$
$M_t = \underline{M}_T = 3$ $D_T = 2.57794$	$M_t = \underline{M}_T = 3$ $D_T = 3.14167$	$M_t = \underline{M}_T = 3$ $D_T = 2.64898$
$M_t = \underline{M}_T = 4$ $D_T = 2.54011$	$M_t = \underline{M}_T = 4$ $D_T = 3.31583$	$M_t = \underline{M}_T = 4$ $D_T = 2.76677$
$M_t = \underline{M}_T = 5$ $D_T = 2.61843$	$M_t = \underline{M}_T = 5$ $D_T = 3.34996$	$M_t = \underline{M}_T = 5$ $D_T = 2.83734$
$i=1, \underline{i}=-1$	**$i=1, \underline{i}=0$**	**$i=1, \underline{i}=1$**
$M_t = \underline{M}_T = 1$ $D_T = 6.33657$	$M_t = \underline{M}_T = 1$ $D_T = 62.78472$	$M_t = \underline{M}_T = 1$ $D_T = 17.98810$
$M_t = \underline{M}_T = 2$ $D_T = 8.69866$	$M_t = \underline{M}_T = 2$ $D_T = 51.30495$	$M_t = \underline{M}_T = 2$ $D_T = 25.37109$
$M_t = \underline{M}_T = 3$ $D_T = 8.83193$	$M_t = \underline{M}_T = 3$ $D_T = 49.97434$	$M_t = \underline{M}_T = 3$ $D_T = 26.22965$
$M_t = \underline{M}_T = 4$ $D_T = 8.69632$	$M_t = \underline{M}_T = 4$ $D_T = 49.52840$	$M_t = \underline{M}_T = 4$ $D_T = 26.48095$
$M_t = \underline{M}_T = 5$ $D_T = 8.71696$	$M_t = \underline{M}_T = 5$ $D_T = 49.24850$	$M_t = \underline{M}_T = 5$ $D_T = 26.54343$
$i=2, \underline{i}=-1$	**$i=2, \underline{i}=0$**	**$i=2, \underline{i}=1$**
$M_t = \underline{M}_T = 1$ $D_T = 0.00000$	$M_t = \underline{M}_T = 1$ $D_T = 0.00000$	$M_t = \underline{M}_T = 1$ $D_T = 0.00000$
$M_t = \underline{M}_T = 2$ $D_T = 0.00000$	$M_t = \underline{M}_T = 2$ $D_T = 1.00812$	$M_t = \underline{M}_T = 2$ $D_T = 0.00000$
$M_t = \underline{M}_T = 3$ $D_T = 0.00000$	$M_t = \underline{M}_T = 3$ $D_T = 1.03808$	$M_t = \underline{M}_T = 3$ $D_T = 0.00000$
$M_t = \underline{M}_T = 4$ $D_T = 0.00000$	$M_t = \underline{M}_T = 4$ $D_T = 1.05363$	$M_t = \underline{M}_T = 4$ $D_T = 0.00000$
$M_t = \underline{M}_T = 5$ $D_T = 0.00000$	$M_t = \underline{M}_T = 5$ $D_T = 1.04436$	$M_t = \underline{M}_T = 5$ $D_T = 0.00000$

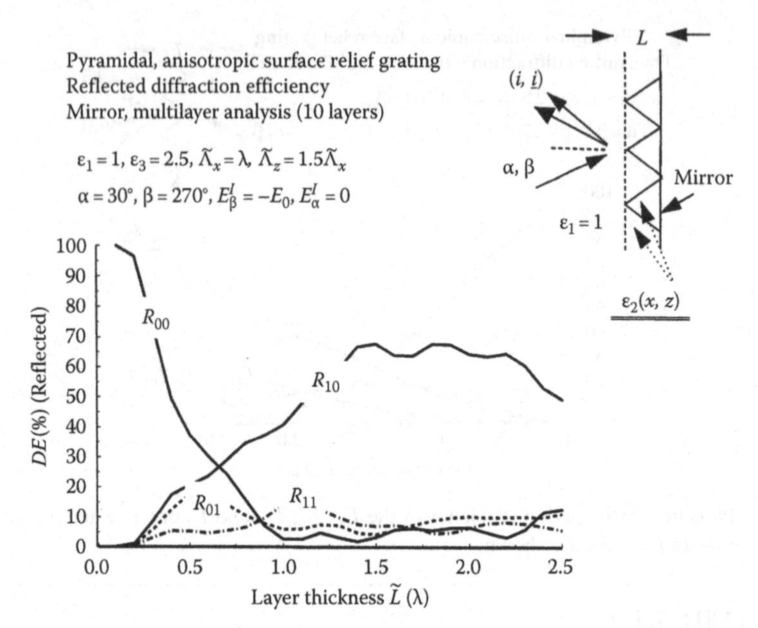

FIGURE 5.29 Pyramidal, anisotropic surface relief grating reflected, diffraction efficiency mirror, multilayer analysis (10 layers).

and $\tilde{L} = 1.7\lambda$ and all other parameters the same as those in Figure 5.27) that results for 15 orders (taking all pair combinations of $i = -2, -1, 0, 1, 2$ and $\underline{i} = -1, 0, 1$) when 5 different matrix truncations $M_T = \underline{M}_T = 1, 2, 3, 4, 5$ are used.

In the final example, a crossed, pyramidal diffraction grating is again studied (same pyramid geometry as that in Figure 5.28), but with a mirror (or a perfectly conducting short circuit plate) placed at the Region 2–3 interface at $\tilde{y} = -\tilde{L}$ (see Figure 5.29). In this case, just the reflected diffraction efficiency was studied (the transmitted diffraction efficiency in Region 3 is zero). The overall EM analysis in this case requires that the tangential EM fields at $\tilde{y} = -\tilde{L}$ be zero. Imposing this condition (see Section 5.3 for a similar one-dimensional analysis) leads to a set of multilayer matrix equations from which the EM fields in the diffraction grating system may be found. Figure 5.29 shows plots of the diffraction efficiency of the R_{00}, R_{10}, R_{01}, and R_{11} orders of the mirror-grating system when the grating thickness is varied from $\tilde{L} = 0\lambda$ to 2.5λ; when $\tilde{\Lambda}_x = \lambda$ and $\tilde{\Lambda}_z = 1.5\tilde{\Lambda}_x$; when $\alpha = 30°$, $\beta = 270°$, $E^I_\beta = -E_0$, and $E^I_\alpha = 0$; when $M_T = \underline{M}_T = 2$; when the values of the parameters n_o^2, n_e^2 were taken to be $n_o^2 = 2$, $n_e^2 = 3$; and when 10 layers ($N_\ell = 10$) were used to carry out the two-dimensional multilayer analysis. As can be seen from Figure 5.29 for the grating under study, power is again diffracted out of the R_{00} pump order and into higher orders. Conservation of power was observed to a high degree of accuracy.

PROBLEMS

5.1 In Figure 5.2, Bragg diffraction from an index-matched grating was studied. This problem concerns studying the diffraction efficiency that results when the incident angle deviates from the Bragg angle.

(a) Using either the RCWA full-field formulation of Section 5.2.1 or the RCWA wave equation method (or second-order differential equation method) of Section 5.2.2 reproduce numerically the order $i = -1, 0, 1$ diffraction efficiencies of Figure 5.2 for the geometry and numerical values used in this figure. The Bragg angle corresponding to this figure was $\theta_i = \theta_{iB} = 10°$.

(b) Calculate using either of the RCWA formulations of Section 5.2, the order $i = 0, 1$ diffraction efficiencies for an incidence of angle $\theta_i = 5°$ (non-Bragg angle). Make comparison plots of the $i = 0, 1$ diffraction efficiencies for $\theta_i = 5°$ and $\theta_i = \theta_{iB} = 10°$.

(c) Comment on the effect that deviation from the Bragg angle has on diffraction from the grating of Figure 5.2.
(d) In your solution of (a), be sure to check numerically that the conservation of power holds to a high degree of accuracy.
(e) Check power conservation for (b).

5.2 H-mode, Bragg diffraction from a surface relief, square wave (50% duty cycle) grating was studied in Figure 5.10 (first presented in [19, Fig. 5]). This problem concerns studying the diffraction efficiency that results when the incident angle deviates from the Bragg angle. The Bragg angle corresponding to this figure was $\theta_i = \theta_{iB} = 30°$.
(a) Using the RCWA full-field formulation of Section 5.2.1, calculate numerically for the geometry and numerical values of Figure 5.10 (1) the $i = -1, 0, 1$ order, transmitted diffraction efficiencies of this figure and (2) the $i = 0$ reflected diffraction efficiency. These calculations are to be carried out to ensure that your diffraction efficiency algorithm is working correctly. The $i = 0$ reflected diffraction efficiency is presented in Ref. [19, Fig. 5].
(b) Calculate using the RCWA full-field formulation of Section 5.2.1, the order $i = 0, 1$ transmitted diffraction efficiencies for an incidence of angle $\theta_i = 20°$ (non-Bragg angle). Make comparison plots of the $i = 0, 1$ transmitted diffraction efficiencies for $\theta_i = 20°$ and $\theta_i = \theta_{iB} = 30°$.
(c) Comment on the effect that deviation from the Bragg angle has on reflected and transmitted diffraction from the grating.
(d) In your solution of (a), (b), and (c), be sure to check numerically that the conservation of power holds to a high degree of accuracy.

5.3 H-mode, Bragg diffraction from a surface relief, square wave grating whose groove width occupies two-thirds of the grating period is to be analyzed. Figure 5.10 illustrates a surface relief, square wave grating whose groove width occupies half of the grating period. This problem concerns studying the effect of changing groove width on diffraction efficiency when the square wave grating is illuminated at the Bragg angle. The Bragg angle for this problem is $\theta_i = \theta_{iB} = 30°$.
(a) Using the RCWA full-field formulation of Section 5.2.1, calculate numerically for the geometry and numerical values of Figure 5.10 (1) the $i = -1, 0, 1$ order, transmitted diffraction efficiencies of this figure and (2) the $i = 0$ reflected diffraction efficiency. This calculation is to be carried out to ensure that your diffraction efficiency algorithm is working correctly. The $i = 0$ reflected diffraction efficiency is presented in Ref. [19, Fig. 5].
(b) Calculate using the RCWA full-field formulation of Section 5.3.1, the order $i = 0,1$ transmitted diffraction efficiencies for an incidence of angle $\theta_i = 30°$ when the groove width occupies 60% of the grating period.
(c) Make comparison plots of the transmitted diffraction efficiencies as calculated in (a) and (b).
(d) Comment on the effect that groove width has on diffraction from the gratings described in (a) and (b).
(e) In your solution of (a) and (b), be sure to check numerically that the conservation of power holds to a high degree of accuracy.

5.4 This problem concerns studying H-mode, Bragg diffraction from a sinusoidal surface relief grating whose grating period is equal to the free-space wavelength. The grating is assumed to be nonmagnetic ($\mu_G = 1$), have a relative dielectric permittivity of $\varepsilon_G = 2.5$, and have a vacuum on the incident side. The Bragg angle for this problem is $\theta_i = \theta_{iB} = 30°$. The geometry of the problem may be found in the Figure 5.20 inset and in Ref. [19, Fig. 4].
(a) Using a multilayer RCWA full-field formulation, including $i = -2, -1, 0, 1, 2$ Fourier harmonics, calculate numerically the $i = -1, 0, 1$ order, transmitted diffraction efficiencies for this problem. Make $i = -1, 0, 1$ order, transmitted diffraction efficiency plots versus grating thickness (peak-to-peak distance of the sinusoid making up the surface relief grating). Plot also the $i = 0$ reflected diffraction efficiency.
(b) Compare your solution plots to those found in Ref. [19, Fig. 4].

(c) Comment on the effect that the grating thickness has on diffraction from the gratings described in (a) and (b).
(d) In your solution of (a) and (b), be sure to check numerically that conservation of power holds to a high degree of accuracy.

5.5 This problem concerns studying H-mode, Bragg diffraction from a symmetric blaze surface relief grating whose grating period is equal to the free-space wavelength. The grating is assumed to be nonmagnetic ($\mu_G = 1$), have a relative dielectric permittivity of $\varepsilon_G = 2.5$, and to have a vacuum on the incident side. The Bragg angle for this problem is $\theta_i = \theta_{iB} = 30°$. The geometry of the problem may be seen in Figure 5.18 and Ref. [19, Fig. 6].
(a) Using a multilayer RCWA full-field formulation, including $i = -2, -1, 0, 1, 2$ Fourier harmonics, calculate numerically the $i = -1, 0, 1$ order, transmitted diffraction efficiencies for this problem. Make $i = -1, 0, 1$ order, transmitted diffraction efficiency plots versus grating thickness (peak-to-peak distance of the blaze grating). Plot also the $i = 0$ reflected diffraction efficiency.
(b) Compare your solution plots to those found in Ref. [19, Fig. 6].
(c) Comment on the effect that the grating thickness has on diffraction from the gratings described in (a) and (b).
(d) In your solution of (a) and (b), be sure to check numerically that conservation of power holds to a high degree of accuracy.

5.6 This problem concerns studying E-mode, Bragg diffraction from a sinusoidal surface relief grating whose relative dielectric permittivity of ε_G assumes different values and whose grating period is equal to the free-space wavelength. The grating is assumed to be nonmagnetic ($\mu_G = 1$) and have a grating period equal to the free-space wavelength. The Bragg angle for this problem is $\theta_i = \theta_{iB} = 30°$. The geometry of the problem is shown in the inset of Figure 5.21.
(a) Using a multilayer RCWA full-field formulation, including $i = -4, -3, -2, -1, 0, 1, 2, 3, 4$ Fourier harmonics, calculate numerically the $i = -1, 0, 1$ order, transmitted diffraction efficiencies for this problem when $\varepsilon_G = 4$. Diffraction efficiency results for this case are displayed in Figure 5.21 and have been calculated in Ref. [53, Fig. 6]. Make $i = -1, 0, 1$ order transmitted diffraction efficiency plots versus grating thickness (peak-to-peak distance of the sinusoid making up the surface relief grating) to ensure that your diffraction grating algorithm is working correctly. Plot the $i = 0$ reflected diffraction efficiency also.
(b) Repeat (a), but let the value of ε_G assume the values $\varepsilon_G = 2, 4, 8, 16$ and plot the $i = 0, 1$ reflected and transmitted diffraction efficiencies for these values.
(c) Observe and comment on what effect changing the dielectric value ε_G of the grating has on the reflected and transmitted diffraction efficiencies of the different ε_G cases when compared to one another.
(d) In your solution of (a) and (b), be sure to check numerically that conservation of power holds to a high degree of accuracy.

5.7 In Section 5.2.2, the wave equation method (or second-order, differential equation method) was used to study H-mode diffraction from a cosine diffraction whose grating vector was $\vec{K} = (2\pi/\Lambda)\hat{a}_x$, where Λ is the period of the grating (see Figure 5.1 and Figure 5.2 inset). M.G. Moharam and T.K. Gaylord [32] studied H-mode and E-mode diffraction from a slanted, cosine grating whose grating vector in the geometry of Figure 5.1 had the form $\vec{K} = (2\pi/\Lambda)\hat{a}_G$, where $\hat{a}_G$ is a unit vector parallel to the xy plane of Figure 5.1. Implement numerically the H-mode and E-mode diffraction grating algorithm of Ref. [32] and compare your numerical results to their results. Solve numerical examples of your own choosing.

5.8 Investigate the numerical stability of the RCWA algorithm for the study of diffraction from gratings of different layer thickness, layer profile, grating period, frequency, and material composition using the algorithm developed by M.G. Moharam, E.B. Grann, D.A. Pommet, and T.K. Gaylord [22,23] who have studied this problem extensively.

REFERENCES

1. R. Petit, ed., *Electromagnetic Theory of Gratings*, Springer-Verlag, Berlin, Germany, 1980.
2. T.K. Gaylord and M.G. Moharam, Analysis and applications of optical diffraction by gratings, *Proc. IEEE* 73, 894–937, 1985.
3. L. Solymar and D.J. Cooke, *Volume Holography and Volume Gratings*, Academic Press, London, U.K., 1981.
4. D. Maystre, ed., *Selected Papers on Diffraction Gratings,* SPIE Milestone Series, Vol. MS 83, 1993.
5. H. Kogelnik, Coupled wave theory for thick hologram gratings, *Bell Syst. Tech. J.* 48, 2909–2947, 1969.
6. J.A. Kong, Second-order coupled-mode equations for spatially periodic media, *J. Opt. Soc. Am.* 67, 825–829, 1977.
7. J. Prost and P.S. Pershan, Flexoelectricity in nematic and smectic-A liquid crystals, *J. Appl. Phys.* 47(6), 2298–2312, 1978.
8. S.K. Case, Coupled-wave theory for multiply exposed thick holographic gratings, *J. Opt. Soc. Am.* 65, 724–729, 1975.
9. P. Yeh, *Introduction to Photorefractive, Nonlinear Optics*, John Wiley & Sons, Inc., New York, 1993.
10. C.B. Burckhardt, Diffraction of a plane wave at a sinusoidally stratified dielectric grating, *J. Opt. Soc. Am.* 56, 1502–1509, 1966.
11. F.G. Kaspar, Diffraction by thick, periodically stratified gratings with complex dielectric constant, *J. Opt. Soc. Am.* 63, 37–45, 1973.
12. K. Knop, Rigorous diffraction theory for transmission phase gratings with deep rectangular grooves, *J. Opt. Soc. Am.* 68, 1206–1210, 1978.
13. R.S. Chu and J.A. Kong, Modal theory of spatially periodic media, *IEEE Trans. Microwave Theory Tech.* MTT-25, 18–24, 1977.
14. D. Marcuse, Exact theory of TE-wave scattering from blazed dielectric gratings, *Bell Syst. Tech. J.* 55, 1295–1317, 1976.
15. K.C. Chang, V. Shah, and T. Tamir, Scattering and guiding of waves by dielectric gratings with arbitrary profiles, *J. Opt. Soc. Am.* 70, 804–812, 1980.
16. M.G. Moharam and T.K. Gaylord, Rigorous coupled-wave analysis of planar grating diffraction, *J. Opt. Soc. Am.* 71, 811–818, 1981.
17. K. Rokushima and J. Yamakita, Analysis of anisotropic dielectric gratings, *J. Opt. Soc. Am.* 73(7), 901–908, 1983.
18. M.G. Moharam and T.K. Gaylord, Three-dimensional vector coupled-wave analysis of planar-grating diffraction, *J. Opt. Soc. Am.* 73, 1105–112, 1983.
19. M.G. Moharam and T.K. Gaylord, Diffraction analysis of dielectric surface-relief gratings, *J. Opt. Soc. Am.* 72, 1385–1392, 1987.
20. K. Rokushima, J. Yamakita, S. Mori, and K. Tominaga, Unified approach to wave diffraction by space-time periodic anisotropic media, *IEEE Trans. Microwave Theory Tech.* 35, 937–945, 1987.
21. E.N. Glytsis and T.K. Gaylord, Rigorous three-dimensional coupled-wave diffraction analysis of single cascaded anisotropic gratings, *J. Opt. Soc. Am. B* 4, 2061–2080, 1987.
22. M.G. Moharam, E.B. Grann, D.A. Pommet, and T.K. Gaylord, Formulation for stable and efficient implementation of rigorous coupled-wave analysis of binary gratings, *J. Opt. Soc. Am. A* 12(5), 1068–1076, 1995.
23. M.G. Moharam, D.A. Pommet, E.B. Grann, and T.K. Gaylord, Stable implementation of the rigorous coupled-wave analysis for surface-relief gratings: Enhanced transmittance approach, *J. Opt. Soc. Am. A* 12(5), 1077–1086, 1995.
24. M.G. Moharam and T.K. Gaylord, Coupled-wave analysis of reflection gratings, *Appl. Opt.* 20(2), 240–244, 1981.
25. Z. Zylberberg and E. Marom, Rigorous coupled-wave analysis of pure reflection gratings, *J. Opt. Soc. Am.* 73(3), 392–398, 1983.
26. M.G. Moharam and T.K. Gaylord, Comments on analyses of reflection gratings, *JOSA Lett.* 73, 399–401, 1983.
27. D. McCartney, The analysis of volume reflection gratings using optical thin-film techniques, *Opt. Quantum Electron.* 21, 93–107, 1989.
28. M.G. Moharam and T.K. Gaylord, Chain matrix analysis of arbitrary-thickness dielectric reflection gratings, *J. Opt. Soc. Am.* 72, 187–190, 1982.
29. P. Lalanne and G.M. Morris, Highly improved convergence of the coupled-wave method for TM polarization, *J. Opt. Soc. Am. A* 13, 779–784, 1996.

30. N. Chateau and J.-P. Hugonin, Algorithm for the rigorous coupled-wave analysis of grating diffraction, *J. Opt Soc. Am. A* 11, 1321–1331, 1994.
31. E.N. Glytsis and T.K. Gaylord, Three-dimensional (vector) rigorous coupled-wave analysis of anisotropic grating diffraction, *J. Opt. Soc. Am. A* 7(8), 1399–1420, 1990.
32. M.G. Moharam and T.K. Gaylord, Rigorous coupled-wave analysis of grating diffraction-E-mode polarization and losses, *J. Opt. Soc. Am.* 73(4), 451–455, 1983.
33. E.N. Glytsis and T.K. Gaylord, Rigorous 3-D coupled wave analysis of multiple superposed gratings in anisotropic media, *Appl. Opt.* 28(19), 2401–2421, 1989.
34. R. Magnusson and T.K. Gaylord, Equivalence of multiwave coupled wave theory and modal theory for periodic-media diffraction, *JOSA Lett.* 68, 1777–1779, 1978.
35. R. Magnusson and T.K. Gaylord, Analysis of multiwave diffraction by thick gratings, *J. Opt. Soc. Am.* 67, 1165–1170, 1977.
36. E.N. Glytsis and T.K. Gaylord, Antireflection surface structure: Dielectric layer(s) over a high spatial-frequency surface-relief grating on a lossy substrate, *Appl. Opt.* 27(20), 4288–4303, 1988.
37. T.K. Gaylord, M.G. Moharam, and W.E. Baird, Zero-reflectivity highspatial-frequency rectangular-groove dielectric surface relief gratings, *Appl. Opt.* 25(24), 4562–4567, 1986.
38. N.F. Hartman and T.K. Gaylord, Antireflection gold surface-relief gratings: Experimental characteristics, *Appl. Opt.* 27(17), 3738–3743, 1988.
39. M.G. Moharam and T.K. Gaylord, Rigorous coupled-wave analysis of metallic surface-relief gratings, *J. Opt. Soc. Am. A* 3, 1780–1787, 1986.
40. T.K. Gaylord, E.N. Glytsis, and M.G. Moharam, Zero-reflectivity homogeneous layers and high spatial-frequency surface-relief gratings on lossy materials, *Appl. Opt.* 26(15), 3123–3134, 1987.
41. L. Li and C.W. Haggans, Convergence of the coupled-wave method for metallic lamellar diffraction gratings, *J. Opt Soc. Am. A* 10, 1184–1189, 1993.
42. W.E. Baird, M.G. Moharam, and T.K. Gaylord, Diffraction characteristics of planar absorption gratings, *Appl. Phys. B* 32, 15–20, 1983.
43. M.G. Moharam and T.K. Gaylord, Rigorous coupled-wave analysis of metallic surface-relief gratings, *J. Opt Soc. Am. A* 3, 1780–1796, 1986.
44. M.G. Moharam, Coupled-wave analysis of two-dimensional dielectric gratings, *Holographic Optics: Design and Applications, SPIE*, Vol. 883, pp. 8–11, 1988.
45. S.T. Han, Y.-L. Tsao, R.M. Walser, and M.F. Becker, Electromagnetic scattering of two-dimensional surface-relief dielectric gratings, *Appl. Opt.* 31, 2343–2352, 1992.
46. Y.L. Kok, Electromagnetic scattering from a doubly-periodic grating corrugated by cubical cavities, *Proceedings: The Twenty-First Southeastern Symposium on System Theory*, Tallahassee, FL, March 26–29, pp. 493–499, 1989.
47. C. Wu, T. Makino, J. Glinski, R. Maciejko, and S.I. Najafi, Self-consistent coupled-wave theory for circular gratings on planar dielectric waveguides, *J. Lightwave Technol.* 9(10), 1264–1276, 1991.
48. A. Vasara, E. Noponen, J. Turunen, J.M. Miller, and M.R. Taghizadeh, Rigorous diffraction analysis of Dammann gratings, *Opt. Commun.* 81, 337–342, 1991.
49. G. Granet and G. Guizal, Efficient implementation of the coupled-wave method for metallic lamellar gratings in TM polarization, *J. Opt Soc. Am. A* 13, 1019–1023, 1996.
50. M.G. Moharam, T.K. Gaylord, and R. Magnusson, Criteria for Bragg regime diffraction by phase gratings, *Opt. Commun.* 32, 14–18, 1980.
51. L. Li, Use of Fourier series in the analysis of discontinuous periodic structures, *J. Opt. Soc. Am. A* 13, 1870–1876, 1996.
52. J. Yamakita and K. Rokushima, Modal expansion method for dielectric gratings with rectangular grooves, *Proceedings of SPIE*, Vol. 503, pp. 239–243, 1984.
53. J. Yamakita, K. Rokushima, and S. Mori, Numerical analysis of multistep dielectric gratings, *Application and Theory of Periodic Structures, Diffraction Gratings, and Moire Phenomena III, SPIE*, Vol. 815, pp. 153–157, 1987.
54. J. Yamakita and K. Rokushima, Scattering of plane wave waves from dielectric gratings with deep grooves, *Trans. Inst. Electron. Commun. Eng., Jpn* J66-B, 375, 1983.
55. C. Botten, M.S. Craig, R.C. McPhedran, J.L. Adams, and J.R. Andrewarthar, The dielectric lamellar diffraction grating, *Opt. Acta* 28(3), 413–428, 1981.
56. C. Botten, M.S. Craig, R.C. McPhedran, J.L. Adams, and J.R. Andrewarthar, The finitely conducting lamellar diffraction grating, *Opt. Acta* 28(8), 1087–1102, 1981.
57. C. Botten, M.S. Craig, and R.C. McPhedran, Highly conducting lamellar diffraction grating, *Opt. Acta* 28(8), 1103–1106, 1981.

58. L. Li, Multilayer model method for diffraction gratings of arbitrary profile, depth, and permittivity, *J. Opt Soc. Am. A* 10, 2581–2591, 1993.
59. D.M. Pai and K.A. Awada, Analysis of dielectric gratings of arbitrary profiles and thickness, *J. Opt Soc. Am. A* 8, 755–762, 1991.
60. M. Nevievre, Diffraction of light by gratings studied with the differential method, *SPIE Periodic Structures, Grating, Moire Patterns and Diffraction Phenomena*, Vol. 240, pp. 90–96, SPIE Society, 1980.
61. P. Vincent, New improvement of the differential formalism for high-modulated gratings, *SPIE Periodic Structures, Grating, Moire Patterns and Diffraction Phenomena*, Vol. 240, pp. 147–154, SPIE Society, 1980.
62. R. Petit and G. Tayeb, Theoretical and numerical study of gratings consisting of periodic arrays of thin and lossy strips, *J. Opt Soc. Am. A* 7, 1686–1692, 1990.
63. J. Chandezon, M.T. Dupuis, G. Cornet, and D. Maystre, Multicoated gratings: A differential formalism applicable in the entire optical region, *J. Opt. Soc. Am.* 72, 839–846, 1982.
64. D. Maystre, A new general integral theory for dielectric coated gratings, *J. Opt. Soc. of Am.* 64(4), 490–495, 1978.
65. M.K. Moaveni, Plane wave diffraction by dielectric gratings, finite-difference formulation, *IEEE Trans. Antennas Propag.* 37(8), 1026–1031, 1989.
66. H.A. Kalhor and M.K. Moaveni, Analysis of diffraction gratings finite-difference coupling technique, *J. Opt. Soc. Am.* 63, 1584–1588, 1973.
67. M.K. Moaveni, Application of finite differences to the analysis of diffraction gratings embedded in an inhomogeneous and lossy dielectric, *Int. J. Electron.* 61(4), 465–476, 1986.
68. Y. Nakata and M. Koshiba, Boundary-element analysis of plane-wave diffraction from groove-type dielectric and metallic gratings, *J. Opt. Soc. Am. A* 7, 1494–1502, 1990.
69. D.E. Tremain and K.K. Mei, Application of the unimoment method to scattering from periodic dielectric structures, *J. Opt. Soc. Am.* 68, 775–783, 1978.
70. A.K. Cousins and S.C. Gottschalk, Application of the impedance formalism to diffraction gratings with multiple coating layers, *Appl. Opt.* 29, 4268–4271, 1990.
71. R.A. Depine, C.E. Gerber, and V. Brundy, Lossy gratings with a finite number of grooves: A canonical model, *J. Opt. Soc. Am. A* 9, 573–577, 1992.
72. P. St. J. Russell, Power conservation and field structures in uniform dielectric gratings, *J. Opt. Soc. Am. A* 1(3), 293–298, 1984.
73. R. Petit and G. Tayeb, On the use of the energy balance criteria as a check of validity of computations in grating theory, *SPIE Application and Theory of Periodic Structures, Grating, Moire Patterns and Diffraction Phenomena III*, Vol. 815, pp. 2–10, SPIE Society, 1987.
74. R.F. Harrington, *Time Harmonic Electromagnetic Fields*, Section 1–10, Complex Power, McGraw-Hill Book Company, New York, 1961.
75. E.N. Glytsis and T.K. Gaylord, Anisotropic guided-wave diffraction by interdigitated, electrode-induced phase gratings, *Appl. Opt.* 27, 5035–5050, 1988.
76. P.M. Van Den Berg, W.J. Ghijsen, and A. Venema, The electric field problem of an interdigital transducer in a multilayer structure, *IEEE Trans. Microwave Theory Tech.* MTT-33(2), 121–129, 1985.
77. D. Quak and G. Den Boon, Electric input admittance of an interdigital transducer in a layered, anisotropic, semiconducting structure, *IEEE Trans. Sonic Ultrason.* SU-25(1), 45–50, 1978.
78. E.N. Glytsis, T.K. Gaylord, and M.G. Moharam, Electric field, permittivity, and strain distributions induced by interdigitated electrodes on electro-optic waveguides, *IEEE J. Lightwave Technol.* LT-5, 668–683, 1987.
79. K. Rokusshima and J. Yamakita, Analysis of diffraction in periodic liquid crystals: The optics of the chiral smetic C phase, *J. Opt. Soc. Am. A* 4(1), 27–33, 1987.
80. D.Y.K. Ko and J.R. Sambles, Scattering matrix method for propagation of radiation in stratified media: Attenuated reflection studies of liquid crystals, *J. Opt. Soc. Am. A* 55, 1863–1866, 1988.
81. C. Schwartz and L.F. DeSandre, New calculational approach for multilayer stacks, *Appl. Opt.* 26, 3140–3144, 1987.
82. S.L. Chuang and J.A. Kong, Wave scattering from periodic dielectric surface for a general angle of incidence, *Radio Sci.* 17, 545–557, 1982.
83. T.K. Gaylord and M.G. Moharam, Thin and thick gratings: Terminology clarification, Symbols, units, nomenclature, *Appl. Opt.* 20(19), 3271–3273, 1981.
84. T. Jaaskelainen and M. Kuittinen, Inverse grating diffraction problems, *SPIE International Colloquium on Diffractive Optical Elements*, Vol. 1574, pp. 272–281, 1991.

85. E. Gluch, H. Haidner, P. Kipfer, J.T. Sheridan, and N. Streibl, Form birefringence of surface-relief gratings and its angular dependence, *Opt. Commun.* 89, 173–177, 1992.
86. J.M. Elson, L.F. DeSandre, and J.L. Stanford, Analysis of anomalous resonance effects in multilayer-overcoated, low-efficiency gratings, *J. Opt. Soc. Am. A* 5, 74–88, 1988.
87. T. Tamir and H.C. Wang, Scattering of electromagnetic waves by a sinusoidally stratified half space: I. Formal solution and analysis approximations, *Can. J. Phys.* 44, 2073–2094, 1966.
88. T. Tamir, Scattering of electromagnetic waves by a sinusoidally stratified half space: II. Diffraction aspects at the Rayleigh and Bragg wavelengths, *Can. J. Phys.* 44, 2461–2494, 1966.
89. S.T. Peng, T. Tamir, and H.L. Bertoni, Theory of periodic dielectric waveguides, *IEEE Trans. Microwave Theory Tech.* MTT-23, 123–133, 1975.
90. T. Tamir, H.C. Wang, and A.A. Oliner, Wave propagation in sinusoidally stratified dielectric media, *IEEE Trans. Microwave Theory Tech.* MTT-12, 323–335, 1964.
91. R.V. Johnson and R. Tanguay, Optical beam propagation method for birefringent phase grating diffraction, *Opt. Eng.* 25, 235–249, 1986.
92. H. Haus, *Waves and Fields in Optoelectronics*, Section 11.1, Prentice-Hall, New York, 1984.
93. S. Samson, A. Korpel, and H.S. Snyder, Conversion of evanescent waves into propagating waves by vibrating knife edge, *Int. J. Imag. Sys. Tech.* 7, 48–83, 1996.
94. P.N. Butcher and D. Cotter, *The Elements of Nonlinear Optics*, Cambridge, Cambridge, U.K., 1990, p. 127.
95. T.L. Zinenko, A.I. Nosich, and Y. Okuno, Plane wave scattering and absorption by resistive-strip and dielectric strip periodic gratings, *IEEE Trans. Antennas Propag.* 46(10), 1498–1505, 1998.
96. E. Popov, L. Tsonev, and D. Maystre, Gratings-general properties of the Littrow mounting and energy flow distribution, *J. Mod. Opt.* 37(3), 367–377, 1990.
97. E. Popov, L. Tsonev, and D. Maystre, Losses of plasmon surface waves on metallic grating, *J. Mod. Opt.* 37(3), 379–387, 1990.
98. E. Popov and L. Tsonev, Total absorption of light by metallic gratings and energy flow distribution, *Surf. Sci.* 230, 290–294, 1990.
99. E. Popov, Light diffraction by relief gratings: A macroscopic and microscopic view, *Progress in Optics*, 31, 141–187, 1993.
100. B.W. Shore, L. Li, and M.D. Feit, Poynting vectors and electric field distributions in simple dielectric gratings, *J. Mod. Opt.* 44(1), 69–81, 1997.
101. J.M. Jarem and P.P. Banerjee, Application of the complex Poynting theorem to diffraction gratings, *J. Opt. Soc. Am. A* 16(5), 819–831, 1999.

6 Application of RCWA to Analysis of Induced Photorefractive Gratings

6.1 INTRODUCTION TO PHOTOREFRACTIVE MATERIALS

Since its first discovery by Ashkin et al. [1] in the mid-1960s, a tremendous amount of research has been carried out to study the photorefractive (PR) effect and apply it to real-time image processing [2], beam amplification [3], self-pumped phase conjugation [4], four-wave mixing [5], and optical computing [6], to name a few applications. When two coherent plane waves of light intersect in a PR material (see Figure 6.1), they form an intensity interference pattern comprising bright and dark regions. Assuming that the PR material is predominantly n-doped, the electrons migrate from bright to dark regions, thus creating an approximately sinusoidal charge distribution. This diffusion-controlled PR effect in turn creates an electrostatic field that is ideally 90° phase shifted from the intensity pattern and modulates the refractive index of the crystal via the electro-optic effect. The incident plane waves are, in turn, scattered by the grating in a way that one wave may have constructive recombination, while the other may encounter a destructive recombination. This effect leads to energy coupling between the beams through what is commonly referred to as the two-beam-coupling effect [3].

A steady-state nonlinear coupling theory of the two-beam coupling phenomenon, using participating plane waves has been derived and the results reconciled with numerical simulations of the coupling between two, in general, focused, Gaussian beams as shown in Chapter 2 and more rigorously in Ref. [7]. The results indicate that the two-beam coupling parameter Γ [8] strongly depends on the initial intensity ratio of the plane waves (or power ratio of the input beams), an effect that has been experimentally observed before [9,10], but for which only an empirical theory existed [11]. The empirical theory, however, was an improvement over linearized time-dependent theories that predict that Γ is independent of the initial intensity ratio [8]. The difficulty in providing an analytical nonlinear time-dependent theory stems from the fact that it is impossible to exactly decouple the Kukhtarev equations [9] except in the steady state, and numerical simulations of these coupled equations are also rather formidable.

In Chapter 2, we employed the beam propagation method (BPM) to analyze Gaussian beam propagation through PR materials, as well as two-beam coupling in the presence of induced transmission and reflection gratings. Furthermore, the results of z- and P-scan of PR materials was also discussed. The explanations and underlying theories were derived on the basis of a simplified induced refractive index, derived from the more exact Kukhtarev equations, since they are computationally challenging.

Over and above the computational difficulty with the Kukhtarev equations is the fact that in a typical experimental setup, the PR crystal is finite and often has a large linear refractive index (typically greater than 2), prompting reflections from boundaries. This introduces the added complication of taking into account forward and backward traveling waves during the two-beam coupling process, and, in fact, extend the two-beam problem to a degenerate four-wave mixing problem in a PR material. We have shown in Chapter 2 how the BPM can be extended to include

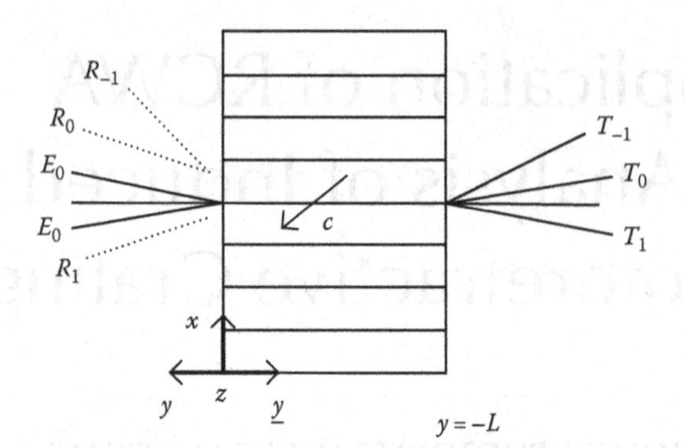

FIGURE 6.1 Geometry of the PR grating. (Reprinted from *Opt. Commun.*, 123, Jarem, J.M. and Banerjee, P.P., A nonlinear, transient analysis of two and multi-wave mixing in a photorefractive material using rigorous coupled-mode diffraction theory, 825–842, Copyright 1996, with permission from Elsevier.)

contradirectional beam propagation and coupling in PR materials. Recently, a "unified" method for solution of four-wave mixing problems in PR crystals in both transmission and reflection geometries has been proposed, and applied to problems of double phase conjugation and two-wave mixing with crossed polarization in cubic crystals [12]. However, the methodology implies linearized interaction equations between the PR grating and the intensity grating, and, furthermore, has been developed for the steady state.

A rigorous analysis of wave interactions, particularly between ordinary and extraordinary polarizations, coupled through induced material anisotropy, has been developed by Wilson et al. [13]. The material analyzed is $LiNbO_3$: Fe where the PR effect is primarily photovoltaic. Arbitrary-direction two-wave coupling is studied, and extended to multi-wave coupling. The analysis is rigorous from an electromagnetics standpoint since it employs the rigorous coupled wave approach [14–16]. Once again, the analysis is restricted to the steady state.

In our approach, we assume the PR effect to be diffusion-dominated, and first write down the expression for the electrostatic field (and hence the induced refractive index profile) as a nonlinear function of the optical intensity in the PR material in the steady state. The time dependent Kukhtarev equations are then decoupled, first under a set of approximations, to yield a differential equation in time for the induced refractive index profile. Rigorous coupled wave analysis (RCWA) is then used to analyze the diffraction of each of the incident plane waves from the slowly induced grating, including the effect of multiple reflections from the interfaces as well as from dielectric mismatches along the direction of propagation created through the changing intensity pattern due to energy transfer. We have performed our simulations first for the case of two coherent incident plane waves to check our results with those in Ref. [7] (derived for the steady state and unidirectional propagation), and then extended our approach to include the case(s) when the incident waves are mutually partially incoherent. We have also investigated the difference in the time responses of two-wave coupling taking the time constant to be intensity-dependent and intensity-independent. Furthermore, we have repeated our calculations with different incident polarizations for different materials with varying photorefractive gains using the exact Kukhtarev model.

The relevance of our generalized approach using a nonlinear time-dependent theory and RCWA may be summarized in the context of the extensive amount of work on the nature of wave-mixing in PR materials. It has been pointed out that spatiotemporal instabilities, possibly due to internal reflections, may cause self-pulsations, irregular fluctuations in time, and optical chaos in self-pumped $BaTiO_3$ phase conjugate mirrors and during naturally-pumped phase conjugation [17–20]. Light waves at the interface of linear and PR media has also been analyzed [21], and it has been experimentally shown that optical bistability may be realized through optical feedback of the signal beam in a two-beam coupling geometry [22]. The effect of a position-dependent time constant during two-wave coupling in both nominally co-propagating and counter-propagating geometries has also been analyzed [23]. The rationale for this calculation is that the PR time constant is, in general, intensity-dependent. Furthermore, PR two-beam coupling with beams of partial spatiotemporal coherence has been theoretically and experimentally investigated [24,25]. The results indicate weaker coupling and beam-profile deformation, along with spatial coherence enhancement and deterioration for the amplified and de-amplified beams, respectively. Finally, recording kinetics in PR materials in a two-beam coupling geometry, resulting in multiple two-wave mixing and fanning, and "kinky" beam paths have been theoretically predicted [26–28]. The results show that at least one higher order may develop with a high diffraction efficiency.

As shown in this chapter, our algorithm can be effectively used to analyze diffraction from induced nonlinear, transient transmission and reflection gratings, and in materials with arbitrary complex permittivities (including absorption) at optical frequencies. The rigorous coupled wave algorithm is a straight forward, easy to understand, diffraction analysis method, which allows in a simple, complete, and elegant way the electromagnetic fields of the diffraction grating and the surrounding medium to be accurately determined. RCWA techniques are routinely used in a wide variety of diffraction grating problems arising in optics and electromagnetics. The electromagnetic solution includes in its analysis the optical of the anisotropy medium as well as all multiply reflected waves that occur from the boundary interfaces. The RCWA method is a generalization of the BPM, and reduces to the BPM when only forward traveling waves are allowed in the analysis. To the best of our knowledge, there exists no other optical analysis method that could even approximately analyze the mismatched boundary case that we present in this chapter (see Figure 6.1). The BPM cannot be applied in the present case because it does not account for multiply reflected waves from the boundary interfaces. We believe that it is useful to use a rigorous electromagnetic field theory even if the material model is simple, because the use of a rigorous field theory at least removes the uncertainty than an inaccurate electromagnetic solution is the source of discrepancy between the overall model and experiment. All discrepancies between theory and experiment can then be ascribed to the materials model.

While the RCWA approach has its advantages over the BPM, in that it can facilitate exact (and time dependent) solutions to the Kukhtarev equations and wave mixing resulting from the induced time-dependent refractive index change, it is almost impossible to analyze beam propagation using this technique. Therefore, to analyze propagation of focused Gaussian beams through PR materials, the BPM or its bidirectional extension is still probably the only tractable method.

6.2 DYNAMIC NONLINEAR MODEL FOR DIFFUSION-CONTROLLED PR MATERIALS

The time-dependent Kukhtarev equations [8] for a PR material for nominal propagation of the optical field along y and assuming transverse field components ($\vec{J} = J\hat{a}_x$, $\vec{E}_s = E_s\hat{a}_x$) and transverse coordinate x are

$$\frac{\partial(N_D^+ - n)}{\partial t} = -\frac{1}{e}\frac{\partial J}{\partial x} \tag{6.1}$$

$$\frac{\partial N_D^+}{\partial t} = (N_D - N_D^+)(sI + \beta) - \gamma_R N_D^+ n \tag{6.2}$$

$$J = eD_s\frac{\partial n}{\partial x} + e\mu n E_s \tag{6.3}$$

$$\varepsilon_s \frac{\partial E_s}{\partial x} = e(N_D^+ - N_A - n) \tag{6.4}$$

where
$n(x, y, t)$ is the free electron density
$N_D^+(x, y, t)$ is the ionized donor concentration
$J(x, y, t)$ is the current density
$E_s(x, y, t)$ is the electrostatic space-charge field
$I(x, y, t)$ is the intensity distribution
N_D is the donor concentration
e is the electronic charge
s is the ionization cross-section
β is the thermal excitation rate (proportional to dark current)
γ_R is the recombination rate
D_s is the diffusion constant
ε_s is the effective quasi-static permittivity $= \underline{K} \cdot \underline{\underline{\varepsilon}}_s \cdot \underline{K}/K^2$
μ is the effective static mobility $= \underline{K} \cdot \underline{\underline{\mu}} \cdot \underline{K}/K^2$
N_A is the acceptor concentration
$K = K\hat{a}_x = [K, 0, 0]$ is the grating vector $K = 2\pi/\Lambda$
Λ is the grating period

The effective values of ε_s and μ are given by Yeh [8, p. 88]. In the following text, we will determine how the dielectric permittivity modulation $\Delta\varepsilon(x, y, t)$ is related to the optical intensity I.

6.3 APPROXIMATE ANALYSIS

To start the analysis, we differentiate Equation 6.4 with respect to time, and after noting that $\partial N_A/\partial t = 0$, we find

$$\frac{\partial(N_D^+ - n)}{\partial t} = \frac{\varepsilon_s}{e}\frac{\partial^2 E}{\partial x \partial t} = -\frac{1}{e}\frac{\partial J}{\partial x} \tag{6.5}$$

If the last two equation statements of Equation 6.5 are integrated with respect to x (the constant of integration can be shown to be zero) and J of Equation 6.3 is substituted, we find

$$\frac{\varepsilon_s}{e}\frac{\partial E_s}{\partial t} = -\mu n E_s - D_s \frac{\partial n}{\partial x} \tag{6.6}$$

For a particular choice of c-axis of the PR crystal, the index of refraction modulation may be taken to be $\Delta n = -r n_o^3 E_s$ where n_o is the ordinary index of refraction of the bulk dielectric of the PR material and where r is the effective electro-optic coefficient. The dielectric permittivity modulation $\Delta\varepsilon$ is related to Δn by $\Delta n = \Delta\varepsilon/(2n_o)$. Thus

$$\Delta\varepsilon(x, y, t) = -E_s r n_o^4 \tag{6.7}$$

Substituting E_s from Equation 6.7, we find the following equation for $\Delta\varepsilon$:

$$\frac{\partial \Delta\varepsilon}{\partial t} + \frac{\mu e n}{\varepsilon_s}\Delta\varepsilon = \frac{D_s e}{\varepsilon_s} r n_o^4 \frac{\partial n}{\partial x} \tag{6.8}$$

This equation shows that the dielectric modulation satisfies a first order, relaxation type of differential equation, which is driven by the spatial gradient of the electron density of n. The equation is nonlinear since the electron density n depends on $\Delta\varepsilon$ through the optical diffraction, reflection, and refraction which occur in the PR media.

At this point, no approximations of Kukhtarev's equations have been made. In order to derive an expression that explicitly shows the dependence of n on I in Equation 6.8, in this chapter we will study the commonly occurring case when $N_A \ll N_D$ and $n \approx N_D^+ \approx N_A$. In this case, the N_D^+ term in the first term of the right-hand side (RHS) of Equation 6.2 may be dropped, and $\gamma_R n$ may be solved to give

$$\gamma_R n = \frac{(sI+\beta)N_D}{N_D^+} - \frac{1}{N_D^+}\frac{\partial N_D^+}{\partial t} \tag{6.9}$$

The second term on RHS of Equation 6.9 is the derivative of a logarithmic term $((1/N_D^+)(\partial N_D^+/\partial t) = (\partial/\partial t)\ln(N_D^+/N_A))$, and may be shown to small relative to the first term on the RHS of Equation 6.9. It has been ignored for now in this approximate analysis. If N_D^+ is approximated by N_A in the denominator of Equation 6.9, we may finally relate the electron density n to the optical intensity I through the linear relations

$$n = \frac{(sI+\beta)N_D}{\gamma_R N_A}, \quad \frac{\partial n}{\partial x} = \frac{sN_D}{\gamma_R N_A}\frac{\partial I}{\partial x} \tag{6.10}$$

Substituting Equation 6.10 in Equation 6.9 we find

$$\tau(I)\frac{\partial \Delta\varepsilon}{\partial t} = -\Delta\varepsilon + \frac{rn_o^4 D_s}{\mu(I+\beta/s)}\frac{\partial I}{\partial x} \tag{6.11}$$

where

$$\tau(I) = \frac{\gamma_R N_A \varepsilon_s}{\mu e N_D(sI+\beta)} \tag{6.12}$$

The dielectric perturbation $\Delta\varepsilon$ may be put in the normalized form

$$\tilde{\tau}(I)\frac{\partial \Delta\varepsilon}{\partial \tilde{t}} = -\Delta\varepsilon + \alpha\frac{\partial I/\partial \tilde{x}}{(I+C)} \tag{6.13}$$

where

$$\tilde{x} = k_0 x, \quad \tilde{t} = \frac{t}{\tau(I_0)}, \quad C = \frac{\beta}{s}, \quad \alpha = \frac{rn_o^4 D_s k_0}{\mu}, \quad \tilde{\tau}(I) = \frac{\tau(I)}{\tau(I_0)} \tag{6.14}$$

and I_0 is the peak intensity of the incident optical field. In Equation 6.14, $\tau(I_0)$ denotes a normalizing constant which makes $\tau(I)$ and t dimensionless. A typical range for values of α is from ~10^{-5} (e.g., in BSO) to ~10^{-2} ($BaTiO_3$). Also, the constant C is of the order of 10^4–10^6 (W/m^2) in most PR materials. We remark also that for small α(~10^{-5}), we have found that the intensity profile I is nearly constant in time; thus Equation 6.13 may be directly integrated w.r.t. $\tilde{t}$ to yield an explicit expression for $\Delta\varepsilon$:

$$\Delta\varepsilon = \alpha\frac{\partial I/\partial\tilde{x}}{C+I}[1-e^{-\tilde{\tau}(I)t}] \tag{6.15}$$

The dielectric tensor of a $BaTiO_3$ crystal is given by

$$\varepsilon_{xx} = n_{CO}^2\cos^2\theta_c + n_{CE}^2\sin^2\theta_c + \Delta\varepsilon(x,y,t)F_{xx}$$

$$\varepsilon_{xy} = (n_{CO}^2 - n_{CE}^2)\sin\theta_c\cos\theta_c + \Delta\varepsilon(x,y,t)F_{xy}$$

$$\varepsilon_{yx} = \varepsilon_{xy}$$

$$\varepsilon_{yy} = n_{CO}^2\sin^2\theta_c + n_{CE}^2\cos^2\theta_c + \Delta\varepsilon(x,y,t)F_{yy} \tag{6.16}$$

where

$$F_{xx} = -F_{OE}\left\{\frac{r_{13}n_{CO}^2}{r_{42}n_{CE}^2}\sin\theta_c\cos^2\theta_c + 2\sin\theta_c\cos^2\theta_c + \frac{r_{33}n_{CE}^2}{r_{42}n_{CO}^2}\sin^3\theta_c\right\}$$

$$F_{xy} = -F_{OE}\left\{\frac{r_{13}n_{CO}^2}{r_{42}n_{CE}^2}\sin^2\theta_c\cos\theta_c + \cos\theta_c[-\cos^2\theta_c + \sin^2\theta_c] - \frac{r_{33}n_{CE}^2}{r_{42}n_{CO}^2}\sin^2\theta_c\cos\theta_c\right\}$$

$$F_{yy} = -F_{OE}\left\{\frac{r_{13}n_{CO}^2}{r_{42}n_{CE}^2}\sin^3\theta_c - 2\sin\theta_c\cos^2\theta_c + \frac{r_{33}n_{CE}^2}{r_{42}n_{CO}^2}\cos^2\theta_c\sin\theta_c\right\}$$

$$F_{OE} = \frac{n_{CO}^2 n_{CE}^2}{n_O^2 n_E^2}$$

$$n_{CO}^2 = n_O^2 - j\varepsilon_O''$$

$$n_{CE}^2 = n_E^2 - j\varepsilon_E'' \tag{6.17}$$

The quantity $\Delta\varepsilon$ in Equation 6.13 is found by solving

$$\tilde{\tau}(I)\frac{\partial\Delta\varepsilon}{\partial\tilde{t}} = -\Delta\varepsilon + \alpha\frac{\partial I/\partial\tilde{x}}{C+I}, \quad \alpha = \frac{D_s}{\mu}n_O^2 n_E^2 k_0 r_{42} \tag{6.18}$$

$\tilde{\tau}(I)$ is given in Equation 6.14. The optical power intensity $I(x, y, t)$ in this problem is assumed proportional to $|E_x|^2 + |E_y|^2$. The terms n_{CO}^2 and n_{CE}^2 are the relative complex dielectric permittivities of the bulk crystal. The terms n_O^2 and n_E^2 are real parts of the relative bulk dielectric permittivity and

ε''_O and ε''_E are the lossy parts. The E_{xsc}, electric space-charge field, is linearly related to the dielectric perturbation or modulation function by the equation

$$\Delta\varepsilon(x,y,t) = n_O^2 n_E^2 r_{42} E_{xsc}(x,y,t) \tag{6.19}$$

The electro-optic constants r_{13}, r_{33}, and r_{42} are specified in [8, Table 1.2, pp. 26–29].

6.3.1 Numerical Algorithm

We will now describe an algorithm based on the RCWA of Chapter 5 that can be used to determine mode coupling and diffraction from a slowly varying PR medium. The temporal variation of the PR medium is specified by Equation 6.18. The algorithm proceeds temporally as follows. At $t = 0$ (first time step) a signal ($\theta = -\theta_i$) and pump wave ($\theta = \theta_i$) is incident on a dark, uniform PR slab of material. The pump and signal wave interfere, creating a periodic optical intensity pattern in the PR slab. The periodic optical intensity modifies the PR slab and produces a small periodic modulation $\Delta\varepsilon$. During this time step, no diffraction has occurred, as the dielectric modulation was zero at the beginning of the time step. At the second time step, the $\Delta\varepsilon$ generated by the optical interference of the first time step will begin to diffract light. The algorithm calculates the total optical intensity in the medium (1) by calculating the EM fields and diffraction from $\Delta\varepsilon$ in the PR slab from the pump wave by itself, (2) by calculating the EM fields and diffraction from the signal wave from $\Delta\varepsilon$ in the PR slab by itself, and (3) by adding these two EM fields together to find the overall EM field in the PR medium, and to find the overall transmitted and reflected EM fields in free space.

The algorithm proceeds by substituting the just described total optical intensity into Equation 6.18 to find a new $\Delta\varepsilon$ modulation, which can be used for the third time step. The algorithm repeats the above process for as many time steps as a solution is required for a solution.

The new value of $\Delta\varepsilon$ for any time step is calculated approximately using the finite difference formula

$$\frac{\partial\Delta\varepsilon}{\partial t} \cong \frac{\Delta\varepsilon(t+\Delta t)-\Delta\varepsilon(t)}{\Delta t} = F(t,I) \tag{6.20}$$

or

$$\Delta\varepsilon(t+\Delta t) = \Delta\varepsilon(t) + F(t,I)\Delta t \tag{6.21}$$

where $F(t, I)$ is the RHS of Equation 6.18, and I is the optical intensity of the light at time t.

The time steps are chosen to be sufficiently small so that only a very small change in $\Delta\varepsilon$ is observed at any given time step.

A numerical simulation of the nonlinear photorefractive dielectric equations of Section 6.2 was made in order to study the interaction of the TE and TM optical electric field with the photorefractive material. The simulation was performed for a photorefractive crystal which was located in free space, and was illuminated by two incident plane waves whose separation angle with the normal to the crystal was 5.71° and −5.71°, respectively, and whose complex amplitudes were, respectively, $E_0 e^{j0°}$ and $E_1 e^{j\psi_{rand}}$. These two plane waves caused an electric field interference pattern whose nulls were $\Lambda = 5\lambda$ apart.

6.3.2 TE Numerical Simulation Results

This section will describe the TE simulation results that arise from the theory of the previous section. For the TE case of this section, the photorefractive crystal was assumed to have an isotropic bulk dielectric of $\varepsilon_{2r} = \varepsilon'_{2r} - j\varepsilon''_{2r}$, was assumed to have a thickness of 1.5 cm, and was assumed to have a wavelength $\lambda = 0.5\,\mu$m. For the TE case, the ratio of the magnitudes of the incident plane

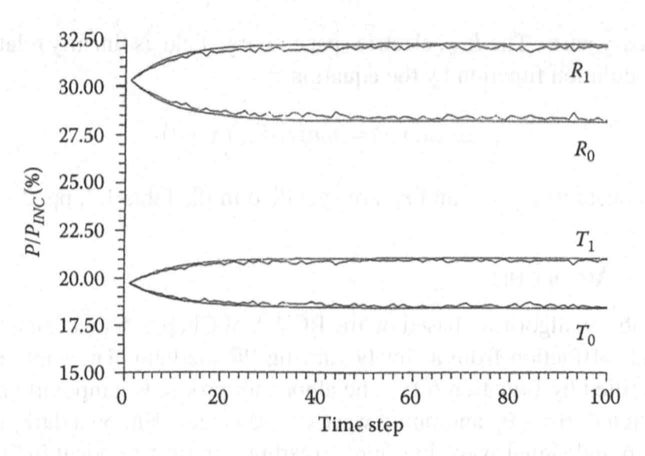

FIGURE 6.2 Plots of the normalized power in the directions T_1, T_0, and R_1, R_0 as a function of time that is transmitted and reflected from a slab of PR material when the incident waves are coherent ($\theta_{rand} = 0°$) and partially coherent ($-30° \leq \theta_{rand} \leq 30°$). The angle of incidence of the pump wave is 5.7°, and angle of incidence of the signal wave is −5.7°. These values are used for all figures. $\Lambda = 5\lambda$, $I_{Inc} = 9.949 \times 10^4$ (W/m²), $\lambda = 0.5\,\mu$m, $E_1 = pE_0\exp(-j\theta_{rand})$, $\alpha = 2 \times 10^{-5}$, $C = 10^4$ (W/m²), $p = 1$. (Reprinted from *Opt. Commun.*, 123, Jarem, J.M. and Banerjee, P.P., A nonlinear, transient analysis of two and multi-wave mixing in a photorefractive material using rigorous coupled-mode diffraction theory, 825–842, Copyright 1996, with permission from Elsevier.)

wave $p = |E_1|/|E_0|$ was taken to have a large range of values, which extended from 0.1 to 10. The parameters were varied in such a way as to keep the total power from both waves I_{inc} transmitted into the photorefractive material constant and equal to $9.949{\times}10^4$ W/m². Concerning the phases of the incident waves, in our simulation, we studied wave coupling when the phase difference ψ_{rand} between the interfering waves was equal to zero, which corresponds to the case of coherent illumination; and we studied the case where ψ_{rand} was uniformly distributed random phase (the limits of the random phase was taken to be ±15° and ±30°), which corresponds to a partially incoherent illumination. Numerically, the random phase was generated by a random number generator.

Figure 6.2 shows the transmitted and reflected normalized power in the directions T_1, T_0, R_1, R_0 as a function of time that result when equal amplitude plane waves ($p = 1.0$) are incident on an initially dark photorefractive material bulk dielectric $\varepsilon_{2r} = 8$. The smooth line shows the result in the case of coherent illumination ($\psi_{rand} = 0$), while the nonsmooth line depicts the response for partially incoherent illumination when the uniform probability density function, generating the random phase data, is bounded to ±30° ($-30° \leq \psi_{rand} \leq 30°$). In our case, as evident from the figure, there is energy transfer between the beams even though the amplitude incident beams are equal.

For partially incoherent illumination, despite a moderate level of phase incoherence, steady state results which are similar to the coherent case are achieved. It is interesting to note that a lower level of energy transfer occurs in the incoherent case as compared to the coherent case, in agreement with the findings in Ref. [24].

Figure 6.3 shows the dielectric constant perturbation $\Delta\varepsilon$ (total dielectric minus average bulk dielectric) that results from the simulation at time step 100 of Figure 6.2. This case corresponds to incoherent illumination by two equal-amplitude interfering plane waves. The x-axis of the figure extends over one grating period Λ (or interference period) of the photorefractive material. The y-axis extends over the length of the crystal from the input plane $y = 0$, to the exit plane at $y = 1.5$ cm. The y coordinate used to plot numerical results is opposite in direction to that used in Section 6.3 and in Figure 6.1 (RCWA coordinate system). The coordinate system of Section 6.3 was chosen to coincide with the coordinate system used by Gaylord [16]. As can be seen from the figure, the dielectric permittivity perturbation is very uniform in the direction of propagation (y) and blaze like in the transverse grating vector direction (x). Thirty layers were used to model the possible longitudinal inhomogeneity of the grating.

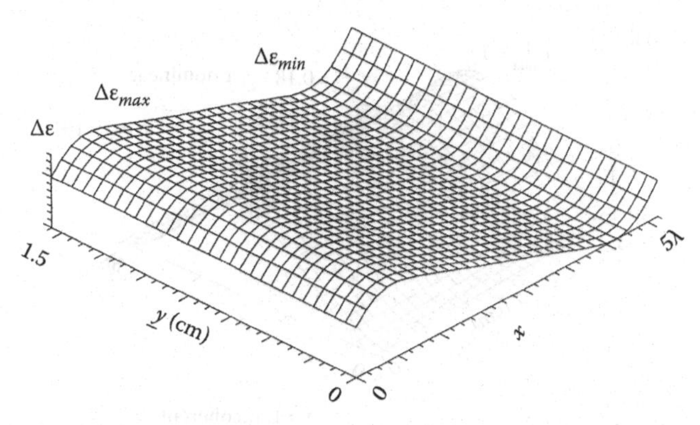

FIGURE 6.3 Plots of the relative dielectric perturbation $\Delta\varepsilon$ when $p = 1$, $\alpha = 2 \times 10^{-5}$, $\varepsilon'_{2r} = 8$, at the time step 100. The $\Delta\varepsilon$ shown correspond to the random case of Figure 6.4, $y = 0$ is the incident side and $y = 1.5$ cm is the exit plane. All other parameters are the same as in Figure 6.2 $\Delta\varepsilon_{min} = -3.8 \times 10^{-6}$ and $\Delta\varepsilon_{max} = -3.51 \times 10^{-6}$ $\Delta\varepsilon$. (Reprinted from *Opt. Commun.*, 123, Jarem, J.M. and Banerjee, P.P., A nonlinear, transient analysis of two and multi-wave mixing in a photorefractive material using rigorous coupled-mode diffraction theory, 825–842, Copyright 1996, with permission from Elsevier.)

The $\Delta\varepsilon$ variation seen in Figure 6.3 can be understood in the following way. At time step 100 of Figure 6.2, $\Delta\varepsilon$ of the PR material has reached steady state. Thus in Equation 6.18, $\tau(\partial\Delta\varepsilon/\partial t) = 0$ and

$$\Delta\varepsilon = \alpha\frac{\partial I/\partial\tilde{x}}{C+I} \tag{6.22}$$

Because the PR material is weakly modulated, the electric field intensity inside the PR material is very little different than that of the unmodulated sample, and thus the intensity I of Equation 6.18 is

$$I(z) = I_0\cos^2\kappa\tilde{x} \tag{6.23}$$

where $I_0 = 4.443 \times 10^4$ (W/m²), $C = 10^4$ (W/m²), $\alpha = 2 \times 10^{-5}$, and $\kappa = 0.1$.

Substitution of Equation 6.23 in Equation 6.22 shows

$$\Delta\varepsilon = \frac{-\alpha I_0\kappa\sin^2\kappa\tilde{x}}{C+I_0\cos^2\kappa\tilde{x}} \tag{6.24}$$

A numerical comparison of Equation 6.24 with the numerical data of Figure 6.3 shows almost exact equality between the two sets of data.

Figure 6.4a shows a plot of the position-dependent [23] nonlinear τ^{-1} function which results from using the more exact space-charge, dielectric constant equation, Equation 6.18. The nonlinear τ^{-1} function was chosen to be unity at the center of the grating fringe. Figure 6.4b shows the normalized power in, for instance, the R_1 direction (see inset, Figure 6.1) when $\tau = 1$ for coherent and incoherent illumination ($-30° \le \psi_{rand} \le 30°$) and compares with the case when the nonlinear τ function of Figure 6.4a is used in the numerical simulation. As can be predicted from Figure 6.4b, the net effect of the nonlinear τ function is, in general, to reduce and slow down the transfer of modal energy in the R_1 direction. Similar results occur for the normalized power in the R_0, T_1, and T_0 directions. It appears that the nonlinear τ does not have a strong effect on the mode coupling as regards the final steady state (although the position-dependent case yields a slightly lower steady state value [23]); for this reason, a constant τ has been chosen for all subsequent calculations in the interest of computation time.

Figure 6.5a through c shows plots of the dielectric modulation $\Delta\varepsilon$ as a function of time t (time step) and transverse dimension (fringing field direction) x when sampled at the center of the photorefractive medium $y = 0.75$ cm. Because of the longitudinal uniformity of the $\Delta\varepsilon$ plots (see Figure 6.3), the $y = 0.75$ cm plane is representative of $\Delta\varepsilon$ at other longitudinal planes. Figure 6.5a shows the $\Delta\varepsilon$

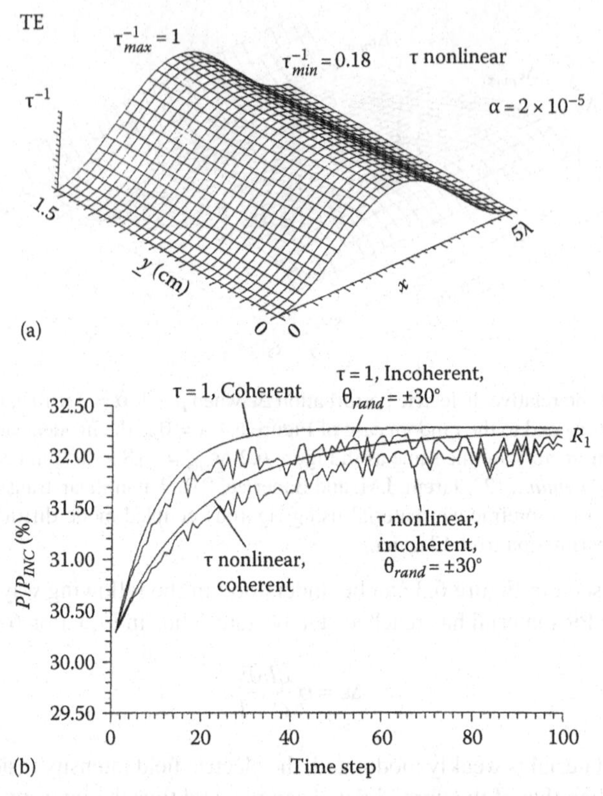

FIGURE 6.4 (a) Plot of the nonlinear time factor τ^{-1} as a function of transverse distance x and longitudinal distance y when $p = 1$, $\alpha = 2 \times 10^{-5}$, $\varepsilon_{2r} = 8$, $\lambda = 0.5\,\mu\text{m}$, $\theta_{rand} = 0°$ (coherent case). (b) Comparison of the reflected normalized power R_1 for the case of Figure 6.2, when the τ factor was taken to be constant ($\tau = 1$) and nonlinear (a) in the cases when the incident waves were coherent ($\theta_{rand} = 0°$) and incoherent ($-30° \leq \theta_{rand} \leq 30°$). As can be seen, the nonlinear τ factor increases the time necessary for the system to reach steady state. (Reprinted from *Opt. Commun.*, 123, Jarem, J.M. and Banerjee, P.P., A nonlinear, transient analysis of two and multi-wave mixing in a photorefractive material using rigorous coupled-mode diffraction theory, 825–842, Copyright 1996, with permission from Elsevier.)

growth when $\tau = 1$ and coherent illumination was used; Figure 6.5b shows the $\Delta\varepsilon$ growth when τ was nonlinear (see Figure 6.4a) and coherent illumination was used; and Figure 6.5c shows the $\Delta\varepsilon$ growth when τ was nonlinear (see Figure 6.4a) and incoherent illumination was used. As can be seen from a comparison of Figure 6.5a and b, the nonlinear τ function reduces the rate of the $\Delta\varepsilon$ growth relative to the $\tau = 1$ case. A comparison of Figure 6.5b and c shows that for the same τ function (the nonlinear τ was used), incoherent illumination produces a lower $\Delta\varepsilon$ modulation ($\Delta\varepsilon_{max} \cong 3.38 \times 10^{-6}$, $\Delta\varepsilon_{min} \cong 3.36 \times 10^{-6}$) than does coherent illumination ($\Delta\varepsilon_{max} = 3.77 \times 10^{-6}$, $\Delta\varepsilon_{min} = -3.77 \times 10^{-6}$), and hence, reduced energy transfer.

Figure 6.6 shows a plot of the two-wave coupling coefficient Γ, which results when a coherent wave ($\psi_{rand} = 0°$) and when two incoherent waves ($-15° \leq \psi_{rand} \leq 15°$, $-30° \leq \psi_{rand} \leq 30°$) are incident on the photorefractive slab. The Γ [7] is defined by

$$\Gamma_t = 10\log\left\{\frac{\left|\dfrac{E_1^t(L)}{E_{1inc}^t(0)}\right|^2}{\left|\dfrac{E_0^t(L)}{E_{0inc}^t(0)}\right|^2}\right\},\quad \Gamma_r = 10\log\left\{\frac{\left|\dfrac{E_1^r(0)}{E_{1inc}^r(0)}\right|^2}{\left|\dfrac{E_0^r(0)}{E_{0inc}^r(0)}\right|^2}\right\} \tag{6.25}$$

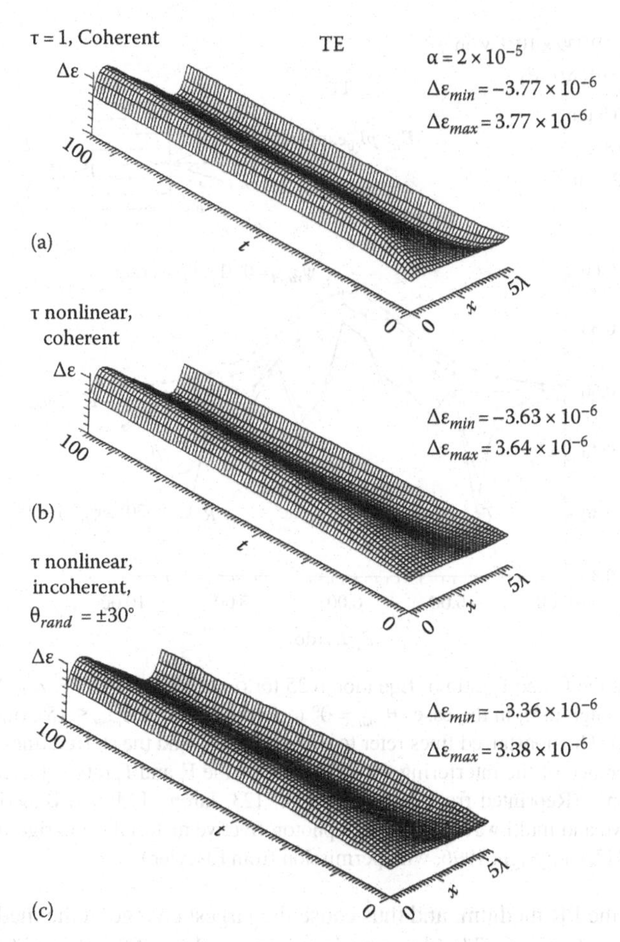

FIGURE 6.5 Plots of $\Delta\varepsilon$ versus transverse dimension x and time step as measured at a plane midpoint in the PR slab ($y = 0.75$ cm, slab length 1.5 cm) for the parameters of Figures 6.2 through 6.4. (a) $\Delta\varepsilon$ coherent illumination was used and $\tau = 1$, (b) $\Delta\varepsilon$ coherent illumination was used and the nonlinear factor τ of Figure 6.4a was used, (c) $\Delta\varepsilon$ when nonlinear factor τ of Figure 6.4a was used and when incoherent illumination was used $\alpha = 2 \times 10^{-5}$. (Reprinted from *Opt. Commun.*, 123, Jarem, J.M. and Banerjee, P.P., A nonlinear, transient analysis of two and multi-wave mixing in a photorefractive material using rigorous coupled-mode diffraction theory, 825–842, Copyright 1996, with permission from Elsevier.)

In these expressions $E^t_{i,inc}(0)$ $(i = 0, 1)$ is the incident electric field in the ith order direction, $E^t_i(L)$ $(i = 0, 1)$ is the transmitted electric field at the final time step which is transmitted out of the slab in the ith order direction, $E^r_{i,inc}(0)$ $(i = 0, 1)$ is the reflected incident electric field in the ith order direction, and $E^r_i(0)$ $(i = 0, 1)$ is the reflected electric field at the final time step which is reflected out of the slab in the ith order direction. The gamma mode ratio is a measure that shows to what degree the photorefractive medium has been able to convert power from one mode to another.

For the plots of Figure 6.6, the bulk dielectric was taken to $\varepsilon_{2r} = 8$, $\alpha = 2 \times 10^{-5}$, and the incident wave's refraction ratio $p = |E_1|/|E_0|$ was varied from $p = 0.1$ (−10 db) to 10 (10 db). The value of E_0 and E_1 was adjusted to keep the incident power $I_{inc} = 9.949 \times 10^{-4}$ (W/m²) the same for all values of p used.

The results of Figure 6.6 show that the PR medium transfers the most energy when the incident waves are coherent, and that a gradual decrease in coupling efficiency occurs as the waves become more incoherent as seen by the drop in Γ_t and Γ_r when $-15° \leq \psi_{rand} \leq 15°$ and $-30° \leq \psi_{rand} \leq 30°$. The results of Figure 6.6 also show that the maximum mode coupling occurs when the wave amplitudes are equal $p = 1 = |E_1|/|E_0|$. This is to be expected since the equal amplitude waves cause the greatest

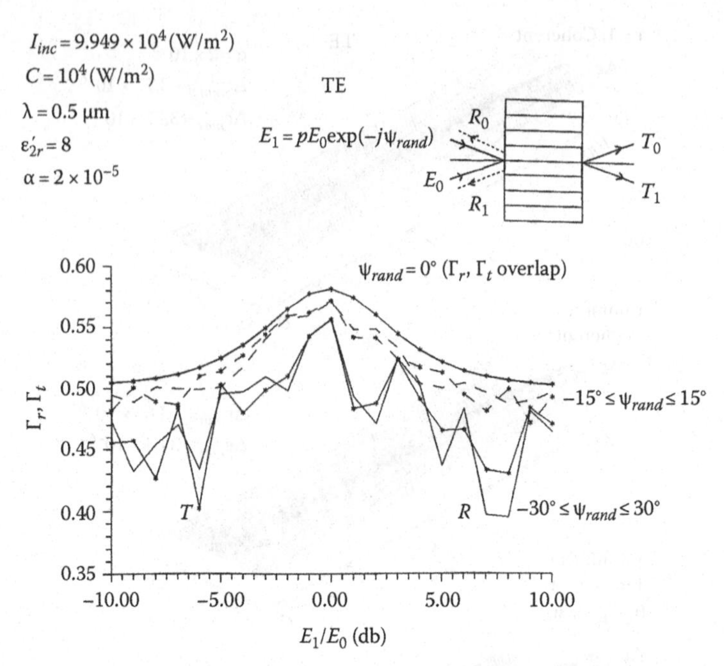

FIGURE 6.6 Plots of the Γ_r and Γ_t ratio of Equation 6.25 for $\alpha = 2 \times 10^{-5}$, $\varepsilon'_{2r} = 8$, $\lambda = 0.5\,\mu$m and slab length 1.5 cm (Figures 6.2 through 6.5), in the cases $\theta_{rand} = 0°$ (coherent), $-15° \le \theta_{rand} \le 15°$, (incoherent), and $-30° \le \theta_{rand} \le 30°$ (incoherent). The nonstarred lines refer to the reflected Γ_r and the starred lines to the Γ_t ratio. As can be seen, as the incoherence of the interfering waves increases, the Γ_r and Γ_t ratios decrease. $I_{inc} = 9.949 \times 10^4$ (W/m²), $C = 10^4$ (W/m²). (Reprinted from *Opt. Commun.*, 123, Jarem, J.M. and Banerjee, P.P., A nonlinear, transient analysis of two and multi-wave mixing in a photorefractive material using rigorous coupled-mode diffraction theory, 825–842, Copyright 1996, with permission from Elsevier.)

inference pattern in the PR medium, and thus cause the largest change in the medium that causes the most change in the mode power. The decrease in mode coupling efficiency with increasing incoherence is expected since increasing incoherence implies less interference of the waves, and therefore less change in the PR medium and thus less mode coupling. Note also that the fluctuations in Γ increase with ψ_{rand}, indicating possibly reduced coherence of the signal at the output, in agreement with [25].

Figure 6.7 shows the power efficiency that results when the $\alpha/\sqrt{\varepsilon'_{2r}}$ ratio is significantly increased over the value used in Figures 6.2 through 6.6 in the case when the bulk index nearly matched the free space ($\varepsilon_{2r} = 1 - j10^{-6}$) and in the case when the PR medium is mismatched to free space ($\varepsilon_{2r} = 8 - j2.82 \times 10^{-7}$). In Figure 6.7, the $\alpha/\sqrt{\varepsilon'_{2r}}$ has been taken to be 3.18×10^{-4}, 3.88×10^{-4} and 4.59×10^{-4}, and fractional ratio of the incident beams has been taken to $p = |E_1|/|E_0| = 0.01$.

The waves marked T_{0a} and T_{1a}, T_{0b} and T_{1b}, T_{0c} and T_{1c}, show the diffraction efficiencies for the three values of $\alpha/\sqrt{\varepsilon'_{2r}}$ as indicated in the figure inset. Because of index matching (for the cases of Figure 6.7a), the reflected waves from the slab were nearly zero and thus are not shown. As can be seen from Figure 6.7a, the main effect that is observed from Figure 6.7a using the values of $\alpha/\sqrt{\varepsilon'_{2r}}$ that were given is the effect that a great deal of energy is transferred from the (order 0) transmitted E_0 wave (which is large at $t = 0$) to the (order 1) transmitted E_1 wave (which is small at $t = 0$). As can be seen from Figure 6.7a, the conversion of the modal energy depends very strongly on the value $\alpha/\sqrt{\varepsilon'_{2r}}$ used, and affects the value of power which is diffracted in the T_1 direction and the speed with which the mode power transfer reaches steady state. In Figure 6.7a, the presence of absorption ($\varepsilon''_{2r} = 1. \times 10^{-6}$) seemed to have a minimal effect on the diffraction except to, of course, attenuate the T_0 and T_1 propagating waves.

Figure 6.7b shows the diffraction that occurs when the real part of the bulk dielectric is $\varepsilon'_{r2} = 8$ and $\alpha/\sqrt{\varepsilon'_{2r}} = 4.59 \times 10^{-4}$. The value of $\alpha/\sqrt{\varepsilon'_{2r}}$ used in Figure 6.7b is the same as in "c" case of

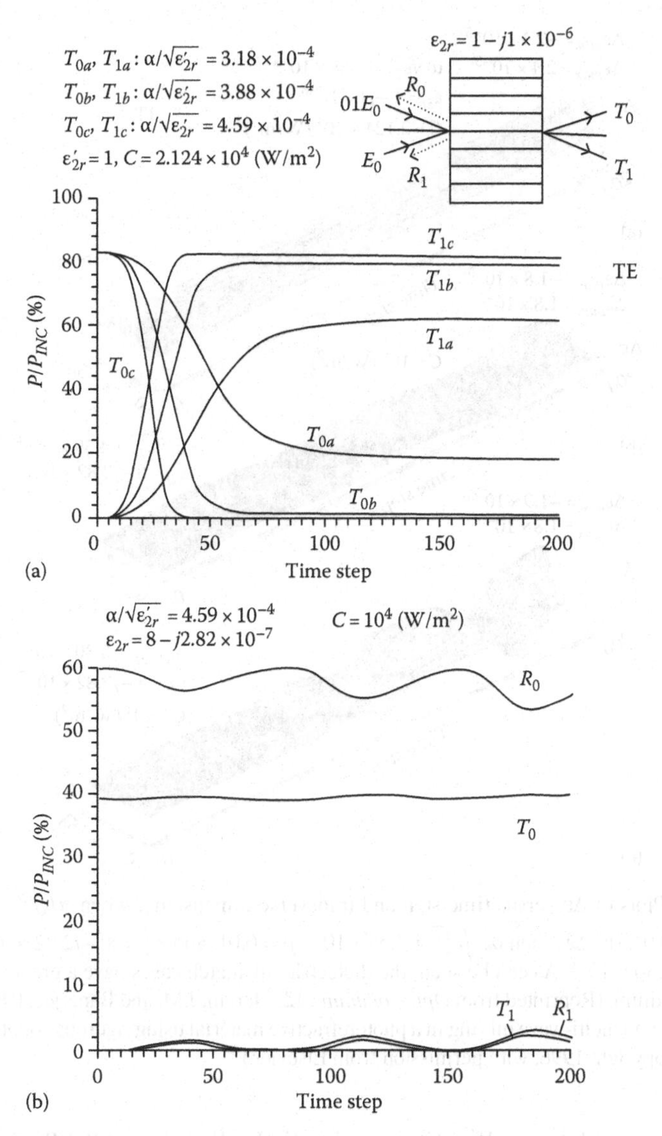

FIGURE 6.7 (a) Plot of the power transmitted in the plots of the T_0 and T_1 directions as functions of the time step in the nearly matched case when $\varepsilon_{2r} = 1 - j1 \times 10^{-4}$ and $p = 0.01$ for the values of $\alpha = 3.18 \times 10^{-4}$, $\alpha = 3.88 \times 10^{-4}$, and $\alpha = 4.59 \times 10^{-4}$ when the PR slab length was 1.5 cm. Also $C = 2.124 \times 10^4$ (W/m²). As α increases, the speed and completeness of the mode conversion increases. (b) Plot of the normalized power reflected and transmitted in the directions R_0, R_1 and T_0, T_1 when $\alpha/\sqrt{\varepsilon'_{2r}} = 4.59 \times 10^{-4}$, $\varepsilon_{2r} = 8 - j2.92 \times 10^{-7}$, $C = 2.124 \times 10^4$ (W/m²), and $p = 0.01$. Because of the dielectric mismatch, aperiodic variation of the normalized power results. (Reprinted from *Opt. Commun.*, 123, Jarem, J.M. and Banerjee, P.P., A nonlinear, transient analysis of two and multi-wave mixing in a photorefractive material using rigorous coupled-mode diffraction theory, 825–842, Copyright 1996, with permission from Elsevier.)

Figure 6.7a. In the simulation case of Figure 6.7b, because there is a large bulk dielectric mismatch between free space and the PR medium, the incident and diffracted optical energy is multiply reflected from the dielectric boundaries on the incident and transmitted sides in a complicated way, causing a standing wave pattern to arise in the PR medium. The electric field power intensity of this standing wave pattern changes slowly in time, modifies the PR medium according to Equation 6.14, and thus further diffracts the incident and reflected light. The net effect over time is a highly inhomogeneous PR medium. The simulations, shown in Figure 6.7b, predict that the

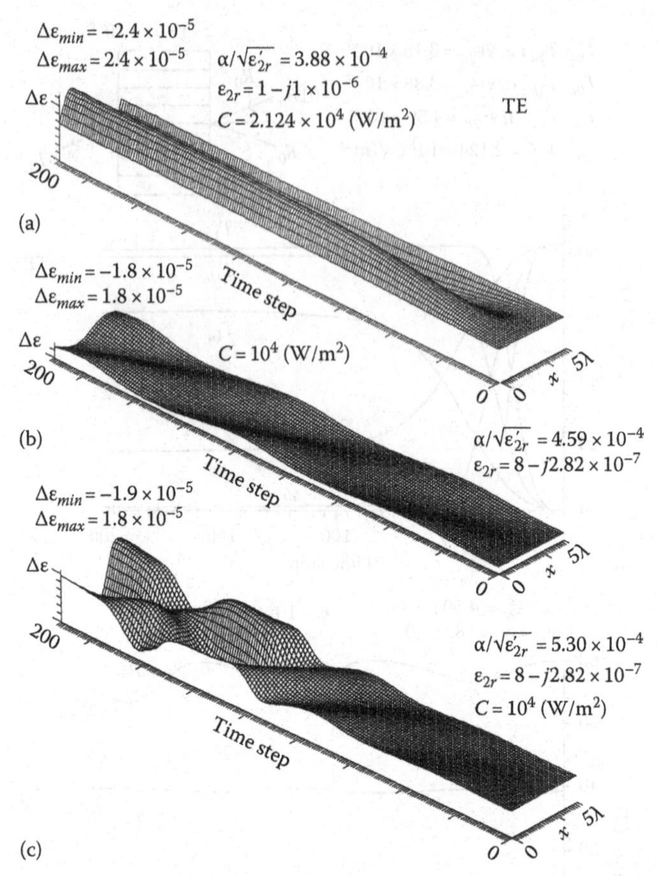

FIGURE 6.8 (a) Plots of $\Delta\varepsilon$ versus time step and transverse dimension x when $\alpha/\sqrt{\varepsilon'_{2r}} = 3.88\times10^{-4}$, $\varepsilon_{2r} = 1 - j1 \times 10^{-6}$, $p = 0.01$. (b) $\Delta\varepsilon$ when $\alpha/\sqrt{\varepsilon'_{2r}} = 4.59\times10^{-4}$, $p = 0.01$, and $\varepsilon_{2r} = 8 - j2.92 \times 10^{-7}$. (c) $\alpha/\sqrt{\varepsilon'_{2r}}$ has increased to $\varepsilon_{2r} = 5.30 \times 10^{-4}$. As can be seen, the dielectric mismatch cases have a great effect on the $\Delta\varepsilon$ that forms in the PR medium. (Reprinted from *Opt. Commun.*, 123, Jarem, J.M. and Banerjee, P.P., A nonlinear, transient analysis of two and multi-wave mixing in a photorefractive material using rigorous coupled-mode diffraction theory, 825–842, Copyright 1996, with permission from Elsevier.)

optical energy may oscillate or self-pulsate quasi-periodically between the R_0, T_0, T_1, and R_1 directions, transferring energy back and forth between the different orders. This is due to the temporal, longitudinally inhomogeneous PR medium which has been formed by the interfering waves. Similar results have been reported during self-pumped and mutually pumped phase conjugation in PR materials [17–20].

Figure 6.8 shows a plot of the dielectric perturbation $\Delta\varepsilon$ that occurs at the line $y = 0.75$ cm (this line is midpoint between the PR medium boundaries) as a function of time step and transverse distance x (wavelength) in the case (1) when the bulk dielectric is matched ($\varepsilon'_{2r} = 1$) to free space (Figure 6.8a, ($\alpha/\sqrt{\varepsilon'_{2r}} = 3.88\times10^{-4}$)), (2) when the bulk dielectric is mismatched ($\varepsilon'_{2r} = 8$) to free space (Figure 6.8b, $\alpha/\sqrt{\varepsilon'_{2r}} = 4.59\times10^{-4}$) and (3) when the bulk dielectric is mismatched ($\varepsilon'_{r2} = 8$) to free space (Figure 6.8c, $\alpha/\sqrt{\varepsilon'_{2r}} = 5.30\times10^{-4}$). As can be seen from these figures, when the boundary is matched, the dielectric modulation $\Delta\varepsilon$ approaches steady state quickly (Figure 6.8a), whereas when the bulk dielectric is mismatched, the dielectric modulation doesn't approach a final steady state, but oscillates in the diagonal ripple pattern shown in Figure 6.8b. The $\Delta\varepsilon$ temporal pattern of Figure 6.8b, as mentioned earlier, is a result of the optical standing wave pattern slowly changing the PR medium, thus further causing a time change in the PR medium.

Figure 6.8c shows the same case as Figure 6.8b (this is a mismatched case) except that $\alpha/\sqrt{\varepsilon'_{2r}}$ has been increased to a value $\alpha/\sqrt{\varepsilon'_{2r}} = 5.30\times10^{-4}$. Because of the higher $\alpha/\sqrt{\varepsilon'_{2r}}$ ratio (stronger PR medium), the diffracted waves more strongly influence the standing wave pattern of the PR medium, and thus more severely change the PR medium. As can be seen from Figure 6.8c, the diffracted waves build up to such a point that almost chaotic behavior occurs as time proceeds [17]. A simulation was run for the same parameters as given in Figure 6.8 except that $\varepsilon''_{2r} = 0$ (no absorption). The resulting $\Delta\varepsilon$ was nearly identical to that as seen in Figure 6.8c. For this case conservation of power was observed numerically almost exactly.

Figure 6.9 shows the PR medium as a function of x and y that results in the matched case at the time step 200 (last time step of Figures 6.7 and 6.8) when (1) $\alpha/\sqrt{\varepsilon'_{2r}} = 3.18\times10^{-4}$ (Figure 6.9a), (2) $\alpha/\sqrt{\varepsilon'_{2r}} = 3.88\times10^{-4}$ (Figure 6.9b), and when (3) $\alpha/\sqrt{\varepsilon'_{2r}} = 4.59\times10^{-4}$ (Figure 6.9c). The three values of $\alpha/\sqrt{\varepsilon'_{2r}}$ used in Figure 6.9a through c correspond to values of $\alpha/\sqrt{\varepsilon'_{2r}}$ used in Figure 6.7 (cases 6.7a and b, respectively). As can be seen from these plots, the dielectric modulation $\Delta\varepsilon$ is highly inhomogeneous longitudinally, with a peak maximum and minimum value occurring

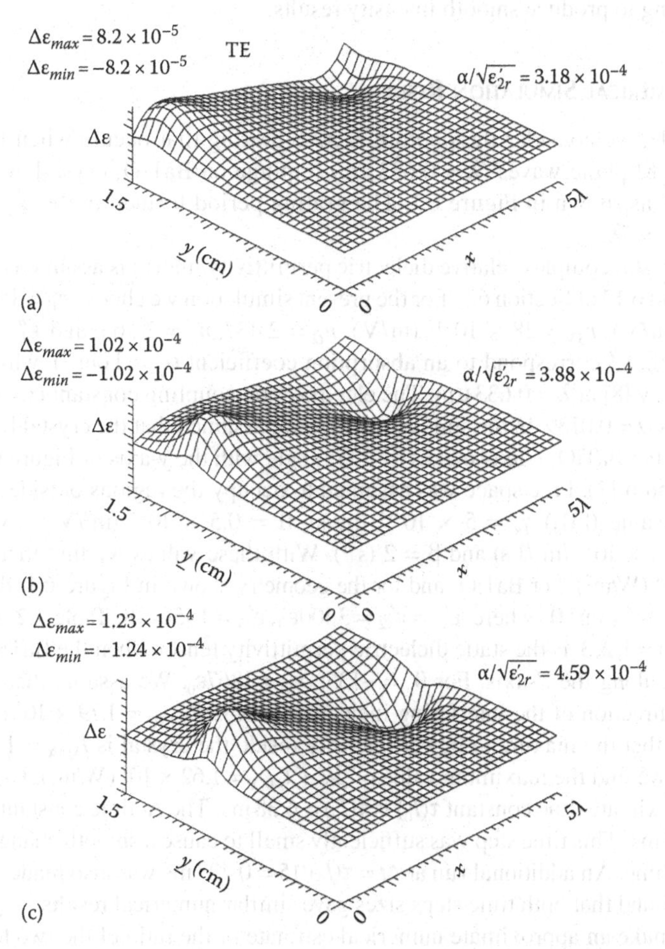

FIGURE 6.9 $\Delta\varepsilon$ at time step 200, $p = 0.01$, $\varepsilon_{2r} = 1 - j1 \times 10^{-6}$, slab length equal $L = 1.5$ (cm). (a) $\alpha/\sqrt{\varepsilon'_{2r}} = 3.18\times10^{-4}$, (b) $\alpha/\sqrt{\varepsilon'_{2r}} = 3.88\times10^{-4}$, and (c) $\alpha/\sqrt{\varepsilon'_{2r}} = 4.59\times10^{-4}$. Higher values of $\alpha/\sqrt{\varepsilon'_{2r}}$ cause mode conversion and interference in the PR medium to occur closer to the incident side as the $\alpha/\sqrt{\varepsilon'_{2r}}$ value increases. (Reprinted from *Opt. Commun.*, 123, Jarem, J.M. and Banerjee, P.P., A nonlinear, transient analysis of two and multi-wave mixing in a photorefractive material using rigorous coupled-mode diffraction theory, 825–842, Copyright 1996, with permission from Elsevier.)

in the PR medium. The position of the peak maximum and minimum values is very dependent on the $\alpha/\sqrt{\varepsilon'_{2r}}$ value used and tends to occur closer to the incident side as the value of $\alpha/\sqrt{\varepsilon'_{2r}}$ increases. This occurs because the larger value of $\alpha/\sqrt{\varepsilon'_{2r}}$ causes the weak signal to be amplified over a shorter distance in the PR medium, thus causing the peaks in a maximum and minimum to form more quickly. Notice that the higher values of $\alpha/\sqrt{\varepsilon'_{2r}}$ caused a much more complete transfer of optical energy as can be seen in Figure 6.7 (cases 6.7a, 6.7b, and 6.7c). Note that the diffraction numerical calculation can only be made correctly using a longitudinally cascaded diffraction analysis algorithm. The plots shown in Figure 6.9 are steady state plots. This was verified by noting that there was no change in the plots shown with plots recorded at time step 100.

In all our numerical simulations for the TE case, 30 layers were used to model the inhomogeneity of the PR medium. Because each individual layer consisted of many wavelengths ($L = 1.5\,\text{cm}/30$, $\lambda = 0.5\,\mu\text{m}$), it was necessary when calculating the intensity $I(x, y)$, to average this over a number of y points distributed over a one-wavelength interval in the center of each layer. If this were not performed, random widely varying samples of the standing wave pattern would be obtained, leading to samples which would not be representative of the average intensity. Three points provided sufficient averaging to produce smooth intensity results.

6.3.3 TM Numerical Simulation Results

As a TM example, we consider the photorefractive grating that occurs when two equal amplitudes, in phase TM plane waves ($\lambda = 0.633\,\mu\text{m}$) impinge on $BaTiO_3$ crystal at an angle of incidence $\theta_i = \pm 5.71°$ as shown in Figure 6.1. The grating period formed in the crystal at this angle of incidence is $\Lambda = 5\lambda$.

The $BaTiO_3$ crystal complex relative dielectric permittivity matrix is assumed to be described by Equations 6.16 and 6.17 of Section 6.3. For the present simulation we choose $r_{42} = 1640 \times 10^{-12}$ (m/V) $r_{13} = 8 \times 10^{-12}$ (m/V), $r_{33} = 28 \times 10^{-12}$ (m/V), $n_O = 2.437$, $n_E = 2.365$, and $\varepsilon''_E = \varepsilon''_O = 2.42 \times 10^{-6}$ (these values of ε''_O, ε''_E correspond to an absorption coefficient $\alpha_p = 1\,\text{cm}^{-1}$), which are the values given for $BaTiO_3$ by [8] at $\lambda = 0.633\,\mu\text{m}$. The electro-optic coupling constant $\alpha = n_O^2 n_E^2 k_0 r_{42} D_s/\mu$ for the values given is $\alpha = 0.0139$. In the simulation, we have assumed that the crystal length $L = 1500\lambda = 0.94\,\text{mm}$ and that the $BaTiO_3$ c-axis makes a $-135°$ angle with the y-axis of Figure 6.1 ($\theta_c = -135°$ in Equations 6.16 and 6.17). Free space was assumed to occupy the regions outside the crystal.

We further assume ([31]) $\gamma_R = 5 \times 10^{-14}$ (m^3/s), $\mu = 0.5 \times 10^{-4}$ (m^2/V s), $N_A/N_D = 0.01$, and assume ([32]) $s = 1 \times 10^{-5}$ (m^2/J s) and $\beta = 2$ (s^{-1}). With these values we find that the dark current $C = (\beta/s) = 2 \times 10^5$ (W/m^2). For $BaTiO_3$ and for the geometry shown in Figure 6.1, the effective value of $\varepsilon_s = \varepsilon'_{s2}\cos^2\theta_c + \varepsilon'_{s3}\sin^2\theta_c$ where $\varepsilon'_{s1} = \varepsilon'_{s2} = 3600\varepsilon_0$, $\varepsilon'_{s3} = 135\varepsilon_0$ [8, Table 1.2, p. 28], and where $\varepsilon'_s = [\varepsilon'_{si}\delta_{i,i'}]$, $(i,i') = 1,2,3$ is the static dielectric permittivity tensor when the $BaTiO_3$ PR crystal has its c-axis aligned along the z'-axis. For $\theta_c = -135°$, $\varepsilon_s = 1867\varepsilon_0$. We assume that the total incident power in the $-y$ direction of the interfering incident waves is $P_{T\,INC} = 1.79 \times 10^7$ (W/m^2). With this value, it is found that the maximum power intensity inside the crystal is $I_{MAX} = 1.60 \times 10^7$ (W/m^2). Using this value, we find the maximum value of $\beta/s + I_{MAX} = 1.62 \times 10^7$ (W/m^2). Using this value, we find that the approximate time constant $\tau(I_0)$ is $\tau(I_0) = 6.38\,\text{ms}$. The Δt time constant was chosen to be $\Delta t = \tau(I_0)/5 = 1.27\,\text{ms}$. This time step was sufficiently small to cause a smooth change in the dielectric modulation with time. An additional run at $\Delta t = \tau(I_0)/15 = 0.423\,\text{ms}$ was also made, to check the time step size. It was found that both time steps sizes gave similar numerical results.

We will now make an approximate numerical estimate of the ratio of the two terms on the RHS of Equation 6.13 for the present $BaTiO_3$ example. We assume that the ionized donor density N_D^+ varies on a timescale of about $\tau(I_0)$ which equals 6.38 ms. We approximate

$$\frac{1}{N_D^+}\frac{\partial N_D^+}{\partial t} \approx \frac{1}{N_D^+}\left(\frac{N_D^+}{\tau(I_0)}\right) = \frac{1}{\tau(I_0)} = 1.56\times 10^2\ \text{s}^{-1}$$

For the first term of the RHS of Equation 6.9 we have,

$$\frac{(sI+\beta)N_D}{N_D^+} \cong \frac{s(I+C)N_D}{N_A} = (10^{-5})(1.62\times10^{7})(10^{2})\ \text{s}^{-1} = 1.62\times10^{4}\ \text{s}^{-1}$$

The ratio of the second term to the first term of RHS of Equation 6.13 is 0.96×10^{-2}. We see that the second term is roughly a hundred times smaller than the first that justifies the approximation used.

The numerical simulation for this case was performed by using Equation 6.18 to calculate $\Delta\varepsilon$ at every time step and using RCWA to calculate the grating diffraction at every time step. The RCWA calculation was performed using $N_L = 160$ layers which gave each layer a length of $\Delta L = 9.375\lambda$. The optical power intensity $I_{po\,int}(x,y,t) \propto (E_x E_x^* + E_y E_y^*)$ (E_x and E_y are the optical electric fields) was calculated at 10 equally spaced points over the layer length $\Delta L = L/N_L = 9.375\lambda$ ($I_{po\,int}(x, y, t)$ was sampled every 0.9375λ). The values of $I_{po\,int}(x, y, t)$ were then averaged. This value was used as the average optical power intensity in Equation 6.22.

Averaging the intensity over the layer length ΔL represents an important part of the interaction of the incident optical light with the PR medium. Physically averaging the optical intensity over ΔL represents the way that partially coherent light or light with a finite frequency spectrum (or both) would interact with the PR medium. When perfectly coherent, monochromatic light enters the PR slab, the energy is multiply reflected at the interfaces, and a strong standing wave pattern is formed in the PR slab. The distance from peak to peak longitudinally (y direction) is a few free space wavelengths λ. When the frequency of the light is changed by a small amount, the peaks and nulls change position by a few wavelengths, and thus in the crystal, a standing wave pattern is formed whose peaks are in an entirely different position than the original monochromatic wave from which the frequency was changed. When many different frequency waves are added together, the peaks and nulls will tend to average out and a longitudinally average field will result. Thus spatial averaging simulates the frequency spread of real optical energy.

Figure 6.10 shows the dielectric modulation function $\Delta\varepsilon(x, y, t)$ that results for the $BaTiO_3$ simulation at time step = 100. At this time the photorefractive crystal is nearly in steady state. The minimum and maximum dielectric modulation that occur are $\Delta\varepsilon_{min} = -0.00605$ and $\Delta\varepsilon_{max} = 0.00565$. Using the relation $\Delta\varepsilon = r_{42} n_O^2 n_E^2 E_{xsc}$, this corresponds to a minimum and maximum space-charge electric field of $E_{xsc}^{min} = -11.1 \times 10^4$ (V/m) and 9.26×10^4 (V/m). These figures are roughly in line (to a factor of 2) with typical values given in [8, Fig. 3.3b, p. 91]. The simulation shown in Figure 6.10 predicts an interesting feature of the dielectric modulation that two different dielectric slanted gratings have formed as a result of the photorefractive-optical interaction. The first grating extends from about zero to 300λ and has a grating period of $\Lambda_1 = 120\lambda$ with grating vector given by

$$\vec{K}_1 = \frac{2\pi}{\Lambda_1}[\sin\theta_1 \hat{a}_x + \cos\theta_1 \hat{a}_y], \quad \theta_1 = 1.5° \tag{6.26}$$

in the geometry of Figure 6.1. The second grating extends from about $y = 300\lambda$ to 1500λ. This grating is more clearly defined and has a grating period of about $\Lambda_2 = 75\lambda$:

$$\vec{K}_2 = \frac{2\pi}{\Lambda_2}[-\sin\theta_2 \hat{a}_x + \cos\theta_2 \hat{a}_y], \quad \theta_2 = 4.5° \tag{6.27}$$

Figure 6.10 shows, respectively, the power which is transmitted and reflected in different diffraction directions as a function of time. As can be seen from the power transmission curves of Figure 6.10, the $BaTiO_3$ PR crystal at the first time step is uniform, and power is transmitted equally in the T_0 and T_1 directions. Within a few time steps, a modulation grating forms and power is rapidly depleted from the T_1 direction and transferred to the T_0 direction. As can be seen from Figure 6.10b the T_0 modal direction is completely depleted of power. As time progresses over a period of approximately

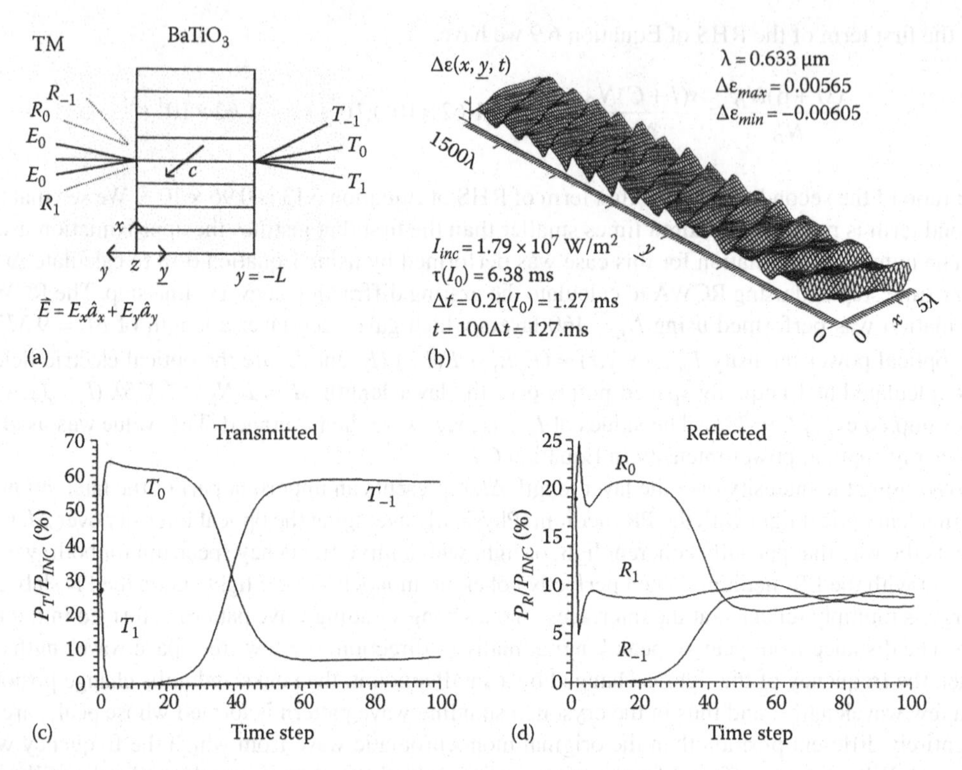

FIGURE 6.10 The geometry (a), dielectric modulation (b), diffracted power (c, d), that results when TM optical energy illuminates $BaTiO_3$ are shown. $I_{inc} = 1.79 \times 10^7$ (W/m²), $\tau(I_0) = 6.38$ ms, $t = 100\Delta t$, and $\lambda = 0.633$ μm. (Reprinted from *Opt. Commun.*, 123, Jarem, J.M. and Banerjee, P.P., A nonlinear, transient analysis of two and multi-wave mixing in a photorefractive material using rigorous coupled-mode diffraction theory, 825–842, Copyright 1996, with permission from Elsevier.)

30 time steps, diffracted waves in the T_{-1} direction build up, and finally a strong mode conversion from the T_0 to the T_{-1} order occurs and the grating goes into a quasi-steady state form. The T_0 mode drops to about 10% of the total diffracted power and the T_{-1} modal power builds from 0% to 60%. The figures are similar to those in Ref. [26] that also illustrate the growth of a higher order. The reflected power in the R_0, R_1, and R_{-1} directions follows a similar time history as did the transmitted except that the R_1 order is not depleted to a zero value by drops from 18% to above 10% as time increases. In contrast to the results in Refs. [26,28], we predict that the +1 order in the transmitted case virtually decays to zero. This may be due to the fact that in the unmatched case (a true experimental possibility), the reflected orders also interact. This example can also be viewed as a case of six-wave coupling in PR materials, such as in $KNbO_3$, which leads to phase conjugation in a two-beam coupling geometry [33].

The $BaTiO_3$ simulation was performed for four other PR crystal lengths which were $L_{i_L} = 1500\lambda + i_L 18.75\lambda, i_L = 1,2,3,4$. An interesting result of the analysis was that for the lengths of 1518.75λ ($i_L = 1$) and 1556.25λ ($i_L = 3$) that simulation showed that diffraction in the T_{-1} and R_1 directions did not occur at all where they did occur for the values of $i_L = 2$ and 4. Evidently the growth of the T_{-1} and R_1 mode perturbations depends on a resonant length of the crystal.

6.3.4 Discussion of Results from Approximate Analysis

By using rigorous coupled wave diffraction theory along with a time-dependent nonlinear formulation, we have analyzed two and multi-wave coupling in a PR material. The two-beam coupling gain has been plotted as a function of the incident intensity ratio for both transmitted and reflected beams. Coherent and partially incoherent cases have been analyzed, and the results show that reasonable beam coupling occurs up to a certain degree of incoherence. Computations performed with both uniform and intensity-dependent time constants show that although the overall time needed to achieve steady state somewhat differs, the final steady state values are unaffected. We have also examined wave coupling in PR media with different gain coefficients and for the first time, cases where there exists significant linear refractive index mismatch between the material and the surrounding medium. The analysis has been extended to analyze an inhomogeneous, anisotropic PR material, e.g., $BaTiO_3$, that includes a nearly exact solution of the Kukhtarev equations. Our simulations thus far predict periodic and nonperiodic oscillations during two-beam coupling, as well as generation of higher transmitted and reflected orders. We believe our analysis is the first that studies the anisotropy of the diffracting region, multiple reflections, and the time dependence of beam coupling simultaneously in a rigorous way.

In our work, we have studied the following new phenomena. We have studied the effect that an intensity-dependent timescale has on beam coupling in a PR medium Figure 6.4b and have studied the effect of partial coherence on beam coupling (Figures 6.5 and 6.6). We have also studied in detail six wave coupling example in $BaTiO_3$ that is shown in Figure 6.10. In the $BaTiO_3$ example, we have shown numerically that when mismatched boundaries occur that the PR coupling seems to be resonant and very sensitive to the overall length L of the grating. For $BaTiO_3$ we haven't carried out experimental verification because we do not have the resources to do so at the present time. However, we would like to point out that the $BaTiO_3$ case which we have analyzed numerically is a very realistic one and one for which experiments could readily be performed. We feel that the case we have analyzed numerically would be a very interesting one to do experimentally. Any discrepancies between theory and experiment would be due to the PR model and not the optical diffraction analysis. The mismatched boundaries could serve as partial Fabry–Perot effect and perhaps make it possible to determine from the optical diffraction data some of the Kukhtarev PR model parameters (mobility constant, N_A, N_D value etc.), which pertain to the $BaTiO_3$ crystal under consideration. Our developed model can thus be effectively used to study wave mixing in PR materials with induced transmission and reflection gratings.

6.4 EXACT ANALYSIS [34]

In Section 6.3, two significant approximations were made to the Kukhtarev analysis. Both approximations were made to the electron rate production equation [29,30]. The first approximation consists of assuming that the donor density N_D was much greater than the acceptor density N_A, the electron density n, and the ion density N_D^+. The effect of the approximation is that the production of electrons is not limited as the electron donors N_D are depleted. The second approximation consists of assuming that the term $\partial N_D^+/\partial t$ is small compared to the source term in Equation 6.2 that depends on the optical intensity. The effect of this approximation is that the full temporal behavior of the system is not included in the analysis. An important impact that the above approximations have, over and above the one already mentioned, is the fact that indirectly, both approximations lead to the omission of second order transverse spatial derivative terms in the analysis. This has an important impact for a few reasons. First, any analysis for which a higher order derivative is ignored limits the analysis to cases where changes in the electric field, the electron density, etc in the transverse direction are small. Thus, for example, if higher order diffraction (which varies rapidly in the transverse direction) should be excited by the PR system, the effect of this cannot be in general studied for arbitrary hologram wavenumbers because the second order derivative terms may contribute more to the analysis than the first order derivatives. Thus, the analysis could at best be only valid over a range of hologram wavenumbers [29,30]. The purpose of this section is to study an exact nonlinear time-dependent solution of the Kukhtarev equations in conjunction with RCWA.

To start the analysis, we substitute Equation 6.3 into Equation 6.1 and find

$$\frac{\partial n}{\partial t} = \frac{\partial N_D^+}{\partial t} + \mu \frac{\partial n E_s}{\partial x} + D_s \frac{\partial^2 n}{\partial x^2} \tag{6.28}$$

Using Equation 6.4, we find

$$N_D^+ = \frac{\varepsilon_s}{e} \frac{\partial E_s}{\partial x} + N_A + n \tag{6.29}$$

or

$$\frac{\partial N_D^+}{\partial t} = \frac{\varepsilon_s}{e} \frac{\partial^2 E_s}{\partial x \partial t} + \frac{\partial n}{\partial t} \tag{6.30}$$

Equation 6.30 results after taking the fact that $(\partial N_A/\partial t) = 0$. If the $(\partial N_D^+/\partial t)$ of Equation 6.30 is substituted in Equation 6.28 and $(\partial n/\partial t)$ is cancelled on the right and left-hand side of the resulting equation, we find

$$\frac{\varepsilon_s}{e} \frac{\partial^2 E_s}{\partial x \partial t} + \mu \frac{\partial (n E_s)}{\partial x} + D_s \frac{\partial^2 n}{\partial x^2} = 0 \tag{6.31}$$

or, after integration with respect to x,

$$\frac{\partial E_s}{\partial t} + \frac{\mu e}{\varepsilon_s} n E_s = -D_s \frac{e}{\varepsilon_s} \frac{\partial n}{\partial x} \tag{6.32}$$

In Equation 6.32 we have set the integration constant to zero, assuming all dependent variables and their derivatives tend to zero as $|x| \to \infty$. For a particular choice of c-axis of the PR crystal, the dielectric permittivity modulation $\Delta\varepsilon$ is related to the electrostatic electric field by

$$\Delta\varepsilon(x, y, t) = n_o^2 n_e^2 r_{42} E_s(x, y, t) \tag{6.33}$$

Substituting E_s from Equation 6.33, we find the following equation for $\Delta\varepsilon$:

$$\frac{\partial \Delta\varepsilon}{\partial t} + \frac{\mu e}{\varepsilon_s} (n \Delta\varepsilon) = -\frac{D_s e}{\varepsilon_s} r_{42} n_o^2 n_e^2 \frac{\partial n}{\partial x} \tag{6.34}$$

Up to this point no use has been made of Equation 6.29, the electron rate production equation. If $(\partial N_D^+/\partial t)$ of Equation 6.30 is substituted in Equation 6.29, we find

$$\frac{\varepsilon_s}{e} \frac{\partial^2 E_s}{\partial x \partial t} + \frac{\partial n}{\partial t} = (sI + \beta)(N_D - N_D^+) - \gamma_R n N_D^+ \tag{6.35}$$

If Equation 6.31 is used to eliminate the $(\varepsilon_s/e)(\partial^2 E_s/\partial x \partial t)$ and N_D^+ of Equation 6.29 is substituted, we find

$$-\mu \frac{\partial n E_s}{\partial x} - D_s \frac{\partial^2 n}{\partial x^2} + \frac{\partial n}{\partial t} = (sI + \beta)\left(N_D - \frac{\varepsilon_s}{e} \frac{\partial E_s}{\partial x} - N_A - n \right) - \gamma_R n \left(\frac{\varepsilon_s}{e} \frac{\partial E_s}{\partial x} + N_A + n \right) \tag{6.36}$$

At this point, it is useful to introduce normalized coordinates and variables and also to perform a small amount of algebraic manipulation of Equation 6.36. Letting $C = \beta/s$, $(sI + \beta) = sC(1 + I/C)$, $\tilde{x} = k_0 x$, $\tilde{t} = \beta t$, $\tilde{n} = n/N_A$, and using Equation 6.33 to express E_s in terms of $\Delta\varepsilon$, we find

$$\Gamma_2 \frac{\partial^2 \tilde{n}}{\partial \tilde{x}^2} + \Gamma_1 \Delta\varepsilon \frac{\partial \tilde{n}}{\partial \tilde{x}} + \left\{ -\Gamma_3 \left(1 + \frac{I}{C}\right) + (\Gamma_1 - \Gamma_4) \frac{\partial \Delta\varepsilon}{\partial \tilde{x}} - \Gamma_3 \Gamma_4 (1 + \tilde{n}) \right\} \tilde{n}$$
$$= \Gamma_3 \left\{ \frac{\partial \tilde{n}}{\partial \tilde{t}} - \left(1 + \frac{I}{C}\right) \left(\frac{N_D}{N_A} - \frac{1}{\Gamma_3} \frac{\partial \Delta\varepsilon}{\partial x} \right) \right\} \quad (6.37)$$

where

$$\Gamma_1 = \frac{\mu e N_A}{\beta \varepsilon_s}, \quad \Gamma_2 = \frac{e k_0 D_s N_A n_o^2 n_e^2 r_{42}}{\beta \varepsilon_s}, \quad \Gamma_3 = \frac{e N_A n_o^2 n_e^2 r_{42}}{\varepsilon_s k_0}, \quad \Gamma_4 = \frac{\gamma_R N_A}{\beta} \quad (6.38)$$

Also Equation 6.34 in normalized form can be written as

$$\frac{\partial \Delta\varepsilon}{\partial \tilde{t}} + \Gamma_1 \tilde{n} \Delta\varepsilon = -\Gamma_2 \frac{\partial \tilde{n}}{\partial \tilde{x}} \quad (6.39)$$

Equations 6.37 and 6.39 are a pair of coupled, nonlinear equations for the electron density $\tilde{n}$ and dielectric modulation function $\Delta\varepsilon$. The form of both of these equations for $\tilde{n}$ and $\Delta\varepsilon$ at any given time and at any given point in the PR medium depends on the value of the optical intensity I at that point in space and time. The value of I itself in the PR medium depends on the incident optical field and depends on the optical energy which has been transmitted, reflected, and diffracted by the dielectric modulation function $\Delta\varepsilon$ that exists in the PR medium at a given time. These transmitted, reflected, and diffracted fields may be found, as mentioned in the introduction, through the use of a diffraction algorithm called rigorous coupled wave theory which is the subject of the next section.

6.4.1 Finite Difference Kukhtarev Analysis

We will now describe an algorithm based on the exact Kukhtarev equations and RCWA, which can be used to determine mode coupling and diffraction from a slowly varying PR medium. The temporal variation of the PR medium is specified by Equations 6.37 through 6.39. The algorithm proceeds temporally as follows. At $t = 0$ (first time step) a signal ($\theta = -\theta_i$) and pump wave ($\theta = \theta_i$) is incident on a dark, uniform PR slab of material. The pump and signal wave interfere, creating a periodic optical intensity pattern in the PR slab. The periodic optical intensity modifies the PR slab and produces a small periodic dielectric modulation $\Delta\varepsilon$. During this time step, no diffraction has occurred, as the dielectric modulation $\Delta\varepsilon$ was zero at the beginning of the time step. At the second time step, the $\Delta\varepsilon$ generated by the optical interference of the first time step will begin to diffract light. The algorithm calculates the total optical intensity in the medium (1) by calculating the EM fields and diffraction from $\Delta\varepsilon$ in the PR slab from the pump wave by itself, (2) by calculating the EM fields and diffraction from the signal wave from $\Delta\varepsilon$ in the PR slab by itself, and (3) by adding these two EM fields together to find the overall EM field in the PR medium, and thus find the overall transmitted and reflected EM fields in free space.

The algorithm proceeds as follows. The time derivative $(\partial \Delta\varepsilon / \partial \tilde{t})$ is approximated as

$$\frac{\partial \Delta\varepsilon}{\partial \tilde{t}} \cong \frac{\Delta\varepsilon(\tilde{x}, y, \tilde{t} + \delta\tilde{t}) - \Delta\varepsilon(\tilde{x}, y, \tilde{t})}{\delta\tilde{t}}$$

and after substitution in Equation 6.39, and cross multiplication by $\delta\tilde{t}$ we find

$$\Delta\varepsilon(\tilde{x},y,\tilde{t}+\Delta\tilde{t})=\Delta\varepsilon(\tilde{x},y,\tilde{t})+\left\{-\Gamma_1\tilde{n}\Delta\varepsilon-\Gamma_2\frac{\partial\tilde{n}}{\partial\tilde{x}}\right\}\partial\tilde{t} \tag{6.40}$$

This equation is used to advance the $\Delta\varepsilon$ dielectric modulation function in time.

Equation 6.37 is used to the find the electron density $\tilde{n}$ for Equation 6.40. Its determination from Equation 6.37 proceeds as follows. Forming an $\tilde{x}$-grid system of N_p divisions, $\delta\tilde{x}=\tilde{\Lambda}/N_p$, $\tilde{\Lambda}=k_0\Lambda$, we let $\tilde{x}_p=(p-0.5)\delta\tilde{x}$, $p=0,1,\ldots,N_p,N_p+1$, be sampled values of $\tilde{x}$ over the grating period of the PR material at a longitudinal distance y. The points $p=0$ and N_p+1 extend one point outside the grating period. These need to be included in order to specify periodic boundary conditions. We also let $V(p)=V(\tilde{x}_p,y,\tilde{t})$ be any sampled dependent variable (for example, $\tilde{n}(p)=\tilde{n}(\tilde{x}_p,y,\tilde{t})$, $\Delta\varepsilon(p)=\Delta\varepsilon(\tilde{x}_p,y,\tilde{t})$, etc.) of Equations 6.37 or 6.39 in the grating period of the PR material at a longitudinal distance y. Using these definitions we may approximate the first and second spatial derivatives of Equations 6.37 and 6.39 by the well-known finite difference formulas

$$\left.\frac{\partial^2\tilde{n}}{\partial\tilde{x}^2}\right|_{\tilde{x}_p}\cong\frac{\tilde{n}(p+1)-2\tilde{n}(p)+\tilde{n}(p-1)}{(\delta\tilde{x})^2},\quad\left.\frac{\partial\tilde{n}}{\partial\tilde{x}}\right|_{\tilde{x}_p}\cong\frac{\tilde{n}(p+1)-\tilde{n}(p-1)}{2\delta\tilde{x}} \tag{6.41}$$

Letting

$$\begin{aligned}P(p)&=\Gamma_1\Delta\varepsilon(p)\\H(p)&=-\Gamma_3\left(1+\frac{I(p)}{C}\right)+(\Gamma_1-\Gamma_4)\left.\frac{\partial\Delta\varepsilon}{\partial\tilde{x}}\right|_{\tilde{x}=\tilde{x}_p}-\Gamma_3\Gamma_4(1+\tilde{n}(p))\\S(p)&=\Gamma_3\left\{\frac{\partial\tilde{n}(p)}{\partial t}-\left(1+\frac{I(p)}{C}\right)\left(\frac{N_D}{N_A}-1\right)\right\}\end{aligned} \tag{6.42}$$

we find for $p=1,\ldots,N_p$

$$\Gamma_2\left\{\frac{\tilde{n}(p+1)-2\tilde{n}(p)+\tilde{n}(p-1)}{(\delta\tilde{x})^2}\right\}+P(p)\left\{\frac{\tilde{n}(p+1)-\tilde{n}(p-1)}{2\delta\tilde{x}}\right\}+H(p)\tilde{n}(p)=S(p) \tag{6.43}$$

If coefficients of $\tilde{n}(p+1)$, $\tilde{n}(p)$, $\tilde{n}(p-1)$ are collected, Equation 6.43 may be put in the following form

$$R_+(p)\tilde{n}(p+1)+R(p)\tilde{n}(p)+R_-(p)\tilde{n}(p-1)=S(p) \tag{6.44}$$

where

$$R_+(p)=\frac{P(p)}{2\delta\tilde{x}}+\frac{\Gamma_2}{\delta\tilde{x}^2},\quad R(p)=H(p)-\frac{2\Gamma_2}{\delta\tilde{x}^2},\quad R_-(p)=-\frac{P(p)}{2\delta\tilde{x}}+\frac{\Gamma_2}{\delta\tilde{x}^2}$$

and where $p=1,2,\ldots,N_p$.

Because the diffraction grating is periodic, we have the important boundary conditions on the variable that $\tilde{n}(0) = \tilde{n}(N_p)$ and $\tilde{n}(N_p + 1) = \tilde{n}(1)$. These equations may be used to eliminate the variables $\tilde{n}(0)$ and $\tilde{n}(N_p + 1)$ in Equation 6.43 and thus give an equation which depends only on the unknowns $\tilde{n}(p)$, $p = 1,\ldots, N_p$. We thus observe that Equation 6.44 represents a system of N_p equations in N_p unknowns $\tilde{n}(p)$, $p = 1,\ldots, N_p$.

The system of finite difference equations given by Equation 6.44 may be conveniently expressed in terms of a matrix equation:

$$\underline{L} = \begin{bmatrix} R(1) & R_+(1) & 0 & 0 & 0 & 0 & 0 & 0 & 0 & R_-(1) \\ R_-(2) & R(2) & R_+(2) & 0 & 0 & 0 & 0 & 0 & 0 & 0 \\ 0 & R_-(3) & R(3) & R_+(3) & 0 & 0 & 0 & 0 & 0 & 0 \\ 0 & 0 & \cdot & \cdot & \cdot & 0 & 0 & 0 & 0 & 0 \\ 0 & 0 & 0 & \cdot & \cdot & \cdot & 0 & 0 & 0 & 0 \\ 0 & 0 & 0 & 0 & \cdot & \cdot & \cdot & 0 & 0 & 0 \\ 0 & 0 & 0 & 0 & 0 & \cdot & \cdot & \cdot & 0 & 0 \\ 0 & 0 & 0 & 0 & 0 & 0 & R_-(N_p-2) & R(N_p-2) & R_+(N_p-2) & 0 \\ 0 & 0 & 0 & 0 & 0 & 0 & 0 & R_-(N_p-1) & R(N_p-1) & R_+(N_p-1) \\ R_+(N_p) & 0 & 0 & 0 & 0 & 0 & 0 & 0 & R_-(N_p) & R(N_p) \end{bmatrix}$$

$$\underline{S} = [S(1) \quad S(2) \quad \cdots \quad S(N_p)]^T, \quad \underline{n} = [n(1) \quad n(2) \quad \cdots \quad n(N_p)]^T$$

$$\underline{L}\,\underline{\tilde{n}} = \underline{S} \tag{6.45}$$

Inverting this matrix equation gives

$$\underline{\tilde{n}} = \underline{L}^{-1}\underline{S} \tag{6.46}$$

which determines the electron density profile for a given value of y in the PR medium.

The algorithm proceeds as follows. Once $\tilde{n}(\tilde{x}, y, \tilde{t})$ is determined, ($\underline{\tilde{n}}(p)$ is determined from matrix inversion and $\underline{\tilde{n}}(p)$ specifies $\tilde{n}(\tilde{x}, y, \tilde{t})$) and its derivative $(\partial\tilde{n}(\tilde{x}, y, \tilde{t})/\partial\tilde{x})$ is calculated, $\tilde{n}(\tilde{x}, y, \tilde{t})$ and $(\partial\tilde{n}(\tilde{x}, y, \tilde{t})/\partial\tilde{x})$ are then substituted back into Equation 6.40. Once a new value of $\Delta\varepsilon(\tilde{x}, y, \tilde{t} + \delta\tilde{t})$ is found from Equation 6.40, this new value of $\Delta\varepsilon(\tilde{x}, y, \tilde{t} + \delta\tilde{t})$ is calculated for all values of y (all discrete layers of the PR slab). Once this step is completed, RCWA is used to study diffraction from the new value of $\Delta\varepsilon(\tilde{x}, y, \tilde{t} + \delta\tilde{t})$, and thus a new optical intensity value $I(\tilde{x}, y, \tilde{t} + \delta\tilde{t})$ is found. The new intensity $I(\tilde{x}, y, \tilde{t} + \delta\tilde{t})$ along with $\Delta\varepsilon(\tilde{x}, y, \tilde{t} + \delta\tilde{t})$ and its x derivative, is substituted into Equations 6.44 and 6.45 and a new value of $\tilde{n}(\tilde{x}, y, \tilde{t} + \delta\tilde{t})$ is found. By repeating the above steps for many iterations, the time evolution of the PR material and the optical diffracted intensity may be found.

6.4.2 TM Numerical Simulation Results

We consider the photorefractive grating that occurs when two in-phase TM plane waves (λ = 0.633 μm) of amplitudes E_0 and E_1 impinge on $BaTiO_3$ crystal at an angle of incidence θ_i as shown in Figure 6.11. The $BaTiO_3$ crystal complex relative dielectric permittivity matrix is assumed to be described by Equations 6.16 and 6.17. For the present simulation, we choose $r_{42} = 1640 \times 10^{-12}$ (m/V), $r_{13} = 8 \times 10^{-12}$ (m/V), $r_{33} = 28 \times 10^{-12}$ (m/V), $n_O = 2.437$, $n_E = 2.365$, and $\varepsilon_E'' = \varepsilon_O'' = 2.42 \times 10^{-6}$ (these values of ε_O'', ε_E'' correspond to an absorption coefficient $\alpha_p = 1\ \mathrm{cm}^{-1}$), which are the values given for $BaTiO_3$ by [8] at $\lambda = 0.633$ μm. In the simulation, we have assumed that the $BaTiO_3$ c-axis makes a 45° angle with the y-axis of Figure 6.1. We further assume [31,32] $\gamma_R = 5 \times 10^{-14}$ (m³/s), $\mu = 0.5 \times 10^{-4}$ (m²/V s), $N_A = 3 \times 10^{22}$, $N_D = 200N_A$, and assume $s = 1 \times 10^{-5}$ (m²/J s) and $\beta = 2$ (s⁻¹). With these values, we find that the dark current $C = (\beta/s) = 2 \times 10^5$ (W/m²). For $BaTiO_3$ and for the geometry shown in

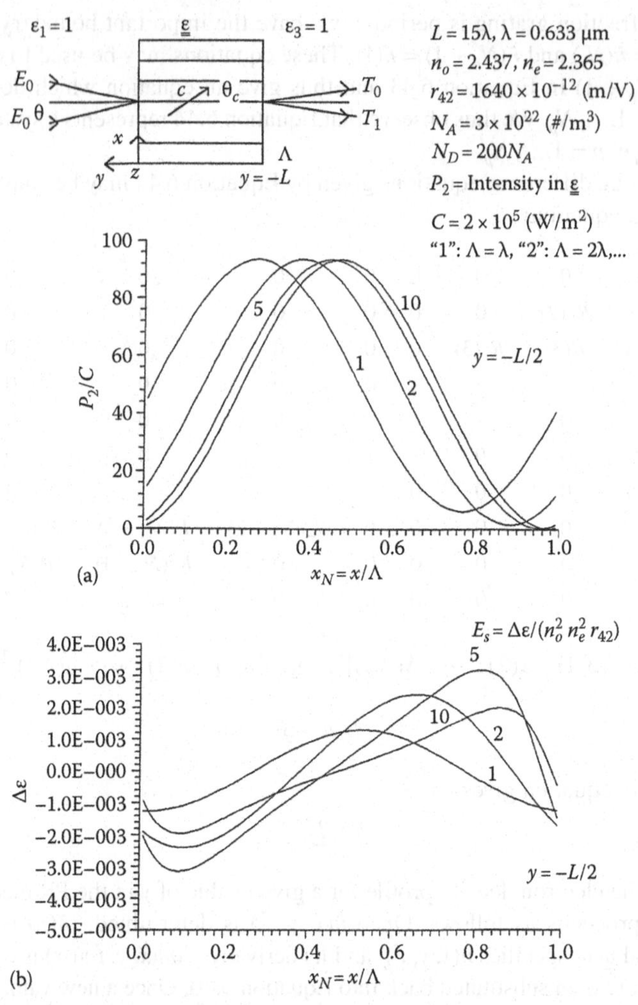

FIGURE 6.11 (a) The optical power intensity (normalized to the dark current C) in the grating when the grating period is $\Lambda = 1\lambda$, 2λ, 5λ, and 10λ as a function of the normalized grating distance $x_N = x/\Lambda$. The incident power (evaluated at $y = -L/2$) was adjusted in order that the intensity profile for each different size grating period would have the same peak intensity. (b) The steady state dielectric modulation function $\Delta\varepsilon$ (also evaluated at $y = -L/2$) that results when the intensity profiles of (a) were used to determine $\Delta\varepsilon$. Because the PR grating was so thin, the intensity profiles of (a) were not assumed to change with time as the $\Delta\varepsilon$ profiles reached steady state. All grating parameters used in the simulation not listed on the figure are given in Section 6.3. (From Jarem, J.M. and Banerjee, P.P., An exact, dynamical analysis of the Kukhtarev equations in photorefractive, barium titanate using rigorous coupled wave diffraction theory, *J. Opt. Soc. Am. A*, 819–831, 1996. With permission of Optical Society of America.)

Figure 6.1, the effective value of $\varepsilon_s = \varepsilon'_{s2}\cos^2\theta_c + \varepsilon'_{s3}\sin^2\theta_c$ where $\varepsilon'_{s1} = \varepsilon'_{s2} = 3600\varepsilon_0$, $\varepsilon'_{s3} = 135\varepsilon_0$ [8, Table 1.2, p. 28], and where $\underline{\varepsilon}'_s = [\varepsilon'_{si}\delta_{i,i'}]$, $(i,i') = 1,2,3$ is the static dielectric permittivity tensor when the $BaTiO_3$ PR crystal has its c-axis aligned along the z'-axis. For $\theta_c = 45°$, $\varepsilon_s = 1867\varepsilon_0$.

Figure 6.11 shows the intensity profile and steady state dielectric modulation function $\Delta\varepsilon$ that results from that profile when two equal amplitude plane waves impinge on a $BaTiO_3$ grating whose thickness is $L = 15\lambda$. The grating thickness has chosen to be so thin that no appreciable diffraction occurs within the grating. Figure 6.11a shows the normalized intensity profile for $\Lambda = 1\lambda$, 2λ, 5λ, 10λ at $y = -L/2$. The power intensity has the approximate intensity profile of a squared sinusoidal wave. The intensity profile peak is shifted to the left as the grating period becomes smaller. This is an effect of the anisotropy of the PR crystal. The incident amplitudes of the interfering waves have been chosen to keep the peak intensity the same for the different grating periods shown in Figure 6.11a. The dielectric modulation function $\Delta\varepsilon$ that results from the intensity profiles of Figure 6.12a is shown in Figure 6.11b. In Figure 6.11b, the curve marked "1" corresponds to a grating period $\Lambda = 1\lambda$, the curve marked "2" corresponds to a grating period $\Lambda = 2\lambda$, and so on. The $\Delta\varepsilon$ of each of the curve has been plotted as a function of the normalized variable $x_N = x/\Lambda$ as was the power intensity.

As can be seen in Figure 6.11b, the size of the grating period Λ has a large effect on the value of $\Delta\varepsilon$ that results. When the grating period is small (on the order of $\Lambda = 1\lambda$) the magnitude of $\Delta\varepsilon$ is small. As the grating period increases in size, the maximum magnitude value of the $\Delta\varepsilon$ profile increases in size until the grating period is about $\Lambda \cong 5\lambda$. During this range of $1\lambda \le \Lambda \le 5\lambda$, the $\Delta\varepsilon$ profile gradually changes shape with the $\Delta\varepsilon$ peak (maximum and minimum) rising more sharply as the grating period increases. When Λ is increased to the range $5\lambda \le \Lambda \le 10\lambda$, the value the maximum magnitude value of the $\Delta\varepsilon$ profile decreases in size. The $\Delta\varepsilon$ profile further gradually changes shape. The increase and then decrease of the maximum magnitude value of the $\Delta\varepsilon$ profile results because the equilibrium equation for n and $\Delta\varepsilon$, namely, Equation 6.37, contains zero, first, and second order x derivatives. Thus when Λ is small ($K = 2\pi/\Lambda$ is large) the second order x derivatives terms are large, and when Λ is large the zero order derivative term tends to be large. Intermediate to this ($\Lambda \cong 4\lambda$) to 5λ, the largest magnitude $\Delta\varepsilon$ profiles occur, reminiscent of the dependence of the linearized two-beam coupling coefficient on Λ. We note that for all values of the grating size Λ that the $\Delta\varepsilon$ function is shifted to the right of the intensity profile, as can be seen by inspecting Figure 6.12a and b.

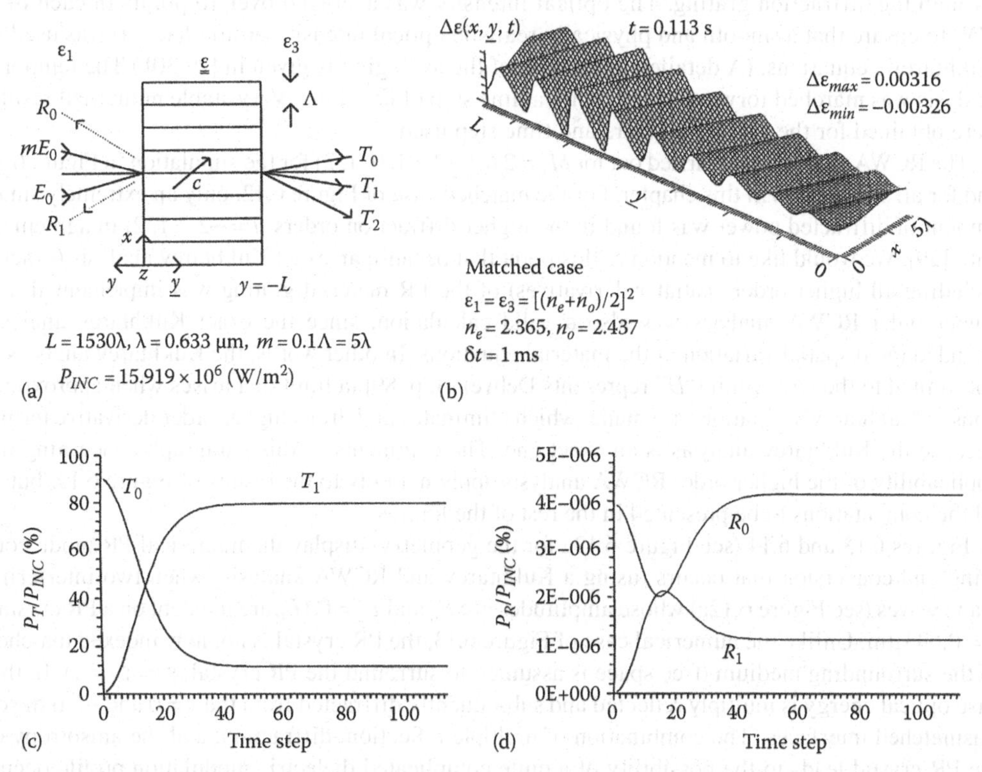

FIGURE 6.12 The geometry (a), the dielectric perturbation function $\Delta\varepsilon$ (b) and the power transmitted in the T_1, T_0 (c) R_1, R_0 (d) directions is shown when the regions (Regions 1 and 3) bounding the PR crystal are indexed matched to the PR crystal. All grating parameters used in the simulation not listed on the figure are given in Section 6.3. (From Jarem, J.M. and Banerjee, P.P., An exact, dynamical analysis of the Kukhtarev equations in photorefractive, barium titanate using rigorous coupled wave diffraction theory, *J. Opt. Soc. Am. A*, 819–831, 1996. With permission of Optical Society of America.)

Figure 6.12a through d displays the numerical PR mode diffraction, coupling, and conversion that occurs (using the Kukhtarev equations and RCWA) when two interfering plane waves (see Figure 6.12a) whose amplitudes are E_0 and $E_1 = 0.1E_0$ are incident on an index matched PR crystal of length $L = 1530\lambda$, $\lambda = 0.633\,\mu m$. The angle of incidence is such to make the grating period $\Lambda = 5\lambda$. Figure 6.12c shows the normalized power which is transmitted through T_0 and T_1 directions as a function of time step, and Figure 6.12d shows the normalized power which is reflected in the R_0 and R_1 directions as a function of time step ($\delta t = 1\,ms$). As can be seen from Figure 6.12d, because of the index matching almost zero power ($\cong 10^{-6}$) is reflected from the PR grating in either the R_0 and R_1 directions. Figure 6.12c shows, for the geometry and material parameters of the case under consideration, that a large amount of energy is transferred from the T_0 to the T_1 in a period of about 35 ms at which time the grating dynamics rapidly approaches steady state. The power in the T_0 and T_1 directions add to about 90% of the incident power. Because the grating is assumed to be lossy, the other 10% of the incident optical power is absorbed as heat in the grating. Figure 6.12b shows the dielectric modulation function $\Delta\varepsilon$ that results at $t = 113\,ms$ when the grating has been in steady state for a long time. The profile has the form of a slanted sinusoidal grating which grows steadily from a small value at $y = 0$ to a peak to peak value of $\Delta\varepsilon_{max} - \Delta\varepsilon_{min} = 0.00642$, which occurs at about $\underline{y} \cong 1000\lambda$. It may be noticed that the grating profile is slightly skewed at $\underline{y} \cong 1000\lambda$. This may be a slight nonlinearity effect.

The simulation grid used 34 δx divisions to solve Kukhtarev's equations (Equation 6.37) for each layer (thus the matrix equation which was inverted in Equation 6.45 was 34 × 34 size) and the grid used $N_L = 160$ layers to describe and simulate the optical wave (RCWA was used to determine the electromagnetic or optical fields of the system) as it propagated and diffracted through the diffraction grating. The optical intensity was averaged over 10 points in each $\delta y = L/N_L$ to ensure that a smooth and physically realistic optical intensity profile $I(x, y, t)$ was used in Kukhtarev's equations. (A detailed discussion of the averaging is given in [29,30].) The temporal analysis was marched forward in time with a time step of $\delta t = 1\,ms$. Very stable numerical results were obtained for the grid parameters and time step used.

The RCWA analysis was carried out for $M_T = 2$ ($i = -2, -1, 0, 1, 2$) for the simulation in Figure 6.12 and for all simulations in this chapter. For the matched case of Figure 6.12, only an extremely small amount of diffracted power was found in the higher diffraction orders ($i = -2, -1, 2$) in agreement with [29]. We would like to mention at this point that because an exact Kukhtarev analysis (exactly including all higher order spatial x-derivatives) of the PR material grating was implemented, the higher order RCWA analysis was also a valid calculation, since the exact Kukhtarev analysis included rapid spatial variation of the material equations. In other words, the Kukhtarev analysis is not limited to the "E_D" limit ("D" represents Debye) [8, p. 89] (a limit that arises when approximations to Kukhtarev's equations are made, which eliminates and drops higher order derivative terms) because the Kukhtarev analysis is an exact one. The comments of this paragraph concerning the applicability of the higher order RCWA analysis apply not only to the results of Figure 6.12, but to all the computations to be presented in the rest of the figures.

Figures 6.13 and 6.14 (see Figure 6.12a for the geometry) display the numerical PR mode coupling and conversion that occurs (using a Kukhtarev and RCWA analysis) when two interfering plane waves (see Figure 6.12a) whose amplitudes are E_0 and $E_1 = 0.1E_0$ are incident on a PR crystal, $\lambda = 0.633\,\mu m$. Unlike the numerical case of Figure 6.13, the PR crystal is not now indexed matched to the surrounding medium (free space is assumed to surround the PR crystal, $\varepsilon_1 = \varepsilon_3 = 1$). In this case optical energy is multiply reflected and subsequently diffracted from the $y = 0$ and $-L$ indexed mismatched interfaces. The combination of multiple reflection, diffraction, and the anisotropy of the PR crystal leads to the possibility of a quite complicated dielectric modulation profile occurring in the PR grating region. For this reason, the numerical simulation of the mismatched grating case was made for several, closely spaced crystal lengths L in order that the effect of crystal length on grating formation could be fully studied. For the values of $L_p = 1482.1875\lambda + p\Delta L$, $\Delta L = 9.5625\lambda$, $p = 1, 2, 3, 4, 5, 6$, Figure 6.13a and b shows the power which is transmitted in the T_0 and T_1 directions, respectively, of Figure 6.12a. Figure 6.13c and d shows the power which is reflected

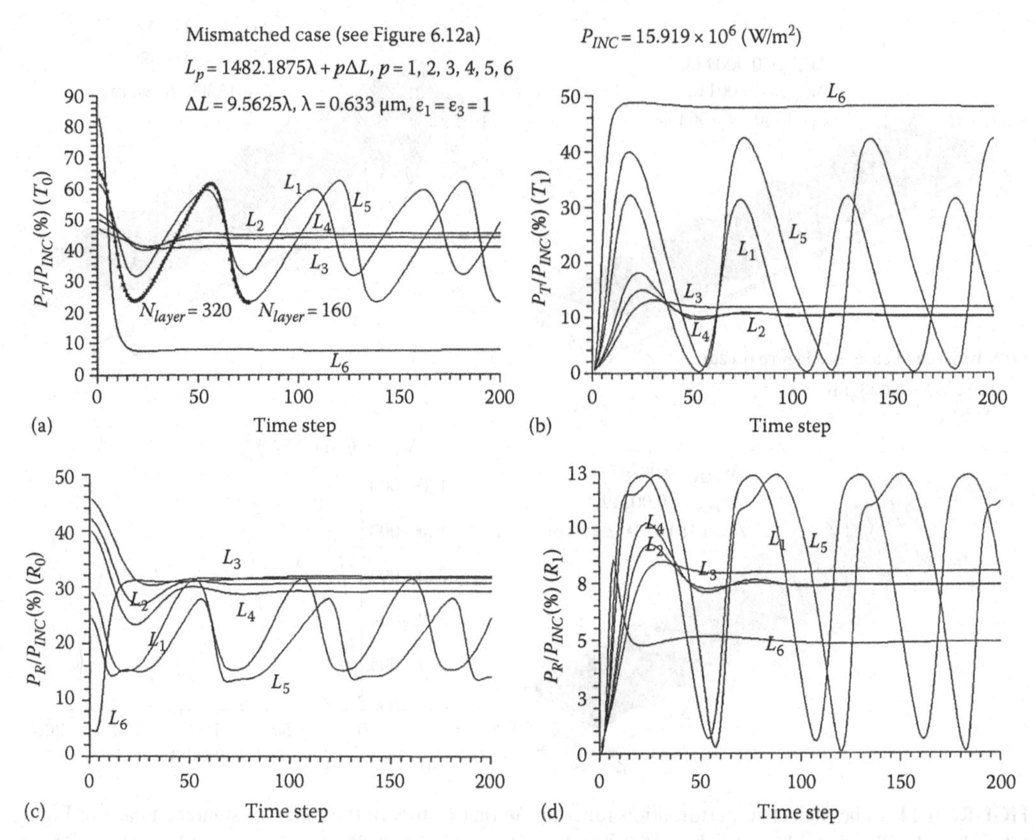

FIGURE 6.13 The power transmitted in the T_0, T_1, R_0, and R_1 directions is shown when the regions (Regions 1 and 3) bounding the PR crystal are not indexed matched to the PR crystal (see Figure 6.12) (air was assumed $\varepsilon_1 = \varepsilon_3 = 1$) is shown in (a), (b), (c), and (d), respectively, for six slightly different PR crystal lengths L. All grating parameters used in the simulation not listed on the figure are given in Section 6.3. The *starred* line on the L_5 curve of (a) used 320 longitudinal layers, whereas all other simulation runs used 160 layers. Note that the number of longitudinal divisions made virtually no difference in the simulation. (From Jarem, J.M. and Banerjee, P.P., An exact, dynamical analysis of the Kukhtarev equations in photorefractive, barium titanate using rigorous coupled wave diffraction theory, *J. Opt. Soc. Am. A*, 819–831, 1996. With permission of Optical Society of America.)

in the R_0 and R_1 directions, respectively, of Figure 6.12a. Again no appreciable higher order diffraction was observed.

There are several interesting features of the plots for this case. The first is that a relatively small change in the overall length L of the PR crystal can make a very large change in the power that is transmitted and reflected in the different directions from the crystal. For example in Figure 6.13b, the power transmitted in the T_1 direction for the L_4 length is small (about 12% in steady state) whereas when the length is increased to L_6 the transmitted power jumps to a large value (about 50%). A length change of only about $2\Delta L \cong 19\lambda$ has occurred. A second interesting feature of the plots is the fact that depending on the length L, the power transmitted (T_0 and T_1 directions) or reflected (R_0 and R_1 directions) may go into an oscillatory steady state or may go into a nonoscillatory steady state. In Figure 6.13b, it is observed that the L_1 and L_5 lengths form oscillatory steady states, whereas the lengths L_2, L_3, L_4, and L_6 form nonoscillatory steady states. It is also interesting to note that the oscillatory steady state periods depend on the length L. For example, the period of the length L_1 is about 50 ms whereas the length L_5 is about 65 ms.

We would like to point out that the results of Figure 6.13 did not change when the number of layers was changed from 160 to 320 layers. The *starred line* of length L_5 in Figure 6.13a was calculated using 320 layers (the power intensity was averaged over 10 points for each of the 320 longitudinal

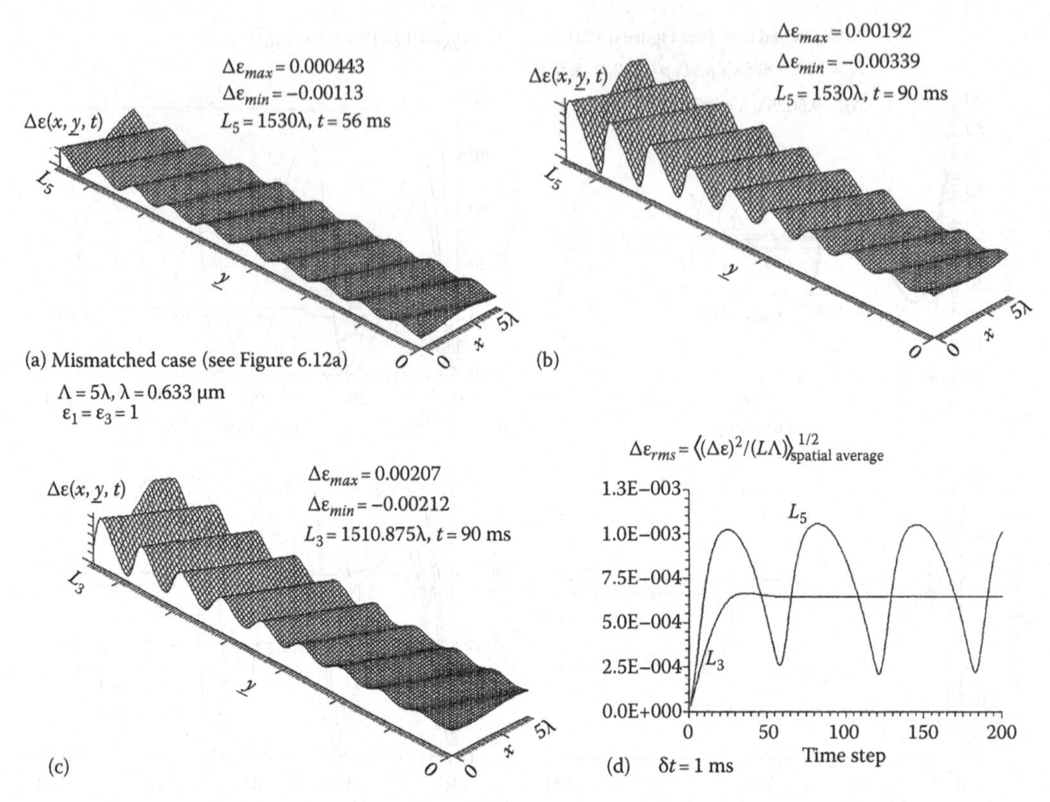

FIGURE 6.14 The dielectric perturbation function $\Delta\varepsilon$ that results in the index mismatched case of Figure 6.13 when the PR crystal length is $L_5 = 1530\lambda$ and $t = 56$ (a) and 90 ms (b) (oscillatory steady state, see Figure 6.13, $L_5 = 1530\lambda$). (c) The dielectric perturbation function $\Delta\varepsilon$ that results in the index mismatched case of Figure 6.14 when the PR crystal length is $L_3 = 1510.875\lambda$ and $t = 90$ ms (nonoscillatory steady state, see Figure 6.14, $L_3 = 1510.875\lambda$). (a–c) are drawn to the same scale. (d) The RMS $\Delta\varepsilon$ (proportional to the electrostatic energy in a stored in a grating period) as a function of time step when the crystal length is $L_3 = 1510.875\lambda$ and when the crystal length is $L_5 = 1530\lambda$ is shown. (From Jarem, J.M. and Banerjee, P.P., An exact, dynamical analysis of the Kukhtarev equations in photorefractive, barium titanate using rigorous coupled wave diffraction theory, *J. Opt. Soc. Am. A*, 819–831, 1996. With permission of Optical Society of America.)

divisions) and all other plots were made using 160 layers (the power intensity was averaged over 10 points for each of the 160 longitudinal divisions). As can be seen from Figure 6.13a, there was no difference in the numerical results which were obtained.

Figure 6.14a and b shows, respectively, the dielectric modulation function $\Delta\varepsilon$ that results when $L = L_5 = 1530\lambda$ (the grating system for this length is in a oscillatory steady state, see Figure 6.13a through d) at times $t = 56$ and 90 ms. Figure 6.14c shows the $\Delta\varepsilon$ dielectric modulation function that results when $L = L_3 = 1510.875\lambda$ at time $t = 90$ ms (the grating system for this length is in nonoscillatory steady state, see Figure 6.13a through d). From Figure 6.14a and b, one notices an interesting property of the oscillatory steady state; namely, that in the oscillatory steady state, that the shape of the dielectric modulation function profile $\Delta\varepsilon$ doesn't change shape with time, but that the peak to peak amplitude of the $\Delta\varepsilon$ profile changes periodically in time with the same period as the diffracted powers T_0, T_1, R_0, and R_1. In Figure 6.14a at time $t = 56$ ms ($L = L_5 = 1530\lambda$), we notice $\Delta\varepsilon_{max} - \Delta\varepsilon_{min} = 1.57 \times 10^{-3}$, whereas in Figure 6.14b $t = 90$ ms ($L = L_5 = 1530\lambda$) and $\Delta\varepsilon_{max} - \Delta\varepsilon_{min} = 5.31 \times 10^{-3}$. Figure 6.14a and b (and Figure 6.14c also) are drawn on the same scale. They have almost an identical shape, but the peak to peak amplitude of Figure 6.14a is about 25% that of Figure 6.14b. Figure 6.14d shows a plot of the root mean square (RMS) amplitude of the dielectric modulation function profile $\Delta\varepsilon$ (the RMS formula is given in Figure 6.14d) as a function of time step ($\delta t = 1$ ms) when the PR crystal length is $L = L_3 = 1510.875\lambda$ and when the PR crystal length is $L = L_5 = 1530\lambda$. As can be seen from the plots of Figure 6.14d, the $L = L_5 = 1530\lambda$ curve shows that in the oscillatory steady state case, that the RMS value (and therefore peak to peak value) of the dielectric modulation function profile $\Delta\varepsilon$ does go through maximum and minimum values whereas the $L = L_3 = 1510.875\lambda$ curve in the nonoscillatory steady state reaches a steady state RMS value of $\Delta\varepsilon$. It is interesting and reasonable that the nonoscillatory RMS value of $\Delta\varepsilon$ is almost exactly the average of the oscillatory RMS $\Delta\varepsilon$ value. We would like to note that the RMS $\Delta\varepsilon$ values shown in Figure 6.14d is proportional to the electrostatic energy (Equation 6.33) that is stored over one grating width Λ and length L of the PR crystal. Thus another way of viewing the oscillatory steady state is that the electrostatic (or quasi-static) energy of the PR crystal is in a continuous state of gaining and losing electrostatic energy as time progresses. We also note from Figure 6.14d that the rise and fall of the RMS value of $\Delta\varepsilon$ (also electrostatic energy) is not symmetric in time, but builds up from a minimum to maximum in about 30 ms and falls from a maximum to a minimum in about 40 ms.

Figure 6.15 displays the numerical PR mode coupling and diffraction that occurs (using the Kukhtarev and RCWA analysis) when two interfering plane waves whose amplitudes are E_0 and

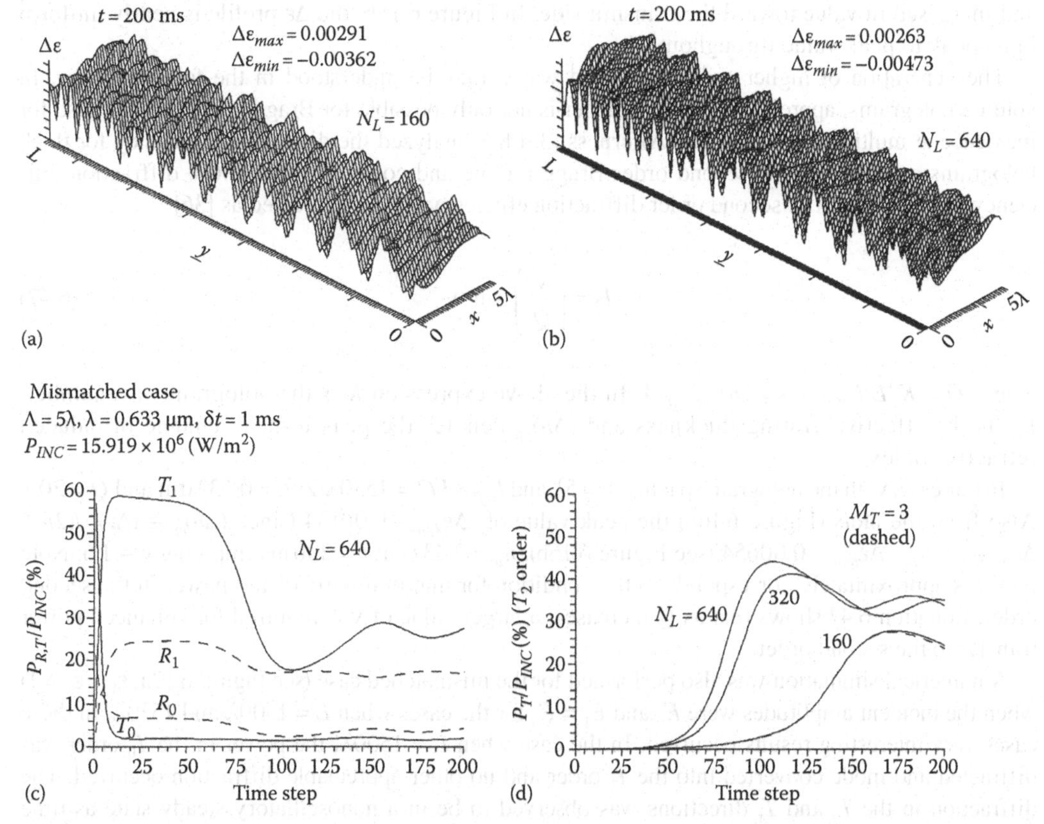

FIGURE 6.15 Numerical PR mode coupling and diffraction that occur (using Kukhtarev equations and RCWA, Section 6.3) when two interfering plane waves whose amplitudes are E_0 and $E_1 = 0.4E_0$ are incident on a PR crystal, ($\lambda = 0.633\,\mu$m, $L = 1530\lambda = L_5$) that is not index matched to the surrounding medium (free space is assumed to surround the PR crystal, $\varepsilon_1 = \varepsilon_3 = 1$). (a) The dielectric modulation function $\Delta\varepsilon$ that results when $N_L = 160$ layers is used. (b) The results when $N_L = 640$ layers is used. (c) The transmitted and reflected power diffracted in the zero and first orders when $N_L = 640$ layers is used. (d) The power transmitted in the second order when $N_L = 160$, 320, and 640 layers are used. The dashed line shown in (d) shows the transmitted power that is diffracted in the second order when $N_L = 160$ layers and $M_T = 3$. (From Jarem, J.M. and Banerjee, P.P., An exact, dynamical analysis of the Kukhtarev equations in photorefractive, barium titanate using rigorous coupled wave diffraction theory, *J. Opt. Soc. Am. A*, 819–831, 1996. With permission of Optical Society of America.)

$E_1 = 0.4E_0$ are incident on a PR crystal ($\lambda = 0.633\,\mu\text{m}$, $L = 1530\lambda = L_5$), which is not index matched to the surrounding medium (free space is assumed to surround the PR crystal, $\varepsilon_1 = \varepsilon_3 = 1$). The peak to peak value of $\Delta\varepsilon$ is approximately 0.00654 as evident from Figure 6.15b. The diffraction response in this case is very different than in the Figures 6.13 and 6.14 case, although this case differs only from the case of Figures 6.13 and 6.14 ($L = L_5 = 1530\lambda$ curves) only in the fact that the signal amplitude E_1 is $E_1 = 0.4E_0$ rather $E_1 = 0.1E_0$. In this case as can be seen from Figure 6.15c and d, power is initially diffracted into the first order diffraction mode directions T_1 and R_1 and then later at about a $t = 100\,\text{ms}$, diffracted into the T_2 and R_2 directions. In the Figures 6.13 and 6.14, diffraction into higher orders was not observed at all. Thus the diffraction in Figures 6.13 and 6.14 constitutes the case of four wave coupling, whereas that of Figure 6.15 represents six wave coupling. Another large difference with the results of Figures 6.13 and 6.14 is that in the case of Figure 6.15 the pronounced oscillation that occurred in Figures 6.13 and 6.14 did not occur in Figure 6.15c and d. The response of Figure 6.15c and d seems to indicate that T_0 attains a quasi-steady state while there is power exchange between T_1 and T_2. A third major difference in the PR response of Figures 6.13, 6.14, and 6.15 was in the shape of the dielectric modulation function $\Delta\varepsilon$ that was formed. In Figure 6.14a through c, the $\Delta\varepsilon$ profile was small at the incidence side and increased in value toward the transmit side. In Figure 6.15b, the $\Delta\varepsilon$ profile is nearly uniform in its peak to peak value throughout.

The generation of higher order diffracted waves may be understood in the following way. In volume holograms, appreciable power transfer is not only possible for Bragg incidence, but also for incidence at multiple-Bragg angles. Alferness [35] has analyzed the diffraction efficiency for thick holograms operating in the second order Bragg regime and concluded that 100% diffraction efficiency was possible. The second order diffraction efficiency can be expressed as [36]

$$I_2 = \left(\frac{\nu}{Q}\right)^2 \sin^2 \nu \tag{6.47}$$

where $Q = K^2L/k_0$, $\nu = k_0(\Delta n)_{pp}L_{eff}/4$. In the above expression K is the hologram wavenumber, L_{eff} is the effective grating thickness and $(\Delta n)_{pp}$ denotes the peak-to-peak change in induced refractive index.

In our case, with the hologram spacing $\Lambda = 5\lambda$ and $L_{eff} \approx L/2 = 1530\lambda/2$, $\lambda = 0.633\,\mu\text{m}$ and $Q = 200$. Also from the plots (Figure 6.16b) the peak value of $(\Delta n)_{pp} \cong 0.00134$ (since $(\Delta n)_{pp} \cong (\Delta\varepsilon)_{pp}/(2n_o)$, $\Delta\varepsilon_{pp} \cong \Delta\varepsilon_{max} - \Delta\varepsilon_{min} = 0.00654$ (see Figure 6.16b), $n_o = 2.437$) at $t = 200\,\text{ms}$ implying $\nu \cong 1.6$. Note that this approximately corresponds to the condition for maximum diffracted power in the second order. Equation 6.47 shows that as Q increases, a larger value of ν is required for enhanced power transfer to the second order.

A numerical simulation was also performed for the mismatched case (see Figure 6.12a, $\varepsilon_1 = \varepsilon_3 = 1$) when the incident amplitudes were E_0 and $E_1 = E_0$ for the cases when $L = 1500\lambda$ and 1530λ. In these cases very interesting results occurred. In the case when $L = 1500\lambda$, the power in the T_0 order was diffracted and mode converted into the T_1 order and no other appreciable diffraction occurred. The diffraction in the T_0 and T_1 directions was observed to be in a nonoscillatory steady state as time increased. For $L = 1500\lambda$, the peak to peak $(\Delta\varepsilon)_{pp} \cong \Delta\varepsilon_{max} - \Delta\varepsilon_{min} = 3.86 \times 10^{-3}$, which made $(\Delta n)_p \cong 0.79 \times 10^{-3}$. Further it was observed that the peak to peak dielectric modulation function $(\Delta\varepsilon)_{pp}$ decreased in value, nearly linearly, from the incidence side to the transmission side, assuming a very small value at the transmission side. Thus the effective length of the grating was about $L_{eff} \cong L/2 = 1500\lambda/2$. In the case when $L = 1530\lambda$, power was initially diffracted from the T_0 order to the T_1 order and then subsequently at about $t = 75\,\text{ms}$ was diffracted from the T_1 order to the T_2 second order. For this case, the peak to peak dielectric modulation function $(\Delta\varepsilon)_{pp} \cong \Delta\varepsilon_{max} - \Delta\varepsilon_{min} = 5.33 \times 10^{-3}$ which made $(\Delta n)_{pp} \cong 1.09 \times 10^{-3}$. From this simulation, we find the interesting result that the index mismatched PR crystal for certain lengths appears to be resonant in the sense that for certain lengths

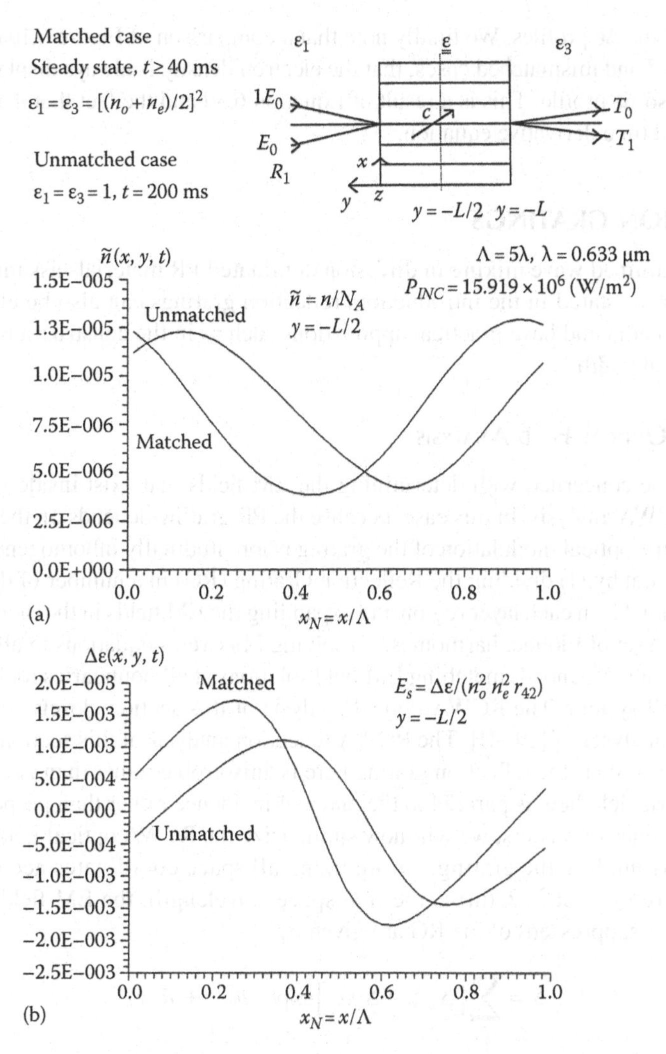

FIGURE 6.16 (a) Plot of the normalized electron density $\tilde{n} = n/N_A$ that result from the simulation shown in the matched case ($t = 113$ ms, $y = -L/2$, $L = 1530\lambda$, $\lambda = 0.633$ μm, $\Lambda = 5\lambda$) and the mismatched case ($t = 200$ ms, $y = -L/2$, $L = L_5 = 1530\lambda$, $\lambda = 0.633$ μm, $\Lambda = 5\lambda$). (b) Plot of the dielectric modulation function $\Delta\varepsilon$ obtained in the same location as the electron densities of (a). (From Jarem, J.M. and Banerjee, P.P., An exact, dynamical analysis of the Kukhtarev equations in photorefractive, barium titanate using rigorous coupled wave diffraction theory, *J. Opt. Soc. Am. A*, 819–831, 1996. With permission of Optical Society of America.)

(in this case $L = 1530\lambda$), the optical energy in the PR crystal can interact with the crystal in such a way that the dielectric modulation function $\Delta\varepsilon$ can build up to relatively large values in the crystal.

Figure 6.16a shows plots of the normalized electron density $\tilde{n} = n/N_A$ which result from the simulation shown in the matched case of Figure 6.13 ($t = 113$ ms, $y = -L/2$, $L = 1530\lambda$, $\lambda = 0.633$ μm, $\Lambda = 5\lambda$) and the mismatched case of Figures 6.14 and 6.15 ($t = 200$ ms, $y = -L/2$, $L = L_5 = 1530\lambda$, $\lambda = 0.633$ μm, $\Lambda = 5\lambda$). As can be seen from these plots, we first notice that the results of the simulation show that the normalized electron density $\tilde{n} = n/N_A$ assumes a very small value of $\tilde{n} \approx 10^{-5}$ or less. Secondly, we note that the presence of matched or mismatched boundaries makes a significant difference as to where the peaks of the electron density occur and the peak to peak size of the electron density. Figure 6.16b shows a plot of the dielectric modulation function $\Delta\varepsilon$ obtained at the same location as the electron densities of Figure 6.16a were obtained. As we can see from these plots the presence of matched or mismatched boundaries makes a significant difference as to position of the peaks and the peak to peak amplitude of the $\Delta\varepsilon$ profiles. We finally note that a comparison of Figure 6.16a and b, show that in both the matched and mismatched cases, that the electron density is always displaced from and out of phase with the so $\Delta\varepsilon$ profile. This is a result of Equation 6.39 and the fact that $\tilde{n}$ and $\Delta\varepsilon$ are related by a first space and time derivative equation.

6.5 REFLECTION GRATINGS

So far we have examined wave mixing in diffusion dominated PR materials assuming transmission gratings. However, as stated in the introduction, reflection gratings can also be efficiently induced and stored in PR media and have practical applications such as in the construction of tunable filters of very low spectral width.

6.5.1 RCWA Optical Field Analysis

This section will be concerned with determining the EM fields that exist inside of a PR reflection grating using a RCWA analysis. In this case, because the PR grating depends on the optical intensity, the magnitude of the optical modulation of the grating is longitudinally inhomogeneous. The RCWA analysis is carried out by (1) dividing the Reflection Grating (RG) in a number of discrete layers, (2) expanding the EM fields in each layer region and expanding the EM fields in the incident and transmit sides of the RG in a set of Floquet harmonics, (3) solving Maxwell's equations in all regions in terms of the Floquet harmonics, and (4) matching EM field solutions at all boundaries to determine the EM fields of the overall system. The RCWA optical analysis of this section closely follows the RCWA reflection grating analyses of [39–41]. The RCWA reflection analysis of this section differs from that of [39–41] in two ways: (1) the reflection grating here is anisotropic rather than isotropic, and (2) the EM incident electric field here is parallel to the plane of incidence rather than perpendicular.

For the convenience of readers, we will now summarize the RCWA optical field equations in the RG and space surrounding the grating. Normalizing all space coordinates according to $x = k_0\tilde{x}$, $y = k_0\tilde{y}$, etc., where $k_0 = 2\pi/\lambda$, λ (m) is the free space wavelength, the EM fields $\vec{E}$, $\vec{H}$ in the ℓth layer (the ℓ index is suppressed) of the RG are given by

$$\vec{S} = \sum_i \left[S_{xi}\hat{a}_x + S_{yi}\hat{a}_y\right]\exp(-jk_{xi}x + j\xi_i y)$$

$$\vec{U} = \sum_i \left[U_{zi}\hat{a}_z\right]\exp(-jk_{xi}x + j\xi_i y) \tag{6.48}$$

where $\vec{S} = \vec{E}$, $\vec{U} = \eta_0\vec{H}$, $\eta_0 = 377\ (\Omega)$, $k_{xi} = \sqrt{\varepsilon_1}\ \sin\theta_I - i(\lambda/\Lambda)\sin(\varphi)$, $\xi_i = -i(\lambda/\Lambda)\cos(\varphi)$, and $\varphi = 0°$. The angle φ [39–41] is the tilt angle of the DG with respect to the planar interfaces and may be taken to be zero for a pure RG analysis [41]. In terms of $\vec{S}$, $\vec{U}$, Maxwell's equations are

$$\nabla\times\vec{S} = -j\vec{U},\quad \nabla\times\vec{U} = j\underline{\varepsilon}\vec{S},\quad \text{where } \underline{\varepsilon} = \begin{bmatrix} \varepsilon_{xx} & \varepsilon_{xy} & 0 \\ \varepsilon_{yx} & \varepsilon_{yy} & 0 \\ 0 & 0 & \varepsilon_{zz} \end{bmatrix} \tag{6.49}$$

In this section, the anisotropic permittivity tensor is assumed to have its c-axis at $\theta_C = 45$ to the crystal interfaces as shown in Figure 6.17. The permittivity tensor elements, including the RG modulation due to the presence of a nonzero optical intensity inside of photorefractive $BaTiO_3$, and a nonzero longitudinal electrostatic field for the c-axis shown in Figure 6.17 have been derived and specified in [5]. After substituting Equation 6.48 in Maxwell's Equation 6.49 following [39–41], the following state variable equations arise:

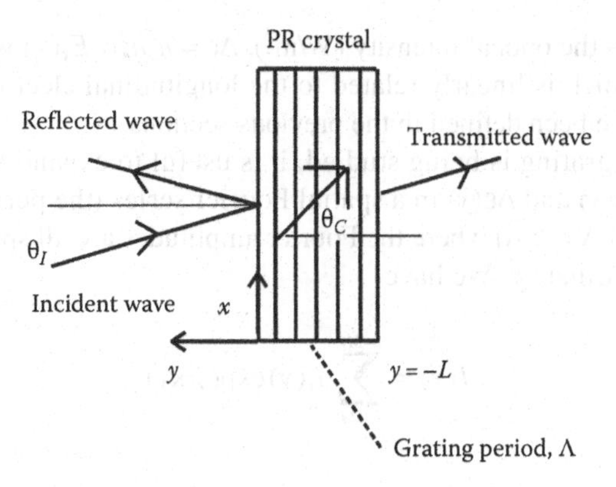

FIGURE 6.17 Problem geometry.

$$\frac{\partial \underline{V}}{\partial y} = \begin{bmatrix} \underline{\underline{a_{11}}} & \underline{\underline{a_{12}}} \\ \underline{\underline{a_{21}}} & \underline{\underline{a_{22}}} \end{bmatrix} \underline{V} \tag{6.50}$$

where $\underline{V} = \left[\underline{S_x}^T \underline{U_z}^T\right]^T$, $\underline{S_x} = [S_{xi}]$, $\underline{U_z} = [U_{zi}]$, $i = -M_T,\ldots, M_T$,

$$\underline{\underline{a_{11}}} = j\left[\underline{\underline{K}}\,\underline{\underline{\varepsilon_{yy}^{-1}}}\,\underline{\underline{\varepsilon_{yx}}} - \underline{\underline{\xi}}\right], \quad \underline{\underline{a_{12}}} = j\left[-\underline{\underline{K}}\,\underline{\underline{\varepsilon_{yy}^{-1}}}\,\underline{\underline{K}} + \underline{\underline{I}}\right],$$

$$\underline{\underline{a_{21}}} = j\left[\underline{\underline{\varepsilon_{xx}}} - \underline{\underline{\varepsilon_{xy}}}\,\underline{\underline{\varepsilon_{yy}^{-1}}}\,\underline{\underline{\varepsilon_{yx}}}\right], \quad \underline{\underline{a_{22}}} = j\left[-\underline{\underline{\xi}} + \underline{\underline{\varepsilon_{xy}}}\,\underline{\underline{\varepsilon_{yy}^{-1}}}\,\underline{\underline{K}}\right]$$

$$\underline{\underline{K}} = \left[k_{xi}\delta_{i,i'}\right], \quad \underline{\underline{\varepsilon_{\alpha\beta}}} = \left[\varepsilon_{\alpha\beta i-i'}\right], \quad (\alpha,\beta) = (x,y), \quad \underline{\underline{\xi}} = \left[\xi_i\delta_{i,i'}\right], \quad \delta_{i,i'} = \begin{cases} 1, & i = i' \\ 0, & i \neq i' \end{cases}$$

Solution of the state variable equations gives the propagating and nonpropagating eigenmodes fields in each thin layer. Summing over these eigenmodes and matching boundary conditions at each thin layer and in Regions 1 and 3, the full EM solution in all regions may be found.

6.5.2 Material Analysis

In this section, we will present a material analysis using Kukhtarev's equations that is applicable to the reflection grating geometry. For the present case, it is assumed that no variation occurs in the transverse x-direction, and that a PR reflection grating is formed in y-longitudinal direction. Following the analysis of transmission gratings above, but taking care to use the longitudinal spatial variable y (the direction of the reflection grating vector), we find that Kukhtarev's material equations may be reduced to the following differential equations

$$\Gamma_2\frac{\partial^2 n}{\partial y^2} + \Gamma_1\Delta\varepsilon\frac{\partial n}{\partial y} + \left\{-\Gamma_3\left[1+\frac{I(y)}{C}\right] + [\Gamma_1 - \Gamma_4]\frac{\partial\Delta\varepsilon}{\partial y} - \Gamma_3\Gamma_4[1+n]\right\}n$$

$$= \Gamma_3\left\{\frac{\partial n}{\partial t} - \left[1+\frac{I(y)}{C}\right]\left[\frac{N_D}{N_A} - 1\right]\right\} + \left[1+\frac{I(y)}{C}\right]\frac{\partial\Delta\varepsilon}{\partial y} \tag{6.51}$$

$$\frac{\partial\Delta\varepsilon}{\partial t} = -\Gamma_1 n\Delta\varepsilon - \Gamma_2\frac{\partial n}{\partial y} \tag{6.52}$$

In these equations, $I(y)$ is the optical intensity (W/m²), $\Delta\varepsilon = n_o^2 n_e^2 r_{42} E_s(y)$ is a normalized dielectric modulation function, which is linearly related to the longitudinal electrostatic field $E_s(y)$ (V/m). All other parameters have been defined in the previous sections.

Because a reflection grating is being studied, it is useful to expand the optical field $I(y)$ and the material variables $n(y)$ and $\Delta\varepsilon(y)$ in a spatial Fourier series (the period of the Fourier series is the grating wavelength $\Lambda = k_0\tilde{\Lambda}$) where the Fourier amplitudes are all spatially varying functions of the longitudinal coordinate y. We have

$$I(y) = \sum_{i=-M_c}^{i=M_c} I_i(y)\exp(ji\kappa y) \tag{6.53}$$

$$n(y) = \sum_{i=-M_c}^{i=M_c} n_i(y)\exp(ji\kappa y) \tag{6.54}$$

$$\Delta\varepsilon(y) = \sum_{i=-M_c}^{i=M_c} \Delta\varepsilon_i(y)\exp(ji\kappa y) \tag{6.55}$$

where $\kappa = 2\pi/\Lambda$ and M_c equals the number of Fourier components. If the Equations 6.53 through 6.55 are substituted into Equation 6.51, and the coefficients of $\exp(ji\kappa y)$ are equated, we finally obtain

$$\frac{\partial^2 n_i(y)}{\partial y^2} + 2ji\kappa\frac{\partial n_i(y)}{\partial y} - \left[\alpha^2 + (i\kappa)^2\right]n_i(y) = H_i(y), \quad i = -M_C, \ldots, M_C \tag{6.56}$$

where

$$\alpha = \left(\frac{\Gamma_3}{\Gamma_2}(1+\Gamma_4)\right)^{1/2}$$

and

$$H_i = \sum_{i'} T_{i-i'} n_{i'} + F_i \tag{6.57}$$

and where

$$T_i = -ji\kappa\frac{\Gamma_1}{\Gamma_2}\Delta\varepsilon_i + \frac{\Gamma_3}{\Gamma_2}\left(\frac{I_i}{C}\right) - \frac{\Gamma_1 - \Gamma_4}{\Gamma_2}\left[\frac{\partial\Delta\varepsilon_i}{\partial y} + ji\kappa\Delta\varepsilon_i\right] + \frac{\Gamma_3\Gamma_4}{\Gamma_2}n_i \tag{6.58}$$

$$F_i = -\frac{\Gamma_1}{\Gamma_2}\sum_{i'}\Delta\varepsilon_{i-i'}\frac{\partial n_{i'}}{\partial y} + \frac{\Gamma_3}{\Gamma_2}\left[\frac{\partial n_i}{\partial t} - \left(\delta_{i,0} + \frac{I_i}{C}\right)\left(\frac{N_D}{N_A} - 1\right)\right]$$
$$+\frac{1}{\Gamma_2}\sum_{i'}\left[\frac{\partial\Delta\varepsilon_{i-i'}}{\partial y} + j(i-i')\kappa\Delta\varepsilon_{i-i'}\right]\left(\delta_{i',0} + \frac{I_i}{C}\right) \tag{6.59}$$

Equations 6.52 to 6.59 for the interval $-L \le y \le 0$ ($L = k_0\tilde{L}$, $\tilde{L}$ (m) is the crystal layer thickness) represents a set of $2M_C + 1$ spatially varying equations from whose solution all material variables may be determined. In this section, we impose the boundary conditions that the normalized electron density $n(y)$ vanish at the crystal interfaces $y = 0$ and $-L$. This boundary condition further imposes the boundary conditions on Equation 6.59 that

$$n_i(0) = n_i(-L) = 0, \quad i = -M_C, \ldots, M_C \tag{6.60}$$

To proceed further, we now for the moment, regard the RHS of Equation 6.59 as a known function y. Equation 6.59 along with its boundary conditions, for each i, is classified as a linear, second order, nonhomogeneous differential equation. The solution to this type of equation is well known and can be found by using a Green's function approach. The Green's function approach consists of: (1) setting the RHS of Equation 6.59 to a Dirac delta function $\delta(y - y')$; (2) solving the resulting differential equation

$$\frac{\partial^2 g_i(y|y')}{\partial y^2} + 2ji\kappa\frac{\partial g_i(y|y')}{\partial y} - \left[\alpha^2 + (i\kappa)^2\right] g_i(y|y') = \delta(y - y') \tag{6.61}$$

with the boundary conditions and continuity condition, respectively

$$g_i(0|y') = g_i(-L|y') = 0, \quad g_i(y|y')\big|_{y=y'^-} = g_i(y|y')\big|_{y=y'^+} \tag{6.62}$$

where y'^- and y'^+ represent locations an infinitesimal to the left and right of y'; and (3) superposing the Green function solutions times the nonhomogeneous RHS H_i, to find the overall response of the system. Regarding $H_i(y)$ as a known function, the solution for $n_i(y)$ is given by

$$n_i(y) = \int_{-L}^{0} g_i(y|y')H_i(y')dy' \tag{6.63}$$

Although Equation 6.63 is an exact integral for the differential equation (Equation 6.59), it is an unnecessarily complicated one to carry for the current analysis. For typical PR parameters used in this section, the constant α is on the same order of magnitude as $\kappa = 2\pi/\Lambda$. Investigation of the Green's function $g_i(y|y')$ for the just described values of α and κ, shows that this Green's function has a significant nonzero value only within a few Λ of the point $y = y'$ in the interval. Investigation has further shown that most of the exponential terms are exponentially small. After analysis, it is found that $g_i(y|y')$ is well approximated by [37]

$$g_i(y|y') = -\frac{1}{s_1 - s_2}\begin{cases} \exp[s_1(y - y')], & -L \le y < y' \\ \exp[s_2(y - y')], & y' < y \le 0 \end{cases} \tag{6.64}$$

where $s_{1,2} = \pm\alpha - ji\kappa$. The function $g_i(y|y')$ is significantly nonzero only when $|y - y'| \le 5\Lambda$. A basic assumption that has been made in this section is that all of the spatially varying amplitudes of Equations 6.53 through 6.55 vary over a much greater length than the $\exp(ji\kappa y)$ functions. Using this assumption, one notices that in the integral of Equation 6.63, the $H_i(y')$ function may be approximately taken to a constant over the range, where $g_i(y|y')$ of Equation 6.64 is nonzero. With the approximations that (1) $H_i(y')$ is taken constant and evaluated at $y = y'$ where the Green's function is maximum, (2) the approximate Green's function of Equation 6.64 is used, and (3) that the limits of the integral of Equation 6.63 be taken to be $y' = -\infty$ to ∞, it turns out that the integral of Equation 6.63 may be evaluated in closed form. The result of the integration is

$$n_i(y) = -G_i H_i(y) \tag{6.65}$$

where

$$G_i = \frac{1}{\alpha^2 + (i\kappa)^2} \tag{6.66}$$

Substituting (6.57) into (6.65), and collecting terms on $n_i(y)$, we derive a matrix equation for the normalized electron density $n_i(y)$:

$$\sum_{i'} \left[\delta_{i,i'} + G_i T_{i-i'}\right] n_{i'}(y) = \sum_{i'} L_{i,i'} n_{i'}(y) = -G_i F_i \tag{6.67}$$

To proceed further, we substitute the spatially varying approximations Equations 6.53 through 6.55 for $n(y)$ and $\Delta\varepsilon(y)$ into Equation 6.52 and find after equating like coefficients of $\exp(ji\kappa y)$ to each other we find

$$\frac{\partial \Delta\varepsilon_i}{\partial t} = -\Gamma_1 \sum_{i'} \left[n_{i-i'} \Delta\varepsilon_{i'}\right] - \Gamma_2 \left[\frac{\partial n_i}{\partial y} + ji\kappa n_i\right] \tag{6.68}$$

If Equation 6.67 is substituted for $n_i(y)$ and matrix terms common to $\Delta\varepsilon_i$ are collected in Equation 6.68, it may placed in the usual state variable form:

$$\frac{\partial \Delta\varepsilon_i}{\partial t} + \sum_{i'} \left[A_{i,i'} \Delta\varepsilon_{i'}\right] = f_i \tag{6.69}$$

where $A_{i,i'}$ and f_i are defined in [37]. The complete temporal numerical solution is detailed in [37].

6.5.3 Numerical Results

This section will be concerned with presenting a numerical simulation of a reflection grating based on the above model. For the present simulation we choose, we assume a lossless grating and assume $r_{42} = 1640 \times 10^{-12}$ (m/V), $r_{13} = 8 \times 10^{-12}$ (m/V), $n_o = 2.437$, $n_e = 2.365$, which are the given crystal values for $BaTiO_3$ by [8] at $\lambda = 0.633\,\mu$m. For these values, the grating wavelength had the value $\tilde{\Lambda} = 0.132\,\mu$m. In the simulation we have assumed that the $BaTiO_3$ c-axis makes a 45° angle with the $-y$-axis of Figure 6.18. We further assume $\gamma_R = 5 \times 10^{-14}$ (m³/s), $\mu = 0.25 \times 10^{-4}$ (m²/Vs), $N_A = 3 \times 10^{22}$ (1/m³), $N_D = 200N_A$, assume $s = 1 \times 10^{-5}$ (m²/J s), and $\beta = 2$ (s⁻¹). With these values we find that the dark current is $C = 2 \times 10^5$ (W/m²). For $BaTiO_3$ and for the geometry shown in Figure 6.17, the effective value of $\varepsilon_s = \varepsilon'_{s2}\cos^2(\theta_c) + \varepsilon'_{s3}\sin^2(\theta_c)$ where $\varepsilon'_{s1} = \varepsilon'_{s2} = 3600\varepsilon_0$ $\varepsilon'_{s3} = 135\varepsilon_0$ [8, Table 1.2, p. 28] and where $\underline{\underline{\varepsilon'_s}} = [\varepsilon_{si}\delta_{i,i'}]$, $(i,i') = 1,2,3$ is the static dielectric permittivity tensor when $BaTiO_3$ PR crystal has its c-axis aligned along the z'-axis. The optical and material analysis was carried out using values of $M_T = 5$ and $M_C = 10$. Also $\tilde{L} = 1000\tilde{\Lambda} = 0.132$ mm, $\kappa = \tilde{\kappa}/k_0 = 4.795$, $\alpha = 4.858$, $\theta_I = 6°$.

In Ref. [37], we have shown the plots of the dielectric modulation function $\Delta\varepsilon(y, t)$ (proportional to the electrostatic space charge $E_s(y, t)$) as calculated over a range of two grating wavelengths $\tilde{\Lambda}$(m), $\Lambda = k_0\tilde{\Lambda}$, from a position $123\tilde{\Lambda}$ to $125\tilde{\Lambda}$ from the incident side of the grating at the times of $\tilde{t} = 0.059, 0.477, 4.97, 30.1, 69.0$ (ms). Also plotted was the reflected diffraction efficiency

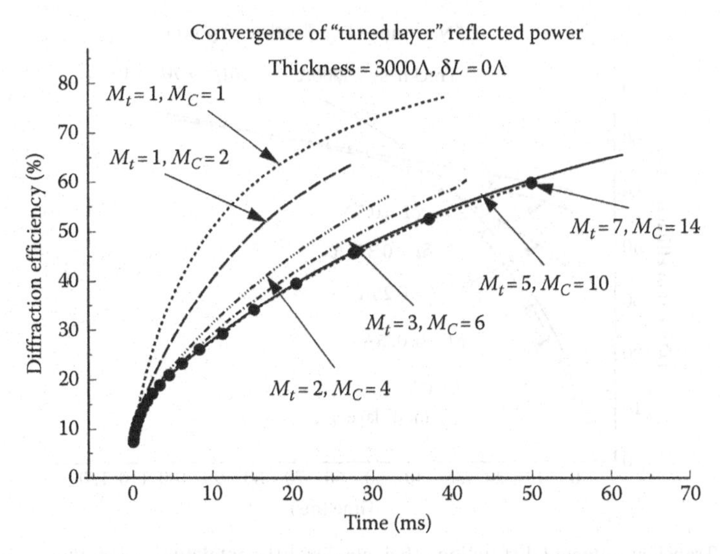

FIGURE 6.18 Transient reflected diffraction efficiency for different optical and material harmonics.

$DE_R(\%)$ as a function of time for the simulation under consideration. We have also shown plots of the terms of the electron density balance equation (Equation 6.67) at a given time $\tilde{t}$ = 30.1 (ms). These plots were all done *only* for a tuned layer and *only* for a thickness of 1000 Λ. Also in [37], no study was done of the convergence of the numerical solution with different spatial harmonics M_C (number of harmonics in the material equations) and M_t (the number of harmonics used in RCWA analysis).

We will now present additional numerical simulations for the diffraction efficiency and the various terms of the electron density balance equation for different lengths of the photorefractive material, and for tuned and detuned cases, respectively. In Figure 6.18, we study the convergence of the solution (diffraction efficiency) for a tuned layer of thickness 3000 Λ by varying the values of M_C from M_C = 1 to 14 and M_t = 1 to 7. As can be seen from Figure 6.18, reasonable convergence is achieved for values of M_C = 6 and M_t = 3 or higher. A nominal combination of M_C = 10 and M_t = 5 was chosen for all following calculations.

Figure 6.19 shows the reflected diffraction efficiency results for tuned layers of different thickness as a function of time. As expected, the diffraction efficiency increases with layer

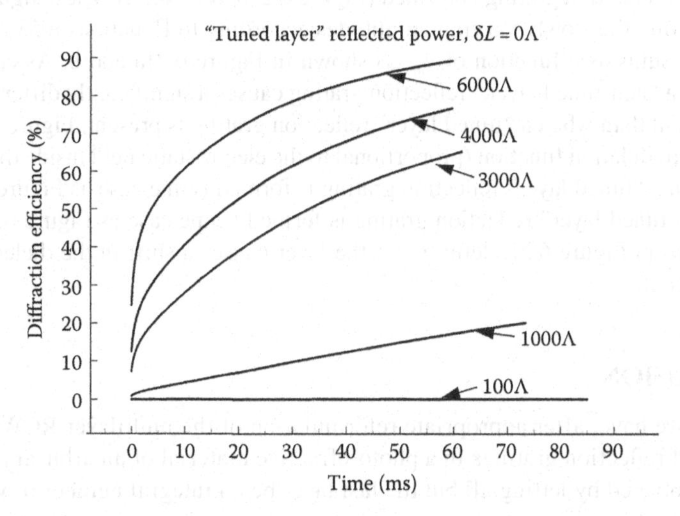

FIGURE 6.19 Transient reflected diffraction efficiency for different tuned layer thicknesses.

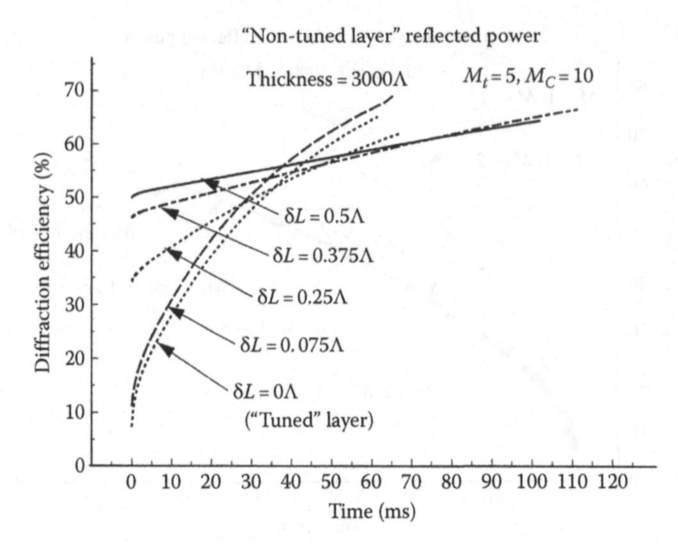

FIGURE 6.20 Transient reflected diffraction efficiency for different detuning lengths.

thickness. Note that the time evolution is non-exponential, and the initial buildup of the diffraction efficiency is more rapid as the thickness increases. Conservation of transmitted and reflected power was observed for all computations.

In Figure 6.20, we show the time evolution of the diffraction efficiencies for a tuned layer of thickness of 3000Λ and several other layers detuned from this value by 0.075Λ, 0.25Λ, 0.375Λ, and 0.5Λ. As can be seen, the amount of detuning has a profound effect on the diffraction efficiency. For a slightly detuned layer, the initial reflection is small, but increases in time rapidly, similar to the tuned case. On the other hand, a severely detuned layer has an initially large reflection, due to boundary mismatch, and the rate of increase of the diffraction efficiency is much slower due to the fact that a smaller percentage of the incident illumination actually enters into and interacts within the photorefractive material to form the reflection grating.

Figure 6.21a and b shows plots of several terms which make up the electron density matrix equation as given by Equation 6.67 (the terms making up Equation 6.67 are specified in Equations 6.56 through 6.59), when a "tuned layer" reflection grating is formed (Figure 6.21a, $\delta L = 0\Lambda$) and when a "non-tuned layer" reflection grating is formed (Figure 6.21b, $\delta L = 0.5\Lambda$). These figures were formed by Fourier summing the slowly varying amplitudes according to Equations 6.53 through 6.55 and then plotting the sums as a function of y_{loc} as shown in Figure 6.21a and b. As can be seen when comparing them a "non-tuned layer," reflection grating causes a significantly different electron density profile to result than when a "tuned layer" reflection grating is present. Figure 6.22 shows plots of the dielectric modulation function (proportional to the electrostatic field inside the reflection grating) when a when a "tuned layer" reflection grating is formed (same case as Figure 6.21a, $\delta L = 0\Lambda$) and when a "non-tuned layer" reflection grating is formed (same case as Figure 6.21b, $\delta L = 0.5\Lambda$). As can be seen from Figure 6.21, detuning of the layer causes a shift in the dielectric modulation function $\Delta\varepsilon$ to occur.

6.6 CONCLUSION

In this section, we have, after appropriate reformulation of the multilayer RCWA, analyzed the time evolution of reflection gratings in a photorefractive material of an arbitrary thickness. This has been accomplished by letting all but the last layer be an integral number of wavelengths and

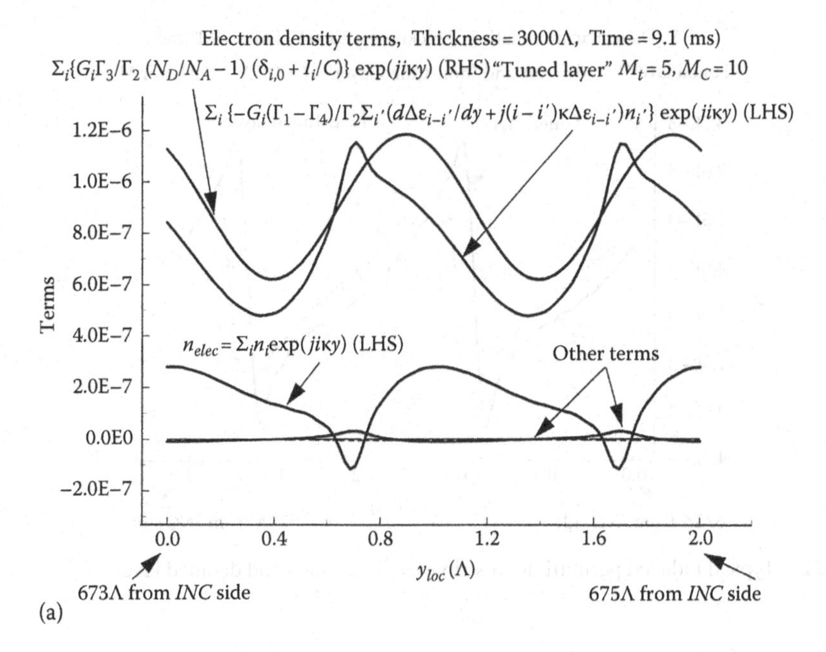

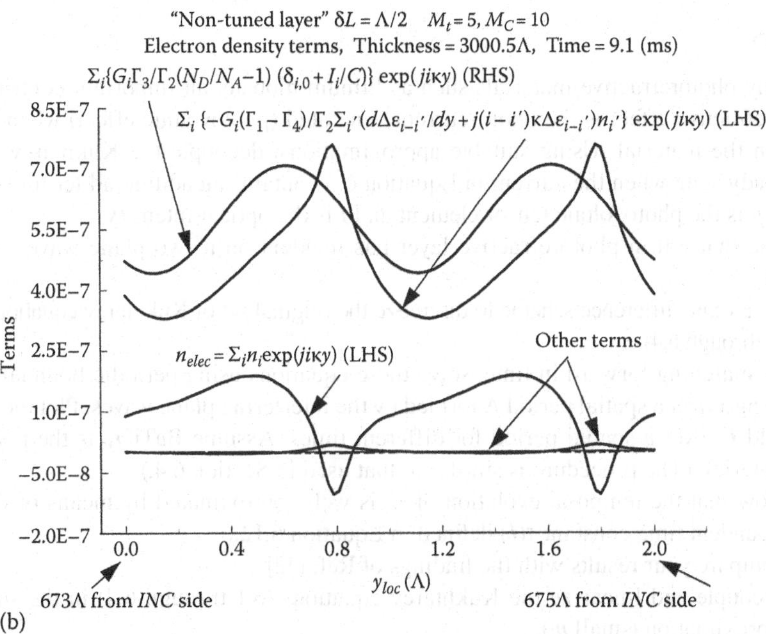

FIGURE 6.21 Spatial distribution of different terms of Equation 6.67 for (a) a tuned layer and (b) a detuned layer.

then letting it have an arbitrary length. The reflected diffraction efficiency has been plotted as a function of time for various layer thicknesses, viz., from 100Λ to 6000Λ. The evolution of the diffraction efficiency has also been shown for various detunings δL ranging from 0Λ to 0.5Λ. Numerical convergence was also analyzed. We have found that a minimum number of spatial harmonics (viz., $M_C = 6$ and $M_t = 3$) should be used for reasonable accuracy. Examples of the spatial distribution of various physical quantities such as the electron density, induced permittivity etc., have been provided.

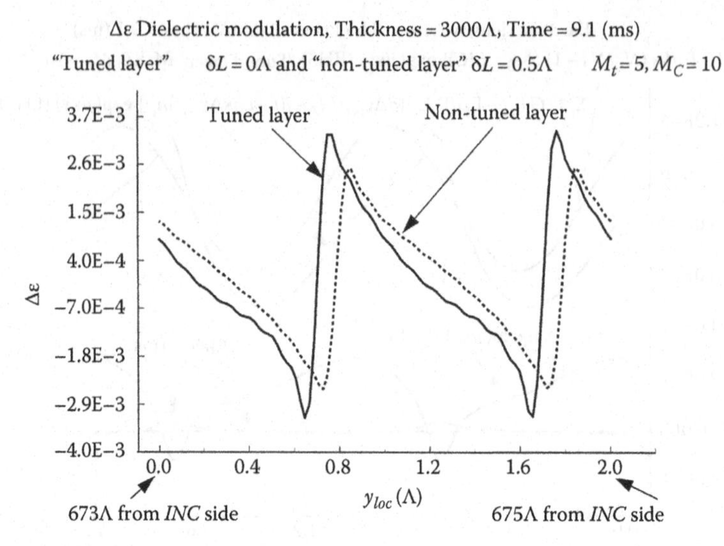

FIGURE 6.22 Typical induced permittivity distribution for a tuned and detuned layer.

PROBLEMS

6.1 In many photorefractive materials such as lithium niobate, an important contribution to the photorefractive effect is the photo-current generated (photovoltaic effect) when light is incident on the material. Using suitable approximations, decouple the Kukhtarev equations in the steady state when the current in Equation 6.3 contains an additional term pI on the RHS, where p is the photovoltaic tensor element and I is the optical intensity.

6.2 Assume that a thin photorefractive layer has incident on it two plane waves of amplitude ratio p.

(a) Use a finite difference scheme to discretize the original set of Kukhtarev equations (Equations 6.1 through 6.4).

(b) By marching forward in time, solve these equations using periodic boundary conditions along x over a spatial period Λ formed by the interfering plane waves. Plot the electrostatic field E_s over a spatial period for different times. Assume $BaTiO_3$ as the photorefractive material. (The procedure is similar to that used in Section 6.4.)

(c) Show that the temporal evolution of E_s is well approximated by means of the intensity-dependent time constant $\tau(I)$ defined in Equation 6.12.

(d) Compare your results with the findings of Ref. [42].

6.3 (a) Decouple and linearize the Kukhtarev equations (6.1 through 6.4) in the small contrast approximation (small m).

(b) Derive the set of differential equations coupling the two incident electric fields during interaction in the photorefractive material.

(c) Exactly solve these equations in the steady state to determine the longitudinal variation of the two electric fields [8].

(d) Solve the differential equations derived in (b) in the transient case. Take the limit as $t \to \infty$ and compare your answer with (c).

6.4 Repeat Problem 6.3, but now including a photovoltaic term in the Kukhtarev equations (see Problem 6.1).

6.5 Assume that two optical waves which are incident on a photorefractive material are frequency offset from one another by Ω. This sets up a running interference pattern proportional to $\cos(Kx-\Omega t)$. Decouple and linearize the Kukhtarev equations (6.1 through 6.4) in the small contrast approximation (small m). Analyze the resulting equations and find an expression for the space-charge field in terms of the intensity and the frequency offset [8].

6.6 Assume that two optical waves which are frequency offset from one another by Ω, as in Problem 6.5, are incident on a photorefractive material at $t = 0$.
(a) Reformulate the numerical algorithm for the exact Kukhtarev equations to account for the traveling wave nature of the intensity grating.
(b) Obtain a numerical solution for the photorefractive and optical response of $BaTiO_3$. What is the effect of varying Ω?

6.7 Investigate two-wave mixing in a photovoltaic material for arbitrary modulation m, and for the case where the photovoltaic contribution is represented in the current equation in the set of Kukhtarev equations as $p(N_D - N_D^+)I$. Plot $|E_{1s}|/m$, Imag$[E_{1s}]/m$ versus E_0 where E_0 is the dc term in the Fourier series expansion of the space-charge field and E_1s represents the amplitude of the fundamental spatial harmonic. (*Hint:* refer to Lee and Chan [43] for description of the numerical scheme, and to Cook et al. [44] for experimental determination of the effect of the photovoltaic field.)

6.8 Investigate multiwave interaction in a diffusion dominated photorefractive material when two incident optical fields, one a plane wave (reference) and one the optical field originating from an object $t(x, y)$ and spatially Fourier transformed by a converging lens of focal length f interact in the material. Assume that the nominal directions of propagation of the two input fields make an angle θ w.r.t. each other in air. Find approximate expressions for the 0, 1, 2, and −1 diffraction orders. Show that the −1 diffraction order contains the forward traveling phase conjugate of the object [33]. What information is there in the second order?

6.9 In the presence of drift and diffusion, show that the space-charge field from the Kukhtarev equations can be expressed in the form $(E_{sn}+(d/d\zeta))[(I_n/1+dE_{sn}/d\zeta)]=g$, where E_{sn}, I_n, and ζ are normalized values for the electrostatic field, intensity and distance, respectively [45]. Using this equation, determine approximate expressions for the variation of the space-charge field with intensity for (a) the drift dominated and (b) diffusion dominated cases.

6.10 Using the form of the ODE for the space-charge field in Problem 6.9, and assuming drift only, find the spatial evolution equations for two-wave mixing in the case of (a) reflection (b) transmission geometry. Solve these equations using appropriate numerical techniques and plot the variation of the pump and the signal waves during energy exchange in drift dominated photorefractive materials.

6.11 Six-wave mixing in photorefractive materials is often observed, when a wave incident at a small angle to the normal to the photorefractive material travels into the material and is reflected off the photorefractive–air interface at the back of the material. Light, also amplified from "noise," travels along the normal in both directions; furthermore, there is higher order diffraction (see Problem 6.8) traveling in the forward and backward directions [46]. (a) Sketch the six participating waves. (b) Using the approximate model of a photovoltaic material, write down the evolution equations for the six waves. (c) By solving these equations or otherwise, predict the behavior of the forward and backward traveling higher order diffracted waves.

REFERENCES

1. A. Ashkin, G.D. Boyd, J.M. Dziedzic, R.G. Smith, A.A. Ballman, J.J. Levinstein, and K. Nassau, Optically induced refractive index inhomogeneities in $LiNbO_3$ and $LiTaO_3$, *Appl. Phys. Lett.* 9, 72–74, 1966.
2. P. Gunter and J.-P. Huignard, eds., *Photorefractive Materials and Their Applications, I & II*, Springer, Berlin, Germany, 1989.
3. J.-P. Huignard and A. Marrakchi, Coherent signal beam amplification in two-wave mixing experiments with photorefractive $Bi_{12}SiO_{20}$ crystals, *Opt. Commun.* 38, 249–258, 1981.
4. J. Feinberg, Self-pumped continuous-wave conjugator using internal reflections, *Opt. Lett.* 7, 486–488, 1982.

5. J.O. White and A. Yariv, Real-time image processing via four-wave mixing in a photorefractive medium, *Appl. Phys. Lett.* 37, 5–7, 1980.
6. P. Yeh and A.E.T. Chiou, Real-time contrast reversal via four-wave mixing in nonlinear media, *Opt. Commun.* 64, 160–162, 1987.
7. K. Ratnam and P.P. Banerjee, Nonlinear theory of two-beam coupling in a photorefractive material, *Opt. Commun.* 107, 522–530, 1994.
8. P. Yeh, *Introduction to Photorefractive Nonlinear Optics*, Wiley Series in Pure and Applied Optics, John Wiley & Sons, Inc., New York, 1993.
9. N.V. Kukhtarev, M.B. Markov, S.G. Odulov, M.S. Soskin, and V.L. Vinetsky, Holographic storage in electrooptic crystals, *Ferroelectrics* 22, 949–964, 1979.
10. R.A. Vazquez, F.R. Vachss, R.R. Neurgaonkar, and D. Ewbank, Large photorefractive coupling coefficient in a thin cerbium-doped strontium barium niobate crystal, *J. Opt. Soc. Am. B* 9, 1932–1941, 1992.
11. J.E. Millerd, E.M. Garmire, M.B. Klein, B.A. Wechsler, F.P. Strohkendl, and G.A. Brost, Photorefractive response at high modulation depths in $Bi_{12}TiO_{20}$, *J. Opt. Soc. Am. B* 8, 1449–1453, 1991.
12. M.R. Belic and M. Petrovic, Unified method for solution of wave equations in photorefractive media, *J. Opt. Soc. Am. B* 11, 481–485, 1994.
13. D.W. Wilson, E.N. Glytsis, N.F. Hartman, and T.K. Gaylord, Beam diameter threshold for polarization conversion photoinduced by spatially oscillating bulk photovoltaic currents in $LiNbO_3$:Fe, *J. Opt. Soc. Am. B* 9, 1714–1725, 1992.
14. M.G. Moharam and T.K. Gaylord, Rigorous coupled-wave analysis of planar grating diffraction, *J. Opt. Soc. Am.* 71, 811–818, 1981.
15. K. Rokushima, J. Yamakita, S. Mori, and K. Tominaga, Unified approach to wave diffraction by space-time periodic anisotropic media, *IEEE Trans. Microwave Theory Tech.* 35, 937–945, 1987.
16. E.N. Glytsis and T.K. Gaylord, Rigorous three-dimensional coupled-wave diffraction analysis of single cascaded anisotropic gratings, *J. Opt. Soc. Am. B* 4, 2061–2080, 1987.
17. P. Gunter, E. Voit, M.Z. Zha, and J. Albers, Self-pulsation and optical chaos in self-pumped photorefractive $BaTiO_3$, *Opt. Commun.* 55, 210–214, 1985.
18. A. Nowak, T.R. Moore, and R.A. Fisher, Observation of internal beam production in $BaTiO_3$ phase conjugator, *J. Opt. Soc. Am. B* 5, 1864–1878, 1988.
19. P.M. Jeffrey and R.W. Eason, Lyapunov exponent analysis of irregular fluctuations in a self-pumped $BaTiO_3$ phase conjugate mirror, *J. Opt. Soc. Am. B* 11, 476–480, 1994.
20. D. Wang, Z. Zhang, X. Wu, and P. Ye, Instabilities in a mutually pumped phase conjugator of $BaTiO_3$, *J. Opt. Soc. Am. B* 7, 2289–2293, 1990.
21. R. Daisy and B. Fischer, Light waves at the interface of linear and photorefractive media, *J. Opt. Soc. Am. B* 11, 1059–1063, 1994.
22. E.D. Baraban, H.-Z. Zhang, and R.W. Boyd, Optical bistability by two-wave mixing in photorefractive crystals, *J. Opt. Soc. Am. B* 9, 1689–1692, 1992.
23. L.-K. Dai, C. Gu, and P. Yeh, Effect of position-dependent time-constant on photorefractive two-wave mixing, *J. Opt. Soc. Am. B* 9, 1693–1697, 1992.
24. H. Kong, C. Wu, and M. Cronin-Golomb, Photorefractive two-beam coupling with reduced spatial coherence, *Opt. Lett.* 16, 1183–1185, 1991.
25. M. Cronin-Golomb, H. Kong, and W. Krolikowski, Photorefractive two-bean coupling with light of partial spatio-temporal coherence, *J. Opt. Soc. Am. B* 9, 1698–1703, 1992.
26. M. Snowbell, M. Horowitz, and B. Fischer, Dynamics of multiple two-wave mixing and fanning in photorefractive materials, *J. Opt. Soc. Am. B* 9, 1972–1982, 1994.
27. E. Serrano, V. Lopez, M. Carrascosa, and F. Agrillo-Lopez, Recording and erasure kinetics in photorefractive materials at large modulation depths, *J. Opt. Soc. Am. B* 11, 670–675, 1994.
28. W.P. Brown and G.C. Valley, Kinky beam paths inside photorefractive crystals, *J. Opt. Soc. Am. B* 10, 1901–1906, 1993.
29. J.M. Jarem and P.P. Banerjee, A nonlinear, transient analysis of two and multi-wave mixing in a photorefractive material using rigorous coupled-mode diffraction theory, *Opt. Commun.* 123, 825–842, 1996.
30. P.P. Banerjee and J. Jarem, Transient wave mixing and recording kinetics in photorefractive barium titanate: A nonlinear coupled mode approach, *Opt. Eng.* 34, 2254–2260, 1995.
31. P.P. Banerjee and J.J. Liu, Perturbational analysis of steady-state and transient-beam fanning in thin and thick photorefractive media, *J. Opt. Soc. B* 10(8), 1417–1423, 1993.
32. P.P. Banerjee and R.M. Misra, Dependence of photorefractive beam fanning on beam parameters, *Opt. Commun.* 100, 166–172, 1993.

33. P.P. Banerjee, H.-L. Yu, D. Gregory, and N. Kukhtarev, Phase conjugation, edge detection and image broadcasting using two-beam coupling in photorefractive potassium niobate, *Opt. Photon. Technol. Lett.* 28, 89–92, 1996.
34. J.M. Jarem and P.P. Banerjee, An exact, dynamical analysis of the Kukhtarev equations in photorefractive, barium titanate using rigorous coupled wave diffraction theory, *J. Opt. Soc. Am. A* 13, 819–831, 1996.
35. R. Alferness, Analysis of propagation at the second-order Bragg angle of a thick holographic grating, *J. Opt. Soc. Am.* 66, 353–362, 1976.
36. A. Korpel, *Acousto-Optics*, Marcel-Dekker, New York, 1988.
37. J.M. Jarem and P.P. Banerjee, Rigorous coupled-wave analysis of photorefractive reflection gratings, *J. Opt. Soc. Am. B*, 15(7), 2099–2106, 1998.
38. J.M. Jarem and P.P. Banerjee, Time domain state variable analysis of induced reflection gratings in photorefractive materials, In *Proceedings SPIE Helographic Materials IV*, 3924, 161–170, San Joe, CA, 1998.
39. M.G. Moharam and T.K. Gaylord, Coupled-wave analysis of reflection gratings, *Appl. Opt.*, 20, 240–244, 1981.
40. M.G. Moharam and T.K. Gaylord, Chain-matrix analysis of arbitrary-thickness dielectric reflection gratings, *J. Opt. Soc. Am.* 72, 187–190, 1982.
41. Z. Zylberberg and E. Marom, Rigorous coupled-wave analysis of pure reflection gratings, *J. Opt. Soc. Am.* 73, 392–401, 1983.
42. N. Singh, S.P. Nader, and P.P. Banerjee, Time dependent nonlinear photorefractive response to sinusoidal gratings, *Opt. Commun.* 136, 487–495, 1997.
43. W.S. Lee and T.S. Chan, Photorefractive hologram writing with high modulation depth in photovoltaic media under different boundary conditions, *Opt. Commun.* 281, 5884–5888, 2008.
44. G. Cook, J. Duignan, and D.C. Jones, Photovoltaic contribution to counter-propagating two beam coupling in photorefractive lithium niobate, *Opt. Commun.* 192, 393–398, 2001.
45. E. DelRe, A. Ciattoni, B. Crosignani, and M. Tamburrini, Approach to space-charge field description in photorefractive crystals, *J. Opt. Soc. Am. B* 15, 1469–1475, 1998.
46. M.A. Saleh, P.P. Banerjee, J. Carns, G. Cook, and D.R. Evans, Stimulated photorefractive backscatter leading to six-wave mixing and phase conjugation in iron-doped lithium niobate, *Appl. Opt.* 46, 6151–6160, 2007.

7 Rigorous Coupled Wave Analysis of Inhomogeneous Cylindrical and Spherical Systems

7.1 INTRODUCTION

In Chapters 4 through 6, the rigorous coupled wave analysis (RCWA) method and the spectral domain techniques were used extensively to treat the solution of Maxwell's equations for planar dielectric systems that were isotropic, anisotropic, and bianisotropic. Chapter 4 concentrated on the case when the dielectric layers were transversely homogeneous and the source and the EM fields of the system could be effectively represented as a Fourier k-space integral (i.e., waveguide slot, dipole antenna, etc.). Chapter 5 concentrated on the case when the dielectric layers were periodic diffraction gratings and the source of the system was Rayleigh plane waves. In Chapter 6, the RCWA method was used to study diffraction from photorefractive gratings.

In this chapter, we will deal with the problem of using the RCWA or exponential matrix method to solve Maxwell's equations in circular and spherical systems that may be inhomogeneous in the radial and angular coordinates and that may be isotropic or anisotropic. In cylindrical or spherical coordinates, the RCWA method is applied by expanding all EM field and source quantities in Floquet harmonics that are periodic in the angles φ or θ. The method is similar to that used when applying the RCWA method to planar diffraction gratings systems where all EM field and source quantities are expanded in Floquet harmonics, which are periodic in the grating periods Λ_x and Λ_z.

Two major differences exist between the cylindrical and spherical RCWA formulation presented in this chapter and the RCWA formulation presented in earlier chapters. The first difference is that in cylindrical or spherical systems, when Maxwell's equations are being reduced to state variable form (a set of first order, partial differential equations), the scale factors associated with Maxwell's equations vary as $1/\rho$, for example, in cylindrical coordinates or vary as $1/r$ in spherical coordinates, whereas in homogeneous planar diffraction grating systems, this variation does not occur. This difference does not cause significant trouble, however, and may be overcome by simply dividing the cylindrical or spherical system into a set of thin layers where the scale factors are nearly homogeneous, and then using a multilayer analysis as was performed in Chapters 5 and 6 to solve the cascaded system. The second major difference is that the field solutions that exist in the uniform space region that bound the inhomogeneous scatterer have Maxwell equation solutions that consist of Hankel and Bessel functions for cylindrical systems and Tesseral harmonics (i.e., half-order Bessel and Legendre polynomial solutions) for spherical systems. When using the RCWA planar diffraction grating method, the Maxwell equation solutions consist of Rayleigh plane waves. Having to use Hankel and Bessel and Tesseral harmonic functions in the boundary matching procedure causes the overall solution to be more complicated than in the planar diffraction grating case. A detailed description of the differences of the RCWA cylindrical and planar diffraction grating formulations will be given, along with several numerical examples.

7.2 RIGOROUS COUPLED WAVE ANALYSIS CIRCULAR CYLINDRICAL SYSTEMS

An important and well-known problem in electromagnetics is the problem of determining the scattering that occurs when an electromagnetic wave is incident on circular cylindrical object. This problem has been extensively studied in the cases where (1) the EM incident wave is an oblique or nonoblique plane wave, (2) the incident EM wave has been generated by a line source, (3) the circular cylindrical scattering object is an inhomogeneous dielectric, and (4) the circular cylindrical object is a dielectric-coated metallic object [1–7]. The problem of determining plane wave and line source scattering from eccentric circular dielectric systems (circular dielectric cylinders of varying dielectric value whose axes are not centered on a single line) has also been studied. Recently, Kishk et al. [8] has obtained a complete solution to this problem. Ref. [8] also gives a complete literature survey of scattering from eccentric and centered circular cylindrical dielectric systems.

A problem concerning circular cylindrical object scattering that has not received a great deal of attention is the problem of determining the scattering and radiation that occurs when the circular cylindrical dielectric system contains a region whose permittivity is inhomogeneous and periodic in the phi(φ) direction. Figure 7.1 shows two examples of such a system. In this section, we treat the cases (1) where the φ-inhomogeneity and its excitation are only periodic over the φ-period of $\Lambda_\varphi = 2\pi$ (e.g., plane wave scattering off a φ-inhomogeneous dielectric cylinder, see Figure 7.1a) and (2) where the φ-inhomogeneity and its excitation possess a higher symmetry than the first case (e.g., Figure 7.1b with a centered line source) in which the φ-period may be taken to be $\Lambda_\varphi = 2\pi/p$, $p \geq 2$ where p is an integer. The φ-inhomogeneous cylindrical dielectric system, which is being studied in this paper, may also be viewed as a circular diffraction grating that has been placed in a circular region.

The solution of the just stated problem may have applications, for example, to scattering from circular shaped, frequency-selective surfaces and scattering from circular surfaces covered with periodically spaced radar absorbing material (RAM). It may also be used as a numerical cross check of other numerical algorithms that study scattering from dielectric systems.

The solution method to be proposed in this section to solve the circular diffraction grating problem will be based on an algorithm called rigorous coupled wave theory, which has been used extensively to determine diffraction from a planar dielectric diffraction grating [9–14]. This algorithm calculates the diffraction from planar grating in four basic steps, which are (1) expand all electric, magnetic field, and dielectric permittivity tensor components in a set of Floquet harmonics (this is an exponential Fourier series whose period is the diffraction grating period); (2) solve Maxwell's equations in the nondiffractive regions on the incident and transmit sides of the diffraction grating; (3) solve Maxwell's equations in the diffractive region of space using a state variable approach (the EM fields in the diffraction gratings consist of an infinite number of forward and backward, propagating and nonpropagating state variable eigenmodes); and (4) use the solutions of Steps (2) and (3) to match EM boundary conditions at the front and back boundaries of the diffraction grating. The solution of Maxwell's equations in the incident and transmit sides of the planar diffraction grating

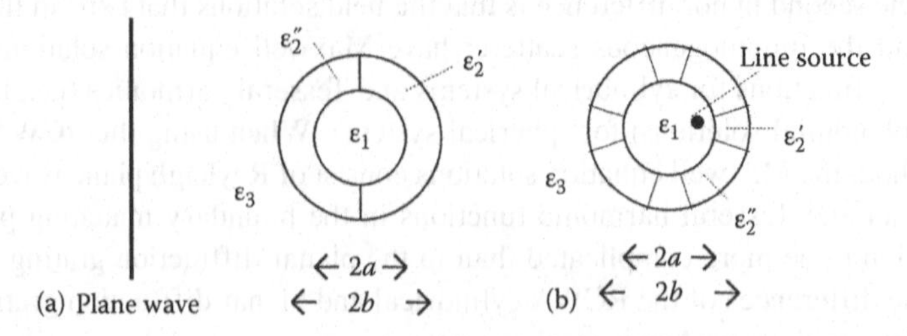

FIGURE 7.1 (a) The geometry of a φ-periodic system when Region 2 is a semicylindrical half shell and a plane wave is incident on the cylindrical system. (b) An eight section four grating period shell system, which is excited by a line source located at $\vec{\rho}_s = \rho_s \hat{a}_x$, $0 \leq \rho_s < a$. (From Jarem, J.M., *J. Electromagn. Waves Appl.*, 11, 197, 1997. With permission of Brill Publishing.)

consists of an infinite number of propagating and evanescent plane waves whose x-propagation factors are $e^{-j\tilde{k}_{xi}\tilde{x}}$ ($\tilde{x}$ is a coordinate along the grating interface) where $\tilde{k}_{xi} = \tilde{k}_{x0} - i\tilde{K}_x$, where $\tilde{k}_{x0} = k_0\sqrt{\varepsilon_I}\sin\theta_I$, $k_0 = 2\pi/\lambda$ where λ is the free space wavelength, $i = -\infty,\ldots,-1, 0, 1,\ldots\infty$, $\tilde{K}_x = 2\pi/\tilde{\Lambda}_x$, $\tilde{\Lambda}_x$ is the grating period, ε_I is the relative permittivity on the incident side, and θ_I is incident angle of the plane wave incident on the diffraction grating. This algorithm is the subject of Chapters 5 and 6.

The solution method described for planar diffraction gratings will now be used to determine the EM fields of the circular cylindrical φ-periodic problem (Section 7.3). In the regions bounding the φ-periodic region, (see Figure 7.1), the Maxwell equation solution will consist of an infinite sum of Bessel and Hankel function solutions $J_i e^{ji\varphi}$, $H_i^{(1)} e^{ji\varphi}$, and $H_i^{(2)} e^{ji\varphi}$ where $i = -\infty, \ldots, -1, 0, 1, \ldots, \infty$. The $e^{ji\varphi}$ factor that makes up the cylindrical solutions is analogous to the $e^{-j\tilde{k}_{xi}\tilde{x}}$ Floquet harmonic x-propagation factor used in the planar diffraction analysis. In the φ-periodic cylindrical region, all electric field, magnetic field, and dielectric permittivity tensor components are expanded in a set of $e^{ji\varphi}$ exponential Fourier series harmonics (Floquet harmonics), and Maxwell's equations are then cast in state variable form and then solved numerically. Because of the radial inhomogeneous nature of the state variable equations in cylindrical coordinates, the state variable equations are solved by dividing the φ-periodic cylindrical region into a series of thin layers (thin enough where the radial coordinate is approximately constant in the layer), solving Maxwell state variable equations in each layer, and the matching EM boundary conditions from one layer to the next to obtain an overall solution in the φ-periodic cylindrical region. This ladder approach is identical to the approach used by [12] to solve planar surface relief diffraction gratings.

In Section 7.3, the RCWA method is used to study EM scattering from spatially inhomogenous cylindrical systems and next used to determine the radiation and scattering that arises from inhomogeneous, anisotropic cylindrical dielectric and permeable systems, which have arbitrary radial and azimuthal spatial variation (Section 7.4). The RCWA state matrix equations and the associated boundary matrix equations (derived from a multilayer ladder analysis) are presented and solved for the cases when a plane wave (TM polarization, electric field parallel to the cylinder axis) or electric line source is incident on a cylinder that possesses an inhomogeneous permittivity profile $\varepsilon(\rho, \varphi)$ and possesses inhomogeneous, anisotropic permeability profiles $\mu_{\rho\rho}(\rho, \varphi)$, $\mu_{\rho\varphi}(\rho, \varphi)$, $\mu_{\varphi\rho}(\rho, \varphi)$, and $\mu_{\varphi\varphi}(\rho, \varphi)$. In this paper, radiation and scattering from three inhomogeneous examples were studied using the cylindrical RCWA method.

Finally, we present a rigorous coupled wave analysis of the electromagnetic radiation that occurs when a centered electric dipole excites power and energy in a general three dimensional inhomogeneous spherical system (Section 7.5). The formulation consists of a multilayer state variable (SV) analysis of Maxwell's equations in spherical coordinates (the SV analysis used transverse-to-r spherical EM field components) as well as a presentation of the EM fields, which exist in the interior and exterior regions that bound the inhomogeneous spherical system. A detailed description of the matrix processing, which is involved with finding the final EM fields of the overall system, is given.

7.3 RIGOROUS COUPLED WAVE ANALYSIS MATHEMATICAL FORMULATION

7.3.1 Introduction

This section is concerned with the problem of determining the EM fields that arise when a plane wave (see Figure 7.1a) and an off-center-interior line source (see Figure 7.1b) excite EM fields in a circular cylindrical dielectric system as shown in Figure 7.1. The EM analysis will be carried out by (1) solving Maxwell's equation in the interior and exterior regions of Figure 7.1 in terms of cylindrical Bessel functions, (2) solving Maxwell's equation in the dielectric shell region by using a multilayer state variable approach, and (3) matching EM boundary conditions at the interfaces. It is convenient to introduce normalized coordinates. We let $a = k_0\tilde{a}$, $b = k_0\tilde{b}$, $\rho = k_0\tilde{\rho}$, etc., where unnormalized coordinates are in *meters* and $k_0 = 2\pi/\lambda$ is the free space wave number (1/m).

7.3.2 Basic Equations

It is assumed that all fields and the medium are z independent and that the relative dielectric permittivity in Region 2 is given by

$$\varepsilon_2(\rho,\varphi)=\sum_{i=-\infty}^{\infty}\breve{\varepsilon}_i(\rho)e^{ji\varphi},\quad 0\le\varphi\le 2\pi,\quad a\le\rho\le b \tag{7.1}$$

where $\breve{\varepsilon}_i(\rho)$ represents φ-exponential Fourier coefficients in Region 2 (see Figure 7.1). The permeability is assumed to be that of free space $\mu = \mu_0$.

To begin the analysis, we determine the EM fields in the regions interior and exterior to the Region 2 dielectric shell. In the interior region, the EM fields of an off-center line source and the general electric scattered fields are given by [5,6]

$$E_z^{(1)}=\left[\sum_{i=-\infty}^{\infty}c_i^{(1)}J_i(X_1)e^{ji\varphi}\right]+c_0^I H_0^{(2)}\left(\sqrt{\varepsilon_1}\left|\vec{\rho}-\vec{\rho}_s\right|\right) \tag{7.2a}$$

$$H_\varphi^{(1)}=\frac{-j\sqrt{\varepsilon_1}}{\eta_0}\frac{\partial E_z^{(1)}}{\partial X_1} \tag{7.2b}$$

where $c_0^I=-\omega\mu_0 I/4$, $\vec{\rho}_s=\rho_s\hat{a}_x$, $\eta_0=\sqrt{\mu_0/\varepsilon_0}$, $X_1=\sqrt{\varepsilon_1}k_0\tilde{\rho}=\sqrt{\varepsilon_1}\,\rho$, $0\le\rho\le a$, and I is the electric current line source. The EM fields in the region $\rho_s<\rho\le a$ may be expressed as [5,6]

$$E_z^{(1)}=\sum_{i=-\infty}^{\infty}\left[c_i^{(1)}J_i(X_1)+c_0^I J_i(X_{1s})H_i^{(2)}(X_1)\right]e^{ji\varphi}=\sum_{i=-\infty}^{\infty}s_{zi}^{(1)}(X_1)e^{ji\varphi} \tag{7.3a}$$

$$H_\varphi^{(1)}=\frac{-j}{\eta_0}\sum_{i=-\infty}^{\infty}\sqrt{\varepsilon_1}\left[c_i^{(1)}J_i'(X_1)+c_0^I J_i(X_{1s})H_i^{(2)\prime}(X_1)\right]e^{ji\varphi}=\frac{-j}{\eta_0}\sum_{i=-\infty}^{\infty}u_{\varphi i}^{(1)}(X_1)e^{ji\varphi} \tag{7.3b}$$

where $J'(X)=dJ(X)/dX$, etc., $X_{1s}=\sqrt{\varepsilon_1}k_0\tilde{\rho}_s=\sqrt{\varepsilon_1}\rho_s$, and ε_1 is the relative permittivity of Region 1. In the exterior region, the EM fields are a sum of an incident plane wave (electric field given by $\vec{E}^I=E_0^I e^{-j\sqrt{\varepsilon_3}x}\hat{a}_z$, $x=k_0\tilde{x}$) and a general EM scattered wave. The exterior EM fields in Region 3 are given by

$$E_z^{(3)}=\sum_{i=-\infty}^{\infty}\left[c_i^{(3)}H_i^{(2)}(X_3)+E_i^I J_i(X_3)\right]e^{ji\varphi}=\sum_{i=-\infty}^{\infty}s_{zi}^{(3)}(X_3)e^{ji\varphi} \tag{7.4a}$$

$$H_\varphi^{(3)}=\frac{-j}{\eta_0}\sum_{i=-\infty}^{\infty}\sqrt{\varepsilon_3}\left[c_i^{(3)}H_i^{(2)\prime}(X_3)+E_i^I J_i'(X_3)\right]e^{ji\varphi}=\frac{-j}{\eta_0}\sum_{i=-\infty}^{\infty}u_{\varphi i}^{(3)}(X_3)e^{ji\varphi} \tag{7.4b}$$

where $E_i^I=E_0^I j^{-i}$, $X_3=\sqrt{\varepsilon_3}k_0\tilde{\rho}=\sqrt{\varepsilon_3}\,\rho$, and ε_3 is the relative permittivity in Region 3.

In Region 2, the middle cylindrical dielectric region, we divide the dielectric region into L thin shell layers of thickness d_ℓ, $b-a=\sum_{\ell=1}^{L} d_\ell$ and solve Maxwell's equations in cylindrical coordinates by a state variable approach in each thin layer. The layers are assumed to be thin enough in order that the ρ dependence of $\varepsilon_2(\rho, \varphi)$ and the ρ scale factors may be treated as a constant in each layer. Letting $\rho = k_0\tilde{\rho}$, we find that Maxwell's equations in a cylindrical shell of radius ρ are given by

$$\frac{\partial E_z}{\partial \rho} = j\eta_0 H_\varphi \tag{7.5a}$$

$$\frac{\partial[\eta_0 \rho H_\varphi]}{\partial \rho} + \frac{1}{j\rho}\frac{\partial^2 E_z}{\partial \varphi^2} = j\varepsilon_2 \rho E_z \tag{7.5b}$$

To solve Equation 7.5a and b, we expand E_z, $\varepsilon_2(\rho, \varphi)$, and $\eta_0 \rho H_\varphi$ in the Floquet harmonics:

$$E_z = \sum_{i=-\infty}^{\infty} s_{zi}(\rho) e^{ji\varphi} \tag{7.6a}$$

$$\eta_0 \rho H_\varphi = \sum_{i=-\infty}^{\infty} u_{\varphi i}(\rho) e^{ji\varphi} \tag{7.6b}$$

$$\varepsilon_2(\rho,\varphi) E_z = \sum_{i=-\infty}^{\infty}\left[\sum_{i'=-\infty}^{\infty} \breve{\varepsilon}_{i-i'}\, s_{zi'}\right] e^{ji\varphi} \tag{7.6c}$$

If these expansions are substituted in Equation 7.5a and b, and after letting $\underline{s}_z(\rho) = [s_{zi}(\rho)]$, $\underline{u}_\varphi(\rho) = [u_{\varphi i}(\rho)]$ be column matrices and $\underline{\underline{\varepsilon}} = [\breve{\varepsilon}_{i-i'}]$, $\underline{\underline{K}} = [iK\delta_{i,i'}]$, $K = 2\pi/\psi_0$ (ψ_0 is the circular grating period and $\delta_{i,i'}$ is the Kronecker delta) be square matrices we find after a small amount of manipulation

$$\frac{\partial \underline{V}}{\partial \rho} = \underline{\underline{A}}\,\underline{V}, \quad \underline{V} = \begin{bmatrix} \underline{s}_z \\ \underline{u}_\varphi \end{bmatrix}$$

$$\underline{\underline{A}} = \begin{bmatrix} \underline{\underline{0}} & \dfrac{(\underline{\underline{I}})}{\rho} \\ \dfrac{-(\underline{\underline{K}}\,\underline{\underline{K}})}{\rho} + \rho\underline{\underline{\varepsilon}} & \underline{\underline{0}} \end{bmatrix} \tag{7.7}$$

In the ℓth cylindrical shell, it is convenient to introduce the local coordinates $s_1 = \rho - b$ for $b - d_1 \le \rho \le b$, $s_2 = \rho - (b - d_1)$ for $b - d_1 - d_2 \le \rho \le b - d_1, \ldots,$ and $s_L = \rho - [b - d_1 - d_2 \cdots - d_{L-1}]$ for $b - d_1 - d_2 \cdots - d_L \le \rho \le b - d_1 - d_2 \cdots - d_{L-1}$.

The state variable equation (Equation 7.7) in each cylindrical shell may be expressed in the local coordinates. Further, if the thickness of each cylindrical shell is chosen to be sufficiently thin so that the ρ variation in $\underline{\underline{A}}$ is negligible, the $\underline{\underline{A}}(\rho)$ matrix of Equation 7.7 may be approximated by the thin shell's midpoint value where $\rho_1^{mid} = b - d_1/2$, $\rho_2^{mid} = b - d_1 - d_2/2, \ldots,$ $\rho_L^{mid} = b - d_1 -, \cdots d_{L-1} - d_L/2$.

Letting $\underline{A}_\ell = \underline{A}\big|_{\rho_\ell^{mid}}$, we have the approximate state variable equation in each thin shell given by

$$\frac{\partial \underline{V}_\ell(s_\ell)}{\partial s_\ell} = \underline{A}_\ell \underline{V}_\ell(s_\ell), \quad \ell = 1, \ldots, L \tag{7.8}$$

If Equation 7.8 is truncated at order M_T ($i = -M_T, \ldots, -1, 0, 1, \ldots, M_T$), Equation 7.8 represents a $N_T = 2(2M_T + 1)$ state variable equation (with matrix $(\underline{A}_\ell)_{N_T \times N_T}$). The solution of this equation is given by

$$\underline{V}_{\ell n}(s_\ell) = \underline{V}_{\ell n}\, e^{q_{n\ell} s_\ell} \tag{7.9}$$

where $q_{n\ell}$ and $\underline{V}_{\ell n}$ are the nth eigenvalue and eigenvector of the constant matrix $\underline{A}_\ell$ ($\underline{A}_\ell \underline{V}_{\ell n} = q_{n\ell} \underline{V}_{\ell n}$). Using the $\underline{V}_{\ell n}(s_\ell)$ eigenvector solution, the general EM fields in the ℓth thin shell region are given by

$$E_{z\ell} = \sum_{i=-M_T}^{M_T} \sum_{n=1}^{N_T} c_{n\ell} s_{zin\ell} e^{q_{n\ell} s_\ell} \tag{7.10a}$$

$$\eta_0 \rho H_{\varphi\ell} = \sum_{i=-M_T}^{M_T} \sum_{n=1}^{N_T} c_{n\ell} u_{\varphi in\ell} e^{q_{n\ell} s_\ell} \tag{7.10b}$$

$$\underline{V}_{\ell n}^t = [\underline{s}_{zn\ell}^t \, \underline{u}_{\varphi n\ell}^t] \tag{7.10c}$$

where t represents matrix transpose. For $M_T = 1$, for example, by suppressing the $n\ell$ subscripts, we have

$$\underline{V}^t = [V_1, V_2, V_3, V_4, V_5, V_6]$$

$$\underline{s}_z^t = [s_{z,-1}, s_{z,0}, s_{z,1}] = [V_1, V_2, V_3]$$

$$\underline{u}_\varphi^t = [u_{\varphi,-1}, u_{\varphi,0}, u_{\varphi,1}] = [V_4, V_5, V_6] \tag{7.11}$$

To proceed further, it is necessary to match EM boundary conditions at all interfaces to determine all the unknowns of the system. If the series in Regions 1 and 3 are truncated for $|i| \le M_T$, then there are $2M_T + 1$ unknowns in Region 1, $2M_T + 1$ unknowns in Region 3, and $LN_T = L[2(2M_T + 1)]$ unknowns in Regions $\ell = 1, \ldots, L$. There are $L + 1$ interfaces, and $2(2M_T + 1)$ equations to be matched at each interface ($(2M_T + 1)$ equations for each $e^{ji\varphi}$ coefficient of E_z and $(2M_T + 1)$ equations for each $e^{ji\varphi}$ coefficient of H_φ). Thus for every truncation order M_T, there are an equal number of equations and unknowns from which the EM solution of the overall system may be obtained.

Although a large matrix equation exists from which the overall solution of the problem may be obtained, a more efficient solution method is to use a ladder approach [12] (i.e., successively relate unknown coefficients from one layer to the next) to express the $c_{n\ell}$ coefficients of the Lth last layer in terms of the $c_{n\ell}$ coefficients of the first layer and then match boundary conditions at $\rho = a$ and b

interfaces to obtain the final unknowns of the system. At the ℓth and $(\ell + 1)$th interface, by matching the $E_{z\ell}$ and $\eta_0\rho H_{\varphi\ell}$ fields to the $E_{z,\ell+1}$ and $\eta_0\rho H_{\varphi\ell+1}$ fields, we have

$$\sum_{n=1}^{N_T} c_{n\ell} s_{zin\ell} e^{-q_{n\ell} d_\ell} = \sum_{n=1}^{N_T} c_{n,\ell+1} s_{zin,\ell+1} \tag{7.12a}$$

$$\sum_{n=1}^{N_T} c_{n\ell}\, u_{\varphi in\ell} e^{-q_{n\ell} d_\ell} = \sum_{n=1}^{N_T} c_{n,\ell+1} u_{\varphi in,\ell+1} \qquad \begin{matrix} i = M_T, \ldots, M_T \\ \ell = 1, \ldots, L-1 \end{matrix} \tag{7.12b}$$

Letting $\underline{C_\ell^t} = [c_{1\ell}, \ldots, c_{N_T\ell}]$, these equations may be written as

$$\underline{\underline{D}}_\ell \underline{C}_\ell = \underline{\underline{E}}_\ell \underline{C}_{\ell+1} \tag{7.13a}$$

or

$$\underline{C}_{\ell+1} = \underline{\underline{E}}_\ell^{-1} \underline{\underline{D}}_\ell \underline{C}_\ell = \underline{\underline{F}}_\ell \underline{C}_\ell \quad \ell = 1, \ldots, L-1 \tag{7.13b}$$

where the −1 superscript denotes matrix inverse. Substituting successively, we have

$$\underline{C}_L = \underline{\underline{F}}_{L-1}(\underline{\underline{F}}_{L-2} \cdots (\underline{\underline{F}}_1 \underline{C}_1)) = \underline{\underline{M}}\underline{C}_1 \tag{7.14}$$

At the $\rho = a$ boundary, if we match the $E_z^{(1)}$ solution with the E_{zL} solution (the $\ell = L$ thin layer is assumed adjacent to Region 1 and the $\ell = 1$ thin layer is assumed adjacent to Region 3) and solve for the Region 1 $c_i^{(1)}$ coefficient, we find

$$c_i^{(1)} = \frac{\left[-c_0^I J_i(X_{1s}) H_i^{(2)}(X_{1a}) + \sum_{n=1}^{N_T} c_{nL} s_{zinL} e^{-q_{nL} d_L}\right]}{J_i(X_{1a})} \tag{7.15}$$

where $X_{1a} = \sqrt{\varepsilon_1}\, a$.

If the $\eta_0 a H_\varphi^{(1)}$ solution is matched with the $\eta_0 a H_{\varphi L}$ solution, the $c_i^{(1)}$ coefficient is substituted, and when the well-known Wronskian equation for Bessel functions is used, it is found that

$$c_0^I J_i(X_{1s}) = \sum_{n=1}^{N_T} c_{nL} e^{-q_{nL} d_L} \left[-\frac{\pi J_i(X_{1a})}{2}\right] \left\{ ja\sqrt{\varepsilon_1}\, \frac{J_i'(X_{1a})}{J_i(X_{1a})} s_{zinL} + u_{\varphi inL} \right\}, \quad i = -M_T, \ldots, 0, \ldots, M_T \tag{7.16}$$

At the $\rho = b$ boundary after matching the tangential electric field $E_z^{(3)}$ from Region (3) to the electric field E_{z1} from the $\ell = 1$ layer and solving for the Region 3 $c_i^{(3)}$ coefficient, we have

$$c_i^{(3)} = \frac{\left[-E_i^I J_i(X_{3b}) + \sum_{n=1}^{N_T} c_{n1} s_{zin1}\right]}{H_i^{(2)}(X_{3b})}, \quad i = -M_T, \ldots, 0, \ldots, M_T \tag{7.17}$$

where $X_{3b} = \sqrt{\varepsilon_3}b$. If the $\eta_0 bH_\varphi^{(3)}$ is matched with the $\eta_0 bH_{\varphi 1}$ field solution of the first layer, the $c_i^{(3)}$ coefficient is substituted and again when a Wronskian Bessel function relation is used, it is found

$$E_i^I = \sum_{n=1}^{N_T} c_{n1}\left[\frac{\pi H_i^{(2)}(X_{3b})}{2}\right]\left\{jb\sqrt{\varepsilon_3}\frac{H_i^{(2)\prime}(X_{3b})}{H_i^{(2)}(X_{3b})}s_{zin1} + u_{\varphi in1}\right\},\quad i = -M_T,\ldots,0,\ldots,M_T \tag{7.18}$$

Equation 7.16 represents a set of $2M_T + 1$ equations, Equation 7.18 represents a set of $2M_T + 1$ equations, and the matrix equation (Equation 7.14) represents a set of $N_T = 2(2M_T + 1)$ equations. Thus, Equations 7.14, 7.16, and 7.18 represent a set of $2N_T = 4(2M_T + 1)$ equations to calculate the $2N_T$ set of unknowns represented by $\underline{C}_1$ and $\underline{C}_L$. Once these quantities are known, all other unknown coefficients in the system may be found.

An important quantity to calculate is the normalized power of each order. We consider the important cases when the power is either radiated from the line source in Region 1 ($c_0^I \neq 0$, $E_0^I = 0$) or the power is scattered by a plane wave from Region 3 ($c_0^I = 0$, $E_0^I \neq 0$). In the case, when $c_0^I \neq 0$, $E_0^I = 0$, it is useful to calculate the normalized power of each order radiated at three different places, namely, at $\rho = a$, b, and ∞. These are useful points to calculate the normalized power because, according to the laws of conservation of power, the sum of the power over all orders at each interface should be equal (or conserved) if there are no losses in the system. Thus, a check of the power conservation is a check of the numerical consistency of the solution. It is useful secondly because it gives information on how the different power levels of the orders change as the EM waves propagate through the system. After the substitution of the electric and magnetic fields into the Poynting real power formula, and carrying out the φ-integrals, the normalized power in each order ($P_{Ni} = P_i^{RAD}/P^{INC}$, i is the order) is given by at $\rho = a$ by

$$P_{Ni}(a) = \frac{\pi a\sqrt{\varepsilon_1}}{2\left|c_0^I\right|^2}\mathrm{Re}\left\{-s_{zi}^{(1)}(X_{1a})u_{\varphi i}^{(1)*}(X_{1a})\right\} \tag{7.19}$$

where $s_{zi}^{(1)}(X_{1a})$, $u_{\varphi i}^{(1)}(X_{1a})$ are defined in Equation 7.3 and the * represents the complex conjugate. At $\rho = b$, the normalized power radiated by the ith order is given by

$$P_{Ni}(b) = \frac{\pi b\sqrt{\varepsilon_3}}{2\left|c_0^I\right|^2}\mathrm{Re}\left\{-j\left|c_i^{(3)}\right|^2 H_i^{(2)}(X_{3b})H_i^{(2)\prime *}(X_{3b})\right\} \tag{7.20}$$

At $\rho = \infty$, the normalized power radiated by the ith order is given by

$$P_{Ni}(\infty) = \frac{\left|c_i^{(3)}\right|^2}{\left|c_0^I\right|^2} \tag{7.21}$$

For the plane wave scattering case ($c_0^I = 0$, $E_0^I \neq 0$), it is useful to calculate the normalized scattered power ({(power/meter)/wavelength}/{Poynting power intensity (W/m^2)}) in each order. The normalized scattered power at $\rho = b$ is given by

$$P_{Ni}^{Scat}(b) = \frac{(P_i^{Scat}(b)/\lambda)}{S_{INC}} = b\sqrt{\varepsilon_3}\,\mathrm{Re}\left\{-j\frac{\left|c_i^{(3)}\right|^2}{\left|E_0^I\right|^2}H_i^{(2)}(X_{3b})H_i^{(2)\prime *}(X_{3b})\right\} \tag{7.22}$$

The normalized scattered power at $\rho = \infty$ is given by

$$P_{Ni}^{Scat}(\infty) = \frac{(P_i^{Scat}(\infty)/\lambda)}{S_{INC}} = \frac{2}{\pi}\frac{\left|c_i^{(3)}\right|^2}{\left|E_0^I\right|^2} \tag{7.23}$$

In Equations 7.22 and 7.23, S_{INC} is the power per unit area (W/m^2) of the incident plane wave, and P_i^{scat} is the scattered power per unit length (W/m) of the ith order.

7.3.3 Numerical Results

In this section, two numerical examples of scattering from φ-periodic cylindrical systems are presented. In the first example, we study the scattering that occurs when a plane wave is incident on the φ-periodic dielectric system shown in Figure 7.1a. In this cylindrical system, the inner cylinder has a relative dielectric value of $\varepsilon_1 = 1.5$, Region 2 consists of two semicircular dielectric regions where $\varepsilon_2' = 2.8$ for the right half dielectric region ($-90° \leq \phi \leq 90°$) and $\varepsilon_2'' = 2.3$ for the left half dielectric region, and Region 3 consists of free space $\varepsilon_3 = 1$. The ϕ-Fourier coefficients (square wave Fourier coefficients) are given by

$$\breve{\varepsilon}_0 = (\varepsilon_2' - \varepsilon_2'')\left(\frac{\psi_1}{\psi_2}\right) + \varepsilon_2''$$
$$\breve{\varepsilon}_i = (\varepsilon_2' - \varepsilon_2'')\left(\frac{\psi_1}{\psi_2}\right)\frac{\sin(\pi i)}{\pi i}, \quad |i| \geq 1 \tag{7.24}$$

where $\psi_0 = 2\pi$ and $\psi_1 = \pi$.

Figure 7.2 shows a comparison of the φ-periodic semicylindrical shell plane wave scattered power in each order (calculated at $\tilde{\rho} = \tilde{b}$, normalized according to Equation 7.22, using $M_T = 20$ and $L = 30$ layers) with that of a uniform dielectric shell as a function of order i when $b = k_0\tilde{b} = 10$. In Region 2, the uniform shell dielectric value was taken to be $\varepsilon_2 = 2.5$. (This is value as the average or bulk dielectric value used for the semicylindrical shell.) The scattering from the uniform dielectric shell (*dashed curves* Figure 7.2) was calculated by both the current state variable algorithm (using $M_T = 20$ and $L = 30$ layers) and by solving Maxwell's equations in Regions 1–3 in terms of Bessel and Hankel functions and then matching electromagnetic boundary conditions at the interfaces.

The numerical results obtained by both methods were so close that the two dashed curves shown in Figure 7.2 (both labeled *uniform*) cannot be distinguished. As can be seen from Figure 7.2, the half cylinder shell causes an increased oscillation of the order power over that of the uniform shell. We note that the order power is symmetric for positive and negative values of the order as is expected. We note that the order $m = \pm 5$ produces the largest change in diffracted power from the uniform shell case for the value of $b = k_0\tilde{b} = 10$ that was used.

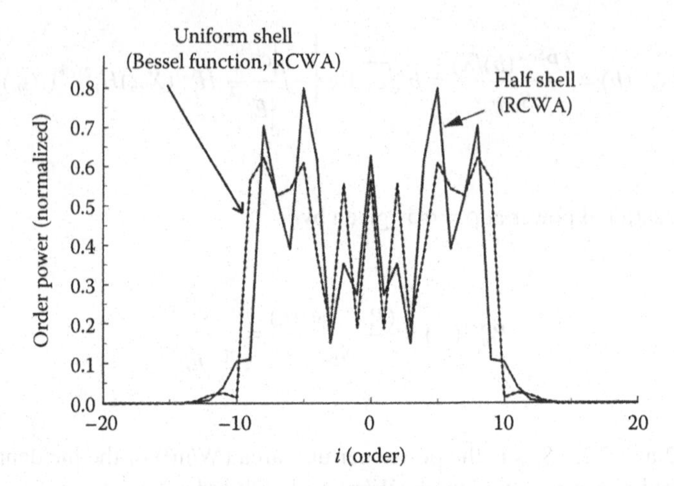

FIGURE 7.2 A comparison of the φ-periodic semicylindrical shell plane wave scattered power in each order (see Figure 7.1a) (calculated at $\tilde{\rho} = \tilde{b}$, normalized according to Equation 7.22) with that of a uniform dielectric shell as a function of order i when $b = k_0\tilde{b} = 10$. In Region 2, the uniform shell dielectric value was taken to be $\varepsilon_2 = 2.5$. (From Jarem, J.M., *J. Electromagn. Waves Appl.*, 11, 197, 1997. With permission of Brill Publishing.)

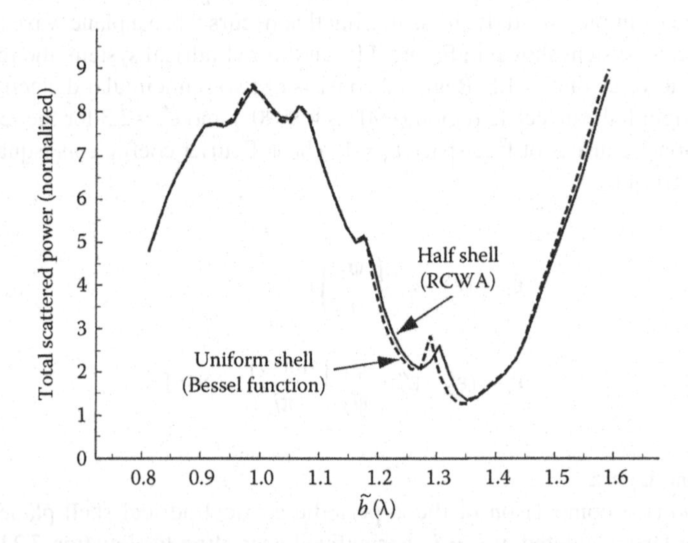

FIGURE 7.3 The total plane wave scattered power from the semicylindrical half shell (see Figure 7.1) (calculated at $\tilde{\rho} = \tilde{b}$, *solid line curve*, normalized, see Equation 7.23, labeled periodic) when $a = k_0\tilde{a} = 5$ and where $\tilde{b}$ ranges from $\tilde{b} = 0.8\lambda$ to 1.6λ. The value of $M_T = 10$ was used to make this plot. For comparison, Figure 7.3 also shows the total plane wave scattered power (*dashed curve* Figure 7.3) that results when a plane wave is incident on the same dielectric system, which has already been described in Figure 7.3, except that Region 2 is now taken to be a uniform dielectric shell whose dielectric value is $\varepsilon_2 = 2.5$. (From Jarem, J.M., *J. Electromagn. Waves Appl.*, 11, 197, 1997. With permission of Brill Publishing.)

Figure 7.3 shows the total plane wave scattered power from the semicylindrical half shell described at the beginning of this section (see Figure 7.1a) (calculated at $\tilde{\rho} = \tilde{b}$, *solid line curve*, normalized, see Equation 7.22) when $a = k_0\tilde{a} = 5$ and where $\tilde{b}$ ranges from $\tilde{b} = 0.8\lambda$ to 1.6λ. The values of $M_T = 10$ and $L = 30$ layers (for each Region 2 shell of inner radius a and outer radius b) were used to make this plot. For comparison, Figure 7.3 also shows the total plane wave scattered power (*dashed curve* Figure 7.3) that results when a plane wave is incident on the same dielectric system, which has already been described in Figure 7.3, except that Region 2 is now taken to be a

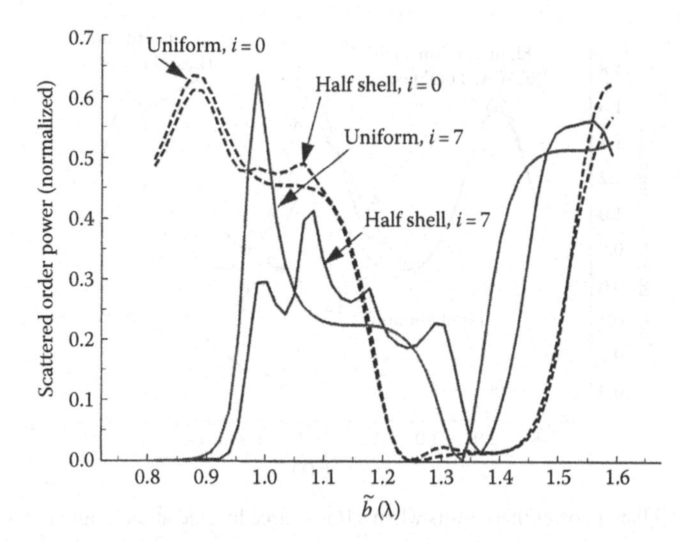

FIGURE 7.4 A comparison of the uniform and φ-periodic half shell dielectric systems (same case as described in Figures 7.2 and 7.3) for the orders of $i = 0$ and 7 when $a = k_0\tilde{a} = 5$ and $b = k_0\tilde{b}$ is varied from values of $b = 5$ to 10. (From Jarem, J.M., *J. Electromagn. Waves Appl.*, 11, 197, 1997. With permission of Brill Publishing.)

uniform dielectric shell whose dielectric value is $\varepsilon_2 = 2.5$. The scattering from the uniform shell (*dashed curve* Figure 7.3) was calculated by both the RCWA algorithm and by solving Maxwell's equations in Regions 1–3 in terms of Bessel and Hankel functions and then matching electromagnetic boundary conditions at the interfaces. Nearly identical numerical results were obtained by both methods. The *dashed curve* shown in Figure 7.3 was calculated by using Bessel and Hankel functions in all three regions and then matching boundary conditions at the interfaces. As can be seen from Figure 7.3, the presence of the semicylindrical shell results in a small but perceptible difference in the power that is scattered when compared to the scattered power from the uniform shell system.

Figure 7.4 shows a comparison of the uniform and φ-periodic half shell dielectric systems (same case as described in Figures 7.2 and 7.3, normalized by Equation 7.22) for the orders of $i = 0$ and 7 when $a = k_0\tilde{a} = 5$ and $b = k_0\tilde{b}$ is varied from values of $b = 5$ to 10 ($M_T = 20$ and $L = 30$ layers). As can be seen from Figure 7.4 for the $i = 0$ order that the uniform and half shell case have only a small difference in the scattered power, whereas for the $i = 7$ order, a large difference in the scattered order power occurs.

Figure 7.5 shows the radiated power (normalized) that results when a line source located at the origin radiates into the quarter φ-periodic shell shown in Figure 7.1b. The values of the relative dielectric in the Region 2 quarter shell centered at $\varphi = 0°(|\varphi| \le 45°)$ are $\varepsilon_2' = 3.25$ for $-22.5° \le \phi \le 22.5°$ and $\varepsilon_2'' = 1.75$ for $22.5° < |\varphi| \le 45°$. The Region 2 quarter shell regions centered at $\varphi = 90°$, 180°, and 270° repeat the $\varphi = 0°$ centered pattern. The Regions 1 and 3 dielectric values are the same as those used in the first example. In this case, because of the centered location of the line source, the grating period of the system may be taken to be $\psi_0 = \pi/2$. Using this grating period ($\psi_0 = \pi/2$) and the Region 2 dielectric values already given, the Fourier coefficients are given by Equation 5.182 with $\psi_1 = 45°$. Also shown in Figure 7.5, for comparison, is the radiation that results when a line source radiates through a uniform dielectric shell (Regions 1 and 3 have the same values as previous examples), and Region 2 has a bulk dielectric value of $\varepsilon_2 = 2.5$. In Figure 7.5, $a = k_0\tilde{a} = 5$, and the radiated power (normalized in Equation 7.20) is plotted versus the outer radius $b = k_0\tilde{b}$ with $b = k_0\tilde{b}$ varying from 5 to 10 ($M_T = 4$ and $L = 50$ layers). The *solid lines* of Figure 7.5 show a comparison of the total power (normalized) radiated by the uniform and quarter shell periodic dielectric systems. For the uniform shell, the $m = 0$ order also represents the total power radiated by the system since

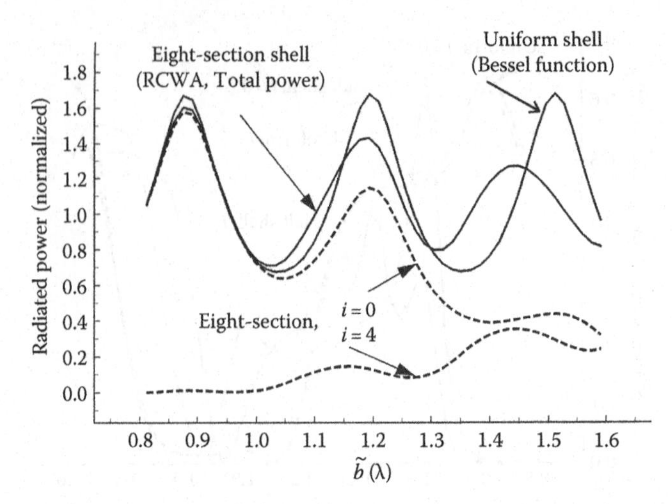

FIGURE 7.5 The radiated power that results when a line source located at the origin radiates into the quarter φ-periodic shell shown in Figure 7.1b. Here $a = k_0\tilde{a} = 5$ and the radiated power is plotted versus the outer radius $b = k_0\tilde{b}$ with $b = k_0\tilde{b}$ varying from 5 to 10. (From Jarem, J.M., *J. Electromagn. Waves Appl.*, 11, 197, 1997. With permission of Brill Publishing.)

the line source is centered on the cylinder axis. Figure 7.5 also shows the $i = 0$ and 4 order powers. The $i = 4$ order shown in Figure 7.5 is based on a $\psi_0 = 2\pi$, full circle grating period. It is the $i = 1$ order that is based on a $\psi_0 = \pi/2$, quarter circle grating period. The uniform shell radiated power was calculated both by the current algorithm and by solving Maxwell's equations in Regions 1–3 in terms of Bessel and Hankel functions and matching EM boundary conditions at the interfaces. Both methods gave nearly identical results. The Bessel function matching method was used to make the plot of Figure 7.4. As can be seen, the quarter shell dielectric system causes significantly different radiation than did the uniform shell system, although both systems had the same bulk dielectric value in Region 2.

The total radiated power shown in Figure 7.5 was determined by calculating the *i*th order power at $\tilde{\rho} = \tilde{b}$ (Equation 5.127) and summing these order powers to obtain the total scattered power. The power was calculated at $\tilde{\rho} = \infty$ and found to almost exactly equal that found at $\tilde{\rho} = \tilde{b}$. The *i*th order power was also calculated at $\tilde{\rho} = \tilde{a}$ (Equation 7.19). It was found that the order power of the higher orders at $\tilde{\rho} = \tilde{a}$ was almost exactly zero for $|i| \geq 1$ and that the $i = 0$ power at $\tilde{\rho} = \tilde{a}$ almost exactly equaled the total radiated power calculated at $\tilde{\rho} = \tilde{b}$. We thus see that conservation of power was obeyed to a high degree of accuracy. It is interesting that in the interior region that almost no power was radiated and diffracted into higher orders in the interior region of the cylindrical system.

7.4 ANISOTROPIC CYLINDRICAL SCATTERING

7.4.1 INTRODUCTION

A problem concerning circular cylindrical object scattering, which has been studied, is the problem of determining the scattering and radiation that occurs when a circular cylindrical dielectric system contains a region whose permittivity is inhomogeneous and periodic in the phi (φ) azimuthal direction [16–18]. Elsherbeni and Hamid [16] study EM and transverse magnetic (TM, electric field parallel to the cylinder axis) scattering from the inhomogeneous radial dielectric shell permittivity profile $\varepsilon(\rho, \phi) = \varepsilon_a(\rho_0/\rho)^2(\eta - \delta\cos(2\varphi))$ where ε_a, ρ_0, η, and δ are constants defined in [16] and ρ and φ are cylindrical coordinates. Mathieu functions are used to solve for the EM fields in the inhomogeneous shell region. The choice of $\varepsilon(\rho, \varphi)$ used by [16–18] was necessary in order that the Region 2 solution could be expressed in terms of Mathieu functions. A limitation of the solution of

[16–18] is the fact that their solution does not apply to an arbitrary $\varepsilon(\rho, \varphi)$ profile but only the one to which a Mathieu function solution can be found.

In the previous section, we generalized the work of [16–18] and presented an EM cylindrical solution algorithm to analyze radiation and scattering from isotropic dielectric cylindrical systems, which have an arbitrary radial and azimuthal $\varepsilon(\rho, \varphi)$ profile rather than the $\varepsilon(\rho, \varphi)$ profile used by [16–18]. The solution algorithm above and in [15] was based on a recently developed EM planar diffraction grating algorithm called rigorous coupled wave analysis [12,13]. The purpose of this section will be to extend the RCWA cylindrical algorithm of [15] to handle the analysis of anisotropic, inhomogeneous dielectric, and permeable material cylinders. Other research on uniform anisotropic cylinder scattering may be found in [19,20].

Specifically the algorithm of this section [21,23] will study the case when (1) the electric field is polarized parallel to the material cylindrical axis (TM case), (2) the cylindrical scattering object has an arbitrary, isotropic, inhomogeneous, dielectric permittivity profile $\varepsilon(\rho, \varphi)$, and (3) the cylindrical scattering object has arbitrary, anisotropic inhomogeneous relative permeability tensor profiles $\mu_{xx}(x, y)$, $\mu_{xy}(x, y)$, $\mu_{yx}(x, y)$, $\mu_{yy}(x, y)$ (μ_{xz}, μ_{zx}, μ_{zy}, and μ_{yz} are taken to be zero). Equations 7.26 and 7.27 of this section and [19,20] express the tensor elements in cylindrical components. The analysis of this section also applies to the case when (1) the magnetic field is polarized parallel to the cylindrical axis (TE case), (2) the cylindrical scattering object has an arbitrary, isotropic, inhomogeneous permeable profile $\mu(\rho, \varphi)$, and (3) the cylindrical scattering object has arbitrary, anisotropic inhomogeneous relative permittivity tensor profiles $\varepsilon_{xx}(x, y)$, $\varepsilon_{xy}(x, y)$, $\varepsilon_{yx}(x, y)$, and $\varepsilon_{yy}(x, y)$ (ε_{xz}, ε_{zx}, ε_{zy}, and ε_{yz} are taken to be zero). This follows since the TE and TM cases just described are dual to one another.

The solution of this problem is of great interest in several areas of EM research. In the area of cylindrical aperture antenna theory, radial and azimuthal dielectric loading in front of a cylindrical aperture antenna can greatly alter, and therefore possibly enhance, the radiation characteristics of cylindrical aperture antennas [16–18]. Other EM applications include (1) scattering from circular shaped, frequency-selective surfaces, (2) scattering from cylindrical surfaces covered with periodically spaced, inhomogeneous, anisotropic, RAM, and (3) use as a cross check of other numerical algorithms (FD–TD or FE) that study scattering from inhomogeneous, anisotropic systems.

7.4.2 State Variable Analysis

This section is concerned with the problem of determining the EM fields that arise when a plane wave and an off-center, interior line source excite EM fields in a circular cylindrical dielectric, anisotropic, permeability system as shown in Figures 7.6 through 7.8 by using the RCWA method. The EM

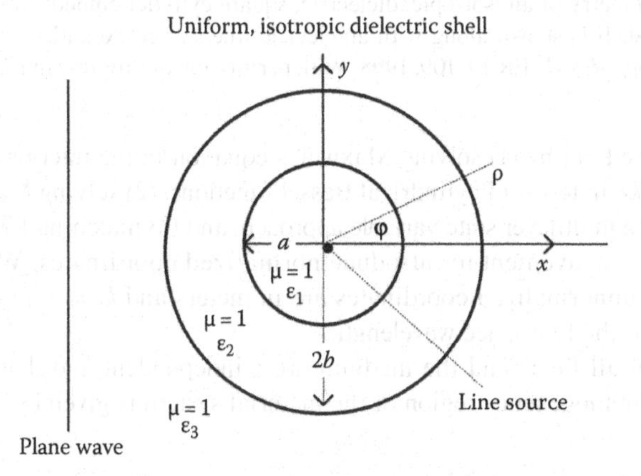

FIGURE 7.6 The geometry of uniform cylindrical shell system when a plane wave is incident on the cylindrical system and when an electric line source excites EM fields in the system is shown. The polarization of the electric field of the plane wave is parallel to the cylinder axis. (Reproduced from Jarem, J.M., *Prog. Electromagn. Res.*, PIER 19, 109, 1998. With permission of Progress in Electromagnetic Research.)

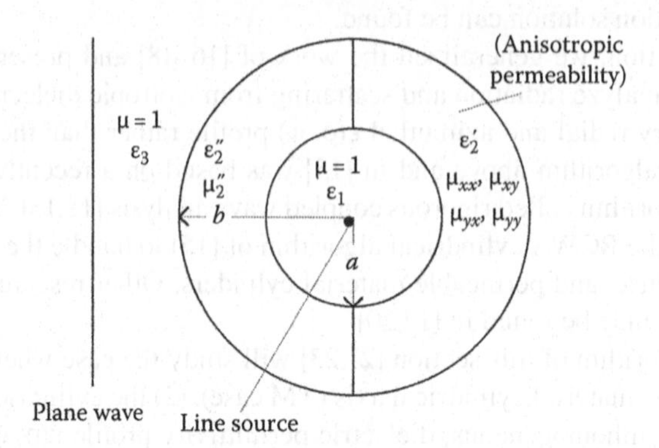

FIGURE 7.7 The geometry of an anisotropic, permeable, cylindrical half shell is shown along with a plane wave (electric field polarized parallel to the cylinder axis) and electric line source excitation. (Reproduced from Jarem, J.M., *Prog. Electromagn. Res.*, PIER 19, 109, 1998. With permission of Progress in Electromagnetic Research.)

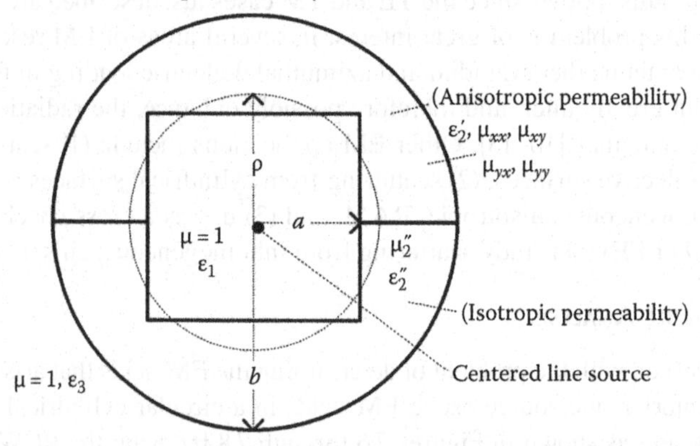

FIGURE 7.8 The geometry of an isotropic, dielectric, square cylinder embedded in an anisotropic, permeable, cylindrical half shell is shown along with an electric line source excitation. (Reproduced from Jarem, J.M., *Prog. Electromagn. Res.*, PIER 19, 109, 1998. With permission of Progress in Electromagnetic Research.)

analysis will be carried out by (1) solving Maxwell's equation in the interior and exterior regions of Figures 7.6 through 7.8 in terms of cylindrical Bessel functions, (2) solving Maxwell's equation in the shell region by using a multilayer state variable approach, and (3) matching EM boundary conditions at the interfaces. It is convenient to introduce normalized coordinates. We let $a = k_0\tilde{a}$, $b = k_0\tilde{b}$, $\rho = k_0\tilde{\rho}$, etc., where unnormalized coordinates are in meters and $k_0 = 2\pi/\lambda$ is the free space wave number (1/m) and λ is the free space wavelength.

It is assumed that all fields and the medium are z independent and that the relative dielectric permittivity in an inhomogeneous region of the material system is given by

$$\varepsilon(\rho,\varphi) = \sum_{i=-\infty}^{\infty} \breve{\varepsilon}(\rho)e^{ji\varphi}, \quad 0 \le \varphi \le 2\pi \tag{7.25}$$

where $\breve{\varepsilon}_i(\rho)$ represent φ-exponential Fourier coefficients. The anisotropic permeability tensor is assumed to be given in rectangular and cylindrical coordinates by [19,20]

$$\underline{\underline{\mu}} = \begin{bmatrix} \mu_{xx} & \mu_{xy} & 0 \\ \mu_{yx} & \mu_{yy} & 0 \\ 0 & 0 & \mu_{zz} \end{bmatrix}, \quad \underline{\underline{\mu}} = \begin{bmatrix} \mu_{\rho\rho} & \mu_{\rho\varphi} & 0 \\ \mu_{\varphi\rho} & \mu_{\varphi\varphi} & 0 \\ 0 & 0 & \mu_{zz} \end{bmatrix} \tag{7.26}$$

where

$$\mu_{\rho\rho} = \mu_{xx}\cos^2(\varphi) + (\mu_{xy} + \mu_{yx})\sin(\varphi)\cos(\varphi) + \mu_{yy}\sin^2(\varphi)$$

$$\mu_{\rho\varphi} = \mu_{xy}\cos^2(\varphi) + (-\mu_{xx} + \mu_{yy})\sin(\varphi)\cos(\varphi) - \mu_{yx}\sin^2(\varphi)$$

$$\mu_{\varphi\rho} = \mu_{yx}\cos^2(\varphi) + (-\mu_{xx} + \mu_{yy})\sin(\varphi)\cos(\varphi) - \mu_{xy}\sin^2(\varphi)$$

$$\mu_{\varphi\varphi} = \mu_{yy}\cos^2(\varphi) + (-\mu_{xy} - \mu_{yx})\sin(\varphi)\cos(\varphi) + \mu_{xx}\sin^2(\varphi) \tag{7.27}$$

The cylindrical permeability tensor components are assumed to be expanded in the exponential Fourier series

$$\mu_{rs}(\rho,\varphi) = \sum_{i=-\infty}^{\infty} \breve{\mu}_{rs}(\rho)e^{ji\varphi} \quad 0 \le \varphi \le 2\pi, \quad (r,s) = (\rho,\varphi) \tag{7.28}$$

where $\breve{\mu}_{rs}(\rho)$ represents φ-exponential Fourier coefficients.

The EM fields interior and exterior (Regions 1 and 3 of Figures 7.6 through 7.8) when a line source (Region 1) and a plane wave (Region 3) excite EM radiation in a cylindrical system are well known to be an infinite expansion of the Fourier–Bessel functions $H_n^{(2)}e^{jn\varphi}$, $J_n e^{jn\varphi}$, $Y_n e^{jn\varphi}$.

In Region 2, the middle cylindrical dielectric region, we divide the dielectric region into L thin shell layers of thickness d_ℓ, $b - a = \sum_{\ell=1}^{L} d_\ell$ ($\ell = 1$ is adjacent to $\rho = b$ and $\ell = L$ is adjacent to $\rho = a$) and solve Maxwell's equations in cylindrical coordinates by a state variable approach in each thin layer. The layers are assumed to be thin enough in order that the ρ dependence of $\varepsilon(\rho, \varphi)$, $\mu_{\rho\rho}(\rho, \varphi)$, $\mu_{\rho\varphi}(\rho, \varphi)$, $\mu_{\rho\varphi}(\rho, \varphi)$, and $\mu_{\varphi\varphi}(\rho, \varphi)$ and the ρ scale factors may be treated as a constant in each layer. Making the substitutions $S_z = E_z$, $U_\rho = \eta_0 H_\rho$, and $U_\varphi = \eta_0 \rho H_\varphi$ where E_z, H_ρ, and H_φ represent the electric and magnetic fields in the thin shell region and $\eta_0 = 377\ \Omega$ is the intrinsic impedance of free space, we find that Maxwell's equations in a cylindrical shell of radius ρ are given by

$$\frac{\partial S_z}{\partial \varphi} = -j\rho\mu_{\rho\rho}U_\rho - j\mu_{\rho\varphi}U_\varphi \tag{7.29}$$

$$\frac{\partial S_z}{\partial \rho} = j\mu_{\varphi\rho}U_\rho + j\left(\frac{\mu_{\varphi\varphi}}{\rho}\right)U_\varphi \tag{7.30}$$

$$\frac{\partial U_\varphi}{\partial \rho} - \frac{\partial U_\rho}{\partial \varphi} = j\rho\varepsilon S_z \tag{7.31}$$

To solve Equations 7.29 through 7.31, we expand $S_z(\rho, \varphi)$, $U_\rho(\rho, \varphi)$, $U_\varphi(\rho, \varphi)$, $\varepsilon(\rho, \varphi)$, and $\mu_{rs}(\rho, \varphi)$, $(r, s) = (\rho, \varphi)$ in the Floquet harmonics:

$$S_z(\rho,\varphi) = \sum_{i=-\infty}^{\infty} s_{zi}(\rho)\, e^{ji\varphi}, \quad U_\rho(\rho,\varphi) = \sum_{i=-\infty}^{\infty} u_{\rho i}(\rho)\, e^{ji\varphi}$$

$$U_\varphi(\rho,\varphi) = \sum_{i=-\infty}^{\infty} u_{\varphi i}(\rho)\, e^{ji\varphi}, \quad \varepsilon(\rho,\varphi)E_z = \sum_{i=-\infty}^{\infty}\left[\sum_{i'=-\infty}^{\infty} \breve{\varepsilon}_{i-i'} s_{zi'}\right] e^{ji\varphi}$$

$$\mu_{rs}(\rho,\varphi)F(\rho,\varphi) = \sum_{i=-\infty}^{\infty}\left[\sum_{i'=-\infty}^{\infty} \breve{\mu}_{rs_{i-i'}} f_{i'}\right] e^{ji\varphi}, \quad (r,s) = (\rho,\varphi) \tag{7.32}$$

where $F(\rho, \varphi)$ represents either $U_\rho(\rho, \varphi)$ or $U_\varphi(\rho, \varphi)$ in Equation 7.32. If these expansions are substituted in Equations 7.29 through 7.31, and after letting $\underline{s}_z(\rho) = [s_{zi}(\rho)]$, $\underline{u}_\rho(\rho) = [u_{\rho i}(\rho)]$, and $\underline{u}_\varphi(\rho) = [u_{\varphi i}(\rho)]$ be column matrices and $\underline{\varepsilon}(\rho) = [\breve{\varepsilon}_{i-i'}(\rho)]$, $\underline{\mu}_{rs}(\rho) = [\breve{\mu}_{rs_{i-i'}}(\rho)]$, $(r, s) = (\rho, \varphi)$, $\underline{K} = [iK\delta_{i,i'}]$, and $K = 2\pi/\Lambda_\varphi$ (Λ_φ is the circular grating period and $\delta_{i,i'}$ is the Kronecker delta) be square matrices, we find after manipulation that

$$\frac{\partial \underline{V}}{\partial \rho} = \underline{A}\,\underline{V}, \quad \underline{V} = \begin{bmatrix} \underline{s}_z \\ \underline{u}_\varphi \end{bmatrix}, \quad \underline{A} = \begin{bmatrix} \underline{A}_{11} & \underline{A}_{12} \\ \underline{A}_{21} & \underline{A}_{22} \end{bmatrix} \tag{7.33}$$

where

$$\underline{A}_{11} = \frac{-j}{\rho}\underline{\mu}_{\varphi\rho}\,\underline{\mu}_{\rho\rho}^{-1}\underline{K}, \quad \underline{A}_{12} = \frac{j}{\rho}\left(-\underline{\mu}_{\varphi\rho}\,\underline{\mu}_{\rho\rho}^{-1}\underline{\mu}_{\rho\varphi} + \underline{\mu}_{\varphi\varphi}\right)$$

$$\underline{A}_{21} = j\left(-\frac{1}{\rho}\underline{K}\,\underline{\mu}_{\rho\rho}^{-1}\underline{K} + \rho\underline{\varepsilon}\right), \quad \underline{A}_{22} = \frac{-j}{\rho}\underline{K}\,\underline{\mu}_{\rho\rho}^{-1}\underline{\mu}_{\rho\varphi}$$

In these equations, u_ρ was eliminated by finding the matrix inverse of $\underline{\mu}_{\rho\rho}$, namely, $\underline{\mu}_{\rho\rho}^{-1}$, and then carrying out appropriate matrix multiplications. If Equation 7.33 is truncated at order M_T ($i = -M_T, \ldots, -1, 0, 1, \ldots, M_T$), Equation 7.33 represents a $N_T = 2(2M_T + 1)$ state variable equation (with matrix $(\underline{A}_\ell)_{N_T \times N_T}$) The solution of this equation is given by $\underline{V}_n(\rho) = \underline{V}_n e^{q_n\rho}$ where q_n and $\underline{V}_n$ are the nth eigenvalue and eigenvector of the constant matrix $\underline{A}_n$. The quantities $\underline{A}_n$, $\underline{V}_n$, and q_n satisfy $\underline{A}_n\underline{V}_n = q_n\underline{V}_n$. The general EM fields in the ℓth thin shell region are given

$$E_z = \sum_{i=-M_T}^{M_T}\sum_{n=1}^{N_T} c_n s_{zin} e^{q_n\rho}, \quad \eta_0 \rho H_\varphi = \sum_{i=-M_T}^{M_T}\sum_{n=1}^{N_T} c_n u_{\varphi in} e^{q_n\rho} \tag{7.34}$$

where $\underline{V}_n^t = \left[\underline{s}_n^t, \underline{u}_n^t\right]$ and where t represents matrix transpose.

Although a large matrix equation exists from which the overall solution of the problem may be obtained, a more efficient solution method is to use a ladder approach [12] (i.e., successively relate unknown coefficients from one layer to the next) to express the c_{nL} coefficients of the Lth last layer in terms of the c_{n1} coefficients of the first layer and then match boundary conditions at the $\rho = a$ and b interfaces to obtain the final unknowns of the system. Using the thin layer coordinate system and latter analysis presented in Section 7.3, we obtain the following overall matrix equation:

$$\underline{C}_L = \underline{F}_{L-1}(\underline{F}_{L-2}\cdots(\underline{F}_1\,\underline{C}_1)) = \underline{M}\,\underline{C}_1 \tag{7.35}$$

$$c_0^I J_i(X_{1s}) = \sum_{n=1}^{N_T} c_{nL} e^{-q_{nL} d_L} \left[-\frac{\pi J_i(X_{1a})}{2} \right] \left\{ ja\sqrt{\varepsilon_1} \frac{J_i'(X_{1a})}{J_i(X_{1a})} s_{zinL} + u_{\varphi inL} \right\} \tag{7.36}$$

$$E_i^I = \sum_{n=1}^{N_T} c_{n1} \left[\frac{\pi H_i^{(2)}(X_{3b})}{2} \right] \left\{ jb\sqrt{\varepsilon_3} \frac{H_i^{(2)'}(X_{3b})}{H_i^{(2)}(X_{3b})} s_{zin1} + u_{\varphi in1} \right\} \tag{7.37}$$

where $i = -M_T, \ldots, 0, \ldots, M_T$, $J'(X) = dJ(X)/dX$, etc., $X_{1s} = \sqrt{\varepsilon_1}\rho_s$, $X_{1a} = \sqrt{\varepsilon_1}a$, and $X_{3b} = \sqrt{\varepsilon_3}b$. Equation 7.36 represents a set of $2M_T + 1$ equations, Equation 7.37 represents a set of $2M_T + 1$ equations, and the matrix equation (Equation 7.35) represents a set of $N_T = 2(2M_T + 1)$ equations. Thus, Equations 7.35 through 7.37) represent a set of $2N_T = 4(2M_T + 1)$ equations to calculate the $2N_T$ set of unknowns represented by $\underline{C}_1$ and $\underline{C}_L$. Once these quantities are known, all other unknown coefficients in the system may be found.

An important quantity to calculate is the normalized power of each order. We consider the important cases when the power is either radiated from the line source in Region 1 ($c_0^I \neq 0, E_0^I = 0$) or the power is scattered by a plane wave from Region 3 ($c_0^I = 0, E_0^I \neq 0$). In the case when $c_0^I \neq 0, E_0^I = 0$, the normalized power in each order is given by $P_{Ni} = P_i^{RAD}/P^{INC}$ where P^{INC} is the incident power of the line source and P_i^{RAD} is the radiation at a radial distance ρ. For the plane wave scattering case ($c_0^I = 0, E_0^I \neq 0$), it is useful to calculate the normalized scattered at $\rho = \infty$

$$P_{Ni}^{Scat}(\infty) = \frac{(P_i^{Scat}(\infty)/\lambda)}{S_{INC}} = \frac{2}{\pi} \frac{\left|c_i^{(3)}\right|^2}{\left|E_0^I\right|^2} \tag{7.38}$$

In Equation 7.38, S_{INC} is the power per unit area (W/m^2) of the incident plane wave, and P_i^{scat} is the scattered power per unit length (W/m) of the ith order.

7.4.3 Numerical Results

In this section, we will study line source radiation and plane wave scattering using the RCWA method for three different material system examples.

The first example consists of studying line source radiation and plane wave scattering from a uniform dielectric shell. In this example, all of space was taken to have a permeability $\mu = 1$ and the permittivity in Regions 1–3 was taken to be respectively $\varepsilon_1 = 1.5$, $\varepsilon_2 = 2.5$, and $\varepsilon_3 = 1$. The inner radius was taken to be $a = k_0\tilde{a} = 5$ ($\tilde{a} = 0.795\lambda$), and the outer shell radius was taken to range from $b = a = 5$ to $b = 10$. Using a centered line source excitation only (see Figure 7.6), Figure 7.9 shows a comparison of the normalized radiated power (all normalized power in this section are assumed normalized either to the incident dipole or incident plane wave amplitude), as determined by the RCWA method (using $L = 10$ layers, $M_T = 1$) and by a Bessel function matching solution method (based on matching Bessel function solutions in Regions 1, 2, and 3) when the outer radius was varied from $b = a = 5$ to $b = 10$. As can be seen from Figure 7.9, excellent agreement exists between the Bessel function matching algorithm and the RCWA method.

Figure 7.10 shows the total radiated power that results when a centered line source radiates through an anisotropic, permeable half shell (see Figure 7.7, $\varepsilon_1 = 1.5$, $\varepsilon_3 = 1$, $\varepsilon_2'' = 1.75$, $\mu_2'' = 1.5$, $\varepsilon_2' = 3.25$, $\mu_{xx} = 1.5$, $\mu_{xy} = 0.3$, $\mu_{yx} = 0.3$, $\mu_{yy} = 1.7$, $M_T = 10$, $L = 10$ layers) when the inner radius is $a = k_0\tilde{a} = 5$ and when the outer radius is varied from $b = a = 5$ to $b = 10$. As can be seen, almost

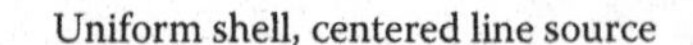

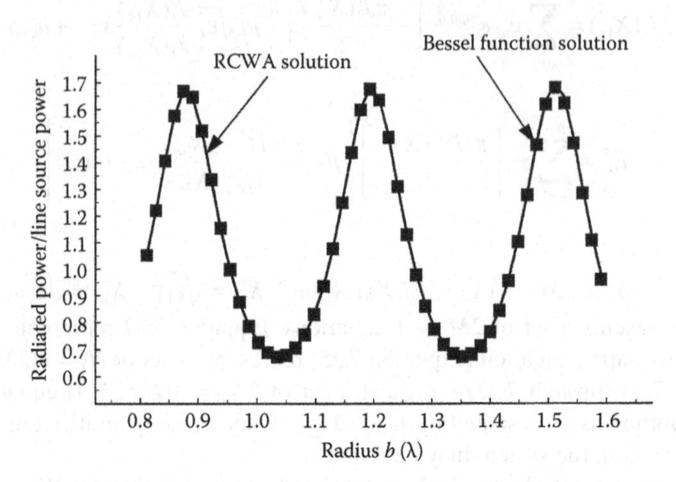

FIGURE 7.9 The normalized radiated power, which results when a centered line source excites a uniform dielectric shell (see Figure 7.6, $\varepsilon_1 = 1.5$, $\varepsilon_2 = 2.5$, $\varepsilon_3 = 1$, $\mu = 1$), is shown when determined by RCWA and when determined by a Bessel function matching solution. (Reproduced from Jarem, J.M., *Prog. Electromagn. Res.*, PIER 19, 109, 1998. With permission of Progress in Electromagnetic Research.)

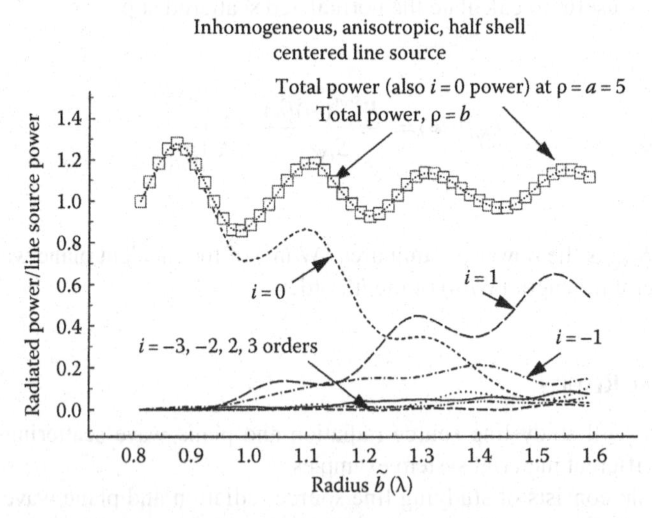

FIGURE 7.10 The total radiated power that results when a centered line source radiates through an anisotropic, permeable half shell (see Figure 7.7, $\varepsilon_1 = 1.5$, $\varepsilon_3 = 1$, $\varepsilon_2'' = 1.75$, $\mu_2'' = 1.5$, $\varepsilon_2' = 3.25$, $\mu_{xx} = 1.5$, $\mu_{xy} = 0.3$, $\mu_{yx} = 0.3$, $\mu_{yy} = 1.7$, $M_T = 10$, $L = 10$ layers) when the inner radius is $a = k_0\tilde{a} = 5$ and when the outer radius is varied from $b = a = 5$ to $b = 10$ is shown. (Reproduced from Jarem, J.M., *Prog. Electromagn. Res.*, PIER 19, 109, 1998. With permission of Progress in Electromagnetic Research.)

exact conservation of power at the inner and outer radius is observed. At $\rho = a$ (inner radius), no power was calculated to be diffracted into higher orders. This is why the total power at $\rho = a$ also equals the $i = 0$ power at $\rho = a$. Also shown in Figure 7.10 are the $i = -1$, 0, and 1 orders radiated at $\rho = b$ (outer radius) and the higher orders $i = -3, -2, 2, 3$. As $\tilde{b}$ is increased from $\tilde{b} = 0.8\lambda$ to 1.6λ in Figure 7.10, one clearly observes that as the outer radius is increased, power is depleted out of the $i = 0$ order and is diffracted into higher orders. One also observes that unequal order power is radiated into the $i = -1$ and 1 orders. This is to be expected and is a result of the anisotropy of the permeable half shell.

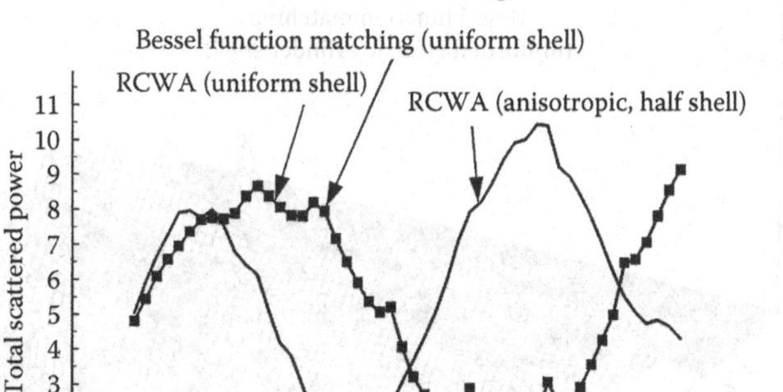

FIGURE 7.11 A comparison of the total plane wave power scattered by the same uniform dielectric shell example as considered in Figure 7.9 (see Figure 7.6, centered line source not present) as determined by the Bessel function matching solution ($M_T = 15$) and as determined by the RCWA method (using $L = 15$ layers, $M_T = 15$) is shown. Plane wave scattering from an anisotropic cylinder is also shown. (Reproduced from Jarem, J.M., *Prog. Electromagn. Res.*, PIER 19, 109, 1998. With permission of Progress in Electromagnetic Research.)

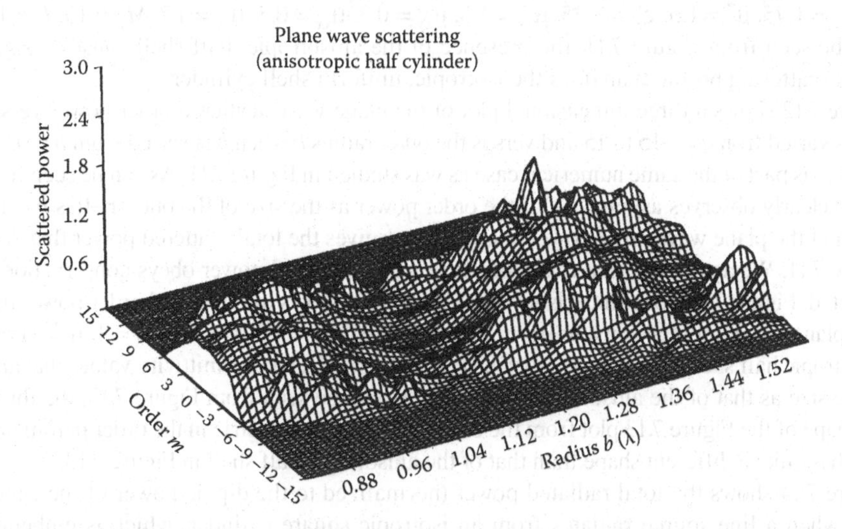

FIGURE 7.12 A three dimensional plot of the plane wave scattered order power verses order *i* when *i* is varied from $i = -15$ to 15 and versus the outer radius $\tilde{b}$ when $\tilde{b}$ is varied from $\tilde{b} = 0.8\lambda$ to 1.6λ is shown. This is part of the same numerical case as was studied in Figure 7.11. (Reproduced from Jarem, J.M., *Prog. Electromagn. Res.*, PIER 19, 109, 1998. With permission of Progress in Electromagnetic Research.)

Figures 7.11 through 7.13 display scattering results when a plane wave is incident on the cylindrical system. Figure 7.11 (*solid line* and *square*) shows a comparison of the total plane wave power scattered by the same uniform dielectric shell example as considered in Figure 7.10 (see Figure 7.6, centered line source not present) as determined by the Bessel function matching solution (*square*, $M_T = 15$) and by the RCWA method (*solid line*, using $L = 15$ layers, $M_T = 15$). As can be seen from Figure 7.11, excellent agreement was obtained between the two methods. As can be seen from Figure 7.11, the RCWA method was able to accurately reproduce even the small resonance peaks that arise in the scattering solution. Figure 7.11 (*solid line* labeled *RCWA [anisotropic, half shell]*) shows the total plane wave scattered power (as a function of the outer radius $\tilde{b}$) that results when a plane wave is incident on an anisotropic, permeable cylindrical half shell (see Figure 7.7, $\varepsilon_1 = 1.5$,

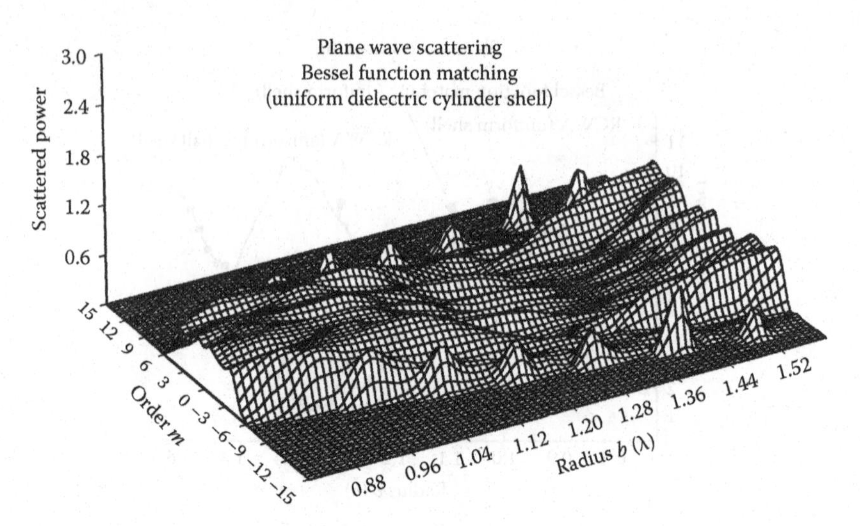

FIGURE 7.13 The scattered order power that occurs when a plane wave impinges on a uniform dielectric shell (see Figure 7.6, $\varepsilon_1 = 1.5$, $\varepsilon_2 = 2.5$, $\varepsilon_3 = 1$, $\mu = 1$) rather than an anisotropic half shell is shown. (Reproduced from Jarem, J.M., *Prog. Electromagn. Res.*, PIER 19, 109, 1998. With permission of Progress in Electromagnetic Research.)

$\varepsilon_3 = 1$, $\varepsilon_2'' = 1.75$, $\mu_2'' = 1.5$, $\varepsilon_2' = 3.25$, $\mu_{xx} = 1.5$, $\mu_{xy} = 0.3$, $\mu_{yx} = 0.3$, $\mu_{yy} = 1.7$, $M_T = 15$, $L = 15$ layers). As can be seen from Figure 7.11, the presence of the anisotropic, half shell causes a significantly different scattering profile than does the isotropic, uniform shell cylinder.

Figure 7.12 shows a three dimensional plot of the plane wave scattered order power versus order i when i is varied from $i = -15$ to 15 and versus the outer radius $\tilde{b}$ when $\tilde{b}$ is varied from $\tilde{b} = 0.8\lambda$ to 1.6λ. Figure 7.12 is part of the same numerical case as was studied in Figure 7.11. As can be seen from Figure 7.12, one clearly observes asymmetry of the order power as the size of the outer radius $\tilde{b}$ is increased. The sum of the plane wave order power at any given $\tilde{b}$ gives the total scattered power that is displayed in Figure 7.11. We again note that this total plane wave scattered power obeys conservation of power as expected. Figure 7.13, for comparison with Figure 7.12, shows the scattered order power that occurs when a plane wave impinges on a uniform dielectric shell ($\varepsilon_1 = 1.5$, $\varepsilon_2 = 2.5$, $\varepsilon_3 = 1$, $\mu = 1$) rather than an anisotropic half shell. The uniform dielectric shell has dielectric permittivity values that are roughly the same size as that of the anisotropic half shell. As can be seen from Figure 7.13, the three dimensional shape of the Figure 7.13 plot from the uniform shell is symmetric in the order parameter i and in general has quite a different shape than that of the anisotropic half shell in Figure 7.13.

Figure 7.14 shows the total radiated power (normalized to the dipole power of the centered line source) when a line source radiates from an isotropic square cylinder, which is embedded in an anisotropic permeable half shell (see Figure 7.8, $\varepsilon_1 = 1.5$, $\varepsilon_3 = 1$, $\varepsilon_2'' = 3.5$, $\mu_2'' = 1$, $\varepsilon_2' = 3.25$, $\mu_{xx} = 1.5$, $\mu_{xy} = 0.3$, $\mu_{yx} = 0.3$, $\mu_{yy} = 1.7$, $M_T = 20$, $L = 25$ layers). The radiated power was calculated at $\rho = a$ ($\tilde{a} = 1\lambda$), which is a circle inscribed in the square cylinder of Region 1, and was calculated at $\rho = b$ ($\tilde{b} = 2.5\lambda$), which is the outer radius of the anisotropic half cylinder. The outer radius $\tilde{b}$ was varied from $\tilde{b} = \sqrt{2}\,\tilde{a} = 1.414\lambda$ to $\tilde{b} = 2.5\lambda$. As can be seen from Figure 7.18, extremely good power conservation was observed at $\rho = a$ ($\tilde{a} = 1\lambda$) and at $\tilde{\rho} = \tilde{b}$. Despite the square shape of the cylinder, no power was observed to be diffracted into higher orders at $\rho = a$ ($\tilde{a} = 1\lambda$). Also shown in Figure 7.14 are the $i = -1$, 0, and 1 orders radiated at $\rho = b$ (outer radius). As in Figure 7.12, one observes that power is depleted from the $i = 0$ order and radiated into higher orders. Figure 7.14 shows the increase in the $i = -1$ and 1 orders, for example, which occurs when $\tilde{b}$ is increased. One also observes in Figure 7.14 that the order power is radiated asymmetrically into the $i = -1$ and 1 orders. As in Figure 7.14, this is expected and is due to the anisotropy of the permeable half shell.

Figure 7.15 shows a plot (*dotted line*) of the relative dielectric permittivity function $\varepsilon(\rho, \varphi)$ when $\tilde{\rho} = 1.241\lambda$ for the square cylinder-anisotropic, half shell case displayed in Figure 7.8. The circular dashed

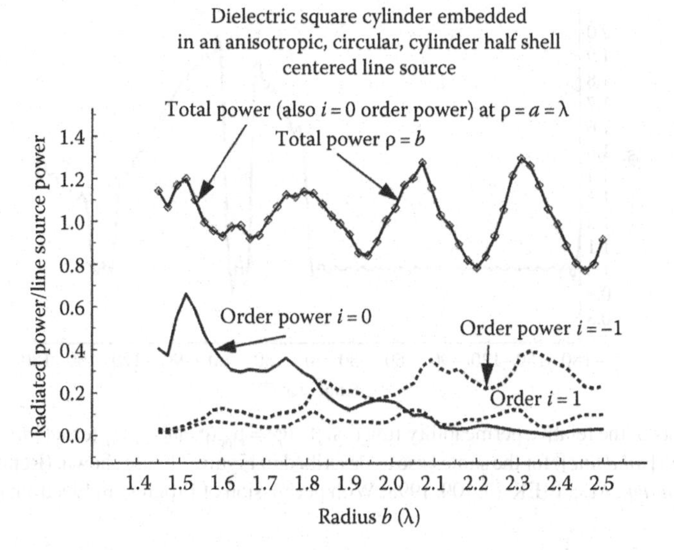

FIGURE 7.14 The total radiated power (normalized to the dipole power of the centered line source) when a line source radiates from an isotropic square cylinder, which is embedded in an anisotropic permeable half shell (see Figure 7.8, $\varepsilon_1 = 1.5$, $\varepsilon_3 = 1$, $\varepsilon_2'' = 3.5$, $\mu_2'' = 1$, $\varepsilon_2' = 3.25$, $\mu_{xx} = 1.5$, $\mu_{xy} = 0.3$, $\mu_{yx} = 0.3$, $\mu_{yy} = 1.7$, $M_T = 20$, $L = 25$ layers), is shown. (Reproduced from Jarem, J.M., *Prog. Electromagn. Res.*, PIER 19, 109, 1998. With permission of Progress in Electromagnetic Research.)

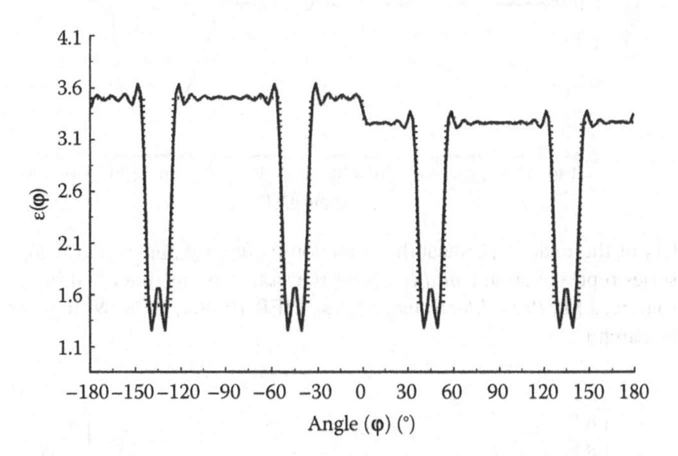

FIGURE 7.15 A plot (*dotted line*) of the relative dielectric permittivity function $\varepsilon(\varphi) = \varepsilon(\rho, \varphi)|_{\rho = 1.241\lambda}$ for the square cylinder-anisotropic, half shell case displayed in Figure 7.8 is shown. The circular dashed line of Figure 7.8 represents the approximate placement of the $\tilde{\rho} = 1.241\lambda$ parameter used to make the plots here. Also shown (*solid line*) is the Fourier series representation of the $\varepsilon(\rho, \varphi)$ profile when $\tilde{\rho} = 1.241\lambda$ and $M_T = 20$. (Reproduced from Jarem, J.M., *Prog. Electromagn. Res.*, PIER 19, 109, 1998. With permission of Progress in Electromagnetic Research.)

line of Figure 7.8 represents the approximate placement of $\tilde{\rho} = 1.241\lambda$ parameter used to make the Figure 7.15 $\varepsilon(\rho, \varphi)$ plots. Also shown in Figure 7.15 (*solid line*) is the Fourier series representation of the $\varepsilon(\rho, \varphi)$ profile when $\tilde{\rho} = 1.241\lambda$ and $M_T = 20$. ($M_T = 20$ was used to make the RCWA method of Figure 7.14.) As can be seen from Figure 7.14, enough Fourier terms ($-40 = -2M_t \le i \le 2M_t = 40$) were used in order to correctly model the inhomogeneous region as defined by the square cylinder. (*Note:* The convolution matrix of Equation 7.32 requires $2M_t = 40$ terms.) Figures 7.16 through 7.18 show the relative permeability tensor profiles $\mu_{\rho\rho}(\rho, \varphi)$, $\mu_{\rho\varphi}(\rho, \varphi)$, and $\mu_{\varphi\rho}(\rho, \varphi)$ $\mu_{\varphi\varphi}(\rho, \varphi)$ for the same case and parameters as shown in Figure 7.15. The numerical example of Figure 7.14 was chosen such that $\mu_{\rho\varphi}(\rho, \varphi) = \mu_{\varphi\rho}(\rho, \varphi)$.

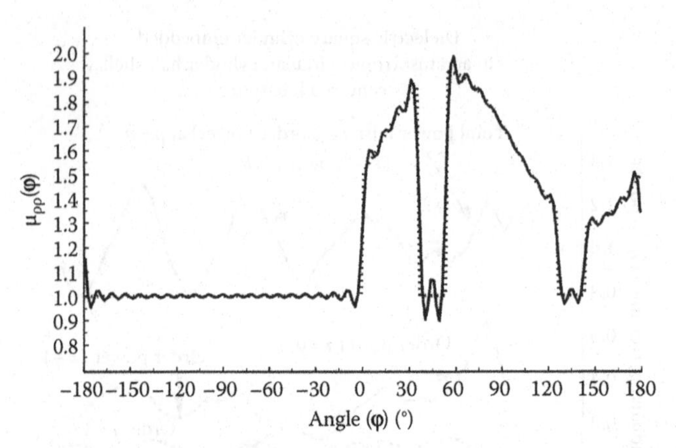

FIGURE 7.16 Plots of the relative permeability function $\mu_{\rho\rho}(\varphi) = \mu_{\rho\rho}(\rho, \varphi)|_{\rho=1.241\lambda}$ (exact [*dotted line*] and Fourier series representation [*solid line*]) for the same case as described in Figure 7.15 are shown. (Reproduced from Jarem, J.M., *Prog. Electromagn. Res.*, PIER 19, 109, 1998. With permission of Progress in Electromagnetic Research.)

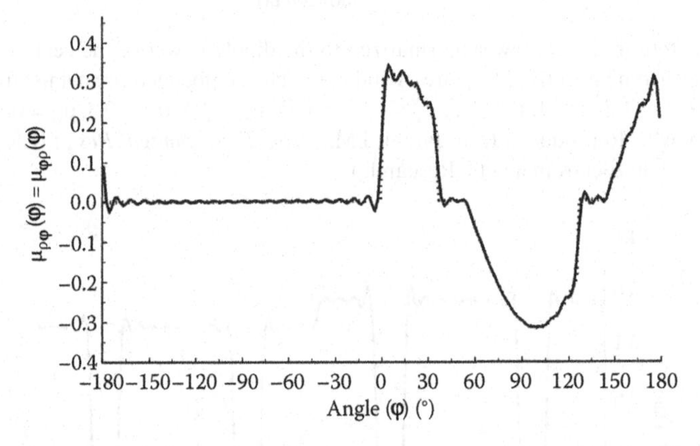

FIGURE 7.17 Plots of the relative permeability function $\mu_{\rho\varphi}(\varphi) = \mu_{\varphi\rho}(\varphi) = \mu_{\rho\varphi}(\rho, \varphi)|_{\rho=1.241\lambda}$ (exact [*dotted line*] and Fourier series representation [*solid line*]) for the same case as described in Figure 7.15 are shown. (Reproduced from Jarem, J.M., *Prog. Electromagn. Res.*, PIER 19, 109, 1998. With permission of Progress in Electromagnetic Research.)

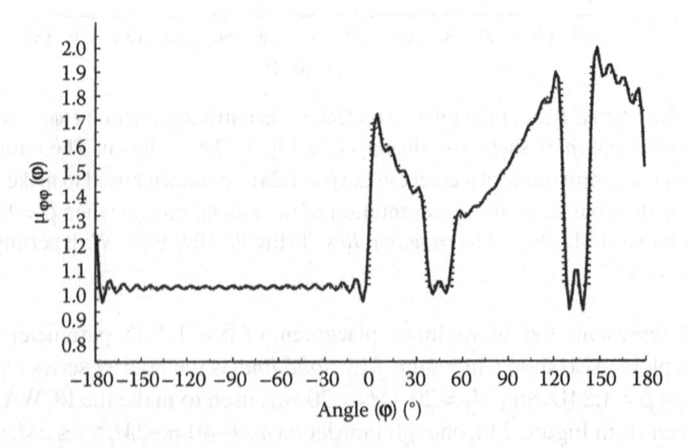

FIGURE 7.18 Plots of the relative permeability function $\mu_{\varphi\varphi}(\varphi) = \mu_{\varphi\varphi}(\rho, \varphi)|_{\rho=1.241\lambda}$ (exact [*dotted line*] and Fourier series representation [*solid line*]) for the same case as described in Figure 7.15 are shown. (Reproduced from Jarem, J.M., *Prog. Electromagn. Res.*, PIER 19, 109, 1998. With permission of Progress in Electromagnetic Research.)

7.5 SPHERICAL INHOMOGENEOUS ANALYSIS

7.5.1 Introduction

Another important and well-known problem in electromagnetic theory is to determine the scattering that occurs when an electromagnetic wave is incident on a spherical object. These problems have been extensively studied in the cases where (1) the EM incident wave is an oblique or nonoblique plane wave, (2) the incident EM wave has been generated by a line source or dipole source, and (3) the circular or spherical object is a dielectric coated metallic object [1–5]. Ren [22] studies scattering from anisotropic, homogeneous spherical systems and Greens' functions associated with anisotropic, homogeneous spherical systems. Ren [22] gives a very complete literature survey of scattering from isotropic and anisotropic spherical systems.

Concerning the problem of EM scattering from inhomogeneous material spherical systems, the RCWA algorithm can be applied to the analysis of radiation and scattering from a spherical inhomogeneous object [23]. Ref. [23] presents the basic spherical equations necessary to analyze an arbitrary 3-D inhomogeneous scatterer by the RCWA method and also presented a simple example of dipole radiation from an inhomogeneous dielectric object, which was azimuthally homogeneous (varied in the θ direction but had no dependence in the φ direction). The spherical RCWA method of this section will extend the results of [23] in the following ways. First, the analysis of this section (following [24]) will study examples where the inhomogeneous scatterer has an inhomogeneous permittivity and permeability profile, which, in addition to varying arbitrarily in the radial direction, also varies arbitrarily in the θ and φ directions. In [23], the inhomogeneity variation is only in the θ direction. This case is numerically much more challenging than [23] because matrix equations for all orders of m and n must be solved rather than a matrix equation for just $m = 0$ and all n. The second way that the results of [23] are extended in this section is that the basic spherical state variable equations and the interior exterior Bessel function equation of [23] will be modified to the general case when the inhomogeneous scatterer and EM source excitation is periodic in the φ coordinate over a region $2\pi/\kappa$ where κ is an integer ($\kappa = 1, 2, 3, \ldots$) rather than being periodic over just 2π as was presented in [23]. This is particularly useful as only a centered line excitation is considered in this section. The third way that the results of [23] are extended is that full radiated power results for radiation from higher order m and n spherical Bessel–Legendre modes are given, whereas in [23], power results were given only for $m = 0$ modes.

This section will be concerned with determining the EM fields that result when a centered electric dipole radiates inside a three dimensionally inhomogeneous material system (see Figure 7.19). This problem may be viewed as either a material shielded antenna source problem or may be viewed as a material microwave cavity problem where the material cavity is formed from the inhomogeneous dielectric and permeable material that surrounds the electric dipole source.

7.5.2 Rigorous Coupled Wave Theory Formulation

This section will be concerned with putting Maxwell's equations in spherical coordinates into a form for which the RCWA formulation can be implemented. We consider the spherical system shown in Figure 7.19. All coordinates will be assumed normalized as $r = k_0\tilde{r}$, $a = k_0\tilde{a}$, etc., where $r = k_0\tilde{r}$, $k_0 = 2\pi/\lambda$, λ is the free space wavelength in meters. In this figure, Region 1 ($0 \le r \le a$) is assumed to be a uniform material with the relative permittivity ε_1 and relative permeability μ_1, Region 3 ($b \le r$) is assumed to be a uniform material with the relative permittivity ε_3 and relative permeability μ_3, and Region 2 with $v = \cos(\theta)$ is assumed to have an arbitrary, inhomogeneous, lossy, relative permittivity $\varepsilon(r, v, \varphi)$ and is assumed to have an inhomogeneous, lossy, relative permeability $\mu(r, v, \varphi)$. For generality, we assume that electromagnetic radiation may impinge on the 3-D object from Region 3 (e.g., a plane wave) or from Region 1 (e.g., a dipole source). We will

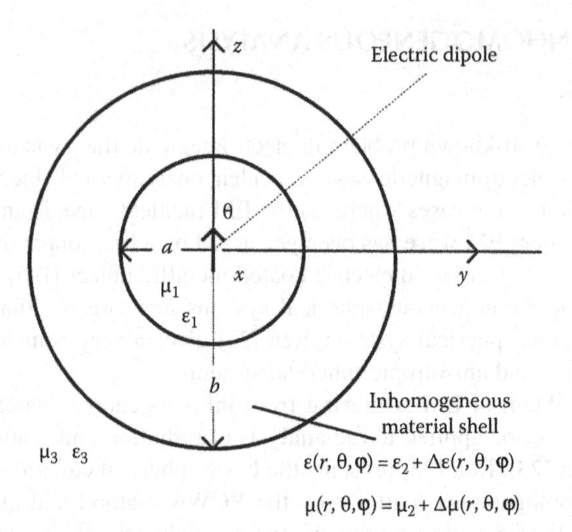

FIGURE 7.19 Geometry of the three dimensionally inhomogeneous spherical system. (Reprinted from Jarem, J.M., Rigorous coupled-wave-theory analysis of dipole scattering from a three-dimensional, inhomogeneous, spherical dielectric and permeable system, *IEEE Microw. Theory Tech.*, 45(8), August 1997. With permission. Copyright 1997 IEEE.)

now put Maxwell's equations of Region 2 into state variable form. If we substitute $\varepsilon(r, v, \varphi)$ and $\mu(r, v, \varphi)$ into the Maxwell's curl equations of Region 2, expand the two Maxwell curl equations into their r, θ, and φ field components, we find that the longitudinal, radial electric and magnetic field components may be expressed in terms of the transverse θ, φ electric and magnetic field components as

$$E_r = -\frac{1}{j\varepsilon(r,v,\varphi)r^2}\left[\frac{\partial U_\varphi}{\partial v} + \frac{1}{\gamma}\frac{\partial U_\theta}{\partial \varphi}\right], \quad H_r = \frac{1}{j\mu(r,v,\varphi)r^2}\left[\frac{\partial S_\varphi}{\partial v} + \frac{1}{\gamma}\frac{\partial S_\theta}{\partial \varphi}\right] \tag{7.39}$$

where $\gamma = (1 - v^2)^{1/2}$, $S_\theta = rE_\theta$, $S_\varphi = r\sin\theta E_\varphi$, $U_\theta = \eta_0 r H_\theta$, $U_\varphi = \eta_0 r \sin\theta H_\varphi$, and $\eta_0 = \sqrt{\mu_0/\varepsilon_0} = 377\,\Omega$. Substituting these equations into the remaining Maxwell curl equations, we find

$$\frac{\partial S_\theta}{\partial r} = j\left[\frac{-\gamma}{r^2}\frac{\partial}{\partial v}\frac{1}{\varepsilon(r,v,\varphi)\gamma}\frac{\partial}{\partial \varphi}\right]U_\theta + j\left[-\frac{\mu(r,v,\varphi)}{\gamma} - \frac{\gamma}{r^2}\frac{\partial}{\partial v}\frac{1}{\varepsilon(r,v,\varphi)}\frac{\partial}{\partial v}\right]U_\varphi \tag{7.40}$$

$$\frac{\partial S_\varphi}{\partial r} = j\left[\mu(r,v,\varphi)\gamma + \frac{1}{r^2\gamma}\frac{\partial}{\partial \varphi}\frac{1}{\varepsilon(r,v,\varphi)}\frac{\partial}{\partial \varphi}\right]U_\theta + j\left[\frac{1}{r^2}\frac{\partial}{\partial \varphi}\frac{1}{\varepsilon(r,v,\varphi)}\frac{\partial}{\partial v}\right]U_\varphi \tag{7.41}$$

$$\frac{\partial U_\theta}{\partial r} = j\left[\frac{\gamma}{r^2}\frac{\partial}{\partial v}\frac{1}{\mu(r,v,\varphi)\gamma}\frac{\partial}{\partial \varphi}\right]S_\theta + j\left[\frac{\varepsilon(r,v,\varphi)}{\gamma} + \frac{\gamma}{r^2}\frac{\partial}{\partial v}\frac{1}{\mu(r,v,\varphi)}\frac{\partial}{\partial v}\right]S_\varphi \tag{7.42}$$

$$\frac{\partial U_\varphi}{\partial r} = j\left[-\varepsilon(r,v,\varphi)\gamma - \frac{1}{r^2\gamma}\frac{\partial}{\partial \varphi}\frac{1}{\mu(r,v,\varphi)}\frac{\partial}{\partial \varphi}\right]S_\theta + j\left[\frac{-1}{r^2}\frac{\partial}{\partial \varphi}\frac{1}{\mu(r,v,\varphi)}\frac{\partial}{\partial v}\right]S_\varphi \tag{7.43}$$

We will now be concerned with developing a multilayer, RCWA method that can be used to solve Equations 7.40 through 7.43 in Region 2. To proceed, we divide Region 2, $a \le r \le b$, into L thin layers of width d_ℓ where $b-a=\sum_{\ell=1}^{L} d_\ell$. We assume that each layer has been made thin enough so that all inhomogeneous functions in the radial coordinate r on the right hand sides of Equations 7.40 through 7.43 may be considered constant in the thin shell region and may be approximated by the midpoint value of r in the thin layer. In each thin spherical shell, it is convenient to introduce the local coordinates $s_1 = r - b$ for $b - d_1 \le r \le b$, $s_2 = r - (b - d_1)$ for $b - d_1 - d_2 \le r \le b - d_1, \ldots$. These local coordinates will be used to express the final state variable equations in each cylindrical shell. In the ℓth thin shell layer, Equations 7.40 through 7.43 are put in state variable form in the local coordinates s_ℓ by expanding all field variables and inhomogeneous factors $\varepsilon(r_\ell^{mid}, \nu, \varphi)$, $\gamma(\nu)$, $1/\mu(r_\ell^{mid}, \nu, \varphi), \ldots$ etc., (these functions are assumed sampled at the ℓth radial midpoint r_ℓ^{mid}) in a two dimensional exponential Fourier series, collecting terms together, which have the same exponential coefficient factors, and forming a set of first order differential equations for the mode amplitudes $S_{\theta im}^{(\ell)}, S_{\varphi im}^{(\ell)}, U_{\theta im}^{(\ell)}, U_{\varphi im}^{(\ell)}$. The mode amplitude expansion for $S_\theta^{(\ell)}(s_\ell, \theta, \varphi)$, for example, is given by $S_\theta^{(\ell)}(s_\ell, \theta, \varphi) = \sum_{i.m} S_{\theta im}^{(\ell)}(s_\ell)\exp[j(i\pi\nu + m\kappa\varphi)]$ where $-\Lambda_\varphi/2 \le \varphi \le \Lambda_\varphi/2$, $-1 \le \nu \le 1$, and $\kappa = 2\pi/\Lambda_\varphi = 1, 2, 3, \ldots$ may be called the azimuthal grating wave number and Λ_φ may be called the azimuthal grating period. The matrix for a general inhomogeneous factor, say $\varepsilon(r_\ell^{mid}, \nu, \varphi)$, for example, is $\underline{\underline{\varepsilon}}^{(\ell)} = [\underline{\underline{\varepsilon}}^{(\ell)}_{(i,m),(i',m')}] = [\breve{\varepsilon}_{i-i',m-m'}]$ where $\breve{\varepsilon}_{i-i',m-m'}$ are the two dimensional Fourier coefficients of $\varepsilon(r_\ell^{mid}, \nu, \varphi)$, and $\underline{\underline{\varepsilon}}^{(\ell)}_{(i,m),(i',m')}$ represents a typical matrix element of the overall matrix $\underline{\underline{\varepsilon}}^{(\ell)}$ (note that (i, m) is an ordered pair representing a single integer in the $\underline{\underline{\varepsilon}}^{(\ell)}$ matrix [same for (i', m')]). The matrices for the differential operators $\partial/\partial\nu$ and $\partial/\partial\varphi$ are given by the diagonal matrices $\underline{\underline{D}}_\nu = [ji\pi\delta_{i,i'}\delta_{m,m'}]$ and $\underline{\underline{D}}_\varphi = [jm\kappa\delta_{i,i'}\delta_{m,m'}]$, respectively, where $\delta_{i,i'}$ is the Kronecker delta, and the matrices describing the modal field amplitudes are given by column matrices (e.g., $\underline{S_\theta^{(\ell)}} = [S_\theta^{(\ell)}{}_{(i,m)}]^t$ [t is transpose]). Replacing each inhomogeneous factor, derivative operator, and field amplitude by the appropriate matrix, the overall system state variable matrix may be found. The first right-hand term of Equations 7.40, for example, is given by $-(j/r_\ell^{mid2})(\underline{\underline{\gamma}}(\underline{\underline{D}}_\nu(\underline{\underline{K}}_{1/\varepsilon\gamma}^{(\ell)}(\underline{\underline{D}}_\varphi\, \underline{U}_\theta)))) = \underline{\underline{A}}_{1,3}^{(\ell)}\, \underline{U}_\theta$ where $\underline{\underline{\gamma}}$ and $\underline{\underline{K}}_{1/\varepsilon\gamma}^{(\ell)}$ matrices represent the factors $\gamma(\nu)$ and $1/(\varepsilon(r_\ell^{mid}, \nu, \varphi)\gamma(\nu))$, respectively. The matrix $\underline{\underline{A}}_{1,3}^{(\ell)}$, which was just formed, represents a square, component, submatrix of the overall state matrix $\underline{\underline{A}}^{(\ell)}$. All component submatrices $\underline{\underline{A}}_{\alpha,\beta}^{(\ell)}, (\alpha,\beta) = (1,4)$ of the overall state matrix $\underline{\underline{A}}^{(\ell)}$ are defined in the same way as was $\underline{\underline{A}}_{1,3}^{(\ell)}$. (Since the component submatrices can be defined by inspection of Equations 7.40 through 7.43, it is not necessary to list the $\underline{\underline{A}}_{\alpha,\beta}^{(\ell)}, (\alpha,\beta) = (1,4)$ matrices specifically.) The overall state variable equations, determined from (7.40 through 7.43) in the ℓth thin shell layer is given by

$$\frac{\partial \underline{V}^{(\ell)}}{\partial s_\ell} = \underline{\underline{A}}^{(\ell)}\, \underline{V}^{(\ell)}, \quad \ell = 1,2,3,\ldots,L \tag{7.44}$$

where

$$\underline{\underline{A}}^{(\ell)} = \begin{bmatrix} 0 & 0 & \underline{\underline{A}}_{1,3}^{(\ell)} & \underline{\underline{A}}_{1,4}^{(\ell)} \\ 0 & 0 & \underline{\underline{A}}_{2,3}^{(\ell)} & \underline{\underline{A}}_{2,4}^{(\ell)} \\ \underline{\underline{A}}_{3,1}^{(\ell)} & \underline{\underline{A}}_{3,2}^{(\ell)} & 0 & 0 \\ \underline{\underline{A}}_{4,1}^{(\ell)} & \underline{\underline{A}}_{4,2}^{(\ell)} & 0 & 0 \end{bmatrix}, \quad \underline{V}^{(\ell)} = \begin{bmatrix} \underline{S_\theta^{(\ell)}} \\ \underline{S_\varphi^{(\ell)}} \\ \underline{U_\theta^{(\ell)}} \\ \underline{U_\varphi^{(\ell)}} \end{bmatrix} \tag{7.45}$$

If the overall state variable equation is truncated with $|i| \le I_T$ and $|m| \le M_T$, then $\underline{\underline{A}}^{(\ell)}$ is a $P_T \times P_T$ square matrix with $P_T = 4(2I_T + 1)(2M_T + 1)$. The solution of the overall state variable matrix

solution is given by $\underline{V^{(\ell)}(s_\ell)}_p = \underline{V^{(\ell)}}_p \exp(q_p^{(\ell)} s_\ell)$ where $q_p^{(\ell)}$ and $\underline{V^{(\ell)}}_p$ $(p = 1, \ldots, P_T)$ are the eigenvalues and eigenvectors, respectively, of the state variable matrix $\underline{\underline{A^{(\ell)}}}$. The overall EM field solution in each thin shell region can be found by adding a linear combination of the P_T eigensolutions. For example, if $|i| \le I_T$, $|m| \le M_T$, $(p = 1, \ldots, P_T)$, then the $S_\theta^{(2,\ell)}(s_\ell, v, \varphi)$ field is given by $S_\theta^{(2,\ell)}(s_\ell, v, \varphi) = \sum_{i,m,p} C_p^{(\ell)} S_{\theta imp}^{(\ell)} \exp[q_p^{(\ell)} s_\ell + j(i\pi v + m\kappa\varphi)]$ where $S_{\theta imp}^{(\ell)}$ is the pth eigenvector component of $\underline{S_\theta^{(\ell)}}_p$ in the overall eigenvector $\underline{V^{(\ell)}}_p$ and $C_p^{(\ell)}$ are unknown EM field expansion coefficients.

Although a large matrix equation could be formed from matching EM boundary conditions at $r = a$, b, and at each thin shell layer interface in the inhomogeneous region, a more efficient solution method is to use a ladder approach similar to that described in Sections 7.3 and 7.4 (i.e., successively relate unknown coefficients from one layer to the next) to express the $C_p^{(L)}$ coefficients of the Lth thin shell layer (located in the layer adjacent to $r = a$) in terms of the $C_p^{(1)}$ coefficients (located in the layer adjacent to $r = b$), and then match boundary conditions at $r = a$ and b interfaces to obtain the final unknowns of the system. At the ℓth and $(\ell + 1)$th interface, matching the tangential magnetic and electric fields, we have

$$\sum_{p=1}^{P_T} C_p^{(\ell)} S_{\theta imp}^{(\ell)} \exp(-q_p^{(\ell)} d_\ell) = \sum_{p=1}^{P_T} C_p^{(\ell+1)} S_{\theta imp}^{(\ell+1)} \tag{7.46}$$

$$\sum_{p=1}^{P_T} C_p^{(\ell)} S_{\varphi imp}^{(\ell)} \exp(-q_p^{(\ell)} d_\ell) = \sum_{p=1}^{P_T} C_p^{(\ell+1)} S_{\varphi imp}^{(\ell+1)} \tag{7.47}$$

$$\sum_{p=1}^{P_T} C_p^{(\ell)} U_{\theta imp}^{(\ell)} \exp(-q_p^{(\ell)} d_\ell) = \sum_{p=1}^{P_T} C_p^{(\ell+1)} U_{\theta imp}^{(\ell+1)} \tag{7.48}$$

$$\sum_{p=1}^{P_T} C_p^{(\ell)} U_{\varphi imp}^{(\ell)} \exp(-q_p^{(\ell)} d_\ell) = \sum_{p=1}^{P_T} C_p^{(\ell+1)} U_{\varphi imp}^{(\ell+1)} \tag{7.49}$$

Letting $\underline{C^{(\ell)}} = [C_1^{(\ell)}, \ldots, C_{P_T}^{(\ell)}]^t$, these equations may be written as

$$\underline{\underline{D_-^{(\ell)}}}\,\underline{C^{(\ell)}} = \underline{\underline{D_+^{(\ell)}}}\,\underline{C^{(\ell+1)}} \tag{7.50}$$

or

$$\underline{C^{(\ell+1)}} = \left[\underline{\underline{D_+^{(\ell)}}}\right]^{-1} \underline{\underline{D_-^{(\ell)}}}\,\underline{C^{(\ell)}} = \underline{\underline{F^{(\ell)}}}\,\underline{C^{(\ell)}} \quad \ell = 1, \ldots, L-1 \tag{7.51}$$

where the −1 superscript denotes matrix inverse. Substituting successively we have

$$\underline{C^{(L)}} = \underline{\underline{F^{(L-1)}}}(\underline{\underline{F^{(L-2)}}}(\cdots(\underline{\underline{F^{(1)}}}\,\underline{C^{(1)}})\cdots) = \underline{\underline{M}}\,\underline{C^{(1)}} \tag{7.52}$$

Another important problem is to relate the fields of Region 1 (interior region) and Region 3 (exterior region) to the fields of Region 2 (inhomogeneous region). The fields in Regions 1 and 3, as is well known, can be expressed in terms of an infinite number of transverse to r electric (TE_r)

and transverse to r magnetic (TM_r) Schelkunoff spherical vector potential modes [5, Chapter 6]. These vector potential modes consist of half order radial Bessel and Hankel functions and consist of Tesseral harmonics (products of Legendre polynomials and φ exponential functions). The scattered field portions of the Regions 1 and 3 Bessel and Hankel function solutions are chosen to satisfy the usual spherical boundary conditions of finiteness at the origin and being an outgoing wave at infinity. In this section, the incident field in Region 1 is the EM field of an infinitesimal dipole. The basic EM boundary matching procedure to be followed in this section is to equate the tangential electric fields at the interfaces $r = a$ and b, eliminate unknown field constants in Regions 1 and 3 in favor of the field constants in Region 2 from these equations, equate the tangential magnetic fields at the interfaces $r = a$ and b, and substitute the electric field matching Region 2 constants into the magnetic field matching equations. This general procedure is precisely the one followed by [9–12] in the analysis of diffraction from planar diffraction gratings. Equating the common terms of $\exp(j\kappa m\varphi)$ of the $S_{\theta m}(v)$ and $S_{\varphi m}(v)$ field components at $r = a$ from Regions 1 and 2, we have:

$$S_{\theta m}^{(1)}(v) = S_{\theta m}^{(1,Scat)}(v) + S_{\theta m}^{(1,INC)}(v) = S_{\theta m}^{(2,a)}(v) \tag{7.53}$$

$$S_{\varphi m}^{(1)}(v) = S_{\varphi m}^{(1,Scat)}(v) + S_{\varphi m}^{(1,INC)}(v) = S_{\varphi m}^{(2,a)}(v) \tag{7.54}$$

where

$$S_{\theta m}^{(1,Scat)}(v) = \sum_{i=-I_T}^{I_T} \left\{ \sum_{n=|\kappa m|+\delta_{m,0}}^{2I_T+|\kappa m|+\delta_{m,0}} \left[E_{Aimn}^{(1)} F_{mn}^{(1)} + E_{Bimn}^{(1)} A_{mn}^{(1)} \right] \right\} \exp(ji\pi v) \tag{7.55}$$

$$S_{\varphi m}^{(1,Scat)}(v) = \sum_{i=-I_T}^{I_T} \left\{ \sum_{n=|\kappa m|+\delta_{m,0}}^{2I_T+|\kappa m|+\delta_{m,0}} \left[E_{Cimn}^{(1)} F_{mn}^{(1)} + E_{Dimn}^{(1)} A_{mn}^{(1)} \right] \right\} \exp(ji\pi v) \tag{7.56}$$

where $E_{Aim}^{(1)} = (-j\kappa m/\varepsilon_1)\hat{J}_n(\beta_1 a) g_{imn}^A$, $E_{Bim}^{(1)} = (-j/\beta_1)\hat{J}_n'(\beta_1 a) g_{imn}^B$, $E_{Cim}^{(1)} = (1/\varepsilon_1)\hat{J}_n(\beta_1 a) g_{imn}^C$, and $E_{Dim}^{(1)} = (\kappa m/\beta_1)\hat{J}_n'(\beta_1 a) g_{imn}^D$.

Letting $g_{Amn} = (1-v^2)^{-1/2} P_n^{|\kappa m|}(v)$, $g_{Bmn} = -(1-v^2)^{1/2} \partial/\partial v (P_n^{|\kappa m|}(v))$, $g_{Cmn} = -(1-v^2)\partial/\partial v (P_n^{|\kappa m|}(v))$, and $g_{Dmn} = P_n^{|\kappa m|}(v)$ where $P_n^{|\kappa m|}(v)$ are associated Legendre functions of order n and $|\kappa m|$, and the coefficients g_{imn}^A, g_{imn}^B, g_{imn}^C, and g_{imn}^D represent the exponential Fourier series expansion coefficients for the terms $g_{Amn}(v)$, $g_{Bmn}(v)$, $g_{Cmn}(v)$, and $g_{Dmn}(v)$, respectively, on the interval $-1 \le v \le 1$ $\left(g_{Amn}(v) = \sum_i g_{imn}^A \exp(ji\pi v) \right)$. In the present analysis, the Fourier coefficients g_{imn}^A, g_{imn}^B, g_{imn}^C, and g_{imn}^D have been determined *exactly* by calculating higher order derivatives of the Bessel function integral representation given in [25, Eq. 9.1.20, p. 360]. The exact calculation of these Fourier coefficients is an important step in order to ensure overall accuracy of the entire RCWA algorithm. The terms $S_{\theta m}^{(1,INC)}(v) = \sum_i S_{\theta im}^{(1,INC)} \exp(ji\pi v)$ and $S_{\varphi m}^{(1,INC)}(v) = \sum_i S_{\varphi im}^{(1,INC)} \exp(ji\pi v)$ represent electric field EM incident waves evaluated at $r = a$, which emanate from Region 1. In this section, it is assumed that a centered electric dipole excites EM radiation in the overall system, and for this source, it is found that $S_{\theta m}^{(1,INC)}(v) = (j/\beta_1) A_{0,1}^I \hat{H}_1'(\beta_1 a)[1-v^2]^{1/2} \delta_{m,0}$, $S_{\varphi m}^{(1,INC)}(v) = 0$, and $A_{0,1}^I$ is the strength of the electric dipole source. In Equations 7.53 through 7.56, $\hat{J}_n(\beta_1 a)$ $\left(\beta_1 = \sqrt{\mu_1 \varepsilon_1}\right)$ represents a spherical Schelkunoff Bessel function [5], and the prime in the expression $\hat{J}_n'(\beta_1 a)$, represents differentiation with respect to the argument of the function $\hat{J}_n(\beta_1 a)$. For $m \ne 0$, the lower n limits start at $|\kappa m|$ since the Legendre polynomials are zero when $|\kappa m| > n$. The terms $S_{\theta m}^{(2,a)}(v)$ and $S_{\varphi m}^{(2,a)}(v)$ represent the state variable solution in Region 2 in the Lth thin shell layer region evaluated at $r = a$ and are given by

$$S_{\theta m}^{(2,a)}(v) = \sum_{i=-I_T}^{I_T} \left\{ \sum_{p=1}^{P_T} C_p^{(L)} S_{\theta imp}^{(L)} \exp(-q_p^{(L)} s_L) \right\} \exp(ji\pi v) = \sum_{i=-I_T}^{I_T} S_{\theta im}^{(2,a)} \exp(ji\pi v) \quad (7.57)$$

$$S_{\varphi m}^{(2,a)}(v) = \sum_{i=-I_T}^{I_T} \left\{ \sum_{p=1}^{P_T} C_p^{(L)} S_{\varphi imp}^{(L)} \exp(-q_p^{(L)} s_L) \right\} \exp(ji\pi v) = \sum_{i=-I_T}^{I_T} S_{\varphi im}^{(2,a)} \exp(ji\pi v) \quad (7.58)$$

When I_T is infinite, the boundary matching equations given by Equations 7.53 through 7.58 are exact. When I_T is truncated at a finite value, and the common coefficients of $\exp(ji\pi v)$ in Equations 7.53 through 7.58 are collected, Equations 7.53 through 7.58 give a $[2(2I_T + 1)] \times [2(2I_T + 1)]$ set of equations from which the $A_{mn}^{(1)}$ and $F_{mn}^{(1)}$ can be expressed in terms of the Region 2 unknown coefficients $C_p^{(L)}$. Letting $\underline{\underline{E_{Am}^{(1)}}}$, $\underline{\underline{E_{Bm}^{(1)}}}$, $\underline{\underline{E_{Cm}^{(1)}}}$, and $\underline{\underline{E_{Dm}^{(1)}}}$ be matrices representing Equations 7.55 and 7.56 (e.g., $\underline{\underline{E_{Am}^{(1)}}} = \left[E_{Aimn}^{(1)} \right]$, where $i = -I_T, \ldots, I_T$ and $n = |\kappa m| + \delta_{m,0}, \ldots, 2I_T + |\kappa m| + \delta_{m,0}$), letting $\underline{S_{\theta m}^{(1,INC)}} = \left[S_{\theta im}^{(1,INC)} \right]$ and $\underline{S_{\varphi m}^{(1,INC)}} = \left[S_{\varphi im}^{(1,INC)} \right]$ for $i = -I_T, \ldots, I_T$, and letting $\underline{\underline{S_{\theta m}^{(2,a)}}} = \left[S_{\theta imp}^{(L)} \exp(-q_p^{(L)} s_L) \right]$ and $\underline{\underline{S_{\varphi m}^{(2,a)}}} = \left[S_{\varphi imp}^{(L)} \exp(-q_p^{(L)} s_L) \right]$ for $i = -I_T, \ldots, I_T$ and $p = 1, \ldots, P_T$, we find the following matrix equation:

$$\underline{\underline{S_m^{(2,a)}}}\,\underline{C^{(L)}} = \begin{bmatrix} \underline{\underline{S_{\theta m}^{(2,a)}}} \\ \underline{\underline{S_{\varphi m}^{(2,a)}}} \end{bmatrix} \underline{C^{(L)}} = \begin{bmatrix} \underline{\underline{E_{Am}^{(1)}}} & \underline{\underline{E_{Bm}^{(1)}}} \\ \underline{\underline{E_{Cm}^{(1)}}} & \underline{\underline{E_{Dm}^{(1)}}} \end{bmatrix} \begin{bmatrix} \underline{F_m^{(1)}} \\ \underline{A_m^{(1)}} \end{bmatrix} + \begin{bmatrix} \underline{S_\theta^{(1,INC)}} \\ \underline{S_\varphi^{(1,INC)}} \end{bmatrix} = \underline{\underline{E_m^{(1)}}} \begin{bmatrix} \underline{F_m^{(1)}} \\ \underline{A_m^{(1)}} \end{bmatrix} + \underline{S_m^{(1,INC)}} \quad (7.59)$$

where

$\underline{\underline{S_m^{(2,a)}}}$ is a $[2(2I_T + 1)] \times P_T$ matrix

$\underline{\underline{E_m^{(1)}}}$ is a $[2(2I_T + 1)] \times [2(2I_T + 1)]$ square matrix

$\underline{S_m^{(1,INC)}}$ is a $2(2I_T + 1)$ column matrix

To proceed further, we match the terms common to $\exp(j\kappa m\varphi)$ of the tangential magnetic field at the Region 1 and Region 2 interfaces at $r = a$ and find

$$U_{\theta m}^{(1)}(v) = U_{\theta m}^{(1,Scat)}(v) + U_{\theta m}^{(1,INC)}(v) = U_{\theta m}^{(2,a)}(v) \quad (7.60)$$

$$U_{\varphi m}^{(1)}(v) = U_{\varphi m}^{(1,Scat)}(v) + U_{\varphi m}^{(1,INC)}(v) = U_{\varphi m}^{(2,a)}(v) \quad (7.61)$$

where

$$U_{\theta m}^{(1,Scat)}(v) = \sum_{i=-I_T}^{I_T} \left\{ \sum_{n=|\kappa m|+\delta_{m,0}}^{2I_T+|\kappa m|+\delta_{m,0}} \left[H_{Bimn}^{(1)} F_{mn}^{(1)} + H_{Aimn}^{(1)} A_{mn}^{(1)} \right] \right\} \exp(ji\pi v) \quad (7.62)$$

$$U_{\varphi m}^{(1,Scat)}(v) = \sum_{i=-I_T}^{I_T} \left\{ \sum_{n=|\kappa m|+\delta_{m,0}}^{2I_T+|\kappa m|+\delta_{m,0}} \left[H_{Dimn}^{(1)} F_{mn}^{(1)} + H_{Cimn}^{(1)} A_{mn}^{(1)} \right] \right\} \exp(ji\pi v) \quad (7.63)$$

where $H^{(1)}_{Aim} = (j\kappa m/\mu_1)\hat{J}_n(\beta_1 a)g^{A}_{imn}$, $H^{(1)}_{Bim} = (-j/\beta_1)\hat{J}'_n(\beta_1 a)g^{B}_{imn}$, $H^{(1)}_{Cim} = (-1/\mu_1)\hat{J}_n(\beta_1 a)g^{C}_{imn}$, and $H^{(1)}_{Dim} = (\kappa m/\beta_1)\hat{J}'_n(\beta_1 a)g^{D}_{imn}$. The terms $U^{(1,INC)}_{\theta m}(v) = \sum_i U^{(1,INC)}_{\theta im}\exp(ji\pi v)$ and $U^{(1,INC)}_{\varphi m}(v) = \sum_i U^{(1,INC)}_{\varphi im}\exp(ji\pi v)$ represent the magnetic field incident waves that may emanate from Region 1. In this section, $U^{(1,INC)}_{\varphi m}(v) = (1/\mu_1)A^{I}_{0,1}\hat{H}_1(\beta_1 a)[1-v^2]\delta_{m,0}$ and $U^{(1,INC)}_{\theta m}(v) = 0$ for a centered electric dipole source. The terms $U^{(2,a)}_{\theta m}(v)$ and $U^{(2,a)}_{\varphi m}(v)$ represent the state variable solution in Region 2 in the Lth thin shell layer region evaluated at $r = a$ and are given by

$$U^{(2,a)}_{\theta m}(v) = \sum_{i=-I_T}^{I_T}\left\{\sum_{p=1}^{P_T} C^{(L)}_p U^{(L)}_{\theta imp}\exp(-q^{(L)}_p s_L)\right\}\exp(ji\pi v) = \sum_{i=-I_T}^{I_T} U^{(2,a)}_{\theta im}\exp(ji\pi v) \tag{7.64}$$

$$U^{(2,a)}_{\varphi m}(v) = \sum_{i=-I_T}^{I_T}\left\{\sum_{p=1}^{P_T} C^{(L)}_p U^{(L)}_{\varphi imp}\exp(-q^{(L)}_p s_L)\right\}\exp(ji\pi v) = \sum_{i=-I_T}^{I_T} U^{(2,a)}_{\varphi im}\exp(ji\pi v) \tag{7.65}$$

Equating common coefficients of $\exp(ji\pi v)$ in Equations 7.60 through 7.65, a similar matrix equation as was formed for the tangential electric field components may be formed for the tangential magnetic field components. We have

$$\underline{U^{(2,a)}_m}\,\underline{C^{(L)}} = \left[\begin{array}{c}\underline{U^{(2,a)}_{\theta m}}\\ \hline \underline{U^{(2,a)}_{\varphi m}}\end{array}\right]\underline{C^{(L)}} = \left[\begin{array}{cc}\underline{H^{(1)}_{Bm}} & \underline{H^{(1)}_{Am}}\\ \hline \underline{H^{(1)}_{Dm}} & \underline{H^{(1)}_{Cm}}\end{array}\right]\left[\begin{array}{c}\underline{F^{(1)}_m}\\ \hline \underline{A^{(1)}_m}\end{array}\right] + \left[\begin{array}{c}\underline{U^{(1,INC)}_{\theta}}\\ \hline \underline{U^{(1,INC)}_{\varphi}}\end{array}\right] = \underline{H^{(1)}_m}\left[\begin{array}{c}\underline{F^{(1)}_m}\\ \hline \underline{A^{(1)}_m}\end{array}\right] + \underline{U^{(1,INC)}_m} \tag{7.66}$$

To proceed further, our objective now is to eliminate the column matrix $\left[\underline{F^{(1)}_m}, \underline{A^{(1)}_m}\right]^t$, from Equation 7.59 and therefore form a single matrix equation for the $\underline{C^{(L)}}$ coefficients alone. By inspecting the matrix equation (Equation 7.59) and their definitions, we notice that two distinct cases arise, namely, the cases when $m \neq 0$ and the case when $m = 0$. In the case of $m \neq 0$, it turns out that the matrices $\underline{E^{(1)}_m}$ and $\underline{H^{(1)}_m}$ are nonsingular; therefore, it is straight forward to invert $\underline{E^{(1)}_m}$ and solve for $\left[\underline{F^{(1)}_m}, \underline{A^{(1)}_m}\right]^t$. For $m \neq 0$, we find

$$\left[\begin{array}{c}\underline{F^{(1)}_m}\\ \hline \underline{A^{(1)}_m}\end{array}\right] = \underline{E^{(1)}_m}^{-1}(\underline{S^{(2,a)}}\ \underline{C^{(L)}}) - \underline{E^{(1)}_m}^{-1}\underline{S^{(1,INC)}_m} = \underline{Z^{(1)}_m}\,\underline{C^{(L)}} - \underline{E^{(1,INC)}_m} \tag{7.67}$$

The determination of $F^{(1)}_{0,n}$ and $A^{(1)}_{0,n}$ ($n = 1, 2, 3, \ldots$) coefficients for the $m = 0$ case requires special matrix processing. We first note for the $m = 0$ case that $E^{(1)}_{Aim} = E^{(1)}_{Dim} = 0$ in Equations 7.55 and 7.56 and thus the matrix equations for $F^{(1)}_{0,n}$ and $A^{(1)}_{0,n}$ are decoupled from one another. One also observes from Equations 7.55 and 7.56 that when solving for either $F^{(1)}_{0,n}$ or $A^{(1)}_{0,n}$, the coefficients of $F^{(1)}_{0,1}, F^{(1)}_{0,3}, F^{(1)}_{0,5}, \ldots$ and $A^{(1)}_{0,1}, A^{(1)}_{0,3}, A^{(1)}_{0,5}, \ldots$ are multiplied by the first derivative Legendre polynomials $\partial/\partial v(P^0_1(v))$, $\partial/\partial v(P^0_3(v)), \ldots$ which are even in v whereas the coefficients of $F^{(1)}_{0,2}, F^{(1)}_{0,4}, F^{(1)}_{0,6}, \ldots$ and $A^{(1)}_{0,2}, A^{(1)}_{0,4}, A^{(1)}_{0,6}, \ldots$ are multiplied by first derivative in Legendre polynomials $\partial/\partial v(P^0_2(v))$, $\partial/\partial v(P^0_4(v)), \ldots$ which are odd in v. This means that when determining the $m = 0$, the Region 1 coefficients $F^{(1)}_{0,n}$ and $A^{(1)}_{0,n}$, the best numerical processing in Equation 7.59 is to decompose $S^{(1,INC)}_{\theta m}(v)$, $S^{(1,INC)}_{\varphi m}(v)$, $S^{(2,a)}_{\theta m}(v)$, and $S^{(2,a)}_{\varphi m}(v)$ for $m = 0$ into a sum of even and odd functions, and from the even functions in Equations 7.55 and 7.56 determine $F^{(1)}_{0,1}, F^{(1)}_{0,3}, F^{(1)}_{0,5}, \ldots$ and $A^{(1)}_{0,1}, A^{(1)}_{0,3}, A^{(1)}_{0,5}, \ldots$ and from the odd functions in Equations 7.55 and 7.56 determine $F^{(1)}_{0,2}, F^{(1)}_{0,4}, F^{(1)}_{0,6}, \ldots$ and $A^{(1)}_{0,2}, A^{(1)}_{0,4}, A^{(1)}_{0,6}, \ldots$. The specific matrix processing that is carried out for say the $F^{(1)}_{0,1}, F^{(1)}_{0,3}, F^{(1)}_{0,5}, \ldots$ coefficients is as follows. After decomposing $S_{\varphi m}^{(1,INC)}(v)$ and $S_{\varphi m}^{(2,a)}(v)$ for $m = 0$ into even and odd functions of v, $F_{0,1}^{(1)}, F_{0,3}^{(1)}, F_{0,5}^{(1)},\ldots$ is determined by (1) expanding the even function part of $S_{\varphi m}^{(1,INC)}(v)$ and $S_{\varphi m}^{(2,a)}(v)$ for $m = 0$ in a $\{\cos(i\pi v)\}_{i=0}^{I_T}$ cosine series (the $\{\cos(i\pi v)\}_{i=0}^{I_T}$ series expansion of $S_{\varphi m}^{(2,a)}(v)$ for $m = 0$ depends on the $C_p^{(L)}$, $p = 1,\ldots,P_T$ coefficients in Region 2 and the $\{\cos(i\pi v)\}_{i=0}^{I_T}$ series expansion of $S_{\varphi m}^{(1,INC)}(v)$ for $m = 0$ depends on the incident EM source waves, which emanate from Region 1); (2) expanding the first derivative Legendre polynomial $\partial/\partial v(P_1^0(v))$, $\partial/\partial v(P_3^0(v))$,… in a $\{\cos(i\pi v)\}_{i=0}^{I_T}$ series; (3) equating common coefficients of the cosine series $\{\cos(i\pi v)\}_{i=0}^{I_T}$; and (4) from these equations, developing an $(I_T + 1) \times (I_T + 1)$ matrix equation, which upon matrix inversion expresses the $F_{0,1}^{(1)}, F_{0,3}^{(1)}, F_{0,5}^{(1)},\ldots$ coefficients in terms of the $C_p^{(L)}$, $p = 1,\ldots,P_T$ coefficients of Region 2 and incident EM wave coefficients of Region 1. The determination of the $F_{0,2}^{(1)}, F_{0,4}^{(1)}, F_{0,6}^{(1)},\ldots$ coefficients is found by (1) expanding the odd function part of $S_{\varphi m}^{(1,INC)}(v)$ and $S_{\varphi m}^{(2,a)}(v)$ for $m = 0$ and the odd derivative Legendre polynomials $\partial/\partial v(P_2^0(v))$, $\partial/\partial v(P_4^0(v))$ functions in a $\{\sin(i\pi v)\}_{i=1}^{I_T}$ series; (2) equating common coefficients of $\{\sin(i\pi v)\}_{i=1}^{I_T}$; and (3) and then forming an $I_T \times I_T$ matrix equation, which, upon matrix inversion, expresses the $F_{0,2}^{(1)}, F_{0,4}^{(1)}, F_{0,6}^{(1)},\ldots$ coefficients in terms of the $C_p^{(L)}$, $p = 1,\ldots,P_T$ coefficients of Region 2 and incident EM wave coefficients of Region 1. After following the above procedure, and combining the even and odd matrix expressions for $F_{0,1}^{(1)}$, $F_{0,3}^{(1)}$, $F_{0,5}^{(1)}$,… and $F_{0,2}^{(1)}$, $F_{0,4}^{(1)}$, $F_{0,6}^{(1)}$,…, a $(2I_T + 1) \times (2I_T + 1)$ matrix relation is found between the overall $F_{0,1}^{(1)}, F_{0,2}^{(1)}, F_{0,3}^{(1)}, F_{0,4}^{(1)},\ldots$ coefficients and the $C_p^{(L)}$, $p = 1,\ldots,P_T$ coefficients and EM incident wave coefficients of Region 1. A similar even and odd analysis allows a $(2I_T + 1) \times (2I_T + 1)$ matrix relation between the $A_{0,n}^{(1)}$ coefficients and the $C_p^{(L)}$, $p = 1,\ldots,P_T$ coefficients of Region 2 and EM incident wave coefficients of Region 1. Altogether, the $F_{0,n}^{(1)}$ and $A_{0,n}^{(1)}$ coefficients for $m = 0$ in matrix form may be expressed as:

$$\begin{bmatrix} \underline{F_0^{(1)}} \\ \underline{A_0^{(1)}} \end{bmatrix} = \begin{bmatrix} \underline{\underline{Z_0^{(1,F)}}} & 0 \\ 0 & \underline{\underline{Z_0^{(1,A)}}} \end{bmatrix} \underline{C^{(L)}} - \underline{E_0^{(1,INC)}} = \underline{\underline{Z_0^{(1)}}}\, \underline{C^{(L)}} - \underline{E_0^{(1,INC)}} \tag{7.68}$$

where the matrix $\underline{\underline{Z_0^{(1)}}}$ is size $[2(2I_T + 1)] \times [2(2I_T + 1)]$.

Substituting $\left[\underline{F_m^{(1)}} \quad \underline{A_m^{(1)}}\right]^t$ from (7.67) $(m \neq 0)$ and $\left[\underline{F_0^{(1)}} \quad \underline{A_0^{(1)}}\right]^t$ of (7.68) $(m = 0)$ into Equation 7.66, we find

$$\left[\underline{\underline{U_m^{(2,a)}}} - \underline{\underline{H_m^{(1)}}}\, \underline{\underline{Z_m^{(1)}}}\right] \underline{C^{(L)}} = \underline{U_m^{(1,INC)}} - \underline{\underline{H_m^{(1)}}}\, \underline{E_m^{(1,INC)}} \tag{7.69}$$

where $m = -M_T,\ldots, -1, 0, 1,\ldots, M_T$. For each m, Equation 7.69 has $2(2I_T + 1)$ rows.

The boundary matching analysis at the $r = b$ interface is identical to that at $r = a$, and the boundary matching equations at the $r = b$ interface are given by Equation 7.53 through 7.69 if (1) one replaces Region "(1)" superscripts with Region "(3)" superscripts, (2) one replaces the spherical Schelkunoff Bessel function $\hat{J}_n(\beta_1 a)$ with the outgoing spherical Schelkunoff Hankel function of the second kind, namely, $\hat{H}_n(\beta_3 b)$, (3) one replaces $S_{\theta m}^{(2,a)}(v)$ and $S_{\varphi m}^{(2,a)}(v)$ in Equations 7.57 and 7.58 with the $\ell = 1$ thin shell, state variable solution evaluated at $r = b$, namely

$$S_{\theta m}^{(2,b)}(v) = \sum_{i=-I_T}^{I_T} \left\{ \sum_{p=1}^{P_T} C_p^{(1)} S_{\theta imp}^{(1)} \right\} \exp(ji\pi v) = \sum_{i=-I_T}^{I_T} S_{\theta im}^{(2,b)} \exp(ji\pi v) \tag{7.70}$$

$$S_{\varphi m}^{(2,b)}(v) = \sum_{i=-I_T}^{I_T} \left\{ \sum_{p=1}^{P_T} C_p^{(1)} S_{\varphi imp}^{(1)} \right\} \exp(ji\pi v) = \sum_{i=-I_T}^{I_T} S_{\varphi im}^{(2,b)} \exp(ji\pi v) \quad (7.71)$$

and (4) one sets all Region 3, incident source terms to zero since EM energy in this paper is assumed to emanate only from a Region 1 centered dipole source. After algebra, it is found that the Region 3 boundary equations are

$$\left[\underline{U}_m^{(2,b)} - \underline{H}_m^{(3)} \, \underline{Z}_m^{(3)} \right] \underline{C}^{(1)} = \underline{U}_m^{(3,INC)} - \underline{H}_m^{(3)} \, \underline{E}_m^{(3,INC)} \quad (7.72)$$

where $m = -M_T, \ldots, -1, 0, 1, \ldots, M_T$ and where the right hand side of Equation 7.72 for the present paper is zero. For each m, Equation 7.72 has $2(2I_T + 1)$ rows. Using $\underline{C}^{(L)} = \underline{M}\,\underline{C}^{(1)}$ from Equation 7.52, we eliminate the $\underline{C}^{(L)}$ column matrix and find

$$\left[\underline{U}_m^{(2,a)} - \underline{H}_m^{(1)} \, \underline{Z}_m^{(1)} \right] \underline{M}\,\underline{C}^{(1)} = \underline{U}_m^{(1,INC)} - \underline{H}_m^{(1)} \, \underline{E}_m^{(1,INC)} \quad (7.73)$$

where $m = -M_T, \ldots, -1, 0, 1, \ldots, M_T$. Including all values of m, Equations 7.72 and 7.73 each represent $[2M_T + 1] \times [2(2I_T + 1)] = P_T/2$ equations for $\underline{C}^{(1)}$. Thus, Equations 7.72 and 7.73 represent together a $P_T \times P_T$ matrix equation from which $C^{(1)}$ may be determined. From knowledge of $\underline{C}^{(1)}$, all other unknown constants of the system may be determined.

Once all of the EM field coefficients are determined, one may calculate the power that is radiated in Regions 1 and 3 of the system. The radiated power associated with a given m and n spherical mode is well known, and the specific formulas may be found in [5, Chapter 6]. In this paper, we will give numerical results in terms of normalized power. The normalized power of a given m and n spherical mode at radial distance r is defined here as the power that is radiated by the m and n spherical mode into a sphere located at a radius r *divided* by the total power radiated by the centered dipole when the centered dipole is in an infinite region whose material parameters are those of Region 1, namely, ε_1 and μ_1.

7.5.3 Numerical Results

In this section, we illustrate the RCWA method of Section 7.5.2 by solving for the radiated and scattered EM fields that result when the Region 2 inhomogeneous material shell is assumed to have, as a specific example, the form

$$\begin{aligned} \varepsilon(r,\theta,\varphi) &= \varepsilon_2 + \Delta\varepsilon_\theta \sin(\theta)\,\mathrm{sgn}(\cos(\theta))[1 + \Delta\varepsilon_\varphi(r)\,\mathrm{sgn}(\varphi)] \\ &= \varepsilon_2 + \Delta\varepsilon_\theta [1 - v^2]^{1/2}\,\mathrm{sgn}(v)[1 + \Delta\varepsilon_\varphi(r)\,\mathrm{sgn}(\varphi)] \\ \mu(r,\theta,\varphi) &= \mu_2 \end{aligned} \quad (7.74)$$

where $\mathrm{sgn}(X) = 1, X > 0$ and $\mathrm{sgn}(X) = -1, X < 0$. For this profile, $\kappa = 1$. This inhomogeneity profile is a convenient one to use, since if it is integrated over a spherical surface of radius r, its average or bulk value is always ε_2, regardless of the value of $\Delta\varepsilon_\theta$ or $\Delta\varepsilon_\varphi$ used. Using this dielectric inhomogeneity profile, three cases will be studied, namely, the cases when: $\Delta\varepsilon_\theta = 0.001$, $\Delta\varepsilon_\varphi(r) = 0.001$ (Case 1); $\Delta\varepsilon_\theta = 2.8$, $\Delta\varepsilon_\varphi(r) = 0.4$ (Case 2); and $\Delta\varepsilon_\theta = 2.8$, $\Delta\varepsilon_\varphi(r) = \alpha_1 r + \alpha_2$ where $\Delta\varepsilon_\varphi(r)\big|_{r=5} = 0.6$ and $\Delta\varepsilon_\varphi(r)\big|_{r=5.5} = 0.15$ for $a \le r \le b$, $a = 5$, and $b = 5.5$ (Case (3)). For all numerical examples of this section, the bulk material parameters will be taken to be $\varepsilon_1 = 1.5$, $\mu_1 = 1$, $\varepsilon_2 = 7$, $\mu_2 = 1.3$, $\varepsilon_3 = 1$, and $\mu_3 = 1$. The first case, which, because of the small values of $\Delta\varepsilon_\theta$ and $\Delta\varepsilon_\varphi$, may be called a homogeneous profile case, represents the application of the RCWA method to the solution of the problem of determining the EM radiation that occurs when a centered dipole radiates through a uniform dielectric shell. This problem of EM radiation through a homogeneous dielectric shell can be solved exactly by matching Bessel function solutions in Regions 1–3; thus, a comparison of the RCWA method with the exact Bessel function matching solution represents a numerical validation of the RCWA method if close numerical results from the two methods occur.

The second case, which may be designated a (θ, φ)-inhomogeneity profile case, represents an inhomogeneous example in which the dielectric shell is homogeneous in the radial r direction but is inhomogeneous in the θ and φ coordinates. This case will be solved by both a single layer RCWA algorithm and by using a multilayer RCWA algorithm. The purpose of solving this second case is to observe in general how much diffraction occurs in higher order spherical modes when a reasonably large θ and φ inhomogeneity material profile is present in the dielectric shell. The purpose of comparing single layer and multilayer RCWA results is to observe the importance that the scale factors of (7.40 through 7.43) have on the overall scattering solution. The purpose also is to study how well the power conservation law is obeyed numerically. Power conservation at different radial distances is a good indication of the accuracy of the numerical solution in a lossless system such as the present one.

The third case, which may be designated a (r, θ, φ)-inhomogeneity profile case, represents a solution of the RCWA method under the most general conditions, namely, when the inhomogeneity variation occurs in the r, θ, and φ coordinates. The radial inhomogeneity variation has been chosen to vary in such a way that the magnitude of the φ-variation in the inhomogeneity profile changes linearly. Single and multilayer analyses have been studied in order to gauge the effect of the radial inhomogeneity variation of the (r, θ, φ)-inhomogeneity profile case relative to that of the (θ, φ)-inhomogeneity profile case. The purpose of studying this case is to see the effect in general that a fully three dimensional inhomogeneity variation has on diffraction and scattering into higher orders. The purpose also of Case (3), as in Case (2), is to study how well the power conservation law is obeyed numerically.

Figure 7.20 shows a comparison of the normalized power radiated through a uniform material shell when calculated by a Bessel function matching algorithm (exact solution), when calculated by a single layer RCWA method, and when calculated by a multilayer RCWA method. Ten layers ($L = 10$)

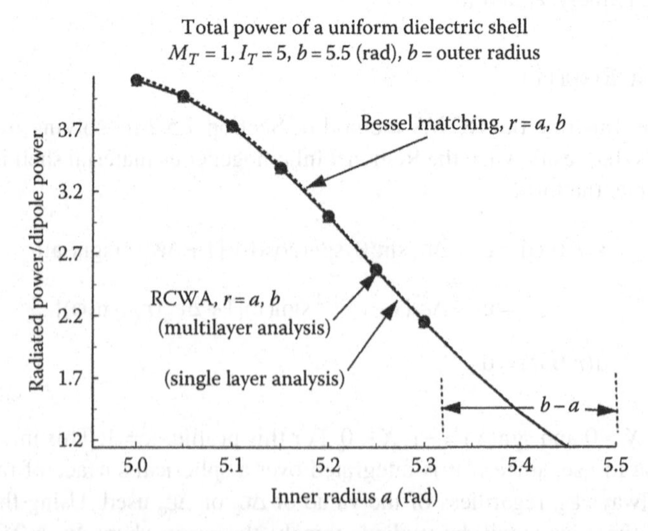

FIGURE 7.20 Normalized total power as obtained by the RCWA method compared to the total normalized power as obtained by matching Bessel function solutions at the interfaces $r = a$ and b. (Reprinted from Jarem, J.M., Rigorous coupled-wave-theory analysis of dipole scattering from a three-dimensional, inhomogeneous, spherical dielectric and permeable system, *IEEE Microw. Theory Tech.*, 45(8), August 1997. With permission. Copyright 1997 IEEE.)

were used to make all multilayer calculations in this section. In Figure 7.20, the normalized power by all methods has been calculated both at $r = a$ and b (The label $r = a, b$ means calculated at $r = a$ and calculated at $r = b$). In Figure 7.20, the outer radius is fixed at $b = 5.5$ (rad), and the inner radius a is varied from 5.0 (rad) to 5.45 (rad). As can be seen from Figure 7.20, there is excellent numerical agreement between the three methods used. We also notice that the $r = a, b$ power results for each of the three methods are so close at $r = a$ and b that the two power curves for each method cannot be distinguished from one another. In Figure 7.20, the RCWA algorithm was calculated using $M_T = 1$ and $I_T = 5$. Because the inhomogeneity factor in this case was very close to that of a perfectly homogeneous shell, the RCWA algorithm could have calculated the power of this case using a value of $M_T = 0$ and $I_T = 5$, which would have meant a significantly smaller matrix equation than would have resulted from $M_T = 1$. A larger matrix equation than necessary was solved for this case in order to test the numerical stability of the algorithm and also to test the sensitivity of the RCWA solution to error in the Fourier coefficients. (Error in the Fourier coefficients would arise for $M_T = 1$ because numerical integration is used to calculate the $\exp(jm\varphi)$, $m = -1, 0, 1$ Fourier coefficients, and thus, instead of the $m = \pm 1$ coefficients being exactly zero, they would have some small value.) The $M_T = 1$ matrix solution showed no ill-conditioned effects from using a larger than needed matrix size and showed no sensitivity to error in the Fourier coefficients. The RCWA algorithm was implemented using $M_T = 1$ and $I_T = 2$. In this case, the RCWA algorithm differed perceptibly from the Bessel matching solution. This indicates that for accurate results, enough Fourier harmonic terms must be included to correctly calculate the state variable solution of Equations 7.40 through 7.43. We finally note that the nearly same case, for the purpose of validation, was also studied in [23] using $M_T = 0$ and $I_T = 5$. The numerical results between Figure 7.20 and Ref. [23] were almost identical.

Figure 7.21 shows a comparison of the total normalized powers that occur when the dielectric shell is taken to be a uniform layer (the power here is calculated at $r = b$ by Bessel function matching) and when the dielectric shell is taken to be a (θ, φ)-inhomogeneity profile with $\Delta\varepsilon_\theta = 2.8$, $\Delta\varepsilon_\varphi(r) = 0.4$ (Case (2)). The power here is calculated at $r = a, b$ by a single layer analysis and a multilayer analysis. As can

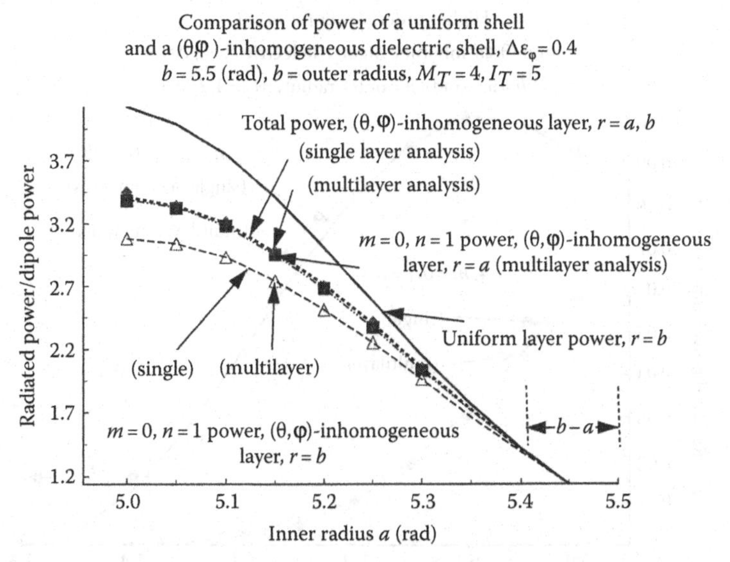

FIGURE 7.21 A comparison of the total normalized powers that occur when the dielectric shell is taken to be a uniform layer (the power here is calculated at $r = a, b$ by Bessel function matching) and when the dielectric shell is taken to be a (θ, φ)-inhomogeneity profile with $\Delta\varepsilon_\theta = 2.8$, $\Delta\varepsilon_\varphi(r) = 0.4$ (Case (2)). The power here is calculated at $r = a, b$ by a single layer analysis and a multilayer analysis. The $m = 0$, $n = 1$ order power is also shown. (Reprinted from Jarem, J.M., Rigorous coupled-wave-theory analysis of dipole scattering from a three-dimensional, inhomogeneous, spherical dielectric and permeable system, *IEEE Microw. Theory Tech.*, 45(8), August 1997. With permission. Copyright 1997 IEEE.)

be seen from Figure 7.21, the presence of the (θ, φ)-inhomogeneity profile causes a marked difference in the total scattered power of the inhomogeneous shell, despite the fact that the bulk dielectric inhomogeneity profile was exactly the same as that of the uniform, homogeneous shell. It is also noticed from Figure 7.21, that for both the single and multilayer analyses, the law of power conservation at $r = a$ and b is obeyed to a reasonable degree of accuracy. Also plotted in Figure 7.21 is the $m = 0, n = 1$ power evaluated at $r = a$. It is noticed that the $m = 0, n = 1$ (θ, φ)-inhomogeneity profile power at $r = a$ almost exactly equals that of the total power at $r = a, b$. This indicates that at $r = a$, no power has been diffracted into higher order modes at the $r = a$ interior boundary shell interface of the system. Figure 7.21 also shows the $m = 0, n = 1$ power as calculated at $r = b$. From this plot, one observes that the $m = 0, n = 1$ power is significantly lower than the $r = a, b$ total power plots. This clearly indicates that as the EM waves have radiated through the dielectric shell, power has been diffracted into the m, n higher order modes of the system. The dielectric shell is acting very much like a planar diffraction grating operating in a transmission mode of operation.

Figure 7.22 shows plots of the $n = 2, m = 0$ and $n = 4, m = 0$ mode order power at $r = b$ for the same (θ, φ)-inhomogeneity profile as was studied in Figure 7.21. The $n = 3, 5$ $(m = 0)$ orders were very small and not plotted. As can be seen from Figure 7.22, as the inhomogeneous shell thickness $b - a$ is increased, the diffracted power is transferred from the $n = 1, m = 0$ lowest order mode (see Figure 7.22) to the $n = 2, m = 0$ mode, to the $n = 4, m = 0$ mode, and to other higher order modes. One also notices the interesting behavior that at about $a = 5.1$ (rad) and $b - a = 0.4$ (rad), the $n = 2, m = 0$ mode power has reached a maximum value and decreases with further increase of the inhomogeneity shell thickness $b - a$. Evidently, the $n = 2, m = 0$ higher order mode is itself transferring energy to other higher order modes. This behavior is very common in planar diffraction gratings [26].

Figure 7.23 shows plots of the $m = 1$ total power (formed by summing all $m = 1, n = 1, 2, 3, \ldots$ mode powers) as computed by using a multilayer analysis (*dotted line-triangle*) and using a single layer analysis (*solid line*). One notices that the single and multilayer analyses give approximately the

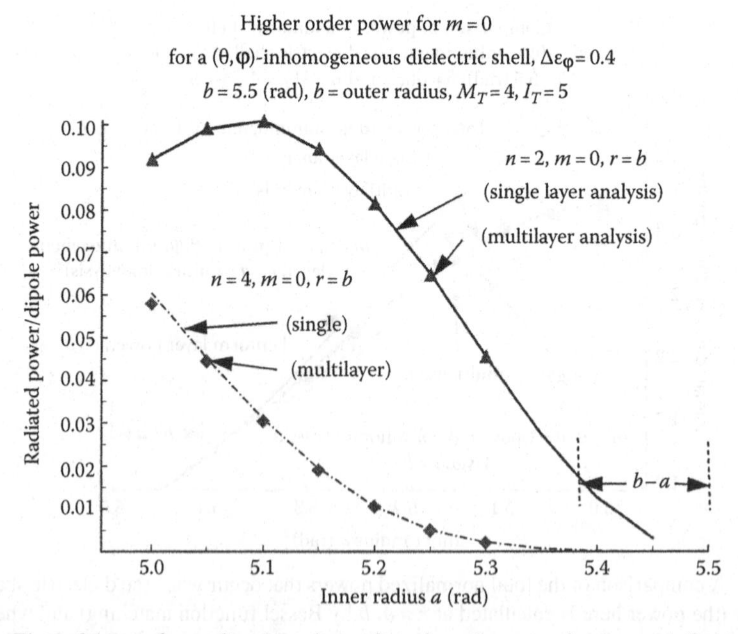

FIGURE 7.22 Plots of the $n = 2, m = 0$ and $n = 4, m = 0$ mode order power for the same (θ, φ)-inhomogeneity profile as was studied in Figure 7.21 are shown. The $n = 3, 5$ $(m = 0)$ orders were very small and not plotted. (Reprinted from Jarem, J.M., Rigorous coupled-wave-theory analysis of dipole scattering from a three-dimensional, inhomogeneous, spherical dielectric and permeable system, *IEEE Microw. Theory Tech.*, 45(8), August 1997. With permission. Copyright 1997 IEEE.)

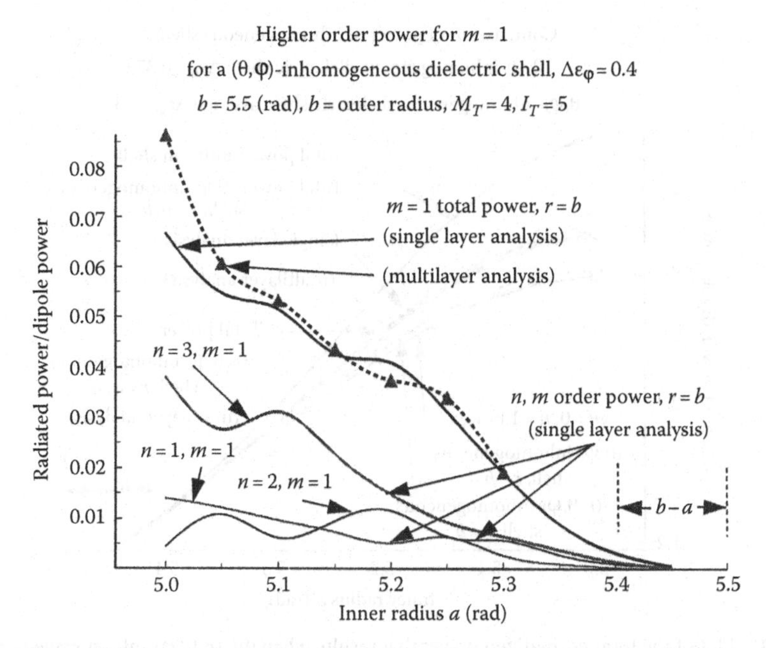

FIGURE 7.23 Plots of the $m = 1$ total power (formed by summing all $m = 1, n = 1, 2, 3, \ldots$ mode powers) as computed by using a multilayer analysis (*dotted line-triangle*) and a single layer analysis (*solid line*) are shown. (Reprinted from Jarem, J.M., Rigorous coupled-wave-theory analysis of dipole scattering from a three-dimensional, inhomogeneous, spherical dielectric and permeable system, *IEEE Microw. Theory Tech.*, 45(8), August 1997. With permission. Copyright 1997 IEEE.)

same result up to about a shell thickness of about $b - a = 0.4$ (rad), but after this value, the multilayer analysis is needed for more accurate results. Figure 7.23 also shows the $m = 1, n = 1, 2, 3$ order power as calculated by a single layer analysis. One observes that as the shell thickness $b - a$ increases, the $m = 1, n = 1, 2, 3$ order power increases.

Figure 7.24 shows a plot of the total normalized power that results when the (r, θ, φ)-inhomogeneity profile of Case (3) was solved using a multilayer RCWA method and using $M_T = 4$ and $I_T = 5$. Also shown for comparison is the total power of a uniform shell system (Case (1) parameters) and the total power that results when a (θ, φ)-inhomogeneity profile was used with $\Delta\varepsilon_\varphi(r)$ set to a constant value of $\Delta\varepsilon_\varphi(r) = 0.375$. This value of $\Delta\varepsilon_\varphi(r)$ exactly equaled the average radial value of the $\Delta\varepsilon_\varphi(r)$ function of Case (3) over the interval $a \le r \le b$, $a = 5$ (rad), and $b = 5.5$ (rad). As can be seen from Figure 7.24, the $\Delta\varepsilon_\varphi(r)$ linear taper causes little difference in total power to be seen between the total power of the (r, θ, φ)-inhomogeneity profile of Case (3) and the total power of the (θ, φ)-inhomogeneity profile that used a constant value of $\Delta\varepsilon_\varphi(r) = 0.375$. We notice from Figure 7.24 that power conservation was observed to hold to a reasonable degree of accuracy. Figure 7.24 also shows a plot of the $m = 0, n = 1$ order power calculated at $r = b$ (rad) for the two inhomogeneity profiles for which the total power was just described. As can be seen from Figure 7.24, a perceptible difference in the plots due to the different inhomogeneity profiles is observed.

Figure 7.25 shows an $m = 1$ (and $m = 3$) total order power comparison between the two inhomogeneity profiles discussed earlier. At $a = 5$ (rad) (shell thickness $b - a = 0.5$ (rad)), the presence of the $\Delta\varepsilon_\varphi(r)$ linear taper for the (r, θ, φ)-inhomogeneity profile of Case (3) causes an observable difference with the $m = 1$ total order power of the (θ, φ)-inhomogeneity profile that used a constant value of $\Delta\varepsilon_\varphi(r) = 0.375$ value. As the same multilayer algorithm was used to calculate the $a = 5$ (rad) $m = 1$ total power plots, with the only difference being that a linear and constant $\Delta\varepsilon_\varphi(r)$ function was used, we conclude that completely correct results can only be achieved in the general (r, θ, φ)-inhomogeneity case by using a multilayer analysis.

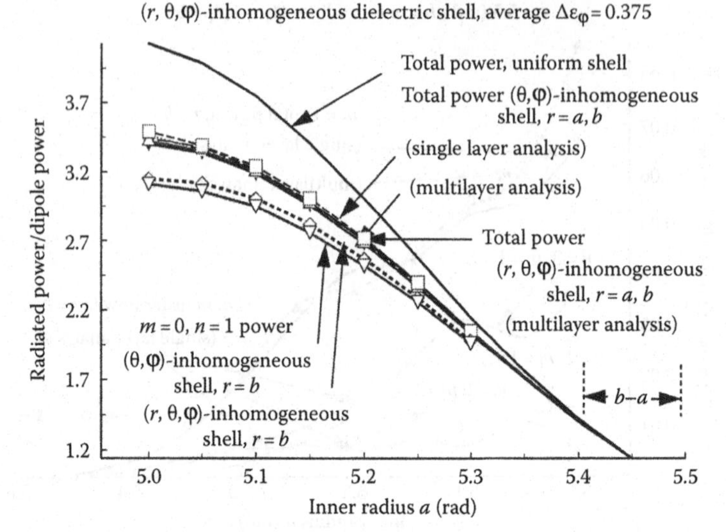

FIGURE 7.24 Plot of the total normalized power that results when the (r, θ, φ)-inhomogeneity profile of Case (3) was solved using a multilayer RCWA method and $M_T = 4$ and $I_T = 5$ is shown. Also shown for comparison is the total power of a uniform shell system (Case (1) parameters) and the total power that when a (θ, φ)-inhomogeneity profile was used with $\Delta\varepsilon_\varphi(r)$ set to a constant value of $\Delta\varepsilon_\varphi(r) = 0.375$. (Reprinted from Jarem, J.M., Rigorous coupled-wave-theory analysis of dipole scattering from a three-dimensional, inhomogeneous, spherical dielectric and permeable system, *IEEE Microw. Theory Tech.*, 45(8), August 1997. With permission. Copyright 1997 IEEE.)

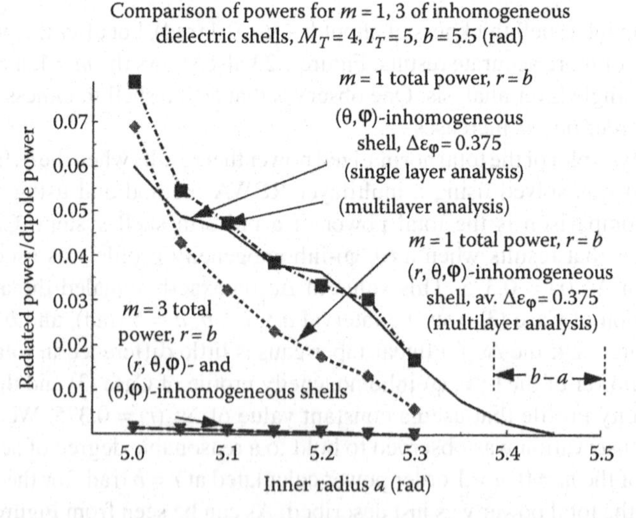

FIGURE 7.25 A $m = 1$ (and $m = 3$) total order power comparison between the two inhomogeneity profiles discussed in Figure 7.24. (Reprinted from Jarem, J.M., Rigorous coupled-wave-theory analysis of dipole scattering from a three-dimensional, inhomogeneous, spherical dielectric and permeable system, *IEEE Microw. Theory Tech.*, 45(8), August 1997. With permission. Copyright 1997 IEEE.)

It is interesting to compare the unit periodic cell formed by the spherical dielectric shell of the present section with the unit cell of a planar, crossed dielectric diffraction grating whose grating dimensions are $\tilde{\Lambda}_x$, $\tilde{\Lambda}_y$. For the spherical system using the inner radius value of $\tilde{r} = \tilde{a}$, the area of the spherical unit cell is $4\pi\tilde{a}^2$. If we chose $\tilde{\Lambda}_x = \tilde{\Lambda}_y = 2\sqrt{\pi}\tilde{a}$, the grating cell areas of the planar system and spherical one are equal. For the present example, the inner surface of the dielectric shell had

$a = k_0\tilde{a} = (2\pi/\lambda)\tilde{a} = 5$ (rad), which thus leads to $\tilde{\Lambda}_x = \tilde{\Lambda}_y = (5/\sqrt{\pi})\lambda = 2.82\lambda$. It is interesting to note that this grating cell size is very typical of many analyses that are made for planar diffraction gratings. For example, in holographic applications, if two interfering plane waves make an angle of 10.21° on opposite sides of a normal to the holographic surface, a one dimensional diffraction grating of width $\tilde{\Lambda}_x = 2.82\lambda$ is formed. We thus see that diffraction from the spherical shell system studied in this section is on the same scale size as diffraction that occurs in many planar diffraction analyses. It is also interesting to note that the spherical shell scattering analysis, which was studied in this section, would, in the area of diffraction grating theory, be classified as a thin grating diffraction analysis. This follows, as the spherical shell thickness is less than the grating period and the percent of modulation for the spherical shell $\Delta\varepsilon_\theta/\varepsilon_2 \cong 2.8/7 = 40\%$, which for planar diffraction grating analysis is large. (Holograms have a depth of modulation on the order of ≈0.03%).

One of the best areas of research concerning the present chapter would be to implement the numerical stability algorithms for the RCWA algorithm [12,13] and greatly increase the size of the spherical and cylindrical scatterer that could be analyzed by the RCWA algorithm.

PROBLEMS

7.1 This problem concerns validating the full-field, circular cylindrical RCWA method presented in Section 7.3 by solving a cylindrical scattering object example by the RCWA method and by a well-known alternate method and then comparing results. For this problem, the validation scattering example to be studied consists of finding the EM fields that arise when a line source centered in a three-region set of concentric, dielectric cylindrical shells excites radiation in the system. Figure 7.1b, with the line source centered and $\varepsilon_2', \varepsilon_2''$ of the figure set equal, shows the geometry of the system.

(a) Referring to Figure 7.1b, numerically calculate the EM fields in the system assuming the system is nonmagnetic, the line source is centered, and that $k_0 a = 5$, $5 \le k_0\tilde{b} \le 10$, $k_0 = 2\pi/\lambda$ (λ is the free space wavelength); $\varepsilon_1 = 1.5$, $\varepsilon_2 \equiv \varepsilon_2' = \varepsilon_2'' = 2.5$, and $\varepsilon_3 = 1$. The numerical solution is to be implemented by finding Bessel function solutions to Maxwell's equations in each region of the system, boundary matching these solutions at the interfaces of the system, and then solving the resulting matrix equation to determine the EM fields of the system. Present details and implement the EM field and matrix solution that you have obtained.

(b) Implement the RCWA method described in Section 7.3 and solve for the EM fields of the system for the scattering system of Part (a).

(c) By using the Bessel function method of Part (a) and the RCWA method of Part (b), assuming $k_0\tilde{a} = 5$, calculate the power defined by Equations 7.19 through 7.21 as a function of the radius $b = k_0\tilde{b}$, $5 \le k_0\tilde{b} \le 10$. Compare your plots to those of Figure 7.5.

(d) Repeat Parts (a), (b), and (c), except take $\varepsilon_2' = \varepsilon_2'' = 10$ rather than $\varepsilon_2 \equiv \varepsilon_2' = \varepsilon_2'' = 2.5$.

(e) Comment on the differences between the Parts (c) and (d) solutions.

7.2 This problem, like Problem 7.1, concerns validating the full-field, circular cylindrical RCWA method presented in Section 7.3 by solving a cylindrical scattering object example by the RCWA method and by a well-known alternate method and then comparing results. For this problem, the validation scattering example to be studied consists of finding the EM fields that arise when a plane wave, incident on a three-region set of concentric, dielectric cylindrical shells, scatters radiation from the system. Figure 7.1a shows the geometry of the system.

(a) Referring to Figure 7.1a, numerically calculate the EM fields in the system assuming the system is nonmagnetic, assuming that a plane wave of amplitude $E_0 = 1$ (V/m) is incident on the cylinder system, assuming $k_0 a = 5$, $5 \le k_0\tilde{b} \le 10$, $k_0 = 2\pi/\lambda$ (λ is the free space wavelength), and assuming $\varepsilon_1 = 1.5$, $\varepsilon_2 \equiv \varepsilon_2' = \varepsilon_2'' = 2.5$, and $\varepsilon_3 = 1$. The numerical solution is to be implemented by finding Bessel function solutions to Maxwell's equations in each region of the system, boundary matching these solutions at the interfaces of the system, and then solving the resulting matrix equation to determine the EM fields of the system. Present details and implement the EM field and matrix solution that you have obtained.

(b) Implement the RCWA method described in Section 7.3 and solve for the EM fields of the system for the scattering system of Part (a).

(c) By using the Bessel function method of Part (a) and the RCWA method of Part (b), assuming $k_0\tilde{a} = 5$, calculate the power defined by Equations 7.19 through 7.21 as a function of the radius $b = k_0\tilde{b}$, $5 \le k_0\tilde{b} \le 10$. Compare your plots to those of Figure 7.11.

(d) Repeat Parts (a), (b), and (c), except take $\varepsilon_2' = \varepsilon_2'' = 10$ rather than $\varepsilon_2 \equiv \varepsilon_2' = \varepsilon_2'' = 2.5$.

(e) Comment on the differences between the Parts (c) and (d) solutions.

7.3 This problem concerns using the full-field, circular cylindrical RCWA method presented in Section 7.4 to calculate the radiation that arises when a centered line source excites an anisotropic cylindrical scattering (geometry is shown in Figure 7.7) using the parameters listed in Figure 7.10. Several steps are to be carried out to determine the radiation of the system including calculation the Fourier coefficients needed to form the state matrix equations of the system, formulation and eigensolution of the state matrix equations of the system, formulation and solution of the overall system matrix as results from EM boundary matching at the interfaces of the system, and finally calculation of the power radiated from the system.

(a) Calculate the Fourier coefficients of $\varepsilon(\rho, \varphi)$, $\mu_{rs}(\rho, \varphi)$, $(r, s) = (\rho, \varphi)$ of Equations 7.25 through 7.28, needed to form the state matrix equations of the system. Make plots of $\varepsilon(\rho, \varphi)$, $\mu_{rs}(\rho, \varphi)$, $(r, s) = (\rho, \varphi)$, and their Fourier sums to be sure that the Fourier coefficients have been calculated correctly and that the Fourier series have converged properly.

(b) Formulate and solve the state variable and boundary matching system matrix equations to find the overall EM fields that exist in the system.

(c) Assuming $k_0\tilde{a} = 5$, calculate the $i = -1, 0, 1$ order power defined by Equations 7.19 through 7.21 as a function of the radius $b = k_0\tilde{b}$, $5 \le k_0\tilde{b} \le 10$.

(d) Again, assuming $k_0\tilde{a} = 5$, to verify conservation of power, calculate the total power radiated by the line source as a function of the radius $b = k_0\tilde{b}$, $5 \le k_0\tilde{b} \le 10$ at $\rho = a$, $\rho = b$, and $\rho \to \infty$.

(e) Repeat Parts (c) and (d) (i.e., using the parameters of Figure 7.10), except take $\mu_{xy} = \mu_{yx} = 0.6$ rather than $\mu_{xy} = \mu_{yx} = 0.3$ as was done in Figure 7.10.

(f) Compare and comment on the effect that the change in the anisotropy parameters from $\mu_{xy} = \mu_{yx} = 0.3$ to $\mu_{xy} = \mu_{yx} = 0.6$ had on radiated power curves of the two systems.

7.4 This problem concerns using the full-field, circular cylindrical RCWA method presented in Section 7.4 to study the effect that varying the anisotropy material parameters of the square-circular system shown in Figure 7.8 has on the radiated power from that system. Figure 7.14 displays the radiated power of the Figure 7.8 system for the given set of parameters listed in the Figure 7.14 caption. As in Problem 7.3, several steps are to be carried out to determine the radiation of the system including Fourier coefficient calculation, state matrix equation analysis, solution of EM boundary matching matrix equations, and calculation of the power radiated from the system.

(a) Calculate the Fourier coefficients of $\varepsilon(\rho, \varphi)$, $\mu_{rs}(\rho, \varphi)$, $(r, s) = (\rho, \varphi)$ of Equations 7.25 through 7.28, needed to form the state matrix equations of the system. Using the material parameters of Figure 7.14, make plots of $\varepsilon(\rho, \varphi)$, $\mu_{rs}(\rho, \varphi)$, $(r, s) = (\rho, \varphi)$, and their Fourier sums to be sure that the Fourier coefficients have been calculated correctly and that the Fourier series have converged properly. Because of the complicated circular-square geometry of the system, the Fourier coefficients of the system need to be calculated separately in the regions $\tilde{a} \le \tilde{\rho} \le \sqrt{2}\tilde{a}$ and $\sqrt{2}\tilde{a} \le \tilde{\rho} \le \tilde{b}$, $\tilde{a} = 1\lambda$. For the region $\tilde{a} \le \tilde{\rho} \le \sqrt{2}\tilde{a}$, make plots of $\varepsilon(\rho, \varphi)$, $\mu_{rs}(\rho, \varphi)$, $(r, s) = (\rho, \varphi)$ and their associated Fourier sums using $\tilde{\rho} = 1.15\lambda$, and for the region $\sqrt{2}\tilde{a} \le \tilde{\rho} \le \tilde{b}$, make the same plots using $\tilde{\rho} = 1.81\lambda$, assuming $\tilde{b} = 2.5\lambda$.

(b) Formulate and solve the state variable and boundary matching system matrix equations to find the overall EM fields that exist in the system using the material parameters of Figure 7.14.

(c) Using the material parameters of Figure 7.14, assuming $\tilde{a} = 1\lambda$, calculate the $i = 0$ order power defined by Equations 7.19 through 7.21 at $\tilde{\rho} = \tilde{a} = 1\lambda$.

(d) Again, assuming $\tilde{a} = 1\lambda$, to verify conservation of power, calculate the total power radiated by the line source as a function of the radius $\sqrt{2}\tilde{a} = \sqrt{2}\lambda \le \tilde{b} \le 2.5\lambda$, at $\tilde{\rho} = \tilde{a}$, $\tilde{\rho} = \tilde{b}$, and $\tilde{\rho} \to \infty$. Compare your answers to those of Figure 7.14.

(e) Repeat Parts (c) and (d) (i.e., using the parameters of Figure 7.14), except take $\mu_{xy} = \mu_{yx} = 0.6$ rather than $\mu_{xy} = \mu_{yx} = 0.3$ as was done in Figure 7.14.

(f) Compare and comment on the effect that the change in the anisotropy parameters from $\mu_{xy} = \mu_{yx} = 0.3$ to $\mu_{xy} = \mu_{yx} = 0.6$ had on radiated power curves of the two systems.

7.5 (a) Repeat Problem 7.4, using the dielectric permittivity parameters specified in Problem 7.4 (and therefore those specified in the Figure 7.14 caption, also) except assuming that the magnetic anisotropy parameters μ_{xx}, μ_{xy}, μ_{yx}, μ_{yy} for this part assume the values $\mu_{xxa} = 1.5$, $\mu_{yya} = 3$, and $\mu_{xyxa} = 1, \mu_{yxa} = 1$. Make plots of the total and $i = -2, -1, 0, 1, 2$ order radiated power and order of the system.

(b) Repeat Part (a), except now assume that the magnetic anisotropy parameters μ_{xx}, μ_{xy}, μ_{yx}, μ_{yy} assume the values $\mu_{xxb} = \mu_{yyb} = \mu_{Bulka}$ and $\mu_{xyb} = \mu_{yxb} = 0$, where $\mu_{Bulka} = \sqrt{\mu_{xxa}\mu_{yya} - \mu_{xya}\mu_{yxa}}$ $\mu_{xyb} = \mu_{yxb} = 0$.

Make plots of the total and $i = -2, -1, 0, 1, 2$ order radiated power $i = -2, -1, 0, 1, 2$ of the system and compare these radiated power results to those found in Part (a).

(c) The system described in Part (a) is anisotropic whereas the system in Part (b) is isotropic. Note that the region "bulk" value defined in this problem by $\mu_{Bulk} = \sqrt{\mu_{xx}\mu_{yy} - \mu_{xy}\mu_{yx}}$ is the same in both parts. Comment on the differences in power that result from the anisotropic case of Part (a) and the isotropic case of Part (b).

(d) In your solution of Parts (a) and (b), make plots of the isotropic and anisotropic material parameters ε_{zz}, $\mu_{\rho\rho}$, $\mu_{\rho\phi}$, $\mu_{\phi\rho}$, $\mu_{\phi\phi}$ to ensure that the Fourier series have been calculated correctly and converge correctly.

(e) Be sure to show that conservation of power holds in your solutions for Parts (a) and (b).

7.6 The key steps and procedures for developing the RCWA algorithm for circular cylindrical coordinates and spherical coordinates has been described in this chapter and the key steps and procedures for the RCWA algorithm for bipolar cylindrical coordinates is given in Chapters 8 and 9 to follow. Using the general RCWA procedures described in Chapters 7 through 9, develop an RCWA algorithm that may be used to solve scattering problems in the elliptical, cylindrical coordinate system. Suggest applications where an RCWA, elliptical, cylindrical coordinate system algorithm might be useful.

7.7 Discuss, describe, and implement general methods that may be used to study how well numerical EM scattering solutions discussed in this chapter obey the law of conservation of power.

REFERENCES

1. M. Le Blanc and G.Y. Delisle, Comments on "Plane wave excitation of an infinite dielectric rod," *IEEE Microwave Guid. Wave Lett.* 5(6), 233–234, 1995.
2. R.B. Keam, Plane wave excitation of an infinite dielectric rod, *IEEE Micowave Guid. Wave Lett.* 4(10), 326–328, 1994.
3. J.R. Wait, *Electromagnetic Radiation from Cylindrical Structures*, Pergamon, New York, 1959.
4. H.Y. Yee, Scattering of electromagnetic waves by circular dielectric coated conducting cylinders with arbitrary cross sections, *IEEE Trans. Antennas Propag.* AP-13, 822–823, 1965.
5. R.F. Harrington, *Time-Harmonic Electromagnetic Fields*, McGraw-Hill, New York, 1961.
6. C.A. Balanis, *Advanced Engineering Electromagnetics*, Wiley, New York, 1989.
7. P. Bhartia, L. Shafai, and M. Hamid, Scattering by an imperfectly conducting conductor with a radially inhomogeneous dielectric coating, *Int. J. Electron.* 31, 531–535, 1971.

8. A.A. Kishk, R.P. Parrikar, and A.Z. Elsherbeni, Electromagnetic scattering from an eccentric multilayered circular cylinder, *IEEE Trans. Antennas Propag.* 40(3), 295–303, 1992.
9. M.G. Moharam and T.K. Gaylord, Rigorous coupled-wave analysis of planar grating diffraction, *J. Opt. Soc. Am.* 71, 811–818, 1981.
10. K. Rokushima and J. Yamakita, Analysis of anisotropic dielectric gratings, *J. Opt. Soc. Am.* 73, 901–908, 1983.
11. M.G. Moharam and T.K. Gaylord, Three-dimensional vector coupled-wave analysis of planar-grating diffraction, *J. Opt. Soc. Am.* 73, 1105–112, 1983.
12. M.G. Moharam and T.K. Gaylord, Diffraction analysis of dielectric surface-relief gratings, *J. Opt. Soc. Am.* 72, 1385–1392, 1982.
13. K. Rokushima, J. Yamakita, S. Mori, and K. Tominaga, Unified approach to wave diffraction by space-time periodic anisotropic media, *IEEE Trans. Microwave Theory Tech.* 35, 937–945, 1987.
14. E.N. Glytsis and T.K. Gaylord, Rigorous three-dimensional coupled-wave diffraction analysis of single cascaded anisotropic gratings, *J. Opt. Soc. Am. B* 4, 2061–2080, 1987.
15. J.M. Jarem, Rigorous coupled wave theory solution of φ-periodic circular cylindrical dielectric systems, *J. Electromagn. Waves Appl.* 11, 197–213, 1997.
16. A.Z. Elsherbeni and M. Hamid, Scattering by a cylindrical dielectric shell with inhomogeneous permittivity profile, *Int. J. Electron.* 58(6), 949–962, 1985.
17. A.Z. Elsherbeni and M. Hamid, Scattering by a cylindrical dielectric shell with radial and azimuthal permittivity profiles, *Proceedings of the 1985 Symposium of Microwave Technology in Industrial Development*, Brazil, July 22–25, 1985, pp. 77–80 (Invited).
18. A.Z. Elsherbeni and M. Tew, Electromagnetic scattering from a circular cylinder of homogeneous dielectric coated by a dielectric shell with a permittivity profile in the radial and azimuthal directions-even TM case, *IEEE Proceedings-1990 Southeastcon*, Session 11A1, 1990, pp. 996–1001.
19. X.B. Wu, Scattering from an anisotropic cylindrical dielectric shell, *Int. J. Infrared Millimeter Waves* 15(10), 1733–1743, 1994.
20. X.B. Wu, An alternative solution of the scattering from an anisotropic cylindrical dielectric shell, *Int. J. Infrared Millimeter Waves* 15(10), 1745–1754, 1994.
21. J.M. Jarem, Rigorous coupled wave theory of anisotropic, azimuthally-inhomogeneous, cylindrical systems, *Prog. Electromagn. Res.* PIER 19, 109–127, 1998.
22. W. Ren, Contributions to the electromagnetic wave theory of bounded homogeneous anisotropic media, The American Physical Society, *Phys. Rev. E* 47(1), 664–673, 1993.
23. J.M. Jarem, A rigorous coupled-wave theory and crossed-diffraction grating analysis of radiation and scattering from three-dimensional inhomogeneous objects, a letter, *IEEE Trans. Antennas Propag.* 46(5), 740–741, 1998.
24. J.M. Jarem, Rigorous coupled-wave-theory analysis of dipole scattering from a three-dimensional, inhomogeneous, spherical dielectric and permeable system, *IEEE Micowave Theory Tech.* 45(8), 1193–1203, 1997.
25. M. Abramowitz and I. Stegum, *Handbook of Mathematical Functions*, Eq. 9.1.20, Dover, New York, 1972, p. 360.
26. J.M. Jarem and P. Banerjee, An exact, dynamical analysis of the Kukhtarev equations in photorefractive barium titanate using rigorous wave coupled wave diffraction theory, *J. Opt. Soc. Am. A* 13(4), 819–831, 1996.

8 Rigorous Coupled Wave Analysis of Inhomogeneous Bipolar Cylindrical Systems

8.1 INTRODUCTION

An important problem in the area of EM spectral domain analysis is the problem of using rigorous coupled wave analysis (RCWA) to determine the EM fields that result when an EM wave is incident on an inhomogeneous material object, that is, one whose dielectric permittivity and magnetic permeability parameters are functions of position. The RCWA algorithm, which was originally designed to study scattering from planar diffraction gratings [1–9], has been used to scattering from inhomogeneous objects when the inhomogeneous object was a phi-dependent circular cylinder [10–13], when the circular cylinder was an anisotropic permeability [14], when the object was a lossy biological material [15], when the object was an elliptical inhomogeneous cylinder [16,17], and when the scattering object was an inhomogeneous sphere [18,19]. Chapter 7 of this book and the first edition of this book [20] describes the RCWA method and its application to *scattering* from inhomogeneous objects.

A limitation of work of [13–15] was that it described circular cylindrical materially inhomogeneous systems for which the interior and exterior circular cylinders of the scattering object were concentric. This is a limitation because scattering objects that have a significantly off center material inhomogeneity cannot easily be analyzed in a concentric circular cylindrical system. Although scattering objects with eccentric, cylindrical inhomogeneity can still be theoretically studied as an inhomogeneous object in a centered cylindrical system, if the difference in material values of eccentric material object was large, a very high number of Fourier harmonics would be required to achieve accurate numerical results. The purpose of this chapter therefore is to further extend the RCWA method to describe EM scattering from inhomogeneous material objects that comprise eccentric, circular, multicylinder systems for which spatially nonuniform material (the dielectric permittivity) occupies the regions between the interfaces of the cylinders. This analysis is carried out in bipolar coordinates (see Figures 8.1 and 8.2) since the coordinates u and v and interfocal parameter a of this coordinate system [21] can be chosen to define the off center exterior and interior boundaries of the scattering object.

The research work contained in this chapter involves the mathematics and the study of cylindrical functions expressed cylindrical coordinate functions [22–29], particularly as applied to electromagnetic scattering from eccentric, cylindrical structures [22,26–29].

The study of EM scattering from an object which is composed of N nonconcentric, completely enclosed, circular cylinders of radius $r_1 > r_2 >, \cdots, > r_N$ where the material parameters of dielectric permittivity and magnetic permeability are homogeneous (or non varying with spatial position) between the adjacent circular boundaries $r_j > r_{j+1}, j = 1, \ldots, N - 1$ has been studied and solved by Kishk, Parrikar, and Elsherbeni [22] (called herein the KPE algorithm). By completely enclosed it is meant that cylinder r_1 completely encloses cylinder r_2, which completely encloses cylinder r_3, etc. The solution of this problem by the KPE algorithm [22] is extremely important to the present work for two reasons. First, the study of EM scattering from eccentric, circular, multicylinder systems by the KPE method [22] provides many very good examples for which the bipolar

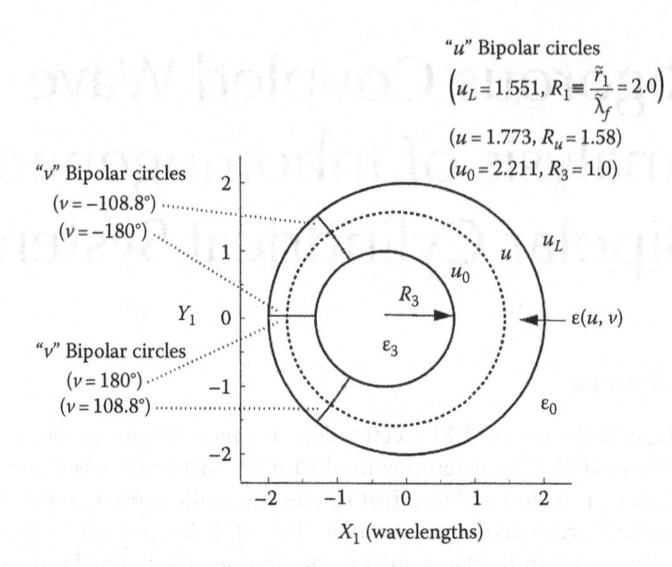

FIGURE 8.1 The geometry of the combined bipolar, cylindrical, rectangular coordinate system of the chapter. The inhomogeneous region, which is described by bipolar coordinates, is divided into L thin layers with interfaces located at $u = u_\ell$, $\ell = 0,\ldots,L$ with the innermost layer at $u = u_0$ and outermost at $u = u_L$. The X_1, Y_1 rectangular coordinate system shown is centered on the outer R_1, $u = u_L$ circle. The figure corresponds to the main scattering example of the chapter (see Figure 8.2) and is drawn to the exact scale of the numerical values displayed. The interfocal distance $a = \tilde{k}_f\tilde{a} = 28.32451318$. X_1(*wavelengths*) here and in Figures 8.2 through 8.9, and Y_1 (*wavelengths*) here and in Figures 8.3 through 8.9 represent the number of wavelengths (positive or negative) from the origin ($X_1 = \tilde{x}_1/\tilde{\lambda}_f$, $Y_1 = \tilde{y}_1/\tilde{\lambda}_f$ are dimensionless). (Reproduced from Jarem, J.M., *Prog. Electromagn. Res.*, PIER 43, 181, 2003. With permission.)

RCWA algorithm may be validated and tested against with respect to numerical accuracy and convergence. Figure 8.1 shows the general bipolar geometry and Figure 8.2 shows a specific scattering example involving three eccentric enclosed cylinders, which are studied extensively in the present work. In Figure 8.2, the interior and exterior cylinders have the same bipolar coordinates as Figure 8.1. One notices from Figure 8.2 that, because of the geometry and conditions stated earlier, the solution of the EM scattering problem displayed in this figure is exactly amenable to an exact solution by the KPE algorithm [22] in the case when the material regions between cylinder interfaces are uniform ($\Delta\varepsilon = 0$ in Figure 8.2). On the other hand, picking the inner and outer circular cylinder boundaries as the interior ($u = u_0 = 2.211$) and exterior ($u = u_L = 1.551$) interfaces of the bipolar scattering geometry (see Figure 8.1), the middle circular cylinder of Figure 8.2 represents an inhomogeneous, uniform-step profile for the RCWA algorithm, if the bulk, relative dielectric permittivities ε_1 and ε_2 are different from one another. To clarify this statement, if the $u = 1.773$ bipolar circle of Figure 8.1 (middle u circle) is placed in Figure 8.2, and one traces the dotted circle from $v = -180°$ to 0°, one would first be in Region 2 ($\varepsilon(u, v) = \varepsilon_2$), then for some v value ($v = -108.8°$ for $u = 1.773$) cross over to Region 1 $\varepsilon(u, v) = \varepsilon_{f1}$. Going from Region 2 ($\varepsilon_2$) to Region 1 ($\varepsilon_{f1}$) represents the step profile for the range $-180° \leq v \leq 0°$. Because it is necessary to use a Fourier series to represent this step profile, this type of inhomogeneity provides a severe test of the RCWA algorithm since the Fourier series of step profiles tends to converge slowly and have a high spectral content.

The second reason is that the study of this problem by the KPE algorithm [22] is useful is because the KPE method [22], after a small amount of mathematical manipulation (Appendix 8.B), can be placed in an algebraic form for which a direct numerical comparison of the system transfer matrices as a result of using the KPE method [22] and the bipolar RCWA method can be made. The transfer matrices were useful for validation purposes because, from the way they

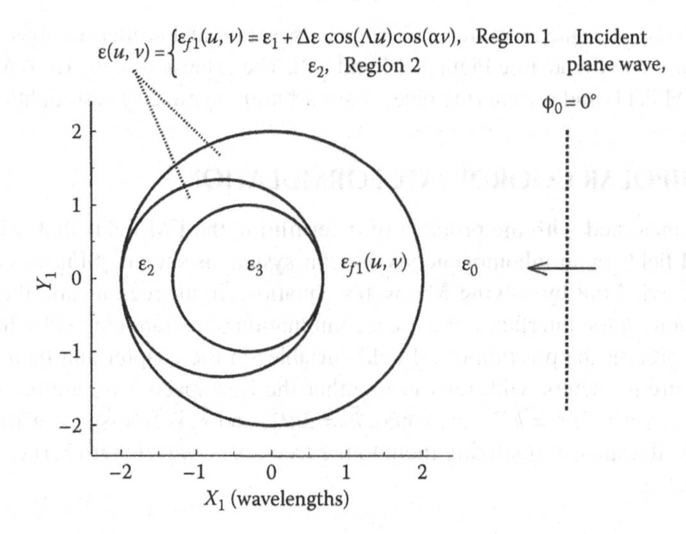

FIGURE 8.2 The geometry of the main scattering example of the chapter. The relative bulk permittivities in Regions 0, 1, 2, and 3, which are displayed in the figure, are assumed to have values $\varepsilon_0 = 1$, $\varepsilon_1 = 2$, $\varepsilon_2 = 3$, and $\varepsilon_3 = 4$, respectively. The direction of the $\phi_0 = 0°$ incident plane wave is shown. The $\phi_0 = 180°$ incident plane wave (not shown) would impinge on the scattering object from the left side of figure. In the case, when $\Delta\varepsilon = 0$, the permittivity inside Region 1 and Region 2 is uniform, and this case thus represents a case for which the KPE algorithm [22] can find an exact solution. (Reproduced from Jarem, J.M., *Prog. Electromagn. Res.*, PIER 43, 181, 2003. With permission.)

were both formulated, they could be meaningfully compared to one another, matrix element to matrix element.

This chapter is based on the research paper of Jarem [26] who has studied the problem of EM scattering from eccentric circular cylinders when one or more of the regions of the eccentric cylinders contain spatially inhomogeneous dielectric material. The analysis of this paper [26] is based as mentioned earlier on using the RCWA algorithm in bipolar coordinates to model the spatial inhomogeneity of the scattering system. Chapter 9 provides many computational examples associated with the bipolar RCWA algorithm presented in this chapter.

The authors of Ref. [27] have studied the problem of scattering from dielectric cylinders when multiple eccentric cylindrical dielectric inclusions are present in the scattering system. The authors of Ref. [28] have further studied the problem when eccentrically stratified, dielectric cylinder with multiple, eccentrically stratified, cylindrical, dielectric inclusions are present in the scattering system. Very recently (November 2009), the author of Ref. [29] studied the problem of EM-wave scattering by an axial slot on a circular perfectly electrical conducting cylinder containing an eccentrically layered inner coating by using a dual-series solution for transverse electric polarization.

The problem of scattering from inhomogeneous eccentric circular, cylindrical composite objects is important, for example, in the areas of; bioelectromagnetics [15], where one might want to know EM field levels in biological materials (for example, EM field penetration in a limb or torso); terrain clutter, where one might want to model EM scattering from eccentric, cylindrical-shaped vegetation; the study of inhomogeneous radar absorbing materials; numerical analysis (validation of mathematical methods other than RCWA); and many other applications as well. It is felt that the work to be presented could be particularly important to the validation of other EM methods (i.e., finite element method [FEM], finite difference–time domain [FD–TD], method of moments [MoM], etc.), since eccentric circular cylindrical systems that possess nonspatially uniform material between the interfaces of the cylinders (for example, see Figures 8.2, 8.3b and c, $\Delta\varepsilon \neq 0$, of this chapter) present a nontrivial scattering geometry (which thus requires nonspatially uniform gridding) for which to test the given algorithm (i.e., FEM, FD–TD, MoM).

The analysis to be presented assumes that source excitation and scattering objects are symmetric with respect to the y coordinate (see Figures 8.1 and 8.2). The extension of the RCWA algorithm to the case where the EM fields and a scattering object have arbitrary symmetry is straightforward.

8.2 RCWA BIPOLAR COORDINATE FORMULATION

This chapter is concerned with the problem of determining the EM fields that arise when a plane wave excites EM fields in an inhomogeneous, bipolar system as shown in Figures 8.1 and 8.2. The EM analysis is carried out by solving Maxwell's equations in all regions and then matching EM boundary conditions at the interfaces. We use a combination of rectangular, cylindrical, and bipolar coordinates to represent the position of all field variables in the chapter and then normalize these coordinates that are in meters, with respect to either the free-space wavenumber $\tilde{k}_f$ when presenting equations (i.e., $x = \tilde{k}_f\tilde{x}$, $y = \tilde{k}_f\tilde{y}$, etc., where $\tilde{k}_f = 2\pi/\tilde{\lambda}_f$ and $\tilde{\lambda}_f$ is free-space wavelength) or when displaying graphical results, normalizing them to the free-space wavelength $\tilde{\lambda}_f$ (i.e., $X \equiv \tilde{x}/\tilde{\lambda}_f = x/2\pi$, $Y \equiv \tilde{y}/\tilde{\lambda}_f = y/2\pi$, etc.).

8.2.1 Bipolar and Eccentric Circular Cylindrical, Scattering Region Coordinate Description

The bipolar coordinates that are used in this chapter to represent the inhomogeneous region of a typical scattering object to be studied herein are shown in Figure 8.1. They are defined by an interfocal distance $a = \tilde{k}_f\tilde{a}$ [21], a "radial" angular coordinate u ($-\infty < u < \infty$), and an angular coordinate v ($-\pi \le v \le \pi$). The rectangular coordinates x, y are related to bipolar coordinates u, v and interfocal distance a by the equations

$$x = \frac{a\sinh(u)}{\cosh(u) - \cos(v)}, \quad y = \frac{-a\sin(v)}{\cosh(u) - \cos(v)} \tag{8.1}$$

These equations have been obtained from Ref. [21], with the minor modification that the angular bipolar coordinate used in Ref. [21] (call it v' in [21] for the moment), was defined on the interval $0 \le v' \le 2\pi$ in Ref. [21], whereas herein the angular coordinate v is defined on the interval $-\pi \le v \le \pi$. Thus, the coordinate v' of Ref. [21] and the coordinate v used here are related by $v = v' - \pi$. Substitution of $v = v' - \pi$ in Equation 8.1 gives the bipolar coordinate formulas used in Ref. [21]. As mentioned in Section 8.1, we assume that the system is symmetric with respect to y coordinate. In bipolar coordinates, the locus of points is defined by setting the "radial" coordinate to a constant value u with $-\pi \le v \le \pi$. This traces out a circle whose center is on the x-axis at $x_{cu} = a/\tanh(u)$ and whose radius is $r_u = a/\sinh(|u|)$. When $u \to \pm\infty$, the circle of constant u's center approaches the interfocal points $x = \pm a$, respectively, and the radius of this circle r_u approaches zero. In bipolar coordinates, the locus of points defined by setting the angular coordinate to a constant value v with $-\infty < u < \infty$, traces out a circle whose center is on the y-axis at $y_{cv} = -a/\tan(v)$ and whose radius is $r_v = a[1 + \cot^2(v)]^{1/2}$. All circles of constant v pass through both interfocal points $x = \pm a$. In bipolar coordinates, the scale factors of the system are for the u, v coordinates defined here are given by

$$h_u(u,v) = h_v(u,v) = h(u,v) \equiv \frac{a}{\cosh(u) - \cos(v)} \tag{8.2}$$

Ref. [21] shows a complete geometry of bipolar coordinates u and v'.

Figure 8.1 shows the rectangular, cylindrical, and bipolar coordinates of a general type of scattering example that one might solve using the bipolar RCWA algorithm, whereas Figure 8.2 shows the details of the particular scattering example that we are most concerned with in this chapter. Figures 8.1 and 8.2 are drawn to the exact scale of the scattering examples that are presented in the chapter. In this chapter, we use bipolar coordinates to describe the inhomogeneous region of the scattering object, and we use the eccentric, circular, cylindrical, coordinate system used by KPE [22] (ρ_j, ϕ_j, $j = 1, 3$ with the origin O_j at the center of each eccentric cylinder in Figure 8.1) to describe the EM fields and geometry outside the inhomogeneous region. The interfaces shown in Figure 8.2 satisfy $\rho_j = r_j$, $r_j = \tilde{k}_f\tilde{r}_j$, $\tilde{r}_j = R_j/2\pi$, $j = 1, 2, 3$. In applying the RCWA algorithm to the geometry of Figure 8.1, it is assumed that the inhomogeneous region displayed there is divided into L thin layers of uniform width and that the interfaces of the thin layers are located on the circles at $u_\ell = u_0 - \ell\Delta u$, $\ell = 0, 1, \ldots, L$, where $\Delta u = (u_0 - u_L)/L > 0$, where $u = u_0$ is located on the circle $\rho_3 = r_3$, and where $u = u_L$ is located on the circle $\rho_1 = r_1$. In Figures 8.1 and 8.2, the rectangular coordinates $X_1 = x_1/2\pi$, $Y_1 = y_1/2\pi$, which are centered on the scattering objects exterior boundary interface $\rho_1 = r_1$ ($u = u_L$), are related to the bipolar rectangular coordinates of Equation 8.1 by the relations $x_1 = x - x_1^c$, $y_1 = y$, where $x_1^c = a/\tanh(u_L)$. In Figures 8.1 and 8.2, the exterior boundary has a radius $r_1 = a/\sinh(u_L)$ and the interior boundary a radius $r_3 = a/\sinh(u_0)$ with $r_3 < r_1$. The rectangular coordinates, call them $x_u(u, v)$, $y_u(u, v)$, whose origin is located at the center of a circle of constant u, call it O_u, are related to the rectangular coordinates $x(u, v)$, $y(u, v)$ of Equation 8.1 by the relations $x_u(u, v) = x(u, v) - x_{cu}$, $y_u(u, v) = y(u, v)$, where $x_{cu} = a/\tanh(u)$. The cylindrical coordinates, call them $\rho_u(u, v)$, $\phi_u(u, v)$, whose origin is located at O_u, are related to $x(u, v)$ and $y(u, v)$ coordinates Equation 8.1 of the overall coordinate system, by the cylindrical coordinate relations

$$\rho_u(u,v) = \sqrt{x_u^2(u,v) + y_u^2(u,v)}, \quad \phi_u(u,v) = \tan^{-1}\left(\frac{y_u(u,v)}{x_u(u,v)}\right) \tag{8.3}$$

Note $\rho_u(u, v) = r_u = a/\sinh(|u|)$.

8.2.2 Bipolar RCWA State Variable Formulation

The EM solution in the inhomogeneous dielectric region, following the procedure in [16,17,20], is obtained by solving Maxwell's equations in bipolar coordinates by a state variable (SV) approach in each thin layer $u_{\ell+1} \le u \le u_\ell$, $\ell = 0, \ldots, L-1$. Making the substitutions $U_{hu}(u, v) = \tilde{\eta}_f h(u, v)H_u(u, v)$ and $U_{hv}(u, v) = \tilde{\eta}_f h(u, v)H_v(u, v)$, where $H_u(u, v)$ and $H_v(u, v)$ represent the magnetic fields in each thin shell region, $\tilde{\eta}_f = 377\ \Omega$, we find that Maxwell's equations in each bipolar, cylindrical shell are given by

$$\frac{\partial E_z(u,v)}{\partial v} = -j\mu U_{hu}(u,v) \tag{8.4}$$

$$\frac{\partial E_z(u,v)}{\partial u} = j\mu U_{hv}(u,v) \tag{8.5}$$

$$\frac{\partial U_{hv}(u,v)}{\partial u} - \frac{\partial U_{hu}(u,v)}{\partial v} = j\varepsilon(u,v)h^2(u,v)E_z(u,v) \tag{8.6}$$

In these equations, $\varepsilon(u,v)$ represents the inhomogeneous relative permittivity in the inhomogeneous region (shown in Figure 8.1).

To solve Equations 8.4 through 8.6, we expand in the Floquet harmonics $-\pi \le v \le \pi$:

$$E_z(u,v) = \sum_{i=-\infty}^{\infty} S_{zi}(u)\exp(jiv), \quad U_{hu}(u,v) = \sum_{i=-\infty}^{\infty} U_{hui}(u)\exp(jiv)$$

$$U_{hv}(u,v) = \sum_{i=-\infty}^{\infty} U_{hvi}(u)\exp(jiv), \quad \varepsilon_h(u,v)E_z(u,v) = \sum_{i=-\infty}^{\infty}\left[\sum_{i'=-\infty}^{\infty} \varepsilon_{h,i-i'} S_{zi'}\right]\exp(jiv) \tag{8.7}$$

$$\varepsilon_h(u,v) \equiv \varepsilon(u,v)h^2(u,v) = \sum_{i=-\infty}^{\infty} \varepsilon_{hi}(u)\exp(jiv)$$

If these expansions are substituted in Equations 8.4 through 8.6, and after letting $\underline{S_z}(u) = [S_{zi}(u)]$, $\underline{U_{hu}}(u) = [U_{hui}(u)]$, and $\underline{U_{hv}}(u) = [U_{hvi}(u)]$ be column matrices and letting $\underline{\underline{\varepsilon_h}}(u) = [\varepsilon_{h,i,i'}(u)]$, $\varepsilon_{h,i,i'}(u) = \varepsilon_{h,i-i'}(u)$, $\underline{I} = [\delta_{i,i'}]$, $\underline{K} = [iK\delta_{i,i'}]$, $K = 2\pi/\Lambda_v$, $\Lambda_v = 2\pi$ (Λ_v may be called the v-angular grating period and $\delta_{i,i'}$ is the Kronecker delta) be square matrices, we find after manipulation [16,17,20,26]

$$\frac{\partial \underline{V}}{\partial u} = \underline{A}\,\underline{V}, \quad \underline{V} = \begin{bmatrix} \underline{S_z^e}(u) \\ \underline{U_{hv}^e}(u) \end{bmatrix}, \quad \underline{A} = \begin{bmatrix} \underline{A_{11}} & \underline{A_{12}} \\ \underline{A_{21}} & \underline{A_{22}} \end{bmatrix} \tag{8.8}$$

where

$$\underline{A_{11}} = 0, \quad \underline{A_{12}} = j\mu\underline{I}, \quad \underline{A_{21}} = j\left[\underline{\underline{\varepsilon_h}} - \frac{1}{\mu}\underline{K}^2\right], \quad \underline{A_{22}} = 0 \tag{8.9}$$

After truncating Equations 8.8 and 8.9 with $i = -I,\ldots,I$, one may determine the eigenvalues and eigenvectors of the matrix $\underline{A}$ and thus obtain a SV solution in each thin layer of the system.

8.2.3 Second-Order Differential Matrix Formulation

An alternate equation for the SV analysis, as presented in Refs. [6,7] (and as reviewed and discussed in Ref. [20]) for planar diffraction gratings, is to reduce Equation 8.8 to a second-order differential matrix equation and perform an eigenanalysis of the resulting equations. Following this procedure we have, for a given thin layer, (with subscript ℓ suppressed)

$$\frac{d}{du}\underline{S_z^e} = \underline{A_{12}}\,\underline{U_{hv}^e}, \quad \frac{d}{du}\underline{U_{hv}^e} = \underline{A_{21}}\,\underline{S_z^e} \tag{8.10}$$

or

$$\frac{d^2}{du^2}\underline{S_z^e} = \underline{A_{12}}\,\underline{A_{21}}\,\underline{S_z^e} \tag{8.11}$$

Letting

$$\underline{S_z^e}(u) = \underline{S_z}(u_\ell)\exp(qu') \tag{8.12}$$

where $u' = u - u_\ell \le 0$, $u_{\ell+1} \le u \le u_\ell$, $\ell = 0,\ldots,L-1$, letting $\underline{C} = -\underline{A_{12}}\,\underline{A_{21}}$, we find after substituting Equation 8.11 into Equation 8.12 and differentiating that

$$\underline{C}\,\underline{S_z^e} = Q\underline{S_z^e} \tag{8.13}$$

where $Q \equiv -q^2$. When taking into account the symmetry with respect to the y-axis as discussed earlier, it was found that $I + 1$ distinct eigenvalues and eigenvectors were associated with the matrix $\underline{C}$, and it was also found that the eigenvalues Q_n were purely real, with some of them assuming positive values and some assuming negative values. In the present work, the eigenvalues (and eigenvectors associated with Q_n) were ordered for each thin layer, such that $Q_1 > Q_2 > \cdots$. Positive values of the eigenvalue Q_n correspond to propagating eigenmodes (since $q_n^2 = -Q_n$), and negative values of Q_n correspond to nonpropagating or evanescent modes.

8.2.4 Thin-Layer, Bipolar Coordinate Eigenfunction Solution

Letting $Q_{n\ell}$ and $\underline{S_{zn\ell}}$ (with the subscript ℓ now included) be the eigenvalues and eigenvectors at $u = u_\ell$, $\ell = 0,\ldots,L-1$ of the matrix $\underline{C_\ell}$ (i.e., $\underline{C_\ell}\,\underline{S_{zn\ell}} = Q_{n\ell}\,\underline{S_{zn\ell}}$), we find that there are two matrix eigensolutions of the original SV matrix $\underline{A}$ of Equation 8.8, which may be expressed in terms of the just mentioned eigenvalues and eigenvectors $Q_{n\ell}$ and $\underline{S_{zn\ell}}$. The electric field and magnetic field portions of these matrix eigensolutions are given by

$$\underline{S_{zn\ell}^{e+}(u')} = \underline{S_{zn\ell}}\exp(-\sqrt{-Q_{n\ell}}\,u') \equiv \underline{S_{zn\ell}}\exp(-q_{n\ell}u')$$

$$\underline{U_{hvn\ell}^{e+}(u')} = \frac{1}{j\mu}\frac{\partial \underline{S_{zn\ell}^{e+}(u')}}{\partial u'} = -Z_{n\ell}\,\underline{S_{zn\ell}^{e+}(u')}$$

$$\underline{S_{zn\ell}^{e-}(u')} = \underline{S_{zn\ell}}\exp(\sqrt{-Q_{n\ell}}\,u') \equiv \underline{S_{zn\ell}}\exp(q_{n\ell}u')$$

$$\underline{U_{hvn\ell}^{e-}(u')} = \frac{1}{j\mu}\frac{\partial \underline{S_{zn\ell}^{e-}(u')}}{\partial u'} = Z_{n\ell}\,\underline{S_{zn\ell}^{e-}(u')} \tag{8.14}$$

where

$$q_{n\ell} = \sqrt{-Q_{n\ell}} = \begin{cases} j\sqrt{Q_{n\ell}}, & Q_{n\ell} \ge 0 \\ \sqrt{-Q_{n\ell}}, & Q_{n\ell} < 0 \end{cases} \tag{8.15}$$

$$Z_{n\ell} = \frac{q_{n\ell}}{j\mu} \tag{8.16}$$

where $u' = u - u_\ell \le 0$, $u_{\ell+1} \le u \le u_\ell$, $\ell = 0,\ldots,L-1$. The eigenmodes (or eigenfunctions) at $u = u_\ell$ (or $u' = 0$) associated with the eigenvalues and eigenvectors $Q_{n\ell}$ and $\underline{S_{zn\ell}}$ of the matrix $\underline{C_\ell}$, respectively, are obtained by summing the Fourier coefficients contained in the matrix $\underline{S_{zn\ell}}$ and are given by

$$S_{zn\ell}(v) = \sum_{i=-I}^{I} S_{zin\ell}\exp(jiv) \tag{8.17}$$

These eigenmodes, which have been normalized to unity, were found numerically to satisfy the orthogonality relation

$$\int_{-\pi}^{\pi} S_{zn\ell}(v) S_{zn'\ell}(v)^*\,dv = \delta_{n,n'} \tag{8.18}$$

where $(n, n') = 1,\ldots,N$ and N is the number of orthogonal eigenmodes under consideration to a high degree of accuracy. This orthogonality relation proved to be very helpful for enforcing EM boundary conditions at different thin-layer interfaces and at the interior and exterior boundaries of the system. We note at this point, that when using the eigensolutions of Equation 8.17 to either expand unknown EM fields or use them to enforce EM boundary conditions at the interfaces of the system, that it is not necessary to always use the full set of $n = 1,\ldots,I+1$ eigenfunctions which are available from the eigen matrix analysis, but if one desires, one may use a smaller set with $N < I + 1$. Using a set of eigensolutions with $N < I + 1$, may be useful as one then avoids using the highest order modes which may suffer truncation error. Or put another way, if it is desired to use N orthogonal modes of a given accuracy, one may use a larger I truncation order to generate those N orthogonal modes.

An important requirement of the bipolar RCWA formulation concerns the fast and accurate numerical calculation of the Fourier series coefficients $\varepsilon_{hi}(u)$ in Equation 8.7. Fast calculations of $\varepsilon_{hi}(u)$ are needed because the bipolar SV solution must be found separately in a possibly large number L of thin layers, and accurate calculation of the Fourier harmonics of $\varepsilon_{hi}(u)$ is needed because these Fourier coefficients are used to form the SV matrices of Equations 8.8 and 8.9. Inaccurate matrix values SV matrices naturally lead to inaccurate and incorrect SV solutions. It turns out that in this chapter Fourier coefficients $\varepsilon_{hi}(u)$ can be calculated almost exactly provided that the Fourier coefficients of the inhomogeneous $\varepsilon(u, v)$ can be calculated exactly (for example, in this chapter, Figures 8.2 and 8.3 corresponds to $\varepsilon(u, v)$ step-cosine profiles which can be done exactly). This follows because the Fourier coefficients of the $h^2(u, v)$ function can be calculated exactly using the residue theorem of complex variable theory [23] and thus the $\varepsilon_{hi}(u)$ Fourier coefficients of the product $\varepsilon_h(u, v) \equiv \varepsilon(u, v)\, h^2(u, v)$ can be calculated from a discrete Fourier convolution of the Fourier coefficients of the two product functions $\varepsilon(u, v)$ and $h^2(u, v)$. Details of the determination of the Fourier coefficients of $h^2(u, v)$ by the residue theorem are given in Appendix 8.A.

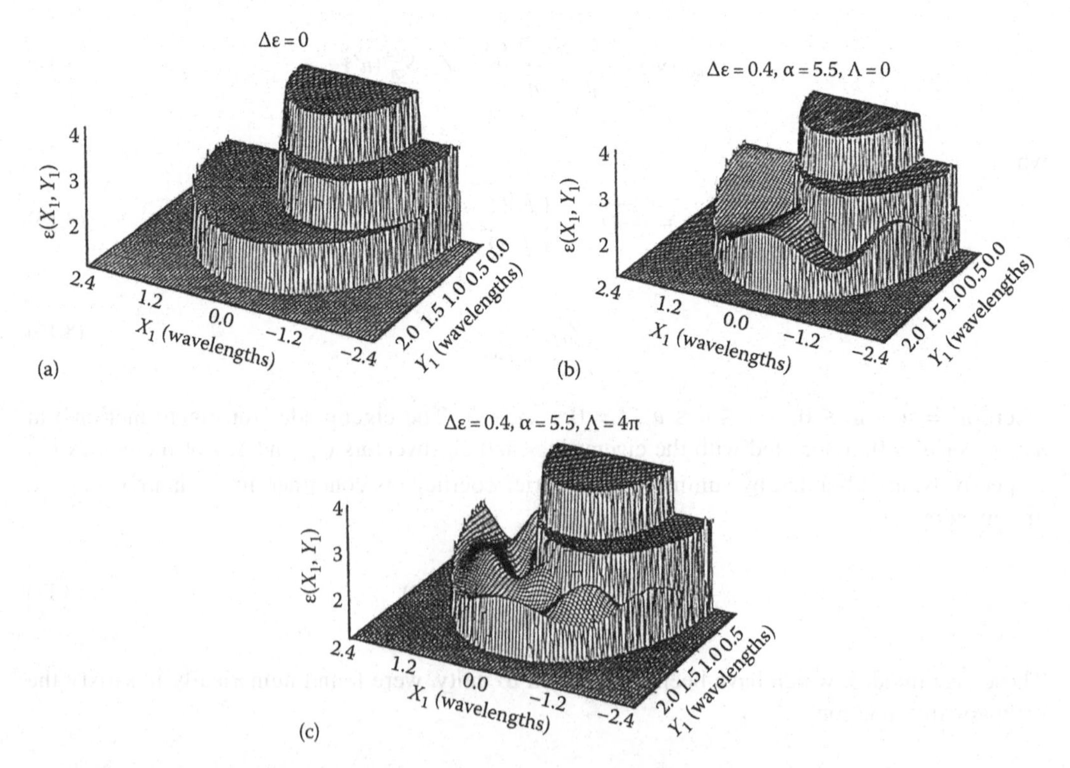

FIGURE 8.3 (a through c) Relative dielectric permittivity function of $\varepsilon(X_1, Y_1)$ (from $\varepsilon(u, v)$ of Equation 8.89) as a function of the rectangular coordinates X_1, Y_1 for the parameters listed in each figure. (Reproduced from Jarem, J.M., *Prog. Electromagn. Res.*, PIER 43, 181, 2003. With permission.)

8.3 BESSEL FUNCTION SOLUTIONS IN HOMOGENEOUS REGIONS OF SCATTERING SYSTEM

We will now present the EM fields in the Regions 0 and 3 (shown in Figures 8.1 and 8.2) which are outside of the inhomogeneous material region of the scattering object and which are assumed to have spatially uniform relative permittivities ε_0 and ε_3, respectively. In these regions, we will use exactly the same coordinates and Bessel functions EM field expansions as were by KPE [22], but specialized to the symmetric problem under consideration. In Region 3, the electric and magnetic fields are given by in general

$$E_z^{(3)}(\rho_3,\phi_3) = \sum_{m'=0}^{N-1}\left\{A_{m'}^{(3)}J_{m'}(k_3\rho_3)+B_{m'}^{I}H_{m'}^{(2)}(k_3\rho_3)\right\}\cos(m'\phi_3) \equiv \sum_{m'=0}^{N-1}E_{m'}^{(3)}(\rho_3)\cos(m'\phi_3) \tag{8.19}$$

$$\begin{aligned}\tilde{\eta}_f H_{\phi_3}^{(3)}(\rho_3,\phi_3) \equiv U_{\phi_3}^{(3)}(\rho_3,\phi_3) &= \sum_{m'=0}^{N-1}\left\{\left[\frac{k_3}{j\mu_3}\right]\left[A_{m'}^{(3)}J_{m'}'(k_3\rho_3)+B_{m'}^{I}H_{m'}^{(2)\prime}(k_3\rho_3)\right]\right\}\cos(m'\phi_3)\\ &\equiv \sum_{m'=0}^{N-1}U_{m'}^{(3)}(\rho_3)\cos(m'\phi_3),\quad \rho_{s3}<\rho_3\le r_3\end{aligned} \tag{8.20}$$

where $\tilde{\eta}_f = 377\ \Omega$, $k_3 = \sqrt{\mu_3\varepsilon_3}$, and $\mu_3 = \mu_0 = 1$ and it is assumed that ρ_{s3} encloses a symmetric source which gives rise to the symmetric incident field which is associated with the $B_{m'}^{I}$ coefficients. In the Region 0, the EM fields are given by

$$E_z^{(1)}(\rho_1,\phi_1) = \sum_{m'=0}^{N-1}\left\{A_{m'}^{I}J_{m'}(k_0\rho_1)+B_{m'}^{(0)}H_{m'}^{(2)}(k_0\rho_1)\right\}\cos(m'(\phi_1-\phi_0)) \equiv \sum_{m'=0}^{N-1}E_{m'}^{(1)}(\rho_1)\cos(m'\phi_1) \tag{8.21}$$

$$\begin{aligned}\tilde{\eta}_f H_{\phi_1}^{(1)}(\rho_1,\phi_1) \equiv U_{\phi_1}^{(1)}(\rho_1,\phi_1) &= \sum_{m'=0}^{N-1}\left\{\left[\frac{k_0}{j\mu_0}\right]\left[A_{m'}^{I}J_{m'}'(k_0\rho_1)+B_{m'}^{(0)}H_{m'}^{(2)\prime}(k_0\rho_1)\right]\right\}\cos(m'(\phi_1-\phi_0))\\ &\equiv \sum_{m'=0}^{N-1}U_{m'}^{(1)}(\rho_1)\cos(m'\phi_1)\end{aligned} \tag{8.22}$$

where $k_0 = \sqrt{\mu_0\varepsilon_0}$, $\mu_1 = \mu_0 = 1$, $r_1 < \rho_1 \le \rho_{s0}$, $\phi_0 = 0$ or $\phi_0 = \pi$ (note $\sin(m'\phi_0) = 0$), and it is assumed that a symmetric source which is exterior to ρ_{s0}, that is, $\rho_1 > \rho_{s0}$, gives rise to a symmetric incident field which is associated with the $A_{m'}^{I}$ coefficients. In this chapter we will be concerned with numerical results for which $B_{m'}^{I}$ will be taken to be zero, for which $\rho_{s0} \to \infty$ and the $A_{m'}^{I}$ coefficients are taken to those coefficients which correspond to an incident plane wave from the direction where $\phi_0 = 0$ or $\phi_0 = \pi$. In Equations 8.19 through 8.22, the number of expansion modes in Regions 0 and 3 has been set equal to the number of SV expansion modes N used in each thin layer.

8.4 THIN-LAYER SV SOLUTION IN THE INHOMOGENEOUS REGION OF THE SCATTERING SYSTEM

The EM field solutions in the inhomogeneous region (see Figure 8.1) in bipolar coordinates in a small range $-\Delta u \le u' \le 0$ ($\Delta u = (u_0 - u_L)/L > 0$, $u' = u - u_\ell \le 0$, $u_{\ell+1} \le u \le u_\ell$, $\ell = 0,\ldots,L-1$) (recall that $u = u_0$ corresponds to the inner boundary r_3 and $u = u_L$ corresponds to the outer boundary r_1

[Figure 8.1]) using the SV eigenvector EM solutions $\underline{S_{zn\ell}^{e+}(u')}$, $\underline{S_{zn\ell}^{e-}(u')}$, $\underline{U_{hvn\ell}^{e+}(u')}$, and $\underline{U_{hvn\ell}^{e-}(u')}$ of Equations 8.14 through 8.18 are given by

$$E_z^{(2)}(u',u_\ell,v) = \sum_{n'=1}^{N} \left\{ C_{n'\ell}^{+} \exp(-q_{n'\ell}u') + C_{n'\ell}^{-} \exp(q_{n'\ell}u') \right\} S_{zn'\ell}(v) \equiv \sum_{n'=1}^{N} E_{n'\ell}(u') S_{zn'\ell}(v) \tag{8.23}$$

$$U_{hv}^{(2)}(u',u_\ell,v) = \sum_{n'=1}^{N} \left\{ -C_{n'\ell}^{+} \exp(-q_{n'\ell}u') + C_{n'\ell}^{-} \exp(q_{n'\ell}u') \right\} Z_{n'\ell} S_{zn'\ell}(v) \equiv \sum_{n'=1}^{N} U_{hn'\ell}(u') S_{zn'\ell}(v) \tag{8.24}$$

where $S_{zn'\ell}(v)$ has been defined in Equation 8.17. In these equations, $C_{n'\ell}^{+}$ and $C_{n'\ell}^{-}$ represent the unknown expansion coefficients in the $u_{\ell+1} \le u \le u_\ell$, $\ell = 0, \ldots, L-1$ thin layer.

8.5 MATCHING OF EM BOUNDARY CONDITIONS AT INTERIOR–EXTERIOR INTERFACES OF THE SCATTERING SYSTEM

This section is concerned with matching EM boundary conditions at all interfaces of the scattering system including matching of layer to layer SV solutions in the inhomogeneous region of the scattering system (Region 2) and matching of Bessel function solutions of Regions 0 and 3 (homogeneous material region) with the SV solution of Region 2. This section discusses the equations which relate bipolar and circular, cylindrical coordinates to one another at a given point in space and also discusses efficient methods of enforcing EM boundary conditions at all interfaces of the system.

8.5.1 Bipolar and Circular Cylindrical Coordinate Relations

Before matching boundary conditions, it is important to relate the cylindrical, bipolar coordinate $\phi_u(u, v)$ defined in Equation 8.3 to the cylindrical coordinates ϕ_1 and ϕ_3 of the Regions 0 and 3 boundaries, respectively. Using Equation 8.3 at $u = u_L$ (Region 0, exterior boundary) and at $u = u_0$ (Region 3, interior boundary), we have, respectively,

$$\phi_1(v) = \phi_u(u_L, v) = \tan^{-1}\left[\frac{y(u_L, v)}{(x(u_L, v) - x_{cu_L})} \right] \tag{8.25}$$

$$\phi_3(v) = \phi_u(u_0, v) = \tan^{-1}\left[\frac{y(u_0, v)}{(x(u_0, v) - x_{cu_0})} \right] \tag{8.26}$$

In matching boundary conditions at the exterior and interior boundaries, respectively, using Equations 8.25 and 8.26, it is important to relate the complicated, cosine functions $\cos(m\phi_1(v))$ and $\cos(m\phi_3(v))$ in Equations 8.19 through 8.22 to the $\exp(jiv)$ Fourier series expansions which occur in Equations 8.23 and 8.24 (after Equation 8.17 has been substituted in Equations 8.23 and 8.24 in a manner which is as accurate and efficient as possible). This may be accomplished by expressing the exponentials $\exp(\pm m\phi_1(v))$ and $\exp(\pm m\phi_3(v))$ which make up the functions $\cos(m\phi_1(v))$ and $\cos(m\phi_3(v))$, respectively, as a complex exponential Fourier series, namely,

$$\exp(jp\phi(u,v)) = \sum_{i=-\infty}^{\infty} \alpha_i^{(p)}(u) \exp(jiv) \tag{8.27}$$

where $p = \pm m$, $u = u_L$, or $u = u_0$, and then numerically determining the exponential Fourier coefficients $\alpha_i^{(p)}(u)$. It turns out that the Fourier coefficients $\alpha_i^{(p)}(u)$ of this series may be calculated exactly using the residue theorem of complex variable theory [23]. Appendix 8.A of this chapter gives details on how these coefficients are calculated and what values these coefficients assume.

8.5.2 Details of Region 2 (Inhomogenous Region)–Region 3 (Homogenous Interior Region) EM Boundary Value Matching

The objective now is to match EM boundary conditions at all interfaces and determine all unknown expansion coefficients of the system as defined in Sections 8.3 and 8.4. Starting at the $u = u_0$ (Region 2–Region 3 interior boundary) we have using the Equations 8.19 through 8.24

$$E_z^{(2)}(u',u_\ell,v)\Big|_{u_\ell=u_0,u'=0,\ell=0} = \left[\sum_{n'=-\infty}^{\infty} E_{n'\ell}(u')S_{zn'\ell}(v)\right]\Bigg|_{u_\ell=u_0,u'=0,\ell=0} = E_z^{(3)}(\rho_3,\phi_3)\Big|_{\rho_3=r_3^-}$$

$$= \left[\sum_{m'=0}^{N-1} E_{m'}^{(3)}(\rho_3)\cos(m'\phi_3)\right]\Bigg|_{\rho_3=r_3^-} \tag{8.28}$$

$$\left[\frac{-1}{h(u,v)}U_{hv}^{(2)}(u',u_\ell,v)\right]\Bigg|_{u_\ell=u_0,u'=0,\ell=0} = \left[\frac{-1}{h(u,v)}\sum_{n'=-\infty}^{\infty} U_{hn'\ell}(u')S_{zn'\ell}(v)\right]\Bigg|_{u_\ell=u_0,u'=0,\ell=0}$$

$$= U_{\phi_3}^{(3)}(\rho_3,\phi_3)\Big|_{\rho_3=r_3^-} = \left[\sum_{m'=0}^{N-1} U_{m'}^{(3)}(\rho_3)\cos(m'\phi_3)\right]\Bigg|_{\rho_3=r_3^-} \tag{8.29}$$

where $\rho_3 = r_3^-$ means just inside the Region 3 interior. The minus sign in Equation 8.29 in the square bracket is present because at $\rho_3 = r_3$, $u = u_0$ interface, at each point on this circle, the unit vectors $\hat{a}_v(u = u_0)$ and $\hat{a}_{\phi_3}(\rho_3 = r_3)$ are in opposite directions or satisfy $\hat{a}_v(u = u_0) = -\hat{a}_{\phi_3}(\rho_3 = r_3)$ and thus opposite signs of the tangential magnetic field components must be included for a correct boundary matching at the interior boundary.

To enforce the electric field boundary condition of Equation 8.28 at the $\rho_3 = r_3$, $u = u_0$ interface, Equation 8.28 is multiplied on both sides by the weighting or testing functions $\{\cos(m\phi_3)\}$, $m = 0,\ldots,N-1$ and then integrated over the range $-\pi \le \phi_3 \le \pi$. This results in the equation

$$\left[\sum_{n'=-\infty}^{\infty} E_{n'\ell}(u')\int_{-\pi}^{\pi} S_{zn'\ell}(v)\cos(m\phi_3(v))d\phi_3(v)\right]\Bigg|_{u=u_0,u'=0,\ell=0} = \left[\sum_{m'=0}^{N-1} E_{m'}^{(3)}(\rho_3)\int_{-\pi}^{\pi}\cos(m\phi_3)\cos(m'\phi_3)d\phi_3\right]\Bigg|_{\rho_3=r_3^-}$$

$$= \sum_{m'=0}^{N-1} E_{m'}^{(3)}(r_3)\pi[1+\delta_{m,0}]\delta_{m,m'} \tag{8.30}$$

The integral on the left hand side, after a change of variables on the interior circle ($\rho_3 = r_3$, $u = u_\ell|_{\ell=0} = u_0$) from integration with respect to the $\phi_3(v)$ variable (Equation 8.23), to integration with respect to the v variable (noting that $\phi_3(v)|_{v=\pi} = -\pi$ and $\phi_3(v)|_{v=-\pi} = \pi$ [see Figure 8.1]), becomes, after substituting $\ell = 0$ in $S_{zn'\ell}(v)$,

$$Z_{m,n'}^{cE3} = \int_{\pi}^{-\pi} \left[S_{zn'0}(v)\cos(m\phi_3(v))\frac{d\phi_3(v)}{dv}\right] dv \tag{8.31}$$

This integral may be evaluated exactly by; (1) substituting $\cos(m\phi_3(v))(d\phi_3(v)/dv)$ in Equation 8.31 the Fourier series ($\rho_3 = r_3$, $u = u_\ell|_{\ell=0} = u_0$)

$$\cos(m\phi_3(v))\frac{d\phi_3(v)}{dv} = \sum_{i=-\infty}^{\infty}\left(\frac{1}{2}\left[\zeta_i^{(m)}(u_0) + \zeta_i^{(-m)}(u_0)\right]\right)\exp(jiv) \tag{8.32}$$

where the Fourier coefficients $\zeta_i^{(p)}(u)$ are defined by

$$\exp(jp\phi_u(u,v))\frac{d\phi_u(u,v)}{dv} = \sum_{i=-\infty}^{\infty}\zeta_i^{(p)}(u)\exp(jiv) \tag{8.33}$$

where $p = \pm m$ and $\phi_3(v)$ is given in Equation 8.26; (2) substituting the exponential Fourier series expansion for $S_{zn'\ell}(v)$, $\ell=0$ as defined by Equation 8.17; and (3) integrating, in a straightforward way, the product of the two just described exponential Fourier series over the limits defined by the integral in Equation 8.32. The result of the integration is

$$Z_{m,n'}^{cE3} = -\pi\sum_{i=-I}^{I}\left(\zeta_i^{(m)}(u_0) + \zeta_i^{(-m)}(u_0)\right)S_{z,-i,n',0} \tag{8.34}$$

In Equations 8.32 through 8.34, the Fourier coefficients $\zeta_i^{(p)}(u)$ of Equation 8.33, like the Fourier coefficients $\alpha_i^{(p)}(u)$ of Equation 8.27, may be evaluated exactly using the residue theorem, and the details are given in Appendix 8.A. The final electric field equation becomes after substituting $\ell = 0$, evaluating the Kronecker delta function $\delta_{m,m'}$ in Equation 8.30 is,

$$\sum_{n'=1}^{N} Z_{m,n'}^{cE3}E_{n'0}(0) \equiv \sum_{n'=1}^{N} Z_{m,n'}^{cE3}E_{n'}^{SV3} = E_m^{(3)}(r_3^-)\pi[1+\delta_{m,0}] \tag{8.35}$$

where the coefficient $E_{n'}^{SV3}$ has been defined to be $E_{n'}^{SV3} \equiv E_{n'\ell}(u')|_{u'=0,\rho_3=r_3^+,\ell=0}$, and where $E_{n'\ell}(u')$ is defined in Equation 8.23. The superscript "SV3" in this equation refers to the evaluation of the SV solution at $\rho_3 = r_3^+$ where $\rho_3 = r_3^+$ means just inside the inhomogeneous scattering object region.

To enforce the magnetic field boundary condition of Equation 8.29 at the $\rho_3 = r_3$, $u = u_0$ interface, Equation 8.29 is multiplied on both sides, just as the electric field Equation 8.28 was, by the weighting or testing functions $\{\cos(m\phi_3)\}$, $m = 0,\ldots,N-1$ and then integrated over the range $-\pi \le \phi_3 \le \pi$. The resulting equation is

$$\left[\sum_{n'=-\infty}^{\infty} U_{hn'\ell}(u') \int_{-\pi}^{\pi} \left[\frac{-1}{h(u,v)}\right] S_{zn'\ell}(v)\cos(m\phi_3(v))\,d\phi_3(v)\right]\Bigg|_{u=u_0,u'=0,\ell=0}$$

$$= \left[\sum_{m'=0}^{N-1} U_{m'}^{(3)}(\rho_3)\int_{-\pi}^{\pi}\cos(m\phi_3)\cos(m'\phi_3)\,d\phi_3\right]\Bigg|_{\rho_3=r_3}$$

$$= \sum_{m'=0}^{N-1} U_{m'}^{(3)}(r_3)\pi[1+\delta_{m,0}]\delta_{m,m'} \tag{8.36}$$

The integral on the left hand side of Equation 8.36 may be evaluated in a similar way as was the integral in Equation 8.30. Calling the integral $Z_{m,n'}^{cU3}$, changing variables from ϕ_3 to v, and letting $\ell = 0$ and $u = u_0$ in Equation 8.36, we find that the integral may be written

$$Z_{m,n'}^{cU3} = \int_{\pi}^{-\pi}\left[\frac{-1}{h(u_0,v)}\right] S_{zn'0}(v)\cos(m\phi_3(v))\frac{d}{dv}\phi_3(v)\,dv \tag{8.37}$$

Carrying out the differentiation of $(d/dv)\phi_3(v)$, a small amount of algebra that shows

$$\frac{1}{h(u_0,v)}\frac{d}{dv}\phi_3(v) = \frac{-\sinh(u_0)}{a} \tag{8.38}$$

Substitution of Equation 8.38 in Equation 8.37, and using the Fourier series expansion of $S_{zn'\ell}(v)$ in Equation 8.17, it turns out that $Z_{m,n'}^{cU3}$ may be evaluated exactly as

$$Z_{m,n'}^{cU3} = -\pi\left[\frac{\sinh(u_0)}{a}\right]\sum_{i=-I}^{I}\left(\alpha_i^{(m)}(u_0)+\alpha_i^{(-m)}(u_0)\right)S_{z,-i,n',0} \tag{8.39}$$

where the Fourier coefficients $\alpha_i^{(m)}(u)$ have been defined earlier. The final magnetic field equation after evaluating the Kronecker delta function $\delta_{m,m'}$ in Equation 8.36 is given by

$$\sum_{n'=1}^{N} Z_{m,n'}^{cU3} U_{hn'}^{SV3} = U_m^{(3)}(r_3^-)\pi[1+\delta_{m,0}] \tag{8.40}$$

where the coefficient $U_{hn'}^{SV3}$ has been defined to be $U_{hn'}^{SV3} \equiv U_{hn'\ell}(u')\big|_{u'=0,\rho_3=r_3^+,\ell=0}$ where $U_{hn'\ell}(u')$ is defined in Equation 8.24.

8.5.3 Region 0 (Homogenous Exterior Region)–Region 2 (Inhomogenous Region) EM Boundary Value Matching

A nearly identical procedure as was used to match EM boundary conditions at the $\rho_3 = r_3$ interface may be used to match EM boundary conditions at the $\rho_1 = r_1$, $u = u_L$ interface. We find that at the $\rho_1 = r_1$, $u = u_L$ interface the electric and magnetic field boundary condition equations are given by

$$\sum_{n'=1}^{N} Z_{m,n'}^{cE1} E_{n'}^{SV1} = E_m^{(1)}(r_1^+)\pi[1+\delta_{m,0}] \tag{8.41}$$

$$\sum_{n'=1}^{N} Z_{m,n'}^{cU1} U_{hn'}^{SV1} = U_m^{(1)}(r_1^+)\pi[1+\delta_{m,0}] \tag{8.42}$$

where

$$E_{n'}^{SV1} \equiv E_{n'\ell}(u')\Big|_{u'=-\Delta u, \rho_1=r_1^-, \ell=L-1} \tag{8.43}$$

$$U_{hn'}^{SV1} \equiv U_{hn'\ell}(u')\Big|_{u'=-\Delta u, \rho_1=r_1^-, \ell=L-1} \tag{8.44}$$

where the weighting or testing functions $\{\cos(m\phi_1)\}$, $m = 0,\ldots,N-1$ have been used to enforce EM boundary conditions over the interval $-\pi \le \phi_1 \le \pi$ in the same way as the $\{\cos(m\phi_3)\}$, $m = 0,\ldots,N-1$ were used to enforce boundary conditions at the $\rho_3 = r_3$, $u = u_0$ boundary.

8.5.4 Details of Layer-to-Layer EM Boundary Value Matching in the Inhomogeneous Region

Matching boundary conditions at the $u = u_\ell$, $\ell = 1,2,\ldots,L-1$ thin-layer interfaces (these interfaces are located entirely inside the inhomogeneous region) we have, using the electric field expression of Equation 8.23 at $u = u_\ell$,

$$E_z^{(2)}(u',u_\ell,v)\Big|_{u'=0} = E_z^{(2)}(u',u_{\ell-1},v)\Big|_{u'=-\Delta u} \tag{8.45}$$

or after evaluation, the left hand side of Equation 8.45 is

$$E_z^{(2)}(u',u_\ell,v)\Big|_{u'=0} = \sum_{n'=1}^{N}\left\{C_{n'\ell}^{+} + C_{n'\ell}^{-}\right\} S_{zn'\ell}(v) \equiv \sum_{n'=1}^{N} E_{n'\ell}(0) S_{zn'\ell}(v), \tag{8.46}$$

the right hand side of Equation 8.45 is

$$\begin{aligned} E_z^{(2)}(u',u_{\ell-1},v)\Big|_{u'=-\Delta u} &= \sum_{n'=1}^{N}\left\{C_{n',\ell-1}^{+}\exp(-q_{n',\ell-1}(-\Delta u)) + C_{n',\ell-1}^{-}\exp(q_{n',\ell-1}(-\Delta u))\right\} S_{zn',\ell-1}(v) \\ &\equiv \sum_{n'=1}^{N} E_{n',\ell-1}(-\Delta u) S_{zn',\ell-1}(v) \end{aligned} \tag{8.47}$$

or altogether

$$\sum_{n'=1}^{N} E_{n'\ell}(0) S_{zn'\ell}(v) = \sum_{n'=1}^{N} E_{n',\ell-1}(-\Delta u) S_{zn',\ell-1}(v) \tag{8.48}$$

where

$$E_{n',\ell-1}(-\Delta u) = C^{+}_{n',\ell-1}\exp(-q_{n',\ell-1}(-\Delta u)) + C^{-}_{n'\ell-1}\exp(q_{n',\ell-1}(-\Delta u)), \tag{8.49}$$

$$E_{n'\ell}(0) = C^{+}_{n'\ell} + C^{-}_{n'\ell} \tag{8.50}$$

To enforce the electric field boundary condition of Equations 8.45 and 8.48, Equation 8.48 is multiplied on both sides by the weighting, enforcing, or testing functions $\{S_{zn\ell}(v)\}$, $n = 1,2,\ldots,N$ of Equation 8.17, and then integrated over the interval $-\pi \le v \le \pi$. Carrying out this operation and using the property that the functions $\{S_{zn\ell}(v)\}$ are orthonormal (Equation 8.18), it is found that

$$E_{n\ell}(0) = \sum_{n'=1}^{N} E_{n',\ell-1}(-\Delta u) S^{nn'\ell} \tag{8.51}$$

where

$$S^{nn'\ell} = \int_{-\pi}^{\pi} S_{zn\ell}(v) S_{zn',\ell-1}(v)\,dv = 2\pi \sum_{i=-I}^{I} S_{zin\ell} S_{z,-i,n,\ell-1} \tag{8.52}$$

where $\ell = 1,2,\ldots,L-1$. We note because the eigenfunctions $S_{zn\ell}(v)$ of Equation 8.17 change as the value of ℓ changes from interface to interface, that the functions $S_{zn\ell}(v)$ and $S_{zn',\ell-1}(v)$ in the integral of Equation 8.52 are not orthogonal, and thus $S^{nn'\ell}$ is nonzero in general when $n \ne n'$.

Evaluating the magnetic field Equation 8.24 at $u = u^{\pm}_{\ell}$ in a similar way as the electric field was evaluated, we find

$$\sum_{n'=1}^{N} U_{hn'\ell}(0) S_{zn'\ell}(v) = \sum_{n'=1}^{N} U_{hn',\ell-1}(-\Delta u) S_{zn',\ell-1}(v) \tag{8.53}$$

where

$$U_{hn',\ell-1}(-\Delta u) = Z_{n',\ell-1}\left\{-C^{+}_{n',\ell-1}\exp(-q_{n',\ell-1}(-\Delta u)) + C^{-}_{n',\ell-1}\exp(q_{n',\ell-1}(-\Delta u))\right\}, \tag{8.54}$$

$$U_{hn'\ell}(0) = Z_{n'\ell}\left\{-C^{+}_{n'\ell} + C^{-}_{n'\ell}\right\} \tag{8.55}$$

If the magnetic field boundary condition of Equation 8.53 is multiplied on both sides by the weighting or testing functions $\{S_{zn\ell}(v)\}$ with $n = 1,2,\ldots,N$ and then integrated over the interval $-\pi \le v \le \pi$ it is found, following the same steps as were used done to enforce the electric field boundary condition,

$$U_{hn\ell}(0) = \sum_{n'=1}^{N} U_{hn',\ell-1}(-\Delta u) S^{nn'\ell} \tag{8.56}$$

where $S^{nn'\ell}$ has been previously defined.

8.5.5 Inhomogeneous Region Ladder-Matrix

The layer-to-layer boundary matching equations presented in Section 8.5.4 may be expressed in such a form from which a ladder-matrix relating the innermost and outermost SV EM field coefficients of the inhomogeneous region may be found. This ladder-matrix is very useful as it allows a very fast and efficient method to compute the final system matrix and thus compute the overall EM fields of the system. This ladder-matrix approach has been used extensively in Chapters 5 and 6 for planar diffraction gratings and Chapter 7 for the study of EM scattering from cylindrical and spherical systems. Details of the ladder-matrix boundary matching equations used in this chapter are given now.

Inspecting Equations 8.50 and 8.55 (after using a value of $\ell-1$ in these equations), we note that the $C^{+}_{n',\ell-1}$ and $C^{-}_{n',\ell-1}$ coefficients are related to the $E_{n',\ell-1}(0)$ and $U_{hn',\ell-1}(0)$ coefficients by a simple 2×2 set of equations. If these 2×2 equations are inverted, we find

$$C^{+}_{n',\ell-1}=\left[E_{n',\ell-1}(0)-\frac{1}{Z_{n',\ell-1}}U_{hn',\ell-1}(0)\right]\Big/2 \tag{8.57}$$

$$C^{-}_{n',\ell-1}=\left[E_{n',\ell-1}(0)+\frac{1}{Z_{n',\ell-1}}U_{hn',\ell-1}(0)\right]\Big/2 \tag{8.58}$$

If $C^{+}_{n',\ell-1}$ and $C^{-}_{n',\ell-1}$ of Equations 8.57 and 8.58 are substituted back into Equations 8.49 and 8.54 and the resulting $E_{n',\ell-1}(-\Delta u)$, $U_{hn',\ell-1}(-\Delta u)$ expressions are further substituted back into Equations 8.51 and 8.56, we find that the coefficients $E_{n'\ell}(0)$ and $U_{hn'\ell}(0)$ may be expressed in terms of $E_{n',\ell-1}(0)$ and $U_{hn',\ell-1}(0)$ by the relations

$$E_{n\ell}(0)=\sum_{n'=1}^{N}\left\{K^{EE}_{nn'\ell}E_{n',\ell-1}(0)+K^{EU}_{nn'\ell}U_{hn',\ell-1}(0)\right\} \tag{8.59}$$

and

$$U_{n\ell}(0)=\sum_{n'=1}^{N}\left\{K^{UE}_{n\ell}E_{n',\ell-1}(0)+K^{UU}_{nn'\ell}U_{hn',\ell-1}(0)\right\} \tag{8.60}$$

where

$$K^{EE}_{nn'\ell}=K^{UU}_{nn'\ell}=S^{nn'\ell}\cosh\left[q_{n',\ell-1}(-\Delta u)\right] \tag{8.61}$$

$$K^{EU}_{nn'\ell}=\left[\frac{S^{nn'\ell}}{Z_{n',\ell-1}}\right]\sinh\left[q_{n',\ell-1}(-\Delta u)\right]$$

$$K^{UE}_{nn'\ell}=\left[S^{nn'\ell}Z_{n',\ell-1}\right]\sinh\left[q_{n',\ell-1}(-\Delta u)\right] \tag{8.62}$$

where $\ell=1,2,\ldots,L-1$ and where cosh(·) and sinh(·) represent the hyperbolic cosine and hyperbolic sine functions, respectively. Defining the $N\times N$ matrices $\underline{K^{EE}_{\ell}}\equiv[K^{EE}_{nn'\ell}]$, $\underline{K^{EU}_{\ell}}\equiv[K^{EU}_{nn'\ell}]$, $\underline{K^{UE}_{\ell}}\equiv[K^{UE}_{nn'\ell}]$, and $\underline{K^{UU}_{\ell}}=[K^{UU}_{nn'\ell}]$ with $(n,n')=1,2,\ldots,N$, defining the column matrices $\underline{E_{\ell}}\equiv[E_{n\ell}(0)]$ and $\underline{U_{h\ell}}\equiv[U_{hn\ell}(0)]$ with $n=1,2,\ldots,N$, defining the $2N\times 2N$ matrix

$$\underline{K}_\ell \equiv \begin{bmatrix} \underline{K}_\ell^{EE} & \underline{K}_\ell^{EU} \\ \underline{K}_\ell^{UE} & \underline{K}_\ell^{UU} \end{bmatrix} \tag{8.63}$$

and the $2N$ column matrix

$$\underline{W}_\ell \equiv \begin{bmatrix} \underline{E}_\ell \\ \underline{U}_{h\ell} \end{bmatrix} \tag{8.64}$$

we find

$$\begin{bmatrix} \underline{E}_\ell \\ \underline{U}_{h\ell} \end{bmatrix} \equiv \begin{bmatrix} \underline{K}_\ell^{EE} & \underline{K}_\ell^{EU} \\ \underline{K}_\ell^{UE} & \underline{K}_\ell^{UU} \end{bmatrix} \begin{bmatrix} \underline{E}_{\ell-1} \\ \underline{U}_{h,\ell-1} \end{bmatrix} \tag{8.65}$$

or

$$\underline{W}_\ell = \underline{K}_\ell \, \underline{W}_{\ell-1} \tag{8.66}$$

where $\ell = 1,\ldots,L-1$. We thus see that the electric and magnetic field coefficients $E_{n\ell}(0)$ and $U_{hn\ell}(0)$, $n = 1,\ldots,N$ of the eigenfunctions $S_{zn\ell}(v)$ which corresponds to the $u = u_\ell$ layer, may be expressed in terms of the electric and magnetic field coefficients $E_{n,\ell-1}(0)$ and $U_{hn,\ell-1}(0)$ of the eigenfunctions $S_{zn',\ell-1}(v)$ at $u = u_{\ell-1}$ layer. By starting at $\ell = 1$ and by successive substitution, it is found that

$$\underline{W}_{L-1} = \underline{K}_{L-1}\,\underline{K}_{L-2} \cdots \underline{K}_2\,\underline{K}_1\,\underline{W}_0 \tag{8.67}$$

Forming the column matrices $\underline{E}^{SV1} \equiv [E_n^{SV1}]$ and $\underline{U}_h^{SV1} \equiv [U_{hn}^{SV1}]$, $n = 1,\ldots,N$, respectively, defined by Equations 8.43 and 8.44, we further find that the electric and magnetic field coefficients $E_{n,L-1}(0)$ and $U_{hn,L-1}(0)$ of the $u = u_{L-1}$ interface are related to the field coefficients E_n^{SV1} and U_{hn}^{SV1} of the $u = u_L$, $\rho_1 = r_1$ interface, after using $\ell = L$ in Equations 8.61 through 8.63, by the matrix relation

$$\underline{W}_L \equiv \begin{bmatrix} \underline{E}^{SV1} \\ \underline{U}_h^{SV1} \end{bmatrix} = \begin{bmatrix} \underline{K}_L^{EE} & \underline{K}_L^{EU} \\ \underline{K}_L^{UE} & \underline{K}_L^{UU} \end{bmatrix} \begin{bmatrix} \underline{E}_{L-1} \\ \underline{U}_{h,L-1} \end{bmatrix} \equiv \underline{K}_L\,\underline{W}_{L-1} \tag{8.68}$$

We further note that at the $\ell = 0$ interface, after letting $\underline{W}_0 \equiv \begin{bmatrix} \underline{E}^{SV3} \\ \underline{U}_h^{SV3} \end{bmatrix}$ where $\underline{E}^{SV3} \equiv [E_n^{SV3}]$ and $\underline{U}_h^{SV3} \equiv [U_{hn}^{SV3}]$, $n = 1,\ldots,N$, that

$$\underline{W}_L = \underline{K}_L\underline{K}_{L-1} \cdots \underline{K}_2\underline{K}_1\underline{W}_0 = \underline{K}_T\,\underline{W}_0 \tag{8.69}$$

We thus see that matrix $\underline{K}_T$ (T = Total) relates the SV coefficients of the inner layer to the SV coefficients outer one.

8.6 REGION 1–REGION 3 BESSEL–FOURIER COEFFICIENT TRANSFER MATRIX

In this section, two different Bessel–Fourier coefficient transfer matrices are developed, the first of which is based on using the SV thin-layer, ladder-matrix formulation described in Section 8.5 [26] and the second of which is based on using the KPE Bessel–Hankel function formulation described in Refs. [22,26]. The SV transfer matrix formulation [26] applies to the general case when inhomogeneous dielectric material is contained in Region 2, whereas the KPE formulation [22,26] holds only in the special case when the nonconcentric (or eccentric) circular cylinders each containing homogeneous dielectric material makes up Region 2. For the special case of homogeneous circular cylinders when both algorithms apply, very close numerical values of the ladder matrices were observed. The two ladder-matrix formulations will now be presented.

Our objective now, as just mentioned, is to develop from the SV solution, an SV transfer matrix which relates the circular Bessel–Fourier coefficients of $\cos(m\phi_3)$, $m = 0, N-1$ associated with the electric and magnetic fields on the inner cylinder $\rho_3 = r_3^-$ to the circular Bessel–Fourier coefficients of $\cos(m\phi_1)$, $m = 0, N-1$ associated with the electric and magnetic fields and the outer cylinder $\rho_1 = r_1^+$. This may be accomplished by (1) expressing the coefficients $E_m^{(1)}(r_1^+)$, $U_m^{(1)}(r_1^+)$ in terms of SV coefficients E_n^{SV1}, U_{hn}^{SV1} by Equations 8.41 and 8.42; (2) expressing the SV coefficients E_n^{SV1}, U_{hn}^{SV1} in terms of E_n^{SV3}, U_{hn}^{SV3} by using the matrix $\underline{K_T}$ of Equation 8.69; (3) expressing the SV coefficients E_n^{SV3}, U_{hn}^{SV3} in terms of the coefficients $E_m^{(3)}(r_3^-)$, $U_m^{(3)}(r_3^-)$ by using Equations 8.28 and 8.29; and (4) substituting successively the coefficient relations mentioned in Steps 1–3 to finally express $E_m^{(1)}(r_1^+)$, $U_m^{(1)}(r_1^+)$ in terms of $E_m^{(3)}(r_3^-)$, $U_m^{(3)}(r_3^-)$.

The SV coefficients E_n^{SV3}, U_{hn}^{SV3} may be found in terms $E_m^{(3)}(r_3^-)$ and $U_m^{(3)}(r_3^-)$, respectively, by multiplying Equation 8.28 by $S_{zn\ell}(v)$, $\ell = 0$, by multiplying Equation 8.29 by $-h(u_0,v)S_{zn\ell}(v)$, $\ell = 0$, and then integrating the resulting equations over the range $-\pi \le v \le \pi$. Performing these operations after using the orthonormal property of the $S_{zn\ell}(v)$, $n = 1,\ldots,N$, $\ell = 0$ eigenfunctions, it is found that

$$E_n^{SV3} \equiv E_{n\ell}(u')\Big|_{u'=0,u=u_0,\ell=0} = \sum_{m=0}^{N-1} E_m^{(3)}(r_3^-)\left[\int_{-\pi}^{\pi} S_{zn0}(v)\cos(m\phi_3(v))\,dv\right] \tag{8.70}$$

$$U_{hn}^{SV3} \equiv U_{hn\ell}(u')\Big|_{u'=0,u=u_0,\ell=0} = \sum_{m=0}^{N-1} U_m^{(3)}(r_3^-)\left[-\int_{-\pi}^{\pi} S_{zn0}(v)h(u_0,v)\cos(m\phi_3(v))\,dv\right] \tag{8.71}$$

If one expands $S_{zn0}(v)$ and the exponentials $\exp(\pm jm\phi_3(v))$ which make up $\cos(m\phi_3(v))$ in the exponential Fourier series in the variable v given by Equation 8.27, one finds after substituting the Fourier series of the two terms and using the orthogonality of the Fourier exponentials $\exp(jiv)$, $i = -I,\ldots,I$, that the integral in Equation 8.70 is given by

$$E_n^{SV3} \equiv \sum_{m=0}^{N-1} Z_{nm}^{SE3} E_m^{(3)}(r_3^-),\quad Z_{nm}^{SE3} = \pi\sum_{i=-I}^{I}\left[\alpha_{-i}^{(m)}(u_0)+\alpha_{-i}^{(-m)}(u_0)\right]S_{zin0} \tag{8.72}$$

The integral in Equation 8.71 may be evaluated by (1) expanding the cosine exponential factors $\exp(jp\phi_3(v))/(\cosh(u_0)-\cos(v)) = (h(u_0,v)/a)\exp(jm\phi_3(v))$, $p = \pm m$ which occur in this integral as an exponential Fourier series

$$\frac{\exp(jp\phi_3(v))}{(\cosh(u_0)-\cos(v))} = \sum_{i=-I_t}^{I_t} \beta_i^{(p)}(u_0)\exp(jiv) \tag{8.73}$$

(details given in Appendix 8.A); (2) substituting the $\exp(jiv)$ Fourier expansion for $S_{zn0}(v)$ of Equation 8.17; and (3) carrying out the resulting integral over $-\pi \le v \le \pi$ with respect to v and using the orthogonality of the $\exp(jiv)$ exponentials. From these steps, it is found that

$$U_{hn}^{SV3} \equiv \sum_{m=0}^{N-1} Z_{nm}^{SU3} U_m^{(3)}(r_3^-), \quad Z_{nm}^{SU3} = -\pi a \sum_{i=-I}^{I} \left[\beta_{-i}^{(m)}(u_0) + \beta_{-i}^{(-m)}(u_0)\right] S_{zin0} \tag{8.74}$$

If Equations 8.72 and 8.74 are put in matrix form with $\underline{\underline{Z}}^{SE3} = [Z_{nm}^{SE3}]$, $\underline{\underline{Z}}^{SU3} = [Z_{nm}^{SU3}]$, $\underline{E}^{(3)} = [E_m^{(3)}(r_3^-)]$, $\underline{U}^{(3)} = [U_m^{(3)}(r_3^-)]$ with $m = 0, \ldots, N-1$ and $n = 1, \ldots, N$, we have

$$\underline{W}_0 \equiv \begin{bmatrix} \underline{E}^{SV3} \\ \underline{U}_h^{SV3} \end{bmatrix} = \begin{bmatrix} \underline{\underline{Z}}^{SE3} & 0 \\ 0 & \underline{\underline{Z}}^{SU3} \end{bmatrix} \begin{bmatrix} \underline{E}^{(3)} \\ \underline{U}^{(3)} \end{bmatrix} \equiv \underline{\underline{Z}}^{(3)} \underline{W}^{(3)} \tag{8.75}$$

The coefficients $E_m^{(1)}(r_1^+)$, $U_m^{(1)}(r_1^+)$ are given by, after using Equations 8.41 and 8.42

$$E_m^{(1)}(r_1^+) = \sum_{n=1}^{N} Z_{mn}^{CE1} E_n^{SV1}, \quad Z_{mn}^{CE1} \equiv \frac{1}{\pi(1+\delta_{m,0})} Z_{mn}^{cE1} \tag{8.76}$$

$$U_m^{(1)}(r_1^+) = \sum_{n=1}^{N} Z_{mn}^{CU1} U_{hn}^{SV1}, \quad Z_{mn}^{CU1} \equiv \frac{1}{\pi(1+\delta_{m,0})} Z_{mn}^{cU1} \tag{8.77}$$

If Equations 8.76 and 8.77 are put in matrix form with $\underline{\underline{Z}}^{CE1} = [Z_{mn}^{CE1}]$, $\underline{\underline{Z}}^{CU1} = [Z_{mn}^{CU1}]$, $\underline{E}^{(1)} = [E_m^{(1)}(r_1^+)]$, $\underline{U}^{(1)} = [U_m^{(1)}(r_1^+)]$ with $m = 0, \ldots, N-1$ and $n = 1, \ldots, N$ we have

$$\begin{bmatrix} \underline{E}^{(1)} \\ \underline{U}^{(1)} \end{bmatrix} = \begin{bmatrix} \underline{\underline{Z}}^{CE1} & 0 \\ 0 & \underline{\underline{Z}}^{CU1} \end{bmatrix} \begin{bmatrix} \underline{E}^{SV1} \\ \underline{U}_h^{SV1} \end{bmatrix} \equiv \underline{\underline{Z}}^{(1)} \begin{bmatrix} \underline{E}^{SV1} \\ \underline{U}_h^{SV1} \end{bmatrix} \tag{8.78}$$

Letting $\underline{W}^{(1)} \equiv \begin{bmatrix} \underline{E}^{(1)} \\ \underline{U}^{(1)} \end{bmatrix}$, using $\underline{W}_L = \begin{bmatrix} \underline{E}^{SV1} \\ \underline{U}^{SV1} \end{bmatrix}$ from Equation 8.68, we have altogether $\underline{W}^{(1)} = \underline{\underline{Z}}^{(1)} \underline{W}_L$, $\underline{W}_L = \underline{\underline{K}}_T \underline{W}_0$ from Equation 8.69, and $\underline{W}_0 = \underline{\underline{Z}}^{(3)} \underline{W}^{(3)}$ from Equation 8.75 or after substitution, $\underline{W}^{(1)} = \underline{\underline{Z}}^{(1)} \underline{\underline{K}}_T \underline{\underline{Z}}^{(3)} \underline{W}^{(3)}$. We have

$$\begin{bmatrix} \underline{E}^{(1)} \\ \underline{U}^{(1)} \end{bmatrix} = \begin{bmatrix} \underline{\underline{K}}_{r_1;r_3}^{SVEE} & \underline{\underline{K}}_{r_1;r_3}^{SVEU} \\ \underline{\underline{K}}_{r_1;r_3}^{SVUE} & \underline{\underline{K}}_{r_1;r_3}^{SVUU} \end{bmatrix} \begin{bmatrix} \underline{E}^{(3)} \\ \underline{U}^{(3)} \end{bmatrix} = \underline{\underline{K}}_{r_1;r_3}^{SV} \begin{bmatrix} \underline{E}^{(3)} \\ \underline{U}^{(3)} \end{bmatrix} = \underline{\underline{Z}}^{(1)} \underline{\underline{K}}_T \underline{\underline{Z}}^{(3)} \begin{bmatrix} \underline{E}^{(3)} \\ \underline{U}^{(3)} \end{bmatrix} \tag{8.79}$$

where we have defined $\underline{\underline{K}}_{r_1;r_3}^{SV} \equiv \underline{\underline{Z}}^{(1)} \underline{\underline{K}}_T \underline{\underline{Z}}^{(3)}$. This relation thus uses SV techniques to express, respectively, the electric and magnetic field coefficients $E_m^{(1)}(r_1^+)$ and $U_m^{(1)}(r_1^+)$ of $\cos(m\phi_1)$ in terms of the electric and magnetic field coefficients $E_m^{(3)}(r_3^-)$ and $U_m^{(3)}(r_3^-)$ of $\cos(m\phi_3)$. The matrix $\underline{\underline{K}}_{r_1;r_3}^{SV}$, may be called a transfer matrix, since it relates or transfers the electric and magnetic field Fourier coefficients from the Region 3 inner boundary to the electric and magnetic field Fourier coefficients of the Region 1 exterior interface. It is useful for defining an overall matrix from which all unknowns of the system may be found.

The transfer matrix of Equation 8.79 is also useful, as mentioned earlier, because in the case when uniform materials occupy the regions between the interfaces of adjacent, eccentric cylinders, it may be compared directly to the exact, Bessel function addition theorem analysis that was developed by KPE [22] after an algebraic manipulation of the form of the KPE algorithm is made. This is very useful because by comparing the matrix elements of the SV transfer matrix $\underline{\underline{K}}^{SV}_{r_1;r_3}$ and the matrix elements of the Bessel function transfer matrix, call it $\underline{\underline{K}}^{B}_{r_1;r_3}$, based on the KPE method [22], one can gain insight into how well the SV method is converging with respect to the number of modes used, the number of layers, the Fourier matrix truncation size, etc. Appendix 8.B specifies and gives a derivation of the Bessel function transfer matrix $\underline{\underline{K}}^{B}_{r_1;r_2}$ that results from the KPE algorithm [22] for a single layer between two adjacent interfaces containing a uniform material, and Appendix 8.B also presents the theory of how to cascade together single layer, uniform material, Bessel, transfer matrices to form the overall transfer matrix of a multiple, eccentric layer system. For example, for the two-layer, three-interface composite cylinder shown in Figure 8.2, the cascaded Bessel transfer matrix $\underline{\underline{K}}^{B}_{r_1;r_3}$ would be given by $\underline{\underline{K}}^{B}_{r_1;r_3} = \underline{\underline{K}}^{B}_{r_1;r_2}\underline{\underline{K}}^{B}_{r_2;r_3}$ where $\underline{\underline{K}}^{B}_{r_1;r_2}$ and $\underline{\underline{K}}^{B}_{r_2;r_3}$ are single-layer transfer matrices.

As a numerical example of a comparison, the SV and Bessel transfer matrices, we again consider the scattering object shown in Figure 8.2, and we consider the numerical case when the SV matrix was determined using $L = 4200$ layers and $N = 30$ modes. It was numerically found that for $m = 4$, $m' = 8$ the SV and Bessel transfer EE submatrix elements had, respectively, the values

$$K^{SVEE}_{r_1;r_3}(4,8) = 8.0877\times10^{-3}$$

$$K^{BEE}_{r_1;r_3}(4,8) = 8.0901\times10^{-3}$$

and the SV and Bessel transfer UE submatrix elements had, respectively, the values

$$K^{SVUE}_{r_1;r_3}(4,8) = -j2.4427\times10^{-2}$$

$$K^{BUE}_{r_1;r_3}(4,8) = -j2.4533\times10^{-2}$$

This represents relatively good agreement between the methods.

So far the SV transfer matrix has been derived to express the electric and magnetic field coefficient column matrix evaluated at $\rho_1 = r_1$ in terms of electric and magnetic field coefficient column matrix evaluated at $\rho_3 = r_3$. We would like to mention at this point that the SV transfer matrix, call it $\underline{\underline{K}}^{SV}_{r_u;r_3}$ which expresses the electric and magnetic field coefficient column matrix evaluated at an intermediate layer in the inhomogeneous region, say $\rho_u = r_u$, $u = u_\ell$ in terms of electric and magnetic field coefficient column matrix evaluated at $\rho_3 = r_3$, may be defined by simply cascade multiplying the matrices of Equation 8.69 through all layers between $\rho_3 = r_3$ and the intermediate layer $\rho_u = r_u$, $u = u_\ell$, call this matrix $\underline{\underline{K}}_{u_\ell;u_0}$, and then forming $\underline{\underline{K}}^{SV}_{r_u;r_3} = \underline{\underline{Z}}^{(u)}\underline{\underline{K}}_{u_\ell;u_0}\underline{\underline{Z}}^{(3)}$ where $\underline{\underline{Z}}^{(u)}$ is defined at the intermediate layer $\rho_u = r_u$, $u = u_\ell$ rather than at $\rho_1 = r_1$ as was $\underline{\underline{Z}}^{(1)}$ of Equation 8.78. The transfer matrix $\underline{\underline{K}}^{SV}_{r_u;r_3}$ is very helpful for postprocessing because it may be used to find the cylindrical Fourier, electric and magnetic coefficients $E_m(r_u)$, $U_m(r_u)$ of $\cos(m\phi_u)$, $m = 0, \ldots, N-1$ at any desired internal layer interface $\rho_u = r_u$, $u = u_\ell$, once the coefficients $E^{(3)}_m(r_3^-)$, $U^{(3)}_m(r_3^-)$ are found. All calculations of the EM fields inside the inhomogeneous region of the scattering object were made in this chapter by finding the coefficients $E_m(r_u)$, $U_m(r_u)$ at all intermediate interfaces using the intermediate transfer matrix $\underline{\underline{K}}^{SV}_{r_u;r_3}$ and then summing the appropriate Fourier cosine series.

8.7 OVERALL SYSTEM MATRIX

We now are in a position to define an overall matrix from which all unknowns of the system may be determined. We use the outward SV transfer matrix of Equation 8.79. At $\rho_3 = r_3^-$, the electric and magnetic field Fourier coefficients from Equations 8.19 and 8.20 are given by $m = 0, 1, \ldots, N-1$

$$E_m^{(3)}(r_3^-) = A_m^{(3)} J_m(k_3 r_3^-) + B_m^I H_m^{(2)}(k_3 r_3^-) \tag{8.80}$$

$$U_m^{(3)}(r_3^-) = \left[\frac{k_3}{j\mu_3}\right]\left[A_m^{(3)} J_m'(k_3 r_3^-) + B_m^I H_m^{(2)\prime}(k_3 r_3^-)\right] \tag{8.81}$$

If the $A_m^{(3)}$ coefficients of Equation 8.80 is substituted into Equation 8.81, it turns out that

$$\frac{J_m'(k_3 r_3^-)}{J_m(k_3 r_3^-)} E_m^{(3)}(r_3^-) - \frac{j\mu_3}{k_3} U_m^{(3)}(r_3^-) = \frac{2jB_m^I}{(\pi k_3 r_3^-) J_m(k_3 r_3^-)} \tag{8.82}$$

In deriving Equation 8.82, the Wronskian relation $J_m'(X)H_m^{(2)}(X) - J_m(X)H_m^{(2)\prime}(X) = (2j/\pi)X$ was used. (Use of the Wronskian eliminated the Hankel function and its derivative from Equation 8.82.) At $\rho_1 = r_1^+$, a similar analysis as was performed at $\rho_3 = r_3^-$, shows for $m = 0, 1, \ldots, N-1$

$$\frac{H_m^{(2)\prime}(k_0 r_1^+)}{H_m^{(2)}(k_0 r_1^+)} E_m^{(1)}(r_1^+) - \frac{j\mu_0}{k_0} U_m^{(1)}(r_1^+) = \frac{-2jA_m^I \cos(m\phi_0)}{(\pi k_0 r_1^+) H_m^{(2)}(k_0 r_1^+)} \tag{8.83}$$

In this chapter, it is assumed that there are no internal sources in Region 3, thus $B_m^I = 0$, and that a plane wave of amplitude $E_0^P = 1.0$ (V/m) is incident of the scattering object (see Figure 8.2) and thus $A_m^I = E_0^P(2-\delta_{m,0})j^m$. Equations 8.82 and 8.83, along with the outward transfer matrix of Equation 8.79, form an overall $4N \times 4N$ system matrix system from which the unknowns, namely, $E_m^{(3)}(r_3^-)$, $U_m^{(3)}(r_3^-)$, $E_m^{(1)}(r_1^+)$, $U_m^{(1)}(r_1^+)$, $m = 0, 1, \ldots, N-1$, may be determined. Once these unknowns are found, all other unknowns of the system may be determined.

8.8 ALTERNATE FORMS OF THE BESSEL–FOURIER COEFFICIENT TRANSFER MATRIX

As mentioned earlier, the SV transfer matrix of Equation 8.79 has been derived to express the $\rho_1 = r_1$ (or $u = u_L$) electric and magnetic field column matrices in terms of $\rho_3 = r_3$ (or $u = u_0$) electric and magnetic field column matrices. This may be referred to as an outward transfer matrix as one starts at the inner boundary $\rho_3 = r_3$ and cascade multiplies the layer to layer matrices $\underline{\underline{K_\ell}}$ in Equation 8.69 in an outward direction from $\rho_3 = r_3$ to $\rho_1 = r_1$. We would like to note at this point that the formulation that has been presented in Equations 8.45 through 8.79 can also be used to derive a transfer matrix which does the reverse of what the present transfer matrix $\underline{\underline{K_{r1;r3}^{SV}}}$ does, namely, express the $\rho_3 = r_3$ electric and magnetic field column matrices in terms of $\rho_1 = r_1$ electric and magnetic field column matrices. This may be accomplished by simply using Δu ($\Delta u > 0$) in Equations 8.45 through 8.79 rather than ($-\Delta u$) as was used earlier, and then cascade multiply the resulting equations in an inward direction from $\rho_1 = r_1$, $u = u_L$ to $\rho_3 = r_3$, $u = u_0$ rather than in the outward direction. The resulting transfer matrix $\underline{\underline{K_{r3;r1}^{SV}}}$ may be called an inward transfer matrix as it is formed by cascade multiplying the resulting equations in an inward direction from $\rho_1 = r_1$, $u = u_L$ to $\rho_3 = r_3$, $u = u_0$. If the inward transfer matrix $\underline{\underline{K_{r3;r1}^{SV}}}$ is used to formulate the overall system matrix equations, one then forms a different overall system matrix than is formed by using the outward $\underline{\underline{K_{r1;r3}^{SV}}}$ transfer matrix. In addition to overall system matrix formulations based on pure outward or inward transfer matrices, one may also develop a mixed overall system matrix formulation where one uses a combination of an inward and outward transfer matrices. In this mixed formulation, with the use of the outward and inward transfer matrices, EM boundary conditions are enforced at an interface inside the inhomogeneous scattering object. The ability to develop overall system matrices based on outward, inward, or mixed sets of transfer matrices is very useful as it allows one to implement different overall system matrices, which thus provides a method to at least partially cross-check numerical results by comparing the numerical results of the different overall system matrix solutions. Numerical testing has shown that in all cases tested, that the pure outward transfer matrix produced the most accurate results when compared to the KPE algorithm [22], but that acceptable numerical solutions were also found using the mixed and backward transfer matrix solutions.

8.9 BISTATIC SCATTERING WIDTH

Once the EM fields of the system are known, an important scattering quantity to determine is the bistatic scattering width per unit wavelength (the wavelength is taken in the medium in which the scattering object is located [free space in this chapter]). The scattering width (also called radar cross section per unit length) in units of meters is defined by Ref. [25]

$$\sigma_{2-D}(\phi,\phi_0) = \lim_{\tilde{\rho}_1\to\infty} 2\pi\tilde{\rho}_1 \frac{\left|E_z^s\right|^2}{\left|E_z^{INC}\right|^2} \tag{8.84}$$

thus bistatic scattering width per unit wavelength is defined by

$$\sigma(\phi,\phi_0) \equiv \frac{\sigma_{2-D}(\phi,\phi_0)}{\tilde{\lambda}_0} \tag{8.85}$$

where here we take $\tilde{\lambda}_0 = 1$ (m). In Equation 8.84, E_z^{INC} is the incident plane wave of the system assumed to have an amplitude value $E_0^P = 1.0$ (V/m), ϕ is the scattering angle (measured from the X_1 axis of Figures 8.1 and 8.2), ϕ_0 is the angle of incidence of the plane wave, and E_z^s is the scattered electric field of the system. When E_z^s is substituted in Equation 8.84 it is found, after the symmetry of the present scattering case is taken into account that

$$E_z^s = \sum_{m=0}^{\infty} B_m^{(0)} H_m^{(2)}(\tilde{k}_0\tilde{\rho}_1)\cos(m\phi_0)\cos(m\phi) \tag{8.86}$$

When the Hankel functions $H_m^{(2)}(\tilde{k}_0\tilde{\rho}_1)$ which make up the E_z^s scattered fields are expanded asymptotically as $\tilde{\rho}_1 \to \infty$, it is found

$$\sigma(\phi,\phi_0) = \frac{2}{\pi}\left|F^s(\phi,\phi_0)\right|^2 \tag{8.87}$$

where

$$F^s(\phi,\phi_0) = \sum_{m=0}^{\infty} j^m B_m^{(0)} \cos(m\phi_0)\cos(m\phi) \tag{8.88}$$

Equation 8.87 is the scattering cross section equation used by KPE [22, Eq. 19]. (Keep in mind that KPE [22] studied scattering from composite cylinders which were not in general symmetric so that $F^s(\phi, \phi_0)$ given in Ref. [22] was expressed as an exponential series from $m = -\infty,\ldots,\infty$ rather than the cosine series given in Equation 8.88.)

8.10 VALIDATION OF NUMERICAL RESULTS

An important aspect of the RCWA algorithm that is being developed herein is to properly validate the numerical results of the algorithm. In this chapter, this may be accomplished in the following ways. The first way is to compare numerical results from the RCWA algorithm with the numerical results KPE algorithm as presented in Ref. [22] for cases for which the KPE applies, namely, multiple eccentric cylinders containing uniform materials between the cylinder interfaces. To accomplish this, the author of Ref. [26] has programmed the KPE algorithm as presented in Ref. [22], has programmed the exact matrix equations presented in Ref. [22], and except for a minor computational change (which was to invert layer to layer matrices of Ref. [22] to obtain a reduced system matrix as opposed to solving a large system as was done in Ref. [22]), the exact KPE algorithm was used to compute comparison validation results. A second way to validate is the RCWA method to use the Bessel, cascaded, transfer matrix, $\underline{\underline{K}}_{r_1;r_3}^{B}$ (which is based on the KPE algorithm) which was derived in Appendix 8.B to formulate the overall matrix of the system rather than use the outward SV transfer matrix. Numerical calculations for several different examples showed that both the original KPE algorithm as presented in Ref. [22] and the KPE algorithm based on the Bessel transfer matrix $\underline{\underline{K}}_{r_1;r_3}^{B}$, gave virtually identical results. A third way of validating results when the KPE or Bessel transfer matrix formulation did not apply (i.e., spatially nonuniform materials between the cylinder interfaces), was to compare matrix results using the outward SV transfer matrix formulation with matrix results using a mixed, outward–inward transfer matrix formulation. In the next section, many validation results are presented. We mention that for validation cases involving the KPE algorithm to be presented, that only the KPE algorithm as originally presented in Ref. [22] (referred as the first way) was used, as this seemed to be the most independent way to compare numerical results of the two algorithms. The author of Ref. [26] has programmed the KPE algorithm as presented in Ref. [22] and has used this algorithm to validate all examples which may be analyzed using the KPE algorithm.

8.11 NUMERICAL RESULTS—EXAMPLES OF SCATTERING FROM HOMOGENEOUS AND INHOMOGENEOUS MATERIAL OBJECTS

This section presents several examples to numerically validate the RCWA and also presents several examples where the scattering object of the system contains an inhomogeneous dielectric material. In the following examples, the permeability is assumed to be that of free space everywhere.

As a first example, we refer to KPE [22, Fig. 6, middle curve] (*TM* case, $E_z \neq 0$) which displays the backscatter width $\sigma_b \equiv \sigma(\phi_0, \phi_0)$ as a function of the angle incidence ϕ_0. For this example, the scattering object consists of a one-layer eccentric circular cylinder where the inner radius is $\tilde{r}_2 = 0.3183\lambda_f$, the outer radius $\tilde{r}_1 = 2\tilde{r}_2$, and the inner cylinder center is displaced a distance $\tilde{e}_{12} = 0.1\lambda_f$ to the right of the center of the outer cylinder. The relative permittivity of the inner cylinder is $\varepsilon_2 = 4.0$, of the layer between the inner and outer cylinders is $\varepsilon_1 = 2.0$, and of the region exterior to the outer cylinder is $\varepsilon_0 = 1.0$. Two data points in Ref. [22, Fig. 6] that can be compared directly to the present symmetry case under analysis in this chapter, correspond to the [22, Fig. 6] data points when the angle of incidence is 0° and 180°. When the angle of incidence in KPE [22, Fig. 6] was 0°, the KPE algorithm written by the author of Ref. [26], calculated numerically a backscatter width value of $\sigma_b^{KPE} = 0.21628129000863$, whereas the RCWA method for the same orientation of the plane wave to the scattering object as was used in Ref. [22], gave a value of $\sigma_b^{RCWA} = 0.21628129000806$. Direct visual inspection of the KPE graph [22, Fig. 6] showed a value approximately of $\sigma_b^{KPE} \cong 0.22$ for this case. When the angle of incidence in KPE [22, Fig. 6] was 180°, the KPE algorithm written by the author [26] calculated numerically a backscatter width value of $\sigma_b^{KPE} = 2.625669624816382$, whereas the RCWA method for the same orientation of the plane wave to the scattering object gave a value of $\sigma_b^{RCWA} = 2.625669624816197$. Direct visual inspection of the KPE graph [22, Fig. 6] showed a backscatter width value approximately of $\sigma_b^{KPE} \cong 2.63$. Extremely close agreement for this example is seen between the KPE and RCWA methods.

The second example to be presented corresponds to the two-layer, three-interface, eccentric cylinder system shown in Figure 8.2. Recalling the original definitions $\rho_j \equiv \tilde{k}_f\tilde{\rho}_j$, $r_j \equiv \tilde{k}_f\tilde{r}_j$, $R_j \equiv (r_j/2\pi)$, $j = 1,2,3$ where $\tilde{k}_f \equiv 2\pi/\tilde{\lambda}_f$, $\tilde{\rho}_j, \tilde{r}_j$ are in units of meters, Figure 8.1 shows dimensions and coordinates of the interior region ($u > u_0$, $\rho_3 < r_3$), the inhomogeneous region ($u_L \le u \le u_0$), and the exterior region ($u < u_l$, $\rho_1 > r_1$), whereas Figure 8.2 shows the geometry of relative material inhomogeneity $\varepsilon(u, v)$. In this example, the inner cylinder is defined by a radius of $\tilde{r}_3 = 1.0\tilde{\lambda}_f$ and its center is displaced a distance $\tilde{e}_{13} = 0.31415926\tilde{\lambda}_f$ to the left of the outer cylinder center, the outer cylinder is defined by a radius of $\tilde{r}_1 = 2.0\tilde{\lambda}_f$, and the cylinder, circumscribed between the inner and outer cylinders as shown in Figure 8.2, is an intermediate cylinder of radius $\tilde{r}_2 = 1.34292\tilde{\lambda}_f$ which is offset a distance $\tilde{e}_{12} = 0.6577079\tilde{\lambda}_f$ to the left of the outer cylinder center. The bipolar parameter had a value of $a = \tilde{k}_f\tilde{a} = 28.32451318$ for the system shown in Figures 8.1 and 8.2. As can be seen from Figure 8.2, the intermediate cylinder contains the inner cylinder, and the intermediate cylinder is contained by the outer cylinder. For this example, the inhomogeneity profile, which is studied (see Figure 8.2), is given by

$$\varepsilon(u,v) = \begin{cases} \varepsilon_1 + \Delta\varepsilon\cos(\Lambda u)\cos(\alpha v), & \text{when } (u,v) \text{ is in Region 1} \\ \varepsilon_2, & \text{when } (u,v) \text{ is in Region 2} \end{cases} \tag{8.89}$$

where Region 1 is the region bounded by $\rho_1 = r_1$ on the outside and $\rho_2 = r_2$ on the inside, and Region 2 is the region bounded by $\rho_2 = r_2$ on the outside and $\rho_3 = r_3$ on the inside. Figure 8.3a through c shows, respectively, plots of $\varepsilon(X_1, Y_1)$ (found from function $\varepsilon(u, v)$ of Equation 8.89 when the relative bulk permittivities are $\varepsilon_0 = 1.0$, $\varepsilon_1 = 2.0$, $\varepsilon_2 = 3.0$, and $\varepsilon_3 = 4.0$ and when $\Delta\varepsilon = 0$ (Figure 8.3a); when $\Delta\varepsilon = 0.4$, $\alpha = 5.5$, $\Lambda = 0$ (Figure 8.3b); and when $\Delta\varepsilon = 0.4$, $\alpha = 5.5$, $\Lambda = 4\pi$ (Figure 8.3c). We note that considerable algebra involving the bipolar coordinates of Equation 8.1 is needed (details not given) to mathematically implement the $\varepsilon(u, v)$ function given in Equation 8.89 and shown in Figure 8.3a through c. The $\varepsilon(u, v)$ profile of Equation 8.89 on any bipolar circle of constant u, $-\pi \le v \le \pi$, may be called a combined uniform step-cosine profile, since part of the profile is constant for those values of v which are in Region 2 and a cosine profile for those values of v which are in Region 1.

We now present the electromagnetic fields that result for the second example which is just described for the case when a plane wave is incident (assuming incidence angles either $\phi_0 = 0°$ or $180°$) on the system (see Figure 8.2) and when the parameter $\Delta\varepsilon$ in Equation 8.89 is taken to have a value $\Delta\varepsilon = 0$ (see Figure 8.3a). We note for this case that the material in Regions 1 and 2 are uniform and thus this case represents a case when the KPE [22] method may be used to directly validate the RCWA algorithm. Figure 8.4a and b shows, respectively, as a function of the coordinates X_1, Y_1 for $\phi_0 = 0°$, the real part of the electric field E_z, namely, $E_{zR} = \mathrm{Re}(E_z)$, when the KPE method [22] is used to calculate E_{zR} (Figure 8.4a) and when the RCWA method was used to calculate E_{zR} (Figure 8.4b). In viewing these plots, visually no difference can be seen between the two plots. The KPE plot of

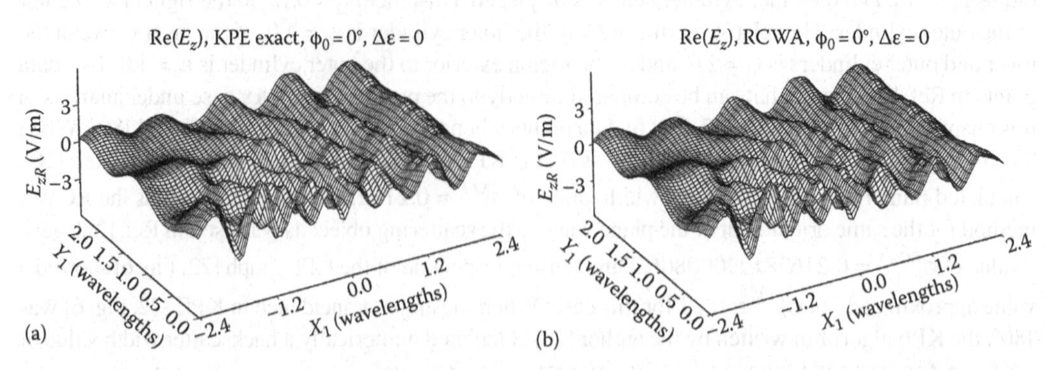

FIGURE 8.4 For the scattering model described in Figures 8.1, 8.2, and 8.3a, for $\phi_0 = 0°$ the real part of the electric field E_z, E_{zR}, that results when the KPE method [22] is used to calculate E_{zR} (a) and when the RCWA method was used to calculate E_{zR} (b). (Reproduced from Jarem, J.M., *Prog. Electromagn. Res.*, PIER 43, 181, 2003. With permission.)

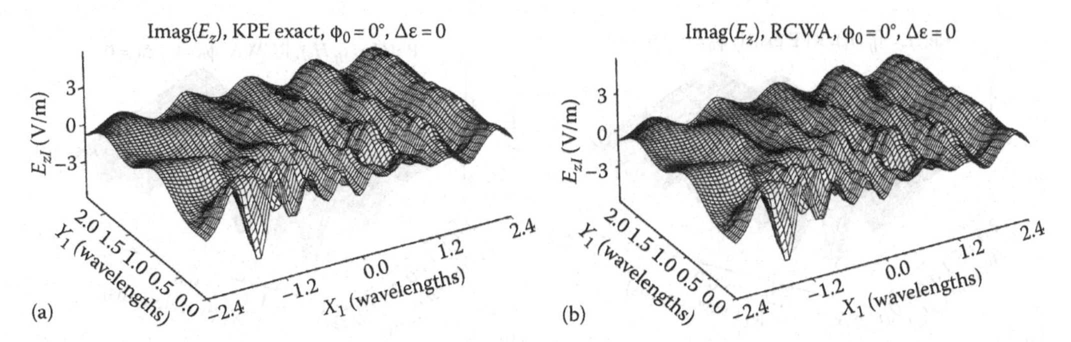

FIGURE 8.5 For the same case described in Figure 8.4 the imaginary part of the electric field E_z, E_{zI}, that results when the KPE method [22] is used to calculate E_{zI} (a) and when the RCWA method was used to calculate E_{zI} (b). (Reproduced from Jarem, J.M., *Prog. Electromagn. Res.*, PIER 43, 181, 2003. With permission.)

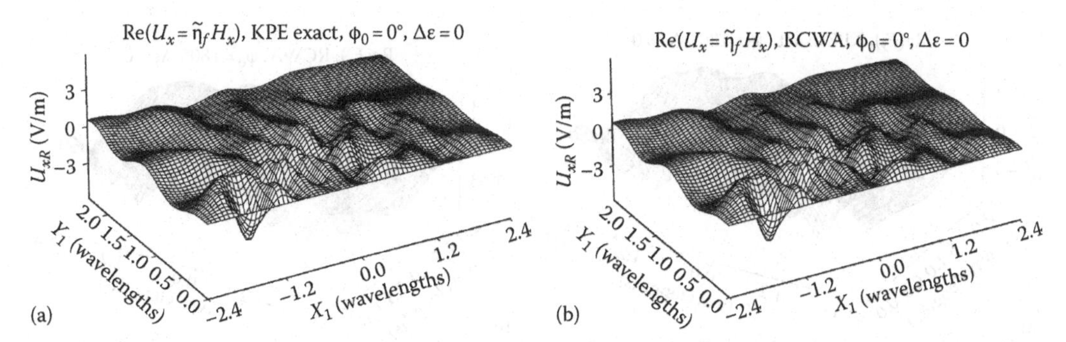

FIGURE 8.6 For the same case described in Figure 8.4 the real part of the magnetic field $U_x = \tilde{\eta}_f H_x$, U_{xR}, that results when the KPE method [22] is used to calculate U_{xR} (a) and when the RCWA method was used to calculate U_{xR} (b). (Reproduced from Jarem, J.M., *Prog. Electromagn. Res.*, PIER 43, 181, 2003. With permission.)

Figure 8.4a was computed by using $N = 40$ ($m = 0,\ldots, 39$) Hankel–Bessel functions modes and was labeled "Exact" because it was determined through numerical testing that the KPE algorithm had completely converged for this number of modes. Figure 8.4b was computed using $N = 30$ SV modes and using $L = 4200$ layers. Figure 8.5a and b shows the imaginary part of the electric field E_z, namely, $E_{zI} = \text{Imag}(E_z)$, for the same parameters as were shown in Figure 8.4a and b. Again in viewing these plots, visually no difference can be seen between the two plots. Figures 8.6 and 8.7 show the real part of the magnetic fields $U_{xR} = \text{Re}(\tilde{\eta}_f H_x)$ and $U_{yR} = \text{Re}(\tilde{\eta}_f H_y)$, respectively, that are associated with the electric field plots of Figures 8.4 and 8.5. Again in viewing these magnetic field plots, visually no difference can be seen between the KPE and RCWA U_{xR} plots and U_{yR} plots. Rectangular components of the magnetic field have been plotted because the geometry of the scattering systems is a mixture of cylindrical, bipolar, and rectangular coordinate systems, thus rectangular components and coordinates are the most convenient ones to use to represent the fields in all regions. In viewing these magnetic field plots of Figures 8.6a and b and 8.7a and b, one notices that for both the U_{xR} and U_{yR} plots the magnetic field component is continuous through the plots. This is to be expected since a uniform value of the magnetic permeability was assumed in all regions of space. One also notices in Figure 8.6a and b that the U_{xR} field is zero on the line $Y_1 = 0$. This is to be expected because the scattering system is symmetric with respect to the Y_1 axis as mentioned earlier.

Figure 8.8a and b shows the real part of the electric field E_{zR} for the same parameters as were used to make the plots of Figures 8.4 through 8.7 except that the angle of incidence was taken to be $\phi_0 = 180°$ rather than $\phi_0 = 0°$. In viewing these electric field plots, visually no difference can be seen between the KPE (Figure 8.8a) and RCWA (Figure 8.8b) plots. It is interesting to note in comparing the E_{zR} field patterns of Figure 8.4 ($\phi_0 = 0°$) with those of Figure 8.8 ($\phi_0 = 180°$), that totally different

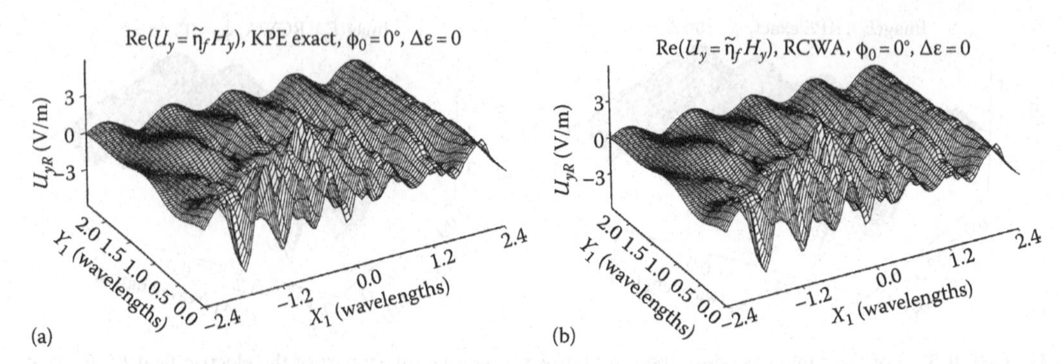

FIGURE 8.7 For the same case described in Figure 8.4 the real part of the magnetic field $U_y = \tilde{\eta}_f H_y$, U_{yR}, that results when the KPE method [22] is used to calculate U_{yR} (a) and when the RCWA method was used to calculate U_{yR} (b). (Reproduced from Jarem, J.M., *Prog. Electromagn. Res.*, PIER 43, 181, 2003. With permission.)

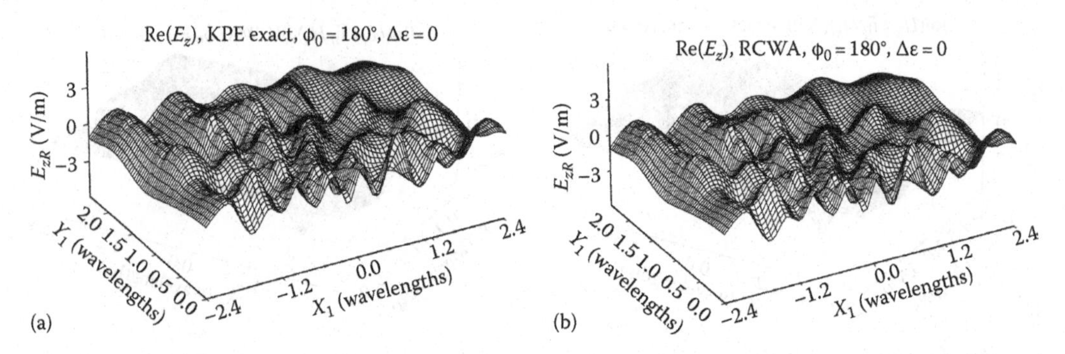

FIGURE 8.8 For the scattering example described in Figures 8.1 through 8.3a for $\phi_0 = 180°$, the real part of the electric field E_z, E_{zR}, that results when the KPE method [22] is used to calculate E_{zR} (a) and when the RCWA method was used to calculate E_{zR} (b). (Reproduced from Jarem, J.M., *Prog. Electromagn. Res.*, PIER 43, 181, 2003. With permission.)

electric field patterns arise for the two different angles of incidence. In the Figure 8.4 ($\phi_0 = 0°$) case, a more fully developed wave structure is seen, whereas in Figure 8.8 ($\phi_0 = 180°$), a much more ripple-like pattern occurs than did in Figure 8.4 plot, particularly inside the scattering structure.

Figure 8.9a and b for $\phi_0 = 0°$ shows, respectively, the real ($E_{zR} = \mathrm{Re}(E_z)$) and imaginary ($E_{zI} = \mathrm{Imag}(E_z)$) parts of the electric field E_z for the same parameter case as were shown in Figure 8.4 except that $\Delta\varepsilon = 0.4$, $\alpha = 5.5$, and $\Lambda = 0$ rather than $\Delta\varepsilon = 0$. Figure 8.3b shows a plot of $\varepsilon(X_1, Y_1)$ (from $\varepsilon(u, v)$

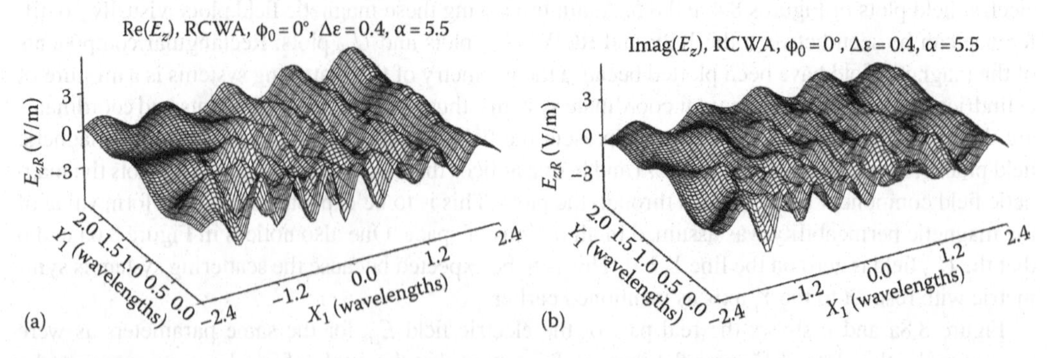

FIGURE 8.9 For $\phi_0 = 0°$, $\Delta\varepsilon = 0.4$, $\alpha = 5.5$, $\Lambda = 0$ ($\varepsilon(X_1, Y_1)$ shown in Figure 8.3b), the real (a) and imaginary (b) parts of the electric field E_z that result when using the RCWA method. Note that because $\Delta\varepsilon \neq 0$ in this case, Region 1 is nonuniform, and thus only the RCWA may be used to determine the EM fields of the system. (Reproduced from Jarem, J.M., *Prog. Electromagn. Res.*, PIER 43, 181, 2003. With permission.)

of Equation 8.89 for this example. In this example, because $\Delta\varepsilon$ is not zero in Equation 8.89, Region 1 is not a uniform material and thus the KPE algorithm cannot be used to calculate the EM fields of the system. The EM fields for the plots of Figure 8.9a and b were made using 30 modes and 4200 layers. A comparison of the Figure 8.9a plot (E_{zR}, $\Delta\varepsilon = 0.4$, $\alpha = 5.5$, $\Lambda = 0$) and Figure 8.4b (E_{zR}, $\Delta\varepsilon = 0$) shows that Figure 8.9a has a very similar field pattern to that shown in Figure 8.4b, but on careful inspection of the two plots, one still notices perceptible difference between the two plots. For example, near the origin of the two plots, one notices that in Figure 8.9a, a higher peak to peak interference pattern of the E_{zR} field occurs than did in the Figure 8.4b plot. A comparison of the Figure 8.9b plot (E_{zI}, $\Delta\varepsilon = 0.4$, $\alpha = 5.5$, $\Lambda = 0$) and Figure 8.5b (plot of E_{zI}, $\Delta\varepsilon = 0$), shows that Figure 8.9b has a very similar field pattern to that shown in Figure 8.5b, but still one may observe perceptible differences between the plots. Overall, the presence of the nonuniform dielectric material in Region 1 of the system causes a definite change in the EM field pattern as to when the material is uniform.

Figure 8.10a shows a comparison of the plots of the db bistatic scattering width $\sigma_{db}(\phi, \phi_0)$ given by $\sigma_{db}(\phi, \phi_0) \equiv 10 \log(\sigma(\phi, \phi_0))$ as a function of ϕ, the scattering angle, for the model described in Figure 8.2 for the case $\phi_0 = 0°$ when $\Delta\varepsilon = 0$ ($\varepsilon(X_1, Y_1)$ shown in Figure 8.3a) as calculated by the KPE method using 40 modes (*dot* in Figure 8.10a) and as calculated by the RCWA method using 30 modes and 4200 layers (*solid line* in Figure 8.10a). As can be seen from the Figure 8.10a plots, extremely close numerical results occur by using the two methods. The close agreement of KPE and RCWA method provides further validation of the RCWA method when the material is uniform. Figure 8.10b presents a plot of $\sigma_{db}(\phi, \phi_0)$ as a function of ϕ for the model described in Figure 8.2 for the case when $\Delta\varepsilon = 0.4$, $\alpha = 5.5$, $\Lambda = 0$ ($\varepsilon(X_1, Y_1)$ shown in Figure 8.3b) for $\phi_0 = 0°$ as calculated by the RCWA method using 30 modes and 4200 layers (*solid line* in Figure 8.10b). Also shown in Figure 8.10b for comparison is the KPE Exact (40 mode) solution (*dashed line*, $\Delta\varepsilon = 0$). As discussed earlier, the $\Delta\varepsilon = 0.4$, $\alpha = 5.5$, $\Lambda = 0$ case is one which cannot be analyzed by the KPE method. In comparing the plots (*solid line* and *dashed line*) of Figure 8.10b, one clearly sees significant differences between the profiles. In Figure 8.10b $\Delta\varepsilon = 0.4$, $\alpha = 5.5$, $\Lambda = 0$ (*solid line*), it almost appears that a modulation feature has arisen in the bistatic profile in comparison with the Figure 8.10b profile ($\Delta\varepsilon = 0$). This would not be an unreasonable result considering the periodic nature ($\alpha = 5.5$) of the $\varepsilon(u, v)$ inhomogeneity profile (see Figure 8.3b) which was used to calculate the plot (*solid line*) of Figure 8.10b. Figure 8.11a and b shows a comparison of the plots of the db bistatic scattering width $\sigma_{db}(\phi, \phi_0)$ as a function of ϕ for the same cases as Figure 8.10a and b except that $\phi_0 = 180°$ rather

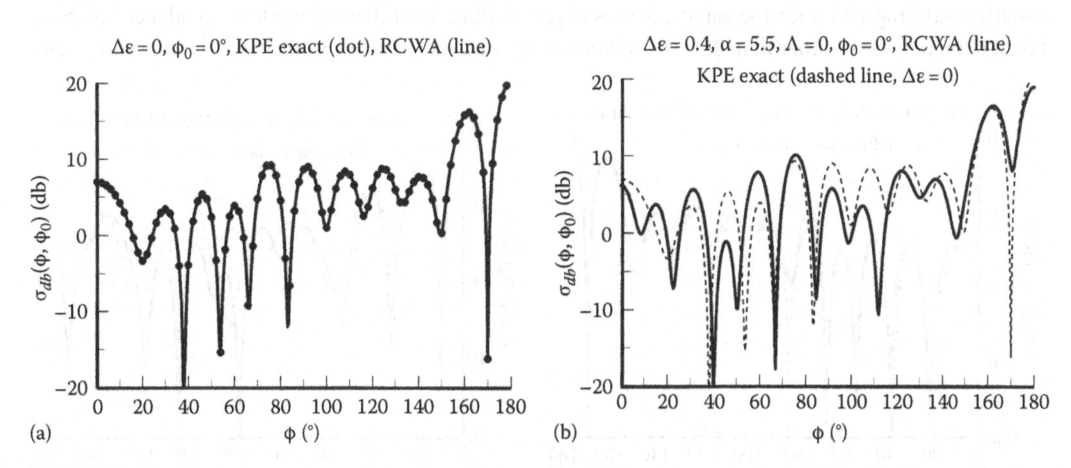

FIGURE 8.10 (a) A comparison of the plots of the $\sigma_{db}(\phi, \phi_0)$ bistatic scattering widths (in db) as a function of ϕ for the scattering model described in Figures 8.1 through 8.3a for the case when $\Delta\varepsilon = 0$ (Figure 8.3a) and when $\phi_0 = 0°$, as calculated by the KPE method [22] using 40 modes (*dot* in (a)) and as calculated by the RCWA method using 30 modes and 4200 layers (*solid line* in (a)). (b) $\sigma_{db}(\phi, \phi_0)$ for the same case as (a), except $\Delta\varepsilon = 0.4$, $\alpha = 5.5$, $\Lambda = 0$ ($\varepsilon(X_1, Y_1)$ shown in Figure 8.3b). Also shown in (b) for comparison is the KPE Exact (40 mode) solution (*dashed line*, $\Delta\varepsilon = 0$). (Reproduced from Jarem, J.M., *Prog. Electromagn. Res.*, PIER 43, 181, 2003. With permission.)

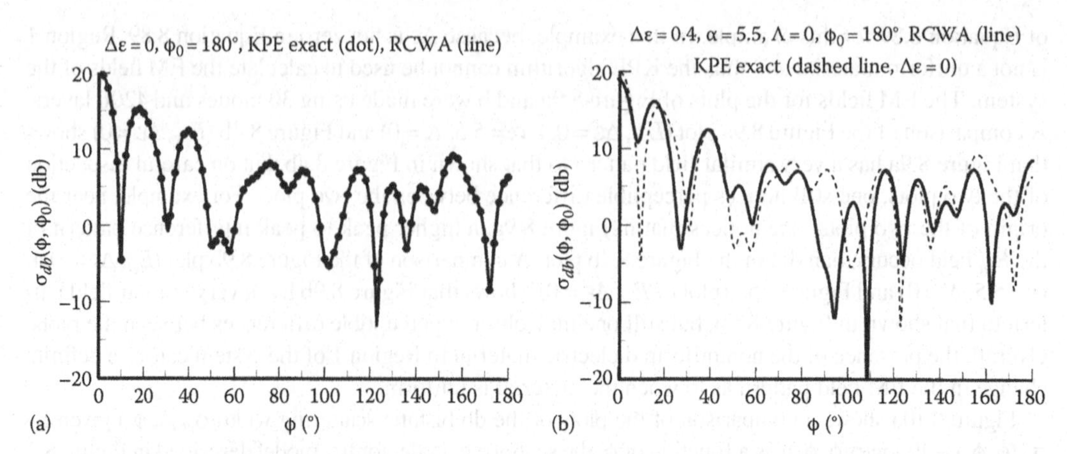

FIGURE 8.11 $\sigma_{db}(\phi, \phi_0)$ for the same cases as displayed in Figure 8.10 except $\phi_0 = 180°$ rather than $\phi_0 = 0°$, (a) $\Delta\varepsilon = 0$ and (b) $\Delta\varepsilon = 0.4$. (Reproduced from Jarem, J.M., *Prog. Electromagn. Res.*, PIER 43, 181, 2003. With permission.)

than $\phi_0 = 0°$. In Figure 8.11a, extremely good validation of the RCWA is seen. In the Figure 8.11b plot, the presence of the nonuniform material in Region 1 in the scattering model causes a significantly different bistatic scattering width profile to occur (*solid line* in Figure 8.11b) as when the material is uniform (*dashed line*, $\Delta\varepsilon = 0$, Figure 8.11b). Overall, for angles of incidences $\phi_0 = 0°$ and 180°, the presence of the nonuniform material in Region 1 (see Figures 8.2 and 8.3b) produces a significantly different bistatic scattering width profile (*solid line*, Figures 8.10b and 8.11b) than when the Region 1 material is uniform (*dashed line*, Figures 8.10b and 8.11b).

Figure 8.12a shows a comparison of the plots of the db bistatic scattering width $\sigma_{db}(\phi, \phi_0)$ as a function of ϕ for the model of the second example described earlier (see Figure 8.2) for the case when $\Delta\varepsilon = 0.4$, $\alpha = 5.5$, $\Lambda = 4\pi$ ($\varepsilon(X_1, Y_1)$ shown in Figure 8.3c), $\phi_0 = 0°$ as calculated by the RCWA method using 30 modes and 4200 layers (*solid line* in Figure 8.12a). Also shown for comparison is the KPE exact solution (*dashed line* in Figure 8.12a). As can be seen from the Figure 8.12a plots, the presence of the material inhomogeneity in Region 1 (see Figure 8.2) causes a perceptible difference in the bistatic scattering width as compared to the case when Region 1 is uniform. Figure 8.12b shows a plot of the bistatic scattering width for the same cases as Figure 8.12a except that the angle of incidence has been taken to be $\phi_0 = 180°$ rather than $\phi_0 = 0°$. Again one notices, the presence of the material inhomogeneity

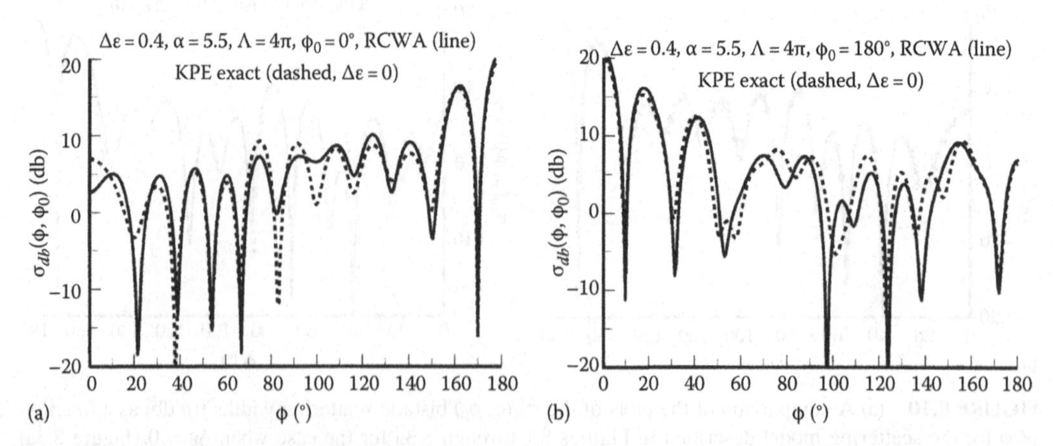

FIGURE 8.12 (a) A comparison of the plots of $\sigma_{db}(\phi, \phi_0)$ given as a function of ϕ for the scattering model described in Figures 8.1, 8.2, and 8.3c for the case when $\Delta\varepsilon = 0.4$, $\alpha = 5.5$, $\Lambda = 4\pi$ (corresponding to Figure 8.3c) when $\phi_0 = 0°$ as calculated by the RCWA method (*solid line* in (a)). Figure (b) shows $\sigma_{db}(\phi, \phi_0)$ for the same case as (a) except $\phi_0 = 180°$. (Reproduced from Jarem, J.M., *Prog. Electromagn. Res.*, PIER 43, 181, 2003. With permission.)

in Region 1 (see Figure 8.2) causes a perceptible difference in the bistatic scattering width as compared to the case when Region 1 is uniform. It is interesting to note that in comparing Figures 8.10b and 8.12a (both $\phi_0 = 0°$), that the inhomogeneity used in Figure 8.10b ($\Delta\varepsilon = 0.4$, $\alpha = 5.5$, $\Lambda = 0$, Figure 8.3b) caused a larger difference with the uniform region cases than did the inhomogeneity ($\Delta\varepsilon = 0.4$, $\alpha = 5.5$, $\Lambda = 4\pi$, Figure 8.3c) used in Figure 8.12a. This is not surprising because although the inhomogeneity used in Figure 8.12a had a more rapid spatial variation overall than the one used in Figure 8.10a, its radial u variation nevertheless tended to average out the EM field, thus producing less marked difference with the uniform region case than did the inhomogeneity used in Figure 8.12a. Overall in Figures 8.10b, 8.11b, 8.12a and b, the differences between the uniform material case *dashed line* and nonuniform material *solid line* are very believable and are of the order one would expect for the inhomogeneity profiles shown in Figure 8.3b and c.

8.12 ERROR AND CONVERGENCE ANALYSIS

We would like to mention that plots similar to those of Figures 8.4 through 8.8 have been made for cases when a lower number of modes and a lower number of layers were used in the RCWA algorithm. It was found in these cases, that visually, just as is seen in Figures 8.4 through 8.8, that virtually no difference between KPE and RCWA methods could be observed. For this reason, in order to study the convergence of the RCWA method in more detail, two error analyses were made, one which studied the peak difference in the electric and magnetic field values between the KPE and RCWA methods over a given X_1, Y_1 region call it R_1, and the second which studied the root mean square (RMS) differences between the methods for the same region R_1 and same cases. The peak difference was studied by defining the relative peak error difference measure

$$E_{rel}^{peak} = 100 \times \frac{\underset{R_1}{Max}\left|F - F_{KPE*}^{exact}\right|}{\underset{R_1}{Max}\left|F_{KPE*}^{exact}\right|} (\%) \tag{8.90}$$

where F is any of complex field values E_z, U_x, U_x which have been computed by the RCWA method or by the KPE method (when the KPE method used a lower number of modes than was used for F_{KPE*}^{exact}), F_{KPE*}^{exact} is any of the complex field values E_z, U_x, U_x which have been computed by the KPE Exact method using 40 modes (called KPE* in Equation 8.90). The symbol "$|\;|$" in Equation 8.90 represents the magnitude of a complex number, and the Max and R_1 in Equation 8.90 means to find the maximum value over the region R_1. The RMS relative difference error measure, E_{rel}^{RMS}, is given by Equation 8.90 when the numerator of the second factor in Equation 8.90 is replaced by the standard RMS formula

$$\left\{\frac{1}{N_s}\sum_{p=1}^{N_s}\left|F_p - F_{p,KPE*}^{p,exact}\right|^2\right\}^{1/2}$$

where the p represents a sample of the complex fields F and F_{KPE*}^{excat} at a point P in region R_1 and N_s represents the number of samples made over the region R_1.

In this chapter, the region R_1, for both the peak and RMS error measures, was taken to be the X_1, Y_1 plane shown in Figures 8.4 through 8.8. For both the peak and RMS error measures, a very fine rectangular, sample grid consisting of $N_s = 240 \times 120$, nearly uniformly spaced, X_1, Y_1 points was used to compute the numerical errors. We say nearly uniformly spaced because when sampling the EM fields inside the inhomogeneous region on an exactly uniform rectangular grid at a point X_{R1}, Y_{R1}, the field point X_1, Y_1 that was actually used was the point X_1, Y_1 which was on a $\rho_u = r_u$, $u = u_\ell \, \ell = 1, \ldots, L-1$ internal bipolar circle which was closest in distance to the X_{R1}, Y_{R1} point. This point was used because, as discussed earlier, the cylindrical Fourier, electric and magnetic coefficients $E_m(r_u)$, $U_m(r_u)$ of $\cos(m\phi_u)$ were computed exactly on the internal layer interfaces $\rho_u = r_u$, $u = u_\ell \, \ell = 1, \ldots, L-1$ by matrix multiplying the intermediate transfer matrix $\underline{\underline{K_{r_u;r_3}^{SV}}}$ times the column matrix $\underline{W^{(3)}}$ to form the Fourier coefficients $E_m(r_3^-), U_m(r_3^-)$. Thus, by choosing this X_1, Y_1 point as described, the most accurate calculation of the internal EM fields possible was made. This was particularly important in the present example because the EM field summations involved a large number of Fourier harmonics and even a very small spatial difference between the point X_{R1}, Y_{R1} of a perfectly uniform spaced rectangular grid and those of a point X_1, Y_1 on the internal layer interfaces $\rho_u = r_u$, $u = u_\ell$ $\ell = 1,\ldots, L-1$ where the RCWA harmonics were computed, could make relatively large difference in the complex values (particularly the complex phase angle) of the EM field solutions which were being evaluated.

Table 8.1 shows for the E_z, H_x, H_y components, the percent (%) Peak Relative Error when compared to the Exact KPE solution [22] (the Exact KPE solution is labeled KPE* in Table 8.1), as results: when the KPE solution is calculated using 22, 26, and 30 modes: when the RCWA method is calculated using; 22 modes, 2800 layers; 26 modes, 2800 layers; and 30 modes, 4200 layers for the cases when the angle of incidence is $\phi_0 = 0°$ and $\phi_0 = 180°$. As can be seen from Table 8.1, for both the KPE and RCWA methods, that the peak relative error with KPE*, E_{rel}^{peak}, decreases rapidly as the number of modes used increases from 22 to 26 to 30 modes for both angles of incidence $\phi_0 = 0°$ and 180°. It is also observed for both $\phi_0 = 0°$ and 180° for the KPE method, that for 30 modes, for all the field components E_z, H_x, H_y, that the peak relative error E_{rel}^{peak}, is extremely small, on the order 0.02%–0.03%. With this extremely small peak error, the present author feels confident that the KPE solution has completely converged when calculated by using 40 modes and that labeling the 40 mode KPE solution as the exact solution is very justified.

It is also interesting to note that for the 22 and 26 mode calculations the KPE and RCWA solutions are showing very similar peak relative error differences (approximately 4%–8% for the 22 mode case and 0.3%–1.5% for the 26 mode case) with the exact KPE solution for both angles of incidence tested and for all field components. It is very reasonable that the number of modes should be an important factor in determining the accuracy of the solutions regardless of whether the KPE or RCWA method is used. It is not surprising that the KPE method is always closer to the exact KPE* solution since for a given number of modes, the Bessel addition formulas used by [22], express exactly the modal coefficients from the inner ($\rho_3 = r_3$) to outer ($\rho_1 = r_1$) layers in terms of Bessel functions, whereas the RCWA method always requires an SV matrix whose matrix elements are constant in each layer, and therefore represent only approximate EM solutions in each thin layer, thus leading to less accurate results than the KPE method can provide.

Table 8.1 in the last column lists the backscatter width, $\sigma_b = \sigma(\phi_0, \phi_0)$, that results for the KPE and RCWA methods for the different computation cases which have been discussed earlier. One notices for the $\phi_0 = 0°$ case that σ_b converges very rapidly to the KPE exact solution (listed in Table 8.1 in parentheses) when using 22, 26, and 30 modes, for both the RCWA and KPE methods. One notices for the $\phi_0 = 180°$ case, however, that KPE method converges rapidly to the exact σ_b value when using 22, 26, and 30 modes, but that significant error in calculating σ_b arises for the 22 mode case when using the RCWA method. More accurate σ_b RCWA results are found for this case when the RCWA method uses 26 and 30 modes. Table 8.2 lists E_{rel}^{RMS}, the % RMS relative error with the exact KPE solution as defined earlier, for same cases that were discussed earlier in Table 8.1. As can be seen from inspecting Table 8.2, all of the trends discussed for Table 8.1 are seen again in Table 8.2. One notices in Table 8.2, the RMS relative error is much smaller than the peak relative errors of Table 8.1. This is not surprising since peak error is a much more severe test of accuracy than an RMS error which tends to average out places where relatively large error might occur.

For the case when the permittivity of the material between the interfaces is nonuniform (i.e., when $\Delta\varepsilon \neq 0$), it is not possible to use the KPE algorithm to validate numerical results. As mentioned earlier, one possible way to at least partially validate the RCWA algorithm for this case is to solve a system matrix equation which is based on a mixed combination of outward and inward transfer matrices, and compare the numerical results of this algorithm with the main RCWA algorithm of the chapter, namely, the algorithm based on a pure outward transfer matrix. Table 8.3 shows the

TABLE 8.1
The % Peak Relative Error with the Exact KPE Solution (Called KPE*), E_{rel}^{peak}, Is Displayed for the E_z, H_x, H_y Components for Several Different Computational Cases

Method ($\Delta\varepsilon = 0$)	# Modes, #Layers	ϕ_0 Angle of Incidence (°)	E_z (% Peak Relative Error with KPE*)	H_x (% Peak Relative Error with KPE*)	H_y (% Peak Relative Error with KPE*)	Backscatter Width $\sigma_b = \sigma(\phi_0, \phi_0)$
KPE* (Exact solution)	40	0	0.0	0.0	0.0	(4.95907)
KPE	22	0	4.4564	5.8828	4.2741	4.95919
RCWA	22, 2800	0	4.4834	5.9464	4.2980	4.95545
KPE	26	0	0.3702	0.7270	0.5074	4.95907
RCWA	26, 2800	0	0.3790	1.2108	0.7035	4.95933
KPE	30	0	0.0128	0.0333	0.0224	4.95907
RCWA	30, 4200	0	0.2256	1.0280	0.4694	4.95901
KPE* (Exact solution)	40	180	0.0	0.0	0.0	(4.38545)
KPE	22	180	4.2371	5.2686	4.6176	4.38527
RCWA	22, 2800	180	6.2794	8.3671	5.1142	4.11560
KPE	26	180	0.3571	0.5722	0.5482	4.38545
RCWA	26, 2800	180	0.4094	1.5942	0.6391	4.35349
KPE	30	180	0.0132	0.0261	0.0244	4.38545
RCWA	30, 4200	180	0.2671	1.2189	0.5668	4.36584

Source: Reproduced from Jarem, J.M., *Prog. Electromagn. Res.*, 43, 181, 2003. With permission.

Note: The Exact KPE solution (40 mode) is labeled KPE*. Also listed is the backscatter width, $\sigma_b = \sigma(\phi_0, \phi_0)$, that results for the KPE and RCWA methods.

TABLE 8.2
The % RMS Relative Error with the Exact KPE Solution (Called KPE*) E_{rel}^{RMS} Is Displayed for the E_z, H_x, H_y Components for the Same Case as Table 8.1

Method ($\Delta\varepsilon = 0$)	# Modes, #Layers	ϕ_0 Angle of Incidence (°)	E_z (% RMS Relative Error with KPE*)	H_x (% RMS Relative Error with KPE*)	H_y (% RMS Relative Error with KPE*)	Backscatter Width $\sigma_b = \sigma(\phi_0, \phi_0)$
KPE* (Exact solution)	40	0	0.0	0.0	0.0	(4.95907)
KPE	22	0	0.4934	0.5726	0.3697	4.95919
RCWA	22, 2800	0	0.5641	0.7126	0.4251	4.95545
KPE	26	0	0.0300	0.0453	0.0291	4.95907
RCWA	26, 2800	0	0.0715	0.1312	0.0738	4.95933
KPE	30	0	0.0008	0.0016	0.0010	4.95907
RCWA	30, 4200	0	0.0426	0.0081	0.0045	4.95901
KPE* (Exact solution)	40	180	0.0	0.0	0.0	(4.38545)
KPE	22	180	0.6189	0.6971	0.4610	4.38527
RCWA	22, 2800	180	1.3355	1.5614	1.0870	4.11560
KPE	26	180	0.0285	0.0357	0.0315	4.38545
RCWA	26, 2800	180	0.1081	0.1520	0.1018	4.35349
KPE	30	180	0.0008	0.0012	0.0011	4.38545
RCWA	30, 4200	180	0.0622	0.0853	0.0569	4.36584

Source: Reproduced from Jarem, J.M., *Prog. Electromagn. Res.*, 43, 181, 2003. With permission.
Note: Also listed is backscatter width, $\sigma_b = \sigma(\phi_0, \phi_0)$, that results for the KPE and RCWA methods.

TABLE 8.3
The % RMS Relative Error E_{rel}^{RMS} between the RCWA* Algorithm (Based on a Pure Outward Transfer Matrix) and the RCWAm Algorithm Solution (Based on the Use of a ¾ Outward and ¼ Inward Transfer Matrix) Is Displayed for the E_z, H_x, H_y Components for Different Computational Cases

Method, $\Delta\varepsilon$, α, Λ	# Modes, #Layers	ϕ_0 Angle of Incidence (°)	E_z (% Peak Relative Error with RCWA*)	H_x (% Peak Relative Error with RCWA*)	H_y (% Peak Relative Error with RCWA*)	Backscatter Width $\sigma_b = \sigma(\phi_0, \phi_0)$
$\Delta\varepsilon = 0$						
KPE* (Exact)	40	0				(4.95907)
RCWA*	22, 2800	0	0.0	0.0	0.0	4.95545
RCWA1	22, 2800	0	1.2193	2.3026	0.7540	4.94900
RCWA*	26, 2800	0	0.0	0.0	0.0	4.95933
RCWA1	26, 2800	0	0.1991	0.2552	0.1785	4.95702
RCWA*	30, 4200	0	0.0	0.0	0.0	4.95901
RCWA1	30, 4200	0	0.1232	0.1541	0.1084	4.95744
KPE* (Exact)	40	180				(4.38545)
RCWA*	22, 2800	180	0.0	0.0	0.0	4.11560
RCWA1	22, 2800	180	3.5397	6.0937	2.3769	4.16989
RCWA*	26, 2800	180	0.0	0.0	0.0	4.39533
RCWA1	26, 2800	180	0.4259	0.7202	0.4219	4.39533
RCWA*	30, 4200	180	0.0	0.0	0.0	4.36584
RCWA1	30, 4200	180	0.2422	0.3034	0.2249	4.39421

(continued)

TABLE 8.3 (continued)
The % RMS Relative Error E_{rel}^{RMS} between the RCWA* Algorithm (Based on a Pure Outward Transfer Matrix) and the RCWAm Algorithm Solution (Based on the Use of a ¾ Outward and ¼ Inward Transfer Matrix) Is Displayed for the E_z, H_x, H_y Components for Different Computational Cases

Method, $\Delta\varepsilon$, α, Λ	# Modes, #Layers	ϕ_0 Angle of Incidence (°)	E_z (% Peak Relative Error with RCWA*)	H_x (% Peak Relative Error with RCWA*)	H_y (% Peak Relative Error with RCWA*)	Backscatter Width $\sigma_b = \sigma(\phi_0, \phi_0)$
$\Delta\varepsilon = 0.4, \alpha = 5.5, \Lambda = 0$						
RCWA*	30, 4200	0	0.0	0.0	0.0	4.09170
RCWA1	30, 4200	0	0.6920	1.1630	0.7738	4.09625
RCWA*	30, 4200	180	0.0	0.0	0.0	7.12139
RCWA1	30, 4200	180	0.7489	1.7432	0.9104	7.16171
$\Delta\varepsilon = 0.4, \alpha = 5.5, \Lambda = 4\pi$						
RCWA*	30, 4200	0	0.0	0.0	0.0	1.82968
RCWA1	30, 4200	0	0.2358	0.3970	0.2542	1.83414
RCWA*	30, 4200	180	0.0	0.0	0.0	4.79632
RCWA1	30, 4200	180	0.2257	0.3625	0.2230	4.82439

Source: Reproduced from Jarem, J.M., *Prog. Electromagn. Res.*, 43, 181, 2003. With permission.

Note: Also listed is backscatter width, $\sigma_b = \sigma(\phi_0, \phi_0)$, that results for the KPE Exact (40 mode), RCWA and RCWAm methods.

error results of such an analysis using a mixed transfer matrix method, entitled RCWAm, which was based on the use of an outward transfer matrix $\underline{\underline{Z}}^{SV}_{r_u;r_3}$ and the use of an inward transfer matrix $\underline{\underline{Z}}^{SV}_{r_1;r_u}$ (r_u is located 0.75 L number (#) of layers outward from r_3 and 0.25 L number (#) of layers inward from r_1) to formulate the overall matrix equation of the overall system. The error measure was obtained by replacing F^{exact}_{KPE*} in Equation 8.90 with $F^{out}_{RCWA}*$ where $F^{out}_{RCWA}*$ was obtained by using any of the complex field values E_z, U_x, U_x computed by the RCWAm method, based on a pure outward transfer method using 30 modes and 4200 layers (called the RCWA* method). All EM fields associated with the RCWAm method were calculated from the electric and magnetic Fourier–Bessel coefficients which corresponded to the $\rho_3 = r_3$ boundary. The region R_1 used for Table 8.3 is the same one used for Tables 8.1 and 8.2. In Table 8.3, the first set of data, corresponding to $\Delta\varepsilon = 0$, shows the relative peak error E^{peak}_{rel} for E_z, H_x, H_y between the RCWA* and RCWAm algorithms for different numbers of modes and layers for $\phi_0 = 0°, 180°$. In this set of data, one clearly sees the peak error between the two RCWA algorithm solutions decreases as the number of modes and layers increases. The rate of error reduction with increase in the number of modes is very similar to that seen in the Table 8.1 relative peak error data when the KPE and RCWA algorithm solutions were compared with the KPE* Exact (40 mode) solution. The second and third sets of data in Table 8.3 show the peak error difference E^{peak}_{rel} for the two inhomogeneous material examples corresponding to Figure 8.3b and c, respectively, for just the case when 30 modes and 4200 layers have been used in both RCWA algorithms for $\phi_0 = 0°, 180°$. In comparing all three data sets for 30 modes and 4200 layers, one notices that relative peak error E^{peak}_{rel} for the second data set ($\Delta\varepsilon = 0.4$, $\alpha = 5.5$, $\Lambda = 0$) is clearly larger ($E^{peak}_{rel} \cong 1\% - 2\%$) than that of the first ($\Delta\varepsilon = 0$) and third ($\Delta\varepsilon = 0.4$, $\alpha = 5.5$, $\Lambda = 4\pi$) data sets where the error is $E^{peak}_{rel} \cong 0.1\% - 0.4\%$. This result seems very much in line with observations made that the difference in the bistatic scattering widths seen in Figures 8.10b and 8.11b (comparison of $\Delta\varepsilon = 0$ [*dashed line*] with $\Delta\varepsilon = 0.4$, $\alpha = 5.5$, $\Lambda = 0$ [*solid line*]) than was the difference seen in Figure 8.12a and b (comparison of $\Delta\varepsilon = 0$ [*dashed line*] with $\Delta\varepsilon = 0.4$, $\alpha = 5.5$, $\Lambda = 4\pi$ [*solid line*]).

8.13 SUMMARY, CONCLUSIONS, AND FUTURE WORK

The chapter has presented a bipolar RCWA algorithm for the calculation of EM fields and EM scattering from an inhomogeneous material, composite eccentric circular cylinder system. The basic RCWA formulation in bipolar coordinates has been presented as well as a complete description of the boundary matching equations that were used. Extensive use of the residue theorem to calculate various bipolar interaction integrals (Appendix 8.A) allowed an extremely accurate and fast numerical implementation of both the RCWA formulation and the boundary matching equations. The RCWA algorithm was extensively validated using an algorithm developed by [22] to study scattering from uniform material eccentric cylinders (termed the KPE algorithm in this chapter). For validation purposes, the chapter presented a slightly altered formulation of the KPE algorithm (Appendix 8.B) which allowed the transfer matrices of the RCWA and KPE algorithms to be compared. Numerical validation results of the RCWA algorithm were presented in Figures 8.4 through 8.12 and in Tables 8.1 through 8.3. Extremely close numerical results between the KPE and RCWA algorithms were observed. In Figures 8.9, 8.10b, 8.11b, and 8.12b numerical examples involving spatially inhomogeneous materials were presented. Very reasonable numerical results were obtained. In this chapter, the RCWA algorithm was applied only to the case where the EM fields and scattering object were symmetric with respect to y coordinate. We would like to mention that the extension of the RCWA algorithm to the case where the EM fields and a scattering object have arbitrary symmetry is straightforward.

There are several areas for which the RCWA algorithm that has been presented could be improved. One area concerns the numerical calculation of the eigenvalues and eigenfunctions of the RCWA algorithm. Numerical testing has shown that in determining the SV eigenmodes that the higher the mode number and the larger the magnitude of the eigenvalue Q_n, that the coefficients of the eigenfunction $S_{vn}(v)$ for that mode have a certain integer index value, call it $i_{n,max}$, for which the magnitude of the coefficient is largest. As one calculates the magnitude of the coefficients of $S_{zin}(v)$ of $S_{vn}(v)$ (Equation 8.17) for integer values above or below this value, the magnitudes of the adjacent coefficients die off more and more rapidly as the magnitude of Q_n increases. Put in spectral domain terms, as the magnitude of the eigenvalue Q_n increases, the coefficients center about $i_{n,max}$ become more and more narrow band. This means that when solving the SV matrix for a given higher order mode, one might be able to perform an eigenanalysis on a much smaller (or greatly reduced) SV matrix which has been truncated around those matrix elements (i.e., associated with $i_{n,max}$) which contribute most to the given higher order mode. This would, of course, save a great deal of computational time, particularly for numerical scattering examples which are much larger than have already been studied. Another area where the RCWA computation time could be greatly reduced is if parallel processing were used to carry out the SV eigenanalysis of the system. The SV eigenanalysis is very amenable to parallel processing since the eigenanalysis in each thin layer may be calculated independently of every other layer. As the eigenanalysis is the most time consuming step in algorithm, parallel process would reduce the processing time approximately by a factor of $1/L$ where L is the number of layers.

In the next chapter, several additional computationally examples are presented using the RCWA algorithm described in this chapter.

PROBLEMS

8.1 Using the fact that

$$\exp(j\phi_u(u,v)) = \cos(\phi_u(u,v)) + j\sin(\phi_u(u,v)) = \frac{(x - x_{cu} + jy)}{r_u} \tag{P8.1.1}$$

show Equation 8.A.5, namely,

$$\exp(j\phi_u(u,v)) = \frac{F}{(\cosh(u) - \cos(v))} \tag{P8.1.2}$$

where

$$F = \cosh(u)\cos(v) - 1 - j\sinh(u)\sin(v) \tag{P8.1.3}$$

8.2 (a) Show that the second part of F in Equation 8.A.5 is given by

$$F = \frac{1}{2}\exp(-jv)\exp(-u)\left[\exp(jv) - \exp(u)\right]^2 \equiv \frac{1}{2}z^{-1}\exp(-u)[z - \exp(u)]^2 \tag{P8.2.1}$$

where $z = \exp(jv)$ and

(b) Show that $\exp(j\phi_u(u,v))$ in Equation 8.A.7 is given by

$$\exp(j\phi_u(u,v)) = -\exp(-u)\left[\frac{\exp(jv)-\exp(u)}{\exp(jv)-\exp(-u)}\right] \tag{P8.2.2}$$

8.3 Show that Equation 8.A.2, namely,

$$\alpha_i^{(-m)}(u) = \frac{1}{2\pi}\int_{-\pi}^{\pi} \exp(-jm\phi_u(u,v) - jiv)dv \tag{P8.3.1}$$

is purely real.

8.4 Equation 8.A.2, namely,

$$\alpha_i^{(-m)}(u) = \frac{1}{2\pi}\int_{-\pi}^{\pi} \exp(-jm\phi_u(u,v) - jiv)dv \tag{P8.4.1}$$

may be calculated for $m \geq 1$, $i \geq 0$ by using Equation 8.A.9, namely,

$$\alpha_i^{(-m)}(u) = (-1)^m \exp(mu)\,\frac{1}{i!}\left\{\frac{d^i}{dz^i}\left[\frac{z-\exp(-u)}{z-\exp(u)}\right]^m\right\}\Bigg|_{z=0} \tag{P8.4.2}$$

and by using $\alpha_i^{(-m)}(u) = 0$ for $i \leq -1$.

(a) Calculate $\alpha_2^{(-1)}(u)$ using Equation P8.4.2 for $u = 1.551778524$. Use Leibniz's theorem for differentiation of a product (see Ref. [24]). Verify your answer to a high degree of accuracy by performing a numerical integration of Equation P8.4.1.

(b) Using Equation P8.4.2, calculate $\alpha_i^{(-m)}(u)$ for $u = 1.551778524$, for $m = 3$, $i = -1, 0, 1, 3, 5, 8$, and for $m = 6$, $i = 8$. Verify your answer to a high degree of accuracy by performing a numerical integration of Equation P8.4.1.

8.5 The Fourier coefficient $\beta_i^{(-m)}$ in the Fourier series of Equation 8.73, namely,

$$\exp\left(\frac{-jm\phi_u(u,v)}{\cosh(u)-\cos(v)}\right) = \sum_{i=-I_t}^{I_t} \beta_i^{(-m)}(u)\exp(jiv) \tag{P8.5.1}$$

is given by

$$\beta_i^{(-m)}(u) = \frac{1}{2\pi}\int_{-\pi}^{\pi}\left[\frac{\exp(-jm\phi_u(u,v))}{\cosh(u)-\cos(v)}\right]\exp(-jiv)dv \tag{P8.5.2}$$

(a) Using complex variable theory, show that this coefficient may be calculated for $m \geq 1$, $i \geq 1$, by using Equation 8.A.12, namely,

$$\beta_i^{(-m)}(u) = (-1)^{m+1}\exp(mu)\frac{2}{(i-1)!}\left\{\frac{d^{i-1}}{dz^{i-1}}\left[\frac{(z-\exp(-u))^{m-1}}{(z-\exp(u))^{m+1}}\right]\right\}\Bigg|_{z=0} \tag{P8.5.3}$$

and show that this coefficient may be calculated for $i \leq 0$ by using $\beta_i^{(-m)}(u) = 0$.

(b) Using Equation P8.5.3, calculate $\beta_i^{(-m)}(u)$ for $u = 1.551778524$, for $m = 3$, $i = -1, 0, 1, 2, 8$, and for $m = 6$, $i = 8$. Verify your answer to a high degree of accuracy by performing a numerical integration of Equation P8.5.2.

8.6 The Fourier coefficient $\beta_i^{(-m)}$ of the Fourier series (Equation 8.73),

$$\frac{\exp(-jm\phi_u(u,v))}{\cosh(u)-\cos(v)} = \sum_{i=-I_t}^{I_t} \beta_i^{(-m)}(u)\exp(jiv) \tag{P8.6.1}$$

for $m = 0$ is given by

$$\beta_i^{(0)}(u) = \frac{1}{2\pi}\int_{-\pi}^{\pi}\left[\frac{1}{\cosh(u)-\cos(v)}\right]\exp(-jiv)dv \tag{P8.6.2}$$

(a) Using complex variable theory, show that this coefficient may be calculated for $m = 0$, $|i| \geq 0$, by using Equation 8.A.13, namely,

$$\beta_i^{(0)}(u) = \frac{1}{\sinh(u)}\exp\left(-|i|u\right) \tag{P8.6.3}$$

(b) Using Equation P8.6.3, calculate $\beta_i^{(0)}(u)$ for $u = 1.551778524$, for $i = -1, 0, 1, 2, 8$. Verify your answer to a high degree of accuracy by performing a numerical integration of Equation P8.6.2.

8.7 (a) Show that

$$\frac{d\phi_u(u,v)}{dv} = -\frac{\sinh(u)}{\cosh(u)-\cos(v)} \tag{P8.7.1}$$

where $\phi_u(u, v)$ is defined in Equation 8.3.

(b) Show that the Fourier coefficient $\varsigma_i^{(0)}$ of the Fourier series (Equation 8.33),

$$\frac{d\phi_u(u,v)}{dv} = \sum_{i=-I_t}^{I_t} \varsigma_i^{(0)}(u)\exp(jiv) \tag{P8.7.2}$$

is given by $\varsigma_i^{(0)} = -\exp\left(-|i|u\right)$ for all i (see also Equation 8.A.19).

8.8 Find the Fourier coefficients of Equation 8.89, namely,

$$\varepsilon_u(u,v) = \begin{cases} \varepsilon_1 + \delta\varepsilon\cos(\alpha v), & 0 \leq |v| \leq v_1(u) \\ \varepsilon_2, & v_1(u) < |v| \leq \pi \end{cases} \tag{P8.8.1}$$

For the Fourier series

$$\varepsilon_u(u,v) = \sum_{i=-I_t}^{I_t} \varepsilon_i(u)\exp(jiv) \tag{P8.8.2}$$

8.9 The eigenvalue matrix for Equation 8.13 has the general form

$$C_{i,i'} = ri^2\delta_{i,i'} + s\varepsilon_{i,i'} = ri^2\delta_{i,i'} + s\tilde{\varepsilon}_{i-i'} \quad (i,i') = (-I_T,\ldots,-1,0,1,\ldots,I_T) \tag{P8.9.1}$$

where $\tilde{\varepsilon}_k = \tilde{\varepsilon}_{-k}$, where r, s are constants, and where

$$\delta_{i,i'} = \begin{cases} 1, & i = i' \\ 0, & i \neq i' \end{cases} \tag{P8.9.2}$$

(a) Let $\underline{\underline{C}} = [C_{i,j}]$ and $\underline{S} = [S_i]$ and $(i, i') = (-1, 0, 1)$. Show that for the case when $S_0 \neq 0$ and $S_{-1} = S_1 \neq 0$ (which may be called an "even" matrix case) that the matrix equation

$$\frac{d\underline{S}}{du} = \underline{\underline{C}}\,\underline{S} \tag{P8.9.3}$$

may be solved by the reduced matrix equation (E for even)

$$\frac{d\underline{S_E}}{du} = \underline{\underline{C_E}}\,\underline{S_E} \tag{P8.9.4}$$

where

$$\underline{\underline{C_E}} = \begin{bmatrix} C_{0,0} & C_{0,-1} + C_{0,1} \\ C_{1,0} & C_{1,-1} + C_{1,1} \end{bmatrix} \quad \text{and} \quad \underline{S_E} = \begin{bmatrix} S_0 \\ S_1 \end{bmatrix} \tag{P8.9.5}$$

(b) Generalize the results of (a) to find the form of the matrix for the "even" case ($S_0 \neq 0$, $S_{-i} = S_i \neq 0$, $i = 1,\ldots,I_T$) when $I_T \geq 1$.

8.10 The eigenvalue matrix for Equation 8.13 has the general form

$$C_{i,i'} = ri^2\delta_{i,i'} + s\varepsilon_{i,i'} = ri^2\delta_{i,i'} + s\tilde{\varepsilon}_{i-i'} \quad (i,i') = (-I_T,\ldots,-1,0,1,\ldots,I_T) \tag{P8.10.1}$$

where $\tilde{\varepsilon}_k = \tilde{\varepsilon}_{-k}$, where r, s are constants, and where

$$\delta_{i,i'} = \begin{cases} 1, & i = i' \\ 0, & i \neq i' \end{cases} \tag{P8.10.2}$$

(a) Let $\underline{\underline{C}} = [C_{i,j}]$, $S = [S_i]$ and $(i, i') = (-1, 0, 1)$. Show that for the case when $S_0 = 0$ and $S_{-1} = -S_1 \neq 0$ (which may be called an "odd" matrix case) that the matrix equation

$$\frac{d\underline{S}}{du} = \underline{\underline{C}}\,\underline{S} \tag{P8.10.3}$$

may be solved by the reduced matrix equation (O for odd)

$$\frac{d\underline{S_O}}{du} = \underline{\underline{C_O}}\,\underline{S_O} \tag{P8.10.4}$$

where

$$\underline{C_O} = [-C_{1,1} + C_{1,-1}] \quad \text{and} \quad \underline{S_O} = [S_1] \tag{P8.10.5}$$

(b) Generalize the results of (a) to find the form of the matrix for the "odd" case ($S_0 = 0$, $S_{-i} = -S_i \neq 0$, $i = 1, \ldots, I_T$) when $I_T \geq 1$.

8.11 Show that the Fourier series coefficient of the function

$$h^2(u,v) = \frac{a^2}{(\cosh(u) - \cos(v))^2} \tag{P8.11.1}$$

is given by Equation 8.A.21 (*sq* means squared)

$$h_i^{sq}(u) = a^2 \exp\left(-|i|u\right)\left[\frac{\cosh(u) + |i|\sinh(u)}{\sinh^3(u)}\right] \tag{P8.11.2}$$

8.12 (a) Using the Bessel function integral formula of Ref. [24, Eq. 9.1.21, p. 360], namely,

$$J_m(z) = \frac{j^{-m}}{\pi}\int_0^{\pi} \exp(jz\cos(\theta))\cos(m\theta)\,d\theta \tag{P8.12.1}$$

express

$$\vec{E}^{Inc} = E_0 \exp(j\vec{k}_I \cdot \vec{r})\hat{a}_z, \quad \vec{k}_I = k_{Ix}\hat{a}_x + k_{Iy}\hat{a}_y = k_I\cos(\phi_0)\hat{a}_x + k_I\sin(\phi_0)\hat{a}_y,$$
$$\vec{r} = x\hat{a}_x + y\hat{a}_y + z\hat{a}_z \tag{P8.12.2}$$

as a Fourier series of the cylindrical coordinate $\phi = \tan^{-1}(y/x)$. The $\vec{E}^{Inc}$ represents the general form of an incident plane wave solution of Maxwell's equations propagating in a uniform region of space having relative permeability μ and relative permittivity ε. The coordinates x, y, z are dimensionless ($x = \tilde{k}_f\tilde{x}$, etc., where $\tilde{k}_f = 2\pi/\tilde{\lambda}_f$, $\tilde{x}$, $\tilde{\lambda}_f$ (m), $\tilde{\lambda}_f$ (free-space wavelength)).

(b) For the special symmetric case when $\phi_0 = 0$, π express Equation P8.12.2 as a $m = 0$, 1, 2, …, ∞ cosine Fourier series.

8.13 Derive the wavelength normalized scatter width (also called line width) defined by

$$\sigma(\phi,\phi_0) \equiv \frac{\sigma_{2-D}}{\tilde{\lambda}_f} \tag{P8.13.1}$$

where [25]

$$\sigma_{2-D} = \lim_{\tilde{\rho}\to\infty} 2\pi\tilde{\rho}\frac{\left|E_z^s\right|^2}{\left|E_z^{Inc}\right|^2} \tag{P8.13.2}$$

where $\tilde{\lambda}_f$ is the free-space wavelength in meters,

$$\vec{E}^s = E_0 \sum_{m=0}^{\infty} B_m^{(0)} H_m^{(2)}(\tilde{k}_I \tilde{\rho}) \cos(m\phi_0) \cos(m\phi) \hat{a}_z \tag{P8.13.3}$$

where, $\vec{\tilde{k}}_I = \tilde{k}_{Ix}\hat{a}_x + \tilde{k}_{Iy}\hat{a}_y = \tilde{k}_I \cos(\phi_0)\hat{a}_x + \tilde{k}_I \sin(\phi_0)\hat{a}_y$, $\vec{\tilde{r}} = \tilde{x}\hat{a}_x + \tilde{y}\hat{a}_y + \tilde{z}\hat{a}_z$, $\phi_0 = 0, \pi$ is the EM field scattered by the incident field $\vec{E}^{Inc}$ (specified in Equation P8.12.2).

APPENDIX 8.A

This appendix presents the calculation of the exponential Fourier coefficients that are used in this chapter. We first determine the Fourier coefficients of the series

$$\exp(-jm\phi_u(u,v)) = \sum_{i=-\infty}^{\infty} \alpha_i^{(-m)}(u) \exp(jiv) \tag{8.A.1}$$

$$\alpha_i^{(-m)}(u) = \frac{1}{2\pi} \int_{-\pi}^{\pi} \exp(-jm\phi_u(u,v) - jiv) dv \tag{8.A.2}$$

where $m \geq 1$, and $u > 0$, and the angle $\phi_u(u, v)$ defined in Equation 8.3 is the circular, cylindrical coordinate defined at the center of the O_u circle whose radius is $r_u = a/\sinh(u)$ (see Figure 8.1) and whose center is $x_{cu} = a/\tanh(u)$ (see Figure 8.1). We first study the function

$$\exp(j\phi_u(u,v)) = \cos(\phi_u(u,v)) + j\sin(\phi_u(u,v)) = \frac{(x - x_{cu} + jy)}{r_u} \tag{8.A.3}$$

where $x = a\sinh(u)/(\cosh(u) - \cos(v))$ and $y = -a\sin(v)/(\cosh(u) - \cos(v))$ are the bipolar coordinates of Equation 8.1. Substituting x, y, and x_{cu} into Equation 8.A.3 one finds after algebra

$$\exp(j\phi_u(u,v)) = \frac{F}{(\cosh(u) - \cos(v))} \tag{8.A.4}$$

$$F = \cosh(u)\cos(v) - 1 - j\sinh(u)\sin(v) = \frac{1}{2}\exp(-jv)\exp(-u)\left[\exp(jv) - \exp(u)\right]^2 \tag{8.A.5}$$

Algebraic manipulation shows

$$\cosh(u) - \cos(v) = -\frac{1}{2}\exp(-jv)\left[\exp(jv) - \exp(-u)\right]\left[\exp(jv) - \exp(u)\right] \tag{8.A.6}$$

After simplification, it is found that

$$\exp(j\phi_u(u,v)) = -\exp(-u)\left[\frac{\exp(jv) - \exp(u)}{\exp(jv) - \exp(-u)}\right] \tag{8.A.7}$$

If the $\exp(j\phi_u(u, v))$ is raised to the $-m$ power and substituted into Equation 8.A.2 and following the complex variable theory of Ref. [23], a substitution $z = \exp(jv)$ is made, and it is found that the resulting Equation 8.A.2 integral is

$$\alpha_i^{(-m)} = \frac{1}{2\pi j}(-1)^m \exp(mu) \oint_{C;|z|=1} \left[\frac{z-\exp(-u)}{z-\exp(u)}\right]^m \frac{dz}{z^{i+1}} \tag{8.A.8}$$

where the substitutions $\exp(-jiv) = z^{-i}$, $dz = j\exp(jv)dv = jzdv$, $dv = dz/jz$ have been made in Equation 8.A.8, and where C; $|z| = 1$ represents a counter clockwise integration in the complex z plane around the line $|z| = 1$. The factor $[(z-\exp(-u))/(z-\exp(u))]^m$ for $m \geq 1$ is analytic in the region of the complex plane $|z| \leq 1$, and the factor z^{i+1} represents a pole of order $i + 1$ for $i > 0$. We thus see using the residue theorem [23] for $m \geq 1$

$$\alpha_i^{(-m)}(u) = (-1)^m \exp(mu) \frac{1}{i!}\left\{\frac{d^i}{dz^i}\left[\frac{z-\exp(-u)}{z-\exp(u)}\right]^m\right\}\Bigg|_{z=0} \tag{8.A.9}$$

for $i \geq 0$ and $\alpha_i^{(-m)} = 0$ for $i \leq -1$.

For the case when $m = 0$,

$$\alpha_i^{(0)}(u) = \begin{cases} 1, & i = 0 \\ 0, & i \neq 0 \end{cases} \tag{8.A.10}$$

The coefficients $\alpha_i^{(m)}$ for $m \geq 1$ may be found by taking the complex conjugate Equation 8.A.1 and comparing the coefficients of $\exp(jiv)$ of the two series. The result is for $m \geq 1$ and for all i is

$$\alpha_i^{(m)} = \alpha_{-i}^{(-m)*} \tag{8.A.11}$$

Since $\alpha_i^{(-m)}(u)$ is purely real, the complex conjugate may be omitted. The higher order derivatives in Equation 8.A.9 may be readily calculated and evaluated by summing the series specified by the Leibniz derivative product rule given in Ref. [24]. The validity of Equation 8.A.9 has been checked for several cases by direct numerical integration of Equation 8.A.2.

An analysis similar to that performed to calculate the $\alpha_i^{(m)}$ coefficients shows that to calculate the $\beta_i^{(-m)}(u)$ coefficients for $u > 0$ for the Fourier series of Equation 8.73, Section 8.3, are given for $m \geq 1$

$$\beta_i^{(-m)}(u) = (-1)^{m+1} \exp(mu) \frac{2}{(i-1)!}\left\{\frac{d^{i-1}}{dz^{i-1}}\left[\frac{(z-\exp(-u))^{m-1}}{(z-\exp(u))^{m+1}}\right]\right\}\Bigg|_{z=0} \tag{8.A.12}$$

for $i \geq 1$ and $\beta_i^{(-m)}(u) = 0$ for $i \leq 0$. For $m = 0$, it is found

$$\beta_i^{(0)}(u) = \frac{1}{\sinh(u)} \exp\left(-|i|u\right) \tag{8.A.13}$$

for all i. The $\beta_i^{(m)}(u)$ coefficients for $m \geq 1$ are given by for all i by

$$\beta_i^{(m)}(u) = \beta_{-i}^{(-m)}(u)^* \tag{8.A.14}$$

Since $\beta_{-i}^{(-m)}$ is purely real, the complex conjugate may be omitted.

The $\zeta_i^{(m)}(u)$ coefficient of Equation 8.33 for $u > 0$ is given by the integral

$$\zeta_i^{(m)}(u) = \frac{1}{2\pi}\int_{-\pi}^{\pi}\left[\frac{d}{dv}\phi_u(u,v)\right]\exp(jm\phi_u(u,v) - jiv)\,dv \tag{8.A.15}$$

For $m \neq 0$, the $\zeta_i^{(m)}(u)$ coefficient may be calculated from the $\alpha_i^{(m)}(u)$ coefficients as follows. We note for $m \neq 0$,

$$\exp(jm\phi_u(u,v))\frac{d}{dv}\phi_u(u,v) = \frac{1}{jm}\frac{d}{dv}\exp(jm\phi_u(u,v)) = \frac{1}{jm}\frac{d}{dv}\left[\sum_{i=-\infty}^{\infty}\alpha_i^{(m)}(u)\exp(jiv)\right] \tag{8.A.16}$$

$$\exp(ji\phi_u(u,v))\frac{d}{dv}\phi_u(u,v) = \sum_{i=-\infty}^{\infty}\frac{i}{m}\alpha_i^{(m)}(u)\exp(jiv) \tag{8.A.17}$$

For all $m \neq 0$, we find

$$\zeta_i^{(m)}(u) = \frac{i}{m}\alpha_i^{(m)}(u) \tag{8.A.18}$$

For $m = 0$, it turns out for all i

$$\zeta_i^{(0)}(u) = -\exp(-|i|u) \tag{8.A.19}$$

The Fourier series for the function $\varepsilon_h(u, v) \equiv h^2(u, v)\varepsilon(u, v)$, Equation 8.7, is given by

$$\varepsilon_h(u,v) = \sum_{i=-\infty}^{\infty}\left[\sum_{i'=-\infty}^{\infty}h_{i-i'}^{sq}(u)\,\varepsilon_{i'}(u)\right]\exp(jiv) \tag{8.A.20}$$

where $\varepsilon_{i'}(u)$ are the Fourier coefficients of the relative dielectric permittivity function $\varepsilon(u,v)$ and $h_i^{sq}(u)$ are the Fourier coefficient of the squared scale factor function $h^2(u,v)$.

Using the residue theorem, it may be shown that

$$h_i^{sq}(u) = a^2\exp(-|i|u)\left[\frac{\cosh(u) + |i|\sinh(u)}{\sinh^3(u)}\right] \tag{8.A.21}$$

for all for all i.

In addition to Equation 8.A.9, the validity of the other Fourier coefficient formulas presented in this appendix have been checked for several cases by direct numerical integration of the integrals which defined them.

APPENDIX 8.B

This appendix gives a derivation of the single-layer, eccentric circle, and Bessel transfer matrix which was described in the main text. The derivation based on a modification of the KPE algorithm presented in Ref. [22]. We derive, as a representative example, the transfer matrix for the layer enclosed by the $\rho_1 = r_1$ and $\rho_2 = r_2$ circles of Figure 8.2 with the material parameters taken to be $\varepsilon_1 = \varepsilon$ and $\mu_1 = \mu$. Let $E_z^{(p)}$ and $U_\phi^{(p)} = \tilde{\eta}_f H_\phi^{(p)}$, $p = 1, 2$ be the electric and magnetic fields at the $\rho_1 = r_1^-$ and $\rho_2 = r_2^+$ boundaries, respectively, we have KPE [22]

$$E_z^{(p)} = \sum_{m=-\infty}^{\infty}\left[A_{em}^{(p)} J_m(X_p) + B_{em}^{(p)} H_m^{(2)}(X_p)\right]\exp(jm\phi_p) \tag{8.B.1}$$

$$U_\phi^{(p)} \equiv \tilde{\eta}_f H_\phi^{(p)} = \frac{k}{j\mu}\sum_{m=-\infty}^{\infty}\left[A_{em}^{(p)} J_m'(X_p) + B_{em}^{(p)} H_m^{(2)\prime}(X_p)\right]\exp(jm\phi_p) \tag{8.B.2}$$

where $X_p = k\rho_p$, $p = 1, 2$, $k = \sqrt{\mu\varepsilon}$,

$$A_{em}^{(1)} = \sum_{m'=-\infty}^{\infty} Z_{m,m'}^{\text{exp}} A_{em'}^{(2)} \tag{8.B.3}$$

$$B_{em}^{(1)} = \sum_{m'=-\infty}^{\infty} Z_{m,m'}^{\text{exp}} B_{em'}^{(2)} \tag{8.B.4}$$

$$Z_{m,m'}^{\text{exp}} = \exp\left[j(m'-m)\,\phi_{12}\right] J_{m-m'}(ke_{12}) \tag{8.B.5}$$

$$(m, m') = -\infty, \ldots, \infty$$

where $e_{12} = \tilde{k}_f \tilde{e}_{12} > 0$ is the magnitude of the separation distance of the centers of the O_1 and O_2 circles, where $\phi_{12} = 0$ when the O_2 circle center is to the right of the O_1 circle center, and $\phi_{12} = \pi$ when the O_2 circle center is to the left of the O_1 circle center. The subscript "e" in Equations 8.B.1 through 8.B.4 represents "exponential." If we take advantage of the symmetry of the present problem and perform algebra we find, $p = 1, 2$

$$E_z^{(p)} = \sum_{m=0}^{\infty}\left[A_{cm}^{(p)} J_m(X_p) + B_{cm}^{(p)} H_m^{(2)}(X_p)\right]\cos(m\phi_p) \equiv \sum_{m=0}^{\infty} E_m(\rho_p)\cos(m\phi_p) \tag{8.B.6}$$

$$U_\phi^{(p)} \equiv \tilde{\eta}_f H_\phi^{(p)} = \sum_{m=0}^{\infty}\frac{k}{j\mu}\left[A_{cm}^{(p)} J_m'(X_p) + B_{cm}^{(p)} H_m^{(2)\prime}(X_p)\right]\cos(m\phi_p) \equiv \sum_{m=0}^{\infty} U_m(\rho_p)\cos(m\phi_p) \tag{8.B.7}$$

where

$$A_{cm}^{(1)} = \sum_{m'=0}^{\infty} Z_{m,m'}^{\cos} A_{cm'}^{(2)} \tag{8.B.8}$$

$$B_{cm}^{(1)} = \sum_{m'=-\infty}^{\infty} Z_{m,m'}^{\cos} B_{cm'}^{(2)} \tag{8.B.9}$$

and

$$Z_{m,m'}^{\cos} = (2-\delta_{0,m}) \begin{cases} Z_{m,m'}^{\exp}, & m'=0 \\ \frac{1}{2}\left[(-1)^{m'} Z_{m,-m'}^{\exp} + Z_{m,m'}^{\exp}\right], & m' \geq 1 \end{cases} \tag{8.B.10}$$

and where

$$E_m(\rho_p) = A_{cm}^{(p)} J_m(X_p) + B_{cm}^{(p)} H_m^{(2)}(X_p) \tag{8.B.11}$$

$$U_m(\rho_p) = \frac{k}{j\mu}\left[A_{cm}^{(p)} J_m'(X_p) + B_{cm}^{(p)} H_m^{(2)'}(X_p)\right] \tag{8.B.12}$$

where the subscript "c" in Equations 8.B.6 and 8.B.7 represents "cosine" and has been placed there to distinguish the exponential coefficients of Equations 8.B.1 through 8.B.4. For each value of $m = 0, 1, \ldots$, the 2×2 equations in Equations 8.B.11 and 8.B.12 which specify $E_m(\rho_p)$ and $U_m(\rho_p)$ in terms of $A_{cm}^{(p)}$ and $B_{cm}^{(p)}$, may be inverted to express the $A_{cm}^{(p)}$ and $B_{cm}^{(p)}$ coefficients in terms of the $E_m(\rho_p)$ and $U_m(\rho_p)$ coefficients. These resulting expressions may be further simplified by using the Wronskian relation of Equation 8.82 that applies for the $J_m(X_p)$, $H_m^{(2)}(X_p)$ functions. If the $A_{cm}^{(2)}$, $B_{cm}^{(2)}$ coefficients, expressed as 2×2 linear combination of the $E_m(r_2^+)$, $U_m(r_2^+)$ coefficients are substituted in Equations 8.B.8 and 8.B.9, and if the $A_{cm}^{(1)}$, $B_{cm}^{(1)}$ coefficients, expressed as 2×2 linear combination of the $E_m(r_1^-)$, $U_m(r_1^-)$ coefficients are substituted in Equations 8.B.11 and 8.B.12, the coefficients $E_m(r_1^-)$, $U_m(r_1^-)$ may be expressed in terms of $E_m(r_2^+)$, $U_m(r_2^+)$. The final results are

$$E_m(r_1^-) = \sum_{m'=0}^{\infty}\left[Z_{m,m'}^{BEE} E_{m'}(r_2^+) + Z_{m,m'}^{BEU} U_{m'}(r_2^+)\right] \tag{8.B.13}$$

$$U_m(r_1^-) = \sum_{m'=0}^{\infty}\left[Z_{m,m'}^{BUE} E_{m'}(r_2^+) + Z_{m,m'}^{BUU} U_{m'}(r_2^+)\right] \tag{8.B.14}$$

where

$$Z_{m,m'}^{BEE} = \left[\frac{j\pi}{2} X_2\right]\left[J_m(X_1) H_{m'}^{(2)'}(X_2) - H_m^{(2)}(X_1) J_{m'}'(X_2)\right] Z_{m,m'}^{\cos} \tag{8.B.15}$$

$$Z_{m,m'}^{BEU} = \left[\frac{\pi}{2} X_2 \left(\frac{\mu}{k}\right)\right]\left[J_m(X_1) H_{m'}^{(2)}(X_2) - H_m^{(2)}(X_1) J_{m'}(X_2)\right] Z_{m,m'}^{\cos} \tag{8.B.16}$$

$$Z_{m,m'}^{BUE} = \left[\frac{\pi}{2} X_2 \left(\frac{k}{\mu}\right)\right]\left[J_m'(X_1) H_{m'}^{(2)'}(X_2) - H_m^{(2)'}(X_1) J_{m'}'(X_2)\right] Z_{m,m'}^{\cos} \tag{8.B.17}$$

$$Z_{m,m'}^{BUU} = \left[\frac{j\pi}{2} X_2\right]\left[-J_m'(X_1) H_{m'}^{(2)}(X_2) + H_m^{(2)'}(X_1) J_{m'}(X_2)\right] Z_{m,m'}^{\cos} \tag{8.B.18}$$

for $(m, m') = 0, 1, \ldots, \infty$. The overall transfer matrix given by this layer is

$$\underline{\underline{Z_{r_1;r_2}^{B}}} \equiv \begin{bmatrix} \underline{\underline{Z^{BEE}}} & \underline{\underline{Z^{BEU}}} \\ \underline{\underline{Z^{BUE}}} & \underline{\underline{Z^{BUU}}} \end{bmatrix} \tag{8.B.19}$$

where $\underline{\underline{Z^{BEE}}} = \left[Z_{m,m'}^{BEE}\right]$, etc.

A nice feature of the transfer matrix of Equation 8.B.19 is the fact that in the case of multieccentric cylinders, the overall transfer matrix of the system may be found by the proper cascade matrix multiplication of the transfer matrices of each individual layer. For the two-layer example of Figure 8.2, assuming in general different materials in each layer, let the single-layer transfer matrices be $\underline{\underline{Z_{r_2;r_3}^{B}}}$ and $\underline{\underline{Z_{r_1;r_2}^{B}}}$, the electric and magnetic field coefficient column matrices be $\underline{E(r_3^+)} = [E_m(r_3^+)]$, $\underline{U(r_3^+)} = [U_m(r_3^+)]$, etc. We have

$$\begin{bmatrix} \underline{E(r_1^-)} \\ \underline{U(r_1^-)} \end{bmatrix} = \underline{\underline{Z_{r_1;r_2}^{B}}} \begin{bmatrix} \underline{E(r_2^+)} \\ \underline{U(r_2^+)} \end{bmatrix} \tag{8.B.20}$$

$$\begin{bmatrix} \underline{E(r_2^-)} \\ \underline{U(r_2^-)} \end{bmatrix} = \underline{\underline{Z_{r_2;r_3}^{B}}} \begin{bmatrix} \underline{E(r_3^+)} \\ \underline{U(r_3^+)} \end{bmatrix} \tag{8.B.21}$$

Because the tangential electric and fields are continuous at the all interfaces, the electric and magnetic coefficients satisfy $E_m(r_j^-) = E_m(r_j^+)$, $U_m(r_j^-) = U_m(r_j^+)$ for $j = 1, 2, 3$ and thus

$$\begin{bmatrix} \underline{E(r_1^+)} \\ \underline{U(r_1^+)} \end{bmatrix} = \underline{\underline{Z_{r_1;r_3}^{B}}} \begin{bmatrix} \underline{E(r_3^-)} \\ \underline{U(r_3^-)} \end{bmatrix} \tag{8.B.22}$$

where $\underline{\underline{Z_{r_1;r_3}^{B}}} = \underline{\underline{Z_{r_1;r_2}^{B}}}\,\underline{\underline{Z_{r_2;r_3}^{B}}}$. For Figure 8.2, $\underline{\underline{Z_{r_1;r_3}^{B}}}$ is the Bessel transfer matrix which expresses the Bessel coefficients $E_m(r_1^+)$, $U_m(r_1^+)$ in terms of $E_m(r_3^-)$, $U_m(r_3^-)$.

REFERENCES

1. M.G. Moharam and T.K. Gaylord, Rigorous coupled-wave analysis of planar grating diffraction, *J. Opt. Soc. Am.* 71, 811–818, 1981.
2. M.G. Moharam and T.K. Gaylord, Diffraction analysis of dielectric surface-relief gratings, *J. Opt. Soc. Am.* 72, 1385–1392, 1982.
3. K. Rokushima and J. Yamakita, Analysis of anisotropic dielectric gratings, *J. Opt. Soc. Am.* 73, 901–908, 1983.
4. M.G. Moharam and T.K. Gaylord, Three-dimensional vector coupled-wave analysis of planar-grating diffraction, *J. Opt. Soc. Am.* 73, 1105–1112, 1983.
5. E.N. Glytsis and T.K. Gaylord, Rigorous three-dimensional coupled-wave diffraction analysis of single and cascaded anisotropic gratings, *J. Opt. Soc. Am. A* 4, 2061–2080, 1987.
6. M.G. Moharam, E.B. Grann, D.A. Pommet, and T.K. Gaylord, Formulation for stable and efficient implementation of rigorous coupled-wave analysis of binary gratings, *J. Opt. Soc. Am. A* 12(5), 1068–1076, 1995.
7. M.G. Moharam, D.A. Pommet, E.B. Grann, and T.K. Gaylord, Stable implementation of the rigorous coupled-wave analysis for surface-relief gratings: Enhanced transmittance matrix approach, *J. Opt. Soc. Am. A* 12(5), 1077–1086, 1995.

8. F. Frezza, L. Pajewski, and G. Schettini, Characterization and design of two-dimensional electromganetic band gap structures by use of a full-wave method for diffraction gratings, *IEEE Trans. Micowave Theory Tech.* 51(3), 941–951, 2003.
9. J.M. Jarem and P.P. Banerjee, An exact, dynamical analysis of the Kukhtarev equations in photorefractive barium titanate using rigorous wave coupled wave diffraction theory, *J. Opt. Soc. Am. A* 13(4), 819–831, 1996.
10. A.Z. Elsherbeni and M. Hamid, Scattering by a cylindrical dielectric shell with inhomogeneous permittivity profile, *Int. J. Electron.* 58(6), 949–962, 1985.
11. A.Z. Elsherbeni and M. Hamid, Scattering by a cylindrical dielectric shell with radial and azimuthal permittivity profiles, in *Proceedings of the 1985 Symposium on Microwave Technology in Industrial Development*, Brazil (Invited), pp. 77–80, July 22–25, 1985.
12. A.Z. Elsherbeni and M. Tew, Electromagnetic scattering from a circular cylinder of homogeneous dielectric coated by a dielectric shell with a permittivity profile in the radial and azimuthal directions-even TM case, in *IEEE Proceedings-1990 Southeastcon*, Session 11A1, pp. 996–1001, 1990.
13. J.M. Jarem, Rigorous coupled wave theory solution of phi-periodic circular cylindrical dielectric systems, *J. Electromagn. Waves Appl.* 11, 197–213, 1997.
14. J.M. Jarem, Rigorous coupled wave theory of anisotropic, azimuthally-inhomogeneous, cylindrical systems, *Prog. Electromagn. Res.* PIER 19, Chap. 4, 109–127, 1998.
15. J.M. Jarem, and P.P. Banerjee, Bioelectromagnetics: A rigorous coupled wave analysis of cylindrical biological tissue, in *Proceedings of the International Conference on Mathematics and Engineering Techniques in Medicine and Biological Sciences (METMBS'00)*, F. Valafar, ed., Vol. II, pp. 467–472, Las Vegas, NV, June 26–29, 2000.
16. J.M. Jarem, Rigorous coupled wave analysis of radially and azimuthally-inhomogeneous, elliptical, cylindrical systems, *Prog. Electromagn. Res.* PIER 34, 89–115, 2001.
17. J.M. Jarem, Validation and numerical convergence and validation of the Hankel-Bessel and Mathieu rigorous coupled wave analysis algorithms for radially and azimuthally-inhomogeneous, elliptical, cylindrical systems, *Prog. Electromagn. Res.* PIER 36, 153–177, 2002.
18. J.M. Jarem, Rigorous coupled-wave-theory analysis of dipole scattering from a three-dimensional, inhomogeneous, spherical dielectric and permeable system, *IEEE Trans. Micowave Theory Tech.* 45(8), 1193–1203, 1997.
19. J.M. Jarem, A rigorous coupled-wave theory and crossed-diffraction grating analysis of radiation and scattering from three-dimensional inhomogeneous objects, a Letter, *IEEE Trans. Antennas Propag.* 46(5), 740–741, 1998.
20. J.M. Jarem and P.P. Banerjee, *Computational Methods for Electromagnetic and Optical Systems*, Marcel Dekker, Inc., New York, 2000.
21. P.M. Morse and H. Feshbach, *Methods of Theoretical Physics*, Chap. 10, McGraw-Hill Book Company, Inc., New York, 1953, pp. 1210–1215.
22. A.A. Kishk, R.P. Parrikar, and A.Z. Elsherbeni, Electromagnetic scattering from an eccentric multilayered circular cylinder, *IEEE Trans. Antennas Propag.* AP-40(3), 295–303, 1992.
23. R.V. Churchill, *Complex Variables and Applications*, McGraw-Hill Book Company, New York, 1960.
24. M. Abramowitz, Elementary analytical methods, Chap. 3 in *Handbook of Mathematical Functions*, M. Abramowitz and I. Stegum, eds., Dover, New York, 1972, Eq. 3.3.8, p. 12.
25. C.A. Balanis, *Advanced Engineering Electromagnetics*, John Wiley & Sons, New York, 1989, Eq. 11–21b, p. 577.
26. J.M. Jarem, Rigorous coupled wave analysis of bipolar cylindrical systems: Scattering from inhomogeneous dielectric material, eccentric, composite circular cylinders, *Prog. Electromagn. Res.* 43, 181–237, 2003.
27. L.G. Stratigaki, M.P. Ioannidou, and D.P. Chrissoulidis, Scattering from a dielectric cylinder with multiple eccentric cylindrical dielectric inclusions, *IEEE Proc.-Microwave Antennas Propag.* 143(6), 505–511, 1996.
28. M.P. Ioannidou, K.D. Kapsalas, and D.P. Chrissoulidis, Electromagnetic-wave scattering by an eccentrically stratified, dielectric cylinder with multiple, eccentrically, stratified, cylindrical, dielectric inclusions, *J. Electromagn. Waves Appl.* 18(4), 495–516, 2004.
29. M.P. Ioannidou, EM-wave scattering by an axial slot on a circular PEC cylinder with an eccentrically layered inner coating: A dual-series solution for TE polarization, *IEEE Trans. Antennas Propag.* AP-57(11), 3512–3519, 2009.

9 Bipolar Coordinate RCWA Computational Examples and Case Studies

9.1 INTRODUCTION

This chapter, presented in a step-by-step computational example (CE) and case study (CS) solution format, is intended as a mathematical and computer program algorithm supplement to the rigorous coupled wave analysis (RCWA) method (based on state variable (SV) theory) discussed in Chapter 8. The basic mathematical algorithms and equations presented in Chapter 8, many of which are only summarized, require for their detailed derivation complex variable integration in the complex plane (Cauchy's integral and higher-order integral formulas, i.e., residue theorem, for example), the detailed derivation of complicated bipolar coordinate trigonometric equations and relations, detailed derivations of the EM fields as calculated in the bipolar coordinate geometry, and many other computations as well. For reproducing the numerical results presented in Chapter 8 and producing new numerical results for cases not studied herein, many small computer programs and small computer subprograms need to be constructed in order that an overall system program can be constructed. The computer programs that need to be constructed carry out tasks such as (1) the cross-checking of the correctness of complicated trigonometric and hyperbolic trigonometric bipolar coordinate formulas; (2) the cross-checking of the accuracy and correctness of complex variable integrals; (3) the calculation of eigenvalues and eigenfunctions associated with solving Maxwell's equations in bipolar coordinates (and plots of these eigenfunctions in order to check if sensible results have been obtained); (4) the checking, testing, and studying the orthogonality of the eigenfunctions found from solving Maxwell's equations in bipolar coordinates; (5) the calculation of cascade-ladder transfer matrices used to formulate final matrix equations from which the overall unknowns of the scattering system may be determined; (6) the calculation and display of the geometry of the scattering system in bipolar coordinates; (7) the calculation and display of the electric and magnetic fields once an EM field solution has been obtained; (8) the calculation of the bistatic line widths (i.e., also called scattering cross section); and (9) the calculation and cross-checking of many other quantities associated with the bipolar RCWA algorithm as well.

The step-by-step CE and CS format presented in this chapter is felt by the authors to be a very natural way from which small computer programs and algorithms can be constructed by the reader whose aim is to form an overall system computer program to study EM scattering. For example, if a program is to be constructed to calculate bipolar eigenfunctions (resulting from the solutions of Maxwell's equation), giving a suggested listing of the detailed steps to calculate these eigenfunctions for a given set of numerical parameters and then presenting plots of the eigenfunctions and associated numerical results provides the reader with a useful cross-check that the algorithm developed by the reader's algorithm is working correctly. A problem such as this requires not only eigenvalue determination but also careful calculation of the Fourier coefficients of a given dielectric inhomogeneity profile (using complex variable residue theory) and verification (by graphical means) that the calculated Fourier series is converging correctly. Overall, the presentation of computational examples and case studies provides the reader with the opportunity to cross-check the reader's own numerical calculations, thus aiding the reader's computer algorithm construction. Hopefully by working through the computational examples listed in this chapter, constructing small programs and subprograms, and then comparing those numerical results with those presented herein, the reader may construct an overall system computer program from which to study scattering from inhomogeneous eccentric cylinders using the bipolar RCWA algorithm.

To validate bipolar RCWA (i.e., SV analysis) numerical results in this chapter for scattering objects composed of spatially homogeneous dielectric permittivity regions, two different validation algorithms are being used, both of which are based on the eccentric, circular coordinate, Bessel algorithm developed by Kishk, Parrikar, and Elsherbeni (KPE) of Ref. [22] in Chapter 8. The first algorithm (derived in Appendix 8.B) is called herein the KPE Bessel aperture (BA) algorithm (or BA method, for short) and consists in using transfer matrices to relate the Bessel–Fourier series coefficients of the tangential electric and magnetic fields that exist at a given circular interface to a second given interface. It is called a "Bessel aperture" algorithm herein because if a radiating aperture were present on one of the surfaces of the dielectric cylinders, radiation from the aperture would be determined by calculating the radiation from the tangential electric and magnetic fields (expressed in terms of Bessel functions) calculated at the aperture surface located on the dielectric cylinder. The second validation algorithm is called herein the KPE Bessel coefficient (BC) algorithm (or BC method, for short) and consists in determining directly (by solving a large system matrix) the Bessel coefficients (hence its name) of all of the Bessel function modes that exist in each homogeneous region of the scatterer. This algorithm is the one originally derived by KPE (Ref. [22] in Chapter 8) and the details of its derivation are given in Chapter 8. The two algorithms are different because the KPE Bessel Aperture method is based on multiplying cascade matrices to determine the final matrix equations of the system, whereas in the KPE BC method, a large matrix is solved to determine the Bessel function modal coefficients directly form the large system matrix. Through the remaining chapter, the two methods will be referred to as the BA or BC validation methods.

It should be noted that the BC method, as implemented in this chapter, solves for the EM field coefficients of the system *without* taking into account the even symmetry of the problem. Thus, it is entirely reasonable, and to be expected, that ill-conditioned results will occur for the BC method for lower values of the mode truncation than would occur if the system equations had been solved taking into account the even symmetry of the problem, as the solution taking into account even symmetry involves a much smaller system matrix than in the case when symmetry is not accounted for (see Problems 8.9 and 8.10). As the BA method and the BC method are basically the same method, and because the BA method *does* take into account the even symmetry of the problem (see Appendix 8.B), it is thus overall reasonable to expect that the BA method as implemented herein will show better conditioning than does the BC method as implemented here (see Figure 9.28).

The case studies to be presented in this chapter are as follows. CS I (CE9.1 through CE9.5) presents numerical examples of several of the analytic techniques necessary to carry out the RCWA Bessel aperture and BC methods, including (1) numerical examples of Fourier series expansions of the permittivity function $\varepsilon(x, y)$, the scale factors $h(x, y)$, and products $\varepsilon(x, y)h(x, y)$ (using the discrete convolution theorem) (CE9.1); (2) numerical examples of the eigenvalues and eigenfunctions resulting from the solution of Maxwell's equations in a given thin expressed in bipolar coordinates (CE9.2 and CE9.3); (3) numerical examples of the coordinate transformations to implement both the BA and BC validation methods (CE9.4); and (4) numerical comparison examples of the BA and SV transfer matrices (CE9.5) that result (at least partially) from the numerical results of CE9.1 through CE9.4. CS II (CE9.6 and CE9.7) presents numerical examples of the electromagnetic fields and bistatic line widths which are calculated for scattering from a relatively small dielectric object (diameter is $(4/\pi)\tilde{\lambda}_f$), composed of spatially homogeneous regions, whereas Case Studies CS III (CE9.8 through CE9.13), CS IV (CE9.14 through CE9.16), and CS V (CE9.17 and CE9.18) study the electromagnetic fields and bistatic line widths which are calculated for plane wave scattering that results from three different, large dielectric objects (diameter $4\tilde{\lambda}_f$), each composed of spatially homogeneous regions and each containing different dielectric permittivities. In CS II through V (CE9.6 through CE9.18), because the dielectric objects are composed of spatially homogeneous regions, EM field results as calculated by the RCWA method are compared to those obtained by the BA and BC validation methods, methods which represent valid solution techniques for the homogeneous region cases under consideration. This comparison has been done to estimate the correctness of the RCWA method (i.e., SV method). CS VI (CE9.19) and VII (CE9.20 and CE9.21) presents numerical examples of the electromagnetic fields and bistatic line widths, which are calculated for scattering, from spatially inhomogeneous regions dielectric objects (diameter $4\tilde{\lambda}_f$) containing both different bulk dielectrics and different inhomogeneity profiles. In CS III through VIII (CE9.8 through CE9.21) convergence of the electromagnetic BA, BC, and RCWA field solutions verses truncation mode number and number of thin layers (for the RCWA method only) are presented. Also in CS III through VIII, many EM field and bistatic line width comparisons between the different studied cases have been made in order to emphasize the affect that different homogeneous and inhomogeneous dielectric profiles have on the EM field scattering of the system (assuming the same incident field). CS IX presents conservation of power numerical results as determined through use of the optical theorem (also called the forward scattering theorem). Appendix 9.A, given as Computational Example 9.A, provides the details of the interpolation scheme that was used for EM field calculations performed throughout Chapters 8 and 9.

We begin by presenting numerical examples using the analytic techniques discussed in Chapter 8 (used to implement the different EM field methods used in Chapters 8) (CS I), followed by the EM field case studies (CS II through IX) discussed earlier.

9.2 CASE STUDY I: FOURIER SERIES EXPANSION, EIGENVALUE AND EIGENFUNCTION ANALYSIS, AND TRANSFER MATRIX ANALYSIS

Computational Example 9.1

This CE concerns expressing a given dielectric permittivity profile function $\varepsilon_u(u, v)$ (u is a constant), expressing the scale function $h^2(u, v)$, and expressing the product function $\varepsilon_h(u, v) \equiv \varepsilon_u(u, v)h^2(u, v)$ each in a Fourier exponential series in the bipolar coordinate $-\pi < v \leq \pi$ and verifying that the Fourier exponential series for these functions have converged correctly to the function it is meant to represent. The correct numerical calculation of the Fourier exponential coefficients for these functions is very important for the calculation of the thin layer bipolar eigenfunctions (see Section 8.1.4) associated with the solution of Maxwell's equations (see 8.4.2) in bipolar coordinates. For this CE, specifically carry out the following steps for the numerical example and truncation values specified as follows:

(a) Using the values of $a = 9.015972$, $u = 1.815498$, $N = 11$, $I = 4(N - 1)$, $\varepsilon_1 = 2$, $\varepsilon_2 = 3$, $\alpha = 1.5$, $\delta\varepsilon = 1.2$, and $v_1 = 101.3361°$, plot the exact function $\varepsilon_u(u, v)$ given by Problem 8.8

$$\varepsilon_u(u,v) = \begin{cases} \varepsilon_1 + \delta\varepsilon\cos(\alpha v), & 0 \leq |v| \leq v_1(u) \\ \varepsilon_2, & v_1(u) < |v| \leq \pi \end{cases} \tag{CE9.1}$$

plot the Fourier series

$$\varepsilon_{u,A}(u,v) = \sum_{i=-I}^{I} \varepsilon_i(u)\exp(jiv) \tag{CE9.2}$$

(subscript "A" approximate) using the Fourier series coefficients $\varepsilon_i(u)$ calculated in Problem 8.8, and verify that the Fourier series of Equation CE9.2 is approximating the exact function $\varepsilon_u(u, v)$ properly.

(b) Using the values of $a = 9.015972$, $u = 1.815498$, and $N_t = 8(N-1)$, plot the exact function

$$h^2(u,v) = \frac{a^2}{\left(\cosh(u) - \cos(v)\right)^2} \tag{CE9.3}$$

plot the Fourier series summation function

$$h_{I_t}^{sq}(u,v) = \sum_{i=-N_t}^{N_t} h_i^{sq}(u)\exp(jiv) \tag{CE9.4}$$

using the Fourier series coefficients calculated in Problem 8.11, and verify that the Fourier series of Equation CE9.4 is approximating the exact function $h^2(u, v)$ properly.

(c) Using the Fourier series summation functions of parts (a) and (b), use the convolution formula of Equation 8.A.20, namely,

$$\varepsilon_{h,A}(u,v) = \sum_{i=-I}^{I}\left[\sum_{i'=-2I}^{2I} h_{i-i'}^{sq}(u)\varepsilon_{i'}(u)\right]\exp(jiv) \tag{CE9.5}$$

to approximate the exact function

$$\varepsilon_h(u,v) \equiv \varepsilon_u(u,v)h^2(u,v) \tag{CE9.6}$$

Verify that the Fourier series of Equation 8.7 of Chapter 8 is approximating the exact function $\varepsilon_h(u, v) = \varepsilon_u(u, v)h^2(u, v)$ properly.

Computational Example 9.1(a) Solution

Figure 9.1 show plots the original functions for which Fourier plots were to be made and the Fourier plots that resulted after summation of the Fourier series. In the part (a) Fourier plot, the overshoot is the normal Gibbs' phenomenon that is associated with any Fourier series plot when a discontinuous function is being represented in a Fourier series.

Computational Example 9.1(b) Solution

Figure 9.2 shows plots the original function $h^2(u, v)$ function ($h(u, v)$ is the bipolar coordinate scale factor) and its corresponding Fourier series. Because the $h^2(u, v) = a^2/(\cosh(u) - \cos(v))^2$ function of Equation CE9.3 is a continuous function for the period $-\pi < v \leq \pi$, no Gibbs' phenomenon was observed in its Fourier series plot.

Computational Example 9.1(c) Solution

The part (c) plot was carried out using the convolution of Equation CE9.5. Again as in part (a) a small Gibbs' like behavior is seen in the Fourier series plot due to the discontinuity of the $\varepsilon_h(u, v) \equiv \varepsilon_u(u, v)h^2(u, v)$ function (Figure 9.3).

Computational Example 9.2

This CE is closely related to Computational Example 9.1 and concerns using the Fourier exponential series coefficients as specified for the function $\varepsilon_h(u, v) \equiv \varepsilon_u(u, v)h^2(u, v)$ to determine the eigenvalues of the thin layer bipolar eigenfunctions of Section 8.1.4 associated with the solution of Maxwell's

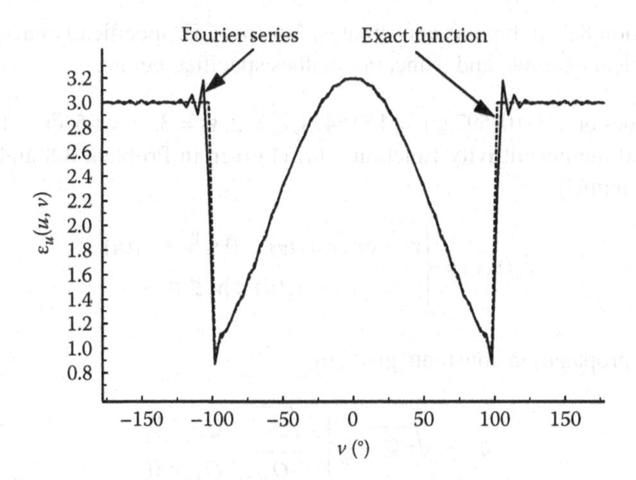

FIGURE 9.1 Plots of relative dielectric permittivity function and its Fourier series representation.

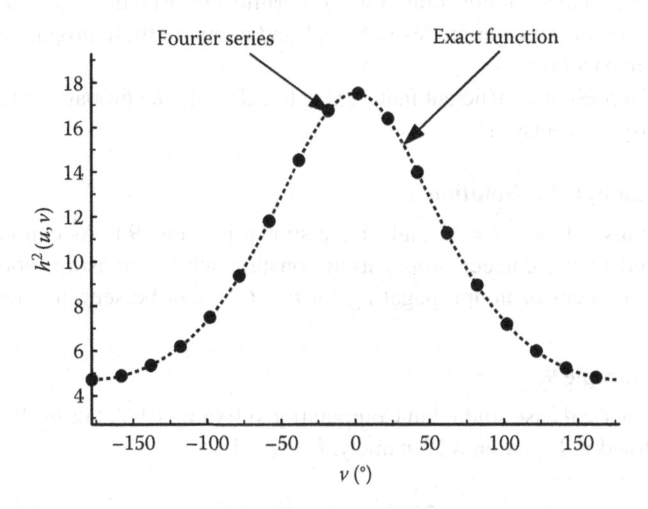

FIGURE 9.2 Plots of the squared scale function and its Fourier series representation.

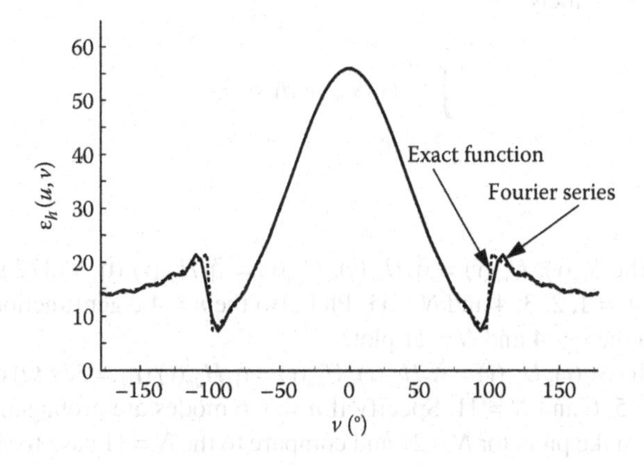

FIGURE 9.3 Plots of $\varepsilon_h(u, v) \equiv \varepsilon_u(u, v)\, h^2(u, v)$ and its Fourier series representation.

equations (see Section 8.2) in bipolar coordinates. For this CE, specifically carry out the following steps for the numerical example and truncation values specified below:

(a) Using the values of $a = 9.015972$, $u = 1.815498$, $\varepsilon_1 = 2$, $\varepsilon_2 = 3$, $\alpha = 1.5$, $\delta\varepsilon = 1.2$, $\delta\varepsilon = 1.2$, $v_1 = 101.3361°$, and the permittivity function $\varepsilon_u(u, v)$ given in Problem 8.8 and Computational Example 9.1, namely,

$$\varepsilon_u(u,v) = \begin{cases} \varepsilon_1 + \delta\varepsilon\cos(\alpha v), & 0 \le |v| \le v_1(u) \\ \varepsilon_2, & v_1(u) < |v| \le \pi \end{cases} \tag{CE9.7}$$

calculate the propagation constants given by

$$q_{n\ell} = \sqrt{-Q_{n\ell}} = \begin{cases} j\sqrt{Q_{n\ell}}, & Q_{n\ell} \ge 0 \\ \sqrt{-Q_{n\ell}}, & Q_{n\ell} < 0 \end{cases} \tag{CE9.8}$$

using Equations 8.7 through 8.17 for the case of $N = 11$ eigenmodes.

(b) Calculate the propagation constants for the permittivity function $\varepsilon_u(u, v)$ defined in part (a) when the number of eigenmodes is $N = 21$ and compare these propagation constants to those found in part (a).

(c) Does $N = 11$ represent a sufficient matrix size to calculate the propagation constants of the system. Justify your answer.

Computational Example 9.2 Solution

The propagation constants for $N = 11$ and 21 are shown in Table 9.1. As can be seen, the $N = 11$ case converges well to the correct propagation constant with increasing mode number N. The modes become evanescent or nonpropagating for $n \ge 6$ as can be seen for the below numerical results.

Computational Example 9.3

Referring to the numerical case studied in Computational Example 9.2, taking $N = 11$, normalize the eigenfunctions defined in Equation 8.17, namely, $I = N - 1$

$$S_{zn}(v) = \sum_{i=-I}^{I} S_{zin} \exp(jiv), \quad n = 1, \ldots, N \tag{CE9.9}$$

by using Equation 8.18, namely,

$$\int_{-\pi}^{\pi} S_{zn}(v) S_{zn'}(v) dv = \delta_{n,n'} \tag{CE9.10}$$

and taking $n = n'$.

(a) Make plots of the $S_{zn}(v)$, $U_{vn}(v) = \tilde{\eta}_f H_{vn}(v)$, $U_{un}(v) = \tilde{\eta}_f H_{un}(v)$ ($\tilde{\eta}_f = 377\ \Omega$) eigenfunctions for the value of $n = 1, 2, 3, 4$ and $N = 11$. Plot also the $n = 4$ eigenfunction $S_{zn}(v)$ for $N = 21$ and compare to the $n = 4$ and $N = 11$ plot.

(b) Make plots of the $S_{zn}(v)$, $U_{vn}(v) = \tilde{\eta}_f H_{vn}(v)$, $U_{un}(v) = \tilde{\eta}_f H_{un}(v)$ ($\tilde{\eta}_f = 377\ \Omega$) eigenfunctions for the value of $n = 5, 6$ and $N = 11$. Specify if $n = 5, 6$ modes are propagating or evanescent. For $n = 6$ mode make plots for $N = 21$ and compare to the $N = 11$ case to check convergence of the eigenfunction.

TABLE 9.1
Propagation Constants for $N = 11$ and $N = 21$

qExp(n) = (0.0000000E+00, 7.031100) $n = 1$ ($N = 21$)
qExp(n) = (0.0000000E+00, 7.031099) $n = 1$ ($N = 11$)

qExp(n) = (0.0000000E+00, 5.111681) $n = 2$ ($N = 21$)
qExp(n) = (0.0000000E+00, 5.111608) $n = 2$ ($N = 11$)

qExp(n) = (0.0000000E+00, 3.901430) $n = 3$ ($N = 21$)
qExp(n) = (0.0000000E+00, 3.900685) $n = 3$ ($N = 11$)

qExp(n) = (0.0000000E+00, 3.308395) $n = 4$ ($N = 21$)
qExp(n) = (0.0000000E+00, 3.307570) $n = 4$ ($N = 11$)

qExp(n) = (0.0000000E+00, 2.681036) $n = 5$ ($N = 21$)
qExp(n) = (0.0000000E+00, 2.678781) $n = 5$ ($N = 11$)

qExp(n) = (1.658647, 0.0000000E+00) $n = 6$ ($N = 21$)
qExp(n) = (1.667689, 0.0000000E+00) $n = 6$ ($N = 11$)

qExp(n) = (3.508559, 0.0000000E+00) $n = 7$ ($N = 21$)
qExp(n) = (3.511534, 0.0000000E+00) $n = 7$ ($N = 11$)

qExp(n) = (5.080310, 0.0000000E+00) $n = 8$ ($N = 21$)
qExp(n) = (5.090958, 0.0000000E+00) $n = 8$ ($N = 11$)

qExp(n) = (6.315886, 0.0000000E+00) $n = 9$ ($N = 21$)
qExp(n) = (6.364629, 0.0000000E+00) $n = 9$ ($N = 11$)

qExp(n) = (7.575145, 0.0000000E+00) $n = 10$ ($N = 21$)
qExp(n) = (7.669509, 0.0000000E+00) $n = 10$ ($N = 11$)

qExp(n) = (8.701915, 0.0000000E+00) $n = 11$ ($N = 21$)
qExp(n) = (8.854873, 0.0000000E+00) $n = 11$ ($N = 11$)

(c) Make plots of the $S_{zn}(v)$, $U_{vn}(v) = \tilde{\eta}_f H_{vn}(v)$, $U_{un}(v) = \tilde{\eta}_f H_{un}(v)$, ($\tilde{\eta}_f = 377\ \Omega$) eigenfunctions for the value of $n = 11$ for $N = 11$ and $N = 21$ and check the convergence of the eigenfunction. Specify if the $n = 11$ mode is propagating or evanescent.

Computational Example 9.3(a) Solution

Plots of the $S_{zn}(v)$, $U_{vn}(v) = \tilde{\eta}_f H_{vn}(v)$, $U_{un}(v) = \tilde{\eta}_f H_{un}(v)$, ($\tilde{\eta}_f = 377\ \Omega$) eigenfunctions for the value of $n = 1, 2, 3, 4$ and $N = 11$ are shown. The $n = 4$ eigenfunction $S_{zn}(v)$ for $N = 21$ (truncation value) is compared to the $n = 4$, $N = 11$ (truncation value) plot, and the two plots are shown to be quite close together showing that the $n = 4$ eigenfunction has converged well for the truncation value of $N = 11$ (Figure 9.4).

Computational Example 9.3(b) Solution

Plots of the $S_{zn}(v)$, $U_{vn}(v) = \tilde{\eta}_f H_{vn}(v)$, $U_{un}(v) = \tilde{\eta}_f H_{un}(v)$, ($\tilde{\eta}_f = 377\ \Omega$) eigenfunctions for the value of $n = 5, 6$ and $N = 11$ are shown. The $n = 5$ mode ids propagating whereas the $n = 6$ is evanescent. Convergence of the $n = 6$ mode is seen for $N = 11$, as the $N = 11$ and $N = 21$ truncation value plots, are quite close together (Figure 9.5).

Computational Example 9.3(c) Solution

Plots of the $S_{zn}(v)$, $U_{vn}(v) = \tilde{\eta}_f H_{vn}(v)$, $U_{un}(v) = \tilde{\eta}_f H_{un}(v)$, ($\tilde{\eta}_f = 377\ \Omega$) eigenfunctions for the value of $n = 11$ for truncation values of $N = 11$ and 21 are shown in Figure 9.6. The mode is evanescent for $n = 11$ and the convergence for $N = 11$ is not as good as was found for smaller mode number values, i.e., $n = 1, 2, 3, 4, 5, 6$ values as can be seen from the difference in the eigenmodes that were found for the truncation values of $N = 11$ and 21.

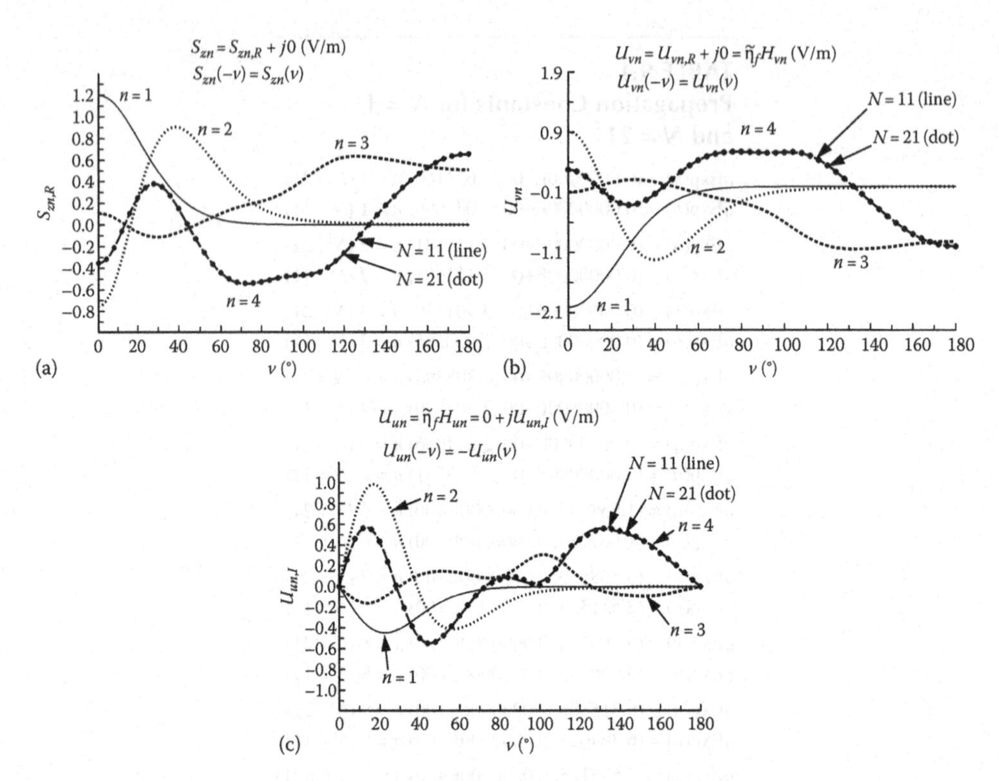

FIGURE 9.4 Plots of the (a) $S_{zn}(\nu)$, (b) $U_{vn}(\nu) = \tilde{\eta}_f H_{vn}(\nu)$, (c) $U_{un}(\nu) = \tilde{\eta}_f H_{un}(\nu)$, ($\tilde{\eta}_f$ = 377 Ω) eigenfunctions for the value of n = 1, 2, 3, 4 and N = 11 are shown.

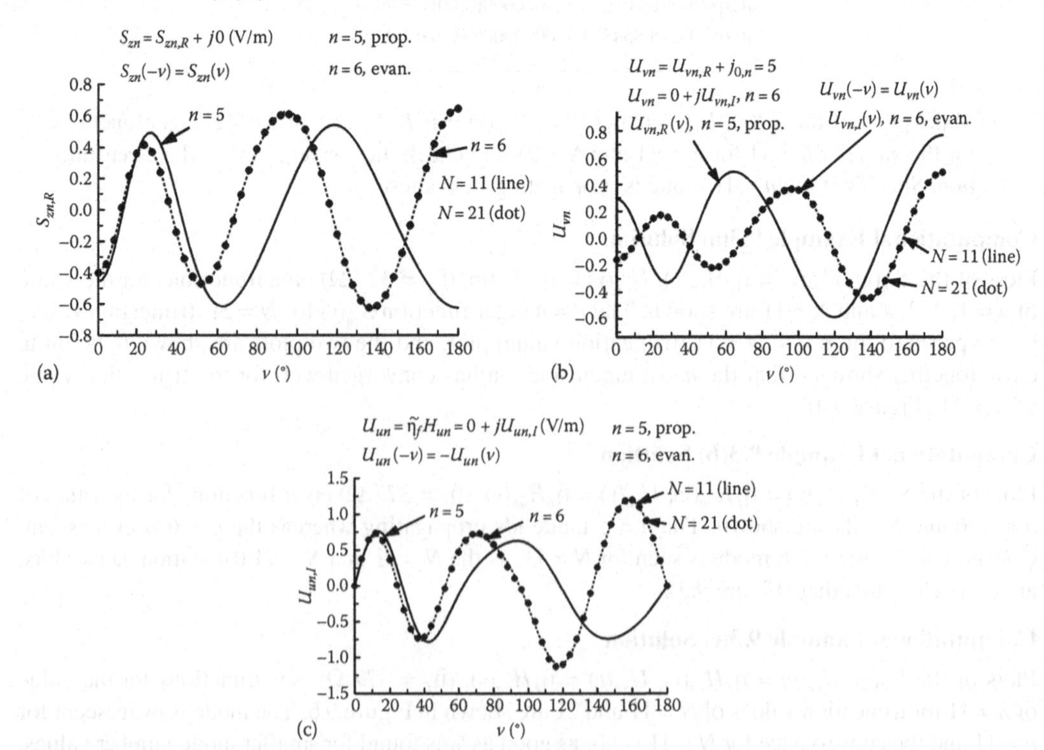

FIGURE 9.5 Plots of the (a) $S_{zn}(\nu)$, (b) $U_{vn}(\nu) = \tilde{\eta}_f H_{vn}(\nu)$, (c) $U_{un}(\nu) = \tilde{\eta}_f H_{un}(\nu)$, ($\tilde{\eta}_f$ = 377 Ω) eigenfunctions for the value of n = 5, 6 and N = 11 are shown.

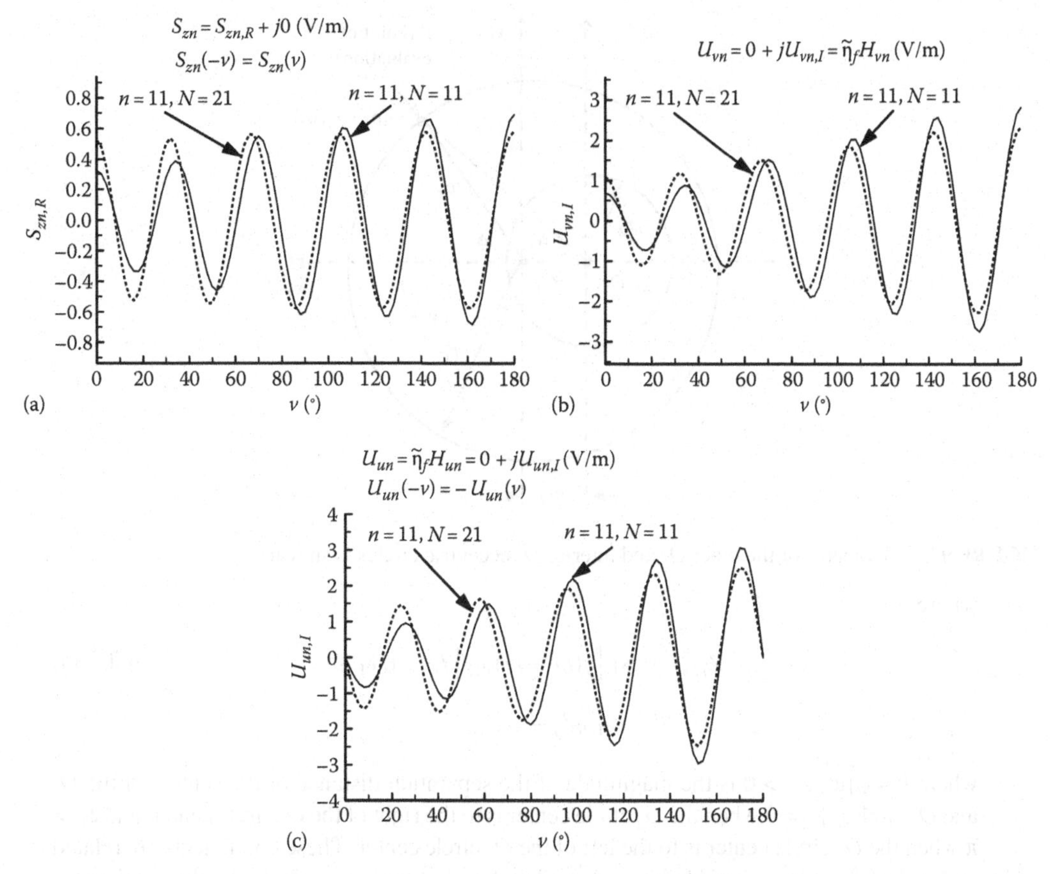

FIGURE 9.6 Plots of the (a) $S_{zn}(\nu)$, (b) $U_{vn}(\nu) = \tilde{\eta}_f H_{vn}(\nu)$, (c) $U_{un}(\nu) = \tilde{\eta}_f H_{un}(\nu)$, ($\tilde{\eta}_f = 377\,\Omega$) eigenfunctions for the value of $n = 11$ for truncation values of $N = 11$ and 21 are shown.

Computational Example 9.4

(a) Verify for the specific numerical case to be detailed below (see Figure 9.7) that the Bessel series coefficients $A_{\text{cm}}^{(1)}, B_{\text{cm}}^{(1)}$, described in Appendix 8.B for circle O_1 (radius r_1) are correctly expressed in terms of the Bessel series coefficients $A_{\text{cm}}^{(2)}, B_{\text{cm}}^{(2)}$ of interior, eccentric circle O_2 (radius r_2) by the relations

$$A_{\text{cm}}^{(1)} = \sum_{m'=0}^{M_t} Z_{m,m'}^{\cos} A_{\text{cm}'}^{(2)} \tag{CE9.11}$$

$$B_{\text{cm}}^{(1)} = \sum_{m'=-\infty}^{M_t} Z_{m,m'}^{\cos} B_{\text{cm}'}^{(2)} \tag{CE9.12}$$

where

$$Z_{m,m'}^{\cos} = (2-\delta_{0,m})\begin{cases} Z_{m,m'}^{\exp}, & m'=0 \\ \dfrac{1}{2}\left[(-1)^{m'} Z_{m,-m'}^{\exp} + Z_{m,m'}^{\exp}\right], & m' \geq 1 \end{cases} \tag{CE9.13}$$

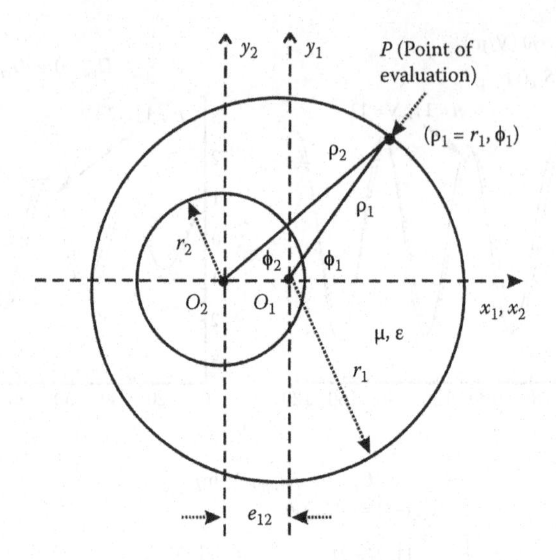

FIGURE 9.7 Geometry of the outer O_1 and interior O_2 eccentric circles is shown.

where

$$Z^{\exp}_{m,m'} = \exp\left[j(m'-m)\phi_{12}\right]J_{m-m'}(ke_{12}) \tag{CE9.14}$$

$$(m,m') = -\infty,\ldots,\infty$$

where $k=\sqrt{\mu\varepsilon}$, $e_{12}>0$ is the magnitude of the separation distance of the centers of the O_1 and O_2 circles, $\phi_{12}=0$ when the O_2 circle center is to the right of the O_1 circle center, and $\phi_{12}=\pi$ when the O_2 circle center is to the left of the O_1 circle center. These coefficients are related to the electric and magnetic fields evaluated at the inner and outer circles by the relations

$$E_z(\rho_p,\phi_p)=\sum_{m=0}^{M_t}\left[A^{(p)}_{\text{cm}}J_m(k\rho_p)+B^{(p)}_{\text{cm}}H^{(2)}_m(k\rho_p)\right]\cos(m\phi_p)\equiv\sum_{m=0}^{\infty}E_m(\rho_p)\cos(m\phi_p) \tag{CE9.15}$$

$$U_\phi(\rho_p,\phi_p)\equiv\tilde{\eta}_f H_\phi=\sum_{m=0}^{M_t}\frac{k}{j\mu}\left[A^{(p)}_{\text{cm}}J'_m(k\rho_p)+B^{(p)}_{\text{cm}}H^{(2)'}_m(k\rho_p)\right]\cos(m\phi_p)$$

$$\equiv\sum_{m=0}^{\infty}U_m(k\rho_p)\cos(m\phi_p) \tag{CE9.16}$$

where $p = 1, 2$ and $M_t \to \infty$.

(b) Verify the relations of Equations CE9.11 through CE9.16 numerically, for finite truncation values of the infinite series, namely, $m = M_t = 8, 10, 16$, using the point P shown in Figure 9.7 and by taking

$$r_2=2,\quad e_{12}=0.2\pi,\quad \phi_{12}=\pi,\quad 0^\circ<\phi_2<180^\circ$$

$$\varepsilon_0=1,\quad \varepsilon_1=2,\quad \varepsilon_2=3,\quad \text{and}\quad \varepsilon_3=4$$

$$A^{(2)}_{\text{cm}}=\frac{(-1+j)^{m+1}}{1000},\quad m\le 8$$

$$B_{cm}^{(2)} = \frac{(-2-j)^{m+1}}{1000}, \quad m \leq 8$$

$$A_{cm}^{(2)} = B_{cm}^{(2)} = 0, \quad m > 8 \tag{CE9.17}$$

and by following the steps listed below:

1. Calculating $A_{cm}^{(1)}$, $B_{cm}^{(1)}$ using Equations CE9.11 through CE9.14 using $A_{cm}^{(2)}$, $B_{cm}^{(2)}$ of Equation CE9.17.
2. Using the point P (shown in Figure 9.7, expressed in O_1 cylindrical coordinates, evaluate $E_z(\rho_1, \phi_1)$ of Equation CE9.15, using the numerical coefficients defined in part 1).
3. Using the point P (shown in Figure 9.7), expressed in O_2 cylindrical coordinates, evaluate $E_z(\rho_2, \phi_2)$ of Equations CE9.14 and CE9.15, using the $A_{cm}^{(2)}, B_{cm}^{(2)}$ numerical coefficients by Equation CE9.17.
4. Calculating and then comparing the values of $E_z(\rho_1, \phi_1)$ and $E_z(\rho_2, \phi_2)$ of Equation CE9.15 for the numerical case and the truncation values $m = M_t = 8$, 10, and 16 under consideration. The comparison is to be made by making plots of $E_z(\rho_1, \phi_1)$ and $E_z(\rho_2, \phi_2)$ (real and imaginary parts) versus the coordinate ϕ_2 ($0° < \phi_2 < 180°$) for the series truncation values of $M_t = 8$, 10, and 16. (*Note:* The ϕ_1 coordinate is a function of the ϕ_2 coordinate.)

(c) From the plots of part (b(4)), comment on if $M_t = 16$ is a sufficiently large infinite series truncation value for reliable numerical results using Equations CE9.11 through CE9.16.

Computational Example 9.4(a) Solution

Plots of the real ($E_{zR}^{(p)} = \text{Re}(E_z(\rho_p,\phi_p))$) and imaginary part ($E_{zI}^{(p)} = \text{Imag}(E_z(\rho_p,\phi_p))$) of Equation CE9.15 for $p = 1, 2$

$$E_z(\rho_p,\phi_p) = \sum_{m=0}^{\infty}\left[A_{cm}^{(p)}J_m(k\rho_p) + B_{cm}^{(p)}H_m^{(2)}(k\rho_p)\right]\cos(m\phi_p) \equiv \sum_{m=0}^{\infty} E_m(\rho_p)\cos(m\phi_p)$$

namely $E_z(\rho_1,\phi_1)$, $E_z(\rho_2,\phi_2)$ are shown in Figure 9.8 (the point of evaluation is shown in Figure 9.7). The $E_{zR}^{(1)} = \text{Re}(E_z(\rho_1,\phi_1))$ electric field is shown for the series truncation values of $M_t = 8$, 10, and 16.

Computational Example 9.4(b) and 9.4(c) Solutions

As can be seen from Figure 9.8a and b, for both the real and imaginary plots, that the values of $M_t = 8$ and 10 (dashed lines) are not sufficiently large enough values for the $E_z(\rho_1,\phi_1)$ field to converge to the $E_z(\rho_2,\phi_2)$ field for $0° < \phi_2 < 180°$. However, the $M_t = 16$ plots (solid line and dot plots of Figure 9.8a

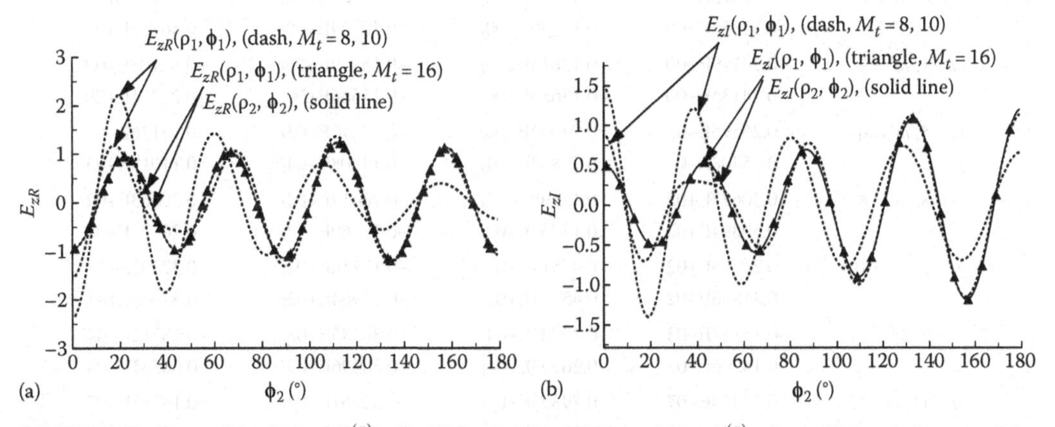

FIGURE 9.8 Plots of the (a) real $E_{zR}^{(p)} = \text{Re}(E_z(\rho_p,\phi_p))$ and (b) imaginary $E_{zI}^{(p)} = \text{Imag}(E_z(\rho_p,\phi_p))$ parts of the electric field are shown.

and b shown) clearly are equal to one another, showing that the below $M_t = 16$ is a large value enough to achieve convergence of the two different Bessel function expansion as given in Equation CE9.15.

Computational Example 9.5

For scattering geometry of Figure 9.7 taking,

$$r_1 = \tilde{k}_f \tilde{r}_1 = 1, \quad r_2 = \tilde{k}_f \tilde{r}_2 = 2, \quad e_{12} = \tilde{k}_f \left|\tilde{e}_{12}\right| = 0.2\pi$$

and taking the relative dielectric values of each region to be $\varepsilon_0 = 1$, $\varepsilon_1 = 2$, $\varepsilon_2 = 3$, and $\varepsilon_3 = 4$; calculate and compare a few matrix elements of the exact BA transfer matrices of (8.B.19), namely,

$$\underline{\underline{Z}}^{B}_{r_1;r_2} \equiv \begin{bmatrix} \underline{\underline{Z}}^{BEE} & \underline{\underline{Z}}^{BEU} \\ \underline{\underline{Z}}^{BUE} & \underline{\underline{Z}}^{BUU} \end{bmatrix} \tag{CE9.18}$$

and SV transfer matrices of Equation 8.79 (Ref. [26] in Chapter 8), namely,

$$\underline{\underline{K}}^{SV}_{r_1;r_3} = \begin{bmatrix} \underline{\underline{K}}^{SVEE}_{r_1;r_3} & \underline{\underline{K}}^{SVEU}_{r_1;r_3} \\ \underline{\underline{K}}^{SVUE}_{r_1;r_3} & \underline{\underline{K}}^{SVUU}_{r_1;r_3} \end{bmatrix} \tag{CE9.19}$$

taking the number of layers to be 200 and the Bessel truncation value to be 16. This example represents a homogeneous region case where the KPE BA and SV (part of the RCWA method) solutions may be directly and meaningfully compared. This example is the same as Ref. [22, Figure 6] in Chapter 8.

Computational Example 9.5 Solution

Table 9.2 lists the matrix elements which were obtained for parameter values listed in the example. The upper number in each box represents the BA transfer matrix element (Equation CE9.18) and the lower number the SV transfer matrix element (Equation CE9.19). The left column lists the (m, m')

TABLE 9.2
Comparison of KPE (Ref. [22] in Chapter 8) Bessel Aperture and State Variable Transfer (Ref. [26] in Chapter 8) Matrices (Bessel $Z_{m,m'}$ [*Top*], SV $Z_{m,m'}$ [*Bottom*])

Mode Numbers	$ZEE_{m,m'}$	$ZEU_{m,m'}$	$ZUE_{m,m'}$	$ZUU_{m,m'}$
$m = 0, m' = 0$	−0.52412E+00	0.11625E+00j	0.30501E+00j	−0.56179E+00
	−0.52358E+00	0.11850E+00j	0.30924E+00j	−0.56155E+00
$m = 1, m' = 2$	−0.24119E+00	−0.13816E+00j	−0.15879E+00j	−0.22103E+00
	−0.24135E+00	−0.13689E+00j	−0.15566E+00j	−0.22201E+00
$m = 5, m' = 0$	0.25595E−03	0.77042E−04j	0.13577E−03j	0.11136E−03
	0.25408E−03	0.77802E−04j	0.13808E−04j	0.11017E−03
$m = 8, m' = 8$	0.50677E+02	0.13698E+02j	−0.76417E+02j	0.20663E+02
	0.50894E+02	0.13752E+02j	−0.76869E+02j	0.20777E+02
$m = 6, m' = 10$	0.21715E+02	0.45514E+01j	−0.17750E+02j	0.37203E+01
	0.21836E+02	0.45757E+01j	−0.17964E+02j	0.37645E+01
$m = 2, m' = 11$	−0.15357E−03	−0.29001E−04j	0.36385E−03j	−0.68712E−04
	−0.14296E−03	−0.26993E−04j	0.36296E−03j	−0.68531E−04
$m = 11, m' = 2$	−0.14314E−07	0.79393E−07j	−0.32961E−07j	−0.18293E−07
	−0.13866E−07	0.79139E−07j	−0.36286E−07j	−0.18268E−07

modal numbers used to define a given matrix element. As can be seen, quite close numerical values have been found from the two different algorithms.

9.3 CASE STUDY II: COMPARISON OF KPE BA, BC VALIDATION METHODS, AND SV METHODS FOR RELATIVELY SMALL DIAMETER SCATTERING OBJECTS

Computational Example 9.6

(a) Using the SV method, using the scattering geometry shown in Figure 9.9,

$$\text{spatial dimensions } r_{Outer} = \tilde{k}_f \tilde{r}_{Outer} = 4, \quad r_{Inner} = \tilde{k}_f \tilde{r}_{Inner} = 2, \quad e = \tilde{k}_f \left|\tilde{e}\right| = 0.2\pi,$$

$$\text{and relative permittivity values } \varepsilon_{Outer} = 2, \quad \varepsilon_{Inner} = 4, \quad \varepsilon_0 = 1,$$

solve for the scattered E_z^s electric field $\tilde{\rho} \to \infty$ field, assuming that the plane wave defined by $\vec{E}^{Inc} = E_0 \exp(j\vec{k}_I \cdot \vec{r})\hat{a}_z, \vec{k}_I = k_{Ix}\hat{a}_x + k_{Iy}\hat{a}_y$

$$\vec{r} = x\hat{a}_x + y\hat{a}_y + z\hat{a}_z \vec{k}_I = k_{Ix}\hat{a}_x + k_{Iy}\hat{a}_y = k_I \cos(\phi_0)\hat{a}_x + k_I \sin(\phi_0)\hat{a}_y \tag{CE9.20}$$

$k_I = (\mu\varepsilon)^{1/2}$, $\phi_0 = 0°$ $E_0 = 1.0$ (V/m) is incident on the dielectric cylinder. The coordinates are normalized (i.e., $x = \tilde{k}_f\tilde{x}$, $\tilde{k}_f = 2\pi/\tilde{\lambda}_f$, $\tilde{\lambda}_f$ is the free space wavelength). The far electric field has the form (see Problems 8.12, 8.13, and Equations 8.86 through 8.88)

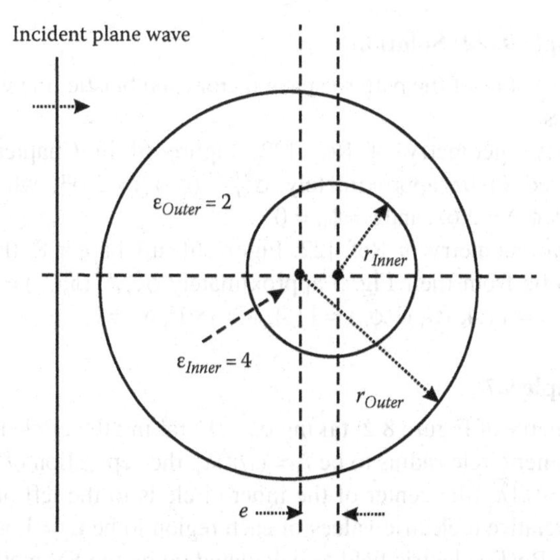

FIGURE 9.9 Geometry for Computational Example 9.6 is shown. Refer to the backscatter geometry of Ref. [22, Figure 6] in Chapter 8.

$$E_z^s \to E_0 \left[\frac{2j}{\pi \tilde{k}_I \tilde{\rho}}\right]^{1/2} \exp(-j\tilde{k}_I\tilde{\rho}) F_{SV}(\phi,\phi_0) \quad \text{(CE9.21)}$$

$$F_{SV}(\phi,\phi_0) = \sum_{m=-\infty}^{\infty} j^m B_{m,SV}^{(0)} \cos(m\phi_0)\cos(m\phi) \quad \text{(CE9.22)}$$

$F_{SV}(\phi, \phi_0)$ is called the pattern space factor of the system. You must implement and solve the matrix Equations 8.80 through 8.83 (formulated using the SV transfer matrices of Computational Example 9.5) to determine the coefficients $B_{m,SV}^{(0)}$.

(b) Repeat part (a) using the exact BA transfer matrices to implement and solve the matrix Equations 8.80 through 8.83. Let the coefficients of this solution be $B_{m,BA}^{(0)}$. The pattern space factor of the system is given by

$$F_{BA}(\phi,\phi_0) = \sum_{m=-\infty}^{\infty} j^m B_{m,BA}^{(0)} \cos(m\phi_0)\cos(m\phi) \quad \text{(CE9.23)}$$

(c) Make a table comparing the coefficients $B_{m,BA}^{(0)}$ and $B_{m,SV}^{(0)}$ for $m = 0, 1, 2, \ldots, 8$.

(d) Plot the two pattern space factors (PSF) $F_{BE}(\phi, \phi_0)$, $F_{SV}(\phi, \phi_0)$ and bistatic line width $\sigma_{BA}(\phi,\phi_0)$, $\sigma_{SV}(\phi,\phi_0)$, and $0° \le \phi \le 180°$, and compare results with that of Ref. [22, Figure 6] in Chapter 8.

Computational Example 9.6, Part (a), (b), and (c) Solutions

Each box in Table 9.3 lists overall system matrix equation solutions for $B_{m,BES}^{(0)}$ (top number, solution based on KPE BA transfer matrices) and $B_{m,SV}^{(0)}$ (bottom number, solution base on SV transfer matrices) for $m = 0, 1, 2, \ldots, 8$. As can be seen from the table below very close values were obtained by the two methods.

Computational Example 9.6(d) Solution

Figure 9.10a and b shows plots of the pattern space factors and biastic line widths respectively using the BA and SV methods.

For the same relative geometry of Ref. [22, Figure 6] in Chapter 8, their back scatter line width was observed to be approximately $\sigma_{KPE}^{b,Fig.6}(\phi,\phi_0) = 2.63$, whereas from Figure 9.10 $\sigma_{BA}(\phi,\phi_0) = 2.63, \sigma_{SV}(\phi,\phi_0) = 2.62$, and $\phi = \phi_0 = 0°$.

For the same relative geometry of Ref. [22, Figure 6] in Chapter 8, their forward scatter line width was observed to be from their Fig. 5 approximately $\sigma_{KPE}^{f,Fig.5}(\phi,\phi_0) = 11.4$, whereas from the figures plotted $\sigma_{BA}^{f}(\phi,\phi_0) = 11.4$, $\sigma_{SV}^{f}(\phi,\phi_0) = 11.4$, $\phi = 180°$, $\phi_0 = 0$.

Computational Example 9.7

Using the general geometry of Figure 8.2; taking $\phi_0 = 0°$; taking the circle radius of the outer circle to be $\tilde{r}_1 = (2/\pi)\tilde{\lambda}_f$, the inner circle radius to be $\tilde{r}_3 = (1/\pi)\tilde{\lambda}_f$, the separation of the centers of inner and outer circles to be $\tilde{e}_{13} = 0.1\tilde{\lambda}_f$ (the center of the inner circle is to the left of the center of the outer circle); and taking the relative dielectric values of each region to be $\varepsilon_0 = 1$, $\varepsilon_1 = 2$, $\varepsilon_2 = 3$, and $\varepsilon_3 = 4$; make plots of the $E_{zR} = \mathrm{Re}(E_z)$ electric field as calculated (a) by the SV method, using 200 thin layers (Ref. [26] in Chapter 8), (b) by the KPE (Ref. [22] in Chapter 8) BA method, and (c) by the KPE

TABLE 9.3
Exact Bessel Function and State Variable Electric Far-Field Coefficients

Mode Order	$B^{(0)}_{m,BA}$ (Top), $B^{(0)}_{m,SV}$ (Bottom)
$m = 0$	(−0.223987E−000, 0.373811E−000)
	(−0.229153E−000, 0.377263E−000)
$m = 1$	(−0.578125E−000, −1.431918E−002)
	(−0.586970E−000, −1.910897E−002)
$m = 2$	(0.712468E−000, −0.189252E−000)
	(0.714336E−000, −0.190040E−000)
$m = 3$	(0.894154E−000, 1.528494E−000)
	(0.891799E−000, 1.538230E−000)
$m = 4$	(−1.703924E−000, −0.988040E−000)
	(−1.673181E−000, −1.005214E−000)
$m = 5$	(0.209672E−000, 2.054497E−002)
	(0.208292E−000, 2.188926E−002)
$m = 6$	(−4.033155E−003, 1.113300E−002)
	(−4.085096E−003, 1.095678E−002)
$m = 7$	(−6.833338E−004, 1.599906E−004)
	(−6.774510E−004, 1.577348E−004)
$m = 8$	(−1.897455E−005, −6.549961E−005)
	(−1.844916E−005, −6.525036E−005)

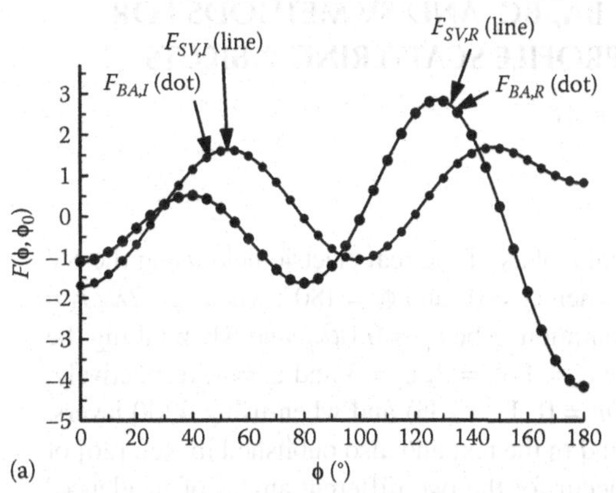

FIGURE 9.10 Plots the two pattern space factors (a) $F_{BE}(\phi, \phi_0)$, $F_{SV}(\phi, \phi_0)$, and bistatic line width (b) $\sigma_{BA}(\phi, \phi_0)$, $\sigma_{SV}(\phi, \phi_0)$, and $0° \leq \phi \leq 180°$ for Computational Example 9.6d are shown.

(Ref. [22] in Chapter 8) BC method. In your solution use 17 Bessel function modes ($m = 0, 1, \ldots, 16$) for all three methods to solve the matrix equations.

Computational Example 9.7, Part (a), (b), and (c) Solutions

In observing the three, E_{zR}, real electric field plots, it can be seen that very close agreement exists between the three calculation methods (Figure 9.11).

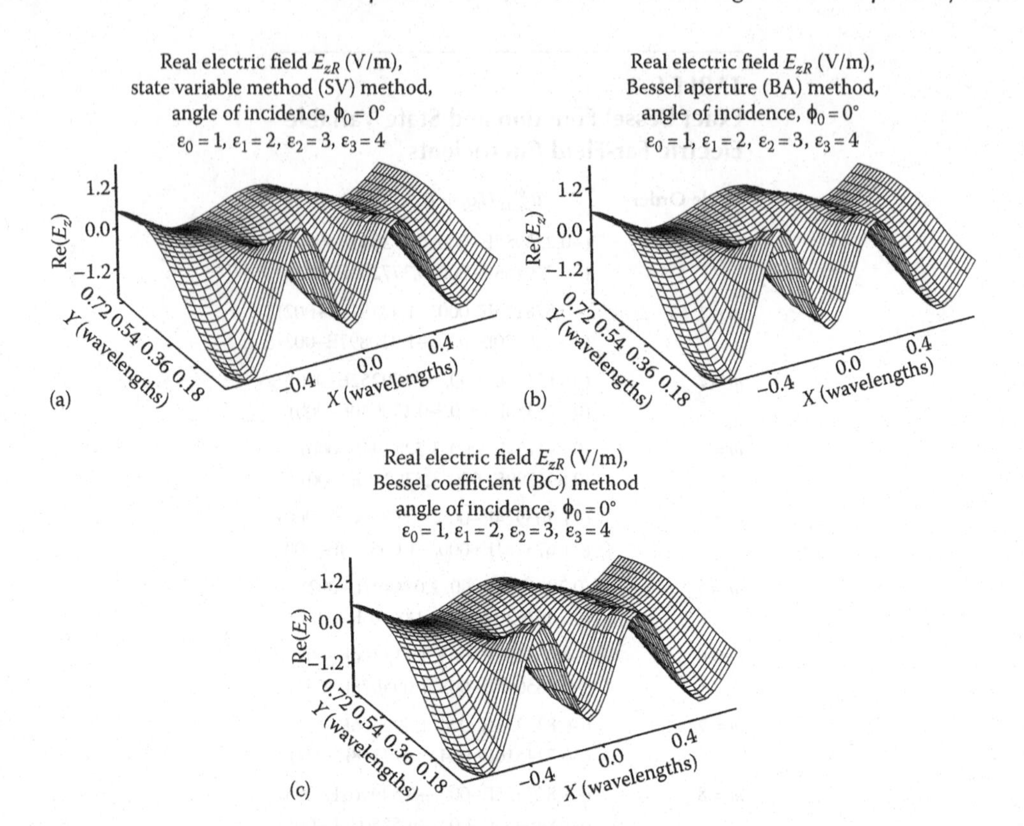

FIGURE 9.11 Plots of the $E_{zR} = \text{Re}(E_z)$ electric field as calculated (a) by the state variable (SV) method, using 200 thin layers (Ref. [26] in Chapter 8), (b) by the KPE (Ref. [22] in Chapter 8) Bessel aperture (BA) method, and (c) by the KPE (Ref. [22] in Chapter 8) Bessel coefficient (BC) method.

9.4 CASE STUDY III: COMPARISON OF BA, BC, AND SV METHODS FOR GRADUALLY, STEPPED-UP, INDEX PROFILE SCATTERING OBJECTS

(Diameter = $2\tilde{r}_1 = 4\tilde{\lambda}_f$, $\varepsilon_0 = 1$, $\varepsilon_1 = 2$, $\varepsilon_2 = 3$, and $\varepsilon_3 = 4$)

Computational Example 9.8

(a) Using the general geometry of Figure 8.2, make plots of the real electric field using the SV method (call it $E_{zR}^{SV} = \text{Re}(E_z^{SV})$) for the cases when $\phi_0 = 0°$ and $\phi_0 = 180°$; when $\tilde{r}_1 = 2\tilde{\lambda}_f$, $\tilde{r}_3 = \tilde{\lambda}_f$, when taking the inner and outer circle separation to be $\tilde{e}_{13} = 0.1\pi\tilde{\lambda}_f$; and when taking the relative dielectric values of each region to be $\varepsilon_0 = 1$, $\varepsilon_1 = 2$, $\varepsilon_2 = 3$ and $\varepsilon_3 = 4$, respectively; and when using 36 Bessel function modes ($m = 0, 1, \ldots, 35$) and when using 4200 layers. Compare your answers and plots to those found in the text and also published in Ref. [26] of Chapter 8. Do different scattering patterns occur for the two different angles of incidence? For visualization purposes, make a plot of the scattering object dielectric function $\varepsilon(x, y)$.

(b) To validate your SV solution, make plots of the BA solution $E_{zR}^{BA} = \text{Re}(E_z^{BA})$ for $\phi_0 = 0°$ and $\phi_0 = 180°$ for the parameter case described in part (a).

(c) To further validate your SV solution also make plots of the difference, real electric field given by $E_{zR}^{diff} = \text{Re}(E_z^{SV} - E_z^{BA})$ for $\phi_0 = 0°$ and $\phi_0 = 180°$ for the parameter case described in part (a).

(d) In this problem, the BA method electric field E_z^{BA} may be considered the "exact" electric field solution to the EM scattering problem and thus the one from which an accurate error analysis may be made. Comment on how well the SV method has calculated the correct solution.

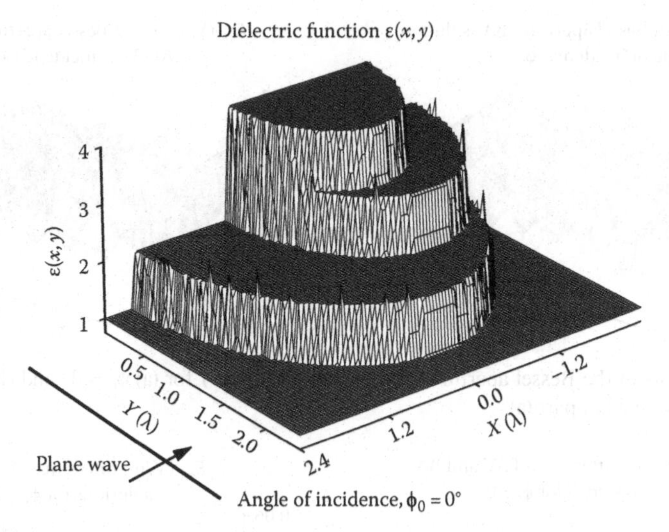

FIGURE 9.12 Plot of the scattering object dielectric function $\varepsilon(x, y)$.

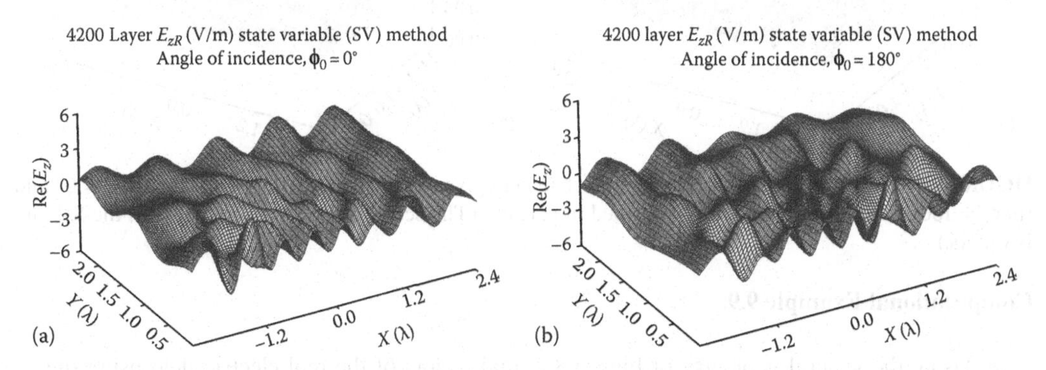

FIGURE 9.13 Plots of the real electric field using the state variable (SV) method (call it $E_{zR}^{SV} = \text{Re}(E_z^{SV})$) for the cases when (a) $\phi_0 = 0°$ and (b) $\phi_0 = 180°$.

Computational Example 9.8(a) Solution

The scattering object dielectric function $\varepsilon(x, y)$ is shown in Figure 9.12 and the 4200 SV layer solution for $\phi_0 = 0°$ and 180° is shown in Figure 9.13. Quite different electric field patterns result for the two different angles of incidence.

Computational Example 9.8(b) Solution

The 4200 BA solution for $\phi_0 = 0°$ and 180° is shown in Figure 9.14.

Computational Example 9.8(c) Solution

The error difference plots for the SV 4200 layer calculation and the BA solution for the $\phi_0 = 0°$ and 180° angles of incidence are shown in Figure 9.15.

Computational Example 9.8(d) Solution

The SV solution has calculated the correct solution fairly accurately as seen from Figures 9.13 through 9.15.

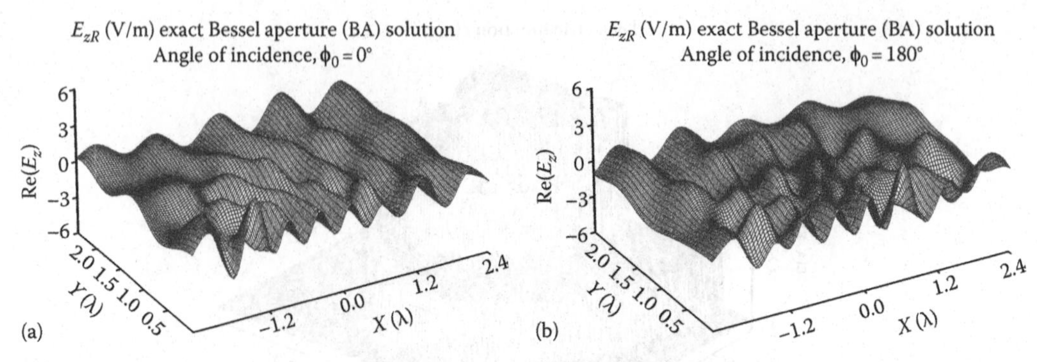

FIGURE 9.14 Plots of the Bessel aperture solution $E_{ZR}^{BA} = \text{Re}(E_Z^{BA})$ for (a) $\phi_0 = 0°$ and (b) $\phi_0 = 180°$ for the parameter case described in part (a).

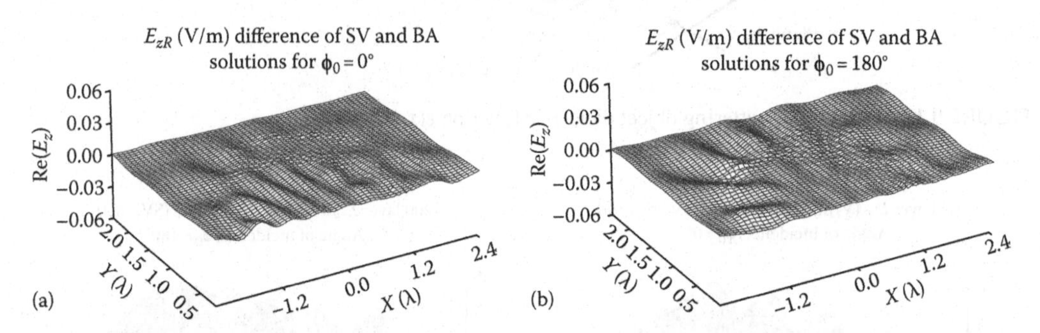

FIGURE 9.15 Plots of the difference, real electric field given by $E_{zR}^{diff} = \text{Re}(E_z^{SV} - E_z^{BA})$ for (a) $\phi_0 = 0°$ and (b) $\phi_0 = 180°$ for the parameter case described in part (a). (The scale is 100 times smaller than the SV and BA plots.)

Computational Example 9.9

(a) Using the general geometry of Figure 8.2, make plots of the real electric field using the BA method (call it $E_{zR}^{BA} = \text{Re}(E_z^{BA})$) and the real electric field using the SV method (call it $E_{zR}^{SV} = \text{Re}(E_z^{SV})$) for the case when $\phi_0 = 0°$; $\tilde{r}_1 = 2\tilde{\lambda}_f$, $\tilde{r}_3 = \tilde{\lambda}_f$, inner and outer circle separation to be $\tilde{e}_{13} = 0.1\pi\tilde{\lambda}_f$; taking the relative dielectric values of each region to be $\varepsilon_0 = 1$, $\varepsilon_1 = 2$, $\varepsilon_2 = 3$, and $\varepsilon_3 = 4$, respectively; and using 36 Bessel function modes $m = 0, 1, \ldots, 35$. The $E_{zR}^{SV} = \text{Re}(E_z^{SV})$ electric field SV plots should be made using 20, 100, and 8400 layers. The dielectric permittivity function $\varepsilon(x, y)$ is shown in Computational Example 9.8.

(b) Make plots of the difference, real electric field given by $E_{zR}^{diff} = \text{Re}(E_z^{SV} - E_z^{BA})$ formed for the cases when 20, 100, and 8400 layers were used to calculate the SV electric field E_z^{SV}. The BA method electric field E_z^{BA} may be considered the "exact" Electric field solution to the EM scattering problem and thus the one from which an accurate error analysis may be made. Comment on the number of layers needed to calculate the electric field by the SV method reliably.

Computational Example 9.9(a) Solution

The 20, 100, and 8400 SV layer solutions and exact BA solution are shown in Figure 9.16 (a through d) using 36 modes. As can be seen, even the 20 and 100 layer SV solutions approximate the BA solution fairly well.

Computational Example 9.9(b) Solution

Error difference plots for the 20, 100, and 8400 layer calculation. From observing the 100 layer plot, very reliable electric field SV plots are being made using 100 or more layers (Figure 9.17).

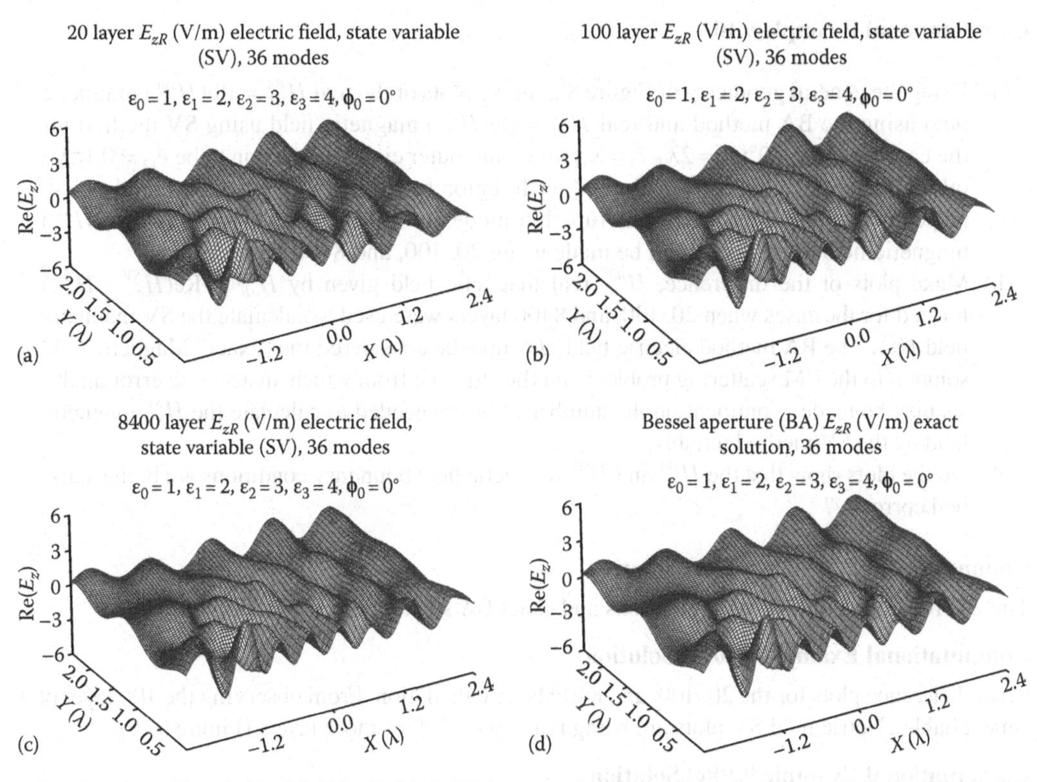

FIGURE 9.16 The (a) 20, (b) 100, (c) 8400 SV layer, and (d) exact Bessel aperture (BA) E_z electric field solutions are shown using 36 modes.

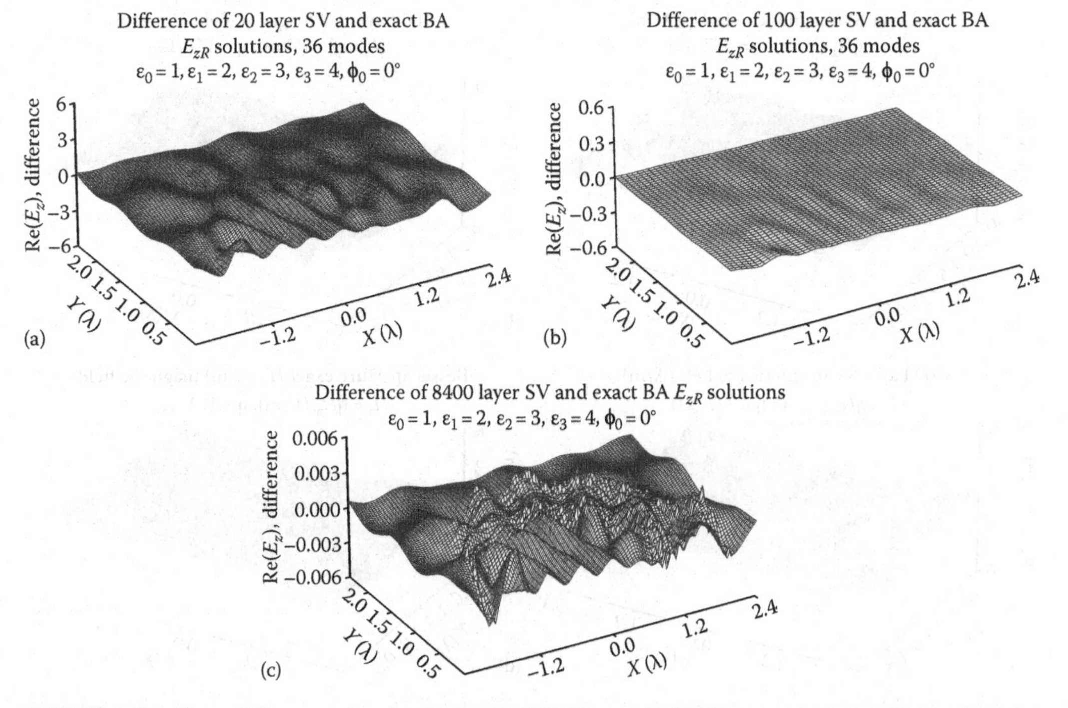

FIGURE 9.17 Error difference plots for the (a) 20, (b) 100, and (c) 8400 layer calculation. From observing the 100 layer plot, very reliable electric field SV plots are being made 100 or more layers.

Computational Example 9.10

(a) Using the general geometry of Figure 8.2, make plots of the real $H_{xR}^{BA} = \text{Re}(H_x^{BA})$ magnetic field using the BA method and real $H_{xR}^{SV} = \text{Re}(H_x^{SV})$ magnetic field using SV method for the case when $\phi_0 = 0°$; $\tilde{r}_1 = 2\tilde{\lambda}_f$, $\tilde{r}_3 = \tilde{\lambda}_f$, inner and outer circle separation to be $\tilde{e}_{13} = 0.1\pi\tilde{\lambda}_f$; taking the relative dielectric values of each region to be $\varepsilon_0 = 1$, $\varepsilon_1 = 2$, $\varepsilon_2 = 3$, and $\varepsilon_3 = 4$, respectively; and using 36 Bessel function modes $m = 0, 1, \ldots, 35$. The $H_{xR}^{SV} = \text{Re}(H_x^{SV})$ magnetic field SV plots should be made using 20, 100, and 8400 layers.

(b) Make plots of the difference, H_{xR}^{diff} real magnetic field given by $H_{xR}^{diff} = \text{Re}(H_x^{SV} - H_x^{BA})$ formed for the cases when 20, 100, and 8400 layers were used to calculate the SV magnetic field H_x^{SV}. The BA method electric field H_x^{BA} may be considered the "exact" Magnetic field solution to the EM scattering problem and thus the one from which an accurate error analysis may be made. Comment on the number of layers needed to calculate the H_x^{SV} magnetic field by the SV method reliably.

(c) Do the plots show that the H_x^{BA} and H_x^{SV} magnetic field boundary conditions are being satisfied correctly?

Computational Example 9.10(a) Solution

The 20, 100, and 8400 SV layer solutions and exact BA solution are shown in Figure 9.18.

Computational Example 9.10(b) Solution

Error difference plots for the 20, 100, and 8400 layer calculation. From observing the 100 layer plot, very reliable electric field SV plots are being made with 100 or more layers (Figure 9.19).

Computational Example 9.10(c) Solution

The plots show that the H_x^{BA}, H_x^{SV} magnetic field boundary conditions are being met properly. Specifically, from observing the plots for the BA and SV methods that $H_x\big|_{y=0} = 0$, is the proper

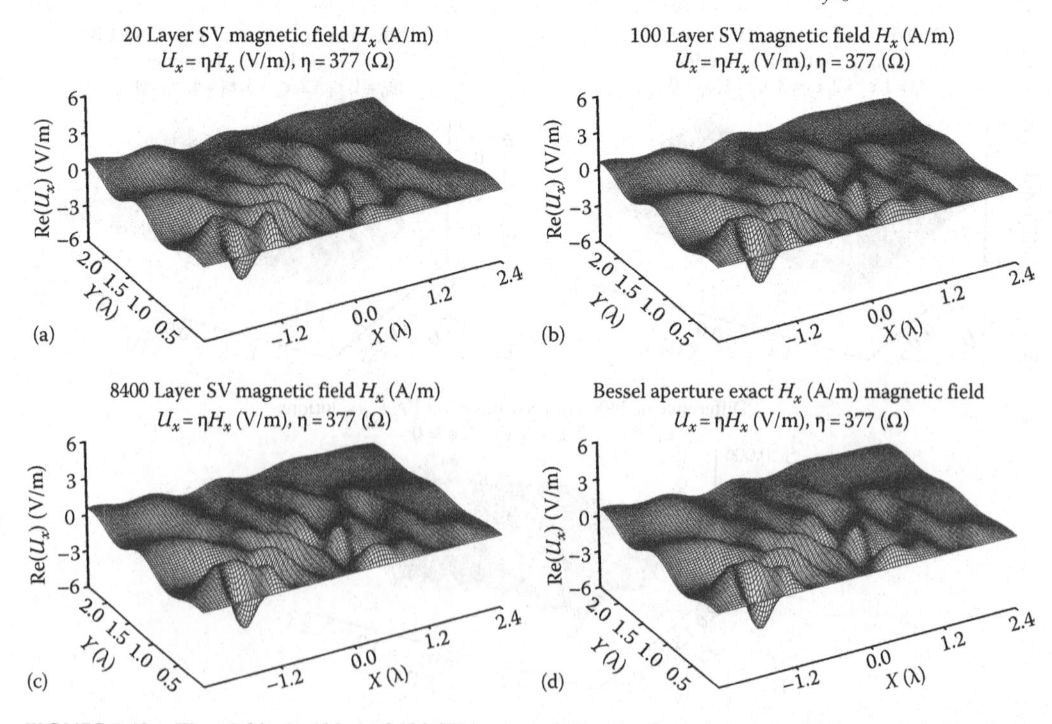

FIGURE 9.18 The (a) 20, (b) 100, (c) 8400 SV layer, and (d) exact Bessel aperture H_x magnetic field solution are shown.

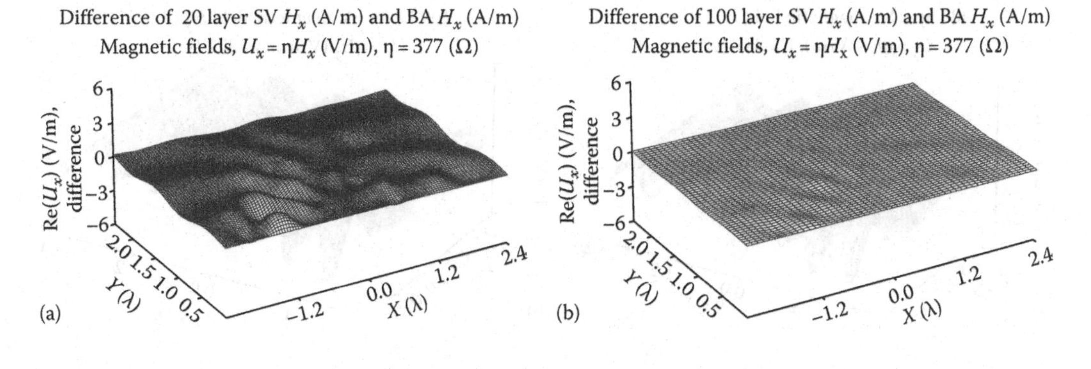

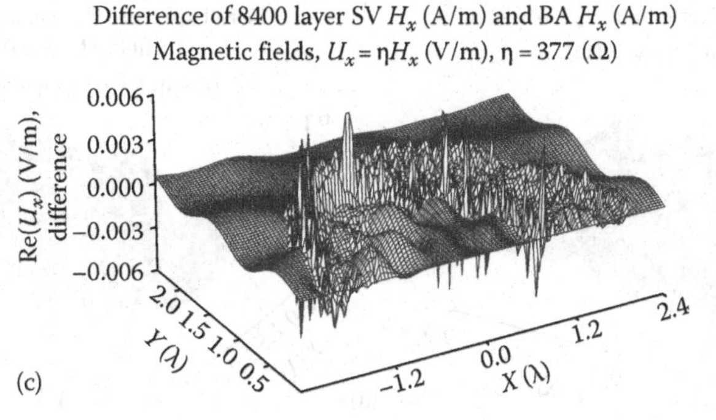

FIGURE 9.19 Magnetic field error difference plots for the (a) 20, (b) 100, and (c) 8400 layer calculations are shown. (The scale is 1000 times smaller than 20 and 100 layer difference plots.)

boundary condition for a perfect magnetic conductor and for the EM symmetry case which is being studied. Also observed throughout the H_x^{BA}, H_x^{SV} plots, it is seen that these magnetic fields are continuous, which is to be expected since the magnetic permeability in all the calculations has been assumed to be that of a vacuum, indicating that both the normal and tangential magnetic field and magnetic flux density field components at all boundaries must be equal, thus leading to continuous magnetic field components at all boundaries.

Computational Example 9.11

(a) Using the general geometry of Figure 8.2, make plots of the real $H_{yR}^{BA} = \text{Re}(H_y^{BA})$ magnetic field using the BA method and real $H_{yR}^{SV} = \text{Re}(H_y^{SV})$ magnetic field using the SV method for the case when $\phi_0 = 0°$; $\tilde{r}_1 = 2\tilde{\lambda}_f$, $\tilde{r}_3 = \tilde{\lambda}_f$, when the inner and outer circle separation is $\tilde{e}_{13} = 0.1\pi\tilde{\lambda}_f$; when the relative dielectric values of each region are taken to be $\varepsilon_0 = 1$, $\varepsilon_1 = 2$, $\varepsilon_2 = 3$, and $\varepsilon_3 = 4$, respectively; and using 36-Bessel function modes $m = 0, 1, \ldots, 35$. The $H_{yR}^{SV} = \text{Re}(H_y^{SV})$ magnetic field SV plots should be made using 20, 100, and 8400 layers.

(b) Make plots of the difference H_{yR}^{diff} real magnetic field given by $H_{yR}^{diff} = \text{Re}(H_y^{SV} - H_y^{BA})$ formed for the cases when 20, 100, and 8400 layers are used to calculate the SV magnetic field H_y^{SV}. The BA method electric field H_y^{BA} may be considered the "exact" magnetic field solution to the EM scattering problem and thus the one from which an accurate error analysis may be made. Comment on the number of layers needed to calculate the H_y^{SV} magnetic field by the SV method reliably.

(c) Make general comments on the 8400 layer and the BA "exact" solution.

(d) Do the plots show that the H_y^{BA}, H_y^{SV} magnetic field boundary conditions are being satisfied correctly?

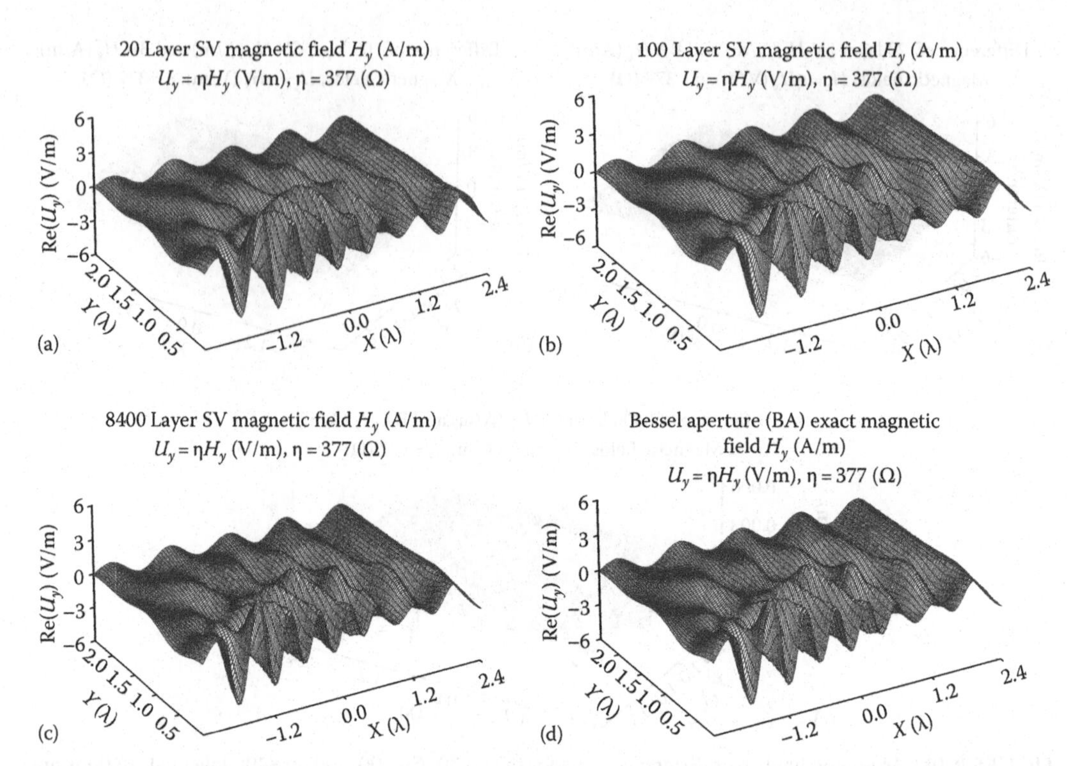

FIGURE 9.20 The (a) 20, (b) 100, (c) 8400 SV layer, and (d) exact Bessel aperture H_y magnetic field solutions are shown.

Computational Example 9.11(a) Solution

The 20, 100, and 8400 SV layer solution and exact BA solution is shown in Figure 9.20.

Computational Example 9.11(b) Solution

Error difference plots for the 20, 100, and 8400 layer calculation. From observing the 100 layer plot, very reliable electric field SV plots are being made for 100 or more layers (Figure 9.21).

Computational Example 9.11(c) Solution

The difference between the 8400 layer and "exact" BA solutions is very small and random. It is difficult to judge whether the exact solution is the SV or BA, since the BA solution itself is based on a finite number of modes. Both the SV and BA solutions are very independent of one another and in a sense both solutions, because of their closeness in value, reinforce each others correctness.

Computational Example 9.11(d) Solution

The plots show that H_y^{BA}, H_y^{SV} magnetic field boundary conditions are being properly met. Specifically, from observing the plots for the BA and SV methods, $H_y\big|_{y=0} \neq 0$ is the proper boundary condition for a perfect magnetic conductor and for the normal magnetic field component. Also observed throughout the H_y^{BA}, H_y^{SV} plots, it is seen that these magnetic fields are continuous, which is to be expected since the magnetic permeability in all the calculations has been assumed to be that of a vacuum, thus indicating both the normal and tangential magnetic field and magnetic flux density field components at all boundaries must be equal, thus leading to continuous magnetic field components at all boundaries.

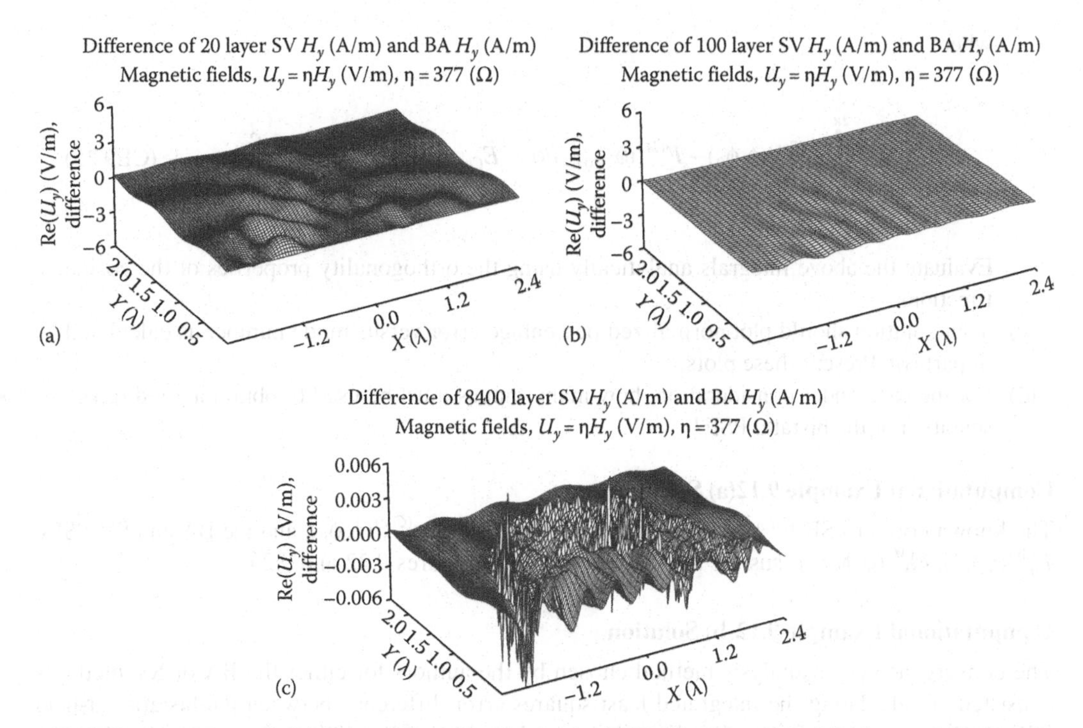

FIGURE 9.21 Magnetic field error difference plots for the (a) 20, (b) 100, and (c) 8400 layer calculations are shown. (The scale is 1000 times smaller than 20 and 100 layer difference plots.)

Computational Example 9.12

Using the general geometry of Figure 8.2, perform a Pattern Space Factor (PSF) error convergence analysis of your design for both the BA and SV methods, based on calculating a known correct PSF $F^{(Cor)}(\phi, \phi_0)$ (see Problems 8.12 and 8.13, and Equations 8.86 through 8.88) to determine the matrix size needed to obtain accurate EM scattering numerical results. Study the following parameter case taking $\phi_0 = 180°$; $\tilde{r}_1 = 2\tilde{\lambda}_f$, $\tilde{r}_3 = \tilde{\lambda}_f$; taking the inner and outer circle separation to be $\tilde{e}_{13} = 0.1\pi\tilde{\lambda}_f$; and taking the relative dielectric values of each region to be $\varepsilon_0 = 1$, $\varepsilon_1 = 2$ $\varepsilon_2 = 3$, and $\varepsilon_3 = 4$, respectively. In the present problem, since $F^{(Cor)}(\phi, \phi_0)$ is not known exactly, take the "correct" PSF. $F^{(Cor)}(\phi, \phi_0)$ is to be estimated and calculated by using the BA method (known to be an accurate method) using a large enough truncation value to obtain accurate numerical results. (Numerical testing for the present case showed that $M = 38$ modes [$m = 0, 1, 2, \ldots, 37$] Bessel function modes was a sufficiently large value to obtain accurate numerical results.)

(a) Calculate the correct PSF $F^{(Cor)}(\phi, \phi_0)$ and use it to calculate and plot the bistatic line width $\sigma^{(Cor)}(\phi, \phi_0)$ versus the angle ϕ. Calculate the BA and SV PSF, $F_{BA}^{(M)}(\phi,\phi_0)$, $F_{SV}^{(M)}(\phi,\phi_0)$, which have been determined by using M Bessel function modes $M < 37$. Use them to calculate and plot the bistatic line widths $\sigma_{BA}^{(M)}(\phi,\phi_0)$ and $\sigma_{SV}^{(M)}(\phi,\phi_0)$. Make one set of plots using 4200 SV layers and taking M, the truncation value to be $M = 15, 17, 35$. Make one set of plots using 100 and 4200 SV layers, taking the truncation value to be $M = 35$ for each plot.

(b) Define the Root Means Square (RMS) normalized integrated error difference between the approximate and correct PSF to be

$$E_{RMS} = \left[\frac{\tilde{E}_M}{\tilde{E}_{Cor}}\right]^{1/2} \tag{CE9.24}$$

where

$$\tilde{E}_M = \frac{2}{\pi}\int_0^{2\pi} \left|F^{(Cor)}(\phi,\phi_0) - F^{(M)}(\phi,\phi_0)\right|^2 d\phi, \quad \tilde{E}_{Cor} = \frac{2}{\pi}\int_0^{2\pi} \left|F^{(Cor)}(\phi,\phi_0)\right|^2 d\phi \tag{CE9.25}$$

Evaluate the above integrals analytically using the orthogonality properties of the cosine functions.

(c) Your solution should plot normalized percentage error versus mode number as calculated in part (b). Present these plots.

(d) For the case under consideration, how many modes must be used to obtain a good representation of the bistatic PSF?

Computational Example 9.12(a) Solution

The known correct PSF $F^{(Cor)}(\phi, \phi_0)$, the bistatic line width $\sigma^{(Cor)}(\phi,\phi_0)$, and the BA and SV PSFs, $F_{BA}^{(M)}(\phi,\phi_0)$, $F_{SV}^{(M)}(\phi,\phi_0)$ versus the angle ϕ are shown in Figures 9.22 and 9.23.

Computational Example 9.12(b) Solution

The convergence error analysis method chosen by the authors for either the BA or SV methods consisted of calculating the integrated least squares error difference between the bistatic far-field PSF of a known correct PSF, call it $F^{(Cor)}(\phi, \phi_0)$, and the PSF, call it $F^{(M)}(\phi, \phi_0)$, as calculated by the BA or SV methods, where $M \equiv M_{trunc} = M_T + 1$ ($m = 0, 1, \ldots, M_T$) is the number of Bessel function modes used to truncate the Bessel function series used to calculate the PSF in a given region (M is called the mode truncation value). The error difference Equation CE9.24 is defined as

$$\tilde{E} = \frac{2}{\pi}\int_0^{2\pi} \left|F^{(Cor)}(\phi,\phi_0) - F^{(M)}(\phi,\phi_0)\right|^2 d\phi \tag{CE9.26}$$

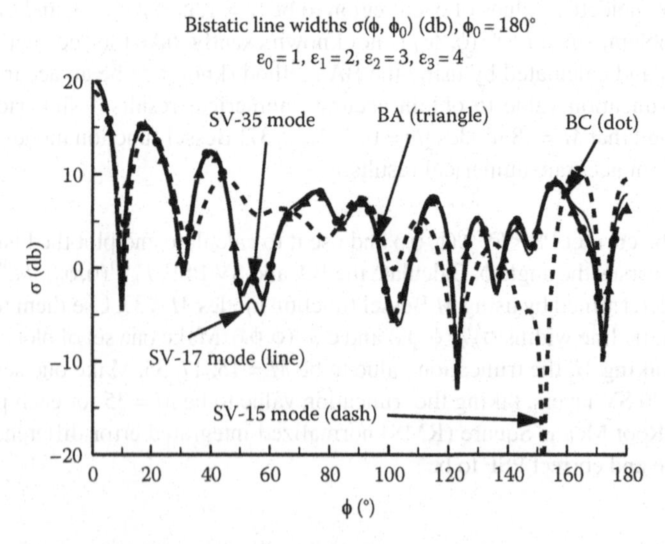

FIGURE 9.22 Plots using 4200 SV layers and taking M the truncation value to be M = 15, 17, and 35 are shown. Comparison of the 32 mode-Bessel coefficient (BC) and the 15, 17, and 35 mode-state variable (SV) (4200 layers) solutions. The 38 mode-Bessel aperture (BA) solution is taken as "correct."

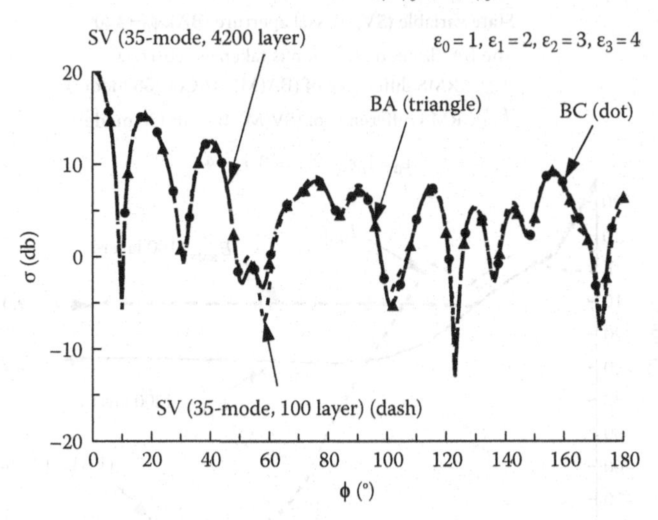

FIGURE 9.23 Plots using 100 and 4200 SV layers and taking the truncation value to be $M = 35$ are shown. Comparison of the 32 mode-Bessel coefficient (BC) and the 35 mode-state variable (SV) (100, 4200 layer) solutions. The 38 mode-Bessel aperture (BA) solution is taken as "correct."

where for the symmetric case being considered $\phi_0 = 0, \pi$ and $F^{(Cor)}(\phi, \phi_0)$, $F^{(M)}(\phi, \phi_0)$ have the general form

$$F(\phi,\phi_0) = \sum_{m=0}^{\infty} B_m \cos(m\phi)\cos(m\phi_0) \tag{CE9.27}$$

Letting $B_m^{(Cor)}$, $B_m^{(M)}$ be the PSF coefficients of $F^{(Cor)}$, $F^{(M)}$ respectively, after carrying out the integration of in the above equation, after using the orthogonality and symmetry of the $\cos(m\phi)$ functions, and after using the fact that $\phi_0 = 0°, 180°$, it is found

$$\tilde{E}_M = 2\sum_{m=0}^{M_T} (1+\delta_{m,0})\left|B_m^{(Cor)} - B_m^{(M)}\right|^2 \tag{CE9.28}$$

where $\delta_{0,0} = 1$ and $\delta_{0,m} = 0$, for $m = 1, 2, \ldots$. The RMS, normalized error used to display error results in this problem is given by

$$E_{RMS} = \left[\frac{\tilde{E}_M}{\tilde{E}_{Cor}}\right]^{1/2} \tag{CE9.29}$$

where

$$\tilde{E}_{Cor} = \frac{2}{\pi}\int_0^{2\pi} \left|F^{(Cor)}(\phi,\phi_0)\right|^2 d\phi \tag{CE9.30}$$

In the present solution, since $F^{(Cor)}(\phi, \phi_0)$ is not known exactly, it has been estimated and calculated by using the BA method (known to be an accurate method) using a large enough truncation value (for this problem 38 modes) to provide accurate numerical results.

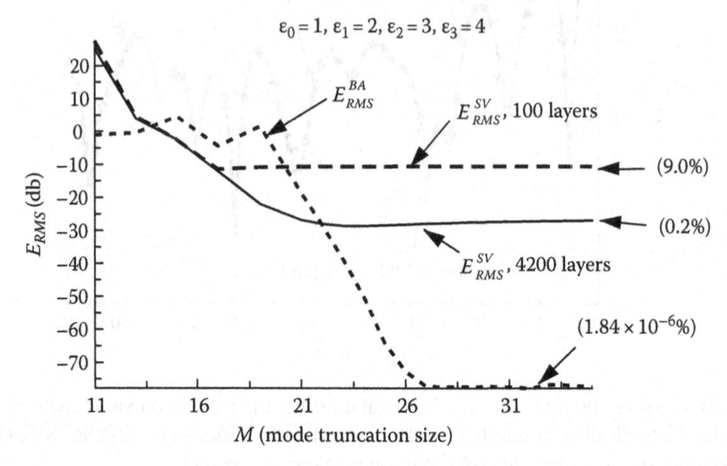

FIGURE 9.24 Plot of normalized error in (db) versus mode number as calculated in part (b). Also listed is percent error at point shown.

Computational Example 9.12(c) Solution

Figure 9.24 shows the RMS integrated, bistatic error E_{RMS} (db) as calculated by Equations CE9.24 through CE9.30, determined by the BA method and the SV methods (using 100 Layers and 4200 layers).

Computational Example 9.12(d) Solution

From viewing the two bistatic line width figures (Figures 9.22 and 9.23) and from viewing the error plots of Figure 9.24, the SV method using 24 modes and 4200 layers, gave excellent bistatic EM line width field results. From the error plot, it's interesting to note that the SV, 4200 Layer solution actually gives a better percentage error for result for $16 \leq M \leq 24$ than did the generally more accurate BA method solution.

Computational Example 9.13

(a) This example concerns studying the convergence of the BA method, the BC method, and the SV method when a different number of Bessel function and SV truncation modes is used to calculate the electric and magnetic fields of the system over a given spatial region (assumed to enclose the scattering object). The electric field convergence is to be studied by using the electric error functions defined respectively by

$$E_{RMS}^{elec}(M) = \frac{\left\{1/(2L_xL_y)\int_0^{L_y}\int_{-L_x}^{L_x}\left|E_z^{(M)}(x,y) - E_z^{exact}(x,y)\right|^2 dx\,dy\right\}^{1/2}}{E_{z,RMS}^{exact}} \tag{CE9.31}$$

$$E_{z,RMS}^{exact} = \left\{\frac{1}{(2L_xL_y)}\int_0^{L_y}\int_{-L_x}^{L_x}\left|E_z^{exact}(x,y)\right|^2 dx\,dy\right\}^{1/2} \tag{CE9.32}$$

where $M \equiv M_{trunc} = M_T + 1$ ($m = 0, 1, \ldots, M_T$) is the number of Bessel function modes used to truncate the Bessel function series used to calculate the EM fields in a given region (M is called the mode truncation value); $E_z^{exact}(x,y)$ represents the "exact" (or "correct") EM solution (The 38-Mode BA solution (i.e., $M = 38$) was taken to be this solution); $E_z^{(M)}(x,y)$ represents either the BA, BC, or SV solutions (using M Bessel function or SV modes respectively); and L_x, L_y defines the x, y region size where the $E_z(x, y)$ electric field difference is to be calculated.

(b) The magnetic field convergence is to be studied by using a magnetic error function defined similarly to that of the electric field, namely,

$$E_{RMS}^{mag}(M) = \frac{\left\{1/(2L_xL_y)\int_0^{L_y}\int_{-L_x}^{L_x}\left[\left|H_x^{(M)} - H_x^{exact}\right|^2 + \left|H_y^{(M)} - H_y^{exact}\right|^2\right]dxdy\right\}^{1/2}}{H_{RMS}^{exact}} \quad \text{(CE9.33)}$$

$$H_{RMS}^{exact} = \left\{\frac{1}{(2L_xL_y)}\int_0^{L_y}\int_{-L_x}^{L_x}\left[\left|H_x^{exact}\right|^2 + \left|H_y^{exact}\right|^2\right]dxdy\right\}^{1/2} \quad \text{(CE9.34)}$$

Using the error functions defined in Equations CE9.31 through CE9.34, make plots of $E_{RMS}^{elec}(M)$ and $E_{RMS}^{mag}(M)$ (expressed as [db]) versus the truncation mode number M. These error plots should be made using the general geometry of Figure 8.2, for the case when $\phi_0 = 180°$; $\tilde{r}_1 = 2\tilde{\lambda}_f$, $\tilde{r}_3 = \tilde{\lambda}_f$, where the inner and outer circle separation is $\tilde{e}_{13} = 0.1\pi\tilde{\lambda}_f$; taking the relative dielectric values of each region to be $\varepsilon_0 = 1$, $\varepsilon_1 = 2$, $\varepsilon_2 = 3$, and $\varepsilon_3 = 4$, respectively; and where L_x, L_y each represent widths of $2.4\tilde{\lambda}_f$ ($\tilde{\lambda}_f$ is the free space wavelength).

Computational Example 9.13(a) Solution

The electric field RMS error of Equation CE9.31 as determined using a different number of SV layers is shown and a different mode truncation value is shown. Also shown are the BA and BC method solutions in Figure 9.25.

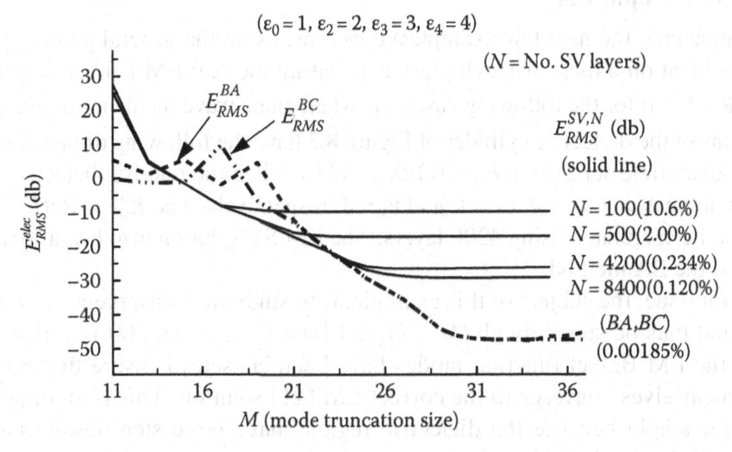

FIGURE 9.25 Plot of normalized electric field error (expressed in db) versus mode number as calculated by Equations CE9.33 and CE9.34. $E_{RMS}^{SV,N}$ in (db) refers to the number of layers N used to make the SV solution.

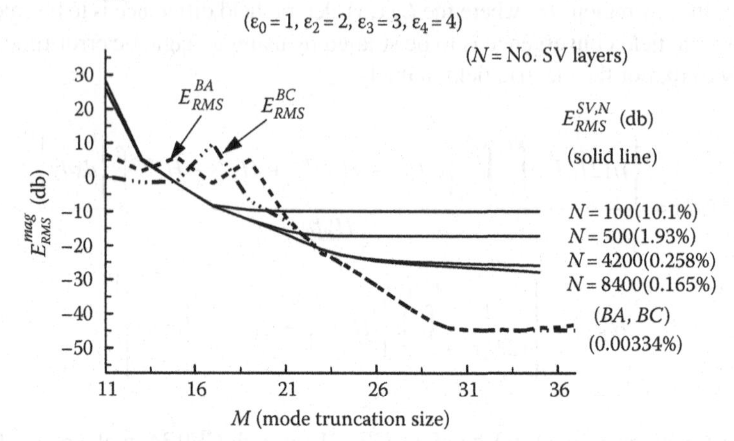

FIGURE 9.26 Plot of normalized magnetic field error (expressed in db) versus mode number as calculated in part (b). $E_{RMS}^{SV,N}$ in (db) refers to the number of layers N used to make the SV solution.

Computational Example 9.13(b) Solution

The magnetic field RMS error of Equation CE9.33 as determined using a different number of SV layers is shown and a different mode truncation value is shown. Also shown are the BA and BC method solutions in Figure 9.26.

9.5 CASE STUDY IV: COMPARISON OF BA, BC, AND SV METHODS FOR MISMATCHED, INDEX PROFILE, SCATTERING OBJECTS

(Diameter $= 2\tilde{r}_1 = 4\tilde{\lambda}_f$, $\varepsilon_0 = 1$, $\varepsilon_1 = 4$, $\varepsilon_2 = 1$ and $\varepsilon_3 = 4$)

Computational Example 9.14

The next example (and the next few examples) concerns, using the general geometry of Figure 8.2 (plane wave incident on a dielectric cylinder), calculating the real EM fields using the SV method (call it $E_{zR}^{SV} = \mathrm{Re}(E_z^{SV})$) for the following cases: (1) when plane wave incidence angle is $\phi_0 = 180°$; (2) when the regions of the dielectric cylinder of Figure 8.2 have the following dimensions: $\tilde{r}_1 = 2\tilde{\lambda}_f$, $\tilde{r}_3 = \tilde{\lambda}_f$, inner and outer circle separation $\tilde{e}_{13} = 0.1\pi\tilde{\lambda}_f$; and (3) when the relative dielectric values of each region is taken to be $\varepsilon_0 = 1$, $\varepsilon_1 = 4$, $\varepsilon_2 = 1$, and $\varepsilon_3 = 4$, respectively. The $E_{zR}^{SV} = \mathrm{Re}(E_z^{SV})$ electric field SV plots should be calculated using 4200 layers. The EM SV solution is to be validated by both the BC method and the BA method.

An important issue, the subject of this example, is to study the convergence of both the BC and BA methods and thus be sure enough ($M = M_T + 1$ [$m = 0, 1,\ldots, M_T$] [M is called the truncation mode value]) the EM Bessel function modes have been chosen to ensure that both the BC and BA methods themselves converge to the correct EM field solution. This is an important question in the present example because the dielectric regions have large step discontinuities and also have sharp discontinuous boundaries, thus making convergence by any method to the exact EM field solution difficult to obtain. A basis for deciding whether enough modes have been chosen for the two validation methods just described to have converged to the correct solution is (1) if both methods show very little change in the EM solution as the number of truncation modes is increased past a certain size and (2) if the two different methods closely agree with each other for a truncation value for which convergence has been achieved. For both the BC and BA matrix methods, the Bessel mode truncation size cannot be made arbitrarily large for a truly complete study, because the modal Bessel function values approach exponentially large values, and thus a meaningful matrix inversion for the numerical solution for this case cannot be carried out, and thus clear numerical instability (i.e., ill-conditioned numerical results) will occur. From the statements just been made, for the two methods under consideration, it is expected that there will be a range of truncation values where the two methods agree closely and where there is little change in the EM solution. For truncation values too much beyond this range of value, large errors in the correct EM solution is expected to occur. For those truncation values where the BC and BA methods closely agree, these solutions can be taken as proper, correct or "exact" validation solutions for the SV method.

To quantitatively study this convergence example, carry out the following steps:

(a) Make a 3-D plot of the dielectric function for this example in order that a proper visualization of the scattering geometry is displayed.
(b) Calculate the EM fields over a given representative region of space by the BC and BA methods, using a number of different $M = M_T + 1$ truncation values, ranging from $M = M_T + 1 = 11, 12, \ldots, 41$. From the resulting field solutions (both electric and magnetic), calculate the integrated EM field difference (see Computational Example 9.13, Equations CE9.31 through CE9.34), call it $E(M, M')$, taking the M, M' value to be $(M, M') = (11, 12, \ldots, 35)$. Make 3-D error plots, and from these plots search for (M, M') regions of convergence.
(c) Once you have determined a reasonable range of truncation convergence values, $M = M_T + 1$, from part (b), plot the electric and magnetic EM fields for the two methods for a few different values of M. Using the results of part (c), plot the difference of the EM fields that result when two different, but close in value, truncation values are used to calculate the EM fields of the system under study. These difference plots should be much smaller than the individual EM field plots used to make the difference plots. These plots are useful, in addition to studying convergence, because they can clearly show regions (near a dielectric step discontinuity, for example) where the EM fields converge to the correct solution slowly with increasing truncation value.
(d) From the results of parts (b,c,d) pick a $M = M_T + 1$ truncation value that will be used to validate the SV solution and also pick the BC or BA validation method which can be expected to be the most accurate to validate SV EM results.

Computational Example 9.14(a) Solution

The dielectric permittivity scattering system and direction of plane wave incidence is shown in Figure 9.27.

Computational Example 9.14(b) Solution

The representative range used to study the EM field solution is that shown in the dielectric function of part (a), namely, $-2.4\tilde{\lambda}_f \le \tilde{x} \le 2.4\tilde{\lambda}_f$, $0 \le \tilde{y} \le 2.4\tilde{\lambda}_f$. From the error analysis plots, see below, it was found that acceptable EM numerical results occurred when $29 \le M \le 38$, $M = M_T + 1$ for the BA method. It is also seen that the BC method clearly started to diverge and give ill-conditioned results at a truncation value of $M = 33$.

It should be noted that the BC method, as implemented in the present chapter, solves for the EM field coefficients of the system *without* taking into account the even symmetry of the problem, as

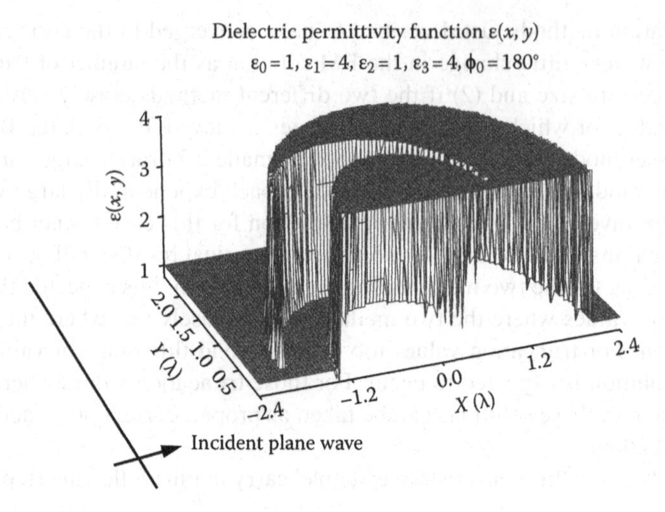

FIGURE 9.27 Plot of the scattering object dielectric function $\varepsilon(x, y)$.

discussed earlier. Thus, it is entirely reasonable that ill-conditioned results will occur for the BC method for lower values of the mode truncation, than would occur if the system equations had been solved taking into account the even symmetry of the problem, as the solution taking into account even symmetry involves a much smaller system matrix than in the case when symmetry is not accounted for (see Problems 8.9 and 8.10). As the BA and the BC methods are basically the same, and because the BA method *does* take into account the even symmetry of the problem, it is thus overall reasonable that the BA method as implemented here does show better conditioning than does the BC method as implemented here. The ill-conditioning of the BC method for values of $M \geq 33$ is clearly seen in Figure 9.28.

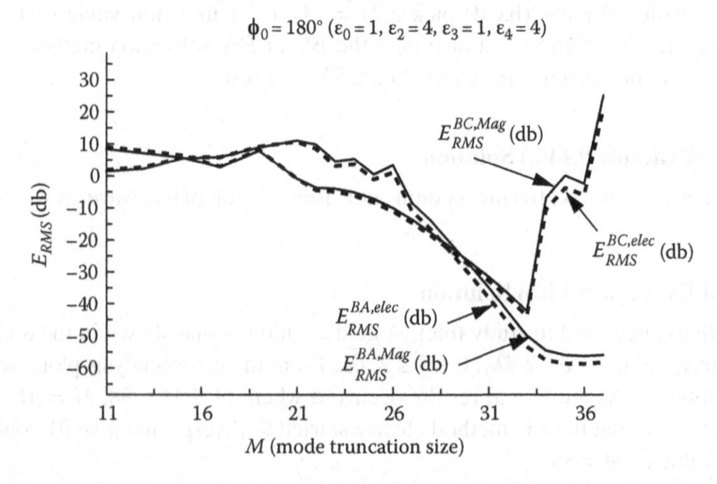

FIGURE 9.28 Plot of normalized electric and field error (expressed in db) error versus mode number as calculated in part (a).

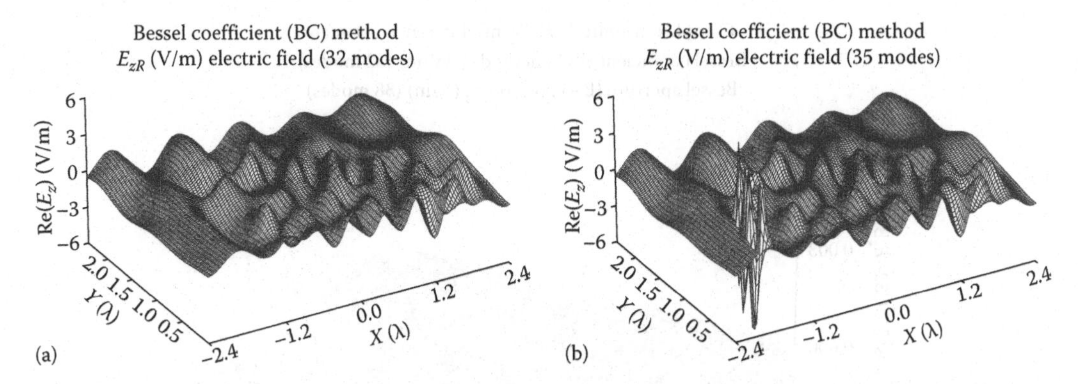

FIGURE 9.29 Electric field as calculated by the Bessel coefficient method, using (a) $M = 32$ and (b) $M = 35$ modes.

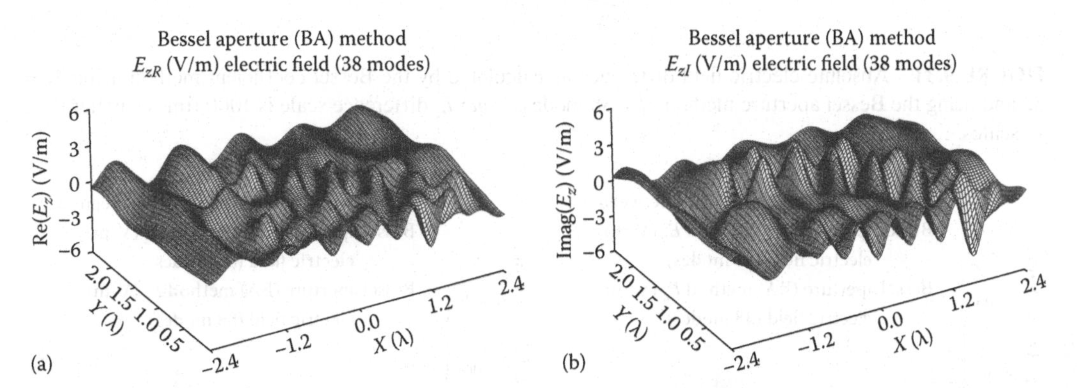

FIGURE 9.30 (a) Real and (b) imaginary electric field as calculated by the Bessel aperture method, using $M = 38$ modes.

Computational Example 9.14(c) Solution

Figure 9.29a and b shows the electric field as calculated by the BC method, using $M = 32$ and 35 modes. As can be seen in the $M = 35$ plot (second graph), numerical ill conditioning has occurred for the BC method.

Figure 9.30a and b shows the real and imaginary parts of the electric field as calculated by the BA method using $M = 38$ modes (call it the 38-mode BA method). As can be seen no ill-conditioned results have occurred despite the larger truncation mode value than used for the BC method.

Figures 9.31 and 9.32a and b show (1) the complex magnitude difference between the BC method electric field using $M = 32$ modes and the BA method electric field using $M = 38$ modes (Figure 9.31); (2) the complex magnitude difference using the BA method between the electric field using $M = 32$ modes and using $M = 38$ modes (Figure 9.32a); and (3) the complex magnitude difference using the BA method between the electric field using $M = 35$ modes and using $M = 38$ modes (Figure 9.32b). From all three difference graphs it is seen that there is close overall agreement between the BC and BA methods, with the BA method clearly giving the better convergence results. The error between the 35-mode BA solution and 38-mode BA electric field solutions (Figure 9.32b) is very small, and gives good confidence that the correct and nearly exact electric field solution to the scattering example has been found. The E_z difference electric field scales on Figure 9.31 and Figure 9.32a are 1000 times smaller that the E_z electric field scale plots of Figures 9.29 and 9.30 presented earlier and 10,000 times smaller on Figure 9.31b than in the E_z electric field scale plots of Figures 9.29 and 9.30.

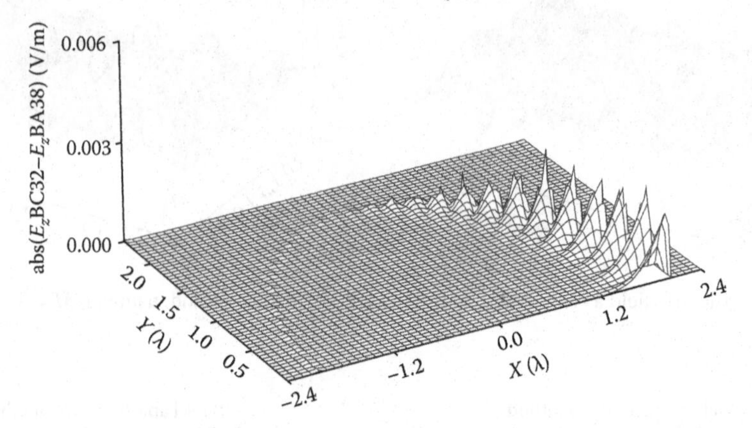

FIGURE 9.31 Absolute electric field difference as calculated by the Bessel coefficient method using M = 32 and using the Bessel aperture method M = 38 modes. *Note*: E_z difference scale is 1000 times smaller than E_z scales.

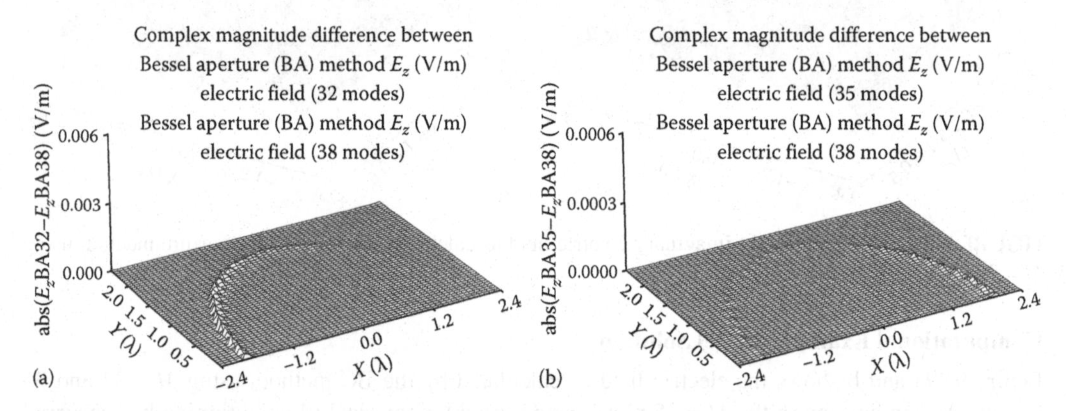

FIGURE 9.32 Absolute electric field difference as calculated between by the 38-mode Bessel aperture method and the (a) 32-mode BA solution (E_z difference scale is 1000 times smaller than E_z scales.) and (b) the 35-mode BA solution. (E_z difference scale is 10,000 times smaller than E_z scales.)

Figure 9.33a through d presents plots of the real and imaginary parts of the H_x, H_y magnetic fields as calculated by the 38-mode BA method, and Figure 9.34a and b following these plots the complex magnitude difference H_x, H_y magnetic fields as calculated between the 35-mode BA method and 38-mode BA method. As seen from Figures 9.33a through d and 9.34a and b, extremely convergent magnetic field results are obtained.

Computational Example 9.14(d) Solution

(d) From the results of part (b through d), the 38-mode BA electric and magnetic field method has been chosen as a correct and nearly exact EM field solution from which to validate SV solutions.

Computational Example 9.15

The example concerns, using the general geometry of Figure 8.2 (plane wave incident on a dielectric cylinder), calculating the real EM fields using the SV method (call it $E_{zR}^{SV} = \text{Re}(E_z^{SV})$) for the following cases: (1) when plane wave incidence angle is $\phi_0 = 180°$; (2) when the regions of the dielectric cylinder of Figure 8.2 have the following dimensions; $\tilde{r}_1 = 2\tilde{\lambda}_f$, $\tilde{r}_3 = \tilde{\lambda}_f$, inner and outer circle

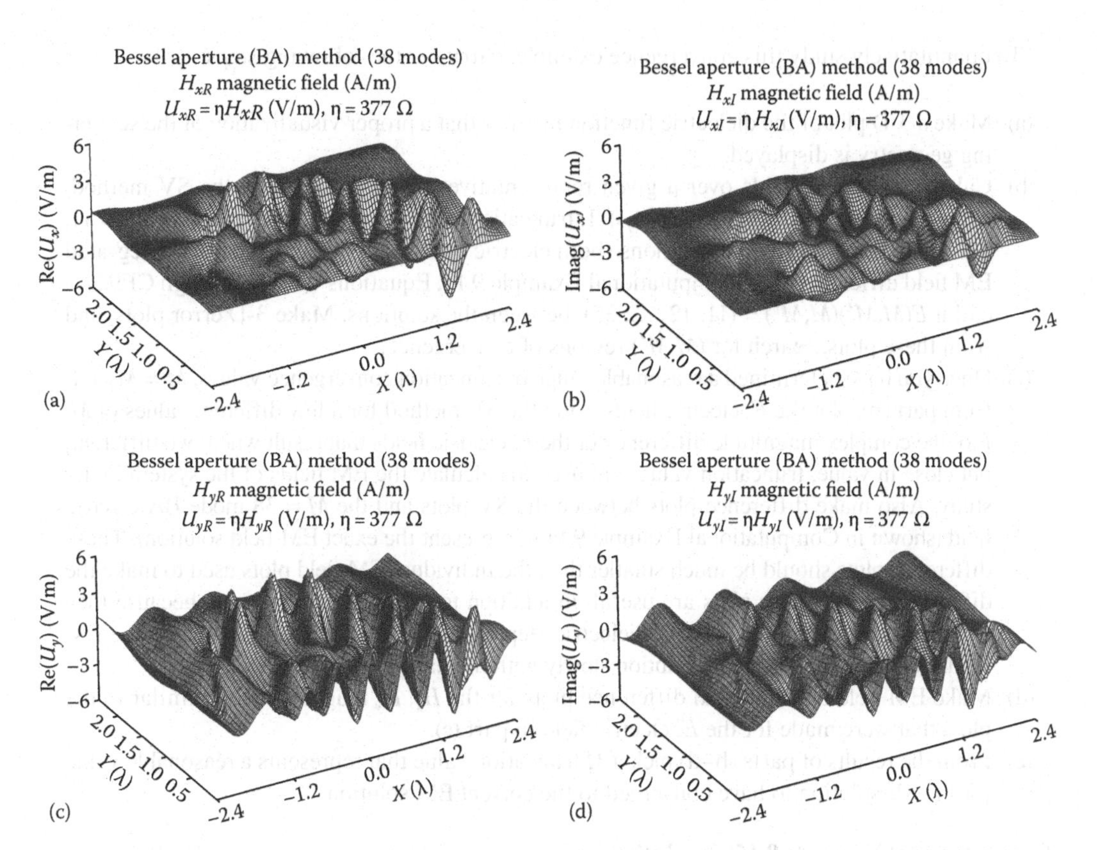

FIGURE 9.33 The (a) $H_{xR} = \text{Re}(H_x)$, (b) $H_{xI} = \text{Imag}(H_x)$, (c) $H_{yR} = \text{Re}(H_y)$ and (d) $H_{yI} = \text{Imag}(H_y)$ magnetic fields as calculated by the Bessel aperture method using $M = 38$ modes is shown.

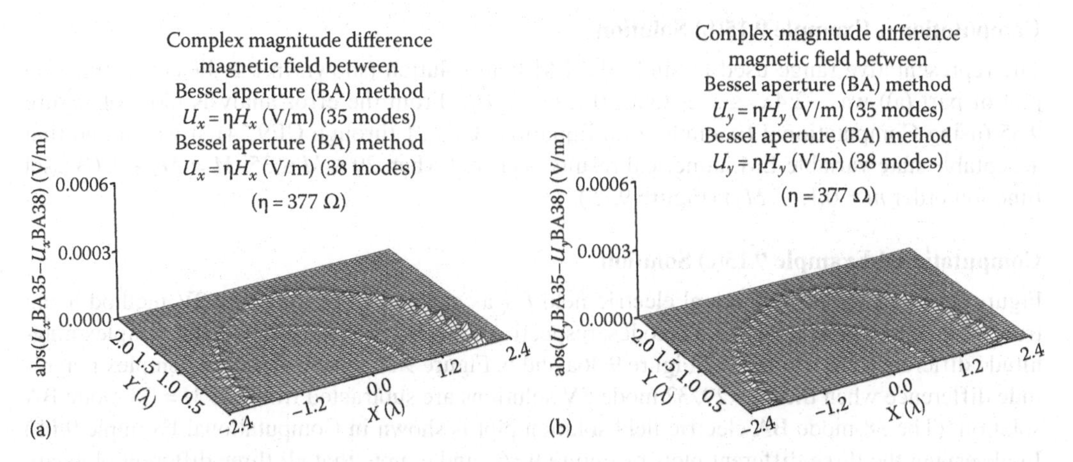

FIGURE 9.34 Complex magnitude field difference between the (a) H_x (U_x difference scale is 10,000 times smaller than U_x scales.) and (b) H_y (U_y difference scale is 10,000 times smaller than U_y scales.) magnetic fields as calculated by the Bessel aperture method using $M = 35$ and $M = 38$ modes is shown.

separation to be $\tilde{e}_{13} = 0.1\pi\tilde{\lambda}_f$; and (3) when the relative dielectric values of each region are taken to be $\varepsilon_0 = 1$, $\varepsilon_1 = 4$, $\varepsilon_2 = 1$, and $\varepsilon_3 = 4$, respectively. The $E_{zR}^{SV} = \text{Re}(E_z^{SV})$ electric field SV plots should be calculated using 4200 layers. The EM SV solution is to be validated by the 38-mode BA method solution shown in Computational Example 9.14 to be an almost exact solution to the EM scattering example as described above.

To quantitatively study this convergence example, carry out the following steps:

(a) Make a 3-D plot of the dielectric function in order that a proper visualization of the scattering geometry is displayed.
(b) Calculate the EM fields over a given representative region of space by the SV method, using a number of different $M = M_T + 1$, truncation values, ranging from $M = M_T + 1 = 11, 12, \ldots, 41$. From these field solutions (both electric and magnetic), calculate the integrated EM field difference (see Computational Example 9.13, Equations CE9.31 through CE9.34), call it $E(M,M')(M,M') = (11, 12, \ldots, 35)$, between the solutions. Make 3-D error plots, and from these plots, search for (M,M') regions of convergence.
(c) Once you have determined a reasonable range of truncation convergence values, $M = M_T + 1$, from part (b), plot the E_z electric fields using the SV method for a few different values of M. Plot the complex magnitude difference of the E_z electric fields that result when two different, but close in value, truncation values are used to calculate the EM fields of the system under study. Also make difference plots between the SV plots and the $M = 38$ mode BA electric field (shown in Computational Example 9.14 to represent the exact EM field solution). These difference plots should be much smaller than the individual EM field plots used to make the difference plots. These plots are useful, in addition to studying convergence, because they can clearly show regions (near a dielectric step discontinuity, for example) where the EM fields converge to the correct solution slowly with increasing truncation value.
(d) Make EM field and EM field difference plots for the H_x, H_y magnetic fields similar to the plots that were made for the E_z electric field in part (c).
(e) From the results of parts (b–d) pick a M truncation value that represents a reasonable value for the SV solution to have converged to the correct EM solution.

Computational Example 9.15(a) Solution

The 3-D dielectric function is shown in Figure 9.27.

Computational Example 9.15(b) Solution

The representative range used to study the EM field solution (shown in the dielectric function plot of part (a)) was $-2.4\tilde{\lambda}_f \leq \tilde{x} \leq 2.4\tilde{\lambda}_f$, $0 \leq \tilde{y} \leq 2.4\tilde{\lambda}_f$. From the error analysis plots of Figure 9.35 (using Computational Example 9.13, Equations CE9.31 through CE9.34), it was found that acceptable state variable EM numerical results occurred when $29 \leq M \leq 35$, $M = M_T + 1$ (Bessel function order $m = 0,1,\cdots,M_T$) (Figure 9.35).

Computational Example 9.15(c) Solution

Figure 9.36a and b shows the real electric field E_{zR} as calculated by using the SV method using truncation values of $M = 29$ and 32 modes, respectively, and Figure 9.36c shows the complex magnitude difference electric field of Figure 9.36a and b. Figure 9.36d and e show the complex magnitude difference when the $M = 29$, 32 mode SV solutions are subtracted from the $M = 38$ mode BA solution. (The 38-mode BA electric field solution plot is shown in Computational Example 9.14.) In observing the three different plots of Figure 9.36c and e, note that all three different plots are all of the same order of magnitude. This indicates that the SV method has converged well for the truncation values of $M = 29$ and 32. *Note:* The complex magnitude difference electric field scales are 100 times smaller than the electric field plots.

Computational Example 9.15(d) Solution

Figure 9.37 shows respectively, the magnetic fields H_x, H_y as calculated by using the SV method using a truncation value of $M = 32$ modes, and Figure 9.37a and b show the complex magnitude difference, respectively, between the $M = 32$ mode, H_x, H_y magnetic field solutions and the $M = 38$

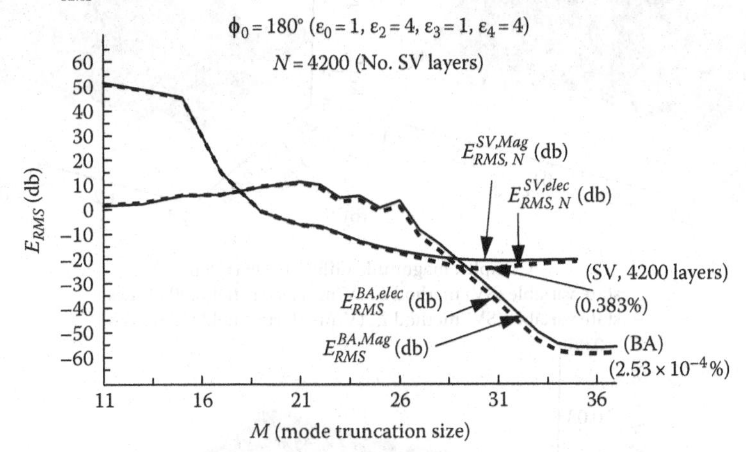

FIGURE 9.35 Plots of state variable root mean square error as a function mode truncation value is shown expressed in db. The 38 mode Bessel aperture EM solution studied in Computational Example 9.14 was taken as the "correct" solution from which error of the state variable method could be calculated.

mode H_x, H_y magnetic field BA solutions (shown in Computational Example 9.14 to represent the exact EM field solution). The 38-mode BA magnetic field solution plots are shown in Computational Example 9.14. *Note:* The H_x, H_y difference field scales are 100 times smaller than that of the first two plots.

Computational Example 9.15(e) Solution

From the results of part (b–d) the 32-mode SV electric and magnetic field solutions have converged well to the correct solution.

Computational Example 9.16

This problem concerns comparing the bistatic line widths that result when the EM scattering is from two different homogeneous dielectric objects being studied.

(a) Using the general geometry of Figure 8.2 (plane wave incident on a dielectric cylinder), calculating the bistatic line width $\sigma(\phi, \phi_0)$ using the SV method for the following cases: (1) when plane wave incidence angle is $\phi_0 = 180°$; (2) when the regions of the dielectric cylinder of Figure 8.2 have the following dimensions; $\tilde{r}_1 = 2\tilde{\lambda}_f$, $\tilde{r}_3 = \tilde{\lambda}_f$, inner and outer circle separation to be $\tilde{e}_{13} = 0.1\pi\tilde{\lambda}_f$; and (3) when the relative dielectric values of each region is taken to be $\varepsilon_0 = 1$, $\varepsilon_1 = 4$, $\varepsilon_2 = 1$, and $\varepsilon_3 = 4$, respectively. The SV EM fields for this problem should be calculated using 4200 layers and using a truncation value of $M = M_T + 1 = 35$. The EM SV solution for this problem is to be validated by using the 38-mode BA method (truncation value $M = 38$) solution shown in Computational Example 9.14, to be very close to the exact solution to the EM scattering problem under consideration. Also validate the solution using the 32-mode BC method (truncation value $M = 32$) described at the beginning of Chapter 9.

(b) Repeat the calculation of part (a) when the geometry and angle of plane wave incidence is identical to that in part (a), except that the relative dielectric values of each region is taken to be $\varepsilon_0 = 1$, $\varepsilon_1 = 2$, $\varepsilon_2 = 3$, and $\varepsilon_3 = 4$, respectively.

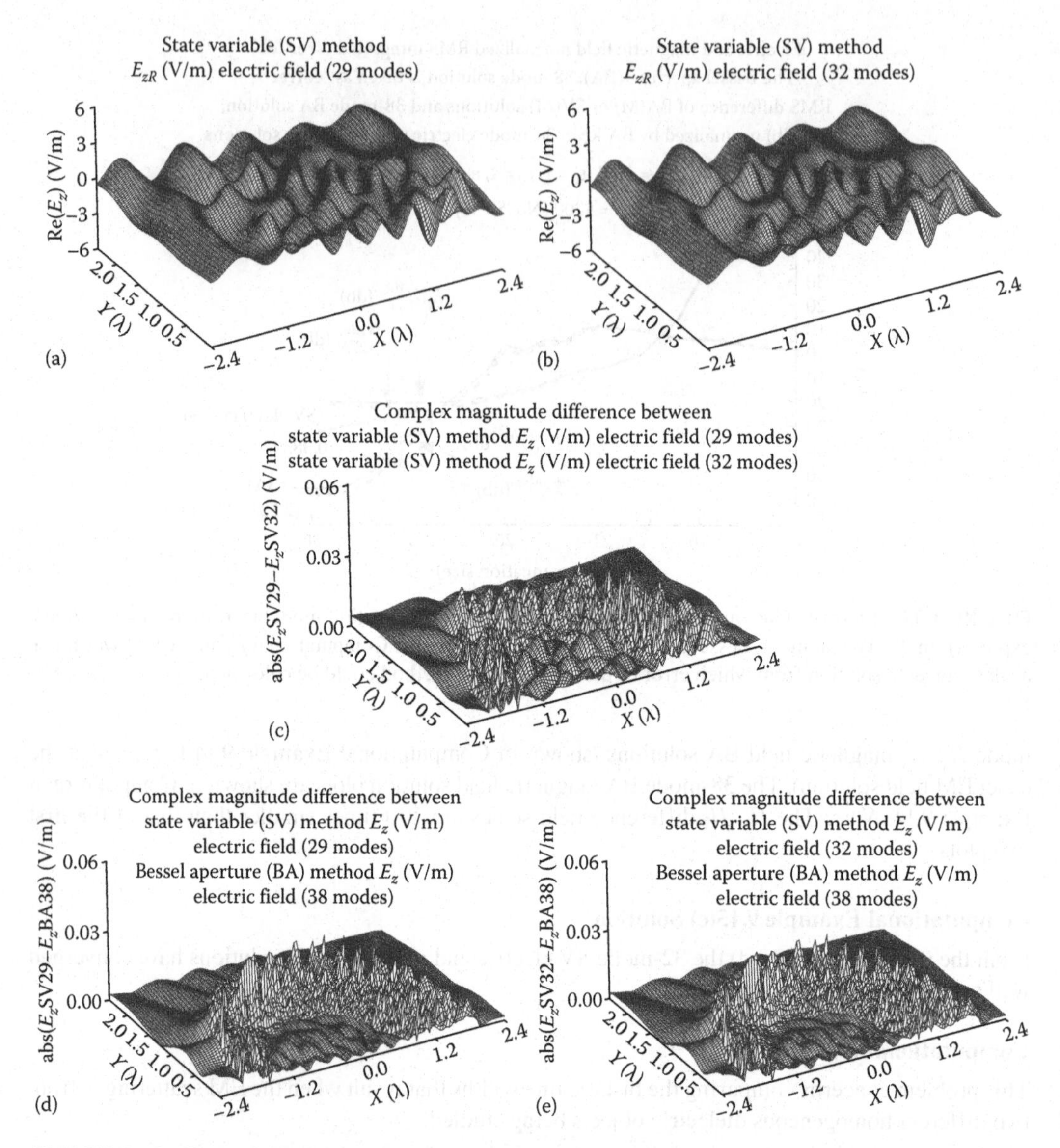

FIGURE 9.36 Plots of the E_z electric field (a and b) and plots of difference electric fields (c through e) as calculated by different methods are shown. (E_z difference scale is 100 times smaller than E_z scales.)

(c) Plot and compare the 35-truncation mode SV bistatic line widths as calculated in parts (a and b) and comment on the effect that the different dielectric composition of two scatterers has on the shape of the bistatic line width.

Computational Example 9.16a, b, and c Solutions

The SV, BA and BC bistatic line widths $\sigma(\phi, \phi_0)$ when the relative dielectric values of each region are taken to be $\varepsilon_0 = 1$, $\varepsilon_1 = 4$, $\varepsilon_2 = 1$, and $\varepsilon_3 = 4$ (Figure 9.38a) and when the material parameters are taken to be $\varepsilon_0 = 1$, $\varepsilon_1 = 2$, $\varepsilon_2 = 3$, and $\varepsilon_3 = 4$ (Figure 9.38b) are shown below. The M truncation values and SV number of layers used to calculate the bistatic line widths are also shown. Figure 9.38c shows that the different dielectric composition of two scatterers results in greatly different bistatic line width shapes.

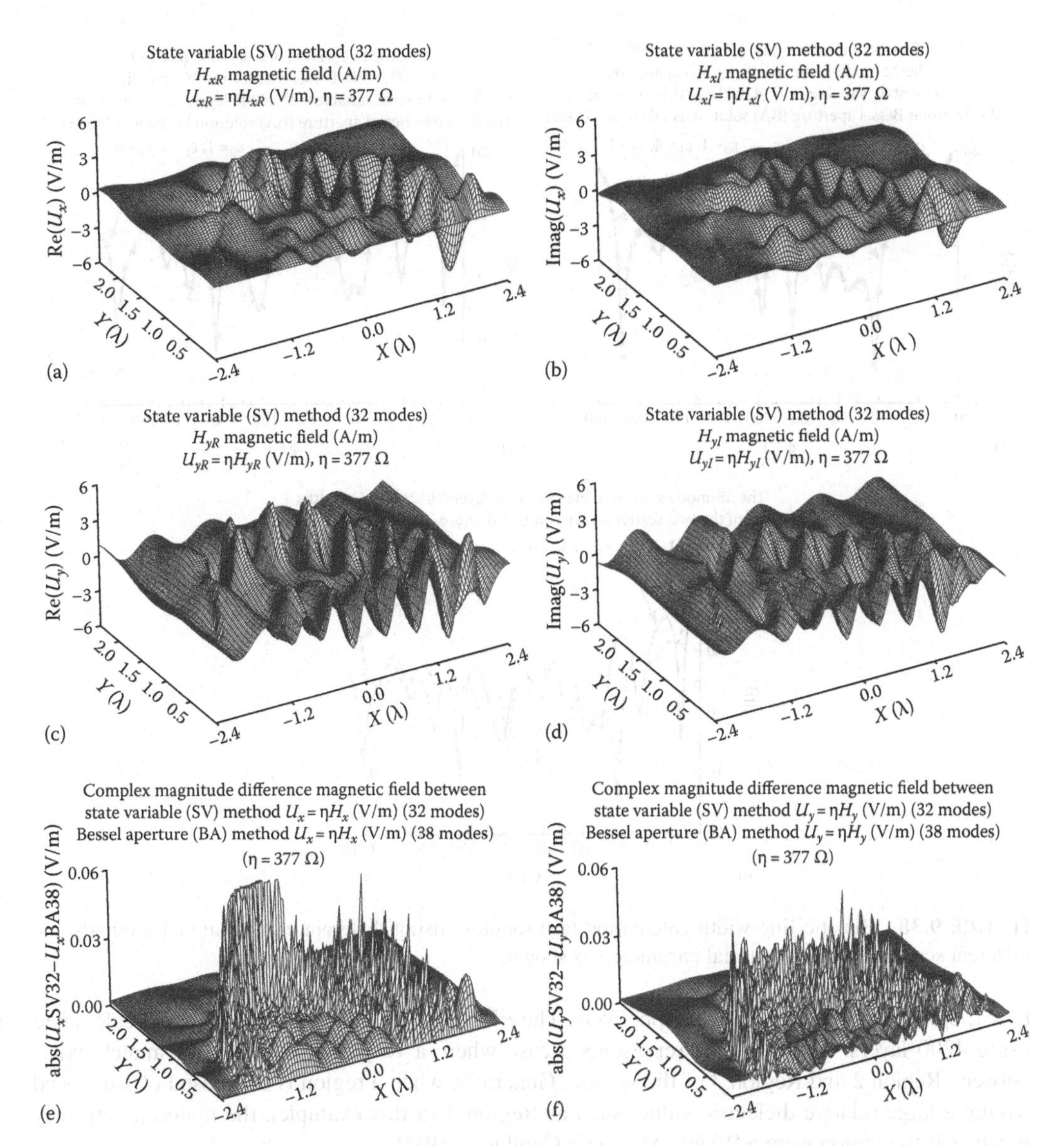

FIGURE 9.37 Plots of the H_x, H_y magnetic fields (a through d) and plots of SV–BA complex magnitude difference magnetic (e and f) are shown. (U_x difference scale is 100 times smaller than U_x scales. U_y difference scale is 100 times smaller than U_y scales.)

9.6 CASE STUDY V: COMPARISON OF BA, BC, AND SV METHODS FOR GRADUALLY, STEPPED-UP, INDEX SCATTERING OBJECTS WITH HIGH INDEX CORE

(Diameter $= 2\tilde{r}_1 = 4\tilde{\lambda}_f$, $\varepsilon_0 = 1$, $\varepsilon_1 = 2$, $\varepsilon_2 = 3$ and $\varepsilon_3 = 32$)

Computational Example 9.17

The example concerns, using the general geometry of Figure 8.2 (plane wave incident on a dielectric cylinder), calculating the real EM fields using the SV method (call it $E_{zR}^{SV} = \mathrm{Re}(E_z^{SV})$) for the following cases: (1) when plane wave incidence angle is $\phi_0 = 180°$; (2) when the regions of the dielectric cylinder of Figure 8.2 have the following dimensions; $\tilde{r}_1 = 2\tilde{\lambda}_f$, $\tilde{r}_3 = \tilde{\lambda}_f$, inner and outer circle separation to be $\tilde{e}_{13} = 0.1\pi\tilde{\lambda}_f$; and (3) when the relative dielectric bulk values of each region is taken to be

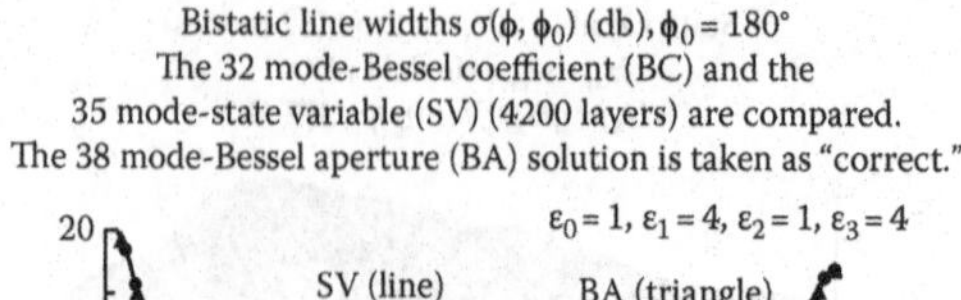

Bistatic line widths σ(ϕ, ϕ0) (db), ϕ0 = 180°
The 32 mode-Bessel coefficient (BC) and the
35 mode-state variable (SV) (4200 layers) are compared.
The 38 mode-Bessel aperture (BA) solution is taken as "correct."

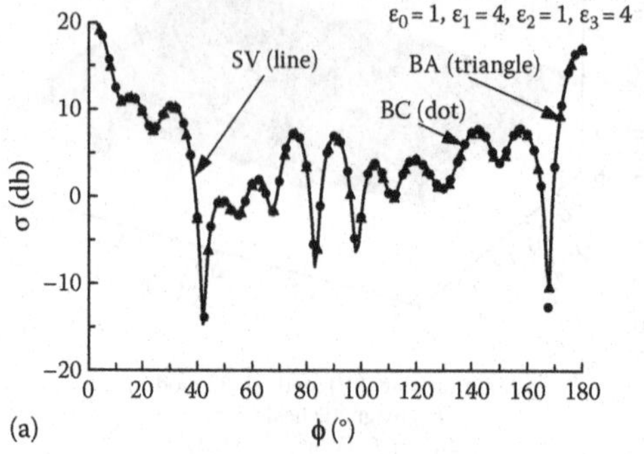

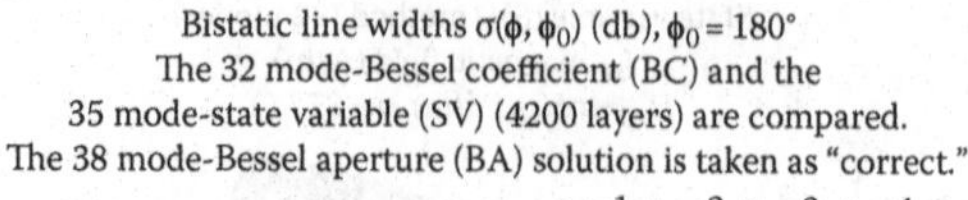

Bistatic line widths σ(ϕ, ϕ0) (db), ϕ0 = 180°
The 32 mode-Bessel coefficient (BC) and the
35 mode-state variable (SV) (4200 layers) are compared.
The 38 mode-Bessel aperture (BA) solution is taken as "correct."

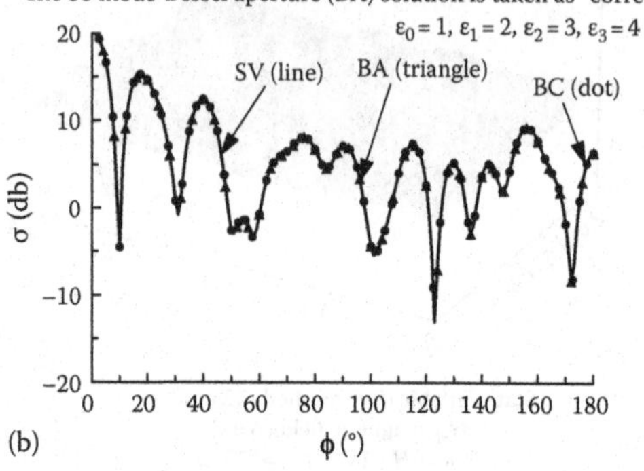

The 35-mode state variable (SV, 4200 layers) bistatic line widths of the two scatterers are compared. σ(ϕ, ϕ0) (db), ϕ0 = 180°

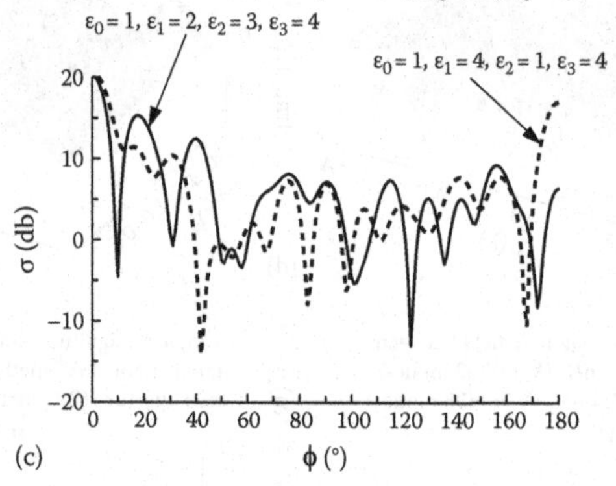

FIGURE 9.38 Bistatic line width calculation (a through c) using different computational methods and different scattering object material parameters is shown.

$\varepsilon_0 = 1$, $\varepsilon_1 = 2$, $\varepsilon_2 = 3$, and $\varepsilon_3 = 32$, respectively. The EM fields for this example should be calculated using 4200 layers. This example represents a case where a very high dielectric mismatch exists between Region 2 and Region 3 of the system. Generally, when a region is composed of a material having a large relative dielectric value (such as Region 3 in this example), the material acts very much as if the region were a Perfect Magnetic Conductor (PMC).

The purpose of this example is twofold, first, to investigate how well the SV method converges to the correct solution (using the BA method to validate SV numerical results), and second, studying both the SV and BA solutions, to see what affect the large relative dielectric value of Region 3 has on the overall EM solution of this example. Does Region 3, composed of a high relative dielectric material, act like a perfect magnetic conductor or not?

To quantitatively study the EM solution of the stated example, carry out the following steps:

(a) Make a 3-D plot of the dielectric function for this example in order that a proper visualization of the scattering geometry is displayed.
(b) Calculate the EM fields over a given representative region of space using the SV and BA methods, using a number of different $M = M_T + 1$, truncation values, ranging from $M = M_T + 1 = 11, 12, \ldots, 41$. From these field solutions (both electric and magnetic), calculate the integrated EM field difference (see Computational Example 9.13, Equations CE9.31 through CE9.34), call it $E(M, M')(M, M') = (11, 12, \ldots, 35)$, (1) between different modal values of the BA solutions; (2) between different modal values of the SV solutions; and

(3) between different modal values of the BA and SV solutions. Make 3-D error plots, and from these plots, search for (M, M') regions of convergence for the BA and SV methods. As the BA method represents the most accurate or almost exact solution (as the scattering regions are homogeneous) for the present example, it can be expected that best convergence and lowest error for this example will be found using the BA method.

(c) Based on the assumption that the BA method will be the most accurate, calculate the RMS difference that results from subtracting the EM fields using the SV method from the EM fields using BA method, call this the RMS error difference E_{RMS}. Make plots of the E_{RMS} difference error versus mode-truncation value M for both the electric and magnetic EM fields. On these plots be sure to identify modal values for which ill-conditioned numerical results start to arise.

(d) Once you have determined a reasonable range of truncation convergence values, $M = M_T + 1$, from parts (b and c), plot the E_z electric fields using the SV method for a few different values of M. Plot also for comparison the E_z electric field using the BA method (which may be considered the almost exact solution as determined from parts (b) and (c)). Plot the complex magnitude difference of the E_z electric fields that result between the SV solution and the almost exact BA solution for different truncation values. These difference plots should be much smaller in magnitude than the individual EM field plots if a correct SV E_z electric field solution has been found. These plots are useful, in addition to studying convergence, because they can clearly show regions (near a dielectric step discontinuity, for example) where the EM fields converge to the correct solution slowly with increasing truncation value.

(e) Make EM field and EM field difference plots for the H_x, H_y magnetic fields similar to the plots that were made for the E_z electric field in part (d).

(f) From the results of parts (b through e) pick a $M = M_T + 1$ truncation value that represents a reasonable value for the SV solution to have converged to the correct EM solution.

(g) Describe the behavior of the EM field solutions found in parts (b through e), particularly in Region 3, a region composed of a material having a large relative dielectric value. Does it act like a perfect magnetic conductor?

Computational Example 9.17(a) Solution (Figure 9.39)

Computational Example 9.17(b) Solution

The representative range used to study the EM field solution is that shown in the dielectric function of part (a), namely, $-2.4\tilde{\lambda}_f \leq \tilde{x} \leq 2.4\tilde{\lambda}_f$, $0 \leq \tilde{y} \leq 2.4\tilde{\lambda}_f$. From the error analysis plots it was found that acceptable EM numerical results occurred when $28 \leq M \leq 32$, $M = M_T + 1$.

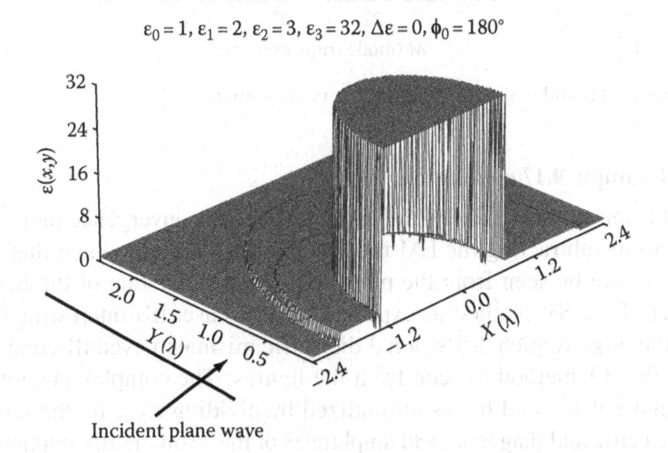

FIGURE 9.39 Dielectric permittivity function and scattering system are shown.

Electric field normalized RMS integrated error (db)
Homogenous $\varepsilon(x, y)$. The BA, 33 mode solution is taken as "correct"
$E_{RMS}^{SV,Hom}$ (db) RMS difference of homogenous SV(M) solutions and 33-mode BA solution,
$E_{RMS}^{BA,Hom}$ (db) RMS difference of BA(M) solutions and 33-mode BA solution,
$\phi_0 = 180°$ ($\varepsilon_0 = 1$, $\varepsilon_1 = 2$, $\varepsilon_2 = 3$, $\varepsilon_3 = 32$, $\Delta\varepsilon = 0$)
$N = 4200$ (No. SV layers)

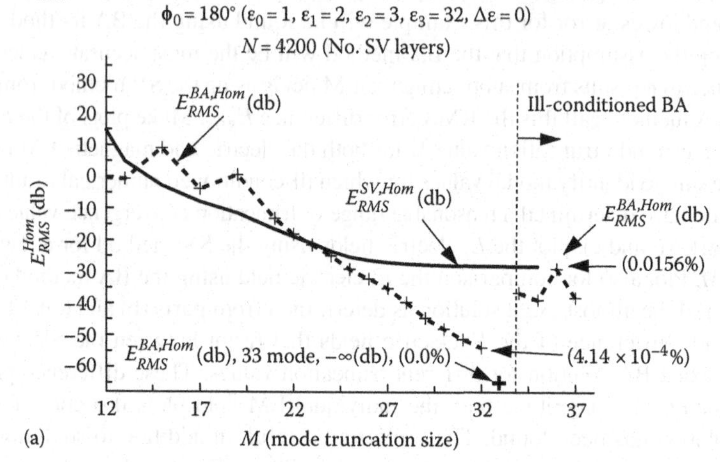

Magnetic field normalized RMS integrated error (db)
Homogenous $\varepsilon(x, y)$. The BA, 33 mode solution is taken as "correct"
$E_{RMS}^{SV,Hom}$ (db) RMS difference of homogenous SV(M) solutions and 33-mode BA solution,
$E_{RMS}^{BA,Hom}$ (db) RMS difference of BA(M) solutions and 33-mode BA solution,
$\phi_0 = 180°$ ($\varepsilon_0 = 1$, $\varepsilon_1 = 2$, $\varepsilon_2 = 3$, $\varepsilon_3 = 32$, $\Delta\varepsilon = 0$)
$N = 4200$ (No. SV layers)

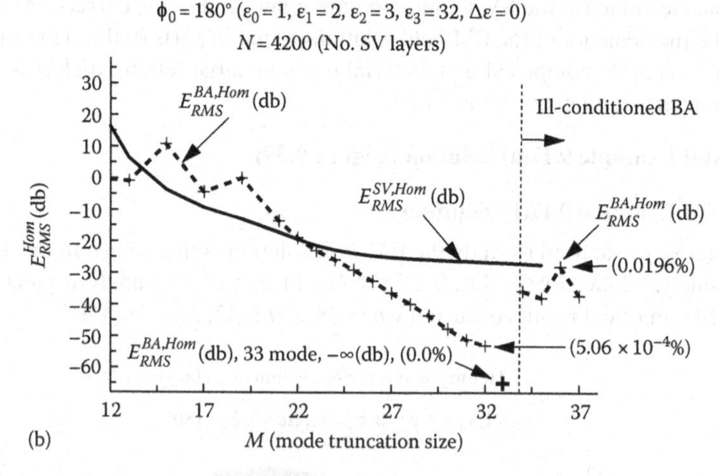

FIGURE 9.40 Electric (a) and magnetic field (b) errors are shown.

Computational Example 9.17(c) Solution

Figure 9.40a and b shows, respectively, the RMS difference convergence plots that result for the present example from subtracting the EM fields using the SV method and that result from using the BA method. As can be seen from the plots, the convergence rate of the BA method is more accurate than that of the SV method as expected for this case. It's interesting to note that probably because of the large Region 2–Region 3 dielectric mismatch even ill-conditioned numerical results occur for the BA method as seen from the figures. The complex magnitude electric field error, E_{RMS}, in Figure 9.40a and b was normalized by dividing E_{RMS} by the integrated, complex magnitude RMS electric and magnetic field amplitudes of the 33-mode BA solution (general formula given in Computational Example 9.13, Equations CE9.31 through CE9.34).

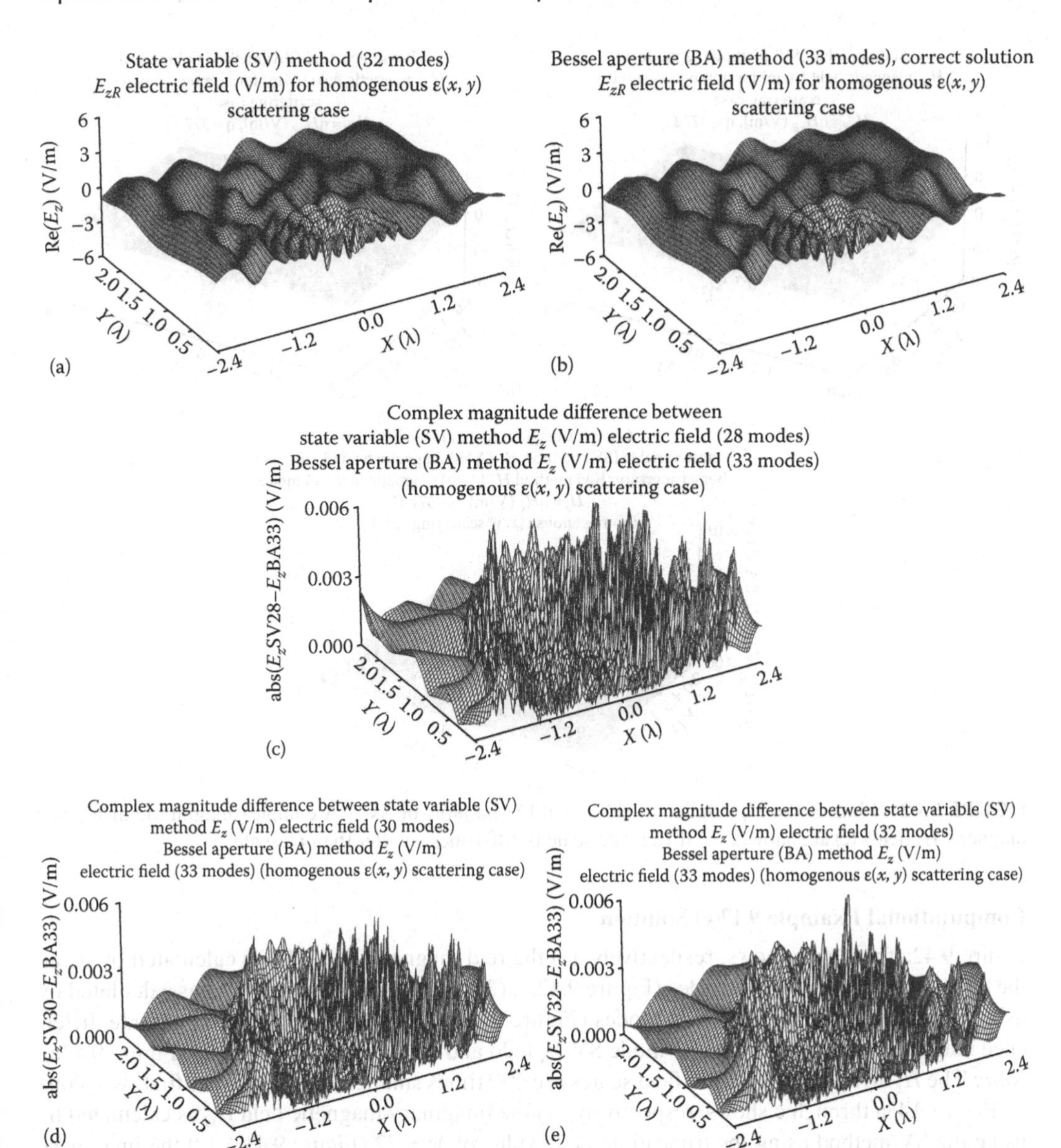

FIGURE 9.41 Plots of the E_z electric field (a and b) and plots of SV–BA complex magnitude difference electric fields (c through e) are shown. (E_z difference scale is 1000 times smaller than E_z scales.)

Computational Example 9.17(d) Solution

Figure 9.41a shows the real electric field E_{zR} as calculated by using the SV method using the $M = 32$ mode truncation value (called the 32-mode SV E_{zR} solution) and Figure 9.41b shows the real electric field E_{zR} as calculated by using the BA method using the $M = 33$ mode truncation value. Figure 9.41c through e shows the complex magnitude difference electric fields between (1) the 28-mode SV E_z and 33-mode BA E_z solutions, (2) the 30-mode SV E_z and 33-mode BA E_z solutions, and (3) the 32-mode SV E_z and 33-mode BA E_z solutions. The 32-mode SV E_z, 33-mode BA complex magnitude difference plot appears smaller in magnitude than the $M = 28$, $M = 30$ difference plots, showing good convergence of the 32-mode SV E_z solution for the truncation value of $M = 32$. *Note:* The E_z difference electric field scales on the Figure 9.41c through e are 1000 times smaller than on Figure 9.41a and b.

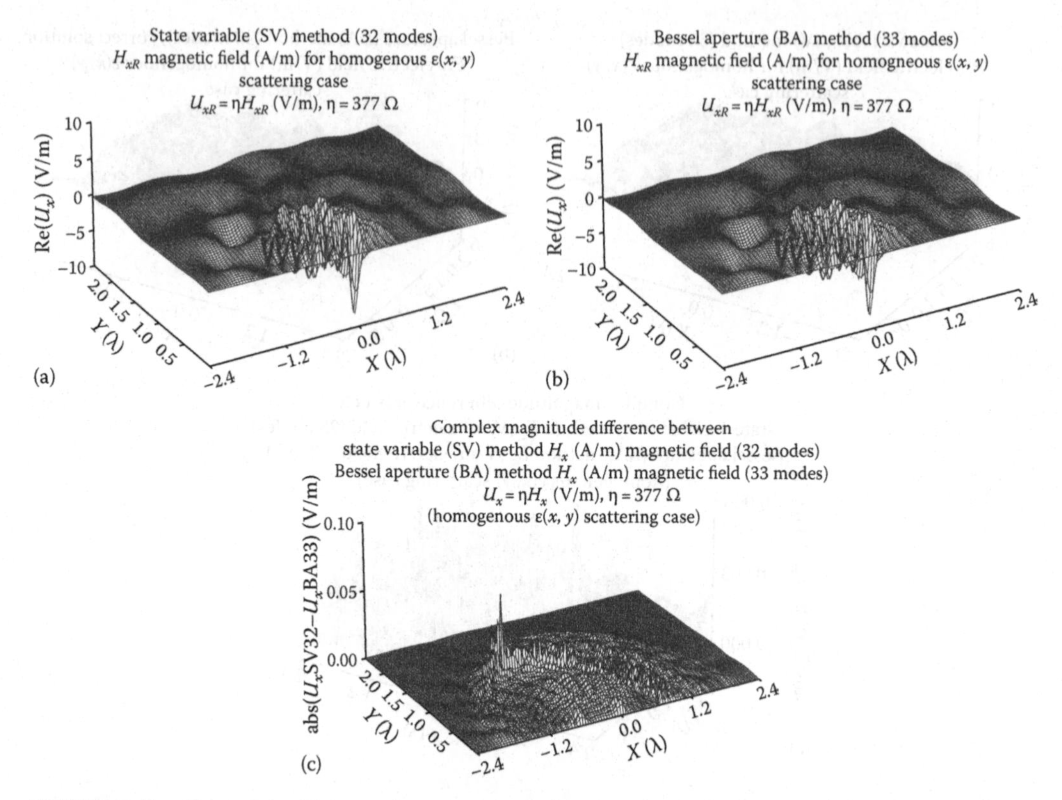

FIGURE 9.42 Plots of the H_x magnetic field (a and b) and plots of SV–BA complex magnitude difference magnetic H_x fields (c) are shown. (U_x difference scale is 100 times smaller than U_x scales.)

Computational Example 9.17(e) Solution

Figure 9.42a through c shows, respectively, (1) the real magnetic field H_{xR} as calculated by using the SV method using $M = 32$ modes (Figure 9.42a), (2) the real magnetic field H_{xR} as calculated by using the BA method using $M = 33$ modes (Figure 9.42b), and (3) the complex magnitude difference magnetic field between the 32-mode SV H_x field and the 33-mode BA H_x field (Figure 9.42c). *Note:* The H_x difference magnetic field scales are 100 times smaller than that of the first two plots.

Figure 9.43a through c shows, respectively, (1) the imaginary magnetic field H_{yI} as calculated by using the SV method using the truncation mode value of $M = 32$ (Figure 9.43a), (2) the imaginary magnetic field H_{yI} as calculated by using the BA method using $M = 33$ (Figure 9.43b), and (3) the complex magnitude difference magnetic field between the 32-mode SV H_y field and the 33-mode BA H_y field (Figure 9.43b). *Note:* The H_y difference magnetic field scales are 100 times smaller than that of the first two plots.

Computational Example 9.17(f) Solution

From the results of part (b through e), the 32-mode SV electric and magnetic field method has converged well to the correct solution.

Computational Example 9.17(g) Solution

As can be seen from both the electric and magnetic field plots, a very short wavelength interference pattern has arisen in the large relative dielectric region (Region 3) of the system. The magnetic field in Region 3 has a large magnitude and is very oscillatory. It is difficult to tell if this magnetic field was spatially averaged in Region 3, if the resulting value would be near zero. If this resulting value

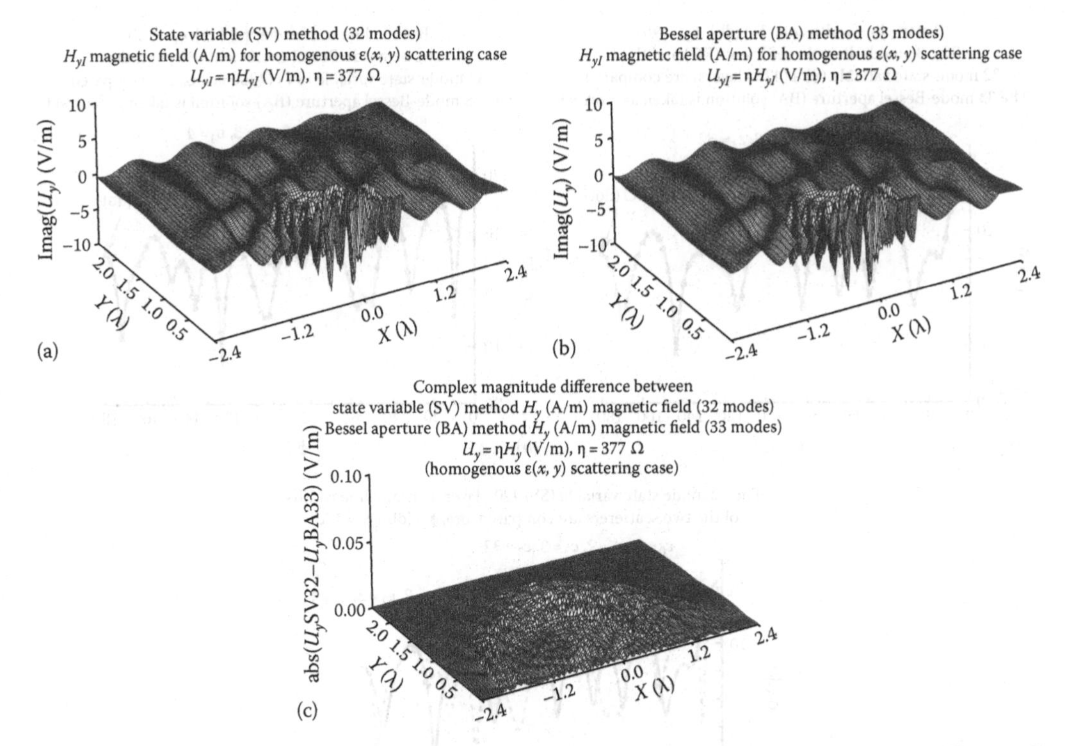

FIGURE 9.43 Plots of the H_y magnetic field (a and b) and plots of SV–BA complex magnitude difference H_y magnetic fields (c) are shown. (U_y difference scale is 100 times smaller than U_y scales.)

was close to zero, then this would indicate that Region 3 was acting like a perfect magnetic conductor in a spatially averaged sense.

Computational Example 9.18

This problem concerns comparing the bistatic line widths that result when the EM scattering is from two different homogeneous dielectric objects being studied.

(a) Using the general geometry of Figure 8.2 (plane wave incident on a dielectric cylinder), calculate the bistatic line width $\sigma(\phi, \phi_0)$ using the SV method for the following cases: (1) when plane wave incidence angle is $\phi_0 = 180°$; (2) when the regions of the dielectric cylinder of Figure 8.2 have the following dimensions; $\tilde{r}_1 = 2\tilde{\lambda}_f$, $\tilde{r}_3 = \tilde{\lambda}_f$, inner and outer circle separation to be $\tilde{e}_{13} = 0.1\pi\tilde{\lambda}_f$; and (3) when the relative dielectric values of each region is taken to be $\varepsilon_0 = 1$, $\varepsilon_1 = 2$, $\varepsilon_2 = 2$, and $\varepsilon_3 = 32$, respectively. The SV EM fields for this problem should be calculated using 4200 layers and using a truncation value of $M = M_T + 1 = 35$. The EM SV solution for this problem is to be validated by using the 33 mode-BA method (truncation value $M = 33$) solution shown in Computational Example 9.17 to be very close to the exact solution to the EM scattering problem described above. Also validate the solution using the 33-mode BC method (truncation value $M = 33$) described at the beginning of Chapter 9.

(b) Repeat the calculation of part (a) when the geometry and angle of plane wave incidence is identical to that in part (a), except that the relative dielectric values of each region is taken to be $\varepsilon_0 = 1$, $\varepsilon_1 = 2$, $\varepsilon_2 = 3$, and $\varepsilon_3 = 4$, respectively. The part (b) analysis was studied in Computational Example 9.12.

(c) Plot and compare the 32-truncation mode SV line widths as calculated in parts (a and b) and comment on the effect that the different dielectric composition of two scatterers has on the shape of the bistatic line width.

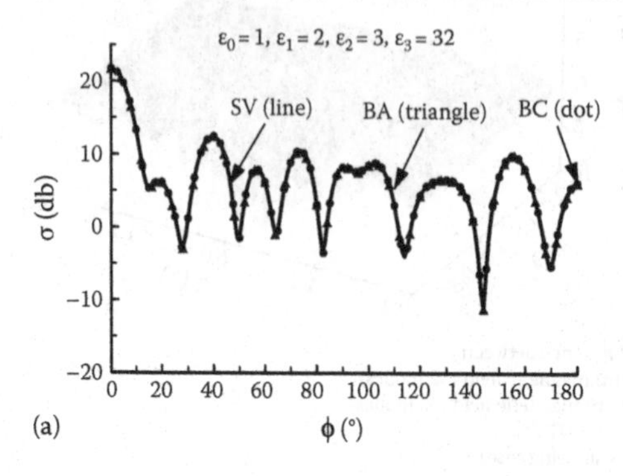

Bistatic line widths σ(ϕ, ϕ0) (db), ϕ0 = 180°
The 33 mode-Bessel coefficient (BC) and the
32 mode-state variable (SV) (4200 layers) are compared.
The 33 mode-Bessel aperture (BA) solution is taken as "correct."
ε0 = 1, ε1 = 2, ε2 = 3, ε3 = 32
SV (line)
BA (triangle)
BC (dot)
σ (db)
ϕ (°)
(a)

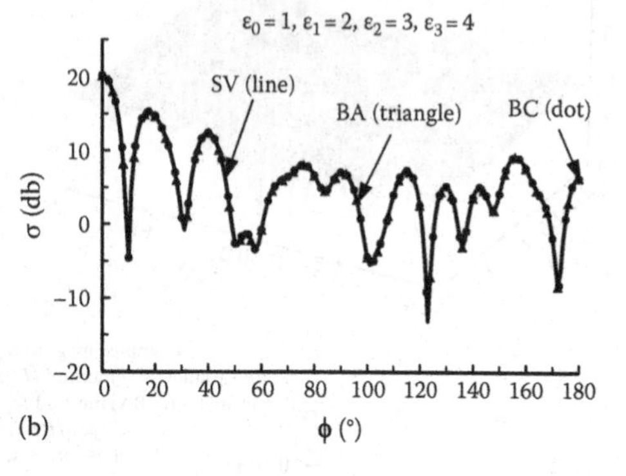

Bistatic line widths σ(ϕ, ϕ0) (db), ϕ0 = 180°
The 32 mode-Bessel coefficient (BC) and the
32 mode-state variable (SV) (4200 layers) are compared.
The 38 mode-Bessel aperture (BA) solution is taken as "correct."
ε0 = 1, ε1 = 2, ε2 = 3, ε3 = 4
SV (line)
BA (triangle)
BC (dot)
σ (db)
ϕ (°)
(b)

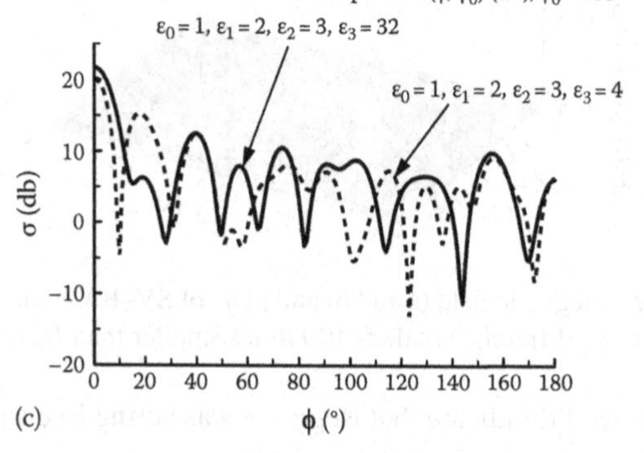

The 32 mode state variable (SV, 4200 layers) bistatic line widths
of the two scatterers are compared. σ(ϕ, ϕ0) (db), ϕ0 = 180°
ε0 = 1, ε1 = 2, ε2 = 3, ε3 = 32
ε0 = 1, ε1 = 2, ε2 = 3, ε3 = 4
σ (db)
ϕ (°)
(c)

FIGURE 9.44 Bistatic line width calculation (a and b) using different computational methods and different scattering object material parameters is shown. (c) The 35-mode SV for the scattering cases when $\varepsilon_0 = 1$, $\varepsilon_1 = 2$, $\varepsilon_2 = 3$, $\varepsilon_3 = 32$ and when $\varepsilon_0 = 1$, $\varepsilon_1 = 2$, $\varepsilon_2 = 3$, $\varepsilon_3 = 4$ are compared.

Computational Example 9.18a, b, and c Solutions

The SV, BA and BC bistatic line widths $\sigma(\phi, \phi_0)$ when the relative dielectric values of each region are taken to be $\varepsilon_0 = 1$, $\varepsilon_1 = 2$, $\varepsilon_2 = 3$, and $\varepsilon_3 = 32$ (Figure 9.44a) and when the material parameters are taken to be; $\varepsilon_0 = 1$, $\varepsilon_1 = 2$, $\varepsilon_2 = 3$, and $\varepsilon_3 = 4$ are shown in Figure 9.44b. The M truncation values and SV number of layers used to calculate the line widths are also shown. Figure 9.44c shows that the different dielectric composition of two scatterers results in greatly different bistatic line width shapes.

9.7 CASE STUDY VI: CALCULATION AND CONVERGENCE ANALYSIS OF EM FIELDS OF AN INHOMOGENEOUS REGION MATERIAL OBJECT USING THE SV METHOD, $\Delta\varepsilon = 1$, $\alpha = 5.5$, $\Lambda = 0$, EXAMPLE

(Diameter = $2\tilde{r}_1 = 4\tilde{\lambda}_f$, for Bulk Dielectric, $\varepsilon_0 = 1$, $\varepsilon_1 = 4$, $\varepsilon_2 = 1$, $\varepsilon_3 = 4$)

Computational Example 9.19

The example concerns, using the general geometry of Figure 8.2 (plane wave incident on a dielectric cylinder), calculating the real EM fields using the SV method (call it $E_{zR}^{SV} = \text{Re}(E_z^{SV})$) for the following cases: (1) when plane wave incidence angle is $\phi_0 = 180°$; (2) when the regions of the dielectric cylinder of Figure 8.2 have the following dimensions; $\tilde{r}_1 = 2\tilde{\lambda}_f$, $\tilde{r}_3 = \tilde{\lambda}_f$, inner and outer circle separation to be $\tilde{e}_{13} = 0.1\pi\tilde{\lambda}_f$; (3) when the relative dielectric bulk values of each region is taken to be $\varepsilon_0 = 1$, $\varepsilon_1 = 4$, $\varepsilon_2 = 1$, and $\varepsilon_3 = 4$, respectively; and (4) when the overall inhomogeneity profile of the scattering object is given by

$$\varepsilon(u,v) = \begin{cases} \varepsilon_1 + \Delta\varepsilon\cos(\Lambda u)\cos(\alpha v), & \text{when } (u,v) \text{ is in Region 1} \\ \varepsilon_2, & \text{when } (u,v) \text{ is in Region 2} \end{cases} \tag{CE9.35}$$

where $\Lambda = 0$, $\alpha = 5.5$, and $\Delta\varepsilon = 1$. The EM fields for this example should be calculated using 4200 layers.

In the following example a BA validation method doesn't apply since the dielectric regions are inhomogeneous as specified by $\varepsilon(u, v)$ given above. Convergence in this example is to be studied by using the SV method to solve for the EM fields of the system when several increasingly large mode truncation values $M = M_T + 1$ have been used to determine the EM fields of the system, and then comparing these solutions to find when little change in them has resulted, indicating convergence. To quantitatively study SV convergence as just described, carry out the following steps:

(a) Make a 3-D plot of the dielectric function $\varepsilon(u, v)$ (resulting in an $\varepsilon(X,Y)$) for this example in order that a proper visualization of the scattering geometry is displayed.
(b) Calculate the EM fields over a given representative region of space by the SV method, using a number of different M, truncation values ranging from $M = M_T + 1 = 11, 12, \ldots, 41$. From these field solutions (both electric and magnetic), (1) calculate the integrated EM field difference (see Computational Example 9.13), call it $E(M,M')(M,M') = (11, 12, \ldots, 35)$, between the solutions and (2) make 3-D error plots, and from these plots, search for (M, M') regions of convergence. Make 2-D plots of SV error $E_{RMS}^{SV,Inh}$ versus the mode truncation values M showing convergence of the SV method for both the electric and magnetic field solutions. Compare the convergence plots for the inhomogeneous case to those made for the homogenous case.
(c) Once you have determined a reasonable range of truncation convergence values M, from part (b), plot the E_z electric fields using the SV method for a few different values of M. Plot the complex magnitude difference of the E_z electric fields that result when two different, but close in value, truncation values are used to calculate the EM fields of the system under study. These difference plots should be much smaller than the individual EM field plots used to make the difference plots. These plots are useful, in addition to studying convergence, because they can clearly show regions (near a dielectric step discontinuity, for example) where the EM fields converge to the correct solution slowly with increasing truncation value.
(d) Make EM field and EM-field difference plots for the H_x, H_y magnetic fields similar to the plots that were made for the E_z electric field in part (c).
(e) Make comparison plots of the M-mode $E_{zR} \equiv \text{Re}(E_z)$ (i.e., M is the truncation value) electric field as calculated when $\Delta\varepsilon = 0$ (homogeneous case) and when $\Delta\varepsilon = 1$ using the bulk dielectric values specified in this example for a truncation value M which you feel reasonable convergence has occurred. Computational Example 9.15 studied the homogeneous case, $\Delta\varepsilon = 0$, for the bulk dielectric values of this example and $M = 32$ was found to be a reasonable value for that example. Comment on any observable difference in the E_{zR} field patterns for the homogeneous ($\Delta\varepsilon = 0$) and inhomogeneous cases ($\Delta\varepsilon = 1$).
(f) From the results of parts (b through e) pick a M truncation value that represents a reasonable value for the SV solution to have converged to the correct EM solution.

Computational Example 9.19(a) Solution

The dielectric permittivity scattering system is shown in two different views in Figure 9.45.

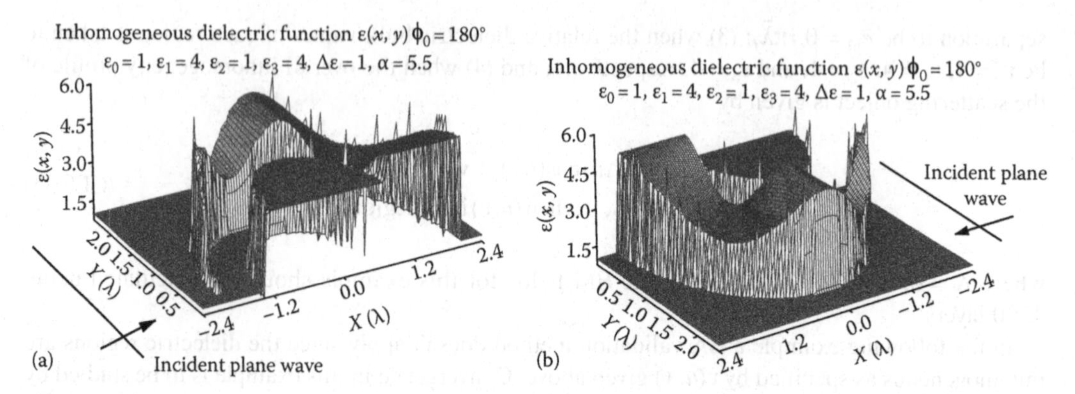

FIGURE 9.45 Relative dielectric permittivity function and geometry of the scatterer in two different views (a and b) are shown.

Computational Example 9.19(b) Solution

The representative range used to study the EM field solution is that shown in the dielectric function of part (a), namely, $-2.4\tilde{\lambda}_f \leq \tilde{x} \leq 2.4\tilde{\lambda}_f$, $0 \leq \tilde{y} \leq 2.4\tilde{\lambda}_f$. From the error analysis plots it was found that acceptable EM numerical results occurred when $28 \leq M \leq 32$, $M = M_T + 1$. Figure 9.46a and b shows, respectively, electric and magnetic field convergence plots for both the SV (inhomogeneous example, $\Delta\varepsilon = 1$, of this example) and the SV (homogeneous example, $\Delta\varepsilon = 0$, of this example ($\Delta\varepsilon = 0$ SV case studied Computational Example 9.14). As can be seen from the plots, the error convergence rates of the SV method for both the inhomogeneous and homogeneous examples studied are very close together for both the electric and magnetic field SV calculations. Also shown for comparison, are error convergence results using the BA method to calculate the EM field scattering from the homogeneous case ($\Delta\varepsilon = 0$). As can be seen from the graphs, the BA method (which only applies to the $\Delta\varepsilon = 0$ homogeneous region case) is much more accurate than the SV method for the $\Delta\varepsilon = 0$ case. Ill-conditioned SV results occurred for the inhomogeneous case when the mode truncation value $M = M_T + 1 \geq 33$. The complex magnitude electric field error $E_{RMS}^{SV,Inh}$ in Figure 9.46a was normalized by the integrated, complex magnitude RMS electric field amplitude of the 32-mode SV solution (Computational Example 9.13, Equations CE9.31 and CE9.32). ($32 = M$ is the mode truncation value) and the complex magnitude magnetic field error $E_{RMS}^{SV,Inh}$ in Figure 9.46b was normalized by the integrated, complex magnitude RMS magnetic field amplitude of the 32-mode SV solution (formula given in Computational Example 9.13, Equations CE9.33 and CE9.34).

Computational Example 9.19(c) Solution

Figure 9.47a through c shows the real electric field E_{zR} as calculated by using the SV method using $M = 28, 30, 32$ modes, and Figure 9.47d and e shows the complex magnitude difference electric field between the $M_{trunc} = 28$ and 32 SV solution and between the $M_{trunc} = 30$ and 32 SV solutions, respectively. The $M = 30, 32$ complex magnitude difference electric field plot (Figure 9.47e) is clearly smaller than the $M = 28, 32$ difference plot (Figure 9.47d), showing good convergence of the SV solution for the value of $M = 32$. *Note:* The E_z difference electric field scales on Figure 9.47d and e graphs are 100 times smaller than those of Figure 9.47a through c.

Computational Example 9.19(d) Solution

Figure 9.48a and b shows respectively, the magnetic fields H_x, H_y as calculated by using the SV method using $M = 32$ modes, and Figure 9.48c and d following these shows the complex magnitude difference, respectively, between the SV $M = 30$ − mode, H_x, H_y magnetic field solutions and the

Electric field normalized RMS integrated error (db)
Inhomogenous $\varepsilon(x, y)$. The SV, 32 mode solution is taken as "correct"
$E_{RMS}^{SV,Inh}$ (db) RMS difference of inhomogenous SV(M) solutions and inhomogenous 32-mode SV solution,
$E_{RMS}^{SV,Hom}$ (db) RMS difference of homogenous SV(M) solutions and 38-mode BA solution,

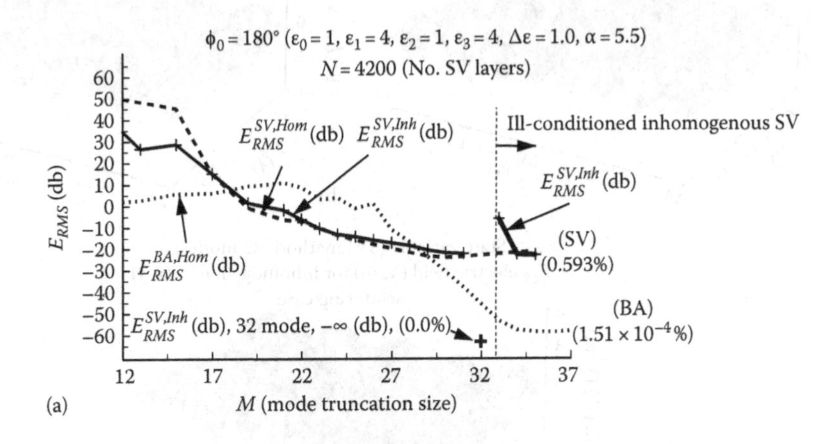

Magnetic field normalized RMS integrated error (db)
Inhomogenous $\varepsilon(x, y)$. The SV, 32 mode solution is taken as "correct"
$E_{RMS}^{SV,Inh}$ (db) RMS difference of inhomogenous SV(M) solutions and inhomogenous 32-mode SV solution,
$E_{RMS}^{SV,Hom}$ (db) RMS difference of homogenous SV(M) solutions and 38-mode BA solution,

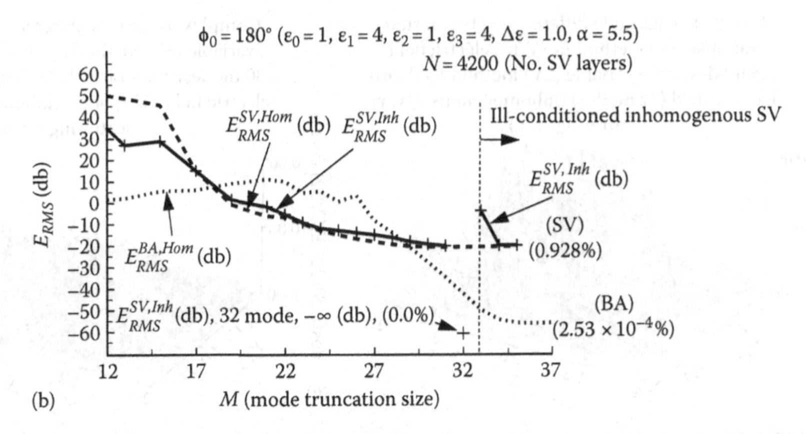

FIGURE 9.46 Electric (a) and magnetic field (b) errors are shown.

SV $M = 32$ – mode H_x, H_y magnetic field solutions. *Note:* The H_x, H_y difference field scales of Figure 9.48c and d are 100 times smaller than the Figure 9.48a and b plots.

Computational Example 9.19(e) Solution

Figure 9.49a and b shows the homogeneous and inhomogeneous permittivity spatial variations of the scatterer. In Figure 9.50, a clearly different electric field pattern is seen in the E_{zR} inhomogeneous electric field (upper) in comparison to the E_{zR} homogeneous electric field pattern, particularly in the regions where the dielectric permittivities of the two scatterers are different.

Computational Example 9.19(f) Solution

From the results of parts (b through e), the 32-mode SV EM field method has produced convergent electric and magnetic field solutions for the inhomogeneous region example under consideration.

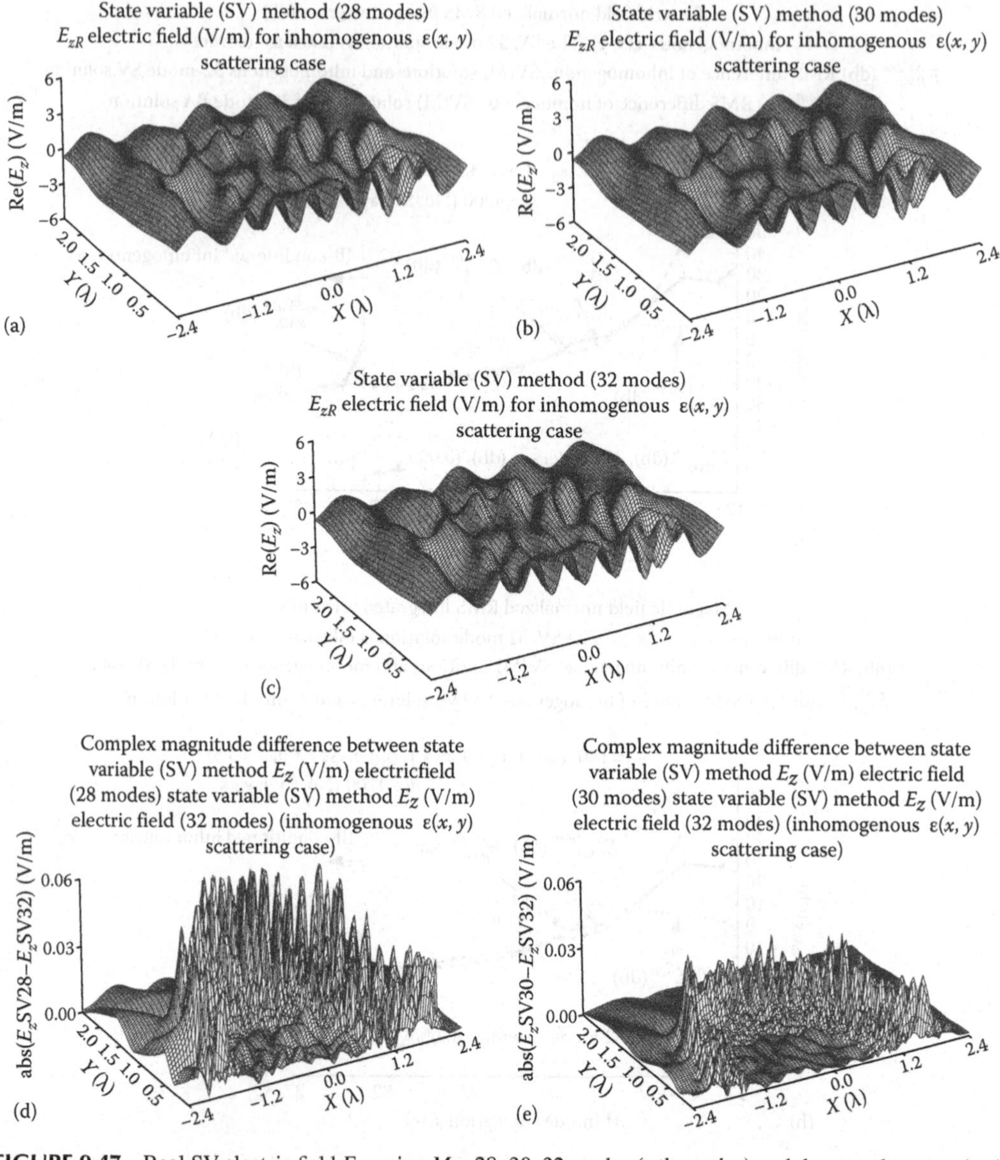

FIGURE 9.47 Real SV electric field E_{zR} using $M = 28, 30, 32$ modes (a through c) and the complex magnitude electric field error difference (d and e) are shown. (E_z difference scale is 100 times smaller than E_z scales.)

9.8 CASE STUDY VII: CALCULATION AND CONVERGENCE ANALYSIS OF EM FIELDS OF AN INHOMOGENEOUS REGION MATERIAL OBJECT USING THE SV METHOD, $\Delta\varepsilon = 0.4$, $\alpha = 5.5$, $\Lambda = 0$, EXAMPLE

(Diameter $= 2\tilde{r}_1 = 4\tilde{\lambda}_f$, for Bulk Dielectric, $\varepsilon_0 = 1$, $\varepsilon_1 = 2$, $\varepsilon_2 = 3$, $\varepsilon_3 = 4$)

Computational Example 9.20

The example concerns, using the general geometry of Figure 8.2 (plane wave incident on a dielectric cylinder), calculating the real EM fields using the SV method (call it $E_{zR}^{SV} = \mathrm{Re}(E_z^{SV})$) for the following cases: (1) when plane wave incidence angle is $\phi_0 = 180°$; (2) when the regions of the dielectric cylinder of Figure 8.2 have the following dimensions; $\tilde{r}_1 = 2\tilde{\lambda}_f$, $\tilde{r}_3 = \tilde{\lambda}_f$, inner and outer

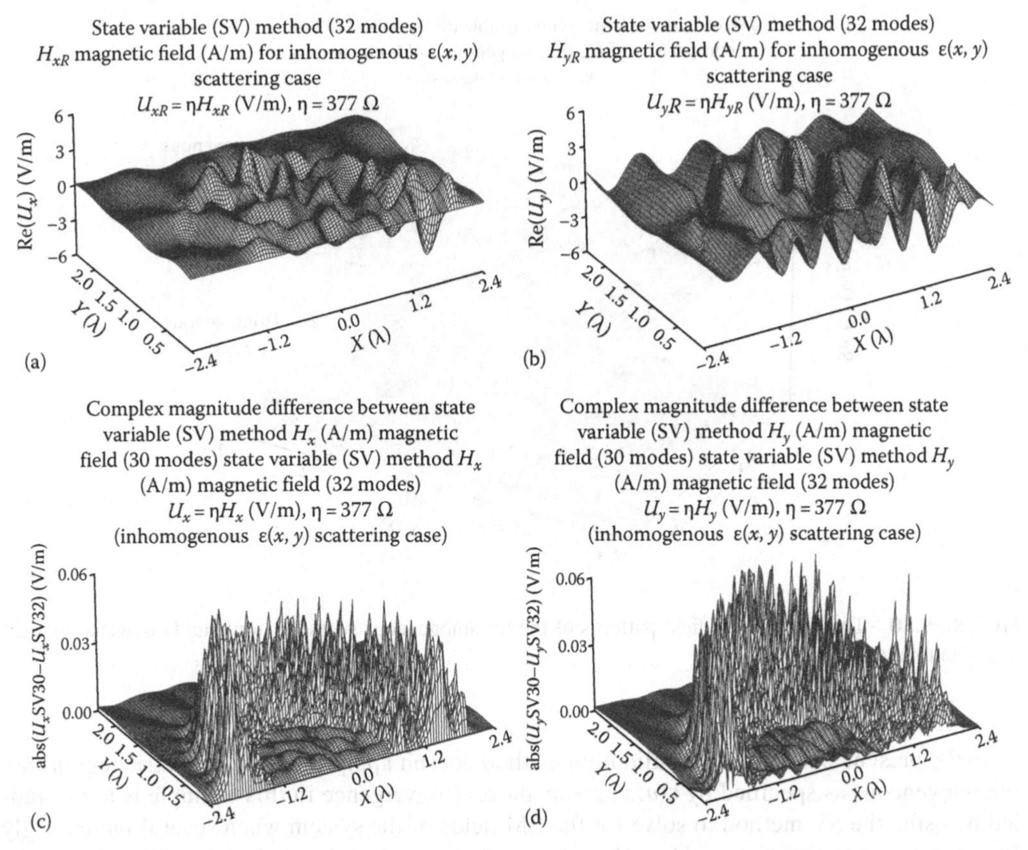

FIGURE 9.48 Real SV magnetic fields H_{xR} (a), H_{yR} (b) using $M = 32$ modes and the complex magnitude magnetic field error difference between the H_{xR}, H_{yR} 30 and 32 SV mode solutions (c and d) are shown. (U_x difference scale is 100 times smaller than U_x scales. U_y difference scale is 100 times smaller than U_y scales.)

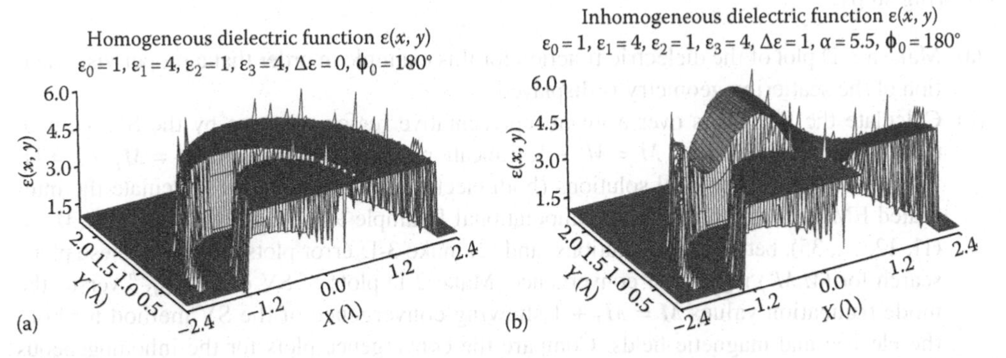

FIGURE 9.49 Dielectric permittivity and scatterer geometric of the homogeneous scatterer (a) and inhomogeneous scatterer (b) are shown.

circle separation to be $\tilde{e}_{13} = 0.1\pi\tilde{\lambda}_f$; (3) when the relative dielectric bulk values of each region is taken to be $\varepsilon_0 = 1$, $\varepsilon_1 = 2$, $\varepsilon_2 = 3$, and $\varepsilon_3 = 4$, respectively; and (4) when the overall inhomogeneity profile of the scattering object is given by

$$\varepsilon(u,v) = \begin{cases} \varepsilon_1 + \Delta\varepsilon\cos(\Lambda u)\cos(\alpha v), & \text{when } (u,v) \text{ is in Region 1} \\ \varepsilon_2, & \text{when } (u,v) \text{ is in Region 2} \end{cases} \quad \text{(CE9.36)}$$

where $\Delta\varepsilon = 0.4$, $\alpha = 5.5$, $\Lambda = 0$. The EM fields for this example should be calculated using 4200 layers.

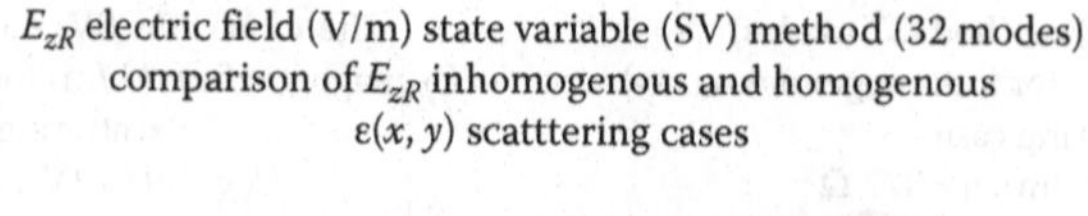

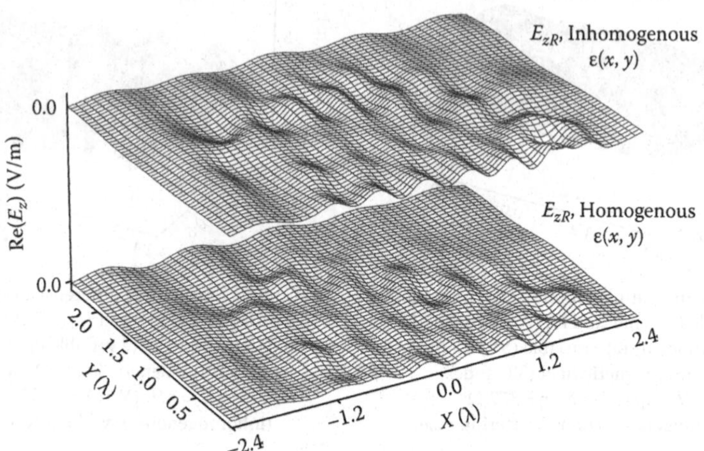

FIGURE 9.50 The real electric field patterns of the inhomogeneous scatterer (top) and homogeneous scatterer (bottom) are shown.

In the present example, a BA validation method doesn't apply since the dielectric regions are inhomogeneous as specified by $\varepsilon(u, v)$ given above. Convergence in this example is to be studied by using the SV method to solve for the EM fields of the system when several increasingly large mode truncation values $M = M_T + 1$ have been used to determine the EM fields of the system, and then by comparing these solutions to find when little change in them has resulted, indicating convergence. To quantitatively study SV convergence as just described, carry out the following steps:

(a) Make a 3-D plot of the dielectric function for this example in order that a proper visualization of the scattering geometry is displayed.
(b) Calculate the EM fields over a given representative region of space by the SV method, using a number of different $M = M_T + 1$, truncation values, ranging from $M = M_T + 1 = 11$, 12, …, 41. From these field solutions (both electric and magnetic), (1) calculate the integrated EM field difference (see Computational Example 9.13), call it $E(M,M')$ $(M,M') =$ (11, 12, …, 35), between the solutions, and (2) make 3-D error plots, and from these plots, search for (M,M') regions of convergence. Make 2-D plots of SV error $E_{RMS}^{SV,Inh}$ versus the mode truncation values $M = M_T + 1$ showing convergence of the SV method for both the electric and magnetic fields. Compare the convergence plots for the inhomogeneous case to those made for the homogenous case.
(c) Once you have determined a reasonable range of truncation convergence values, $M = M_T + 1$, from part (b), plot the E_z electric fields using the SV method for a few different values of M. Plot the complex magnitude difference of the E_z electric fields that result when two different, but close in value, truncation values are used to calculate the EM fields of the system under study. These difference plots should be much smaller than the individual EM field plots used to make the difference plots. These plots are useful, in addition to studying convergence, because they can clearly show regions (near a dielectric step discontinuity, for example) where the EM fields converge to the correct solution slowly with increasing truncation value.

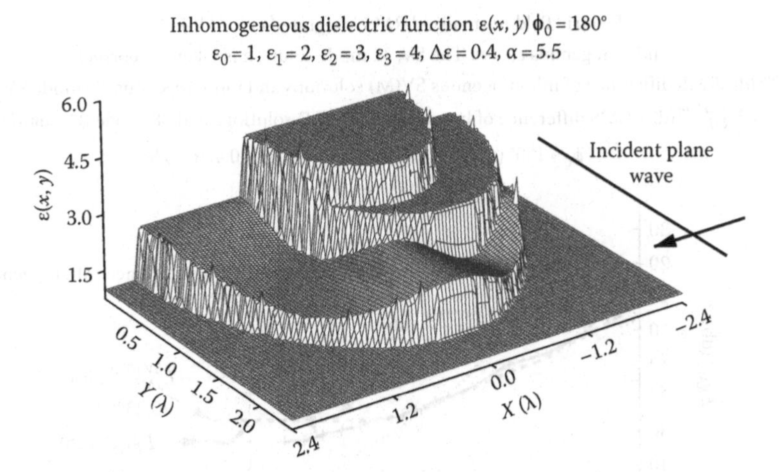

FIGURE 9.51 The dielectric permittivity function and geometry of the scattering system are shown (Λ = 0).

(d) Make EM field and EM field difference plots for the H_x, H_y magnetic fields similar to the plots that were made for the E_z electric field in part (c).
(e) From the results of parts (b through d) pick a $M = M_T + 1$ truncation value that represents a reasonable value for the SV solution to have converged to the correct EM solution.

Computational Example 9.20(a) Solution

The dielectric permittivity function and scattering system is shown in Figure 9.51.

Computational Example 9.20(b) Solution

The representative range used to study the EM field solution is that shown in the dielectric function of part (a), namely, $-2.4\tilde{\lambda}_f \le \tilde{x} \le 2.4\tilde{\lambda}_f$, $0 \le \tilde{y} \le 2.4\tilde{\lambda}_f$, and from the error analysis plots, it was found that acceptable EM numerical results occurred for the truncation mode values of $28 \le M \le 32$, where $M = M_T + 1$. Figure 9.52a and b shows, respectively, electric and magnetic field convergence plots for both the SV (inhomogeneous example, $\Delta\varepsilon = 0.4$, of this example) and the SV (homogeneous example, $\Delta\varepsilon = 0$, of this example ($\Delta\varepsilon = 0$ SV case was studied Computational Example 9.14). As can be seen from the plots, the error convergence rates of the SV method for both the inhomogeneous and homogeneous examples studied are very close together for both the electric and magnetic field SV calculations. Also shown for comparison, are error convergence results using the BA method to calculate the EM field scattering from the homogeneous case ($\Delta\varepsilon = 0$). As can be seen from the graphs, the BA method (which only applies to the $\Delta\varepsilon = 0$ homogeneous region case) is much more accurate than the SV method for the $\Delta\varepsilon = 0$ case. Ill-conditioned SV results occurred for the inhomogeneous case when the mode truncation value $M = M_T + 1 \ge 33$. The complex magnitude electric field error $E_{RMS}^{SV,Inh}$ in Figure 9.52a was normalized by the integrated, complex magnitude RMS electric field amplitude of the 32-mode SV solution (formula given in Computational Example 9.13) ($32 = M$ is the truncation value) and the complex magnitude magnetic field error $E_{RMS}^{SV,Inh}$ in Figure 9.52b was normalized by the integrated, complex magnitude RMS magnetic field amplitude of the 32-mode SV solution (formula given in Computational Example 9.13).

Computational Example 9.20(c) Solution

Figure 9.53a through c shows the real electric field E_{zR} as calculated by using the SV method using $M = 28, 30, 32$ modes, and Figure 9.51d and e shows the complex magnitude difference electric field plots between the $M = 28$ and 32, and the $M = 30$ and 32 field solutions, respectively. The $M = 30$, 32 complex magnitude difference electric field plot of Figure 9.53e is clearly smaller than the $M =$ 28, 32 Figure 9.53d difference plot, thus showing good convergence of the SV solution for the value

Electric field normalized RMS integrated error (db)
inhomogenous $\varepsilon(x, y)$. The SV, 32 mode solution is taken as "correct"
$E_{RMS}^{SV,Inh}$ (db)RMS difference of inhomogenous SV(M) solutions and inhomogenous 32-mode SV solution,
$E_{RMS}^{SV,Hom}$ (db) RMS difference of homogenous SV(M) solutions and 38-mode BA solution,
$\phi_0 = 180°$ ($\varepsilon_0 = 1$, $\varepsilon_1 = 2$, $\varepsilon_2 = 3$, $\varepsilon_3 = 4$, $\Delta\varepsilon = 0.4$, $\alpha = 5.5$)
$N = 4200$ (No. SV layers)

30
20
10
0
−10
−20
−30
−40
−50
E_{RMS} (db)
$E_{RMS}^{BA,Hom}$ (db)
Ill-conditioned inhomogenous SV
$E_{RMS}^{SV,Hom}$ (db)
$E_{RMS}^{SV,Inh}$ (db)
$E_{RMS}^{SV,Hom}$ (db),
(0.234%)
$E_{RMS}^{SV,Inh}$ (db)
(BA)
(1.85 × 10^{-3}%)
$E_{RMS}^{SV,Inh}$ (db), 32 mode, −∞ (db), (0.0%)
12 17 22 27 32 37
(a)
M (mode truncation size)

Magnetic field normalized RMS integrated error (db)
inhomogenous $\varepsilon(x, y)$. The SV, 32 mode solution is taken as "correct"
$E_{RMS}^{SV,Inh}$ (db) RMS difference of inhomogenous SV(M) solutions and inhomogenous 32-mode SV solution,
$E_{RMS}^{SV,Hom}$ (db) RMS difference of homogenous SV(M) solutions and 38-mode BA solution,
$\phi_0 = 180°$ ($\varepsilon_0 = 1$, $\varepsilon_1 = 2$, $\varepsilon_2 = 3$, $\varepsilon_3 = 4$, $\Delta\varepsilon = 0.4$, $\alpha = 5.5$)
$N = 4200$ (No. SV layers)

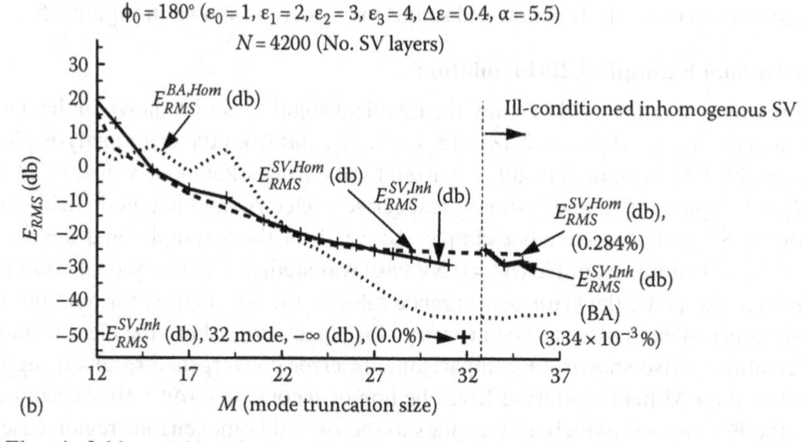

FIGURE 9.52 Electric field error (a) and magnetic field error (b) are shown ($\Lambda = 0$).

of $M = 32$. *Note:* The E_z difference electric field scales of Figure 9.53d and e are 100 times smaller than that of Figure 9.53a through c.

Computational Example 9.20(d) Solution

Figure 9.54a and b shows, respectively, the magnetic fields H_x, H_y as calculated by using the SV method using $M = 32$ modes, and Figure 9.54c and d shows the complex magnitude difference, respectively, between the SV $M = 30$ mode, H_x, H_y magnetic field solutions and the SV $M = 32$, mode H_x, H_y magnetic field solutions. *Note:* The H_x, H_y difference magnetic field scales are 100 times smaller than that of the first two plots.

Computational Example 9.20(e) Solution

From the results of part (b through d) the 32-mode SV electric and magnetic field method has converged well to the correct solution.

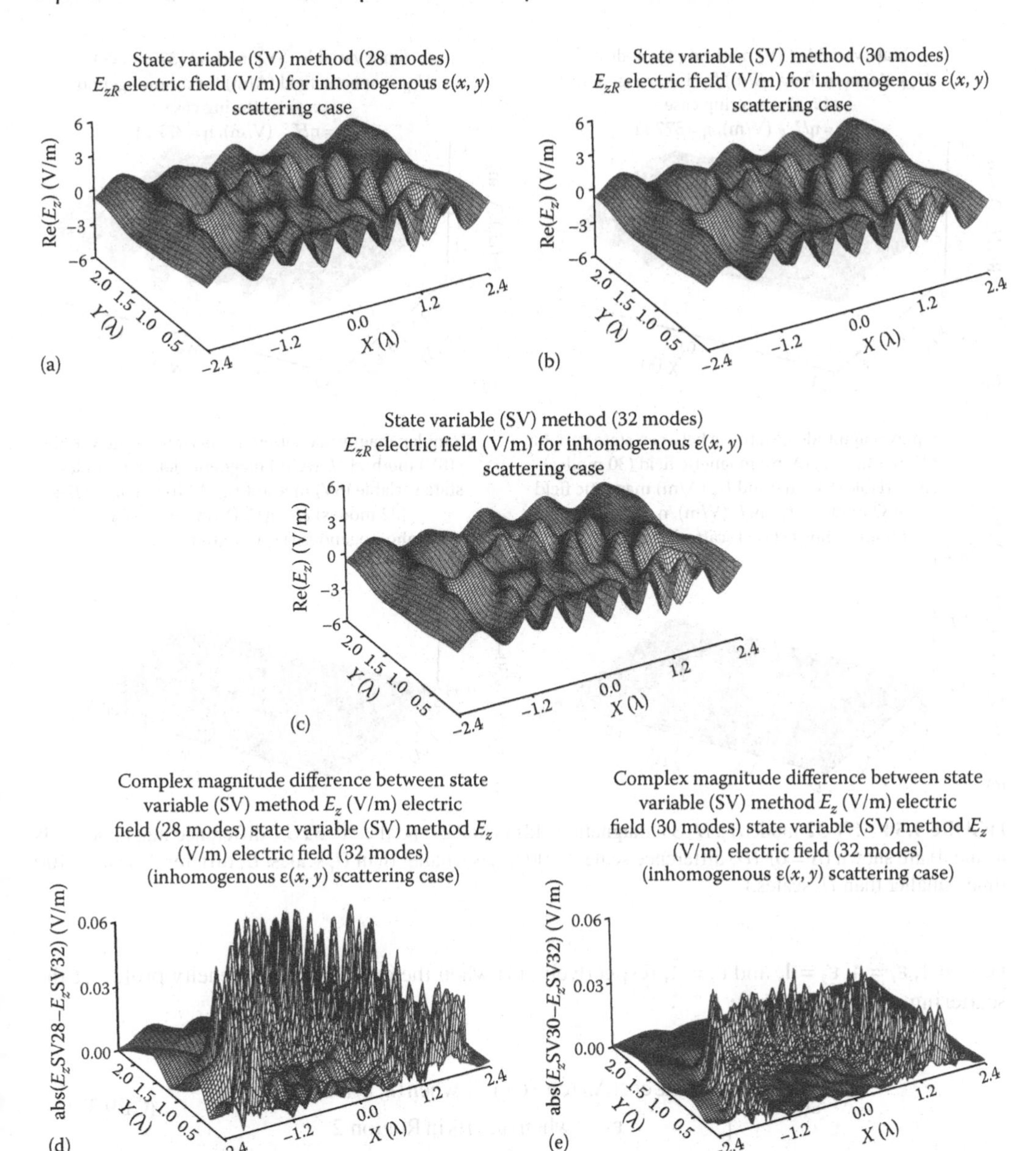

FIGURE 9.53 28, 30, 32 mode real electric field E_{zR} SV solutions (a through c) and electric field difference plots (d and e) are shown ($\Lambda = 0$). (E_z difference scale is 100 times smaller than E_z scales.)

9.9 CASE STUDY VIII: COMPARISON OF HOMOGENEOUS AND INHOMOGENEOUS REGION BISTATIC LINE WIDTHS

Computational Example 9.21

This example concerns, using the general geometry of Figure 8.2 (plane wave incident on a dielectric cylinder), calculating the bistatic line width $\sigma(\phi, \phi_0)$ using the SV method for the following cases: (1) when plane wave incidence angle is $\phi_0 = 180°$; (2) when the regions of the dielectric cylinder of Figure 8.2 have the following dimensions; $\tilde{r}_1 = 2\tilde{\lambda}_f$, $\tilde{r}_3 = \tilde{\lambda}_f$, inner and outer circle separation to be $\tilde{e}_{13} = 0.1\pi\tilde{\lambda}_f$; and (3) when the relative bulk dielectric values of each region is taken to

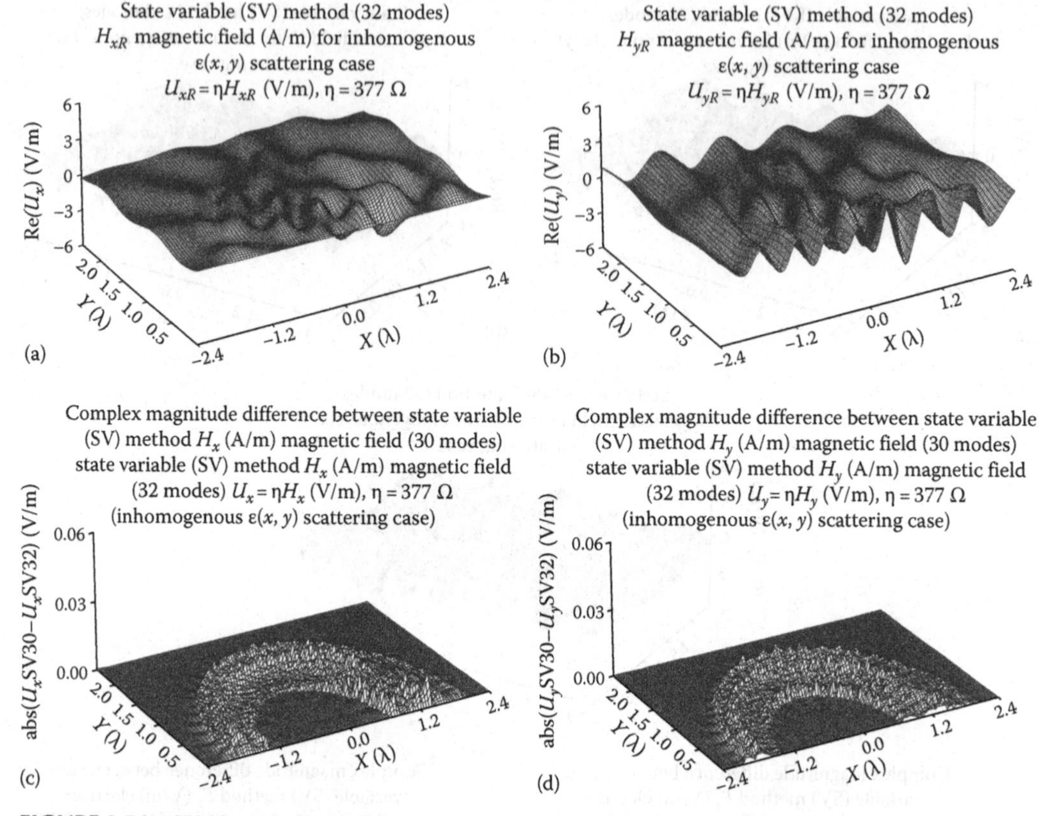

FIGURE 9.54 SV 32-mode H_x, H_y SV magnetic fields (a and b) and the 30–32 SV difference magnetic fields (c and d) are shown ($\Lambda = 0$). (U_x difference scale is 100 times smaller than U_x scales. U_y difference scale is 100 times smaller than U_y scales.)

be $\varepsilon_0 = 1$, $\varepsilon_1 = 4$, $\varepsilon_2 = 1$, and $\varepsilon_3 = 4$, respectively; (4) when the overall inhomogeneity profile of the scattering object is given by

$$\varepsilon(u,v) = \begin{cases} \varepsilon_1 + \Delta\varepsilon\cos(\Lambda u)\cos(\alpha v), & \text{when } (u,v) \text{ is in Region 1} \\ \varepsilon_2, & \text{when } (u,v) \text{ is in Region 2} \end{cases} \tag{CE9.37}$$

where $\Lambda = 0$, $\Delta\varepsilon = 0$ (homogeneous case); and (5) when $\Delta\varepsilon = 1$, $\alpha = 5.5$, $\Lambda = 0$ (inhomogeneous case).

(a) Calculate the SV EM fields for this example using 4200 layers and using a truncation value of $M = M_T + 1 = 35$. The EM SV solution for this example for the $\Delta\varepsilon = 0$ (homogeneous case) is to be validated by using the 38-mode BA method (38 = M is the truncation value) solution shown in Computational Example 9.15 to be very close to the exact solution to the EM scattering example described above and validated by the 32-mode BC method solution. The purpose of this example part is to show what effect the presence of the inhomogeneity in the scatterer has on the bistatic line width $\sigma(\phi, \phi_0)$ of the system.
(b) Repeat part (b) except for taking $\varepsilon_0 = 1$, $\varepsilon_1 = 2$, $\varepsilon_2 = 3$, and $\varepsilon_3 = 4$, and take, $\Delta\varepsilon = 0$ (homogeneous case) and take $\Delta\varepsilon = 0.4$, $\alpha = 5.5$ (inhomogeneous case).
(c) Comment on any effect that inhomogeneity of the scatterers has on the bistatic line widths.

Computational Example 9.21(a) Solution

The SV, BA, and BC bistatic line widths $\sigma(\phi, \phi_0)$ are shown in Figure 9.55a for the homogeneous case ($\Delta\varepsilon = 0$) when the relative dielectric values of each region is taken to be $\varepsilon_0 = 1$, $\varepsilon_1 = 4$, $\varepsilon_2 = 1$, and $\varepsilon_3 = 4$. In Figure 9.55b, the 32-Mode SV solutions for the homogeneous case ($\Delta\varepsilon = 0$) and for the inhomogeneous case ($\Delta\varepsilon = 1$, $\alpha = 5.5$, $\Lambda = 0$) are shown and compared when the relative bulk dielectric values of each region is taken to be $\varepsilon_0 = 1$, $\varepsilon_1 = 4$, $\varepsilon_2 = 1$, and $\varepsilon_3 = 4$.

Computational Example 9.21(b) Solution

The SV, BA, and BC bistatic line widths $\sigma(\phi, \phi_0)$ are shown in Figure 9.56a for the homogeneous case ($\Delta\varepsilon = 0$) when the relative bulk dielectric values of each region is taken to be $\varepsilon_0 = 1$, $\varepsilon_1 = 2$, $\varepsilon_2 = 3$, and $\varepsilon_3 = 4$. In Figure 9.56b, the 32-Mode SV solutions for the homogeneous case ($\Delta\varepsilon = 0$) and for the inhomogeneous case ($\Delta\varepsilon = 1$, $\alpha = 5.5$, $\Lambda = 0$) are shown and compared when the relative bulk dielectric values of each region is taken to be $\varepsilon_0 = 1$, $\varepsilon_1 = 2$, $\varepsilon_2 = 3$, and $\varepsilon_3 = 4$.

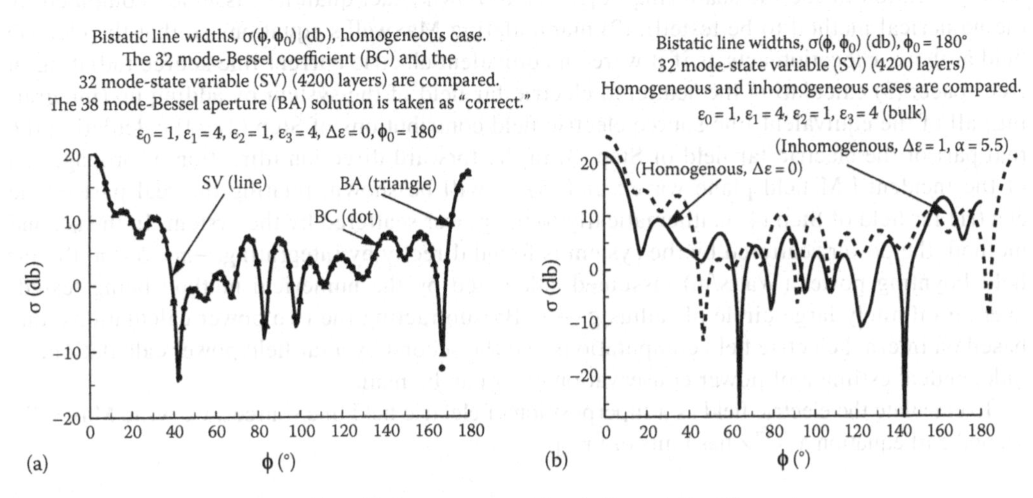

FIGURE 9.55 Homogeneous (a) and inhomogeneous bistatic line widths (b) are shown.

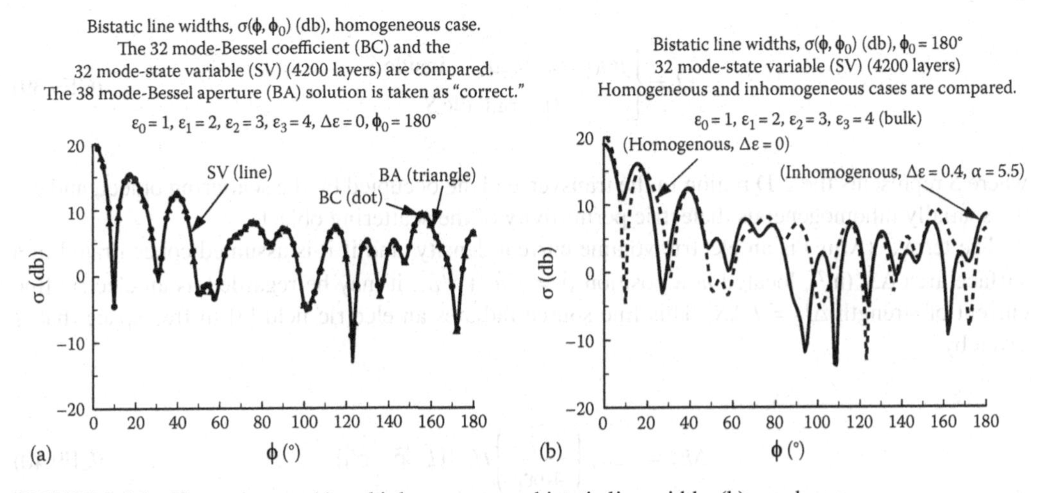

FIGURE 9.56 Homogeneous (a) and inhomogeneous bistatic line widths (b) are shown.

Computational Example 9.21(c) Solution

The different dielectric composition of two scatterers has a great affect on the shape of the bistatic line width as seen from Figures 9.55 and 9.56.

9.10 CASE STUDY IX: CONSERVATION OF POWER ANALYSIS

Computational Example 9.22

Using the Optical Theorem (forward scattering theorem) [1,2] derive an equation for the total scattering line width that may be used to test conservation of power in the scattering system.

Computational Example 9.22 Solution

The optical theorem (also called the forward scattering theorem), which may be used to check total power conservation, is based on calculating the total power scattered by the system (in Watts/m) by two different methods and then comparing the results of the two methods. In the first method, the scattered power of the system is calculated by (1) treating the electric field at an x, y location inside the scattering object as a known exact quantity (assumed computed by the numerical method to be tested); (2) manipulating Maxwell's equations so that the electric field at the x, y location acts as if it were an equivalent electric current line source radiating in free space; (3) calculating the scattered electric far field of the system by adding up (integrating) all of the equivalent line source electric field contributions of Step (2); (4) calculating the real part of the electric far field of Step (3) in the forward direction (direction of propagation of the incident EM field plane wave); and (5) as will be shown, relating the real part of the electric far field of Step (4) mathematically to the power scattered by the system. In the second method, the power scattered by the system is found directly by integrating, $-\pi < \phi < \pi$, the far field Poynting power (Watts/m) (assumed calculated by the numerical method being tested) over an infinitely large circle of radius $\rho \rightarrow \infty$. By subtracting the two power calculations, one based on internal electric field computations and the second by a far field power calculation, an independent estimate of power conservation error can be made.

To calculate the electric field as a superposition of electric field line sources, we write Maxwell's second curl equation as ($\tilde{\nabla}\times$ has units of [m^{-1}])

$$\tilde{\nabla} \times \vec{H} = j\omega\tilde{\varepsilon}\vec{E} = \vec{J} + j\omega\tilde{\varepsilon}_0\vec{E} \tag{CE9.38}$$

where

$$\vec{J} = \begin{cases} j\omega(\tilde{\varepsilon} - \tilde{\varepsilon}_0)E_z\hat{a}_z, & \text{inside } S \\ 0, & \text{outside } S \end{cases} \tag{CE9.39}$$

where S represents the 2-D region in the transverse plane occupied by the scattering object, and $\tilde{\varepsilon}$ is the spatially inhomogeneous dielectric permittivity of the scattering object.

The term $\vec{J}$ (A/m^2) is an electric volume current density, and if it is assumed concentrated in a surface area $\Delta\tilde{S}'$ (m^2), located at a position $\tilde{\vec{\rho}}' = \tilde{x}'\hat{a}_x + \tilde{y}'\hat{a}_y$, it may be regarded as an electric line current of strength $\Delta I_E = J_z\Delta\tilde{S}'$. This line source radiates an electric field [3] in free space that is given by

$$\Delta E_z^s = -\Delta I_E\left(\frac{\tilde{k}_f^2}{4\omega\tilde{\varepsilon}_0}\right)H_0^{(2)}\left(\tilde{k}_f\left|\tilde{\rho} - \tilde{\rho}'\right|\right) \tag{CE9.40}$$

where $H_0^{(2)}$ is a Hankel function of the second kind, $\tilde{k}_f = 2\pi/\tilde{\lambda}_f$, $\tilde{\lambda}_f$ (m) is the free space wavelength. A superposition of the line currents over all points in the scattering object gives the scattered electric field at any point in space. It is given by

$$E_z^s = -j\left(\frac{\tilde{k}_f^2}{4}\right)\iint_S (\varepsilon(\tilde{x}',\tilde{y}')-1)H_0^{(2)}\left(\tilde{k}_f\left|\vec{\tilde{\rho}}-\vec{\tilde{\rho}}'\right|\right)E_z(\tilde{x}',\tilde{y}')d\tilde{x}'d\tilde{y}' \tag{CE9.41}$$

where $\varepsilon(\tilde{x}', \tilde{y}')$ is the relative dielectric permittivity of the system. It is convenient to use normalized coordinates, $x = \tilde{k}_f\tilde{x}$, $\tilde{k}_f = 2\pi/\tilde{\lambda}_f$, etc. and in the present analysis, the incident electric field in Equation CE9.41 which excites the electric field E_z inside the scattering object, is to be taken $\vec{E}^i = E_0\exp(j(x'\cos(\phi_0)+y'\sin(\phi_0))\hat{a}_z$, where E_0 is the plane wave amplitude and ϕ_0 is the angle of incidence (Figure 8.2).

At this point, we are interested in finding E_z^s in Equation CE9.41 in the far-field of the scattering system. We have, $\vec{\rho} = \tilde{k}_f\vec{\tilde{\rho}}$, $(\vec{\rho}-\vec{\rho}')\cdot(\vec{\rho}-\vec{\rho}') = \rho^2 - 2\vec{\rho}\cdot\vec{\rho}' + \rho'^2 \cong \rho^2(1-2\vec{\rho}\cdot\vec{\rho}'/\rho^2)$ as $\rho^2 \gg \rho'^2$ in the far field. Thus $|\vec{\rho}-\vec{\rho}'| \cong \rho(1-2\hat{a}_\rho\cdot\vec{\rho}'/\rho)^{1/2} \cong \rho - \hat{a}_\rho\cdot\vec{\rho}'$ where the approximation $(1 - X)^{1/2} \cong 1 - X/2$ and the expression $\hat{a}_\rho = \vec{\rho}/\rho = \cos(\phi)\hat{a}_x + \sin(\phi)\hat{a}_y$ have been substituted into the last equation for $|\vec{\rho}-\vec{\rho}'|$. In normalized coordinates, substituting the large argument form of the Hankel function $H_0^{(2)}(X) = (2/\pi X)^{1/2}\exp(-j(X-\pi/4))$, $X \to \infty$ in Equation CE9.41, E_z^s becomes

$$E_z^s = \left(\frac{2}{\pi\rho}\right)^{1/2}\exp\left(-j\left(\rho-\frac{\pi}{4}\right)\right)E_0F_J(\phi,\phi_0) \tag{CE9.42a}$$

where

$$F_J(\phi,\phi_0) = -\left(\frac{j}{4}\right)\iint_S (\varepsilon(x',y')-1)E_{zN}(x',y',\phi_0)\exp(j\hat{a}_\rho\cdot\vec{\rho}')dx'dy' \tag{CE9.42b}$$

where $E_{zN}(x', y') = E_z(x', y')/E_0$.

Letting

$$I_J(\phi,\phi_0) = \int_S (\varepsilon(x',y')-1)E_{zN}(x',y',\phi_0)\exp(j(x'\cos\phi + y'\sin\phi))dx'dy' \tag{CE9.43}$$

where $\vec{\rho}' = x'\hat{a}_x + y'\hat{a}_y$ has been substituted in the expression $\hat{a}_\rho\cdot\vec{\rho}'$. We also have

$$F_J(\phi,\phi_0) = -\left(\frac{j}{4}\right)I_J(\phi,\phi_0) \tag{CE9.44}$$

The quantity $F_J(\phi, \phi_0)$ is the pattern space factor of the system as determined by using a superposition of line sources.

The total power per unit length, call it $P_s(\phi_0)$ (Watts/m), scattered by the system may be found using the divergence theorem in conjunction with the assumption that the scattering system is lossless and absorbs no real power. Letting C_∞ be a circle of radius $\rho \to \infty$ in the transverse plane (or x, y plane) occupied by the scattering object, and letting the total fields in the system be

$$\vec{E} = \vec{E}^i + \vec{E}^s = (E_z^i + E_z^s)\hat{a}_z \tag{CE9.45a}$$

$$\vec{H} = \vec{H}^i + \vec{H}^s \tag{CE9.45b}$$

the power per unit length absorbed by the system is given by (using $\tilde{\rho} = \rho/\tilde{k}_f \to \infty$)

$$0 = \oint_{C_\infty} \left(\frac{1}{2\tilde{k}_f}\right) \text{Re}(\vec{E}\times\vec{H}^*)\cdot(-\hat{a}_\rho)\rho d\phi = \oint_{C_\infty} \left(\frac{1}{2\tilde{k}_f}\right) \text{Re}\left((\vec{E}^i+\vec{E}^s)\times(\vec{H}^i+\vec{H}^s)^*\right)\cdot(-\hat{a}_\rho)\rho d\phi \tag{CE9.46}$$

The $\vec{E}^i \times \vec{H}^{i*}$ integral in (CE9.46) is zero, i.e.,

$$0 = \oint_{C_\infty} \left(\frac{1}{2\tilde{k}_f}\right) \text{Re}(\vec{E}^i \times \vec{H}^{i*})\cdot \hat{a}_\rho \rho d\phi \tag{CE9.47}$$

since no power is absorbed by a plane wave located in a vacuum. Thus with $P_s(\phi_0)$ being the power per unit length scattered by the scattering object, we have

$$P_s(\phi_0) \equiv \oint_{C_\infty} \left(\frac{1}{2\tilde{k}_f}\right) \text{Re}(\vec{E}^s \times \vec{H}^{s*})\cdot \hat{a}_\rho \rho d\phi = -\oint_{C_\infty} \left(\frac{1}{2\tilde{k}_f}\right) \text{Re}(\vec{E}^i \times \vec{H}^{s*} + \vec{E}^s \times \vec{H}^{i*})\cdot \hat{a}_\rho \rho d\phi \tag{CE9.48}$$

This also may be expressed as

$$P_s(\phi_0) = -\oint_{C_\infty} \frac{1}{2\tilde{k}_f} \text{Re}(\vec{E}^i \times \vec{H}^* + \vec{E} \times \vec{H}^{i*})\cdot \hat{a}_\rho \rho d\phi \tag{CE9.49a}$$

$$P_s(\phi_0) = -\oint_{C_\infty} \frac{1}{2\tilde{k}_f} \text{Re}(\vec{E}^{i*} \times \vec{H} + \vec{E} \times \vec{H}^{i*})\cdot \hat{a}_\rho \rho d\phi \tag{CE9.49b}$$

$$P_s(\phi_0) = -\text{Re} \int_{C_\infty} \left(\frac{1}{2\tilde{k}_f}\right) (\vec{E}^{i*} \times \vec{H} + \vec{E} \times \vec{H}^{i*})\cdot \hat{a}_\rho \rho d\phi \tag{CE9.49c}$$

Equation CE9.49a may be seen to hold if $\vec{E}$, $\vec{H}$ of Equation CE9.45 are substituted in Equations CE9.49a, and CE9.47 is used for the $\vec{E}^i \times \vec{H}^{i*}$ integral that will now be contained in Equation CE9.49a. Equation CE9.49b and c hold because for each component in the expression $\text{Re}(\vec{E}^i \times \vec{H}^*) = \text{Re}(\vec{E}^{i*} \times \vec{H})$, the real part of a complex number and the real part of it's complex conjugate equal each other.

Use of the 2-D divergence theorem shows

$$\left(\frac{1}{2\tilde{k}_f}\right) \oint_{C_\infty} (\vec{E}^{i*} \times \vec{H})\cdot \hat{a}_\rho \rho d\phi = \left(\frac{1}{2\tilde{k}_f}\right) \int_{S_\infty} \nabla\cdot(\vec{E}^{i*} \times \vec{H}) dS \tag{CE9.50}$$

$$\left(\frac{1}{2\tilde{k}_f}\right) \oint_{C_\infty} (\vec{E} \times \vec{H}^{i*})\cdot \hat{a}_\rho \rho d\phi = \left(\frac{1}{2\tilde{k}_f}\right) \int_{S_\infty} \nabla\cdot(\vec{E} \times \vec{H}^{i*}) dS \tag{CE9.51}$$

where S_∞ be the area enclosed by the circle C_∞. From a vector identity we have

$$\nabla\cdot(\vec{E}^{i*}\times\vec{H}) = \vec{H}\cdot\nabla\times\vec{E}^{i*} - \vec{E}^{i*}\cdot\nabla\times\vec{H} = \vec{H}\cdot j\tilde{\eta}_f\vec{H}^{i*} - \vec{E}^{i*}\cdot\frac{j\varepsilon}{\tilde{\eta}_f}\vec{E} \quad \text{(CE9.52)}$$

$$\nabla\cdot(\vec{E}\times\vec{H}^{i*}) = \vec{H}^{i*}\cdot\nabla\times\vec{E} - \vec{E}\cdot\nabla\times\vec{H}^{i*} = \vec{H}^{i*}\cdot(-j\tilde{\eta}_f)\vec{H} - \vec{E}\cdot\left(\frac{-j\varepsilon_0}{\tilde{\eta}_f}\right)\vec{E}^{i*} \quad \text{(CE9.53)}$$

where $\varepsilon_0 = (\tilde{\varepsilon}_0/\tilde{\varepsilon}_0) = 1$, $\tilde{\eta}_f = (\tilde{\mu}_0/\tilde{\varepsilon}_0) = 377\Omega$, and where Maxwell's equations in normalized form, namely have been substituted for $\nabla\times\vec{E}^{i*}$, $\nabla\times\vec{H}$, $\nabla\times\vec{E}$, and $\nabla\times\vec{H}^{i*}$. Adding Equations CE9.50 and CE9.51 after substitution of Equations CE9.52 and CE9.53 in these equations and canceling terms, after noting $\vec{E}\cdot\vec{E}^{i*} = \vec{E}^{i*}\cdot\vec{E}$ and $\vec{H}^{i*}\cdot\vec{H}^{i} = \vec{H}^{i}\cdot\vec{H}^{i*}$, and and taking the negative real part of the overall resulting equation, we have

$$\oint_{C_\infty}\left(\frac{-1}{2\tilde{k}_f}\right)\mathrm{Re}\,(\vec{E}^{i}\times\vec{H}^{*} + \vec{E}\times\vec{H}^{i*})\cdot\hat{a}_\rho\rho d\phi = \int_{S_\infty}\left(\frac{-1}{2\tilde{k}_f}\right)\mathrm{Re}\left(\left(\frac{-j(\varepsilon-1)}{\tilde{\eta}_f}\right)\vec{E}\cdot\vec{E}^{i*}\right)dS \quad \text{(CE9.54)}$$

In the last integral over the area outside the area S of the scattering object, we have $\varepsilon = \varepsilon_0 = 1$ and thus the integral over this outside region is zero. Using Equations CE9.48 and CE9.54 we have

$$P_s(\phi_0) \equiv \int_{C_\infty}\left(\frac{1}{2\tilde{k}_f}\right)\mathrm{Re}\,(\vec{E}^{s}\times\vec{H}^{s*})\cdot\hat{a}_\rho\rho d\phi = \int_{S}\left(\frac{1}{2\tilde{k}_f}\right)\mathrm{Re}\left(\frac{j(\varepsilon-1)}{\tilde{\eta}_f}\,\vec{E}\cdot\vec{E}^{i*}\right)dx'dy' \quad \text{(CE9.55)}$$

Substituting $\vec{E}_N = (E_z/E_0)\hat{a}_z = E_{zN}\hat{a}_z$ and $\vec{E}^{i} = E_0\exp\,(j(x'\cos(\phi_0) + y'\sin(\phi_0)))\hat{a}_z$ where $\phi_0 = 0, \pi$ for the present problem, we have

$$P_s(\phi_0) = \left(\frac{E_0E_0^{*}}{2\tilde{k}_f\tilde{\eta}_f}\right)\mathrm{Re}\left(j\int_S(\varepsilon(x',y')-1)E_{zN}(x',y',\phi_0)\,\exp(-j(x'\cos(\phi_0)+y'\sin(\phi_0)))\,dx'dy'\right) \quad \text{(CE9.56)}$$

Letting

$$I_D(\phi_0) = \int_S(\varepsilon(x',y')-1)E_{zN}(x',y',\phi_0)\exp(-j(x'\cos(\phi_0)+y'\sin(\phi_0)))\,dx'dy' \quad \text{(CE9.57a)}$$

we have,

$$P_s(\phi_0) = \left(\frac{E_0E_0^{*}}{2\tilde{k}_f\tilde{\eta}_f}\right)\mathrm{Re}(jI_D(\phi_0)) \quad \text{(CE9.57b)}$$

On comparing $I_J(\phi, \phi_0)$ of Equation CE9.43 with $I_D(\phi_0)$ of Equation CE9.57a, we note that if ϕ in Equation CE9.43 is chosen to be $\phi = \phi_f$, where $\phi_f \equiv \phi_0 + \pi$ (ϕ_f is the forward scattering angle), we have

$$I_D(\phi_0) = I_J(\phi_f,\phi_0) = \left(\frac{-4}{j}\right)F_J(\phi_f,\phi_0) \quad \text{(CE9.58)}$$

since $\cos\phi_f = \cos(\phi_0 + \pi) = -\cos(\phi_0)$, $\sin\phi_f = \sin(\phi_0 + \pi) = -\sin(\phi_0)$ and the two integrals are identical. The last equality in Equation CE9.58 results from Equation CE9.44.

We now evaluate the left hand side of Equations CE9.56 and CE9.57b, namely,

$$P_s(\phi_0) = \int_{-\pi}^{\pi} \left(\frac{1}{2\tilde{k}_f}\right) \mathrm{Re}(\vec{E}^s \times \vec{H}^{s*}) \cdot \hat{a}_\rho \rho d\phi \tag{CE9.59}$$

In the far field (see Equation CE9.42a),

$$\vec{E}^s \times \vec{H}^{s*} = \frac{\vec{E}^s \cdot \vec{E}^{s*}}{\tilde{\eta}_f} \hat{a}_\rho = \frac{E_z^s E_z^{s*}}{\tilde{\eta}_f} \hat{a}_\rho = \frac{1}{\tilde{\eta}_f}\left(\frac{2}{\pi\rho}\right) E_0 E_0^* F_J(\phi,\phi_0) F_J^*(\phi,\phi_0) \hat{a}_\rho$$

thus

$$P_s(\phi_0) = \left(\frac{E_0 E_0^*}{2\tilde{k}_f \tilde{\eta}_f}\right)\left(\frac{2}{\pi}\right) \int_{-\pi}^{\pi} F_J(\phi,\phi_0) F_J^*(\phi,\phi_0) d\phi \tag{CE9.60}$$

From Equations CE9.57b and CE9.58 we have ($jI_D(\phi_0) = -4F_J(\phi_f, \phi_0)$)

$$P_s(\phi_0) = \left(\frac{E_0 E_0^*}{2\tilde{k}_f \tilde{\eta}_f}\right) \mathrm{Re}(jI_D(\phi_0)) = \left(\frac{E_0 E_0^*}{2\tilde{k}_f \tilde{\eta}_f}\right) \mathrm{Re}(-4F_J(\phi_f,\phi_0)) \tag{CE9.61}$$

and thus equating Equations CE9.60 and CE9.61, we have

$$P_s(\phi_0) = \left(\frac{E_0 E_0^*}{2\tilde{k}_f \tilde{\eta}_f}\right)\left(\frac{2}{\pi}\right) \int_{-\pi}^{\pi} F_J(\phi,\phi_0) F_J^*(\phi,\phi_0) d\phi = \left(\frac{E_0 E_0^*}{2\tilde{k}_f \tilde{\eta}_f}\right) \mathrm{Re}\left(-4F_J(\phi_f,\phi_0)\right) \tag{CE9.62}$$

Equation CE9.62 represents a statement of the Optical theorem (or forward scattering theorem).

A convenient way to present the conservation of power in the scattering system is in terms the of the total scattering line width. The total scattering line width $\tilde{\sigma}_W$ (*meters*) is defined herein as that width such that the total power $P_{TW} = S_{Inc}\tilde{\sigma}_W \tilde{L}_z = (1/2\tilde{\eta}_f) E_0 E_0^* \tilde{\sigma}_W \tilde{L}_z$ (Watts) transmitted through a flat area of size $\tilde{\sigma}_W \tilde{L}_z$ (the normal of this area is parallel to the propagation direction of the incident plane wave), equals the total power, call it P_{TS}, scattered radially by the scattering object through a cylindrical area whose radius is $\tilde{\rho} \to \infty$ and whose length in the z direction is $\tilde{L}_z$. This amount of power is $P_{TS} = P_s(\phi_0)\tilde{L}_z$. Equating P_{TW} and P_{TS} we have

$$P_{TW} = \left(\frac{1}{2\tilde{k}_f \tilde{\eta}_f}\right) E_0 E_0^* (\tilde{k}_f \tilde{\sigma}_W) \tilde{L}_z = P_s(\phi_0)\tilde{L}_z = P_{TS} \tag{CE9.63}$$

Introducing a dimensionless line width $\sigma_W \equiv \tilde{k}_f \tilde{\sigma}_W$, $\tilde{k}_f = 2\pi/\tilde{\lambda}_f$, canceling $\tilde{L}_z$ from both sides of Equation CE9.63, and after substitution of Equation CE9.62 in Equation CE9.63, we have

$$\left(\frac{1}{2\tilde{k}_f \tilde{\eta}_f}\right) E_0 E_0^* \sigma_W = P_s(\phi_0) = \left(\frac{E_0 E_0^*}{2\tilde{k}_f \tilde{\eta}_f}\right)\left(\frac{2}{\pi}\right) \int_{-\pi}^{\pi} F_J(\phi,\phi_0) F_J^*(\phi,\phi_0) d\phi$$

$$= \left(\frac{E_0 E_0^*}{2\tilde{k}_f \tilde{\eta}_f} \right) \mathrm{Re}(-4F_J(\phi_f, \phi_0)) \tag{CE9.64}$$

Further canceling $E_0 E_0^* / 2\tilde{k}_f \tilde{\eta}_f$ we have

$$\sigma_W(\phi_0) = \left(\frac{2}{\pi} \right) \int_{-\pi}^{\pi} F_J(\phi, \phi_0) F_J^*(\phi, \phi_0) d\phi = \mathrm{Re}(-4F_J(\phi_f, \phi_0)) \tag{CE9.65}$$

This equation represents a dimensionless conservation of power scattering width parameter from which to gauge and study scattering from the dielectric object. The conservation of power as results from a given numerical solution of $F_J(\phi_f, \phi_0)$ may be studied by introducing a normalized percentage error $E(\%)$ defined by

$$E(\%) = \frac{100 |\sigma_{WR} - \sigma_{WL}|}{\sigma_{WR}} (\%) \tag{CE9.66}$$

where

$$\sigma_{WL} = \left(\frac{2}{\pi} \right) \int_{-\pi}^{\pi} F_J(\phi, \phi_0) F_J^*(\phi, \phi_0) d\phi \tag{CE9.67}$$

$$\sigma_{WR} = \mathrm{Re}(-4F_J(\phi_f, \phi_0)) \tag{CE9.68}$$

Numerical examples of the dimensionless line widths σ_{WL}, σ_{WR} and the error $E(\%)$ are given in Computational Examples 9.23 and 9.24 to follow.

For the present problem, the pattern space factor $F_J(\phi, \phi_0)$ for $\phi_0 = 0, \pi$, is given by Equations 8.86 through 8.88

$$F_J(\phi, \phi_0) = \sum_{m=0}^{\infty} B_m^{(0)} j^m \cos(m\phi_0) \cos(m\phi) \tag{CE9.69}$$

Using the orthogonality of the cosine functions it is found

$$\sigma_{WL} = 2 \sum_{m=0}^{\infty} (1 + \delta_{m,0}) \left| B_m^{(0)} \right|^2 \tag{CE9.70}$$

$$\sigma_{WR} = -4\,\mathrm{Re} \left(\sum_{m=0}^{\infty} j^m \cos(m\phi_f) \cos(m\phi_0) B_m^{(0)} \right) \tag{CE9.71}$$

where $\delta_{0,0} = 1$, $\delta_{m,0} = 0$, $m \geq 1$ and $\phi_f \equiv \phi_0 + \pi$.

Numerical examples of the dimensionless line widths σ_{WL}, σ_{WR}, and the error $E(\%)$ are given in Computational Examples 9.23 and 9.24 to follow.

Computational Example 9.23

This example concerns, using the general geometry of Figure 8.2 (plane wave incident on a dielectric cylinder), using the optical theorem (also called the forward scattering theorem) to calculate how accurately the conservation of power in the system holds for the following cases: (1) when plane wave incidence angle is $\phi_0 = 180°$; (2) when the regions of the dielectric cylinder of Figure 8.2 have the following dimensions; $\tilde{r}_1 = 2\tilde{\lambda}_f$, $\tilde{r}_3 = \tilde{\lambda}_f$, inner and outer circle separation to be $\tilde{e}_{13} = 0.1\pi\tilde{\lambda}_f$; (3) when the relative bulk dielectric values of each region is taken to be $\varepsilon_0 = 1$, $\varepsilon_1 = 4$, $\varepsilon_2 = 1$, and $\varepsilon_3 = 4$, respectively; (4) when the overall inhomogeneity profile of the scattering object is given by

$$\varepsilon(u,v) = \begin{cases} \varepsilon_1 + \Delta\varepsilon\cos(\Lambda u)\cos(\alpha v), & \text{when}\,(u,v)\,\text{is in Region 1} \\ \varepsilon_2, & \text{when}\,(u,v)\,\text{is in Region 2} \end{cases} \tag{CE9.72}$$

where $\Lambda = 0$ and $\Delta\varepsilon = 0$ (homogeneous case); and (5) when $\Delta\varepsilon = 1$, $\alpha = 5.5$, and, $\Lambda = 0$ (inhomogeneous case). The SV EM fields for this example should be calculated using 4200 layers and using a truncation value of $M = M_T + 1$.

(a) Using the optical theorem (also called the forward scattering theorem [see Computational Example 9.22]), the EM SV solution for this example for the $\Delta\varepsilon = 0$ (homogeneous case) is to be validated by using the BA method solution shown to be very close to the exact solution (Computational Example 9.15) to the EM scattering example described above and also validated by the BC method solution. Specifically, the accuracy of the conservation of power of the system is to be found by subtracting the total scattering normalized line width σ_W of the scattering object as calculated by the σ_{WL} and σ_{WR} of Equations CE9.70 and CE9.71, and using the size of this difference, to gauge how accurately power conservation holds, and thus ultimately, how accurately numerical computations have been made. Calculate this error difference using $(E(M)(\%) = 100\left(\left|\sigma_{WR} - \sigma_{WL}\right|/\sigma_{WR}\right)(\%))$ (Equation CE9.69), using the BA, BC, and SV methods, and present your results in table form. Choose a mode truncation value $M = M_T + 1$ for which reasonable convergence of the EM field solutions have been found. Plot the error difference $(E(M)(\%) = 100\left(\left|\sigma_{WR} - \sigma_{WL}\right|/\sigma_{WR}\right)(\%))$, expressed in (db) verses $M = M_T + 1$, the mode truncation number, for the homogeneous and inhomogeneous cases given in the problem.

(b) Repeat part (a) for the inhomogeneous case defined by Equation CE9.72 with $\Delta\varepsilon = 1$, $\alpha = 5.5$, and, $\Lambda = 0$.

Computational Example 9.23a and b Solution

The SV, BA, and BC total scattering normalized line width σ_W of the scattering object as calculated by σ_{WL}, σ_{WR}, and $E(\%)$ of Computational Example 9.22 are displayed in Table 9.4 for the homogeneous and inhomogeneous cases specified in the problem definition. The error difference $\left(E(M)(\%) = 100\left|\sigma_{WR} - \sigma_{WL}\right|/\sigma_{WR}(\%)t\right)$ versus M the mode truncation number for the homogeneous and inhomogeneous cases given in the problem are shown in Figure 9.57.

Computational Example 9.24

This example concerns, using the general geometry of Figure 8.2 (plane wave incident on a dielectric cylinder), using the optical theorem (also called the forward scattering theorem) to calculate how accurately the conservation of power in the system holds for the following cases: (1) when plane

TABLE 9.4
Conservation of Power (Optical Theorem), Normalized σ_{WL}, σ_{WR} Line Widths, and Difference Error $E(\%)$, M (Number of Truncated Modes)

Homogeneous	σ_{WL}	σ_{WR}	$E(\%) = 100 \left(\lvert \sigma_{WR} - \sigma_{WL} \rvert / \sigma_{WR} \right)(\%)$
$\varepsilon_0 = 1, \varepsilon_1 = 4, \varepsilon_2 = 1, \varepsilon_3 = 4, \Delta\varepsilon = 0$			
Bessel aperture $M = 35$	53.712	53.712	1.7101×10^{-8}
Bessel coefficient $M = 35$	53.712	53.712	1.7494×10^{-7}
State variable $M = 35$	53.712	53.712	5.4224×10^{-3}
Inhomogeneous			
$\varepsilon_0 = 1, \varepsilon_1 = 4, \varepsilon_2 = 1, \varepsilon_3 = 4, \Delta\varepsilon = 1, \alpha = 5.5$, $\Lambda = 0$ State variable, $M = 29$ (SV only applies)	60.667	60.664	5.1399×10^{-3}

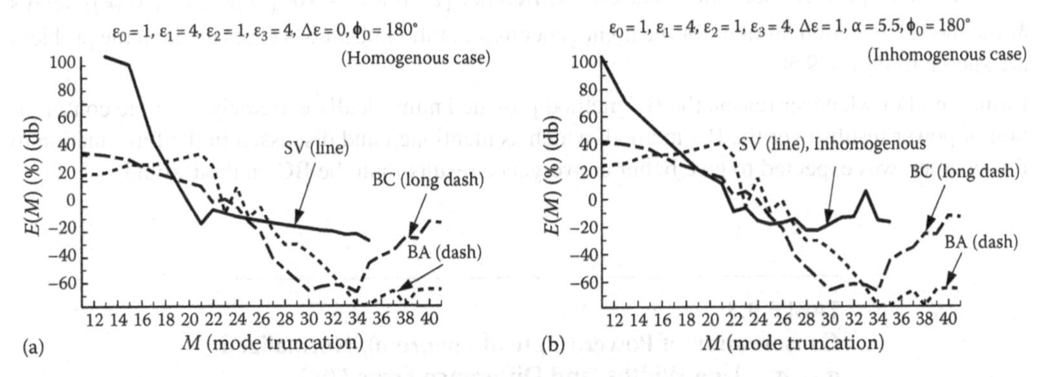

FIGURE 9.57 Conservation errors for the homogenous case (a) and inhomogeneous case (b) are shown. Power conservation scattering line width errors, $E(M)$ (%) (db) of the Bessel coefficient (BC, homogenous), the Bessel aperture (BA, homogenous), and state variable (SV, inhomogeneous) (4200 layers) methods are compared. The BA and BC convergence results only apply to the homogenous case, but have been plotted in the figure to show a comparison of the convergence rates of the different methods.

wave incidence angle is $\phi_0 = 180°$; (2) when the regions of the dielectric cylinder of Figure 8.2 have the following dimensions; $\tilde{r}_1 = 2\tilde{\lambda}_f$, $\tilde{r}_3 = \tilde{\lambda}_f$, inner and outer circle separation to be $\tilde{e}_{13} = 0.1\pi\tilde{\lambda}_f$; and (3) when the relative bulk dielectric values of each region is taken to be $\varepsilon_0 = 1$, $\varepsilon_1 = 2$, $\varepsilon_2 = 3$, and $\varepsilon_3 = 4$, respectively, (4) when the overall inhomogeneity profile of the scattering object is given by

$$\varepsilon(u,v) = \begin{cases} \varepsilon_1 + \Delta\varepsilon\cos(\Lambda u)\cos(\alpha v), & \text{when } (u,v) \text{ is in Region 1} \\ \varepsilon_2, & \text{when } (u,v) \text{ is in Region 2} \end{cases} \tag{CE9.73}$$

where $\Lambda = 0$, $\Delta\varepsilon = 0$ (homogeneous case); and (5) when $\Delta\varepsilon = 0.4$, $\alpha = 5.5$, and $\Lambda = 0$ (inhomogeneous case). The SV EM fields for this example should be calculated using 4200 layers and using a truncation value of $M = M_T + 1$.

(a) Using the optical theorem (also called the forward scattering theorem [see Computational Example 9.22]), the EM SV solution for this example for the $\Delta\varepsilon = 0$ (homogeneous case) is to be validated by using the BA method solution shown to be very close to the exact solution (Computational Example 9.13) to the EM scattering example described above and validated by the BC method solution. Specifically, the accuracy of the conservation of power of the system is to be found by subtracting the total scattering normalized line width σ_W of the scattering object as calculated by the σ_{WL} and σ_{WR} of Equations CE9.70 and CE9.71, and using the size of this difference, to gauge how accurately power conservation holds, and thus ultimately, how accurately the numerical computations have been made. Calculate this error difference using $\left(E(M)(\%) = 100\left(\left|\sigma_{WR} - \sigma_{WL}\right| / \sigma_{WR}\right)(\%)\right)$ (Equation CE9.66), using the BA, BC, and SV methods, and present your results in table form. Choose a mode truncation value $M = M_T + 1$ for which reasonable convergence of the EM field solutions have been found. Plot the error difference $\left(E(M)(\%) = 100\left(\left|\sigma_{WR} - \sigma_{WL}\right| / \sigma_{WR}\right)(\%)\right)$, expressed in (db) verses $M = M_T + 1$, the mode truncation number, for the homogeneous and inhomogeneous cases given in the problem.

(b) Repeat part (a) for the inhomogeneous case defined by Equation CE9.72 with $\Delta\varepsilon = 0.4$, $\alpha = 5.5$, $\Lambda = 0$.

Computational Example 9.24a and b Solution

The SV, BA, and BC total scattering normalized line width σ_W of the scattering object as calculated by σ_{WL}, σ_{WR}, and $E(\%)$ of Computational Example 9.22 are displayed in Table 9.5 for the cases specified in the problem definition. The error difference $\left(E(M)(\%) = 100\left(\left|\sigma_{WR} - \sigma_{WL}\right| / \sigma_{WR}\right)\right)$ versus M the mode truncation number for the homogeneous and inhomogeneous cases given in the problem are shown in Figure 9.58.

Comment: For whatever reason the BC method provided numerically extremely accurate conservation of power results over the BA method, which as mentioned and discussed in the Introduction to this chapter, was expected to give better convergence results than the BC method would.

TABLE 9.5
Conservation of Power (Optical Theorem), Normalized σ_{WL}, σ_{WR} Line Widths, and Difference Error $E(\%)$, M (Number of Truncated Modes)

Homogeneous	σ_{WL}	σ_{WR}	$E(M)(\%) = 100\left(\left\|\sigma_{WR}-\sigma_{WL}\right\| / \sigma_{WR}\right)(\%)$
$\varepsilon_0 = 1, \varepsilon_1 = 2, \varepsilon_2 = 3, \varepsilon_3 = 4, \Delta\varepsilon = 0$			
Bessel aperture			
$M = 35$	48.269	48.269	9.1612×10^{-8}
Bessel coefficient			
$M = 35$	48.269	48.269	4.4161×10^{-14}
State variable			
$M = 35$	48.305	48.305	2.3819×10^{-5}
Inhomogeneous			
$\varepsilon_0 = 1, \varepsilon_1 = 2, \varepsilon_2 = 3, \varepsilon_3 = 4,$	44.379	44.379	8.9027×10^{-5}
$\Delta\varepsilon = 0.4, \alpha = 5.5, \Lambda = 0$			
State variable, $M = 29$			
(SV only applies)			

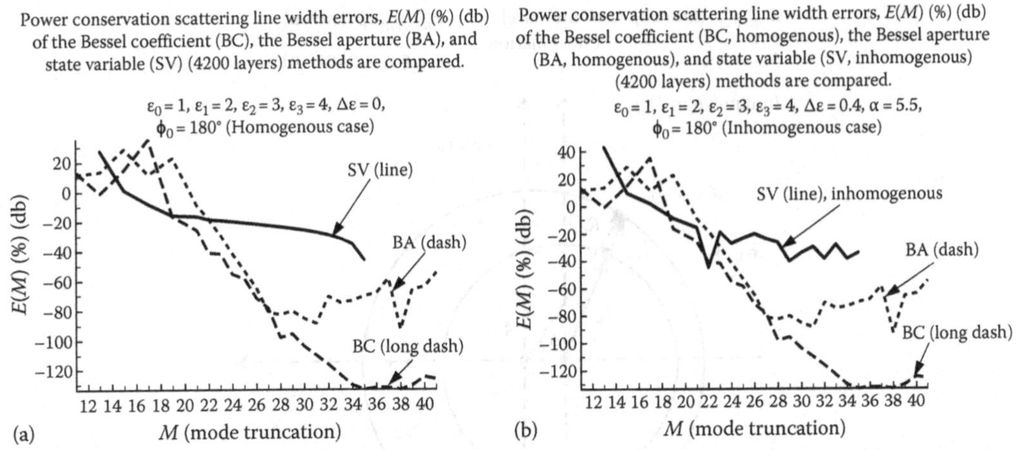

Power conservation scattering line width errors, E(M) (%) (db) of the Bessel coefficient (BC), the Bessel aperture (BA), and state variable (SV) (4200 layers) methods are compared.
ε0 = 1, ε1 = 2, ε2 = 3, ε3 = 4, Δε = 0, ϕ0 = 180° (Homogenous case)
SV (line)
BA (dash)
BC (long dash)
E(M) (%) (db)
M (mode truncation)
(a)
Power conservation scattering line width errors, E(M) (%) (db) of the Bessel coefficient (BC, homogenous), the Bessel aperture (BA, homogenous), and state variable (SV, inhomogenous) (4200 layers) methods are compared.
ε0 = 1, ε1 = 2, ε2 = 3, ε3 = 4, Δε = 0.4, α = 5.5, ϕ0 = 180° (Inhomogenous case)
SV (line), inhomogenous
BA (dash)
BC (long dash)
E(M) (%) (db)
M (mode truncation)
(b)

FIGURE 9.58 Conservation errors for the homogenous case (a) and inhomogeneous case (b) are shown. The BA and BC convergence results only apply to the homogenous case but have been plotted in the figure to show a comparison of the convergence rates of the different methods.

APPENDIX 9.A: INTERPOLATION EQUATIONS

Computational Example 9.A

Develop interpolation equations to calculate the EM field values accurately inside the SV thin layers, given that the EM solution has been calculated on the thin layer boundaries. Draw diagrams illustrating your solution.

Computational Example 9.A Solution

Figure 5.59a through c show the overall interpolation geometry used for plots in the chapter and show the basic equations that need to be solved to implement the overall interpolation scheme.

Basic interpolation problem
y, y1, yu−, yu, yu+
xOu+, yOu+
xOu, yOu
xOu−, yOu−
Ou−
Ou
Ou+
x, x1, xu−, xu, xu+
xOu, yOu
Ou−, Ou, Ou+
Ru−, Ru, Ru+
(a) (known in x1, y1 coordinates)
xOu−, yOu−, xOu+, yOu+
(unknown, to be found in x1, y1 coordinates)

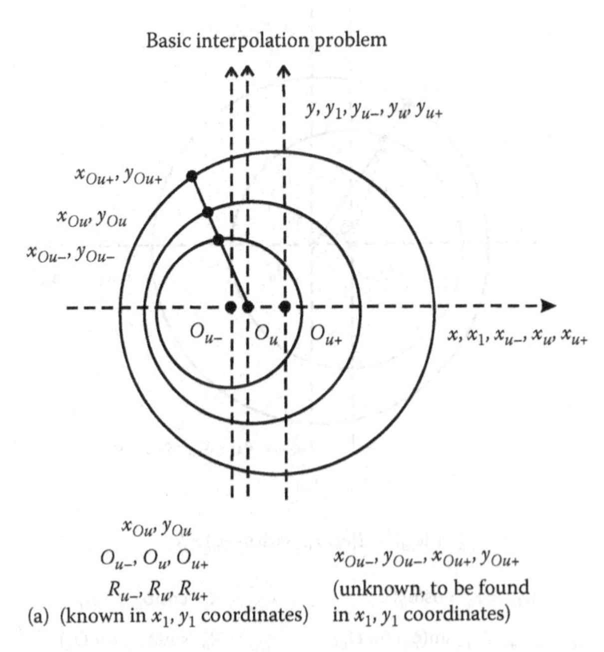

FIGURE 9.59 The geometry of the eccentric cylinder interpolation geometry (a through c) is shown.

(continued)

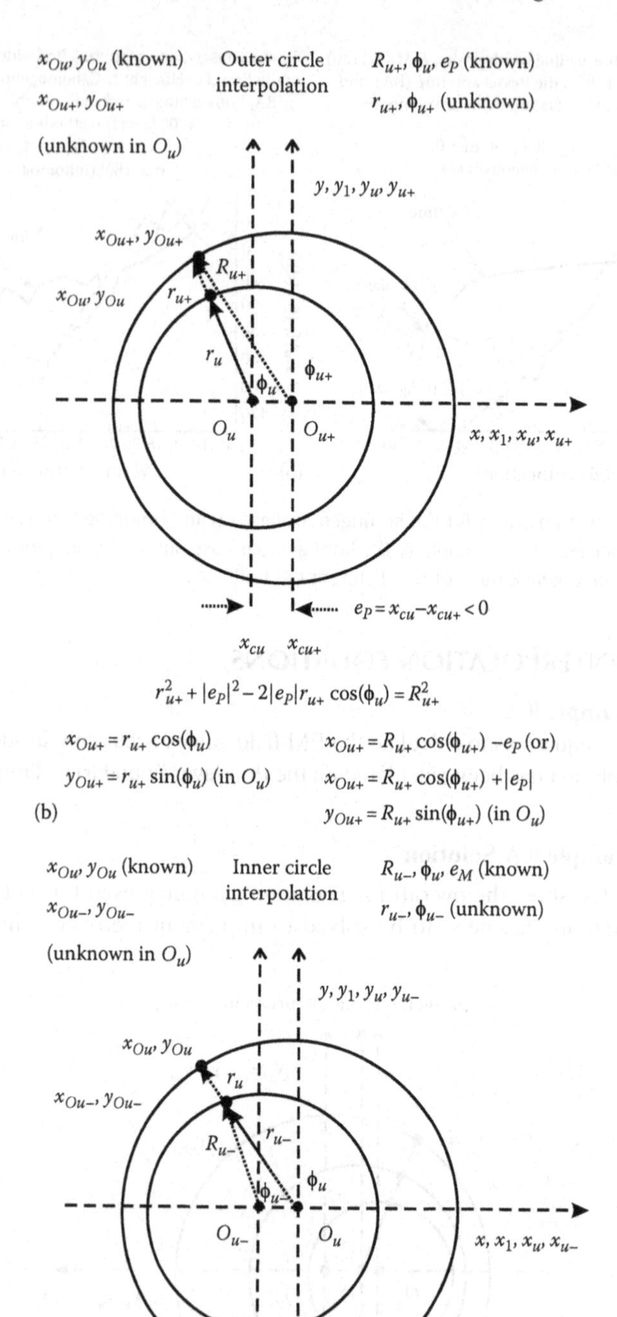

$$r_{u-}^2 + |e_M|^2 - 2|e_M|r_{u-}\cos(\pi - \phi_u) = R_{u-}^2$$

$x_{Ou-} = r_{u-}\cos(\phi_u)$ $\quad$ $x_{Ou-} = R_{u-}\cos(\phi_{u-}) - e_M$

(c) $y_{Ou-} = r_{u-}\sin(\phi_u)$ (in O_u) $\quad$ $y_{Ou-} = R_{u-}\sin(\phi_{u-})$ (in O_u)

FIGURE 9.59 (continued)

In the figures, the EM field values are known anywhere on the R_{Ou+} outer and R_{Ou-} inner circles. The point x_{Ou}, y_{Ou} in the figures represents the point where the EM field is to be evaluated by interpolation using the known EM field values at x_{Ou+}, y_{Ou+} and x_{Ou-}, y_{Ou-}. Figure 9.59a shows the overall geometry and Figure 9.59a and b gives the geometry and equations from which to determine the values of the coordinates x_{Ou+}, y_{Ou+} and x_{Ou-}, y_{Ou-} respectively. Once the values of the x_{Ou+}, y_{Ou+} and x_{Ou-}, y_{Ou-} coordinates are known, the EM fields at these points may be evaluated, and the EM field values at x_{Ou}, y_{Ou} may be found using linear interpolation along the line shown in Figure 9.59a.

REFERENCES

1. A. Ishimaru, *Electromagnetic Wave Propagation, Radiation and Scattering*, Prentice Hall, Upper Saddle River, NJ, 1991.
2. M.P. Ioannidou, EM-wave scattering by an axial slot on a circular PEC cylinder with an eccentrically layered inner coating: A dual-series solution for TE polarization, *IEEE Trans. Antennas Propag*. AP-57(11), 3512–3519, 2009.
3. C.A. Balanis, *Advanced Engineering Electromagnetics*, John Wiley & Sons, New York, 1989.

Index

F

G

H

S

Printed in the United States
by Baker & Taylor Publisher Services